Flora of Thummalapalle Uranium Mining Area, Andhra Pradesh, India

The Authors

Dr. K. Raja Kullayiswamy was born in Anantapuramu district. He completed his M.Sc. in Botany from Sri Krishnadevaraya University, Anantapuramu and obtained Ph.D. from Sri Krishnadevaraya University. At present, he is working as Research Associate in Indian Institute of Science, Bangalore from 2014 till today. He published 16 research papers in various National and International journals and presented papers in 13 National and International conferences and seminars. He acted as reviewer for two international taxonomic journals.

Dr. S. Sandhya Rani did Master degree in Botany with first rank from Sri Venkateswara University and thereafter obtained her Ph.D. degree (1998) from Sri Krishnadevaraya University. At present, she is working as Assistant Professor in the Department of Botany, Sri Krishnadevarya University, Ananatapuramu. She has 16 years of research experience in Plant Taxonomy and 6 years of teaching experience. Her publications include 55 research papers and 6 books. She was Principal Investigator of a research project funded by Department of Atomic Energy, Government of India and Life Member of IAAT, IBS and NESA.

Dr. B. Sadasivaiah obtained M.Sc. and Ph.D. degrees from Sri Krishnadevaraya University, Ananthapuramu, Andhra Pradesh. His expertise lies in the fields of Plant Taxonomy, Ecology and Biodiversity Conservation. He joined as an Assistant Professor in the Department of Botany, Government Degree College (Men), Wanaparthy, Telangana. He published more than 40 research papers and 3 chapters in edited books. He got K.S. Manilal Award from Indian Association for Angiosperm Taxonomy (2008) and Young Scientist Award from Council for Green Revolution (2014). He worked as Research Associate in DAE project under the supervision of Prof. T. Pullaiah. He is not only a botanist but also a Herpetologist, Conservationist and expert in rescuing of wild animals, especially snakes.

Prof. T. Pullaiah obtained M. Sc. (1973) and Ph. D. (1976) degrees in Botany from Andhra University. He was a Post-doctoral Fellow at Moscow State University, Russia, during 1976-1978. He travelled widely in Europe and USA and visited Universities and Botanic Gardens in about 17 countries. He joined Sri Krishnadevaraya University as Lecturer (1979) and became Professor (1993). He held several positions in the University as Dean, Faculty of Life Sciences; Head, Department of Botany; Chairman, Board of Studies in Botany; Head, Department of Sericulture; Co-ordinator and Chairman, Board of Studies in Biotechnology; Principal, Sri Krishnadevaraya University College. He retired from active service in 2011 and was selected by UGC as UGC-BSR Faculty Fellow in the Department of Botany, Sri Krishnadevaraya University. He has published 65 books, 330 research papers and 35 popular articles. His books include Encyclopaedia of World Medicinal Plants (5 volumes), Flora of Andhra Pradesh (5 volumes), Biodiversity in India (8 volumes), Flora of Eastern Ghats (4 volumes), Flora of Telangana (3 volumes), Encyclopaedia of Herbal Antioxidants etc. Apple Academic Press, Canada is bringing out 5 volume edited book on Ethnobotany of India, and they are being distributed by CRC Press. He was the Principal-Investigator of 20 major research projects totalling more than a crore of rupees funded by DBT, DST, CSIR, UGC, BSI, WWF, GCC etc. Under his guidance, 54 students obtained their Ph. D. degrees and 34 students their M. Phil. Degrees. He is the recipient of Best Teacher Award from the Government of Andhra Pradesh, Prof. P. Maheshwari Gold Medal and Dr. G. Panigrahi Memorial Award of Indian Botanical Society and Prof. Y. D. Tiagi Gold Medal of the Indian Association for Angiosperm Taxonomy. He was President of Indian Association for Angiosperm Taxonomy (2013) and President of the Indian Botanical Society (2014). He was a member of Species Survival Commission of International Union for Conservation of Nature and Natural Resources (IUCN).

Flora of Thummalapalle Uranium Mining Area, Andhra Pradesh, India

— *Authors* —

K. Raja Kullayiswamy
S. Sandhya Rani
B. Sadasivaiah
&
T. Pullaiah

Department of Botany,
Sri Krishnadevaraya University
Anantapuramu 515 003
Andhra Pradesh

2017

Regency Publications

A Division of

Astral International Pvt. Ltd.

New Delhi – 110 002

Cataloging in Publication Data--DK
Courtesy: D.K. Agencies (P) Ltd. <docinfo@dkagencies.com>

Raja Kullayiswamy, K., author.
Flora of Thummalapalle uranium mining area, Andhra Pradesh, India / author, K. Raja Kullayiswamy, S. Sandhya Rani, B. Sadasivaiah & T. Pullaiah.
pages cm
Includes indexes.

ISBN 978-93-86071-92-7 (International Edition)

1. Plants--India--Thummalapalle (Kadapa)--Classification. I. Sandhya Rani, S., author. II. Sadasivaiah, B., author. III. Pullaiah, T., author. IV. Title.

QK358.R35 2017 DDC 581.95485 23

Published by : **Regency Publications**
A Division of
Astral International Pvt. Ltd.
– ISO 9001:2015 Certified Company –
4736/23, Ansari Road, Darya Ganj
New Delhi-110 002
Ph. 011-43549197, 23278134
E-mail: info@astralint.com
Website: www.astralint.com

Preface

Natural resources survey like floristic studies, play an important role in the economic development of developing country like India. The alarming rate at which this natural wealth is vanishing becomes the compelling factor for a detailed plant survey of all the plant species for effective conservation and sustainable utilization.

Recently Uranium Mining was launched at Thummalapalle in Kadapa district, Andhra Pradesh, India. Detailed studies are necessary for the steps to be taken to protect and conserve natural resources in the Mining Area. Due to lack of floristic account of the mining area this work has been undertaken to enumerate the plant species in an area of 30 Km radius from the mining site. The purpose of the book is to provide a detailed account on plants found in Thummalapalle Uranium Mining Area. It is hoped that the book will be useful in developing strategies for monitoring and conservation of precious plant resources of Thummalapalle Uranium mining area.

Authors are grateful to Board of Research in Nuclear Sciences, Department of Atomic Energy, Government of India for Financial support to the project (vide reference No. 2008/36/34-BRNS). They are thankful to Dr. A. Vinod Kumar, Head, Environmental Systems and Networking section, Environmental Assessment Division, Bhabha Atomic Research Centre, Mumbai for his guidance, support, help and coordination of the project. They also thank Dr. Debanik Roy, Programme Officer, BRNS, Dr. R.M.Tripathi, Head, RPS (NF), EAD, BARC for suggestions.

Our sincere thanks are due to Teaching faculty of Department of Botany, Sri Krishnadevaraya University, Anantapuramu, Andhra Pradesh, and Sri Krishnadevaraya University authorities for facilities and Research scholars, and well-wishers who have accorded their full cooperation during research work and the preparation of this book. The help and cooperation received from Botanists at Herbaria CAL, MH, BSID and different universities is thankfully acknowledged.

We welcome comments from the readers of the book.

K. Raja Kullayiswamy

S. Sandhya Rani

B. Sadasivaiah

T. Pullaiah

Contents

1

Introduction

The association of man and nature is prehistoric and particularly with plants is very significant. His basic necessities of air, water, food, shelter, clothing and other essential whereabouts led his curiosity to explore the nature and its surroundings. In the process, he gained elementary knowledge of plants and learnt the significance of this association over the years. This unending quest for knowledge in every other discipline, along with socio-economic developments, has further led towards a systematic approach in the field of plant biology, to a concept of documentation and scientific classification of plants which is a backbone of science and technology of future.

Man has been indulging in utilization of forest resources enormously. He is deriving his common needs of food, medicines, fiber, fodder, timber, cosmetics, gums, resins, fragrances, food flavors, dyes, pesticides, industrial enzymes, emulsifiers, leather and animal products from forest. For the overall development of an area, its natural resources need to be utilized judiciously and this is possible only when the resources are identified and evaluated scientifically. Hence, detailed floristic investigation becomes important so that necessary plans can be drawn for its integrated development.

The population of human and livestock have grown tremendously all over the world. In India alone, 3–3.5 times of growth is there from 1951 to 2011 in human population (India census, 2011). Livestock population was 228 million in 1951 and reached 1210 million in 2011. Increased population laid a pressure on biological resource and the balance between production and exploitation is impaired. In addition to this, man encroached the forest areas for settlements, developmental projects and agriculture. Deforestation, habitat loss, habitat fragmentation, over exploitation and invasive alien species lead to the loss of biodiversity. A great deal of the biodiversity is at the verge of extinction and already some species are extinct due to above said activities. The deforestation in the world was 9.4 million hectares per annum during 1990-2000 (FAO, 2000).

Several National, International, Government and Non-Governmental organizations recognized this catastrophe and endeavoring to protect through *in situ* and *ex situ* conservation. In the world's total land area, only 3.7 per cent was under protection as national parks, forest reserves etc. (McNeely *et al.*, 1990). The role of taxonomy is crucial in the task of conservation of the biodiversity but only 1.7 million of species are described (Singh, 2002) which less than 5 per cent of total expected species number. IUCN (2009) was able to assess the threatened status of 2.7 per cent of the described species and their paucity in assessment due to lack of information on the distribution of the described species. A lot of biodiversity is being lost before the description of species and reaching their importance to science.

Mines are the source of mineral supply and conducted explorations for minerals without giving much regards to its impacts on the ecology and environment. Therefore, the mining industry is being placed under the red category, meaning it is in the top bracket of environmental degradation. The chief environmental impacts due to mining are changes in soil stratification, reduced biotic diversity and alteration of structure and functioning of ecosystems. These changes ultimately influence water and nutrient dynamics of trophic interactions. Advanced technology, automation and globalization of mining industries has changed the mining sector around the globe.

India is a developing country; it needs to enhance production of power for growing industries, agriculture and domestic use for the growing population. To satisfy the needs of power, our country is eager to augment the nuclear power production, which it is less than 6 per cent in the total and plans to generate 20,000 megawatts of nuclear power by 2020.

Recently underground uranium mining was launched in the year 2008 at Thummalapalle, near Pulivendula in Kadapa district, Andhra Pradesh. Being underground mining there are no open cast mining or surface activities and hence there is no possibility of biodiversity loss. However, like any other human developmental activity the associated development in the area can lead to changes in biodiversity. To keep a track, a detailed floristic and quantification study of the Uranium Mining Area has been carried out to identify and know the existing baseline data on plants which are common, rare, threatened and endemic. This helps us to implement necessary conservation strategies to preserve the biodiversity.

Hence the need of the hour is intensive exploration of plant diversity in the Mining area like Thummalapalle Uranium Mining Area, Andhra Pradesh. The outcome of the exploration should be known to the public through documentation in the form of scientific publication which showcase the existing plant diversity and its status.

Due to lack of floristic studies in the mining site and surrounding 30 km radius, it has been under taken to enumerate the same. The importance of survey, exploration and exploitation of the floral wealth of the country needs to be reiterated.

2

Topography and General Features

Thummalapalle uranium mining site is located near Thummalapalle village, in Vemula mandal, Kadapa district of Andhra Pradesh. Thummalapalle Uranium Mining Area has been taken as stydy area, 30 km radius from the mining site, which accounts an area of 2828 km^2. The area is spread over in 225 revenue villages under 12 Mandals of two districts Kadapa and Anantapuramu of Andhra Pradesh state. The geographical coordinates of the centre point are NL 14° 19′ 59.3″ and EL 78° 15′ 18.2″. It is bounded on the East by Idupulapaya (14°18′59.77″ N and 78°29′54.91″ E), West by Mudigubba (14°20′23.04″ N and 77°59′50.07″ E), North by Bhadrampalle (14°03′01.98″ N and 78°16′51.20″ E) and South by Nallacheruvu (14°37′44.69″ N and 18°15′12.41″ E). Altitude ranges from 198 m to 875 m.

Palakonda hill ranges in the study area are the prominent and belong to Eastern Ghats. These ranges separate the two districts (Kadapa and Anantapuramu) and extend through Idupulapaya and are connected to Seshachalam (Thirumala) hill ranges and Nallamala hill ranges in the central Eastern Ghats. North-West and North-East corners of the area are mainly dominated by agricultural lands and barren lands. The South, South-West, South-East, West and East corners are with moderate vegetation by scrub. The study area encompasses forest area, agricultural land and barren land. Sagileru and Maddileru are the main rivulets, and annual streams, tanks, reservoirs, springs, waterfalls, ditches are responsible for the wealth of flora and fauna.

The area of 30 km radius is divided in to three zones for study purpose based on mining effect through DAE (Department of Atomic Energy) guidelines which are Core zone, Buffer zone I and Buffer zone II. All three zones are spread in both the districts.

Core zone: This is the first zone in the study area, it starts from the centre of the mining site and end by 5 km radius (78.5718 Km^2). Altitude of this zone ranges from 375 m to 789 m. It covers 9 villages - Thummalapalle, Rachakuntapalle, Mabbuchintalapalle, Bhumaiahgari palle, K. K. Kottala, Kanampalli, Somulavaripalli, Chinnapalli and Velpula in two districts. In this zone Reseved forests Chinta Kanam Bhavi RF and Mallappa Kanuma RF of Kadapa district and Pashuvula Doddi RF, Marrikonda Kattava RF of Anantapuramu district are also located.

Scrub type of forest is the dominant in the core zone followed by waste lands and Agriculture lands. The dominant herbs/grasses are *Heteropogon contortus, Chrysopogon fulvus, Cymbopogon coloratus, Hybanthus enneaspermus, Tridax procumbens, Crotalaria hebecarpa* and *Indigofera cordifolia,* whereas the shrubs include *Catunaregam spinosa, Flueggea leucopyrus* and *Senna montana.* The climbers like *Asparagus racemosus, Gloriosa superba* and *Rivea hypocrateriformis* grow in this zone. *Croton scabiosus,* an endemic tree is very common along with *Dolichandrone atrovirens, Euphorbia tortilis* and *Acacia catechu. Ceropegia spiralis* has been found only in this zone. *Hildegardia populifolia* is the vulnerable tree species, however hundreds of individuals are found near Chinnapalle. Annual streams are more in Rachakuntapalle (Small dam is constructed recently) because of deep valleys and thick vegetation, whereas tanks are found in Velpula, Somulavaripalle and Chinnapalle villages.

Buffer Zone I

This is the second zone in the study area, extending to 15 km radius (628.5718 km^2). Altitude of this zone ranges from 210 m to 875 m. Human habitations in this zone include 39 villages in Kadapa district and 58 villages in Anantapuramu district.

Dry deciduous forest is seen only on south side of this zone while scrub forest spreads as patches in almost all areas along with Agriculture lands in this zone.

Reserved forests like Sakalavani Konda RF, Kokkaramkonda RF, Jerripotuladonakuppa RF, Nigidi RF and Gorivikanuma RF are in Anantapuramu district. This area is dominated by herbs/grasses like *Eriocaulon quinquangulare, Heteropogon contortus, Triumfetta pentandra* and shrubs *Catunaregam spinosa, Solanum pubescens, Abutilon indicum, Senna montana* and *Barleria noctiflora.* The climbers commonly observed are *Asparagus racemosus, Jasminum roxburghianum, Pterolobium hexapetalum, Rhynchosia suaveolens, Grewia flavescens* (straggler) and *Cissus quadrangularis.* The trees seen in this zone are *Balanites aegyptiaca, Dolichandrone atrovirens, Azadirachta indica, Acacia catechu* and *Grewia tiliifolia.*

Buffer Zone II

This is the third zone in the study area extending to 30 km radius (2119.8564 km^2). Altitude ranges from 198 m to 815 m. Human habitations in this zone are 53 villages in Kadapa district and 56 villages in Anantapuramu district.

This zone has good pockets of dry deciduous and scrub forests. Reserved forests in East Side of this zone (Kadapa District) are Idupulapaya RF and Vempalle

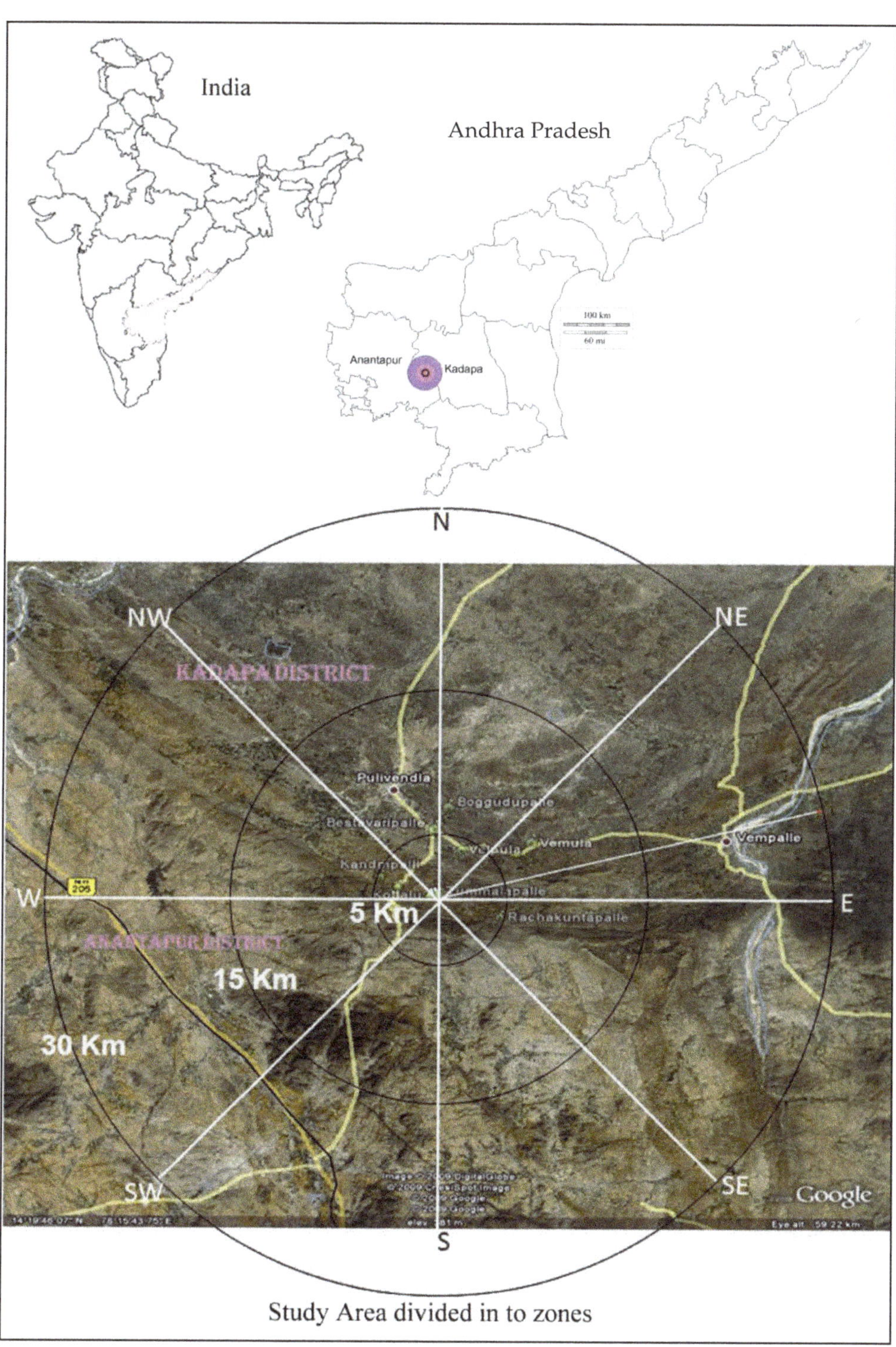

Study Area divided in to zones

RF with dry deciduous forest. South - West corner of this zone (Anantapuramu district) includes Dorigillu RF, Malakavemula RF, Errappakonda RF, Kadiri RF (South) and Godduvelagala RF (West) with deciduous forests. Remaining area is covered by scrub forests, open scrub jungles and rocky hillocks. Grasses *Apluda mutica, Alloteropsis cimicina* and *Heteropogon contortus* are the dominant elements in this zone. Common shrubs are *Catunaregam spinosa, Abutilon indicum* and *Flueggea leucopyrus*. Climbers like *Pterolobium hexapetalum, Jasminum roxburghianum* and *Rhynchosia suaveolens* are found extensivly. *Anogeissus latifolia, Chloroxylon swietenia, Dolichandrone atrovirens, Gyrocarpus americanus, Albizia amara, Wrightia tinctoria* and *Dalbergia lanceolaria* are common tree species in this zone.

Geology

Geological formation of the mining area can broadly be categorized into two distinct and well-marked groups; an older group of Archaen rocks and a younger one of sedimentary rocks towards the north side of the study area. Rocks belonging to the later formation cover the area of south side. Archaen group of rocks are most common and gneissose type though showing considerable local variations. Narrow serrated edges of quartz reefs are also noticed in this area, such as in Kadiri area. At Namalagundu Igneous intrusions are wide spread and this area forms the south western convex flank of Kadapa basin and contact zone between these two rock groups denoting enormous time gap in the evolution of earth's crust. It is called Eparchan unconformity and is indicated by the occurrence of conglomerate. The common rock types of Archaean comprises granite, granodiorite and alkali granite with intrusions of fine grained dolerite. Lithologically Gulcheru Quartzite exhibiting penecontemporaneous deformation in Namalagundu, Intrusives Stromatolitic Dolomite in Vempalli. Namalagundu area comprises variety of rock formations of different geologic times with sharp lithological contacts that can offer a geological feast (Phani, 2011).

Soils

The soil is derived mostly from disintegrated rocks with occasional quartzes formation. It is mostly very shallow and devoid of any humus except in a few sheltered valleys. This mining area of 30 km radius can be grouped into two natural divisions: Of the total mining area 70 per cent has red soil and 30 per cent of black soil. Black cotton soils are found in Peddakudala, Chinnakudala, north of Pulivendula, Yerripalli, Boggudupalli, Peddajuturu etc., while remaining area has red and gravelly soils.

Minerals

Asbestos is mined near Ambakampalle of Kadapa district.

Physiography

Thummalapalle Uranium mining area covers Forests (30 per cent), Water Bodies (4 per cent), Waste land (3 per cent), Agriculture land (57 per cent) and human habitations (6 per cent).

Irrigation

Irrigation facilities in the study area are only 12.6 per cent out of 57 per cent Agriculture land. Remaining agricultural land is rainfed. Out of gross irrigation area tanks account 43 per cent, Tube wells 37 per cent, Wells 7 per cent, Canals 9 per cent and others 4 per cent.

Climate

The climate can be considered as tropical, semi-arid with more number of sunny days. The winter is short, while summer is long and hot. The beginning of November to the later part of February is the coolest season of the year. After February, temperatures begin to rise rapidly and by the second part of May is hottest season. The temperature fluctuates between 44.2°C in summer, to 15.7°C in winter. With the onset of monsoon from June, the temperature slowly falls down and the climate is quite pleasant. During the period of South-West monsoon from June to September very strong and high winds blow across the area. Anantapuramu district is driest district in the State, while Kadapa is much better.

The erratic rainfall is a common phenomenon of the study area, due to the geographical situation, it does not lie definitely within either of the monsoons, being far from the East coast, it does not enjoy the full benefits of North East Monsoon and being cut off by the high Western Ghats, the South West Monsoon also prevented from penetrating and subjected to droughts due to bad seasons. The average rainfall loses its significance due to the abnormal variations from year to year. Most of the rainfall is received from South West monsoon only. Agriculture conditions are more often precarious. The normal rainfall of the two districts in the mining area is 614.0 mm. The normal rainfall of the South West monsoon period is 348.0 mm which forms about 61.2 per cent of the total rainfall in the year. The failure of the rains in this South West monsoon period of June to September leads to drought and failure of crops. The rainfall for North East monsoon period is 266.0 mm only, which forms 29.8 per cent for the total rainfall of the year, from October to November. The other months are almost dry; March, April and May are warm months when the normal daily maximum temperature ranges between 29.1°C to 44.2°C. November, December and January are cooler months when temperature falls to about 15.7°C.

Drainage

There is no major drainage system in the study area. The tributaries of the Penna river, like Sagileru and Maddileru are the main seasonal rivulets that flow in the study area. Papaghni river biannually flows in the Eastern part of the study area in Buffer Zone-II. A good number of annual streams, Tanks, springs and more number of ditches are also present in the area. Yogi Vemana reservoir is in Buffer Zone-II of the study area. Batrepalli waterfall is one of the most attractive, which is located in Buffer Zone-I, near Kadiri in Anantapuramu district and it flows in the months of September to November only. Most of the water sources are located in Buffer Zone-I and II, a few are in Core zone. Rich water sources are mainly in South-East and South-West corners of the study area. All the above water resources enrich the fauna and flora of the area. Details of water bodies and rivulets are given below.

Papaghni River: it is located in Buffer zone II and it flows South to North and connected to Somasila Reservoir.

Maddileru: it is located in Buffer zone II and it flows from South to North and connected to Mylavaram Dam, started with Yogi Vemana Reservoir at Mudigubba.

Moganor yeru: it is located in Buffer zone I and it originates at Kadiri and connected to Yogi Vemana Reservoir.

Dams

Yogi Vemana Reservoir is situated in the Buffer zone II in the mining area towards the West.

Tanks

17 tanks (only 1 in core zone remaining are in Buffer zone I and II) are found in the study area, these all are being used for agriculture irrigation and fishing purpose.

Vegetation

The collective and continuous growth of plants in space is called vegetation. The types of vegetation are determined by climatic, edaphic and biotic factors coupled with altitude.

Out of the total Geographical area of about 2828.571 sq.km in the 30 km radius of study area, nearly 30 per cent (848.5713sq.km) is under forest coverage.

The vegetation of the area is mainly represented by dry deciduous and scrub forests which are protected and reserved. The hillocks in most of the study area are occupied with huge rock boulders and they are common in the West side and hills with vegetation seen in the Eastern part of the study area. The main forest areas are located in Erramalais and Palakondalu. The Erramalais are seen in Kadiri forest range and the Palakondalu in Vempalle forest range.

As per the classification of Champion and Seth (1968) forests of the Thummalapalle Uranium Mining Area can be categorized under two types *viz.* tropical dry deciduous forests and thorny scrub forests.

Dry Deciduous Forests

Dry deciduous type of vegetation is spread over as patches near Mudigubba, Dorigallu, Kalasamudram, Batrepalli, Somulavaripalli and Nigidi of Anantapuramu district, and K. K. Kottala, Thummalapalle, Rachakuntapalli and Idupulapaya of Kadapa district. The Idupulapaya forests are dominating with *Pterocarpus santalinus*, which is highly valuable and endemic species of the Eastern Ghats. The forests of Kadiri range especially Batrepalli, Kalasamudram, Chinnapalli are the main places for *Hildegardia populifolia* which is an endemic and vulnerable taxon. Generally the forests are dominating with *Anogeissus latifolia, Terminalia bellirica, Ficus mollis, Chloroxylon sweitenia, Croton scabiosus, Dolichandrone atrovirens, Gyrocarpus americanus, Albizia amara, Wrightia tinctoria* and *Dalbergia lanceolaria.*

A. Dry deciduous forest, Batrepalli

B. Dry deciduous forest, Idupulapaya

C. Scrub forest, Thummalapalle

D. Ground nut crop, Kalasamudram

E. Bengal gram crop, Peddakudala

F. Thummalapalle village

Water bodies

Stream at Malakavemula (Buffer zone II)

Papaghni river at Gandi (Buffer zone II)

Tank at Batrepalle (Buffer zone I)

Tank at Thalupula (Buffer zone I)

Yogi Vemana reservoir at Mudigubba (Buffer zone II)

Upper Canopy

The upper canopy is usually uneven with a mixture of most of tree species which are practically deciduous during dry summer months consisting of large to middle sized trees like, *Boswellia serrata, Cochlospermum religiosum, Hildegardia populifolia, Kavalama urens* (*Sterculia urens*), *Chloroxylon swietenia, Ailanthus excelsa, Ochna obtusata, Commiphora caudata, Moringa concanensis, Dalbergia latifolia, Dalbergia paniculata, Pterocarpus santalinus, Hardwickia binata, Acacia leucophloea, Albizia amara, A. thompsonii, Anogeissus latifolia, Terminalia arjuna, T. bellirica, Syzygium cumini, Cassia fistula, Gyrocarpus americanus, Mitragyna parvifolia, Morinda angustifolia, Madhuca indica, Diospyros melanoxylon, Wrightia tinctoria, Strychnos nux-vomica, Vitex leucoxylon* etc.

The teak, *Tectona grandis,* occurs in stunted form in Kalasamudram, Batrepalli and Palakonda hill ranges. *Santalum album* has rare occurence in some parts of the mining area. *Dendrocalamus strictus* is occasionally found in the small portions of Kalasamudram and Batrepalli forests.

Lower Canopy

The lower canopy of the forest is dominated by *Annona squamosa, Polyalthia cerasoides, Alangium salvifolium, Benkara malabarica, Butea monosperma, Gardenia resinifera, Capparis grandis, Grewia damine, G. flavescens, Erythroxylum monogynum, Aegle marmelos, Cassine glauca, Sapindus emarginatus, Semecarpus anacardium, Bauhinia racemosa, Cassia fistula, Parkinsonia aculeata, Albizia amara, Prosopis juliflora, Terminalia chebula, T. coriacea, Psydrax umbellata, Gardenia gummifera, G. latifolia, Ixora pavetta, Pavetta tomentosa, Santalum album, Bridelia retusa, Croton scabiosus, Euphorbia antiquorum, E. tortilis, Jatropha curcas, Holoptelea integrifolia, Ficus hispida, Phoenix loureirii, Maerua apetala* etc.

Scrub Forest

These forests were once part of the widely distributed dry deciduous forests but now reduced to such low, degraded forest due to continued, intense biotic interference along with frequent fires. The resistant thorny species ultimately survive and regenerate year after year. Scrub types of forests are more common near the peripheries of dry deciduous forests and small hillocks nearby villages. The forests of Batrepalli, Kalasamudram and Thimmammamarrimanu are with large rocky boulders and poor soils.

The scrub vegetation include many thorny species such as *Euphorbia antiquorum, E. caducifolia, E. tortilis, Acacia catechu, A. horrida, Dichrostachys cinerea, Pterolobium hexapetalum, Flueggea leucopyrus, Carissa spinarum, Ziziphus* spp., *Balanites aegyptiaca, Capparis zeylanica, Catunaregam spinosa* etc. Non spinous shrubs in scrub forest include *Erythroxylum monogynum, Searsia mysorensis, Cadaba fruiticosa, Cissus* spp., *Grewia* spp., *Bridelia montana, Acacia horrida, Dodonaea viscosa, Bauhinia racemosa, Vitex negundo, Ehretia laevis, Phyllanthus reticulatus, Senna montana, Gymnosporia heyneana* etc. The variable climbers and creepers that are found in this area are *Cynanchum viminale, C. sarcomedium, Leptadenia reticulata, Pergularia daemia, Tylophora indica, Cocculus hirsutus, Coccinia grandis, Cardiospermum halicacabum, C. canescens, Abrus precatorius, Cissus quadrangularis, Canavalia gladiata, Jasminum arborescens, Hemidesmus indicus, Gymnema sylvestre, Rivia hypocrateriformis, Ipomoea* spp., *Passiflora foetida, Cryptostegia grandiflora, Holostemma ada-kodien* etc. Non spinous succulents like *Boucerosia lasiantha, Boucerosia umbellata, Caralluma* spp. are met with in this area.

Ground Vegetation

The ground vegetation of the forests, both in Dry deciduous and Scrub is dominated by the grasses like *Cymbopogon coloratus, Cymbopogon caesius, Andropogon pumilus, Apluda mutica, Aristida funiculata, A. hystrix, A. setacea, Brachiaria ramosa, B. reptans, Chloris barbata, Chrysopogon asper, C. fulvus, Cynodon dactylon, Digitaria ciliaris, Heteropogon contortus*. Other herbs that are found in the forests are *Crotalaria hebecarpa, Polycarpaea corymbosa, Sida cordata, S. cordifolia, Pulicaria wightiana, Euphorbia longistyla, E. cristata, E. rosea, Hibiscus lobatus, Malvastrum coromandelianum, Melhania incana, Melochia corchorifolia, Triumfetta pentandra, Tribulus terrestris, Alysicarpus* spp., *Crotalaria medicaginea* var. *neglecta, C. verrucosa, C. retusa, Indigofera glabra, I. prostrata, I. trita, Senna tora, Mollugo* spp., *Hedyotis* spp., *Blumea* spp., *Trichodesma* spp., *Sopubia delphinifolia, Striga* spp., *Barleria acuminata, B. cuspidata, B. stocksii, Indoneesiella echioides, Justicia glauca, J. microntha, J. procumbens, R. vahlii, Priva cordifolia, Anisomeles indica, Leucas aspera, Ocimum americanum, O. tenuiflorum, Boerhavia diffusa, B. erecta, Achyranthes aspera, A. aspera* var. *sicnla, Aerva lanata, Allmania longepedunculata, Allmania nodiflora, Alternanthera sessilis, Gomphrena serrata, Acalypha indica, Amaranthus viridis, Lagascea mollis, Croton bonplandianum* etc. A few members are seen growing among rocks and rock crevices like *Didymocarpus tomentosus, Kohautia aspera, Polycarpaea aurea* etc.

Besides the above two types of forests, floristic components of other habitats are given in following headings.

Disturbed and Fallow Areas Near Villages

Trees like *Prosopis juliflora, Euphorbia tirucallii, Balanites aegyptiaca, Acacia nilotica, A. leucophloea, Ziziphus jujuba, Flueggea leucopyrus, Carissa spinarum* etc are found in disturbed and fallow areas near villages. 70 per cent of the fuel wood of urban and semi-urban population comes from *Prosopis juliflora* only.

The carpet vegetation of plains consists of grasses like *Aristida adscensionis, A. hystrix, Heteropogon contortus, Chloris barbata, Echinochloa colonum, Panicum curviflorum, Brachiaria reptans* etc. Other species that grow among these grasses are *Indigofera cordifolia, Tribulus terrestris, Evolvulus alsinoides, Oldenlandia umbellata, Boerhavia diffusa, Euphorbia hirta, Allmania longepedunculata, Tridax procumbens, Mollugo nudicaulis, Commelina benghalensis, C. attenuata, Cyanotis axillaris, Cyperus rotundus* etc. The common undershrubs of the plains are *Calotropis procera, C. gigantea, Solanum pubescens, S. anguivi, Justicia adhatoda, Senna occidentalis, S. auriculata, Tephrosia purpurea, T. villosa, Cleome gynandra, Corchorus trilocularis, C. aestuans* etc.

Aquatic Habitat

Aquatic vegetation is a rare sight in mining area. They are divided into different classes based on the relationship with water.

Submerged Anchored Hydrophytes

They are entirely or for the most part submerged in water but attached to the soil. Components of this category are *Monochoria vaginalis.*

Floating-leaved Anchored Hydrophytes

They are rooted in the muddy substratum with their leaves lying on water surface. They are represented by *Nymphaea* spp.

Floating-shoots Anchored Hydrophytes

They are rooted in the soil with their roots, shoot and leaves are on the water surface. During dry months, when water recedes, these species become terrestrial and survive in subswampy to nearly dry habitat. Constituents of this category are *Ludwigia octovalvis* and *Ludwigia perennis.*

Emerged Anchored Hydrophytes

The root, the lower part of the stem and sometimes even the lower leaves of these hydrophytes are usually submerged under water. The plants when growing on the marginal belts on the marshy banks of rivers, canals etc., are subjected to periodic flooding and are liable for complete submersion. Plants of this group are *Bacopa monnieri, Limnophila heterophylla, Typha angustata, Cyperus* spp., *Ammannia baccifera* etc.

Wetland Hydrophytes

Here the soil is usually saturated with water at least in the earlier part of life. Their appearance coincides with the gradual drying of water sources. These include *Glinus lotoides, Eclipta prostrata, Melochia corchorifolia, Ipomoea carnea, Cynodon dactylon, Sphaeranthus indicus* etc.

Parasitic Plants

A few parasitic plants have also been recorded from the area. *Dendrophthoe falcata,* a partial stem parasite on *Albizia amara, Acacia* spp., *Zizipus mauritiana, Maerua apetala,* etc. and *Viscum articulatum* on *Ziziphus* spp. and *Croton scabiosus* are seen in dry deciduous forests. *Cassytha filiformis* is a partial stem parasite on a number of species like *Carissa spinarum, Ziziphus jujuba, Hardwickia binata, Azadirachta indica, Erythroxylum monogynum, Acacia chundra, A. horrida, Vitex negundo* and on some orchard trees like orange, mango etc. This species is recorded both in the forest and non-forest areas. *Striga asiatica* is frequent root parasite on grasses including Sorghum and Bajra. *Striga gesnerioides* is a root parasite on *Lepidagathis cristata* and *L. subarmata. Cuscuta reflexa* is a complete stem parasite on *Ziziphus* spp., and *Ipomoea staphylina.*

Avenue Trees

The avenue trees include *Tamarindus indica, Azadirachta indica, Pongamia pinnata, Delonix regia, D. elata, Ficus benghalensis, F. amplissima, F. recemosa, Dalbergia sissoo, Samanea saman* and *Albizia lebbeck. Phoenix sylvestris* grows along the rivers, rivulets and streams.

Cultivated Areas

Weeds

Some of the common weeds in and around cultivated fields of the study area are *Celosia argentea, Cynodon dactylon, Cyperus rotundus, Cyanotis axillaris, Alysicarpus* spp., *Euphorbia hirta, E. indica, E. prostrata, Amaranthus viridis* and other species of *Amaranthus, Aristolochia bracteolata, Digera muricata, Trianthema portulacastrum, Portulaca oleracea, P. quadrifida, Amaranthus viridis* etc.

Crops

Agricultural lands occupy a large part of the study area. Most of these are dry lands and some are irrigated lands. Following different types of crops are grown.

Cereals Crops

Rice is the most important crop in this area. Other cereals like Wheat, Maize, Pearl millet, Jowar are mostly cultivated as dry land crops.

Oil Crops

Most of the Agriculture land is dominated by Ground nut. Castor, Sunflower and Sesame are also grown.

Pulses

Red gram (Pigeon pea), Black gram, Green gram and Horse gram.

Spices

Coriander and Chillies.

Vegetables

Tomato, Brinjal, Onion and Beans.

Fruit Crops

Mango, Orange, Banana, Pomegranate and Sapota

Other crops include Cotton, Betel leaves and Mulberry.

Tectona grandis, Borassus flabellifer and *Phoenix sylvestris* are grown, which also show natural regeneration, along the waste lands around villages and edges of agriculture fields.

Sacred Grove

Sacred groves comprise of patches of forests or natural vegetation, from a few trees forests of several acres – that are usually dedicated to local folk deities. These spaces are protected by local communities because of their religious beliefs and traditional rituals that run through several generations. The degree of sanctity of the sacred forests varies from one grove to another. Sacred groves in Thummalapalle Uranium Mining Area include Mallalamma temple near Batrepalli, Muniswarakunta near Kalasamudram, these two are located in Nigidi forest of Buffer zone I in the study area. Lord Siva and Anjaneyaswamy temples are located at Namalagundu in Buffer zone I in Palakonda hill ranges while Gandi Anjaneya Swamy temple is situated in Buffer zone II, near Idupulapaya in Palakonda hill ranges.

3

Past and Present Work

Past Floristic Studies

Botanists like Roxburgh (1795 - 1810), Wight (1838 - 1853), Beddome (1869 - 1874) and Elliot (1859) conducted intermittent field trips during 19th century and reported on the floristic data of some regions of the state of Andhra Pradesh. Roxburgh made a survey of the Coromandel coast from 1795 - 1820 and published a book on "Plants of the Coromandal coast"; this was a fore-runner to his "Flora of India". Wight in his work "Icones Plantarum Indiae Orientalis" (1838 – 1853) presented illustrations of 2,101 plants in 6 volumes, out of which 1780 species have been reported in "Flora of the Presidency of Madras" by Gamble and Fischer (1915 - 1935). Beddome (1830 – 1811) extensively collected plants from different regions of Andhra Pradesh. He followed the method of Wight in illustrating the plants. His significant contribution is "The Flora Sylvaticae for southern India (1869 - 1874)". Hooker (1872 – 1897) published seven volumes on "Flora of British India". One of the important floristic work on flora of Peninsular India is that of Gamble and Fisher (1915 – 1935) on "Flora of the Presidency of Madras", where they enumerated 4516 species.

Botanical Survey of India is publishing series of volumes on flora of India (Sharma *et al.*, 1993; Sharma and Balakrishnan, 1993; Hajra *et al.*, 1995a, b; 1997; Singh *et al.*, 2000, Balakrishnan *et al.*, 2012). Inventory studies have been done by different authors like Brandis (1906) on Indian trees, Gamble (1881) on Indian timbers, Bor (1960) on grasses of Burma, Ceylon, India and Pakistan, Karthikeyan *et al.* (1989) on monocots of India, Chaudhuri (1967) on Meliaceae of India, Bose *et al.* (1980)

and Mishra (2007) on Orchids of India, Sanjappa (1991) on legumes of India and Balakrishnan and Chakrabarthy (2007) on Euphorbiaceae of India. Monographs based on genus or on special groups include Monograph on Indian *Leucas* (Singh, 2001a), Monograph on Indian subtribe *Cassiinae* (Singh, 2001b), Bladderworts of India (Janarthanam and Henry 1992), genus *Euphorbia* in India (Binojkumar and Balakrishnan, 2010), genus *Acacia* and *Albizia* of India (Chakrabarthy and Gangopadhyay, 1996a, b), genus *Caralluma* in India (Karuppusamy *et al.*, 2013).

Elliot (1859) studied the economic botany of Andhra Pradesh and published "Flora Andhrica" for the plants of Andhra area. Eastern Ghats are situated in 4 states of South-East India, namely Orissa, Andhra Pradesh, Tamil Nadu and small parts of Karnataka. In united Andhra Pradesh, the Ghats covered all the 13 districts. Seshagiri Rao (1998) mentioned three major types of vegetation in the Eastern Ghats eco-region and reported hill top species, medicinal plants, endemic species and plants with horticultural value. Sri Ramamurty *et al.* (1998) published a preliminary checklist of 328 legumes in Eastern Ghats. Pullaiah and Sri Ramamurthy (2001), Pullaiah and Muralidhra Rao (2002), Pullaiah *et al.* (2007, 2011) worked on flora of Eastern Ghats and have given a detailed inventory, brought out 4 volumes on Flora of this Geographical area. Satyanarayana and Sanjappa (1998) listed a total of 25 endemic leguminous taxa occurring in Eastern Ghats eco-region. Sandhya Rani and Pullaiah (2002) reported 560 tree taxa form Eastern Ghats. Sudhakar Reddy *et al.* (2009) prepared a checklist of 510 indigenous and naturalized trees of Eastern Ghats of Andhra Pradesh state. *Utricularia australis, Euphorbia clarkeana, Brachystelma ciliatum* have been reported as new distribution records for Eastern Ghats by Raja Kullayiswamy *et al.* (2013a, b, c).

Floristic works by Kumar and Bhanja (1992) produced a seed manual for forest trees of Andhra Pradesh. While working on Flora of Andhra Pradesh, Pullaiah and Chennaiah (1997), Pullaiah and Ali Moulai (1997), Pullaiah (1997) and Pullaiah and Karuppusamy (2008) reported 2531 species. Sudhakar Reddy *et al.*, (2008) added 272 taxa for the flora of Andhra Pradesh and now the number of species is 2803, covering 1051 genera and 185 families. Pullaiah (2008) worked on Biodiversity and Conservation of Eastern Ghats. Pullaiah *et al.* (2008) presented a brief account on floristic analysis of the districts of Andhra Pradesh and reported Fabaceae and Poaceae as the dominant families of Andhra Pradesh. Murthy *et al.* (2008) mentioned that around 490 taxa have been added the flora of Andhra Pradesh after Gamble. Raju *et al.* (2008) studied and reported 76 Orchids from Andhra Pradesh.

Floristic studies on the flora of Kadapa (earlier known as Cuddapah) district was done by Madhusudhana Rao (1989). Madhava Chetty and Rao (1990) and Rao *et al.* (1993) reported one gymnosperm, six angiosperms as endemic and 25 rare plants for Tirumala hills in the Eastern Ghats of Chittoor district. Pullaiah and Yesoda (1989) explored Anantapur district intensively including Kalasamudram, Batrepalli and Palakonda hills. Chennaiah (1987) worked on flora of Kalasamudram Reserved forest in Anantapur district. B.R.P.Rao *et al.* (2003) reported 96 taxa as additions to the Flora of Anantapur district.

Present Work

The present work has been carried out in a following manner with a view to prepare a detailed floristic account of plants of Thummalapalle Uranium Mining Area.

Field Study

In the present investigation the Mining Area was explored intensively for a period of 4 years covering all the seasons during 2009–2012 and a total 1850 field numbers comprising 7400 specimens were collected. The exploration trips were conducted covering all ranges of the forests, and non-forest areas in the Mining area. Every plant was collected in quadruplicates either in flowering or fruiting with the information of habitat, soil, elevation, vegetation type, frequency of distribution, relative abundance, local names, associates and economic importance etc., of the individual species which was recorded in filed note book with proper care in the field itself. Field numbers have been given for every specimen collected and all the above information was recorded carefully in the field itself. Care was taken in collecting small herbs, aquatic species, bulbs, corms, rhizomes, tubers etc. The photographs were taken with the help of Cannon digital Camera. The major threats in the study area were observed.

Herbarium Preparation

All the collected specimens were poisoned with mercuric chloride and ethyl alcohol solution, pressed, dried and made into herbarium specimens according to the methodology described by Santapau (1958) and Jain and Rao (1977). The specimens were collected in well-tied thick polythene bags. Dry method was adopted in poisoning the specimens with alcohol-mercuric chloride. The whole plants or twigs were dipped in saturated solution of mercuric chloride in ethyl alcohol. Immediately they were placed in between blotting papers with the help of iron plant pressers. After 24 hours, blotting papers were changed and the specimens were spread properly in such a way that their physical appearance is not changed. These spread specimens were changed once again into dried blotting papers and tied in iron pressers. This process of changing was continued until the specimens were completely dried. The dried specimens were pasted with fevicol on a standard herbarium sheet (42 × 28 cm). Only two specimens from the collected quadruplicates were pasted on the herbarium sheets, while the rest were kept as stock. Labels with relevant information such as name of the plant, family, locality, altitude, date of collection and collector's name were affixed on the right hand bottom corner of the mounted sheets.

Identification of Herbarium Specimens

The specimens were examined critically by dissecting the floral parts of duplicate specimens under binocular stereoscope and compound microscope and provisional identification was made with the help of Floras like Gamble's Flora of the Presidency Madras (1915 - 1925), Hooker's Flora of British India (1872 - 1897), Flora of the Tamilnadu Carnatic by Matthew (1981 - 1983), Flora of Andhra Pradesh by Pullaiah and Chennaiah (1997), Pullaiah and Moulali (1997), Pullaiah

(1997), Flora of Eastern Ghats by Pullaiah and Muralidhara Rao (2002), Pullaiah and Sri Rama Murthy (2001) and Pullaiah *et al.* (2007, 2011) and with the help of standard regional and national floras. Ptridophytes were identified with the help of Pteridophytic floras like Pullaiah *et al.* (2003) and Manickam and Irudayaraj (1992). Identifications were further confirmed after matching the specimens with the authentic specimens housed in the Madras Herbarium (MH), Coimbatore and Central National Herbarium (CAL), Kolkata. All these collected specimens after identification were deposited in the herbarium, Department of Botany, Sri Krishnadevaraya University (Acronym SKU), Anantapuramu, Andhra Pradesh and duplicates in Botanical Survey of India (Acronym BSID), Hyderabad and some specimens in Indian Institute of Science, Bangalore (Acronym JCB), Karnataka.

Methodology for Bryophyts

Systematic collection and survey of bryophytes were made from different localities of Thummalapalle Uranium Mining Area. The field data were recorded in the field note book such as habitat, habit, locality, date, altitude, type of vegetation, significant characters, associates etc. Bryophytes have been collected for detailed study and identification. The collected bryophyte specimens were air dried for herbarium preparation. Small portions of fresh specimens were also preserved in 70 per cent alcohol for anatomical studies. The dried specimens were stored in brown paper packets (5″ × 4″) pasted with detailed labeling sheet. Plants of Bryophytes were identified with the help of available literature (Gangulee, 1969–80; Chopra, 1975; Bapna and Kachroo, 2000). The identifications were also confirmed by comparing the herbarium specimens with voucher specimens kept in NBRI, Lucknow and SKU, Ananthapuramu and further confirmation with subject experts. Herbarium specimens were deposited in Botany Department Herbarium (Acronym SKU) of Sri Krishnadevaraya University, Andhra Pradesh, India.

Plan of Systematic Enumeration

Bentham and Hooker (1862-83) system of classification has been followed while presenting the floristic account of study area and with modifications according to the recent works. The endemic and threatened plants were listed with available past literature.

The nomenclature adopted in present work is based on recent taxonomic literature and in accordance with the recommendation made by International Code of Nomenclature for algae, fungi and plants (ICN).

At the beginning, a comprehensive key to the families of Thummalapalle Uranium Mining area has been provided for easy identification of a family. Keys for all the genera of a family, species of particular genus and varieties of species have also been given. These keys are purely artificial and bracketed. The genera within a family and species within genera are arranged in simple alphabetical order to facilitate easy reference.

In the treatment of individual species, the following details are mentioned.

1. **Citation:** Nomenclature has been brought up-to-date as far as possible. Under citation, valid name of the species is given in bold letter which

is followed by its author name, original publication and basionym if necessary and or synonyms mostly to connect with the correct name of the of the taxon for the names published in Flora of British India (1872 - 1897), Flora of Presidency of Madras (1916 - 1935) and more important synonyms in chronological order. Flora of Andhra Pradesh (1997 - 2007) and The Grasses of Burma, Ceylon, India and Pakistan (1960) references are also given at the end of citation.

2. **Local Names:** Vernacular or local names of the species which are used by the local people (Telugu) have been given if available
3. **Plant Description:** The terminology followed is based on Lawrence (1951) and Radford *et al.* (1974). The description of each species contains important features of the taxon as observed in the field and critical study of the specimen in the Herbarium.
4. Habitat, status and phenology of the plant species followed by locality, collector's (abbreviated) name and field number have been given.
5. **Note** has been given for some species wherever necessary.

4

Floristic Analysis

During the present study, a total of 1850 voucher specimens have been collected and studied. Floristic study accounts for a collection of 928 taxa. Of this 892 constitutes Angiosperms, represents 104 families and 455 genera; 14 taxa constitutes

Table 1: Systematic Analysis of Flora of Thummalapalle Uranium Mining Area

Group	*Families*		*Genera*		*Species*		*Taxa*	
	Total	*Per cent*	*Total*	*Per cent*	*Total*	*Per cent*	*Total*	*Per cent*
Bryophytes	8	6.45	14	2.90	21	2.32	22	2.37
Pteridophytes	12	9.67	13	2.69	14	1.54	14	1.5
Gymnosperms	0	0	0	0	0	0	0	0
Angiosperms	104	83.87	455	94.39	869	96.12	892	96.22
Total	**124**		**482**		**904**		**928**	
Systematic Analysis Angiosperms								
Monocotyledons	18	17.3	98	21.53	201	23.13	203	22.75
Dicotyledons	86	82.69	357	78.46	668	76.86	689	77.24
Polypetalae	49	47.11	162	35.60	321	36.93	333	37.33
Gamopetalae	26	25	153	33.62	254	29.22	259	29.03
Monochlamydeae	11	10.57	42	9.23	93	10.70	97	10.87
Total	**104**		**455**		**869**		**892**	

Pteridophytes represent 12 families and 13 Genera; 21 taxa account for Bryophytes representing 8 families belong to 14 Genera. Gymnosperms have not been found in the study area (Table 1).

Table 2: Statistical Analysis of Growth Forms of Angiosperms

Group	*Trees*	*Shrubs*	*Climbers*	*Herbs*
Monocotyledons	3	2	3	196
Dicotyledons				
Polypetalae	87	67	66	113
Gamopetalae	37	46	43	133
Monochlamydae	18	23	4	51
Total	**145**	**138**	**116**	**493**

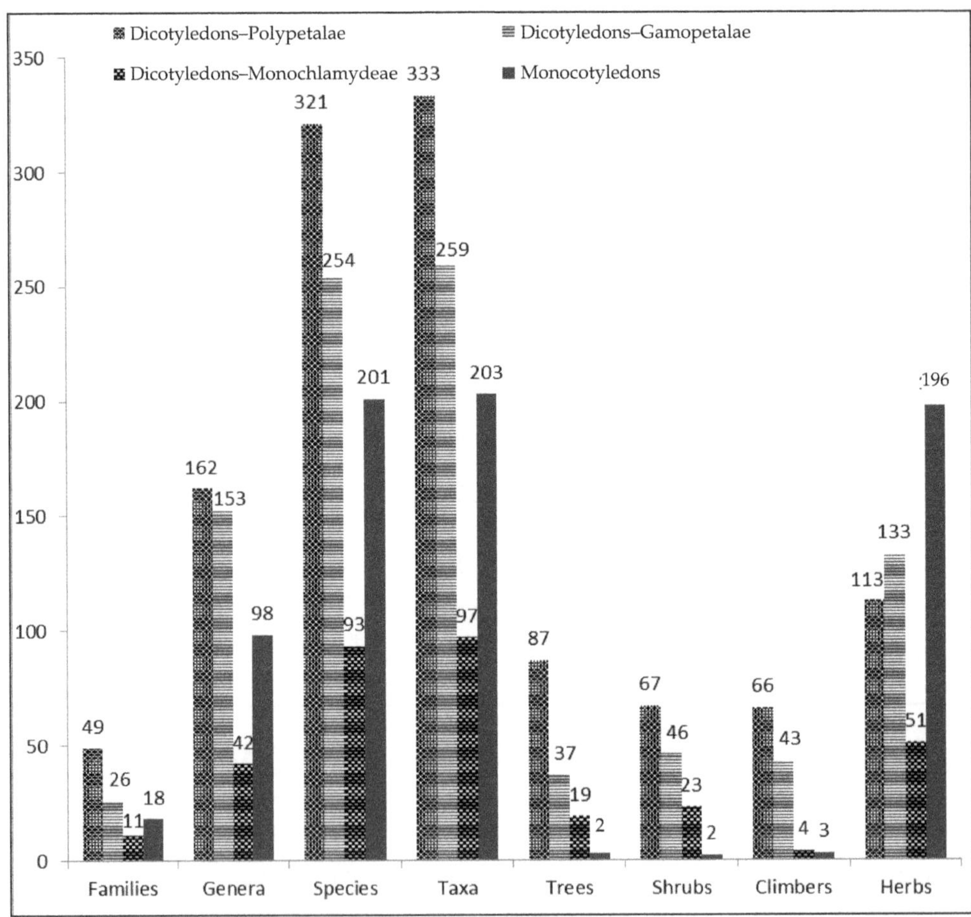

Figure 1: Comparative Analysis of Dicotyledons of Angiosperms by Number.

The Figure 2 clearly demonstrates that maximum representation to the flora is herbaceous community followed by trees, shrubs and climbers (incl. liana and stragglers). Loranthaceae that contributes to the bulk of parasites are mainly confined to deciduous forest. Order of the dominance of families (Figure 3) are as follows. Family Poaceae is the dominant with 112 species followed by Fabaceae (86), Euphorbiaceae (48), Cyperaceae (44), Asteraceae (34), Acanthaceae (33), Rubiaceae (27), Convolvulaceae (25), Mimosaceae (24) and Asclepiadaceae (23). Top 10 dominant genera (Figure 4) include *Euphorbia* with 17 species followed by *Cyperus, Crotalaria, Eragrostis* 14 species each, and *Indigofera* (13), *Fimbristylis* (12), *Senna* (11), *Ipomoea* (10), *Grewia* and *Justicia* 9 species each.

Orchidaceae which is the second largest family in India is represented by only one species in the study area. It is evident that the growth and development of Orchids in open dry deciduous and scrub forests are generally poor.

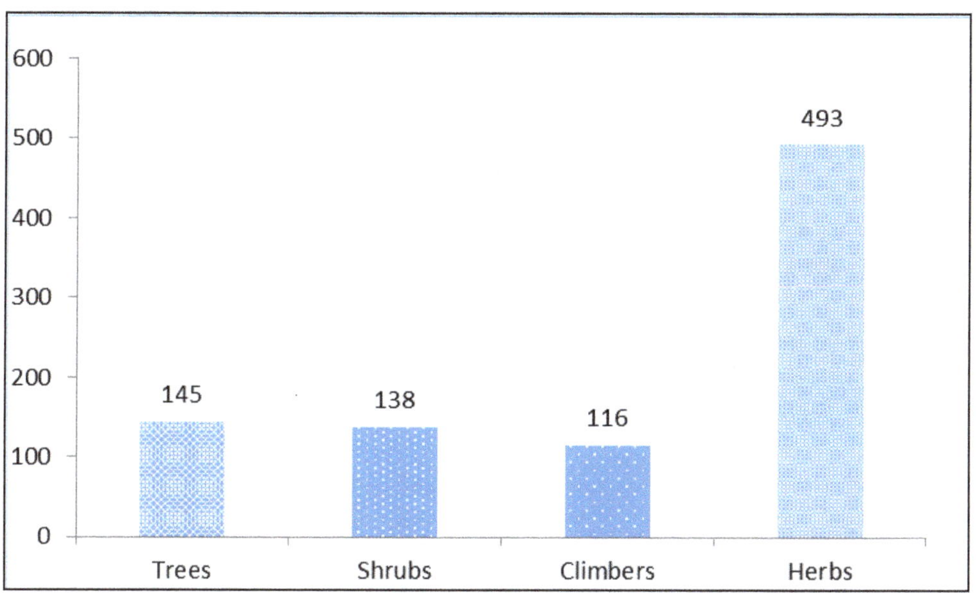

Figure 2: Growth Form Analysis.

Herbs are more dominant in the study area with 495 taxa.

Endemism

Endemism in Indian flora has been discussed by several authors like Chatterjee (1940), Rao (1972), Blasco (1971), Nayar (1980a, b, 1996). Hajra and Mudgal (1997) concentrated on plant diversity of hot spots in India. Ahmedullah and Nayar (1987) who estimated that there are 141 endemic genera within the present political boundaries of India, in which 58 genera belonging to 25 families are confined to peninsular India. Ahmedullah (2000) narrated the endemic flora of India in the introductory volume of India series and reported that India harbors a rich flora with a fairly high degree of endemism due to its varied physiognomy and climate. Being bounded on the north by the Himalayas and with its peninsular part surrounded

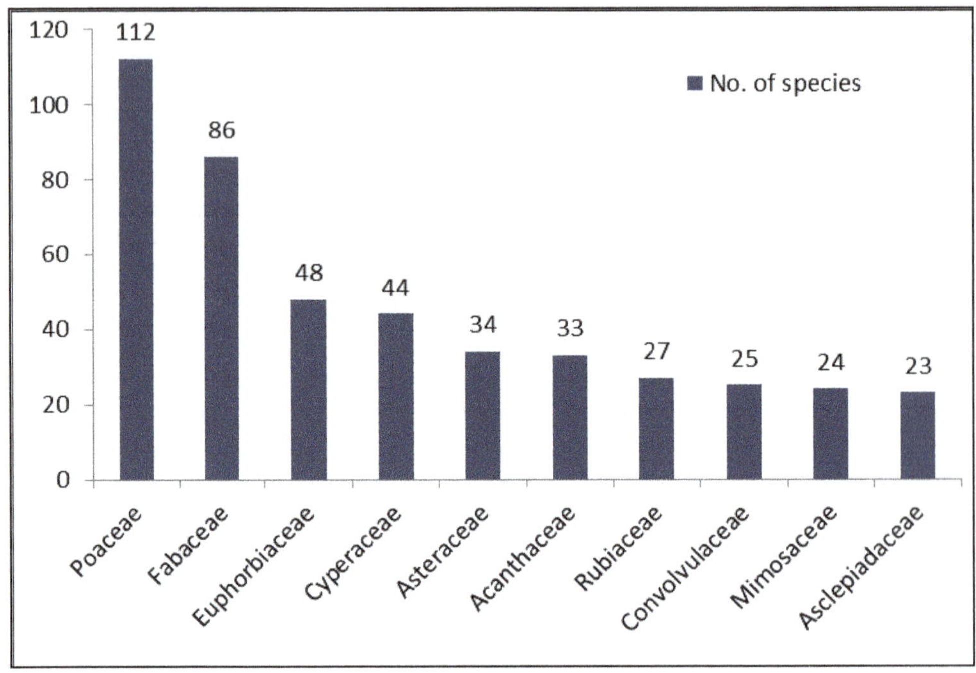

Figure 3: Top 10 Dominant Families Showing in Ascending Order.

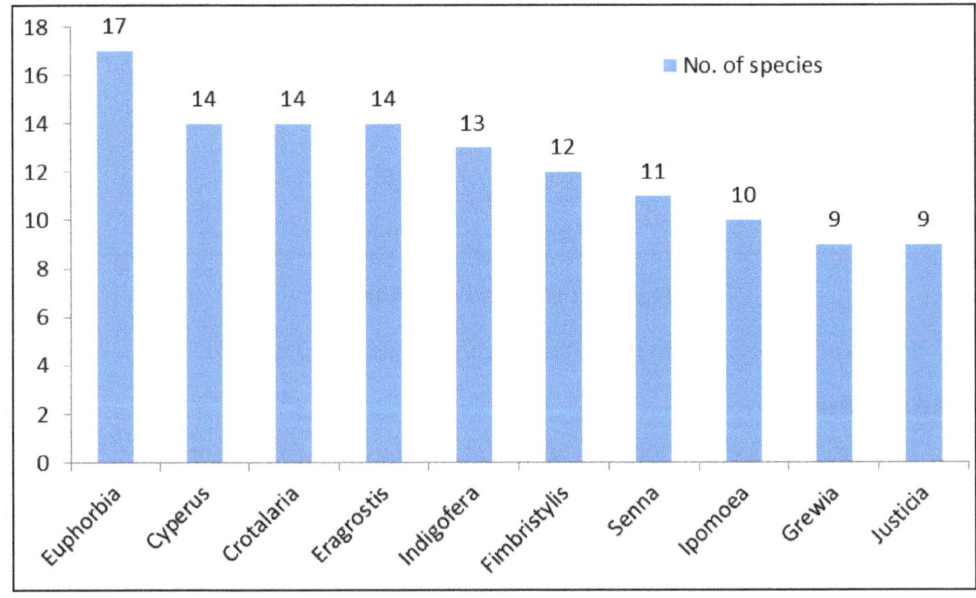

Figure 4: Showing Top 10 Dominant Genera.

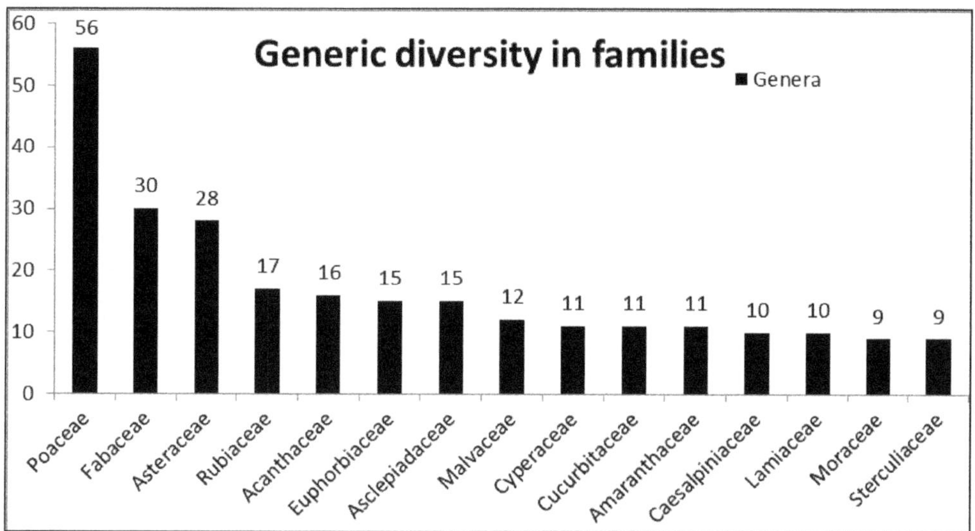

Figure 5: Top 15 Dominant Families Showing Number of Genera with Order of Dominance.

by the different oceans, the isolation of the Indian flora is enough to give rise to endemism.

A total of 45 endemic taxa of Peninsular India (Andhra Pradesh and Eastern Ghats are included) are recorded in the present study, of which six species are endemic to Eastern Ghats, include *Albizia thompsonii, Ceropegia pullaiahii, Decaschistia rufa, Bridelia retusa, Pterocarpus santalinus, Euphorbia longistyla.* Two species namely *Croton scabiosus* and *Hybanthus vatsavayii* are endemic to Andhra Pradesh. 37 taxa are endemic to Peninsular India; these are *Andrographis glandulosa, Andropogon pumilus, Aristida redacta, Arundinella nervosa, Barleria acuminata, Barleria cuspidata, B. stocksii, Brachystelma ciliatum, Caralluma adscendens* var. *attenuata, C. adscendens* var. *fimbriata, Carissa paucinervia, Senna montana, Ceropegia spiralis, Chloris bournei, Chloris quinquesetica, Chrysopogon asper, Cymbopogon coloratus, Decalepis hamiltonii, Didymocarpus tomentosus, Digitaria tomentosa, Dolichandrone atrovirens, Eragrostiella brachyphylla, Eragrostis riparia, Euphorbia corrigioloides, E. fusiformis, Heteropogon fischerianus, Hildegardia populifolia, Iseilema anthephoroides, Lepidagathis subarmata, Lophopogon tridentatus, Maerua apetala, Mimosa prainiana, Panicum fischeri, Plectranthus caninus, Polyalthia cerasoides, Rotala fimbriata, Tephrosia strigosa,*

Out of 892 taxa, only four taxa are categorized under various threat categories. The threatened taxa are *Albizia thompsonii* (Rare), *Ceropegia spiralis* (Rare), *Pterocarpus santalinus* (endangered) and *Hildegardia populifolia* (Vulnerable). Nearly 190 economically important plants were recorded from the study area, some of them are *Carissa spinarum, Erythroxylum monogynum, Grewia flavescens, Hygrophila auriculata, Phoenix sylestris, Syzygium cumini, Tamarindus indica* etc.

List of Abbreviations

ANS: A. Narayana Swamy
ATP: Anantapuramu
BR: B. Ravi Prasad Rao
BSS: B. Sadasivaiah
FAP: Flora of Andhra Pradesh vols. 1–3 (1997)
GK: G. Kavita
GO: G. Obulasu
HJCB-E: Herbarium JCB, Eastern Ghats
JCB: Acronym of Herbarium (Joseph's college, Bangalore)
FBI: Flora of British India by J.D.Hooker (1872 – 1896)
FEG: Flora of Eastern Ghats
FPM: Flora of Presidency of Madras by J.S.Gamble and C.E.C.Fischer (1915-1936)
KDP: Kadapa (formerly known as Cuddapah)
KRKS: K.Raja Kullayi Swamy
KRS: K. Rajasekhar
KSM: K.Sri Ramamurthy
NY: N. Yasodamma
PVP P.V. Prasanna
SKU: Sri Krishnadevaraya University and also acronym of Herbarium
SSR: S. Sandhya Rani
TP: T. Pullaiah

5

Key to Plant Groups

1. Non flowering and non-seed bearing plants .... (2)
1. Flowering and seed bearing plants .... **Angiosperms**
2. Plants gametophytic; conductive tissue not well developed (without lignification), have false roots (rhizoids) and leaves; spores are produced in a capsule .... **Bryophytes**
2. Plants sporophytic; conductive tissue well developed, have true roots and leaves present; spores borne on under surface of the leaf .... Pteridophytes

Angiosperms

KEY TO THE FAMILIES

1a. Stem with vascular tissue forming one or more rings; central pith present or replaced by wood; bark separable; flowers usually 4-5-merous, leaves usually articulated on the stem and with reticulated veins, cotyledons usually two, opposite, the young stem arising between them; usually tap root system .... **(2) (Dicotyledons)**

1b. Stem with scattered vascular tissue and compact periderm; flowers usually 3-merous, leaves usually sheathing at the base and not articulated on the stem and veins usually parallel connected by straight cross-veinlets; cotyledon 1, young stem arising from a lateral cavity; adventitious root system.......................**(101) (Monocotyledons)**

2a. Perianth absent (achlamydous) or of one whorl (monochlamydous) ....**(3)**

2b. Perianth present, biseriate or multiseriate ....**(22)**

3a. Plants aquatic, submerged in still waters, free floating........... **Ceratophyllaceae**

3b. Plants usually terrestrial, few epiphytic with stem parasitic habit ....**(4)**

4a. Perianth absent .... **Euphorbiaceae p.p. (*Euphorbia*)**

4b. Perianth present ....**(5)**

5a. Ovary superior ....**(6)**

5b. Ovary inferior or half inferior ....**(16)**

6a. Plants parasitic, wiry, leafless, anthers with valvular dehiscence...... **Lauraceae**

6b. Plants not parasitic, not wiry, leafy, anthers not as above............**(7)**

7a. Flowers unisexual or polygamo-dioecious (bisexual ad unisexual in same inflorescence) ....**(8)**

7b. Flowers bisexual............**(13)**

8a. Plants without latex or watery sap .... **(9)**

8b. Plants with latex or watery sap .... **(11)**

9a. Stamens on staminal column; fruit follicular, up to 11cm long............**Sterculiaceae p.p.**

(*Firmiana, Hildegardia, Sterculia*)

9b. Stamens free from one another; fruit berry or drupe or capsule, up to 4 cm across............**(10)**

10a. Stamens 5-10 ....**Sapindaceae p.p.** (*Dodonaea, Schleichera*)

10b. Stamens numerous ....**Flacourtiaceae**

11a. Ovary tricarpellary, trilocular with 1-2 ovules in each locule on axile placentation; fruit capsular ....**Euphorbiaceae p.p**

11b.Ovary 1-carpellary, 1-locular with one ovule on pendulous placentation............**(12)**

12a. Plants with milky latex, anthers inflexed in bud, leaves not oblique at base, fruit a sorosis or syconia'............**Moraceae**

12b. Plants with watery sap, anthers erect in bud, leaves oblique at base, fruit a dry, winged samara............**Ulmaceae**

13b. Ovary 1-locular, fruit utricle, albumen floury, fleshy ....**(14)**

13a. Ovary 2-5 locular, fruit capsular, albumen mealy ....**(15)**

14a. Inflorescence usually spike or clustered; bracts and bracteoles present, perianth dry, scarious, hyaline; many ovuled .... **Amaranthaceae**

14b. Inflorescence cymose cluster, bracts and bracteoles absent; perianth membranous, green, fleshy, ovary -1 ovuled .... **Chenopodiaceae**

15a. Tepals free .... **Aizoaceae**

15b. Tepals connate .... **Molluginaceae**

16a. Ovary 6- locular, perennial herbs, flowers zygomorphic ...... **Aristolochiaceae**

16b. Ovary 1-locular.. **(17)**

17a. Plants stem parasites .... **(18)**

17b. Plants autophytic .... **(19)**

18a. Leafy stem without joints; drupes on elongated stalks .... **Loranthaceae**

18b. Leafless stem with joints; drupe solitary or compactly clustered...... **Viscaceae**

19a. Anthers dehiscing by valves, leaves cordifom, 15-neved from base .... **Hernandiaceae**

19b. Anthers not dehiscing by valves, leaves other than cordiform, penninerved .... **(20)**

20a. Stamens twice the number of perianth lobes, fruits winged .... **Combretaceae p.p.** (*Terminalia, Anogeissus*)

20b. Stamens equal to or fewer than perianth lobes .... **(21)**

21a. Herbs; fruit an anthocarp with glands... **Nyctaginaceae**

21b. Trees; fruit drupe without glands .... **Santalaceae**

22a. All or most lobes of inner whorl of perianth (corolla) free from base.......... **(23)**

22b. Inner whorl of perianth (corolla) united, the corolla usually falling away as a whole when shed .... **(74)**

23a. Plants aquatic.. **Nymphaeaceae**

23b. Plants usually terrestrial .... **(24)**

24a. Ovary superior, rarely half inferior .... **(25)**

24b. Ovary inferior .... **(67)**

25a. Stamens more than twice as many as petals .... **(26)**

25b. Stamens twice as many as or as many as petals or fewer .... **(39)**

26a. Ovary apocarpous... **(27)**

26b. Ovary mono or syncarpous .... **(28)**

27a. Trees or shrubs, flowers bisexual.. **Annonaceae**

27b. Climbers, flowers unisexual .... **Menispermaceae p.p.**

28a. Stamens variously connate at base .... **(29)**

28b. Stamens free.. **(31)**

29a. Anthers 1-celled, staminal column adnate to petals .... **Malvaceae**

29b. Anthers 2-celled, staminal column free from petals .... **(30)**

30a. Sepals 5; plants not succulent; placentation axile; capsule loculicidal .... **Sterculiaceae p.p.**

30b. Sepals 2; plants succulent; placentation free central; capsule circumscissile .... **Portulacaceae**

31a. Leaves lobed or compound .... **(32)**

31b. Leaves simple .... **(36)**

32a. Flowers > 10 cm across, seeds with cottony hairs .... **Cochlospermaceae**

32b. Flowers < 5cm across, seeds without cottony hairs .... **(33)**

33a. Trees or straggling shrubs .... **(34)**

33b. Herbs .... **(35)**

34a. Leaves punctate with aromatic glands, ovary without gynophore .... **Rutaceae p.p.** (*Aegle*)

34b. Leaves without glands, ovary borne on gynophore .... **Capparaceae p.p** (*Maerua oblongifolia*)

35a. Plants prickly, with yellow latex; stamens numerous in many series; stigma 5-7-lobed, radiating .... **Papaveraceae**

35b. Plants not prickly, without latex; stamens 6-12; stigma capitate .... **Cleomaceae**

36a. Fruits winged, anthers with appendage connections .... **Dipterocarpaceae**

36b. Fruits not winged, anthers without appendage connectives .... **(37)**

37a. Distinct gynophore present .... **Capparaceae p.p**

37b. Distinct gynophore absent .... **(38)**

38a. Anther dehiscence poricidal; flower disk enlarged in fruit .... **Ochnaceae**

38b. Anther dehiscence longitudinal; flower disk not conspicuous .... **Tiliaceae**

39a. Stamens twice as many as petals or fewer .... **(40)**

39b. Stamens as many as petals or fewer .... **(57)**

40a. Fruit a legume /lomentum, placentation marginal .... **(41)**

40b. Fruit and placentation not as above .... **(43)**

41a. Flowers actinomorphic; calyx and corolla valvate in bud .... **Mimosaceae**

41b. Flowers zygomorphic, perianth segments imbricate in bud .... **(42)**

42a. Corolla papilionaceous, descendingly imbricate, the posterior petals outermost, the 2 anterior petals (keels) often basally connate .... **Fabaceae**

42b. Corolla caesalpinaceous, ascendingly imbricate, the posterior petal inner most, not much variable .... **Ceasalpiniaceae**

43a. Leaves gland dotted, armed or unarmed .... **(44)**

43b. Leaves not gland dotted, unarmed .... **(45)**

44a. Fruit indehiscent, berry; plants armed .... **Rutaceae** (*Atalantia, Naringi*)

44b. Fruit dehiscent, capsular; plants unarmed................................ **Flindersiaceae**

45a. Flowers irregular; petals adnate to staminal tube sheath, inner sepals petaloid; seeds with caruncle.................................... **Polygalaceae**

45b. Flowers regular; petals free from stamens, inner sepals not petaloid; seeds without caruncle .... **(46)**

46a. Leaves compound or pinnatisect.. **(47)**

46b. Leaves simple .... **(55)**

47a. Stamens tetradynamous (4 long 2 short) .... **Brassicaceae**

47b. Stamens other than above.. **(48)**

48a. Stamens connate.. **(49)**

48b. Stamens free.. **(51)**

49a. Herbs; leaf base equal; flowers sub umbellate............................... **Oxalidaceae**

49b. Trees; leaf base oblique; flowers in panicles, polygamodioecious .... **(50)**

50a. Staminal filaments connate in a tube; fruit capsular, baccate or drupe.. **Meliaceae**

50b. Staminal filaments free, inserted at the base of the disk.............. **Ailanthaceae**

51a. Herbs..**Zygophyllaceae**

51b. Trees or climbers .... **(52)**

52a. Ovules on axile placentae .... **(53)**

52b. Ovule on pendulous placentae .... **(54)**

53a. Stamens alternately long and short, bark resinous, stamens 8-10, inserted below or on the disk..................................... **Burseraceae**

53b. Stamens sub equal, bark non resinous, stamens 8, inserted within the disk or excentric.................................... **Sapindaceae p.p.** (*Cardiospermum, Sapindus*)

54a. Plants unarmed; leaves imparipinnate; flowers monoecious or dioecius; ovule pendulous from near the top of the cell ....**Anacardiaceae** (*Lannea*)

54b. Plants armed with strong sharp spines; leaves 2-foliolate, flowers bisexual; ovule pendulous from below the apex ....**Balanitaceae**

55a. Flowers 4-merous; disk absent; stamens 8, free..............**Lythraceae** (*Lawsonia*)

55b. Flowers 5-merous; disk prominent; stamens 10, connate variously .... **(56)**

56a. Scandant shrubs; flowers large, yellow showy; peduncles modified into 2 spiral hooks; petals contorted, fugacious.....................................**Linaceae**

56b. Small trees; flowers small, whitish; peduncles not modified into hooks'; petals with an erect ligule on the inner side......... **Erythroxylaceae**

57a. Flowers irregular; petals unequal.. **(58)**

57b. Flowers regular .... **(59)**

58a. Herbs; leaves simple; flowers coloured, solitary................................ **Violaceae**

58b. Trees; leaves compound; flowers cream in terminal panicles...... **Moringaceae**

59a. Climbers or stragglers .... **(60)**

59b. Non climbers .... **(62)**

60a. Disk not distinct in flower, Ovules numerous on each parietal placentae .... **Passifloraceae**

60b. Disk distinct in flower; ovules 1-2, pendulous in each placentae................ **(61)**

61a. Ovary 1-celled; fruit a drupe .... **Opiliaceae**

61b. Ovary 2-6 celled, usually embedded in the disk; fruit a berry;............ **Vitaceae**

62a. Flowers polygamous, unarmed, stragglers.......... **Celastraceae p.p.** (*Celastrus*)

62b. Flowers only bisexual. .... **(63)**

63a. Plants herbaceous, usually grow in moist or rocky localities; disk absent .... **(64)**

63b. Plants woody, grow in dry localities; disk prominent .... **(66)**

64a. Insectivorous herbs .... **Droseraceae**

64b. Non insectivorous herbs .... **(65)**

65a. Marshy plants; calyx tubular, persistent, 3-6-lobed; ovules on axile placentation ....**Lythraceae** *p.p.* (*Rotala, Ammania*)

65b. Grow nearby rocky areas; calyx not united, sepals 4-5, free, ovules on free central placentation.. **Caryophyllaceae**

66a. Petals cucullate, enclosing stamens; calyx deciduous in fruits.. **Rhamnaceae**

66b. Petals not cucullate, stamens alternate with petals; calyx persistent in fruit................................ **Celastraceae p.p.** (*Cassine, Gymnosporia*)

67a. Flowers umbellate.. **Apiaceae**

67b. Flowers not umbellate.. **(68)**

68a. Plants succulent, spiny; leaves reduced or 0; petals numerous .... **Cactaceae**

68b. Plants not succulent rarely armed; leafy; petals definite .... **(69)**

69a. Ovary 1-celled.. **(70)**

69b. Ovary 2-more-celled.. **(72)**

70a. Anthers dehiscing by apical pores; connectives appendaged; ovary 1-celled .... **Melastomataceae**

70b. Anthers longitudinally dehisced; connectives un-appendaged .... **(71)**

71a. Flowers in clusters; petals 10; ovules solitary .... **Alangiaceae**

71b. Flowers in spikes; petals 4 or 5;
ovules 2-5 .... **Combretaceae p.p.** (*Combretum*)

72a. Plants herbaceous to sub shrubby; stamens 8................................. **Onagraceae**

72b. Plants woody; stamens numerous.. **(73)**

73a. Flowers small (1cm across) in axillary panicles; stamens
inflexed in bud...**Myrtaceae**

73b. Flowers large, showy on woody pendundes in spikes;
stamens erect in bud...**Lecythidaceae**

74a. Ovary inferior or partly so... **(75)**

74b. Ovary superior .... **(78)**

75a. Tendril bearing vines...**Cucurbitaceae**

75b. Non tendrillar plants .... **(76)**

76a. Anthers united in a tube above the style; stipules normal;
leaves usually alternate; flowers actinomorphic or zygomorphic................ **(77)**

76b. Anthers separate from one another; stipules inter or
intra-petiolar; leaves usually opposite; flowers actinomorphic.........**Rubiaceae**

77a. Flowers in heads subtended by an involucre simulating calyx;
ovary 1-celled, 1-ovuled... **Asteraceae**

77b. Flowers in leafy lax terminal racemes; ovary 2-more celled,
many-ovuled.. **Lobeliaceae**

78a. Plants insectivorous or parasitic; leaves inconspicuous or 0......................... **(79)**

78b. Plants autophytic; leaves well developed .... **(80)**

79a. Slender herbs with insectivorous bladders, semiaquatic;
corolla zygomorphic; stamens 2; placentation free-central;
ovules numerous.. **Lentibulariaceae**

79b. Stem wiry, twining, parasitic; corolla actinomorphic;
stamens 5; placentation axile; ovules 4 .... **Cuscutaceae**

80a. Pistils two, nearly or quite separate at maturity .... **(81)**

80b. Pistil one .... **(83)**

81a. Pollen masses free, not agglutinated into pollinia; style one;
anthers sagittate ....**Apocynaceae**

81b. Pollen masses agglutinated into wax like masses or pollinia;
styles 2, united by stigma .... **(82)**

82a. Filaments united, anthers with horny wings.............................**Asclepiadaceae**

82b. Filaments free, anthers without horny wings...............................**Periplocaceae**

83a. Stamens free from corolla lobes .... **(84)**

83b. Stamens inserted on the corolla (epipetalous) .... **(85)**

84a. Ovary 1-celled; placentation free-central; calyx tubular with stalked glands .... **Plumbaginaceae**

84b. Ovary 2-many celled; placentation axile; calyx without glands ...... **Ebenaceae**

85a. Ovary 6-8-celled (carpels 6-8); flowers usually unisexual; corolla 4-8-lobed; stamens as many as corolla lobes and opposite to them .... **Sapotaceae**

85b. Ovary 2-celled (Carpels-2); flowers usually bisexual; corolla usually 5-lobed; stamens as many as corolla lobes or fewer and alternate to them .... **(86)**

86a. Corolla regular or actinomorphic; stamens usually as many as corolla lobes .... **(87)**

86b. Corolla irregular or zygomorphic; stamens usually didynamous, less than corolla lobes .... **(95)**

87a. Inflorescence usually secund or one sided cymes .... **(88)**

87b. Inflorescence not as above .... **(89)**

88a. Herbs or under shrubs; style entire or shortly 2-lobed; boragoid hairs present .... **Boraginaceae**

88b. Trees or shrubs; style once forked or styles 2; boragoid hairs absent .... **Cordiaceae**

89a. Leaves usually alternate .... **(90)**

89b. Leaves opposite or at least lower one opposite .... **(91)**

90a. Plants erect or diffuse, but not twining, without latex; calyx mostly 4-5-lobed; ovary 2-celled; ovules many in each locule in swollen axile placentation .... **Solanaceae**

90b. Plants chiefly twining, at times trailing, diffuse or erect, with latex; ovary 2-celled; Ovules 2 in each loculus or 4-celled with one ovule in each loculus .... **Convolvulaceae**

91a. Stamens 2 .... **(92)**

91b. Stamens 4-5 .... **(93)**

92a. Capsule compressed, fruiting calyx splitting .... **Nyctanthaceae**

92b. Berry globose, fruiting calyx not splitting .... **Oleaceae**

93a. Herbs .... **Gentianaceae**

93b. Shrubs or Trees .... **(94)**

94a. Fruit berry. .... **Loganiaceae**

94b. Fruit drupe .... **Salvadoraceae**

95a. Plants scapose .... **Gesneriaceae**

95b. Plants not scapose .... **(96)**

96a. Leaves pinnately compound; fruit woody, clearly enlarged; seeds winged; stigma dumbbell-shaped, anthers divaricate .... **Bignoniaceae**

96b. Leaves simple, rarely digitate; fruit not woody, hardly enlarged; seed not winged; stigma and anthers otherwise .... **(97)**

97a. Flowers with extra-floral glands at the base of pedicel; ovules on uniseriately and projecting parietal 2-bladed placentation .... **Pedaliaceae**

97b. Flowers without extra floral glands at the base of pedicels; ovules on 1-many on the middle or basal region of placentae........ **(98)**

98a. Fruit capsular; seeds few to numerous; placentation axile........ **(99)**

98b. Fruit drupe, pyrene or nutlet; seeds 4; placenta basal........ **(100)**

99a. Bracts and bracteoles conspicuous; upper lip of corolla crested; anthers bearded; capsule dehiscence elastically; seeds with jointed funicle; placenta never swollen .... **Acanthaceae**

99b. Bracts and bracteoles inconspicuous or absent; upper lip of corolla not crested; anthers not bearded; capsule dehiscence otherwise; seeds with inconspicuous funicle; placenta swollen........ **Scrophulariaceae**

100a. Aromatic; calyx distinctly bilobed; ovary 4-lobed, style gynobasic; fruit of 4 nutlets........ **Lamiaceae**

100b. Non- aromatic; calyx not so; ovary entire; style terminal; fruit drupe or pyrenes .... **Verbenaceae**

101a. Leaves fan-shaped or pinnate; usually very long or very broad, forming a terminal crown, (palm like) fibrous, long petiolate .... **Arecaceae (Palmae)**

101b. Leaves not palm like........ **(102)**

102a. Perianth none or rudimentary .... **(103)**

102b. Perianth present, usually of 2-series, at least the inner petaloid, not of bristes or scales........ **(105)**

103a. Perianth represented by membranous green scales or slender hairs; inflorescence not made up of spikelets; flowers not subtended by dry, chaffy bracts (glumes or scales)........ **Typhaceae**

103b. Perianth 0; flowers subtended by dry, chaffy, usually imbricating bracts (glumes); inflorescence various of spikelets........ **(104)**

104a. Stem obtusely to distinctly trigonous mostly solid; leaves not ligulate, leaf sheaths not split; fruit a compressed or trigonous nut .... **Cyperaceae**

104b. Stem mostly hollow, cylindrical or flattened; leaves ligulate, leaf sheaths split; fruit a caryopsis........ **Poaceae**

105a. Ovary superior .... **(106)**

105b. Ovary inferior .... **(111)**

106a. Plants aquatic .... **(107)**

106b. Plants terrestrial .... **(109)**

107a. Perianth one whorled (only 2), corolla like
(violet coloured) .... **Aponogetonaceae**

107b. Perianth composed of calyx and corolla (3+3) .... **(108)**

108a. Inflorescence subtended by spathe like bract;
perianth deep blue with red spots; pistil one; ovules
on axile placentation .... **Pontederiaceae**

108b. Inflorescence not subtended by spathe like bract; perianth
white; pistils 15-20, free; ovule solitary on basal placentation .. **Alismataceae**

109a. Plants climbing or twining .... **Liliaceae p.p.** (*Asparagus, Gloriosa*)

109b. Plants non climbers .... **(110)**

110a. Flowers in terminal solitary, involucral heads, compact head;
leaves radical .... **Eriocaulaceae**

110b. Flowers and leaves not as above .... **(111)**

111a. Plants usually very xerophytic; leaves thick, sub erect,
rigid, 8 or 9 in tuft .... **Sansevieriaceae**

111b. Plants not as above .... **(113)**

112a. Outer segment of perianth calyx like and different from
the inner corolla segment .... **Commelinaceae**

112b. Outer and inner segments or perianth essentially
like a petaloid .... **Liliaceae p.p.**

113a. Plants usually very xerophytic; leaves mostly fibrous and
thick in a dense tuff at the base of apex of the stem; flowers
mostly in large panicle to 6 m long .... **Agavaceae**

113b. Plants not as above; inflorescence to 50 cm long .... **(114)**

114a. Flowers strongly zygomorphic; corolla spurred, gynostomic
type; pollen often agglutinated into pollinia .... **Orchidaceae**

114b. Flower regular (rarely irregular); stamens 3 or more,
no petaloid staminodia .... **(115)**

115a. Submerged herbs .... **Hydrocharitaceae**

115b. Terrestrial .... **(116)**

116a. Plants twining; flowers unisexual; capsule deeply angular;
seeds winged; leaves 3-5 foliolate, reticulately veined .... **Dioscoreaceae**

116b. Plants not twining; flowers bisexual; fruit and seeds not
angled or winged; leaves simple, large, parallel veined .... **(117)**

117a. Inflorescence raceme, flowers invariably bright to golden-yellow .... **Hypoxidaceae**

117b. Inflorescence umbel, flowers white .... **Amaryllidaceae**

6

Systematic Enumeration

DICOTYLEDONS

ANNONACEAE

Key to Genera

1a. Flowers greenish yellow; connectives concealing anthers .... **Miliusa**

1b. Flowers purple -green or flesh-coloured; connectives not concealing anthers .... **(2)**

2a. Inner petals scaly or 0; carpels connate .... **Annona**

2b. Inner petals not scaly; carpels free .... **Polyalthia**

ANNONA L.

Key to Species

1a. Leaves elliptic or oblong-lanceolate, glabrescent; fruit conical in shape, rough, grooved between seeds .... **A. squamosa**

1b. Leavs broadly ovate, velvety; fruit irregular in shape, smooth, not grooved between seeds .... **A. cherimola**

Annona cherimola Mill., Gard. Dict. ed. 8, no. 5. 1768.

Local name: *Hanuman phalam, Lakshman phalam.*

Moderate deciduous tree, to 3 m high, bark grey. Leaves simple, broadly ovate, 3.5–6.5 cm, velvety-tomentose beneath, base rounded, margin entire, apex obtuse. Flowers purple-green, solitary, extra axillary. Sepals 3. Fruit ovoid, syncarpous, irregular in shape, carpels numerous, not grooved between seeds. Seeds many, black, shiny.

Cultivated for its fruit. Fl. & Fr.: Mar. – Aug. Kalasamudram (ATP), *KRKS & BSS* 39136.

Annona squamosa L., Sp. Pl. 537. 1753; FBI 1: 78. 1872; FPM 1: 20.1915; FAP 1: 68. 1997.

Local name: *Seetha phalam.*

Small deciduous tree to 4 m high, bark brown to black. Leaves simple, elliptic or oblong-lanceolate, 6–10 × 2–4 cm, glabrescent, glaucous beneath, base cuneate, margin entire, apex obtuse. Flowers greenish, solitary on leaf-opposed peduncles or 2–4 in a cluster. Sepals 3, triangular, 3 mm long, pubescent; inner ones minute. Stamens more than 100. Fruit syncarpous, cordate, conical, ovoid, 7 cm across, rough, granular, loosely coherent, grooved between seeds. Seeds many, angular, oblong, blackish brown, shiny.

Occasional, in rocky hills. Fl. & Fr.: Apr. – Nov. Batrepalli (ATP), *KRKS & BSS* 37980.

Fruits are edible

It is a native of tropical America.

MILIUSA Lesch. ex A. DC.

Miliusa tomentosa (Roxb.) Sinclair, Gard. Bull. Straits Settlem. 14: 378. 1955; FAP 1: 69. 1997. *Uvaria tomentosa* Roxb., Pl. Coromandel t. 35. 1795. *Saccopetalum tomentosum* (Roxb.) Hook. f. &Thoms. Fl. Ind. 152. 1855; FBI 1: 88. 1872; FPM 1: 22. 1915.

Local name: *Pedda chilka duduga.*

Large deciduous tree, to 15 m high, bark brown-blackish, deeply longitudinally fissured. Leaves simple, thick-coriaceous, ovate, 7 × 3 cm, tomentose below, base acute, margin entire, apex acute. Flowers greenish, in leaf-opposed cymes or solitary. Sepals 3, ovate or linear-lanceolate, subequal, pilose. Outer petals sepaloid; inner ones saccate below. Fruit berry, monocarps free, dark purple, sub globose-ellipsoid. Seed 1 per carpel.

Rare, in hilly areas of dry deciduous forests. Fl. & Fr.: Feb. – Jun. Kalasamudram (ATP), *KRKS & BSS* 39192.

POLYALTHIA Blume

Polyalthia cerasoides (Roxb.) Bedd., Fl. Sylv. t. 1. 1869; FBI 1: 63. 1872; FPM 1: 17. 1915; FAP 1: 70. 1997. *Uvaria cerasoides* Roxb. Pl. Coromamdel. i. t. 33. 1795.

Local name: *Gutti kaya.*

Moderate evergreen tree to 6 m high, bark grey, rough; young branches densely tomentose. Leaves simple, elliptic-lanceolate or oblong, 9 × 3.5 cm, chartaceous, tomentose, base acute, margin entire, apex acuminate. Flowers purple green, axillary or extra axillary, solitary. Sepals 3, ovate, tomentose. Petals 6, greenish, tomentose. Fruit berry, carpels numerous, free, ovoid, red, pilose, stalks to 2 cm long. Seed 1 per carpel, round, grey.

Occasional, in interior of the dry deciduous forests on hilly slopes. Fl. & Fr.: Jan. – Oct. Kalasamudram (ATP), *KRKS & BSS* 39199.

MENISPERMACEAE

Key to Genera

1a. Leaves ovate-oblong to lanceolate, villous beneath; style simple .... **Cocculus**

1b. Leaves ovate, orbicular, glabrous or pubescent; styles 2–8-partite .... (2)

2a. Inflorescence umbellate or subcorymbose cyme; stigma 3–5-fid.................... (3)

2b. Inflorescence paniculate, racemose; stigma capitate .... (4)

3a. Leaf tomentose, basal nerves 8–10, bracts obovate to orbicular, endocarp curved .... **Cissampelos**

3b. Leaf glabrous, basal nerves 5–7, bracts subulate, endocarp compressed .... **Stephania**

4a. Bark corky, flaking off with age, leaves broadly ovate-cordate, basal nerves 5–7, palmate .... **Tinospora**

4b. Bark not corky; leaves ovate, basal nerves 3–5, sub palmate ....**Pachygone**

CISSAMPELOS L.

Cissampelos pareira L., Sp. Pl. 1031. 1753. var. **hirsuta** (Buch.-Ham. ex DC.) Forman, Kew Bull. 22: 356. 1968; FPM 1: 30.1915; FAP 1:73. 1997. *C. hirsuta* Buch.-Ham. ex DC., Syst. 1: 535. 1817.

Local name: *Adavi bankateega.*

A deciduous climber, branchlets hairy. Leaves ovate, reniform, 3–6 × 3.5–7 cm, tomentose, palmately 5–7-nerved, chartaceous, tomentose, base cordate, margin entire, apex mucronate; petiole 2–10 cm. Flowers unisexual, greenish-yellow, in pendulous, subcorymbose cymes; bracts obovate, bracteoles linear. Sepals and petals 4. Stamens 4. Ovary pilose; style glabrous; stigma 3-fid. Drupe ovoid, red. Seeds curved, white, transverse ridged.

Occasional, in moist and shaded places of forest, twining over hedges and shrubs. Fl. & Fr.: Apr. – Nov. Namalagundu Temple (KDP), *KRKS & BSS* 40164.

Roots are used medicinally especially for cough and snake bites. A strong fibre is extracted from the plant.

COCCULUS DC. *nom. cons.*

Cocculus hirsutus (L.) W. Theob., Burmah ed. 3, 2 : 657. 1860. *Menispermum hirsutum* L., Sp. Pl. 341.1753. *Cocculus hirsutus* (L.) Diels., Engl. Pflanzenr. 46: 236. 1910; FPM 1: 29. 1915; FAP 1: 73.1997. *Cocculus villosus* (Lam.) DC., Syst. 1: 525. 1817; FBI 1: 101. 1872. *Menispermum villosum* Lam., Encycl. 4. 97. 1797.

Local name: *Dusara teega.*

Climber, hirsute hairy; branches densely tomentose. Leaves ovate-oblong, 3–4 × 2–2.5 cm, greyish hairy, base rounded, often cordate, cuneate, rarely cordate, margin entire, apex retuse, mucronate. Flowers unisexual. Male flowers greenish-yellow in slender panicles. Outer sepals oblong to lanceolate; inner ones broadly ovate. Petals 6, free, oblong-ovate, greenish. Stamens 6, free, enclosed by petals. Female flowers in axillary fascicles; styles terete. Drupe curved, obovoid, black. Seed white, curved almost into a ring, furrowed.

Common, in forests and wastelands. Fl. & Fr.: Nov. – Mar. Mabbuchintalapalli (KDP), *KRKS & KRS* 33663.

Leaves and roots are medicinally used.

PACHYGONE Miers

Pachygone ovata (Poir.) Miers ex Hook. f. & Thomson, Fl. Ind. 203. 1855; FBI 1: 105. 1872; FPM 1: 31. 1915. *Cissampelos ovata* Poir. in Lam., Encycl. 5: 10. 1604.

Vern. : *Pedda dusari.*

Climbing shrubs, up to 10 m long, branchlets yellowish pubescent or puberulous when young, later glabrous. Leaves ovate, blunt, 3.5–6 × 2–4 cm, base truncate, rounded or obtuse, margin entire, apex retuse, mucronate, basal nerves 3–5, lateral nerves 1–2 pairs, petioles 2–5 cm long. Flowers minute, white, honey scented in axillary racemes; Male flowers: yellow. Sepals 6–12, outer 3-6 bracteiform, elliptic-rounded, inner 3–6, elliptic, rotund, puberulous outside. Petals 6, oblong; stamens 6, free, clasped by petals, pistillodes 3. Female flowers: Sepals and petals as in male. Staminodes 6, minute. Carpels 3, subovoid, style flattened, stigma minute. Drupes orange to purple when ripe, reniform, 6 × 7 cm, stylar-scar subbasal, endocarp smooth. Seeds horse-shoe-shaped, albumen nil, cotyledons semi-terete, very thick, hard.

Occasional, in dry deciduous forests. Fl. & Fr.: Febr.-Oct. Kalasamudram (ATP), *SSR* & *KSM* 18589.

STEPHANIA Lour.

Stephania japonica (Thunb.) Miers, Contrib. Bot. 3: 213. 1871, *p.p.*; FPM 1: 29. 1915; FAP 1: 75. 1979. *Menispermum japonicum* Thunb., Fl. Jap. 193. 1784. *Stephania hernandiifolia* (Willd.) Walp., Rep. Bot. Syst. 1: 96. 1842; FBI 1: 103. 1872. *Cissampelos hernandiifolia* Willd., Sp. Pl. 4: 861. 1806.

Plate 1

A. *Stephania japonica*, B. *Nymphaea nouchali*, C. *Nymphaea rubra*, D. *Capparis sepiaria*, E. *Maerua apetala*, F. *Hybanthus vatsavayae*

Local name: *Duvvateega.*

Large slender climber, glabrous. Leaves peltate, ovate, triangular, 4.5–6 × 5–6 cm, glabrous, base truncate, round, margin entire, apex acuminate, basally 9-nerved; petiole 3–12 cm long, puberulous. Flowers greenish in axillary capitate umbel, unisexual. Male flowers green-yellow. Sepals 6–8, oblanceolate or spatulate. Petals 3–5, free, orbicular. Stamens 6. Female flowers yellow. Sepals 3–5; carpels solitary, ovoid; style 3–5-fid. Drupe obovoid, red, glabrous. Seeds appressed, fleshy, curved, white.

Rare, along the stream sides of forest. Fl. & Fr.: Aug. – Oct. Namalagundu Temple (KDP), *KRKS & BSS* 39657.

TINOSPORA Miers

Tinospora cordifolia (Willd.) Miers, Ann. Mag. Nat. Hist. Ser. 27 : 35, 38. 1851; FBI 1: 97. 1872; FPM 1: 26. 1915; FAP 1: 76. 1979. *Menispermum cordifolium* Willd., Sp. Pl. 4. 826. 1806.

Local name: *Tippa teega.*

Deciduous woody climber, glabrous, stem green when young, with brown lenticels when mature. Leaves broadly ovate-cordate, 5–6 × 7–8 cm, glabrous, base cordate, margin entire, apex acuminate, basal nerves 5–7; petiole pulvinate, 2–7 cm long. Flowers greenish yellow in axillary pseudoraceme or on leafless branches. Male flowers fascicled, yellow. Outer sepals 3, ovate; inner 3, elliptic. Petals 6, rhombic-unguiculate. Female flowers on pendulons racemes. Petals spatulate; stigma capitate. Drupes red, glabrous, round, with gummy pulp. Seed white, recurved, rounded at both ends, ridged, slightly papillose on surface.

Common, in all wastelands and forest. Fl. & Fr.: Jan. – May & Aug. – Oct. Kalasamudram (ATP), *KRKS & BSS* 39996.

Note: Aerial photosynthetic roots from mature stems appeaer like wires. Stem and roots are medicinal.

NYMPHAEACEAE

NYMPHAEA L. *nom. cons.*

1a. Leaves greenish, glabrous, elliptic-orbicular, base deeply cordate, lobes obtuse; flowers white-blue or purple; calyx acute, thick, glabrous .... **N.nouchali**

1b. Leaves pinkish, pubescent, elliptic, base sagittately lobed, lobes acute; flowers pink; calyx obtuse, thin, pubescent .... **N. rubra**

Nymphaea nouchali Burm. f., Fl. Ind. 120. 1768; FAP 1: 77. 1997. *N. stellata* Willd., Sp. Pl. 2: 1153. 1799; FBI 1: 114.1872; FPM 1: 33. 1915.

Local name: *Kaluva.*

An aquatic, tuberous, perennial herb. Leaves elliptic-orbicular, glabrous, pale purple below, pilose above, 10–11 × 9–10 cm, base deeply cordate, margin dentate, apex rounded, petiole 30–50 cm long. Flowers pale bluish white, axillary, long-peduncled, equal to leaf petiole, hollow. Fruit berry, oblong, 1.5–4 cm across. Seeds numerous, ellipsoid, ribbed.

Occasional in tanks. Fl. & Fr.: Throughout year. Batrepalli (ATP), *KRKS & BSS* 40360.

Nymphaea rubra Roxb. ex Andrews, Bot. Rep. 8 (104): t. 503. 1808; FAP 1: 60. 1997. *N. lotus* auct. non L. (1753); FBI 1:114. 1872.

Local name: *Erra kaluva.*

Aquatic perennial rhizomatous herb. Leaves long-petioled, peltate, sagittate when young, orbicular, elliptic or reniform when mature, 15–25 × 20–30 cm, greenish-pink, pubescent below, margin dentate. Petiole to 1.5 m long, pubescent. Flower crimson red. Sepals 4, thin, pubescent outside. Petals 16–25, oblong-elliptic.

Vary rare, in ponds and tanks. Fl.: Jul. – Jan. Namalagundu (ATP), *KRKS* 42043.

PAPAVARACEAE

ARGEMONE L.

Argemone mexicana L., Sp. Pl. 508. 1753; FBI 1: 117. 1872; FPM 1: 35. 1915; FAP 1: 78. 1997.

Local name: *Yerri kusam.*

Annual, branched, prickly herb, to 60 cm high; branches spinescent with yellow latex and white waxy. Leaves sessile, amplexicaulous, oblong-ovate, prickly on both sides with white colour, 4–7 × 2–2.5 cm, base truncate, margin pinnatifid, apex acute. Flowers yellow, terminal. Capsule, oblong, 2–4 cm, elliptic, prickly. Seeds black, numerous, rugose, shiny.

Common, in wastelands and cultivated lands as a weed. Fl. & Fr.: Throughout the year. Thummalapalle (KDP), *KRKS & KRS* 33674.

Note: This species is a native of Mexico and is widely distributed and naturalised as a weed in India.

CAPPARACEAE

Key to Genera

1a. Sepals fused, forming a distinct tube .... **Maerua**

1b. Sepals mostly free, often slightly connate at base, but never forming a tube .... (2)

2a. Disk large, spatulate, tubular, trumpet-shaped, petaloid; stamens 4–6, inserted halfway on gynophore .... **Cadaba**

2b. Disk small, adaxial, not as above; stamens numerous, inserted at the base of gynophore .... **Capparis**

CADABA Forssk.

Cadaba fruticosa (L.) Druce, Bot. Exch. Club. Soc. Brit. Isles 3: 415. 1914; FAP 1: 83. 1997. *Cleome fruticosa* L., Sp. Pl. 671. 1753. *Cadaba indica* Lam., Encycl. 1: 554. 1785; FPM 1: 43. 1915.

Local name: *Chimurudu chettu, Yenuka-parika.*

Shrub, low straggling among bushes, to 3 m; stem much branched, terete, old twigs glabrescent, smooth. Leaves simple, alternate, elliptic-oblong, 1–2 × 0.6–1 cm, glabrous, base rounded-cuneate, margin entire, apex acute. Flowers greenish white, terminal, corymbose. Sepals greenish, glandular. Petals white, spatulate. Disk prolonged into a 6–9 mm long, trumpet-shaped process, petaloid. Stamens 4–6, spreading. Gynophore 2 cm long; ovary 3–5 mm long. Fruit berry, cylindrical, 5.5 cm. Seeds 2 mm, clothed with red aril and embedded in orange colour pulp.

Common, in wastelands and borders of cultivated lands. Fl. & Fr.: Mar. – June. Thummalapalle (KDP), *KRKS, KRS, SSR & TP* 33655.

CAPPARIS L.

Key to Species

1a. Flowers solitary, greenish, axillary; fruit ribbed .... **C. divaricata**

1b. Flowers not solitary, vary in color, supra axillary or terminal; fruit otherwise .... (2)

2a. Berry up to 8 mm across .... **C. sepiaria**

2b. Berry more than 1 cm across .... (3)

3a. Flowers in supra-axillary vertical series of 2-10 or even more, pink.............................. **C. zeylanica**

3b. Flowers otherwise, never in supra-axillary vertical series, white or creamy.............................. (4)

4a. Inflorescence panicle .... **C. tomentosa**

4b. Inflorescence sub umbels or in corymbs .... (5)

5a. Flowers 6 – 15 per umbel; sepals glabrous.............................. **C. roxburghii**

5b. Flowers 25 – 30 per umbel; sepals sparsely puberulous .... **C. grandis**

Capparis divaricata Lam., Encycl. 1: 606. 1785; FAP 1: 85. 1997. *C. stylosa* DC., Prodr. 1: 246. 1824; FPM 1: 45. 1915.

Local name: *Badreni.*

Small, deciduous tree, to 5 m high; branches slender, zig-zag, stellate-pubescent. Leaves elliptic ovate-lanceolate, 4–5 × 1.5–2 cm, base rounded, margin entire, apex mucronate; stipule thorny, divaricate, 3–4 mm; petiole 3 mm. Flowers greenish, axillary, solitary; buds ovoid, beaked, greyish to brown-pubescent. Calyx lobes elliptic-orbicular, thick, pubescent on both sides. Corolla lobes greenish-yellow, creamy, spatulate. Stamens 45–65. Gynophore 2–2.2 cm long; ovary ovoid. Fruit ovoid-globose, rough surfaced, 5-ribbed, reddish, mesocarp pulpy. Seeds 6–10, black.

Occasional, in forests and foot hills. Fl. & Fr.: May – Aug. Idupulapaya (KDP), *KRKS & BSS* 40314.

Capparis grandis L. f., Suppl. Pl. 263. 1781; FBI 1: 176. 1872; FPM 1: 46. 1915; FAP 1: 85. 1997.

Lacal name: *Nalluppi, Guli, Ragata.*

Deciduous, crooked tree, to 5 m high, bark fissured; young branches olive green, velvety tomentose. Leaves elliptic-oblong, 2–5 × 2–3 cm, pubescent above, pilose below, base cuneate, acute, margin entire, apex obtuse; petiole 1 cm; stipular thorns straight or recurved. Flowers white, in terminal corymbs or clusters. Calyx lobes sparsely puberulous; outer lobes boat-shaped; inner lobes linear-oblong or obovate. Corolla lobes oblong to narrowly obovate. Stamens 35–40. Gynophore 1.3–2.8 cm. Fruit sub globose, somewhat umbonate, 3 cm across, red, pulpy. Seeds black, 2–5, elongate.

Occasional, in foot hills of the forest. Fl. & Fr.: Mar. – Jul. Idupulapaya (KDP), *KRKS & BSS* 39075.

Capparis roxburghii DC., Prodr. 1: 247. 1824; FBI 1: 175. 1872; FPM 1: 46. 1915; FAP 1: 86. 1997.

Large, woody, climbing shrub, to 5 m high; stem glabrous, branches pubescent, early glabrescent; stipular thorns 1.5–3 mm long, on old shoots. Leaves obovate-oblong, 4.5–8 × 2.5–3.5 cm, sub coriaceous, glabrous, base cuneate, margin entire, apex rounded to acute; petiole 1.3–2 cm long. Flowers showy, white, fragrant, 4–5.5 cm across, in terminal, 6–15-flowered corymbs; pedicels 2.5–3.8 cm long. Calyx lobes glabrous, 9–10 × 6–9 mm; outer pair much concave. Corolla lobes obovate, narrowed at base, truncate at apex, 11–15 × 9–11 mm. Stamens 45–60, rosy-purple at mature, 2.5–3 cm long. Gynophore 3.5–5.5 cm long. Ovary ellipsoid or ovoid; stigma capitate. Fruit globose, umbonate, 4.5–6 cm across, pendent. Seeds 35–40, ovoid, 1–1.2 × 1 cm, reddish brown.

Occasional, on road sides, foot hills of dry deciduous forest. Fl. & Fr.: Mar. –Aug. Batrepalli (ATP), *KRKS & BSS* 39613.

Capparis sepiaria L., Syst. Nat. (ed. 10) 2: 1071. 1759; FBI 1: 177. 1872.

Scandent prickly shrubs, rarely trees, 2-6 m high, stems zig-zag, terete, branchlets densely fulvous or grey-pubescent. Leaves oblong, lanceolate or elliptic, rarely obovate, 1.5-3 × 0.8-1.5 cm, thin, coriaceous, base cuneate, margin entire, apex notched. Flowers white in terminal corymbose umbels. Sepals 4, subequal, ovate or suborbicular, inner pair smaller than outer, margin ciliate. Petals 4, obovate or oblong-spatulate, puberulous at base. Stamens 25-40, filaments 10-14 cm long, exserted, anthers brown. Gynophores 5-6 mm long, slender, ovary ovoid. Berry globose, *ca.* 0.8 cm across, blue when ripe. Seeds 2, embedded in a sticky pulp.

Common, in dry forests and scrub jungles. Fl.: Mar. – Jul.; Fr.: Oct. - Nov. Kalasamudram RF (ATP), *EC* 2952.

Capparis tomentosa Lam., Encycl. 1: 606. 1785. *C. floribunda* Wight, Illus. Ind. Bot. 1: 35. 1840; FBI 1: 177. 1872; FPM 1: 46. 1915; FAP 1: 85. 1997.

Woody climber, to 5 m; twigs straight, glabrous, rarely puberulous with simple fulvous hairs. Leaves elliptic-obovate, 3–5 × 2–3.5 cm, pilose below, pubescent above, base round, margin entire, apex obtuse; petiole 0.5 cm; stipular thorns 2, divaricate, recurved. Flowers white, fragrant, terminal, many-flowered sub umbells or paniculate corymbs; buds globose; bracts linear, caducous. Calyx lobes subequal, deflexed, persistent. Gynophore 4–6 mm long, glabrous. Fruit globose, shortly beaked at apex, fleshy, 3 cm in diameter, red. Seeds 2-8, black, ovate.

Occasional, in forest. Fl. & Fr.: Jan. – Apr. Batrepalli (ATP), *KRKS & BSS* 39613.

Capparis zeylanica L., Sp. Pl. ed. 2. 270. 1762; FPM 1: 46. 1915; FAP 1: 86. 1997; *C. horrida* L. f. Suppl. Pl. 264. 1781.

Local name: *Adavi nimma.*

Large, thorny straggler, to 4–6 m long, branches brown or greyish stellate scurf deciduous tomentum at young, densely armed. Leaves elliptic-obovate, 2–6 × 1.5–4 cm, tomentose when young, glabrous when mature, base acute or subcordate, margin entire, apex mucronate; petiole 0.5–0.6 cm; stipular thorns recurved. Flowers white, supra-axillary, 2–6, fading to pink or purple. Calyx lobes subequal; outer pair elliptic or orbicular; inner pair elliptic-oblong, slightly smaller than outer. Corolla lobes rounded at apex, tomentose inside. Disk *ca.* 1 mm in diameter. Stamens many, creamy white, turning to pink. Gynophore glabrous, except at base, *ca.* 7 cm long; ovary glabrous, ellipsoid. Fruit globular or ellipsoid, glabrous, reddish orange to purple, hard, white pulpy. Seeds black, round, up to 25, embedded in pulp.

Common, in scrub and dry deciduous forest. Fl. & Fr.: Apr. – Aug. Thummalapalle (KDP), *KRKS & KRS* 33672; *KRKS & BSS* 39073.

MAERUA Forssk.

Key to Species

1a. Leaves 3–5-foliolate; petals 0; fruit ovoid; tree .... **M. apetala**

1b. Leaves simple; petals 4; fruit elongate, cylindric or moniliform; ascending straggling shrub .... **M. oblongifolia**

Maerua apetala (Roth) M. Jacobs., Blumea 12(2): 207. 1964; FAP 1: 87. 1997. *Capparis apetala* Roth, Nov. Pl. Sp. Ind. Or. 238. 1821. *Niebuhria linearis* DC., Prodr. 1: 244. 1824; FBI 1: 171. 1872. *N. apetala* (Roth) Dunn in Gamble, Flora Madras 1: 41. 1915.

Local name: *Pilli adugu, Nemali-adugu.*

Evergreen tree, to 10 m high, glabrous; branches lenticellate. Leaves 3–5-foliolate, petiole 5–8 cm, leaflets elliptic oblong or elliptic ovate, glabrous, 3–7 × 1–3.5 cm, base acute-attenuate, margin entire, apex acute-mucronate. Flowers greenish yellow, in terminal corymbs or raceme. Calyx tube greenish; lobes ovate, puberulous along margins. Corolla absent. Stamens *ca.* 30. Gynophore 2–2.5 cm. Ovary ovoid, 1-loculed with 2 placentae. Fruit oblong, ovoid, 1–2-ridged, 1.5 cm in diameter, yellowish. Seeds 1–4, globose, pitted, brown.

Occasional, in dry deciduous forest and foot hills of forest. Fl. & Fr.: Apr. – June. Thummalapalle (KDP), *KRKS & KRS* 33695; Batrepalli (ATP), *KRKS & BSS* 39098.

Maerua oblongifolia (Forssk.) A. Rich. in Guill. & Perr., Fl. Seneg. Tent. 1: 32, t. 6. 1847; FAP 1: 88. 1997. *Capparis oblongifolia* Forrsk., Fl. Aegypt.-Arab. 99. 1775. *Maerua arenaria* (DC.) Hook. f. & Thoms. in Hook. f., Fl. Brit. India 1: 171. 1872; FPM 1: 42. 1915. *Niebuhria arenaria* DC., Prodr. 1: 244. 1824.

Local name: *Merupu teega, Puttateega.*

Unarmed straggling shrub, to 3–5 m; branches glabrous; bark smooth, pale brown. Leaves simple, elliptic-oblong or lanceolate, obovate, 3–3.5 × 1.5–2.5 cm, glabrous, base obtuse, margin entire, apex obtuse-mucronate; petioles 6–9 mm long. Flowers greenish white, in axillary corymbs or clusters. Bracts small, ovate, acuminate, rigid. Calyx petaloid, united near base; calyx tube 3–8 mm long; lobes elliptic-oblong. Corolla cup-shaped disk, lobes ovate-lanceolate to obovate. Stamens 20–26. Gynophore 2 cm. Ovary cylindrical; stigma sessile. Fruit berry, moniliform, cylindrical, constricted between seeds, brown, up to 7 cm long. Seeds up to 30, arranged in 3 rows, attached with each other, brown in colour, round, furrowed.

Occasional, in scrub forest and dry deciduous forest. Fl. & Fr.: Jan. – Jun. Thummalapalle (KDP), *KRKS & KRS* 33695; Rachakuntapalle, *KRKS & BSS* 39072.

CLEOMACEAE

CLEOME L.

Key to Species

1a. Leaves simple .... **C. monophylla**

1b. Leaves 3–9-foliolate .... (2)

2a. Flowers white with distinct gynophore........ **C. gynandra**

2b. Flowers coloured without distinct gynophore .... (3)

3a. Leaves more than 3-foliolate; erect herb........ (4)

3b. Leaves 3-foliolate; prostrate herb .... (5)

4a. Flowers pink; stamens above 50; leaves 5–7-foliolate, non glandular .... **C. chelidonii**

4b. Flowers yellow; stamens below 50; leaves 5-foliolate, glandular viscid........ **C. viscosa**

5a. Fruit flattened; stem densely appressed-strigose .... **C. felina**

5b. Fruit cylindrical; stem with scattered, minute prickles .... **C. aspera**

Cleome aspera Koenig ex DC., Prodr. 1: 241. 1824; FBI 1: 169. 1872; FPM 1: 41. 1915; FAP 1: 89. 1997.

Local name: *Chittierretaku, Chitti vaminta.*

Annual spreading herb, to 40 cm high, stem and leaves asperous with minute scattered prickles. Leaves 1–3-foliolate, leaflets unequal, oblong-oblanceolate, 1.5–2.5 × 0.4–0.5 cm, base cuneate, margin entire, apex obtuse or retuse; petiole 2–3 mm. Flowers yellow, axillary, solitary, pedicel 2 cm long. Calyx lobes linear-lanceolate, obtuse, thinly pubescent. Corolla yellow or creamy. Stamens 6. Gynophore 1 mm long; ovary linear; style slender. Fruit dehiscent capsule, terete, narrowed at both ends, glabrous, cylindrical, slightly beaked, 2–3 cm long. Seeds 12–20, curved, 1 mm across, rugose, dark brown.

Occasional, in sandy soil and plains of dry deciduous forest. Fl. & Fr.: Jun. – Nov. Thummalapalle (KDP), *KRKS & SSR* 39017; Kalasamudram (ATP), *KRKS & KRS* 36488.

Cleome chelidonii L. f., Suppl. Pl. 300. 1781; FBI 1: 170. 1872; FPM 1: 41. 1915; FAP 1: 89. 1997.

Annual erect herb, to 20–70 cm high, branches sparsely, bulbous based hairy. Leaves 4–7-foliolate, leaflets obovate, 1–2 × 0.3–0.5 cm, glandular scabrid hairy, base cuneate, margin entire, apex obtuse, petiole 1–4 cm long. Flowers pink or rosy purple, in lax terminal corymbiform, up to 10 cm long racemes, pedicels 2–4 cm. Bracts foliaceous. Calyx lobes narrowly imbricate, appressed, elliptic to obovate, acuminate. Corolla 4–lobed, elliptic, ovate to obovate, narrow at base. Stamens *ca.* 60. Ovary sessile, linear. Fruit cylindrical, glabrous, capsule up to 7 cm. Seeds numerous, 1 mm across, curved, muricate, light yellow or reddish brown.

Common in ponds and village outskirts. Fl. & Fr.: Jul. – Jan. Kalasamudram (ATP), *KRKS* 39525.

Cleome felina L. f., Suppl. Pl. 300. 1781; FBI 1: 170. 1872; FPM 1: 41. 1915; FAP 1: 89. 1997.

Annual prostrate woody herb, to 35 cm, branches radiating from base covered with stiff sharp hairs. Leaves 3-foliolate, leaflets unequal, obovate, 1–1.5 × 0.6 –1 cm, covered with stiff sharp hairs, base cuneate, margin entire-ciliate, apex obtuse; petiole 1.5 cm long. Flowers pink or purple, axillary, solitary, pedicel to 2 cm long. Calyx lobes linear lanceolate, oblong, scabrid. Corolla lobes elliptic-obovate, spatulate. Stamens 25–40. Ovary sessile, 5–6 mm long. Fruit flattened, narrowed at both ends, glabrous, 2–2.5 × 0.4–0.5 cm, beaked. Seeds up to 36, 1 mm across, curved, echinate, with oblique closed cleft, dull white or yellowish-brown.

Common, in wastelands and plains of forest. Fl. & Fr.: July – Nov. Thummala palle (KDP), *KRKS & KRS* 36904; Bhumaiahgari palle (KDP), *KRKS & KRS* 36997.

Cleome gynandra L., Sp. Pl. 671. 1753; FAP 1: 90. 1997. *Gynandropsis pentaphylla* (L.) DC., Prodr. 1: 238. 1824; FBI 1: 171. 1872; FPM 1: 40. 1915.

Local name: *Gabbataku, Vaminta.*

Annual, erect herb, to 1 m high, pubescent with glandular trichomes. Leaves 4–5-foliolate, leaflets unequal, obovate-elliptic, oblanceolate, 2–4 × 1–1.5 cm, pubescent, base cuneate, margin entire, apex obtuse; petiole up to 5 cm long. Flowers white, in terminal raceme, to 25 cm long, bracts foliaceous, trifoliolate, much reduced and sessile upwards. Calyx lobes ovate, elliptic, acuminate. Corolla lobes obovate to oblanceolate, subequal. Androphore 1–25 mm long; stamens 6. Gynophore up to 1.5 cm long; ovary sessile among stamens, oblong-cylindrical. Fruit cylindrical, capsule to 3.5 cm, pubescent, tapering at both ends. Seeds about 40, reniform, 1 × 1 mm, compressed, concentrically striate with ridges, rugulose, brown.

Common, in wastelands and agriculture lands. Fl. & Fr.: Feb. – Nov. Pulivendula (KDP), *KRKS & SSR* 39044.

Cleome monophylla L., Sp. Pl. 672. 1753; FBI 1: 168. 1872; FPM 1: 41. 1915; FAP 1: 90. 1997.

Local name: *Okkaku vaminata, Thikka vaminta.*

Annual erect herb, to 30 cm high, glandular hairy. Leaves simple, ovate – oblong lanceolate, sparsely viscid hairy, 3 – 7 × 0.5 – 1.5 cm, base sub cordate – truncate, margin entire, apex acute; petiole up to 4.5 cm long. Flowers light pink in terminal receme, pedicel to 1 cm. Calyx lobes linear-lanceolate, acuminate. Corolla lobes oblong, obovate or spatulate. Stamens 6. Ovary linear, 5–8 mm long. Fruit cylindrical, oblong, hairy, up to 6 cm. Seeds thick brown, up to 40, ridged with longitudinal striations, curved, rugose, 1 mm diameter, dark brown.

Common, in agriculture lands. Fl. & Fr.: Jun. – Oct. K. K. Kottala (KDP), *KRKS & KRS* 37653; Kalasamudram (ATP), *KRKS & BSS* 39179.

Cleome viscosa L., Sp. Pl. 672. 1753; FBI 1: 170. 1872; FPM 1: 41. 1915; FAP 1: 91. 1997.

Local name: *Kukka vaminta, Errataku.*

Annual erect herb, 0.5–1 m high, glandular viscid hairy. Leaves 3–5-foliolate, leaflets obovate-elliptic, 0.5–5 × 0.4–3.5 cm, viscid hairy, base cuneate, margin entire-glandular ciliate, apex acute; petiole 1–6 cm long. Flowers yellow in axillary and terminal and form a raceme at apices, pedicel 1–3.5 cm. Calyx 4-lobed, oblong, lanceolate, acute at apex, glabrous within. Corolla lobes 4, subequal, obovate, spatulate, cuneate at base. Stamens 12–40. Ovary sessile, oblong, cylindric, beaked. Fruit cylindrical, to 7 cm long, viscid hairy, tapering at both ends. Seeds about 100, curved, rugose, 1 mm diameter, dark brown.

Common, in agriculture and wastelands. Fl. & Fr.: Throughout the year. K.K. Kottala (KDP), *KRKS & KRS* 36442; Kalasamudram (ATP), *KRKS & KRS* 36447.

Note: Leaf is remedy for scorpion sting, which works when crushed and kept at the sting point.

VIOLACEAE

HYBANTHUS Jacq. *nom. cons.*

Key to Species

1a. Stem brownish green to reddish brown; flowers conspicuous; pedicel 6-16 mm long; sepals dark green; stamens 5, all fertile; leaves linear-oblong or obovate; seeds 10–18 .... **H. ennaeaspermus**

1b. Stem pale yellow; flowers inconspicuous; pedicel 2 mm long; sepals yellow; stamens 5, only 2 fertile; leaves linear-lanceolate; seeds 5–9 .... **H. vatsavayae**

Hybanthus ennaeaspermus (L.) F. Muell., Fragm. Phyt. Austr. 10: 81. 1872; FAP 1: 91. 1997. *Viola enneaspermus* L., Sp. Pl. 937. 1753. *Ionidium suffruticosum* (L.) Ging. ex DC. Prodr. 1: 311. 1834; FBI 1: 185. 1872; FPM 1: 49. 1915. *Viola suffruticosa* L., Sp. Pl. 937. 1753.

Local name: *Ratna purusha.*

Annual herb, to 20 cm, much branched, sparsely hairy. Leaves spiral, linear-lanceolate, 2–4 × 0.5–0.6 cm, sparsely hairy, base attenuate, margin distinctly serrate, apex acute, subsessile. Flowers pink, axillary, solitary; pedicel 0.6–1.6 cm long. Calyx subequal; lobes lanceolate, acute, glabrous or ciliate. Corolla lobes unequal; upper two elliptic, symmetric, acuminate, 3–4 mm long; lateral two triangular-oblong, 4–6 mm long, rather expanded and obtuse at apex; lower petal 0.8–1.9 cm long. Stamens 5, *ca.* 2 mm long; anterior stamen with a small recurved fleshy appendage. Ovary globose, glabrous; style thickened towards tip. Fruit capsule, globose, 3-locular, glabrous. Seeds 9, three per locule, white, longitudinally rugose.

Common, in open scrub forest, wastelands. Fl. & Fr.: May – Oct. K. K. Kottala (KDP), *KRKS & KRS* 36419; Kalasamudram (ATP), *KRKS & BSS* 39129.

It usually appears after rains.

Hybanthus vatsavayae C. S. Reddy, J. Econ. Taxon. Bot. 25: 219. 2001.

Annual, erect herb, to 30 cm high; stem angular, pilose; branced from base; root stock woody, pale yellowish. Leaves spiral to whorled, linear to linear-lanceolate, 3–5 × 0.2–0.4 cm, glabrescent, base attenuate, margin entire to distinctly serrate, apex acute-apiculate. Flowers inconspicuous, yellow, axillary, solitary, hairy. Pedicel subsessile, to 1 mm long. Sepals 5-lobed, yellow, 4 mm long, ovate, acuminate. Petals smaller than sepals, caducous. Stamens 5, only 2 fertile. Ovary globose, glabrous. Capsule 5 mm across, 3-lobed, subglobose. Seeds *ca.* 9, three per locule, ellipsoid, 2 mm, longitudinally rugose.

Rare, in forest of Idupulapaya, Fl. & Fr.: July – Oct. Idupulapaya (KDP), *KRKS, BSS & TP* 40318.

COCHLOSPERMACEAE

COCHLOSPERMUM Kunth

Cochlospermum religiosum (L.) Alston in Trimen, Handb. Fl. Ceylon 6 (Suppl.). 14. 1931; FAP 1: 93. 1997. *Bombax religiosum* L., Sp. Pl. 512. 1753. *Cochlospermum gossypium* DC., Prodr. 1: 527. 1824; FBI 1: 190. 1872; FPM 1: 50. 1915.

Local name: *Kondagogu.*

Deciduous tree, 10–15 m high, young branches pilose. Leaves palmately lobed, lobes ovate or lanceolate; midlobe 2.5–10 cm; laterals 2–8 cm, tomentose below, sparsely tomentose above, base cordate, margin entire, apex acute to acuminate; petiole 6–20 cm, tomentose. Flowers yellow in terminal subcorymbs. Sepals 5, obovate, unequal, 1.5–3 × 1–3 cm, pubescent. Corolla 5-lobed, golden yellow, obovate, subequal, 4.5 × 3 cm, deeply emarginated. Stamens numerous in 8–10 clusters. Ovary globose, glabrous, 5-carpellate, 1-celled. Fruit capsule, pear-shaped, 5-valved. Seeds numerous, reniform to cochleate, 6 × 5 mm, brown, woolly.

Common, in open hilly areas of dry deciduous forest. Flowers appear after leaf shedding. Fl. & Fr.: Feb. – Sept. Kalasamudram (ATP), *KRKS & BSS* 39200.

The tree yields a valuable gum.

FLACOURTIACEAE

FLACOURTIA L' Herit.

Flacourtia indica (Burm. f.) Merr., Interpr. Rumph. Herb. Amb. 377. 1917; FAP 1: 95. 1997. *Gmelina indica* Burm. f., Fl. Ind. 132, t. 39. f. 5. 1768. *Flacourtia ramontchi* L' Herit., Stirp. Nov. 3: 59, tt. 30 & 30B. 1785; FBI 1: 193. 1872; FPM 1: 54. 1915. *F. sepiaria* Roxb., Pl. Coromandel. 1: 48, t. 68. 1795; FBI 1: 194. 1872; FPM 1: 54. 1915.

Shrub or small trees, to 8 m high; young branches sparsely brown hairy; stem thick brown. Leaves usually clustered towards apices on older branches, alternate, obovate, 5–9 × 2.5–5 cm, glabrous, base attenuate, margin crenate-serrate, apex acute-emarginate; petiole 1.5 cm. Flowers 3 mm across, pale green, solitary or few in axillary bracteate racemes. Calyx 5-lobed, slightly connate at base, ovate, obtuse, subglabrous outside, hairy inside. Male flowers: Stamens numerous. Female flowers: disk entire; ovary globular, somewhat attenuate at apex with 3–6 radiating structures, *ca.* 1 mm long, terete, styles thick, little or no connate at base, with slightly bilobed recurved stigmas. Berry ellipsoid, subglobose, red, 4–10 mm across. Seeds trigonous, obovoid, 3–5, pale yellow.

Common, in deciduous forests. Fl. & Fr.: Apr. – Aug. Batrepalli (ATP), *KRKS & BSS* 39589.

POLYGALACEAE

POLYGALA L.

Key to Species

1a. Flowers rose; caruncle with broad membranous winged appendages.......... **P. javana**

1b. Flowers blue or yellow or pink; caruncle with shortly toothed or linear to narrowly oblong appendages.......... (2)

2a. Capsules orbicular or suborbicular.......... **P. chinensis**

2b. Capsules oblong, obliquely oblong, elliptic-oblong, ovate-oblong or rhomboid.......... (3)

3a. Wing sepals elliptic-oblong or elliptic-obovate, obtuse to subobtuse at apex; style widened at apex like a knob.......... **P. erioptera**

3b. Wing sepals obliquely ovate or ovate-elliptic, ovate-lanceolate, acute or acuminate at apex; style horse-shoe shaped or hook-like with wings at apex.......... (4)

4a. Petals greenish, tinged with pink, staminal sheath split half way into 3 bundles bearing 3 sessile anthers in each side-bundle and 2 anthers on short free filaments in the mid-bundle; seeds oblong.......... **P. wightiana**

4b. Petals yellow, staminal sheath bearing 8 free filaments; seeds elliptic-obovate or oblong-obovate.......... (5)

5a. Racemes up to 2 cm long; wing sepals glabrous.......... **P. arvensis**

5b. Racemes up to 13 cm long; wing sepals pubescent.......... **P. elongata**

Polygala arvensis Willd. Sp. Pl. 3(2): 876. 1802; FAP 1: 99. 1997. *P. chinensis* auct. non. L. 1753; FBI 1 : 204. 1872; FPM 1: 58. 1915.

Annual, prostrate or erect herb, to 25 cm, sparsely hairy. Leaves elliptic, linear-oblanceolate, 2–2.5 × 0.5–0.6 cm, hairy, base subacute, margin entire, ciliate, apex obtuse- apiculate; petiole 2 mm. Flowers yellow 2 cm across in supra axillary raceme, to 2 cm long, sometimes solitary. Sepals ovate; outer lobes winged, obliquely ovate or falcate, puberulous, acute. Corolla yellow; wing petals broadly obovate, to 2 mm; keels equal to laterals, hooked, appendages fimbriate. Staminal sheath to 1.5 mm; filaments 1.5 mm. Ovary obovoid, turgid; style incurved and apex somewhat hooked. Fruit capsule, obliquely oblong-elliptic. Seeds 2, black, ellipsoid-obovate, villous, caruncle hooded, ¼ covered, 3-lobed.

Common, in scrub forest and wastelands. Fl. & Fr.: July. – Oct. Bhumaiahgari palle (KDP), *KRKS & KRS* 37604; Idupulapaya (KDP), *KRKS & BSS* 40197.

A. *Polygala arvensis*, B. *Polygala wightiana*, C. *Polycarpaea aurea*, D. *Portulaca pilosa*, E. *Portulaca tuberosa*, F. *Abutilon pannosum*

Polygala chinensis L., Sp. Pl. 2: 704. 1753; FBI 1: 204. 1872; FAP 1: 100. 1997. *P. glomerata* Lour., Fl. Cochinch. 426. 1790; FBI 1: 206. 1872.

Erect herb, up to 50 cm high. Leaves elliptic, elliptic-oblanceolate or obovate, 0.6–2.5 × 0.1–0.4 cm, glabrous or sparsely pubescent, base acute, margin entire, apex apiculate; petiole 1–2 mm. Flowers pale blue, in few-flowered, supra-axillary racemes, up to 1.5 cm long. Outer sepals 1.5–3 × 0.8–1 mm, unequal, ovate to ovate-lanceolate, acute, long-mucronate; wing sepals 6 × 3 mm. Lateral petals almost as long as the keel, obovate, keel rounded, hood-shaped with 2 narrowly oblong membranous appendages and a short tooth. Capsule suborbicular, 5–7 mm, pilose, ciliate along narrow marginal wings. Seed obovoid, black, white silky hairy; caruncle hood-shaped, with 2 narrowly oblong membranous appendages and a short tooth, yellowish.

Common, in wastelands and foot hills of forest. Fl. & Fr.: Jun. – Sept. Kalasamudram (ATP), *KRSK & BSS* 39840.

Polygala elongata Klein in Willd., Sp. Pl. 3: 879. 1802; FBI 1: 203. 1872; FPM 1: 58. 1915; FAP 1: 100. 1997.

Annual erect herb, sometimes ascending, up to 40 cm, pubescent on stem. Leaves linear, oblong to lanceolate, 3–4.5 × 0.2–0.4 cm, glabrous above, pubescent below, base attenuate, margin entire, apex obtuse-apiculate. Flowers yellow, 10 × 5 mm, in extra axillary receme to 13 cm long. Calyx lobes elliptic to ovate; wing sepals obliquely ovate or elliptic, glabrescent, margins ciliate, apically acute. Corolla yellow, lateral lobes subequal; keels boat-shaped; crest 2-lobed, each with 4-lobed appendages. Staminal sheath to 3 mm long; jointed at different levels. Capsule elliptic or rhombate, ciliate. Seeds 2, elongate, black, hispid. Caruncle hooded at apex, 3-lobed.

Common, in scrub forests and grass lands. Fl. & Fr.: Aug – Jan. K. K. Kottala (KDP), *KRKS, KRS & TP* 37840.

Polygala erioptera DC., Prodr. 1: 326. 1824; FBI 1: 203. 1872; FPM 1: 59. 1915; FAP 1: 100. 1997.

Annual, erect or ascending herb, to 25 cm high, pubescent. Leaves linear, linear–oblong, 1–2 × 0.2–0.3 cm, pubescent, base narrowly cuneate, margin entire, apex obtuse, apiculate. Flowers pink to purple, 5 mm long in extra axillary, condensed raceme to 1 cm long, 5–8-flowered. Calyx lobes elliptic or obovate, unequal, pubescent; wing sepals oblong or obovate, 5 × 2.5 mm, pilose without, oblique at base, apex rotund, margins often ciliate. Corolla rose-pink; lateral lobes ovate, 2 mm, shorter than keels; crests 2-partite. Staminal sheath to 3 mm, filaments to 1.5 mm. Ovary ovoid, 1 mm long, apex long ciliate; stigma capitate. Capsule oblong, elliptic-oblong. Seeds 2, densely downy-woolly, black, caruncle hooded at tip, 3-lobed.

Common, in grass lands. Fl. & Fr.: Throughout year. Bhumaiahgari palle (KDP), *KRKS & KRS* 37000.

Polygala javana DC., Prodr. 1: 327. 1824; FBI 1: 201. 1872; FPM 1: 58. 1915; FAP 1: 101. 1997.

Annual, erect, glabrescent herb, to 50 cm high. Leaves obovate-oblanceolate, 1–2 × 0.4–0.5 cm, glabrescent-tomentose, base cuneate-attenuate, margin entire, apex obtuse-emarginate, mucronulate; petiole 2 mm. Flowers rose in extra axillary and terminal, to 12 cm long raceme. Calyx 5-lobed, unequal, puberulous, yellow; outer sepals broadly ovate or lanceolate to elliptic-oblong, puberulous; wing sepals broadly ovate, triangular, cordate at base, mucronate at apex. Corolla rose, lobes 5, 3 lobes adnate at base for half the length; lateral petals ovate or elliptic-oblong; keel petal 4–5 mm long, pale yellowish or green with violet multifiliform *ca.* 2 mm long crest at back of apex. Stamens 8, pinkish; sheath 2–4 mm long, encircling the style. Capsule broadly oblong or suborbicular, notched at apex, 4–6 × 3–5 mm, puberulous. Seeds 2, densly stiff-pubescent, oblong to ovoid, ovate, black, caruncle hooded, 2-appendaged, and 1-toothed, appendage apically spongy, basally membranous, covered on tip, 3-armed.

Common, in scrub forest and open dry deciduous forest. Fl. & Fr.: Sept. – Jan. K. K. Kottala (KDP), *KRKS & KRS* 37666.

Polygala wightiana Wall. ex Wight & Arn., Prodr. Fl. Ind. Orient. 38. 1834; FPM 59. 1915. *P. elongata* auct. non Klein 1802; FBI 1: 203. 1872, *p.p.*

Erect herb, to 40 cm high, often dichotomous at apex, glaucous. Leaves subsessile, linear-oblong or narrowly oblanceolate, 3–3.5 × 0.1–0.3 cm, chartaceous, glabrous, base narrowed, margin recurved, apex obtuse, mucronate. Flowers yellow, 3–4 mm long in extra-axillary leaf-opposed, 1–8 cm long racemes. Bracts *ca.* 1 mm long. Calyx persistent; outer lobes unequal, ovate-lanceolate, 2–2.5 × 0.5–1.2 mm, glabrous; wing sepals obliquely ovate, falcate, 5-nerved, glabrous. Corolla lobes 2–2.5 mm long, greenish, tinged with pink; middle lobe keeled, up to 3.5 mm long, with a crest of linear, forked appendages on back; lateral lobes obovate, clawed and hairy at base. Staminal sheath free, *ca.* 1.5 mm long. Ovary ovoid, glabrous, ciliate along margins; style slender, curved, hooked at apex with membranous wings, up to 3.5 mm long; stigma dilated. Capsules overtopped by wing sepals, oblong, emarginate at apex, margined, 2.5–4 × 1.5–2.5 mm, glabrous. Seeds oblongoid, truncate at apex, 2–3 × 0.5–1 mm, black, densely silky; caruncle hood-shaped, with 3 curved small toothed appendages.

Rare, in dry deciduous forest. Fl. & Fr.: Aug. – Dec. Kalasamudram (ATP), *KRKS* 41848.

CARYOPHYLLACEAE

POLYCARPAEA Lam. *nom. cons.*

1a. Herbs woody, whitish pubescent .... **P. aurea**

1b. Herbs herbaceous, pinkish pubescent .... **P. corymbosa**

Polycarpaea aurea Wight & Arn., Ann. Nat. Hist. Series 1, 3: 91. 1839; FPM 1: 65. 1915; FAP 1: 104. 1997. *P. corymbosa* Lam. var. *aurea* (Wight & Arn.) Wight, Ill. 2: 44. t. 110. 1850; FBI 1: 245. 1874.

Local name: *Rathirajuma.*

An erect much branched, perennial woody herb, to 60 cm high, white-pubescent; much branched and suffruiticose; branches terete. Leaves opposite, sessile, linear, 1 × 0.1 cm, pilose, base truncate, margin entire, apex acute, stipules fimbriate. Flowers pink or white in terminal irregular corymbs or cymes. Calyx 5-lobed, free, ovate-lanceolate, entire, acute, *ca.* 2 × 5 mm, reddish brown (calyx longer than corolla, to 1 mm in length). Corolla 5-lobed, free, shining, oblong-obovate, obtuse, *ca.* 1.2 × 0.5 mm. Stamens 5. Capsule, ovate-elliptic, oblong, *ca.* 1.2 × 1 mm, small stiped, 3-valved. Seeds many, reniform, light brown, minute, elongated.

Occasional, in rocky crevices in dry deciduous forest. Fl. & Fr.: Aug. – Feb. Rachakuntapalle (KDP), *SSR & TP* 33624.

Polycarpaea corymbosa (L.) Lam., Tabl. Encycl. 2: 129. 1797; FBI 1: 245. 1874; FPM 1: 65. 1915; FAP 1: 104. 1997. *Achyranthes corymbosa* L., Sp. Pl. 205. 1753.

Local name: *Rajuma, Bommasari.*

Annual, erect, much branched herb, to 30 cm high; stem simple, pinkish pubescent. Leaves whorled in variable sizes, sometimes red, sessile, linear, oblong, flat, 1–2 × 0.1–0.2 cm, base truncate, margin entire, apex acute; stipules membranous, fimbriate. Flowers reddish-pink, in axillary and terminal corymbs or helicoid cymes. Calyx 5-lobed, free, lanceolate, 2–5 mm long (calyx much longer than corolla). Corolla 5-lobed, forming a ring with the stamens at the base, *ca.* 1 mm long. Capsule, oblong, subglobose, glabrous 1 mm across. Seeds numerous, reniform, subglobose, compressed, light brown, 0.3 mm.

Common, in agriculture fields and wastelands in open rocky soils. Fl. & Fr.: Jul. – Dec. Chinnapalli (ATP), *KRKS & KRS* 37813.

PORTULACACEAE

PORTULACA L.

Key to Species

1a. Roots tuberous. – stem woody, light yellow flowers .... **P. tuberosa**

1b. Roots not tuberous .... (2)

2a. Erect or decumbent woody stem; leaves alternate, linear-oblong; flowers rose .... **P. pilosa**

2b. Prostrate or ascending herbs; leaves sub-opposite, oblong or obovate; flowers yellow .... (3)

3a. Nodes with roots and scales or hairs .... **P. quadrifida**

3b. Nodes without roots and scales or hairs .... **P. oleracea**

Portulaca oleracea L., Sp. Pl. 445. 1753; FBI 1: 246. 1874; FPM 1: 66. 1915. *P. suffruticosa* Thwaites Enum. Pl. Zeyl. 24. 1858.

Key to Varieties

1a. Leaves linear; seeds 55–75 per capsule .... var. **linearifolia**

1b. Leaves obovate; seeds 20–25 per capsule .... var. **oleracea**

Portulaca oleracea L. var. **linearifolia** Sivarajan & Manilal, New Botanist. 4: 30. 1977; FAP 1: 106. 1997.

Local name: *Pappukura, Golikura.*

Annual, prostrate, succulent herb, to 20 cm long; stem glabrous, shiny. Leaves alternate to sub-opposite, fleshy, spirally arranged at apex, oblong-obovate, 1–2 × 0.3–0.5 cm, glabrous, base attenuate, margin entire, apex obtuse; petiole 1.5 mm long. Flowers light yellow in terminal clusters, 0.7 cm across. Calyx 8 × 5 mm across. Corolla 4-lobed, 6 × 4 mm. Stamens mumerous, clustered. Fruit capsule, globose, glabrous, oblong-ovoid, 0.3 cm across. Seeds above 50, black or brown, reniform, shiny, 0.5 mm diameter.

Common, in moist localities of agriculture and fallow lands. Fl. & Fr.: Throughout the year. K. K. Kottala (KDP), *KRKS & KRS* 36416.

Portulaca oleracea L., var. **oleracea;** FAP 1: 106. 1997.

Local name: *Godla payalaku, Golikura.*

Annual, erect or decumbent, succulent herb, to 25 cm; stem glabrous, shiny. Leaves basally alternate and terminally opposite, spirally arranged at apex, broadly obovate, 2–2.5 × 0.6–1.5 cm, glabrous, base attenuate, margin entire, apex obtuse; petiole 3 mm long. Flowers yellow, terminal, solitary or in clusters. Calyx 8 × 5 mm across. Corolla lobes 4, 6 × 4 mm. Capsule globose, glabrous, 0.3 cm. Seeds 20–30, black or brown, shiny, 0.5 mm diameter.

Common, in moist localities of agriculture lands and wastelands. Fl. & Fr.: Throughout the year. Kalasamudram (ATP), *KRKS & BSS* 39128.

Portulaca pilosa L., Sp. Pl. 445. 1753; subsp. **pilosa**; FAP 1: 106. 1997. *P. tuberosa* auct. non Roxb., FBI 1: 247. 1874; FPM 1: 66. 1915.

Perennial, erect, fleshy, radially spreading herb, to 30 cm high; stem hard with brown coloured scaly bark, with woody root stock. Leaves alternate, sessile or subsessile, spiral or crowded at apex, linear-lanceolate, 0.5–1.5 × 0.1–0.2 cm, glabrous, base attenuate, margin entire, apex subacute, pilose hairs at leaf base. Flowers pink-red purple at branch apex, solitary, to 0.6 cm across. Calyx 2–5 × 1–4 mm across, lobes ovate. Corolla 3–7 × 2–10 mm across, lobes obovate. Stamens 10–16. Capsule ovoid, globose, 3 mm across. Seeds numerous, dull brown, 0.6 mm, shiny.

Occasional, in sandy soils. Fl. & Fr.: Jan. – Sept. Rachakuntapalli (KDP), *SSR & TP* 33618; Kalasamudram (ATP), *KRKS & BSS* 39181.

Portulaca quadrifida L., Mant. Pl. 1: 73. 1767; FBI 1: 247. 1874; FPM 1: 66. 1915; FAP 1: 106. 1997.

Local name: *Sannapayalaku.*

Annual creeping herb, to 25 cm long, rooting at nodes, stem pink, glabrous, inflated at nodes, succulent. Leaves opposite, succulent, elliptic or oblong-lanceolate, 0.8–1 × 0.1–0.2 cm, glabrous, base attenuate, margin entire, apex acute, white hairy at base. Flowers yellow in terminal clusters or solitary, 2 mm across. Calyx 2-lobed, oblong, to 3 mm long, equal. Corolla 4-lobed, obovate, 5 × 2 mm, basally subconnate. Capsule globose-oblong, ovoid, 2 mm across. Seeds numerous, rugose, black-brown, minute, shiny.

Occasional, in cultivated fields and open forests. Fl. & Fr.: Throughout the year. Thummalapalle (KDP), *KRKS & KRS* 36917.

Portulaca tuberosa Roxb., Fl. Ind. 2: 464. 1832. *P. suffruticosa* Wall. ex Wight & Arn., Prodr. Fl. Ind. Orient. 356. 1834; FBI 1: 246. 1834.

Perennial, erect or decumbent herb, up to 30 cm high, with thick tuberous root. Leaves spiral, oblong, 4-28 × 0.5-5 mm, apex obtuse or rounded, axillary hairs 1-18 mm long. Flowers solitary or 2-4 in capitula, subtended by an involucre or 3-8 cauline leaves. Sepals 2, 2-6 × 1-4 mm. Petals 4-6, bright yellow, 2.5-12 × 1.8-11 mm, obovate, spreading. Stamens 10-25. Styles 3-5-armed. Capsule ovoid-globose, 2-3 mm in diam.

On rocky laterite and sandy soils. Fl. & Fr.: Oct.- Dec. Kalasamudram RF (ATP), *EC* 2900.

MALVACEAE

Key to Genera

1a. Styles as many as carpels or styles undivided and stigmas entire .... (3)

1b. Style twice as many as carpels .... (2)

2a. Mericarp usually glochidiate, rarely smooth; leaves with nectaries on midrib beneath .... **Urena**

2b. Mericarps muricate or often with 1–3 retrorsely barbed awns at the apex, nerves prominent; leaves without nectaries, fruit capsule.- (mericarps vermiform in *P. procumbens*)- seed one per mericarp .... **Pavonia**

3a. Fruit a capsule; staminal column 5-toothed or armed at apex, antheriferous throughout or a major portion of it or tipped with clusters of anthers .... (4)

3b. Fruit a schizocarp; breaking into mericarps at maturity; staminal column without teeth, apex split up into numerous filaments .... (8)

4a. Carpels remain attached to each other and to central axis at maturity; stipules simple .... (5)

4b. Carpels separate from central axis at maturity; stipules laciniate or completely divided into 2–3 segments .... **Decaschistia**

5a. Style distally 5-lobed, with branches spreading at maturity; stigmas more or less capitate or globose .... (6)

5b. Style unbranched; stigmas ribbed or lobed .... **Gossypium**

6a. Calyx irregularly 2–3-lobed, spathaceous, deciduous .... **Abelmoschus**

6b. Calyx regularly 5-lobed, not spathaceous, persistent .... (7)

7a. Capsules winged .... **Fioria**

7b. Capsules not winged .... **Hibiscus**

8a. Seeds 2 or more in each mericarp .... (10)

8a. Seed one in each mericarp .... (9)

9a. Epicalyx present .... **Malvastrum**

9b. Epicalyx absent .... **Sida**

10a. Stems erect, stout; carpels slightly inflated or not; mericarps more or less pointed, mucronate or aristate at apex, wall of mericarps thick .... **Abutilon**

10b. Stems ascending, weak; carpels prominently inflated at maturity; mericarps rounded at apex, wall of mericarps thin and papery .... **Herissantia**

ABELMOSCHUS Medik.

Key to Species

1a. Prostrate under shrub; epicalyx segments 5–6 .... **A. ficulneus**

1b. Erect undershrub; epicalyx segments 10–12 .... **A. angulosus**

Abelmoschus angulosus Wall. ex Wight & Arn. Prodr. Fl. Ind. Orient. 53. 1834. FEG 1: 113. 2002. *Hibiscus angulosus* (Wall. ex Wight & Arn.) Steudel ex Mast. Fl. Brit. India 1: 341. 1874. *Abelmoschus angulosus* var. *grandiflorus* Thwaites Enum. Pl. Zeyl. 26. 1858.

Erect undershrub, up to 2 m tall, hirsute with simple and stellate hairs some what stinging. Leaves palmately 5-lobed, 3–12 cm across, elliptic to orbicular, base cordate, 5–7-nerved, margin dentate to crenate, apex acute; petiole longer than lamina, 15 cm long, villous; leaves gradually reduced towards apex of the stem and as raceme. Flowers light yellow in terminal axils of reduced leaves, up to 7 flowers in each axil. Pedicel 1–3 cm long, pubescent. Epicalyx 10 or 12 segments, hirsute, *ca.* 2 cm long, linear. Calyx lobes 5, *ca.* 3 cm long, ovate, densely villous without, pubescent within. Corolla light yellow with deep purple centre, lobes 8 × 6 cm, glabrous. Capsule 3 – 5 × 1.5 – 2 cm, ovoid to oblong, acute to acuminate, densely hispid.

Very rare, in open dry deciduous forests among rocks. Fl. & Fr.: Oct.- Jan. Kalasamudram RF (ATP), *KRKS* 41872.

Note: This species is a new distributional record for Andhra Pradesh.

Abelmoschus ficulneus (L.) Wight & Arn. ex Wight, Cat. No. 14. 1833; FAP 1: 116. 1997. *Hibiscus ficulneus* L., Sp. Pl. 695. 1753; FBI 1: 340. 1874; FPM 1: 97. 1915.

Prostrate undershrub, up to 2 m long, stem unbranched, simple hairy, rarely prickly with bulbous base. Leaves orbicular, palmately 3–5-lobed, 5–12 cm across, sparsely stiff, simple and stellate hairy on both surfaces, base cordate, lobes 2–8 × 1–4 cm, obovate to spatulate, margin serrate, apex rounded; petioles 1.5–15 cm long; stipules 4–10 mm long. Flowers white becoming pink with deep purple centre, axillary, solitary or in terminal racemes by decrescense of upper leaves. Epicalyx segments 5–6, linear to lanceolate, acute, hirsute, caducous. Calyx in bud lageniform, 3 mm long, lobes linear. Corolla 1.5–2 cm in across; petals 2–3 × 1–2 cm, obovate, rounded at apex, glabrous. Capsule 2–4.5 × 1.5–2 cm, ovoid, 5-angular, obtuse, tomentose; valves hirsute within. Seeds *ca.* 3 mm across, globular, stellate-tomentose in concentric rings, blackish.

Rare in agriculture lands and wastelands, particularly in black soils. Fl. & Fr.: Oct. – Dec. Gollala guduru (KDP), *KRKS* 41968; Chinna kudala (KDP) *KRKS* & *BSS* 40463.

ABUTILON Mill.

Key to Species

1a. Stems, petioles and pedicels with dense long patent simple hairs, few minute stellate hairs and short viscid glandular hairs; corolla orange yellow with a purple centre .... **A. hirtum**

1b. Stems, petioles and pedicels with dense stellate-pubescent, few simple hairs; corolla with or without purple centre .... (2)

2a. Corolla yellow, calyx lobes deltoid to ovate; mericarps densely hairy on dorsal margin; leaves orbicular to rotund-cordate, mericarps rounded at apex .... **A. pannosum**

2b. Corolla yellow with purple centre (ssp. of *A. indicum*); calyx lobes ovate to lanceolate; mericarps dorsally and ventrally stellate-hairy; leaves ovate-cordate, mericarps acute at apex .... **A. indicum**

Abutilon hirtum (Lam.) Sweet, Hort. Brit. ed. 1, 53. 1826; FPM 1: 91. 1915; FAP 1: 118. 1997. *A. graveolens* (Roxb. ex Hornem.) Wight & Arn. ex Wight, Cat. 13. 1833; FBI 1: 327. 1874, incl. var. *hirtum. Sida graveolens* Roxb. [Hort. Beng. 50. 1814, nom. nud.] ex Hort. Hafn. Suppl. 77. 1819.

Erect under shrub, up to 2 m high, much branched, somewhat viscid with unpleasant smell; stems, petioles and pedicels with simple long hairs. Leaves broadly ovate, 5–13 × 5–11 cm, pubescent with minute stellate hairs and with long simple hair below, pubescent above, base cordate, margin irregularly crenate-dentate, apex acute; petiole 3–20 cm long, hirsute; stipule 0.5–1 cm long. Flowers yellow with purple centre, axillary, solitary. Calyx 7–10 mm diam., lobes campanulate, divided from below the middle, lobes ovate, acute to acuminate, 5–10 × 4–6 mm. Corolla yellow with a purple centre; lobes longer than calyx lobes, glabrous without, stellate within, apex rounded, base emarginated, ciliate. Staminal column 5–7 mm, yellow, basal part conical and stellate-hairy. Schizocarp 1–2 cm across, sub-globose, indented at apex; mericarps 15–20, separating, 10–15 mm high, radially 7–10 mm, broadly ovate-reniform, acute at apex, 3-seeded. Seeds *ca.* 2.5 mm across, brownish black, reniform, punctuate with minute stellate hairs,.

Occasional, as a hedge along road sides and waste lands. Fl. & Fr.: Aug. – Dec. Idupulapaya (KDP), *KRKS* 39080.

Note: Leaf and schizocarp of *A. hirtum* and *A. pannosum* are almost similar in appearance but hairy nature, colour of flower and apex of the fruit are the key features to distinguish from *A. pannosum* in the field.

Abutilon indicum (L.) Sweet, Hort. Brit. ed. 1, 54. 1826; FBI 1: 326. 1874; FPM 1: 91. 1915; subsp. **indicum**; Borssum, Blumea 14: 170. 1966; FAP 1: 119. 1997. *Sida indica* L., Cent. Pl. 2: 26. 1756. *A. indicum* var. *populifolium* (Lam.) Wight & Arn. ex Mast. in Hook. f., Fl. Brit. India 1: 326. 1874. *Sida populifolia* Lam., Encycl. 1: 7. 1783.

Local name: *Thuttura benda, Erribenda.*

Perennial shrub to 1.5 m high, young branches with stellate hairs. Leaves simple, cordate, 1–2 × 0.5–2.5 cm, stellate hairy, base cordate, margin crenate-dentate, apex acute, petiole 1–2 cm, simple whitish hairy to 1 mm. Flowers yellow, axillary and terminal in 2-flowered cymes. Calyx 0.7 cm across, base connate, lobes acuminate. Corolla yellow, lobes obovate to 1 cm, apex truncate. Staminal column 0.5 cm long. Schizocarp 1.5–2 cm across; mericarps 10–15, longer than calyx, to 1 cm, shortly acute to mucronate at apex, stellate hairy. Seeds 1–2 per mericarp, 2 × 3 mm, brown, reniform, stellate hairs at microphylar end, simple hairs speading, verrucose.

Common, in open places and road sides. Fl. & Fr.: Mar. – Sept. Thummalapalle (KDP), *KRKS & KRS* 33673; Idupulapaya (KDP), *KRKS* 39080.

Abutilon pannosum (G. Forst.) Schlect., Bot. Zeit. 9: 828. 1851; FAP 1: 120. 1997. *Sida pannosa* G. Forst., Comment. Soc. Reg. Sci. Gotting ser. 2: 9. 1787. *Abutilon glaucum* (Cav.) Sweet, Hort. Brit. ed. 1. 54. 1826; FPM 1: 91. 1915. *Sida glauca* Cav., Icon. 1: 8, t. 11. 1791. *Abutilon muticum* (Del. ex DC.) Sweet, Hort. Brit. ed. 2. 65. 1830; FBI 1: 327. 1874. *Sida mutica* Del. ex DC., Prodr. 1: 470. 1824.

Erect shrub, up to 2 m high, much branched, somewhat viscid with unpleasant smell; stems, petioles and pedicels covered with long patent and short simple, and minute stellate and short glandular hairs. Leaves orbicular to broadly ovate, 2.5–12 × 3.5–13 cm, pubescent with minute stellate hairs and with long simple and glandular hairs, base cordate, margin irregularly crenate-dentate, apex acute; petiole 3–20 cm long, hirsute; stipule 0.5–1 cm long. Flowers yellow, axillary, solitary or cymes. Calyx 7–10 mm diam., lobes 5–10 × 4–6 mm, campanulate, divided to the middle, lobes ovate, acute to acuminate, reflexed in fruit and encloses the fruit. Corolla orange-yellow; lobes longer than calyx lobes, glabrous, apex rounded, base emarginated, ciliate. Staminal column 5–7 mm, yellow, basal part conical and stellate-hairy. Schizocarp 2–3 cm across, globular and stellate hairy, blunted apical margin; mericarps 18–23, separating, 10–15 mm high, radially 7–10 mm, broadly ovate-reniform, somewhat rounded at apex, 2–3-seeded. Seeds *ca.* 2.5 mm across, reniform, punctate by minute stellate hairs, brownish black.

Occasional on road sides, wastelands and scrub forests. Fl. & Fr.: Oct. – Jan. Mabbuchintalapalle (KDP), *KRKS* 41840; *KRKS & BSS* 40063, *KRKS* 269 (HJCB-E).

DECASCHISTIA Wight & Arn.

Decaschistia rufa Craib, Bull. Misc. Inform. Kew 1912: 35 1912; V.V. Sivarajan & A.K. Pradeep, Malvaceae of Southern Peninsular India Mono. 22. 1996. *D. cuddapahensis* T. K. Paul & Nayar, Geobios New rep. 2: 156. 1983; FAP 1: 121. 1997.

Perennial shrub, to 1.3 m high, young stem with brown stellate hairs. Leaves simple, 4–6.5 × 2.2–5.5, upper dark brown velvety, lower light brown or grey velvety, base sub-cordate, margin basally entire, apically distinctly denticulate, serrate, apex acute–obtuse, 5–7-nerved at base; stipules linear-lanceolate, to 0.5 cm; petiole up to 2 cm. Flowers yellow, solitary, axillary, pedicel 2–3 cm. Capsule globose, enclosed with calyx, densely stellate hairy, 5-locular. Seeds one per locule, 2 mm, reniform, sparsely hairy.

Rare in Batrepalli forest. Fl. & Fr.: Jul. – Mar. Batrepalli (ATP), *KRKS & BSS* 39570.

FIORIA Mattei

Fioria vitifolia (L.) Mattei, Bot. Ort. Bot. Palermo n.s. 2: 71. 1916; FAP 1: 122. 1997. *Hibiscus vitifolus* L., Sp. Pl. 696. 1753; FBI 1: 338. 1874; FPM 1: 98. 1915.

Local name: *Keru patti.*

Perennial subshrub, to 1 m high, yellowish stellate hairy. Leaves 3–5-lobed or some times simple, 2–7 × 1.5–5 cm, stellate hairy, base cordate, margin dentate or crenate, apex acute, some times acuminate, 5-nerved at base; petiole 1.5–10 cm long. Flowers yellow with dark purple tinge in centre, axillary, solitary. Epicalyx linear, to 1 cm, calyx longer than the fruit, to 1.5 cm. Calyx campanulate, 5-lobed, lobes 5–15 × 6–10 mm, ovate to deltoid. Corolla yellow; petals 2.5–5 × 1–3 cm, obovate, obtuse or rounded at apex. Staminal column glabrous, antheriferous throughout. Ovary ovoid, obtuse, 5-angular, 5-loculed. Capsule 1.5–2 cm across, 5-winged, beaked, beaks *ca.* 3 mm long, wings bristly, very close at apex, splitting. Seeds 2–5 per locule, brown colour, 2–3 mm, reniform veruculose.

Common, in agriculture lands and foot hills of dry deciduous forests. Fl. & Fr.: Apr. – Nov. Idupulapaya (KDP), *KRKS* & *SSR* 39090; Batrepalli (ATP), *KRKS* 36962.

GOSSYPIUM L.

Gossypium herbaceum L., Sp. Pl. 2: 693. 1753.

A. *Decaschistia rufa*, B. *Hildegardia populifolia*, C. *Melhania hamiltoniana*, D. *Grewia orientalis*, E. *Biophytum reinwardtii*, F. *Naringi crenulata*

Local name: *Patthi.*

Shrub to 1.5 m high. Leaf palmatly lobed. Flowers yellow, axillary, solitary. Friut ellipsoid. Seeds black, ovate, epidermal hairs 1.5–2 cm long.

Cultivated for cotton as crop. Sometimes as an escape. Fl. & Fr.: Through the year. Gollalaguduru (ATP), *KRKS & BSS* 41977.

HERISSANTIA Medik.

Herissantia crispa (L.) Medik., Phil. Bot. 1: 90. 1789. *Sida crispa* L., Sp. Pl. 685. 1753. *Abutilon crispum* (L.) Medik., Malv. 29. 1787; FBI 1: 327. 1874; FPM 1: 91.1915; FAP 1: 117. 1997.

Local name: *Nela benda.*

Annual, often trailing, vine-like, weak-stemmed herb, to 1.5 m long, branchlets densely stellate hairy, pubescent. Leaves ovate, 2–10 × 1.5–8 cm, stellate-hairy, base cordate, margin crenate to serrate, apex acuminate, basally 7–9-nerved. Flowers yellow, 11 mm across, axillary, solitary, pedicel shorter than petiole, to 2 cm long. Calyx *ca.* 8 mm across, campanulate, 5-lobed; lobes ovate to linear-triangular, densely stellate-hairy. Corolla as long as or slightly shorter than calyx, petals broadly obovate, *ca.* 10 mm long, glabrous. Staminal column 2–2.5 mm long, glabrous. Schizocarp, 0.8 cm across, mericarps to 10, globular, carpels inflated, rounded at apex, stellate, 2–3-seeded. Seeds 1.5–2 mm across, curved, reniform, covered with simple hairs, brownish black.

Common, in cultivated lands and foot hills of dry deciduous forests. Fl. & Fr.: Mar. – Jan.Thummalapalle (KDP), *KRKS & KRS* 33690.

HIBISCUS L. *nom. cons.*

Key to Species

1a. Leaves deeply 3-lobed; petiole 1–2 cm long; fruit elliptic-ovoid; herbs .... **H. lobatus**

1b. Leaves shortly lobed or simple; petiole 2–6 cm long; fruit globose; subshrubs .... **H. micranthus**

Hibiscus lobatus (Murray) Kuntze, Revis. Gen. Pl. 3, 2: 19. 1898; FAP 1: 124. 1997. *Solandra lobata* Murr., Commentate Soc. Reg. Sc. Goetting 6: 20, t. f. 1785. *Hibiscus solandra* L., Herit., Stirp. Nov. 1: 103, t. 49. 1788, *nom. illeg.*; FBI 1: 336. 1874; FPM 1: 98. 1915.

Local name: *Pottibenda.*

Annual, erect, pubescent herb, to 70 cm high; stem pubescent with short simple hairs and rarely mixed with stellate hairs. Leaves 3-lobed, 2–9 × 1.5–7.5 cm, upper leaves lanceolate to linear, stellate hairy, lobes 3-5 cm, middle lobe slightly longer than lateral lobes, base cordate, margin crenate or coarsely serrate, apex acute-acuminate; petiole 1–2 cm. Flowers white, axillary, solitary; bud 1 × 0.5 cm; epicalyx connate, ovate, to 1 cm. Calyx campanulate to rotate, 5 – 8 mm across, 5-fid to

5-parted, lobes 5–10 × 1–2 mm, accrescent, up to 1.2 × 0.3 cm, deltoid to lanceolate, hispid without. Corolla white or yellow, 1.3 – 1.8 cm in diam.; petals 10–15 × 6–10 mm, obovate, glabrous. Staminal column *ca.* 6 mm long, antheriferous throughout. Capsule ovoid, oblong 5-loculed, valves smooth inside, locules 3–4-seeded. Seeds 1.5 × 1.3 mm, tetragonous to globose, tubercled or verrucose, glabrous, black.

Occasional, in dry deciduous forest and agriculture lands. Fl. & Fr.: Jul. – Nov. Somulavaripalli (ATP), *KRKS & KRS* 37811.

Hibiscus micranthus L. f., suppl. Pl. 308. 1781; FBI 1: 335. 1874; FPM 1: 97.1915. *H. ovalifolius* (Forssk.) Vahl, Symb. bot. 1:50.1970; FAP 1: 125. 1997; Raja Kullayiswamy *et al.*, J. Econ. Taxon. Bot. 39: 346. 2015.

Local name: *Chittipottikaya, Nityamalli.*

Erect undershrub, to 2 m high, branches with stellate hairs. Leaves ovate to oblong, 0.5–3.5 × 0.3–2 cm, scabrid, stellate hairy, base rounded, margin denticulate-serrate, apex acute, basally 5-nerved; petiole 2–6 mm. Flowers white, axillary, solitary, epicalyx 2 mm, ovate; pedicel 1.2 cm, scabrous appressed hairy. Calyx divided up to the middle, lobes *ca.* 5 mm long, stellate-hairy without and on apical portion within. Corolla 0.6 – 1.2 cm across, purplish-white or pink; petals *ca.*1.2 × 0.4 cm, oblong-obtuse. Capsule globose, dehiscing into 5 valves, smooth without, glossy within. Seeds reniform, 2–4 per locule, black, reniform, silky hairy.

Common, in forests and agriculture lands. Fl. & Fr.: Feb. – Nov. Rachakuntapalle (KDP), *SSR & TP* 33605.

MALVASTRUM A. Gray *nom. cons*

Malvastrum coromandelianum (L.) Garcke, Bonplandia 5: 295. 1857; FPM 1: 88. 1915; FAP 1: 128. 1997. *Malva coromandeliana* L., Sp. P1. 687. 1753. *Malvastrum tricuspidatum* (R. Br.) A. Gray, Pl. Wright 1: 16. 1852; FBI 1: 321. 1874. *Malva tricuspidata* R. Br. in Aiton f., Hort. Kew. ed. 2.4: 210. 1812. *nom. illeg.*

Annual, erect herb or undershrub, to 40 cm, 4-armed appressed stellate hairy, stem woody. Leaves ovate to oblong, 1–3 × 0.4–1.6 cm, simple appressed hairy on both sides, base obtuse-truncate, margin serrate or crenate, apex acute, 5-nerved at base; petiole 0.5–1.5 cm. Flowers yellow, solitary, axillary or in terminal clusters, pedicel *ca.* 3 mm, epicalyx very small, to 1 mm, linear. Calyx campanulate, lobes ovate to 1 cm long, acuminate. Corolla *ca.* 1.5 cm across; petals obliquely obovate, rounded-emarginate at apex, ciliate at base. Staminal column *ca.* 2.5 mm long, conical, glabrous. Schizocarps 2 mm, subglobose, persistent; mericarps 10 – 14, *ca.* 2 mm high, radially 2.5 mm; strongly curved, reniform, apical awns paired. Seeds 1 per mericarp, reniform, 1.5 mm across, brownish-black, glabrous.

Common, in hilly forest areas and weed in cultivated lands and wastelands. Fl. & Fr.: Throughout year. Thummalapalle. (KDP), *KRKS, KRS, SSR & TP* 33670.

Note: It is similar in appearance with *Sida acuta* but, it is distinguished by its epicalyx and hairy nature.

PAVONIA Cav. *nom.cons.*

Key to Species

1a. Erect subshrub, villose or stellate; epicalyx lobes more than 5; flowers whitish-pink .... (2)

1b. Decumbent subshrub, pubescent; epicalyx lobes 5; flowers yellow .... **P. procumbens**

2a. Mericarps slightly winged at angles; calyx connate up to middle .... **P. zeylanica**

2b. Mericarps wingless; calyx connate at base .... **P. odorata**

Pavonia odorata Willd., Sp. P1. 3: 837. 1800; FPM 1: 93. 1915: FAP 1: 129. 1997. *Hibiscus odoratus* Roxb. ex Wight & Arn., Prodr. Fl. Ind. Orient. 47. 1934.

Erect herb, to 50 cm high, with glandular hairs. Leaves simple, orbicular-ovate, 2–6 × 1.5–4 cm, scattered stellate hairy on both surfaces, obscurely 3-lobed, base cordate or less truncate, margin irregularly dentate, apex acuminate, 3–7- nerved at base, petioles 1–8 cm long; stipules *ca.* 2 mm long, linear, deciduous. Flowers white, axillary, solitary; epicalyx 10–12, linear, to 0.7 cm, longer than capsule and flower bud. Calyx lobes ovate connate at base to 0.4 cm. Corolla white to 1 cm, glabrous. Staminal column shorter than petals, glabrous. Ovary globose. Capsule globose-reniform, unarmed, glabrescent, mericarps 5, wingless. Seed 1 per mericarp, blackish-brown, reniform, papillose.

Common, in foot hills of dry deciduous forest. Fl. & Fr. : Aug. – Dec. Batrepalli (ATP), *KRKS & SSR* 36949; Idupulapaya (KDP), *KRKS & BSS* 39062.

Pavonia procumbens (Wallich ex Wight & Arn.) Walp., Rep. Bot. Syst. 1: 301. 1842, non Casaretto 1842; FPM 1: 93. 1915; FAP 1: 129. 1997. *Lebretonia procumbens* Wallich ex Wight & Arn., Prodr. Fl. Ind. Orient. 47. 1834. *Pavonia glechomifolia* (A. Rich.) Garcke ex Schweinf., Beitr. Fl. Aethiop. 1: 54. 1867; FBI 1: 330. 1874. *Lebretonia glechomifolia* A. Rich., Tent. Fl. Abyss. 1: 54. 1847.

Woody herb, to 50 cm long, spreading branches, vine-like, hard, scattered stellate hairy. Leaves often orbicular to ovate, 3-lobed, 1- 4 × 2 – 4 cm, scattered hairy above, villose below, base cordate, margin dentate, apex acute, base cordate, basally 5-nerved; petiole to 2.5 cm long. Flowers yellow, axillary, solitary; pedicel pendulous, up to 5 cm long. Epicalyx lobes 5, elliptic-ovate, hairy, connate at base, acuminate at base and apex. Calyx campanulate, lobes ovate, connate at base. Corolla yellow, 1.5–2 cm across; petals obovate, rounded at apex. Staminal column *ca.* 5 mm long. Fruit schizocarp, mericarps 5 (sometimes 3), triquetrous, obovoid, vermiform, rugose, whitish yellow, hard. Seed 1 per mericarp, 2 mm across, reniform, angular, glabrous, brownish.

Occasional, in dry deciduous forest and near ponds. Fl. & Fr.: May. – Dec. Kalasamudram (ATP), *KRKS & TP* 39143, 39530.

Pavonia zeylanica (L.) Cav., Diss. 3: 134, t. 48 f. 2. 1787; FBI 1: 331. 1874; FPM 1: 93.1915; FAP 1: 129. 1997. *Hibiscus zeylanicus* L., Sp. P1. 697. 1753.

Local name: *Zigata porra.*

Perennial under shrub, to 1.5 m high, stellate hairy, woody branches. Leaves ovate, deeply lobed 1–1.5 × 0.5–1.5 cm, stellate hairy on both surfaces, base cordate, lobes 3, elliptic-oblong, margin irregularly dentate, apex acute; petiole 1–2 cm. Flowers whitish pink, axillary, solitary; pedicels 1–4 cm long. Epicalyx lobes 9–12, linear, 0.5 cm, glandular hairy, longer than flower bud and capsule. Calyx campanulate, 5-lobed, lobes ovate, free to the middle, 3 × 1.5 mm. Corolla white-light pink; petals obovate, glabrous. Staminal column 8 mm long. Capsule 5-locular, subglobose; mericarps 5, 4 × 3 mm, oblong-obtuse, slightly winged at angles, awn less, glabrous. Seed 1 per mericarp, pilose, reniform, 1.5 mm across, black.

Common, in cultivated land and foot hills of forest. Fl. & Fr. Jun. – Dec. K. K. Kottala (KDP), *KRKS & KRS* 36421; Kalasamudram (ATP), *KRKS* 36461; Batrepalli (ATP), *KRKS* 284 (HJCB-E).

SIDA L.

Key to Species

1a. Mericarps 5 .... (2)

1b. Mericarps 6–12 .... (4)

2a. Erect herbs or undershrubs; flowers axillary, solitary initially, ultimately in condensed clusters towards apex .... (3)

2b. Prostrate or diffuse herbs; flowers usually solitary or in few-flowered racemes .... **S. cordata**

3a. Stem and petiole with spiny projections .... **S. spinosa**

3b. Stem and petiole not as above. – viscid .... **S. mysorensis**

4a. Stipules of each pair dissimilar, one linear to lanceolate and other linear to filiform; leaves sparsely hairy, soon glabrescent .... **S. acuta**

4b. Stipules of each pair not dissimilar; leaves densely pubescent .... (5)

5a. Mericarps dehiscent .... (6)

5b. Mericarps indehiscent .... **S. ovata**

6a. Leaves pilose or villous, awns retrorsely hairy .... **S. cordifolia**

6b. Leaves glabrescent or sparsely hairy, awns not retrorse hairy .. **S. rhombifolia**

Sida acuta Burm. f., Fl. Ind. 147. 1768; FPM 1: 90.1915; FAP 1:130. 1997. *S. carpinifolia* auct. non L.f. 1785; FBI 1: 323. 1874.

Local name: *Parkarikampa.*

Annual erect or ascending herb or undershrub, to 50 cm high, pubescent with minute stellate hairs; branches woody and scattered hairy. Leaves simple, lanceolate to linear or ovate-oblong, 2–3 × 0.5–1 cm, sparsely stellate or glabrescent, base obtuse-rounded, margin distinctly serrate-dentate, apex acute, 3-nerved at base; petiole up to 0.6 cm. Flowers yellow, axillary, solitary or 2–4-clusters. Calyx 5–6 mm across, campanulate, 5-lobed, lobes ovate-triangular. Corolla yellow, as long as calyx; petals emarginated at base, sparsely glandular hairy outside. Staminal column *ca.* 4 mm long. Fruit sub globose; mericarps 6–10, reniform, 5 mm across, 2-awned at apex, furrowed. Seeds 1 per mericarp, 2 mm across, brownish black, reniform, verruculose, simple and stellate hairy at apex.

Common, weed in cultivated lands and wastelands also in forest outskirts. Fl. & Fr.: Throughout the year. Rachakuntapalle (KDP), *SSR & TP* 33614

Sida cordata (Burm. f.) Borss. Waalk., Blumea 14: 182. 1966; FAP 1: 131.1997. *Melochia cordata* Burm. f., Fl. Ind. 143. 1768. *Sida veronicifolia* Lam., Encycl. 1: 5. 1783; FPM 1: 89. 1915. *S. humilis* Cav., Diss. 5, t. 134, f. 2. 1788; FBI 1: 322. 1874, incl. var. *veronicifolia*.

Local name: *Gayapuaku*.

Annual, prostrate or ascending herb, to 1 m long; stem with scattered pubescent. Leaves simple, ovate to orbicular, cordate, basal leaves larger than apical, 1–4 × 0.8–3 cm, sparsely stellate and simple hairy, base cordate-truncate, margin crenate-dentate, apex acuminate; petiole up to 2 cm, stellate hairy; 5-nerved at base. Flowers yellow, axillary, solitary or 3–5 clustered. Calyx connate just above the middle, 5-fid; lobes ovate-deltoid to triangular, acuminate, stellate without, glabrous within. Corolla yellow, *ca.* 10 mm across, lobes obovate, ciliate at base; pedicel to 3 cm; peduncle 2.5 cm. Staminal column *ca.* 3 mm long. Schizocarp subglobose, *ca.* 4 × 2 mm, thin, enclosed with calyx, mericarps 5, tetrahedral with rounded angles, beaked. Seeds 1 per mericarp, 3-angular, 2 mm across, dark brown-black, beaked, verrucose.

Common, in wastelands and foot hills of forest. Fl. & Fr.: Sept. – Jan. Rachakuntapalli (KDP), *SSR & TP* 33613; Batrepalli (ATP), *KRKS* 304 (HJCB-E).

Sida cordifolia L., Sp. Pl. 684. 1753; FBI 1: 324. 1874; FPM 1: 89. 1915; FAP 1: 131. 1997.

Local name: *Chiru benda*.

Under shrub, to 1 m high, tomentose or densely pubescent. Leaves simple, ovate, 3–5 × 2–3 cm, tomentose or velutinous on both sides, base cordate, margin crenate-dentate or serrate, apex acute-obtuse, 5–7-nerved at the base, petiole to 4.5 cm, villose hairy. Flowers yellow in axillary and terminal, 3–8-flowered clusters; pedicels 3–10 mm long, accrescent, up to 2 cm. Calyx 0.5 cm across, campanulate; lobes triangular, ovate, acute at apex, silky. Corolla 1 cm across; petals obliquely obovate, truncate at apex; ciliate at base. Staminal column *ca.* 2.5 mm long, simple hairy. Ovary conical, stellate-hairy. Fruit subglobose; mericarps 8–10, *ca.* 3.5 mm long, brown with prominent reticulations, apically 2-awns to 3 mm long. Seeds 2 mm across, 3-angular, flattened, reniform, glabrous, dark brown.

Common, along roadsides and open forests. Fl. & Fr.: Aug. – Dec. Rachakuntapalli (KDP), *SSR & TP* 33613; Thummalapalle mining site (KDP), *KRKS & KRS* 36986; Batrepalli (ATP), *KRKS* 305 (HJCB-E).

Sida mysorensis Wight & Arn., Prodr. Fl. Ind. Orient. 59. 1834; FBI 1: 322. 1874; FAP 1: 132. 1997. *S. glutinosa* Roxb., Fl. Ind. 3: 172. 1832; FPM 1: 89. 1915.

Annual or perennial herb or under shrub to 1 m high, pubescent. Leaves simple, ovate, cordate, upper leaves oblong or orbicular, 6–8 × 4.5–6 cm, pubescent, base cordate, margin serrate to crenate, apex acute or acuminate, basally 7–9-nerved; petiole to 6 cm long, scattered hairy. Flowers yellow, axillary, solitary or 2–4-flowered clusters; pedicels 3–20 mm long, slightly accrescent. Calyx widely campanulate, 5–10 × 3–4 mm; lobes ovate, acuminate, stellate hairy. Corolla yellow 0.8 cm across, slightly exceeding calyx; petals obtriangular, glabrous. Staminal column *ca.* 4 mm long. Ovary *ca.* 1 mm long, brownish black. Fruit globose, brown; mericarps 5, thin, 2.5–3 mm, tetrahedral with rounded angles, acute and short hairy at apex, beaked, 3-angular. Seeds *ca.* 2 mm long, ovoid to obtriangular, dark brown with prominent reticulations, 2 mm across.

Common, in wastelands and road sides. Fl. & Fr.: Oct. – Feb. Rachakuntapalli (KDP), *KRKS & KRS* 37907.

Sida ovata Forssk., Fl. Aegypt.-Arab. 124. 1775; FAP 1: 132. 1997. *S. grewioides* Guill. & Perr., in Guill. *et al.*, Fl. Seneg. Tent. 1: 71. 1830; FBI 1: 323. 1874.

Under shrub to 50 cm high, stellate hairy. Leaves ovate-oblong, 1.5–2.5 × 0.7–1.4 cm, densely stellate-velutinous on both sufaces, base obtuse, margin serrate-crenate, apex obtuse-acute, basally 3-nerved; petiole *ca.* 1 cm long. Flowers yellowish white, axillary, solitary or paired; pedicel 0.6 – 0.7 cm. Calyx 0.6 cm across, campanulate; lobes ovate, accrescent, up to *ca.* 10 × 7 mm, stellate, acute. Corolla yellow, lobes to 0.8 cm long, obliquely truncate, sparsely covered with bulbous base hairs. Staminal column *ca.* 7 mm long. Schizocarp *ca.* 5 mm across, indehiscent, closed by the calyx, nearly black; mericarps 7–8, reniform, protuberant, apically 2 small awned, to 2 mm. Seeds 1 per mericarp, 2 mm diam., reniform, brown with prominent reticulations, simple hairy at apex.

Occasional in scrub forests. Fl. & Fr.: Aug. – Jan. Rachakuntapalli (KDP), *KRKS & KRS* 37627.

Sida rhombifolia L., Sp. Pl. 684. 1753.

Key to Subspecies

1a. Leaves obovate to orbicular, apex rounded, decumbent branching .... ssp. **retusa**

1b. Leaves obovate or rhomboid, apex obtuse or acute, erect branching .... ssp. **rhombifolia**

Sida rhombifolia L., Sp. Pl. 684. 1753; FBI 1: 323. 1874; FPM 1: 90. 1915; FAP 1: 133. 1997; subsp. **rhombifolia** var. **rhombifolia**. Fl. India 3: 290. 1993.

Perennial herbs or under shrubs to 60 cm high, rigid stellate hairy. Leaves obovate, rhomboid, 0.5 – 4 × 0.3–2.7 cm, glabrescent above, pubescent below, basally 3-nerved, base cuneate, margin entire up to middle, dentate towards apex, apex obtuse or acute; petiole to 0.7 cm, stellate hairy. Flowers yellowish-white, axillary, solitary. Calyx 0.5 cm, ovate–rhomboid, acute, accrescent, up to 8 mm in fruit. Corolla yellowish, 0.6 cm. Fruit subglobose; mericarps 9, 3 mm across, brown, reniform, 2-awned at apex, 1.5 mm long. Seeds 1 per locule, reniform, brown, prominently reticulated, 1 mm across.

Occasional as forest undergrowth in dry deciduous forests. Fl. & Fr.: Mar. – Jul. Kalasamudram (ATP), *KRKS & BSS* 39106, 39912.

Sida rhombifolia L., Sp. PI. 684. 1753; FBI 1: 323. 1874; FPM 1: 90. 1915; FAP 1: 133. 1997; subsp. **retusa** (L.) Borss. in Blumea 14: 198. 1966. *Sida retusa* L., Sp. Pl. ed. 2: 961. 1763. *S. rhombifolia* var. *retusa* (L.) Masters FBI 1: 324. 1874.

Undershrubs, ascending or prostrate, up to 50 cm high, branches erect. Leaves 0.5–6 × 0.5–4 cm, obovate to orbicular, upper leaves sometimes lanceolate, base acute to acuminate, Margin dentate, apex mostly retuse rounded or truncate. Pedicels as long as or slightly longer than petioles jointed above or below middle. Calyx 5–6 mm high, accrescent, up to 10 mm. Corolla yellow 2–2.5 cm across. Mericarps with 2 short, up to 1.5 mm long awns. Seeds glabrous, hilum with few stellate hairs.

Occasionla in the rocky areas of dry deciduous forests. Fl. & Fr.: Dec. – Feb. Batrepalli forest (ATP), *KRKS* 296 (HJCB-E).

Sida spinosa L., Sp. Pl. 683. 1753; FAP 1: 133. 1997; FBI 1: 323. 1874; FPM 1: 89. 1915.

Much branched woody erect herb-undershrub; stems and branches rough with spiny, tubercles at the nodes below the petioles. Leaves ovate-elliptic or oblong, 1.5–4 × 1–2.5 cm, basally 3-nerved, minutely stellate-tomentose above, cinereous, whitish beneath, base obtuse-rounded, serrate, apex acute. Flowers yellow with red tinge, axillary solitary or 2 – 5 in clusters. Schizocarp wrinkled, exceeding calyx, mericarps 5; seed ovoid, slightly 3-angled.

Occasional, in dry deciduous forests. Fl. & Fr.: Oct. – Feb. Batrepalli (ATP), *KRKS* 437 (HJCBE).

URENA L.

Urena lobata L. Sp. Pl. 2: 692. 1753; FBI 1: 327. 1874; FPM 1: 92. 1915. subsp. **sinuata** (L.) Borssum, Blumea 14: 142. 1966; FAP 1: 135. 1997. *Urena sinuata* L., Sp. Pl. 692. 1753; FPM 1: 92. 1915.

Suffrutecose undershrub up to 1 m high; stems sparsely to densely stellate-tomentose. Leaves 3–5-palmatilobed with entire to pinnatisect lobes; lamina of lower leaves 4–6 × 5–7 cm, usually suborbicular, base rounded to subcordate, apex shallowly 3-lobed, middle leaves 5–7 × 3–6.5 cm, often more ovate, of upper leaves ovate-elliptic; margin evenly or irregularly serrate; petiole 2–4 cm. long. Stipules *ca.* 3 mm long, lanate. Flowers solitary, axillary; pedicels *ca.* 3 mm long, lanate. Epicalyx *ca.* 6 mm long, connate in the lower 1/3, stellate-pilose. Calyx shorter than the epicalyx, densely lanate at the base of the keels. Petals 10–15 (25) mm long. Fruit *ca.* 10 mm. in diam, 5-lobed.

Common, in wastelands, agriculture borders, roadside, also in forests. Fl. & Fr. Jul. – Dec.

STERCULIACEAE

Key to Genera

1a. Trees .... (2)
1b. Shrubs or herbs .... (6)
2a. Flowers unisexual or polygamous; petals absent .... (3)
2b. Flowers bisexual; petals present .... (5)
3a. Follicles opening long before maturity .... **Firmiana**
3b. Follicles not opening after maturity .... (4)
4a. Follicles membranous; seeds usually 2 .... **Hildegardia**
4b. Follicles coriaceous or woody; seeds many .... **Kavalama**
5a. Flowers actinomorphic; staminal tube narrow, straight; seeds winged .... **Pterospermum**
5b. Flowers zygomorphic; staminal tube dilated; seeds not winged ...... **Helicteres**
6a. Stem herbaceous, trailing, angular; fruit prickly .... **Byttneria**
6b. Stem woody, erect or decumbent, not angular; fruit not prickly .... (7)
7a. Staminodes present .... **Melhania**
7b. Staminodes absent .... (8)
8a. Ovary 5-carpellary; styles central .... **Melochia**
8b. Ovary 1-carpellary; styles subcentric .... **Waltheria**

BYTTNERIA Loefl. *nom. cons.*

Byttneria herbacea Roxb., Pl. Coromandel 1: 28. 1795; FBI 1: 376. 1874; FPM 1: 112. 1915; FAP 1: 138. 1997.

Herbs, prostrate or ascending, sparsely branched with perennial root stock; stems angular when young, stellate-tomentose. Leaves ovate, oblong, ovate-lanceolate, 1.5–7 × 1–3 cm, sparsely hairy, base obtuse or subcordate, margin irregular dentate, 3-nerved at base; petiole 2–3 cm long; stipules 2 mm long. Flowers maroon red, 2–3 in cymes; pedicels *ca.* 4 mm long; bracts 2–6, 1–2 mm long. Calyx up to 4 × 1.5 mm, connate at base; sepals narrowly ovate-lanceolate, acuminate, tomentose. Corolla up to 5 mm long, puberulous, claw slender, limb subulate with hood and 2-fid appendages which enter inside the staminal cup covering the stamens. Stamens in inner series of the staminal cup. Ovary 1 mm in diam., 5-loculed, sparsely hairy. Capsule globose, septicidally 5-valved, shortly prickly, apically horned. Seeds 1 in each locule, 3 × 2 mm, flat, muricate.

Common, in rocky areas of dry deciduous forests. Fl. & Fr.: Aug. – Jan. Namalagundu (KDP), *KRKS & BSS* 39622.

FIRMIANA Marsili

Firmiana colorata (Roxb.) R. Br., Pterocymbium 235. 1844; FPM 1: 107. 1915; FAP 1: 139. 1997. *Sterculia colorata* Roxb., Pl. Coromandel 1: 26, t. 25.795, quoad t. 25. descr. excl. & Fl. Ind. 3: 146. 1832; FBI 1: 350. 1874. *S. rubicunda* Wallich ex Masters in Hook.f., Fl. Brit. India 1: 360. 1874.

Local name: *Karaka, Maraka, Karu boppaja.*

Tree, up to 20 m high; trunk butressed, lenticelled, smooth, greyish green; branches spreading. Leaves ovate, variable, 14–36 × 8–29 cm, palmately 3-lobed, dark geern above, pale green-woolly below, base cordate, margin entire, apex acuminate, basally 7-nerved; petiole 10–20 cm long, pulvinate. Flowers bright red in panicles in the axils of fallen leaves, densely covered with minute stellate hairs and scales. Calyx 1–5 cm long, funnel-shaped, slightly curved, inflated towards apex, teeth acute. Androgynophore 5–10 mm long, exerted; anthers 10. Ovaries 5, flask-shaped, glabrous. Follicles 8–11 × 3–5 cm, oblong, membranous, glabrous, strongly veined, stipitate, opening at early stage. Seeds 2 per follicle, yellow, globose, compressed, wrinkled.

Occasional, in dry deciduous forests. Fl.: Mar. – Apri.; Fr.: Jun. – Aug. Batrepalli (ATP), *KRKS* & *BSS* 39592.

HELICTERES L.

Helicteres isora L., Sp. Pl. 963. 1753; FBI 1: 365. 1874; FPM 1: 107. 1915; FAP 1: 140. 1997.

Local name: *Purikaya.*

Small tree or large shrub, to 5 m high; branches scattered stellate hairy. Leaves broadly elliptic, elliptic-obovate, 7–10 × 3.5–8.5 cm, scattered stellate hairy above, stellate tomentose below, base cordate, margin crenate-serrate, apex acuminate, basally 3-nerved; petiole 0.5–1 cm, stellate hairy. Flowers crimson red, axillary clusters or solitary, tubular. Calyx tubular, 2 cm long, laterally compressed, somewhat 2-lipped, densely stellate, persistent, scattered hairy. Corolla oblong, crimson red, petals 4 – 5 cm long, reflexed, 2 lower shorter and broader than the 3 upper ones, claws winged. Staminal tube 3 – 4 cm long, slightly bent on one side at the tip. Ovary 5-celled, 5-lobed; styles united, as long as the ovary. Fruit follicles, 4 – 8 × 0.5 – 1 cm, cylindrical, spirally twisted with an apical beak, up to 1 cm long, stellate hairy, locules 5, apically beaked. Seeds many, 2 mm long, angular, wrinkled, tomentose.

Common, in dry deciduous forests. Fl. & Fr.: Sept. – Jan. Idupulapaya (KDP), *KRKS & SSR* 39066.

HILDEGARDIA Schott & Endl.

Hildegardia populifolia (Roxb.) Schott & Endl., Melet. Bot. 33. 1832. *Sterculia populifolia* Roxb., Fl. Ind. 3: 148. 1832; FBI 1: 361. 1874; FPM 1: 107. 1915.

Local name: *Galibudaga.*

Deciduous trees, up to 20 m high; bark green, smooth, gum comes out when cut, wood silverish white, soft branches warty. Leaves reniform or deeply cordate, 9.5–11.5 × 8–11 cm, glabrous when mature, brown stellate when young, base cordate, margin entire, apex long-acuminate, basally 7-nerved; petiole 5–10 cm long. Flowers pinkish, polygamous in axillary and terminal, up to 15 cm long panicles. Calyx scarlet red; lobes 6–12 mm long, free nearly to the base, spatulate-linear, downy without. Stamens 10 on *ca.* 2 mm long, staminal column adnate to the base of ovary. Ovary ovoid, hispid, tapering into a short style; stigma 5-lobed. Follicles 5, up to 10 × 5 cm, oblique, strongly veined, obtuse, winged, membranous; Stalk up to 4 cm long. Seed 2, *ca.* 1.3 cm across, ovate-oblong, pale brown.

Occasional, in dry deciduous forests especially in Nigidi forest. Fl.: May – Jun.; Fr.: Oct. – Dec., sometimes through the year. Chinnapalli (ATP), *KRKS & BSS* 37990, *KRKS* 309 (HJCB-E).

Note: Bark fibers used to make ropes and for agriculture usage, bark used as medicine for Animals.

KAVALAMA Raf.

Kavalama urens (Roxb.) Raf., Sylv. Tellur. 72. 1838. *Sterculia urens* Roxb., Pl. Coromandel 1: 25, t. 24. 1795; FBI 1: 355. 1874; FPM 1: 106. 1915; FAP1: 144. 1997.

Local name: *Tabsu.*

Deciduous tree, up to 20 m high; bark creamish white, smooth, papery peeling off; branches speading, marked with large scars; wood reddish. Leaves simple, 11–30 cm in diam., crowded at apex of branches, digitately 3–5-lobed, glabrous or sparsely hairy, base cordate, margin entire, apex caudate-acuminate; petiole 8–10 cm long, tomentose; stipules caducous. Flowers greenish yellow, 5–9 mm across in terminal branched panicles. Calyx *ca.* 5 mm long, campanulate, hoary, tube about as long as the lobes, lobes oblong-lanceolate, acute, glandular hairy inside at base. Male flowers: staminal column *ca.* 3 mm long, anthers 10–15. Female flowers: ovaries 2 mm in diameter, gynandrophore 3 mm long. Follicles 5, spreading, 3–4 × 1–1.5 cm, usually oblong or ovoid-oblong, densely pubescent, often mixed with stinging hairs. Seeds 3–6 in each follicle, *ca.* 7 × 5 mm, oblong, black, glossy.

Common, in dry deciduous forest. Fl.: Dec. – Mar.; Fr.: May – Aug. Kalasamudram (ATP), *KRKS & BSS* 39524.

Note: White Katila gum is extracted from the bark. Seeds are roasted and eaten.

MELHANIA Forssk.

Key to Species

1a. Shrubs; leaves ovate-oblong, densely tomentose on both sides; petiole 2–3 cm .... **M. hamiltoniana**

1b. Herbs; leaves oblong-elliptic, lanceolate, glabrescent above, white tomentose below; petiole 0.2 – 0.7 cm .... **M. incana**

Melhania hamiltoniana Wallich, Pl. Asiat. Rar. 1: 69, t. 77, 1830; FBI 1: 372. 1874; FPM 1: 113. 1915; FAP 1: 141. 1997.

Shrub to 1.5 m high, spreading white tomentose. Leaves ovate-oblong, 7.5–10.5 × 4–7 cm, densely tomentose on both sides, base cordate, rarely truncate, margin serrate-dentate, apex acute, basally 5–7-nerved, petiole 2–3 cm. Flowers yellow, axillary clusters; involucral bracts 4, ovate, stellate hairy, 1 cm across, acuminate. Calyx lobes 5, elliptic-ovate, stellate hairy, 1–1.5 cm long, acute; peduncle 4.5 cm, pedicel 1 cm, stellate hairy. Corolla yellow, 1.2 – 1.8 cm long, obovate, oblique. Staminal tube 2 mm long; staminodes *ca.* 1 cm long. Ovary *ca.* 5 mm in diam., subglobose, hairy, 5-loculed. Fruit capsule, ovoid, villose, stellate hairy, 5-locular, each locule boat-shaped, loculicidally dehiscent. Seeds 5 per locule, 4-angular, reniform, thick brown, 2 mm across, verrucose, beaked.

Occasional, in dry deciduous forest. Fl. & Fr.: Aug. – Sept. Idupulapaya (KDP), *KRKS & BSS* 41984.

Melhania incana Heyne ex Wight & Arn., Prodr. Fl. Ind. Orient. 68. 1834; FBI 1: 372. 1874; FPM 1: 113. 1915; FAP 1: 141. 1997.

Perennial, suffruticose, woody herb, to 30 cm high, stellate tomentose. Leaves oblong-elliptic, lanceolate, 1–3 × 0.4–0.7 cm, glabrescent above, white tomentose below, basally 3- nerved, base obtuse – cordate, margin distinctly serrulate, apex obtuse, petiole 0.2 – 0.7 cm. Flowers yellow in axillary peduncled cymes or clusters; involucral bracts 4, ovate, acute, stellate hairy, 0.4 cm long. Calyx 5-lobed, persistent, 0.7 cm long, ovate, stellate hairy, acuminate. Corolla yellow, 5-lobed, 0.7 cm long, obovate, truncate at apex. Fruit capsule, subglobose, 5-locular, 0.6 cm long, velvety, loculicidally dehiscent. Seeds thick brown, 5 per locule, 3-angular, reniform, verrucose, 2 mm across.

Occasional, in wastelands and agriculture lands. Fl. & Fr.: July - Dec. Thummalapalle (KDP), *KRKS & KRS* 36918.

MELOCHIA L.

Melochia corchorifolia L., Sp. Pl. 675. 1753; FBI 1: 374. 1874; FPM 1: 110. 1915; FAP 1: 142. 1997.

Local name: *Sitakura.*

Stout herb to under shrub, up to 80 cm high, sparsely hairy or glabrous. Leaves variable, ovate, ovate-lanceolate, elliptic-oblong, 1–4 × 0.5–2 cm, glabrous, base cordate-truncate, margin serrate, apex acute-obtuse, basally 5-nerved; petiole 0.5–1.5 cm. Flowers pink or white with yellow tinge near the base, densely crowded in terminal cymes; bracteole 4–5, linear, 0.4 cm long. Calyx ovate, 0.3 cm long, acute. Corolla pink, obovate, spatulate, lobes 0.3 cm long. Staminal cup somewhat spindle-shaped, narrowed at mouth. Fruit capsule, depressed globose, 3–5 mm across, hispid, mucronate, locules 5, reniform, loculicidally 5-valved. Seeds 1 in each locule, dark brown, reniform, 2 mm across, trigonous.

Common, along roadside and open forests especially moist locations. Fl. & Fr.: Jul. – Dec. Batrepalli (ATP), *KRKS & SSR* 36959; Rachakuntapalle (KDP), *KRKS & SSR* 37640.

PTEROSPERMUM Schreb.

Pterospermum xylocarpum (Gaertn.) Santapau & Wagh, Bull. Bot. Surv. India 5: 108. 1963; FAP 1: 143. 1997. *Velaga xylocarpa* Gaertn., Fruct. Sem, Pl. 2: 245, t. 133. f. 2. 1791. *Pterospermum heyneanum* Wallich ex Wight & Arn., Prodr. Fl. Ind. Orient. 69. 1834; FBI 1: 369. 1874.

Local name: *Tada, Polavu, Masapondi.*

Evergreen tree, up to 20 m high; branchlets rusty-tomentose; bark brown, longitudinally striated, wood white. Leaves variable in size and shape, usually oblong or oblong-obovate, 10–19 × 6–12 cm, coriaceous-glabrous above, greyish pubescent below, base cordate-truncate, margin entire at base, coarsely toothed or lobed from middle, apex acuminate; petioles 6–10 mm long, stout. Flowers white, 5 cm across, fragrant, solitary or in 2–3-flowered fascicles; bracteoles imbricate, persistent. Calyx lobes 5, 3–5 × 0.3–0.5 cm, oblong, rusty stellate tomentose outside, silky villous inside. Corolla lobes 5, 2.5–6 × 0.6–1 cm, obovate, stellate-pubescent outside, spreading. Stamens 15, 1–1.3 cm long. Ovary *ca.* 6 × 3 cm, densely rust stellate-hairy. Capsules 5–7.5 × 2–3.6 cm, oblong to pyriform, narrowed at both ends, obtusely 5-angled, rusty stellate-tomentose. Seeds 8–10 in each locule, 10 × 3 mm, orbicular, wing *ca.* 2.2 × 0.8 cm, papery.

Occasional, in hilly areas of dry deciduous forests. Fl.: Oct. – Dec. Fr.: Mar. – May. Rachakuntapalle (KDP), *KRKS & SSR* 39021.

WALTHERIA L.

Waltheria indica L., Sp. Pl. 637. 1753; FBI 1: 375. 1874; FPM 1: 111. 1915; FAP 1: 145. 1997.

Local name: *Nallabenda.*

Perennial undershrub, to 1.5 m high, brown softly stellate hairy. Leaves ovate, elliptic, or oblong, 2–4 × 2–3 cm, velvety tomentose on both sides with stellate hairs, base cordate-truncate, margin crenate-dentate, apex obtuse-acute; petiole to 1.5 cm, stellate hairy; basally 5-nerved. Flowers yellow, axillary clusters or condensed cymes, surrounded by involucral bracts, bracts narrowly lanceolate, stellate hairy. Calyx campanulate, connate up to middle; lobes 5, acute, 0.3 cm, long, stellate. Corolla lobes 5, obovate-spatulate, apex acute, stellate hairs at tip; peduncle 2–3 cm long, pedicels 0.2 cm, sometimes sessile. Staminal cup *ca.* 2 mm high, subconial. Ovary unilocular, pilose. Fruit capsule, one-seeded, subglobose, enclosed with calyx. Seed 1, 2 mm across, dark brown, subglobose, hairy at apex.

Common, in cultivated fields, road sides and wastelands. Fl. & Fr.: Sept. – April. Thummalapalle mining site (KDP), *KRKS, KRS, SSR & TP* 33653.

Note: There are two forms, one with whitish, silky hary, villous, flowers sessile heads in leaf axils, petals light yellow, campanulate and high altitude species and other one maroon, villous, flowers in pedunculate heads in leaf axils, petals yellow, funnel-shaped, low altitude species also grow as weed in cultivated fields (above mentioned speciemen).

TILIACEAE

Key to Genera

1a. Fruit drupe .... **Grewia**

1b. Fruit capsule .... (2)

2a. Fruit prickled, subglobose .... **Triumfetta**

2b. Fruit not prickled, cylindric-elongated .... **Corchorus**

CORCHORUS L.

Key to Species

1a. Capsule terminating into 3, 2-fid each, spreading tip .... (2)

1b. Capsule not terminating into 3 (beak-like), spreading tip .... (3)

2a. Stamens 12–30; capsules 6-angled, 3 angles prominently winged .. **C. aestuans**

2b. Stamens 10–15; capsules cylindric, not winged .... **C. tridens**

3a. Capsules cylindric .... (4)

3b. Capsules 3-angled .... (5)

4a. Capsule > 4 cm long, 10-ribbed, 5-loculed, plant shrubby to 1.5 m high .... **C. olitorius**

4b. Capsule < 2 cm long, 6-ribbed, 3-locular, plant herbaceous to 50 cm high .... **C. fascicularis**

5a. Capsule falcate, strigose hairy, seeds wrinkled, - capsules divaricate each other .... **C. urticifolius**

5b. Capsule neither falcate nor strigose; seeds not wrinkled, - capsule erect .... **C. trilocularis**

Corchorus aestuans L., Syst. Nat. ed. 10, 2: 1079. 1759; FAP 1: 146. 1997. *C. acutangulus* auct. non Forssk. 1755: FBI 1 : 398. 1874; FPM 1: 121. 1915.

Local name: *Kalasa*.

Annual suberect, much branched herbs, to 50 cm high; stems pilose, often purple; branches woody. Leaves ovate-lanceolate or elliptic, 2–3 × 1–2 cm, glabrous above, pilose below, basal appendages absent, base rounded, margin crenate-serrate, apex acute; petiole to 1 cm long, basally 3–5-nerved. Flowers yellow, axillary, solitary or paired; bud obconical; bracts linear. Sepals green without, purple-dotted within, linear-oblong, 3–4 mm long, hooded and apiculate, glabrous. Petals yellow, obovate, 3–5 mm long, obtuse with a glandular claw, hairy at base. Stamens 12–30. Ovary *ca.* 2 mm long, cylindric, carpels 3. Capsules 6-angled, 3-winged, 3-loculed, 2–2.5 cm long, apex 3-beaked, each again divided into 2-fid. Seeds numerous, brown, 3-angled, 1 mm across.

Common, weed in cultivated fields, stream banks, out skirts of forests and roadsides. Fl. & Fr.: Jul. – Jan. Batrepalli (ATP), *KRKS & SSR* 36957.

Corchorus fascicularis Lam., Encycl. 2: 104 1786; FAP 1: 147. 1997; FBI 1: 398. 1874; FPM 1: 122. 1915.

Suffruticose herb to 50 cm high, branches spreading. Leaves ovate or elliptic-ovate, 1.5–3 × 1–2 cm, base rounded, margin dentate, apex obtuse, basally 3-nerved, nerves pominent upper half of the lamina, petiole *ca.* 0.5 cm long. Flowers yellow, axillary or extra axillary 3-(5)-flowered cyme. Sepals green, linear-oblong, 2-3 mm long. Petals yellow, obovate or elliptic-ovate. Stamens 5–10, 3 mm long, anthers yellow. Capsule cylindrical, 1.5–2.5 cm long, 6-ribbed, 3-locular, 3-valved, beak small to 2 mm long, obtuse. Seeds *ca.* 20, 3-gonous, dark brown with white line.

Common, in field bunds and grows after harvesting of crop. Fl. & Fr.: Jul. – Jan. Mabbuchintalapalle (KDP), *KRKS* 430 (HJCB).

Corchorus olitorius L., Sp. Pl. 529. 1753; FBI 1: 397. 1874; FPM 1: 122. 1915; FAP 1: 148. 1997.

Local name: *Parintakura, Parinta.*

Biennial shrub, to 1.2 m high; stem glabrous, woody. Leaves lanceolate, ovate-oblong, 3–9 × 1.2–4 cm, distinctly hairy above nerves, basal appendages 2, prolonged into filiform processes, base obtuse, margin irregular serrate, apex acute, basally 3–5-nerved; petiole 2–3 cm long; stipules 8–12 mm long, subulate, glabrous. Flowers 1–2 in leaf-opposed, shortly pedunculate cyme, bud obconical, apex apiculate. Sepals 5–7 mm long, linear-oblong, apiculate. Petals yellow, 5–7 mm long, oblong, spatulate, obtuse. Stamens many, somewhat united at base. Ovary cylindric, sparsely hairy; carpels 5. Capsule cylindrical, oblong, glabrous, 4.5 cm long, brown, 5-angled, 5-valved, beak 0.5 cm long, 5-fid at apex. Seeds numerous, 3–4 angular, ovate, apex acute, 1 mm across, inconspicuously verrucose, black.

Occasional, on roadsides and agriculture lands. Fl. & Fr.: Apri. – Oct. Rachakuntapalle (KDP), *KRKS & KRS* 37639; Chinnapalli (ATP), *KRKS & BSS* 37992; Kalasamudram (ATP), *KRKS* 40073.

Corchorus tridens L., Mant. Alt. 566. 1771; FBI 1: 398. 1874; FPM 1: 121. 1915; FAP 1: 148. 1997.

Annual, erect herb, to 45 cm high; stem glabrous. Leaf oblong-lanceolate to ovate-lanceolate, 4–10 × 1–2 cm, glabrous except hairy nerves, base truncate or somewhat attenuate, margin serrate, apex acuminate, basally 5-nerved, basal apendages 2, filiform; petiole 1 cm long. Flowers 4, leaf-opposed; bud elliptic-oblong apiculate. Sepals 4–5 mm long, linear-oblong, apiculate. Petals yellow, 3–4 mm long, oblong, obtuse. Stamens 10–15. Ovary cylindric, hairy, carpels 3. Capsule *ca.* 4.5 cm long, 3-angled, trilocular, oblong, glabrous, often curved, terminating into 3-fid, each bifurcate. Seeds numerous, 4-angled, truncate at both ends, 1 mm long, black.

Common, along the damp places, besides agriculture fields and open places. Fl. & Fr.: Through the year. Thummalapalle (KDP), *KRKS & KRS* 36971; Kalasamudram (ATP), *KRKS & TP* 39113.

Corchorus trilocularis L., Mant. Pl. 77. 1767; FBI 1: 397. 1874; FPM 1: 122. 1915; FAP 1: 149. 1997.

Local name: *Bankitutturu.*

Annual, erect, much branched herbs, to 50 cm high; stem pilose hairy. Leaves narrowly ovate-lanceolate or oblong-elliptic, 2–4 × 0.5–2 cm, sparsely hairy on both surfaces, base obtuse, margin irregular crenate, apex obtuse or acute, basal appendages absent or some times minute, basally 3-nerved; petiole 2 cm long, pubescent. Flowers 2, leaf-opposed; bud obconical, apiculate. Sepals 4 – 5 mm long, linear-oblong, acuminate. Petals yellow, 5 – 6.5 mm long, somewhat pandurate, obtuse. Stamens 15 – 20. Ovary cylindrical, hairy, 3-loculed; carpels 3. Capsule cylindrical, oblong, 7 × 2.5 mm, stellate hairy, 3-angular, straight, 3-loculed, transversely septate, thick brown, beak entire. Seeds numerous, 3-angular, 0.5 cm across, truncate at both ends, black.

Common, in wastelands, along the streams and weed in cultivated fields. Fl. & Fr.: Throughout the year. Kalasamudram (ATP), *KRKS & BSS* 39114.

Corchorus urticifolius Wight & Arn., Prodr. Fl. Ind. Orient. 73. 1834. *"urticaefolius"*; FBI 1: 397. 1874; FPM 1: 122. 1915; FAP 1: 149. 1997.

Annual erect herbs, to 30 cm high; stem pubescent. Leaves ovate-lanceolate, 2–5.5 × 1–2 cm, sericeous or thinly pilose, basal appendages absent, base obtuse-attenuate, margin dentate, apex acute-acuminate, basally 5-nerved; petiole 1.5 cm long, sericeous hairy. Flowers leaf-opposed, 2–3; bud elliptic-oblong, apiculate, pubescent. Sepals *ca.* 4 mm long, oblong, acuminate, hairy. Petals yellow, *ca.* 3 mm long, ovate-spatulate, obtuse. Stamens *ca.* 15. Ovary 3-loculed, slightly 3-angled, pubescent. Capsule, slightly 3-angled, to 2.5 cm long, hispid, 3-loculed, brown, beak entire, two capsules divaricate in pseudo-axils. Seed quadrangular, truncate at both ends, 0.5 mm across, wrinkled, black.

Occasional, in forest edges and agriculture fields. Fl. & Fr.: Aug. – Feb. Chinnapalli, (ATP), *KRKS & BSS* 37805; Somulavaripalli (ATP), *KRKS & BSS* 37825; Kalasamudram (ATP), *KRKS & BSS* 39516.

GREWIA L.

Key to Species

1a. Peduncles capillary, 1–2-flowered .... **G. tenax**

1b. Peduncles not capillary, 3–many-flowered .... (2)

2a. Peducles much shorter than or as long as petiole .... (3)

2b. Peduncles much longer than petioles .... (6)

3a. Flower buds to 1 cm or more .... (4)

3b. Flower buds less than 0.8 cm .... (5)

4a. Stem and branches 4-gonous; fruit pilose .... **G. flavescens**

4b. Stem cylindrical; fruit pubescent or villous.-yellow when ripe..... **G. orientalis**

5a. Stipules auriculate .... **G. tiliifolia**

5b. Stipules linear-lanceolate .... **G. villosa**

6a. Leaves 5-nerved at base .... **G. orbiculata**

6b. Leaves 3-nerved at base .... (7)

7a. Leaves glaucous beneath .... **G. damine**

7b. Leaves not glaucous beneath .... (8)

8a. Plant with bisexual and unisexual flowers .... **G. hirsuta**

8b. Plant with unisexual flowers only .... **G. carpinifolia**

Grewia carpinifolia Juss., Ann. Mus. Natl. Hist. Nat. 4: 91. 1804. *G. rhamnifolia* Heyne ex Roth, Nov. Pl. Sp. 244. 1821; FPM 1: 117. 1915; FAP 1: 154. 1997. *G. orientalis* auct. non L. 1753; FBI 1: 384. 1874.

Straggling shrub to 3 m high, stem reddish brown, scabrous hairy. Leaves rhomboid, obovate, ovate-elliptic, 2–11 × 1.2–5 cm, scabrous hairy on nerves of both sides, basally 3-nerved, base obtuse, margin serrate, apex acuminate; petiole to 0.6 cm long; stipules ovate, 0.3 cm long. Flowers in axillary cymes, bud sub-globose, hirsute hairy; peduncle 0.6 cm long, pedicel 0.7 cm long. Sepals 1.2 – 2 cm long, lanceolate, woolly tomentose. Petals 5 – 7 mm long, ovate-lanceolate; glands *ca.* 3 mm long. Androgynophore *ca.* 2 mm long, grooved, woolly. Stamens many. Ovary 1 – 1.5 mm across, globose. Drupes subglobose, hirsute hairy, closely 2-lobed or 4-lobed, 1.1 cm across, mericarp fibrous, apex truncate, apiculate spine with persistent style. Seeds 2, hard, brown, 3-angular, rugose, 0.8 cm across.

Common, on roadsides and field borders. Fl. & Fr.: Jun. – Oct. K. K. Kottala (KDP), *KRKS & KRS* 36434; Batrepalli (ATP), *KRKS* 36948.

Grewia damine Gaertn., Fruct. Sem. Pl. 2: 113. t. 106, f. 9. 1790; FPM 1: 118. 1915 FAP 1: 151. 1997. *G. salvifolia* Heyne ex Roth, Nov. Pl. Sp. 239. 1821; FBI 1: 386. 1874.

Local name: *Adivi pagari, Jara, Konda cipuru, Mnickalupu, Narabudam, Uppidi.*

Small tree or shrub, to 4 m high; branches spreading; bark brown. Leaves ovate-oblong or elliptic, 1.5–4.5 × 0.5–1.5 cm, light brown above, grey hairy beneath, base oblique or obtuse, 3-nerved, margin serrate, apex acute; petiole 0.5 cm long; stipule linear-lanceolate, to 0.5 cm long. Flowers in axillary cymes; peduncle longer than petiole, to 1.2 cm long, pedicel to 1 cm long; flower bud ovoid-oblong, tomentose, 0.5 cm long. Sepals 8–12 mm long, linear to oblong, tomentose. Petals yellow, 3.5–6 mm long, elliptic-oblong, retuse; glands *ca.* 2 mm long, hairy. Stamens many. Ovary *ca.* 1.5 mm across, subglobose, hirsute. Drupes *ca.* 0.8 cm across, distinctly 2-lobed, lobes globose, pubescent, black, pulp sweet, grainy. Seeds brown, round, rugose, 0.4 cm across.

Common, in dry deciduous forests and foot hills of the forest. Fl. & Fr.: May – Sept. Kalasamudram (ATP), *KRKS & BSS* 39182.

Grewia flavescens Juss., Ann. Mus. Natl. Hist. Nat. 4: 91. 1804; FPM 1: 119. 1915; FAP 1: 151. 1997. *G. carpinifolia* sensu Masters in FBI 1: 387. 1874.

Local name: *Cipurutada, Kukka budda, Madekava, Nalli, Tadikamullu.*

Straggler, to 6 m high; stem 4-angular at maturity, stellate hairy, brown, coarsely granulate. Leaves ovate-oblong, ovate-elliptic, 2–6 × 1.5–3 cm, scabrous above, tomentose with stellate hairs, basally 3-nerved, base truncate or rounded, margin irregularly serrate, apex acute–acuminate; petiole 0.3 cm long; stipules linear. Flowers in axillary cymes, bud elliptic-oblong, 1 cm long, stellately hairy; peduncle 0.3 cm long; pedicel 0.5 cm long. Sepals 1.2–1.7 cm long, linear-lanceolate, acute, tomentose outside, glabrous inside. Petals yellow, 5–10 mm long, spatulate, usually 2-fid; glands *ca.* 3 mm long, oblong. Stamens *ca.* 1 cm long. Ovary *ca.* 2 × 0.5 mm, subglobose or ovoid, pilose. Drupes 0.7–1 cm, stellate tomentose, closely 2-lobed, yellow, mericarp fibrous, sweet. Seeds 0.5 cm across, globose, hard, whitish yellow, rugose.

Common, in open hilly areas of dry deciduous and scrub forest. Fl. & Fr.: May – Oct. Rachakuntapalle (KDP), *SSR* & *TP* 33631; Kalasamudram (ATP), *KRKS & BSS* 39501, 36455, 39518.

Grewia hirsuta Vahl, Symb. Bot. 1: 34. 1790; FBI 1: 391. 1874; FPM 1: 119. 1915; FAP 1: 152. 1997.

Local name: *Bidara cipura, Budda cipura, Cittijana, Nuvalcu, Tellajana.*

Shrubs, to 3 m high, stem hirsute hairy. Leaves ovate, lanceolate or ovate-lanceolate, 4–7 × 0.5–2.5 cm, stellate hirsute hairs above, densely tomentose below, basally 3-nerved, base slightly oblique, obtuse, margin irregular serrate-crenate, apex acuminate; petiole 0.5 cm long, stipules linear, to 0.5 cm long. Flowers polygamous in axillary umbellate cymes, bud ovate, stellate tomentose; peduncle 0.5 cm long; pedicel 0.5 cm long. Sepals *ca.* 8 mm long, elliptic-lanceolate, hirsute. Petals white, *ca.* 3 mm long, oblong, rounded; glands *ca.* ½ the length of petals. Stamens more than 40. Ovary *ca.* 2 mm across, globose, densely villose. Drupes subglobose, 0.2 cm across, densely hairy, 4-lobed, fleshy, wrinkled, densely hirsute or hispid. Seeds brown, globose, rugose, 0.3 cm across.

Common, in dry deciduous forests. Fl. & Fr.: Jul. - Oct. Kalasamudram (ATP), *KRKS, BSS* 39188; *KRKS* & *TP* 39517; *KRKS* & *TP* 39822; Batrepalli (ATP), *KRKS* 41898; *KRKS* 308 (HJCB-E).

Grewia orbiculata Rottler, Ges. Naturf. Freunde Berlin Neue Schr. 4: 205. 1803; FBI 1: 386. 1874; FAP 1: 153. 1997. *G. rotundifolia* Juss., Ann. Mus. Natl. Hist. Nat. 4: 92. t. 50, f. 3. 1804; FPM 1: 118. 1915.

Local name: *Jana, Nulitada.*

Tree, to 6 m high, bark brown, longitudinally peeling off. Leaves orbicular, broadly elliptic or obovate, 3–7 × 2.5–5 cm, brown tomentose above, grey tomentose below, base obliquely subcordate, margin distinctly serrate, apex obtuse-acute, 5-nerved at base; stipules subulate; petiole 0.8 cm long. Flowers in axillary cymes; peduncle to 2 cm long; pedicel to 1 cm long, bud elliptic oblong, tomentose. Sepals 6 – 10 mm long, lanceolate, woolly outside. Petals orange-yellow, 4 – 6 mm long, oblong. Stamens *ca.* 4 mm long. Ovary *ca.* 2.5 × 1.5 mm, obovoid, furrowed, pilose. Drupes *ca.* 0.8 cm across, tomentose, subglobose, obscurely 2-lobed, mucronate, not fleshy. Seeds dark brown, 3-angular, 0.4 cm across.

Occasional, on dry deciduous forests. Fl. & Fr.: Mar. – Aug. Kalasamudram (ATP), *KRKS & BSS* 39519.

Grewia orientalis L., Sp. Pl. 964. 1753.; FAP 1: 154. 1997; FPM 1: 117. 1915. *G. columnaris* J.E. Smith in Rees., Cyclop. 17: 5. 1811; FBI 1: 383. 1874.

Shrubs or small trees, up to 2.5 m high; branchlets tomentose, or stellate hairy. Leaves 4–7 × 3–3.5 cm, ovate or obovate, tomentose above, glabrous below, base truncate or subcordate, margin crenulate, apex acuminate; petiole to 8 mm. Cymes axillary or leaf-opposed. Flowers white, polygamous; peduncles more than twice the length of petiole; pedicels divergent. Sepals 5, more than 1 cm, lanceolate, pubescent without stellate hairy. Petals 5, half the length of sepals. Stamens as many as and as long as sepals. Ovary elongate, strigose. Drupe villous, slightly 4-lobed, lobes to 8 mm, subglobose shortly bristly.

Common, in dry deciduous forests and foot hills of forest areas. Fl. & Fr.: Jul. – Dec. Kalasamudram (ATP), *NY* 671.

Grewia tenax (Forssk.) Fiori, Agric. Colon. 5: Suppl. 23. 1912; FAP 1: 156. 1997. *Chadara tenax* Forssk., Fl. Aegypt.-Arab. 105. 1775. *Grewia populifolia* Vahl, Symb. Bot. 1: 33. 1790; FBI 1: 385. 1874. *G. betulaefolia* A. L. Juss., Ann. Mus. Natl. Hist. Nat. 4: 92. t. 2. 1804; FPM 1: 117. 1915.

Local name: *Kadhara kaya, Gundukadira, Kottekalupu.*

Shrub to 3 m high, much branched; stem blackish brown. Leaves elliptic-obovate, broadly ovate, 1.5–3 × 1–2.5 cm, scabrid above, glabrous or sparsely hairy below, basally 5-nerved, base rounded, margin crenate-serrate, apex obtuse-acute; petiole to 0.6 cm long. Flowers, axillary or leaf-opposed 1–2-flowered cymes, bud elliptic-oblong; peduncle *ca.* 1 cm long; pedicel 0.2 cm long. Sepals 1.2–1.8 cm long, linear-oblong, tomentose outside. Petals white, *ca.* 6.5 mm long, linear-oblong, obtuse, usually notched at apex, ciliate at base. Androgynophore *ca.* 0.5 mm long. Stamens many; filaments *ca.* 5 mm long. Ovary *ca.* 2.5 mm across, subglobose, 4-lobed, hirsute. Drupes 6–10 mm across, 2 parted, each half didynamous, sparsely hairy, 1 × 1.5 cm across, each lobe rounded, shiny, orange-yellow; mesocarp fibrous, very sweet. Seeds rounded, brown, rugose, 0.5 cm across.

Common, in scrub forest and field bunds. Fl. & Fr.: Jun. – Sept. Kalasamudram (ATP), *KRKS & KRS* 36459; *KRKS* 41982.

Grewia tiliifolia Vahl, Symb. Bot. 1: 35. 1790; FBI 1: 386. 1874; FPM 1: 118. 1915; FAP 1: 156. 1997. *'tilaefolia'*.

Local name: *Charachi, Nulijana, Tada, Tada-jana.*

Tree, to 6 m high; bark peeling off; stem dark brown; young branches stellate hairy. Leaves elliptic, elliptic-ovate, ovate-rounded 2–4 × 1.5–2.5 cm, glabrescent above, grey pubescent below, basally 4-nerved, base truncate, oblique, obtuse, margin irregular crenate, apex acute; petiole 0.4 cm long, stellate hairy, stipules auriculate, 1 cm. Flowers in axillary umbellate cymes; peduncles 1.5 cm long, stellate hairy; pedicels 0.4 cm long, flower bud subglobose, stellate hairy. Sepals 5 – 8 × 3 mm, elliptic or lanceolate, subacute, tomentose outside. Petals yellow, 3 – 4.5 × 1.5 mm, elliptic-oblong or spatulate, obtuse, notched at apex. Stamens many. Ovary *ca.* 1.2 mm across. Drupes 2.5–5 × 7–10 mm, distinctly 2-lobed; lobes globose, distinctly 2-lobed, stellate hairy, black. Seeds brown, 0.3 cm across.

Common, in dry deciduous forest. Fl. & Fr.: May – Sept. Kalasamudram (ATP), *KRKS & BSS* 39118.

Grewia villosa Willd., Ges. Naturf. Freunde Berlin Neue Schr. 4: 205. 1803; FBI 1: 388. 1874; FPM 1: 119. 1915; FAP 1: 156. 1997.

Local name: *Bantachendla.*

Shrubs or small trees to 4 m high, bark thick brown; branches stellate hairy. Leaves rounded-ovate, orbicular, 2.3–5.5 × 1.7–4.5 cm, brown scabrous above, grey villose beneath, 5–7-nerved at base, base cordate, margin crenate, apex obtuse-acute; petiole to 2 cm long, stellate hairy, stipules linear-lanceolate. Flowers in axillary or leaf-opposed cymes, bud elliptic-oblong, villose; peduncle 0.4 cm long; pedicel 0.2 cm long, villose. Sepals *ca.* 1 cm long, lanceolate, pilose. Petals dull yellow, *ca.* 1 cm long, spatulate, emarginated. Stamens many. Ovary *ca.* 2 mm across, subglobose, densely villous. Drupe subglobose, villose, orange, 1 cm across. Seed hard, subglobose, brown, 0.8 cm across.

Common, in scrub forest and field bunds. Fl. & Fr.: Jun – Oct. K. K. Kottala (KDP), *KRKS & KRS* 36430.

TRIUMFETTA L.

Key to Species

1a. Annual herb or small shrub; leaves palmately 3-lobed, elliptic-lanceolate, margin dentate, tip acuminate; stipule linear; stamens 5–10; spines with a line of simple hairs on capsule .... **T. pentandra**

1b. Perennial under shrub; leaves simple, obovate-rhomboid, margin irregularly serrate, tip acute, stipules lanceolate; stamens 10–25; spines pubescent with stellate hairs on fruit ..... **T. rotundifolia**

Triumfetta pentandra A. Rich. in Guillemin *et al.*, Fl. Seneg. Tent. 93. t. 19. 1831; FPM 1: 120. 1915; FAP 1: 157. 1997. *T. neglecta* Wight & Arn., Prodr. Fl. Ind. Orient. 75. 1834; FBI 1: 396. 1874. *T. rhomboidea* Jacq. var. *pentandra* (A. Rich.) J. L. Ellis, Bull. Bot. Surv. India 24: 209. 1983.

Annual erect herb or undershrub, to 1.2 m high, branches hispid. Leaves palmately 3-lobed, hispid hairy, 3–7 × 1.6– 4 cm, basal leaves 3-lobed, ovate; apical leaves elliptic-lanceolate, 3–6 × 1.5–2 cm, basally 3–5-nerved, base attenuate-acute, margin dentate, apex acuminate; petiole 1–5.5 cm; stipules linear, 0.5 cm. Flowers terminal and leaf-opposed clustered cymes, buds elliptic-oblong, 0.2 cm long; pedicel 0.2 cm. Sepals *ca.* 2.5 mm long, spatulate, stellate hairy. Petals yellow, as long as sepals, spatulate, obtuse at base. Stamens 5 (10). Capsule 4–5 mm across, ovoid, oblong-ellipsoid, tomentose; spines uncinate, 2 mm long, simple hairy. Seeds 4, thick brown, ovate, trigonous, smooth, 1 mm across, beaked.

Common, in open dry deciduous forests. Fl. & Fr.: Sept. – April. Rachakuntapalle (KDP), *SSR & TP* 33607; Somulavaripalle (ATP), *KRKS & KRS* 37821.

Triumfetta rotundifolia Lam., Encycl. 3: 421. 1792; FBI 1: 395. 1874; FPM 1: 120. 1915; FAP 1: 158. 1997.

Under shrub, to 1m high, bushy, branches suffruticose, stellate hairy. Leaves orbicular, rhomboid, 1–3.5 × 0.5 – 2 cm, brown stellate hairy above, grey stellate hairy beneath, basally 3–5-nerved, base obtuse, margin irregularly serrate-crenate, apex acute; petiole 1–2.5 cm long; stipules lanceolate 0.6 cm long. Flowers 5–8, clustered cymes on elongated terminal axis, buds obovate, grey stellate hairy, appendages 5 at apex, black; pedicel 1 mm; bract *ca.* 1.5 mm long, linear. Sepals 5–7 mm long, lorate, apiculate, albido-tomentose. Petals yellow, 4–5 mm long, spatulate, emarginated at apex, ciliate at base. Capsules 0.4–0.6 cm, ovoid-subglobose to globose, pubescent; spines *ca.* 2 mm, straight, stellate hairy. Seeds *ca.* 2, 2.5 mm across, ovate, beaked, blakish brown, smooth.

Occasional, in dry deciduous forest and agriculture lands. Fl. & Fr.: Aug. – Dec. Kalasamudram (ATP), *KRKS & BSS* 39186, 40118.

LINACEAE

HUGONIA L.

Hugonia mystax L., Sp. Pl. 675. 1753 *'myxstrax'*; FBI 1: 413. 1874; FPM 1: 126. 1915; FAP 1: 161. 1997.

Local name: *Tenekaya.*

Climbing shrub or often small tree, to 3 m high, bark whitish-brown; young branches yellowish pubescent. Pair of strong and circinate hooks on branches helps to climb. Leaves 9–12 × 3.5–4.5 cm, alternate, elliptic-lanceolate, obovate, pubescent both sides on nerves, base attenuate, margin distinctly serrate, apex acuminate, lateral veins 14–16 pairs; petiole *ca.* 0.5 cm, pubescent; stipules ovate-lanceolate, 1–1.5 cm long, pubescent. Flowers in axillary cymes. Peduncle hooked, yellow pubescent. Bracteole ovate–lanceolate, 0.6 cm long. Pedicel 0.5 cm long. Sepals 5, ovate-lanceolate, pubescent without. Petals yellow, 5-lobed, ovate, 0.6 cm long. Stamens 10, [8 (3+5) + 2], longer ones *ca.* 3 mm long and shorter ones *ca.* 1 mm long; anthers, dorsifixed, 1 mm long. Ovary globular, glabrous, ovules pendulous; styles 5, (3 free and 2 connate); stigma disk-shaped. Drupes globose, glabrous, apiculate at apex, 1.5 cm diam., surrounded by persistent scarlet sepals, reddish-yellow, shiny, pulp sweet like honey. Seeds pendulous, light brown, 5 mm across.

Rare along the foot hills of dry deciduous forests. Fl. & Fr.: Mar. – Aug. Chinnapalli (ATP), *KRKS & BSS* 37911; Rachakuntapalle (KDP), *KRKS* 41839.

ERYTHROXYLACEAE

ERYTHROXYLUM L.

Erythroxylum monogynum Roxb., Pl. Coromandel 1: t. 88. 1795; FBI 1: 414. 1874; FPM 1: 127. 1915; FAP 1: 162. 1997.

Local name: *Devadaraku, Adavi goranti, Gadara.*

Small tree, to 4–6 m high, glabrous, much branched; bark grey, peeling off, scented. Leaves obovate-elliptic, 1–4.1 × 0.5–2 cm, glabrous, base attenuate, cuneate, margin entire, apex obtuse. Flowers yellowish white, axillary, solitary or 4-flowered fascicles; pedicel to 1 cm long, thickened towards calyx. Calyx campanulate, persistent, glabrous; petals 5, 2 × 2 mm, triangular, acute. Petals yellowish-white, 3 mm long, ligulate, spreading, caducous. Stamens 10 in 2 whorls, equal. Ovary 2 × 1 mm, oblong, longer than staminal tube. Drupes straight, elliptic oblong, subcylindric, 0.8 × 0.4 cm, rose or red, apex mucronate. Seeds 3, oblong, brown, 0.5 × 0.2 cm.

Common, in dry deciduous forests and scrub forests. Fl. & Fr.: Jun. – Nov. Thummalapalle (KDP), *KRKS & KRS* 36910; Batrepalli (ATP), *KRKS* 36956.

Note: Leaves and fruits are edible.

ZYGOPHYLLACEAE

Key to Genera

1a. Erect, glabrescent herb, spiny; leaves 1-3 foliolate, leaflets linear-lanceolate; flowers pink .... **Fagonia**

1b. Prostrate pubescent herb, not spiny; leaves pinnate, leaflets elliptic; flowers yellow .... **Tribulus**

FAGONIA L.

Fagonia indica Burm.f. Fl. Indica 102. 1768. *F. cretica* sensu FAP 1 : 164. 1997. *F. arabica* L., Sp. Pl. 386. 1753; FBI 1 : 425. 1874 *p.p.*; FPM 1 : 130. 1915.

Perennial herb, up to 50 cm high, glandular; internodes terete, 2–3 cm long, striate. Lower leaves trifoliolate, upper ones unifoliate; petioles 3–5 mm long; leaflets linear-oblong, mucronate, 4–12 × 1.5–3 mm; stipular spines 1–2 cm long. Flowers axillary, solitary. Sepals deciduous, oblong-lanceolate, *ca.* 4 mm long, glandular without. Petals bluish violet, obtuse, 5–6 × 3–4 mm. Stamens *ca.* 5 mm long. Capsules *ca.* 5 mm across, glandular-pubescent, 1-seeded. Seeds flat, smooth, glabrous, pale-brown, oblong.

Occasional, on roadsides and foot hills of the forests, especially in scrub and plains. Fl. & Fr.: Aug. – Nov. Mabbuchintalapalle (KDP), *KRKS* 41841; Vemula (KDP), *KRKS & BSS* 41959.

Note : Gamble (in FPM) says that leaflets are linear but Hook.f. (in FBI) says that leaflets are elliptic or linear; there is no specimen from south India with elliptic leaves. One has to see the plants in the field to solve this ambiguty.

TRIBULUS L.

Tribulus terrestris L., Sp. Pl. 387. 1753; FBI 1: 423. 1874; FPM 1: 129. 1915; FAP 1: 165. 1997.

Local name: *Palleru.*

Annual prostrate herb, to 30 cm long, hirsute hairy. Leaves compound, 1–3 cm long, paripinnate; leaflets subsessile, elliptic, 0.5 × 0.3 cm, sparsely hispid above, hispid below, base subcordate, margin entire, apex acute-apiculate; stipules subulate to falcate, 2–5 × 1–1.5 mm. Flowers axillary, solitary; pedicels 1 cm long. Sepals ovate, pubescent. Petals yellow 2.5–5 × 1–3 mm. Stamens 10, 3–4 mm long. Ovary *ca.* 1.5 mm across, with stiff upward spreading hairs. Fruit *ca.* 10 mm across; mericarps 3–5, hispid, 0.4–0.6 × 0.3–0.4 cm across; spines 4, two small to 3 mm long and two large to 5 mm long, divergent. Seeds 2–3, brown, elliptic, 0.2 mm diameter.

Common, weed in open places, cultivated fields and wastelands. Fl. & Fr.: Throughout the year. Thummalapalle (KDP), *SSR & TP* 33637.

OXALIDACEAE

1a. Leaves simple, 3-foliolate; capsule cylindrical .... **Oxalis**

1b. Leaves compound, multifoliolate, capsule elliptic; ovate .... **Biophytum**

BIOPHYTUM A. P. DC.

Biophytum reinwardtii (Zucc.) Klotzsch in Peters Naturw. Reise Mossambique 6 (Bot., 1): 85. 1861; FAP 1: 166. 1997; FBI 1: 437. 1874; FPM 1: 133. 1915. *Oxalis reinwardtii* Zucc., Abh. Math.-Phys. Cl. Königl. Bayer. Akad. Wiss. 1: 274 1831.

Annual, erect, small herb, to 15 cm high, brown hairy. Leaves pinnate, crowded at apex of stem, pinna 4–12 cm long, glabrous; leaflets opposite, paripinnate, subsessile, ovate–elliptic, oblong, 0.5–0.7 × 0.3–0.4 cm, glabrous, base truncate, margin entire, apex apiculate. Flowers in axillary, long-peduncled clusters; peduncle to 10 cm long; pedicel 0.5 cm long; involucral bracts linear, *ca.* 1.5 mm long. Calyx ovate-linear, lanceolate, acute, 4–7 × 0.5–1.5 mm, strigose and glandular hairy, longer than pedicels. Petals yellow with rose lines, obovate, 5–7 × 1–2 mm. Anthers 1.5–2.5 mm long. Ovary 0.5–0.75 mm, ciliate apically. Capsule ovate, ca. 1 cm long, loculicidally dehiscent, 4 seeds per locule. Seeds ellipsoid, 1 × 0.5 mm, obliquely transverse tubercled in spherical lines, light pink, shiny.

Common, in moist places and wastelands. Fl. & Fr.: Jun. – Dec. Bhumaiahgari palle (KDP), *KRKS* 37606; Vemula (KDP), *KRKS* & *BSS* 41953.

OXALIS L.

Oxalis corniculata L., Sp. Pl. 135. 1753; FBI 1: 436. 1874; FPM 1: 132. 1915; FAP 1: 167. 1997.

Perennial procumbent herb, to 45 cm, pubescent, sometimes much thickened and woody; stem pink, branched, puberulous. Leaves 3-foliolate, petiole scattered hairy, 2–4 cm, stipules linear, leaflets obovate, broadly elliptic, 1–1.5 × 1–1.7 cm, glabrous, very thin, base cuneate, margin entire, apex emarginate, subsessile. Flowers umbellate, axillary cymes; peduncle 5 cm long, pubescent; pedicel 2 cm long, pubescent; bracts 3, ovate, bracteoles 3, ovate, hairy. Calyx lobes 5, elliptic oblong. Petals 5, yellow, shortly clawed, oblong, *ca.* 15 × 5 mm, glabrous. Stamens 10 (5+5), alternately short, united below. Ovary ovoid, *ca.* 8 × 4 mm; styles 5, very short, pubescent. Capsule subcylindrical, oblong, 5-angled, 5-locular. Seeds *ca.* 40, 8 per locule, 0.75 mm, flattened.

Occasinal in dry deciduous forests. Fl. & Fr.: Jul. – Dec. Batrepalli (ATP), *KRKS & BSS* 39641.

BALANITACEAE

BALANITES Delile *nom. cons.*

Balanites roxburghii Planch., Ann. Sc. Nat. Bot. ser. 4. 2: 258. 1854; FBI 1: 522. 1874; FPM 1: 164. 1915. *B. aegyptiaca* auct. non (L.) Del.; FAP 1: 181. 1997.

Local name: *Gara.*

Bushy shrub or small tree, 3–4 m high, branches armed; stem green when young, grey when mature; bark whitish, bitter; young parts pubescent; thorns up to 6 cm long, simple or forked or with smaller spines, often bearing leaves and flowers. Leaflets 2, elliptic oblong-obovate, 1.5–2.1 × 0.7–1 cm, glabrous, base acute-attenuate, margin entire, apex obtuse-acute; petiole 0.5–1 cm long; petiolules 1–4 mm long. Flowers greenish-yellow, fragrant, 4 – 12, in axillary or supra-axillary fascicled cymes; pedicels 5 – 15 mm long, pubescent. Sepals 5, imbricate, elliptic-ovate, *ca.* 3 mm long, densely pubescent outside, long silky hairy inside. Petals 5, elliptic-oblong, glabrous outside, silky hairy inside. Stamens 10, 2 – 3 mm long. Ovary globose, *ca.* 2 mm across, surrounded by disk. Drupes subspherical or obovate, 5-grooved, hard, 3–6 × 2–4 cm, puberulous, yellow when ripe with foetid smelling pulp. Seed solitary, embedded in pulp with stony fibrous testa.

Common, in dry deciduous forest and agriculture lands. Fl. & Fr.: Jun. – Nov. K.K. Kottala (KDP), *KRKS & KRS* 36445.

Note: Plant bark used as fish poison, fruit used for medicine.

RUTACEAE

Key to Genera

1a. Large trees; petiole wingless .... **Aegle**

1b. Shrubs or small trees; petiole winged .... (2)

2a. Leaves simple, tip emarginate .... **Atalantia**

2b. Leaves 3–5-foliolate, tip rounded .... **Naringi**

AEGLE Corr., *nom. cons.*

Aegle marmelos (L.) Corr., Trans. Linn. Soc. London 5: 223. 1800; FBI 1: 516. 1875; FPM 1: 161. 1915; FAP 1: 171. 1997. *Crateva marmelos* L., Sp. Pl. 444. 1753. *Aegle marmelos* (L.) Correa var. *mahurensis* Zata, Indian J. Forest. 5: 35. 1982.

Local name: *Bilvamu, Maredu.*

Deciduous tree, to 10 m high, branchlets pubescent, armed, spines axillary, paired to 4 cm long. Leaves alternate, 3–5-foliolate, leaflets ovate-elliptic, lanceolate or oblong-obovate, 2–5 × 1.5–3 cm, middle leaflet bigger than the lateral leaflets, puberulose on both sides, base cuneate, margin subcrenulate, apex obtuse to acute, black when dry; petiole to 6 cm long, petiolule 0.5 cm long. Inflorescence axillary and terminal, racemose or cymose, few-flowered, up to 5 cm long; peduncles 4 cm long, densely puberulent; pedicels 2–4 mm long, densely puberulent. Flowers subglobose in bud, fragrant. Calyx cupular with 5, small, deltate or suborbicular teeth, finely puberulent, caducous. Petals 4 or 5, greenish-white, ovate-obtuse, *ca.* 12 × 6 mm, coriaceous, glandular, glabrous. Stamens numerous, 30–40 in irregular 2 or 3 series, free or irregularly coherent at base, unequal. Ovary ovoid-oblong, glabrous, faintly ridged, greenish, 8–12-locular; locules with many, biseriate ovules. Style very short, glabrous; stigma cylindric or bluntly conical, longitudinally grooved, often sticky. Berry greenish-yellow, subglobose, 5–10 cm diam., epicarp hard with stony cells. Seeds many, embeded in pulp, oblong, oblong-ovoid, flattened, 3 mm across.

Occasional, in dry deciduous forests. Fl. & Fr.: Mar. – Dec. Somulavaripalli (ATP), *KRKS & BSS* 39911.

Note: Fruits ripen after six months. Trees planted in compounds of temples. Fruit yields gum and used for treating chronic diarrhoea and dysentery.

ATALATIA Corr., *nom. cons.*

Atalantia monophylla (L.) DC., Prodr. 1: 535. 1824. var. **monophylla**; FBI 1 : 535. 1874; FPM 1: 159. 1915; FAP 1: 171. 1997. *Limonia monophylla* L., Mant. Pl. 237. 1775.

Local name: *Yerru munukudu, Adavi nimma.*

Large shrub or small trees, to 5 m high; branches glabrous, woody, cylindric, stout spines axillary. Leaves alternate, glabrous, elliptic-oblong, ovate, 2–3 × 1–6 cm, faint above, dotted with oil glands below, base broadly cuneate, margin entire, apex emarginated, petiole 0.5 cm long. Inflorescence axillary, racemose, corymbose or cymose; bracteoles subulate, *ca.* 1.5 mm long, puberulous or glabrous. Flowers ovoid or oblong in bud, widely opening, up to 2 cm across. Calyx cupular, splitting irregularly into 2 lobes at anthesis; lobes obtuse, 0.5–1.5 mm long, densely puberulent to glabrous, persistent. Petals 4 or 5, white, oblong-elliptic, rounded at apex, clawed below, *ca.* 10 × 4 mm. Stamens 8 or 10, alternately short and long. Ovary ovoid, 2.5–4 × 1–1.5 mm, smooth, 3 or 4-locular. Berry globose, yellow, 3–5-locular, 2–8 mm across. Seed 1, ovate, whitish, glabrous.

Occasional, in dry deciduous forest. Fl. & Fr.: Aug. – Feb. Kalasamudram (ATP), *KRKS & BSS* 39561.

NARINGI Adans.

Naringi crenulata (Roxb.) Nicolson, Fl. Hassan Distr. 387. 1976; FAP 1: 176. 1997. *Limonia crenulata* Roxb., Pl. Coromandel t. 86. 1798; FPM 1: 157. 1915. *Limonia acidissima* auct. non. L. 1762; FBI 1: 507. 1875.

Perennial shrub with axillary spines. Stem rounded rough, green when young, light grey-coloured when mature. Leaves alternate, 3–5-foliolate, petiole winged, wing to 0.5 cm wide, leaflets elliptic-oblong, 2–4 × 1–2 cm, gland dotted, base rounded, margin entire, apex rounded. Spines solitary, axillary, straight to 2 cm long. Flowers white in axillary racemes, fragrant, disk yellow, glandular, stamens in bundles. Berry globose, 1–2 cm across, smooth. Seeds 1 or 2, shining, flattened.

Occasional, in deciduous forests. Fl. & Fr. Apr. – Sep. Kalasamudram, Nigidi forests, Anantapur district.

FLINDERSIACEAE

CHLOROXYLON DC. *nom. cons.*

Chloroxylon swietenia DC., Prodr. 1: 625. 1824; FBI 1: 569. 1874; FPM 1: 152. 1915; FAP 1: 192. 1997.

Local name: *Billudu.*

Deciduous tree, to 6 m high, most parts greyish-pubescent; bark grey-brownish, corky, fissured. Leaves up to 30 cm long; petiole slender, terete, 3–4 cm long; leaflets 10–20 pairs, alternate and opposite, ovate-oblong, 2.5–3 × 1.2 cm, green above, pale and glaucous beneth, base cuneate, margin entire, apex obtuse. Inflorescence 10–15 cm long, pyramidal grey-pubescent. Flowers 10 mm across; peduncle *ca.* 5 mm long, puberulent. Sepals 5, minute, deltoid, puberulent. Petals 5, imbricate, spatulate, clawed, acute at apex, 3–5 × 1.2–3 cm, creamy-white, puberulent on both sides. Stamens 10, shorter than petals. Ovary immersed in disk, greenish, *ca.* 2 mm broad, 3-lobed, 3-locular, ovules 2 – 8 in each locule, biseriate. Capsule ovoid-oblong, *ca.* 3 × 1.5 cm, smooth, 3-valved. Seeds 4 in each valve, compressed, 15 × 3 mm, distally alate, brown.

Common, in dry deciduous forests. Fl.: Mar. – Apr.; Fr.: Jul. – Sept. Idupulapaya (KDP) *KRKS & BSS* 41983.

SIMAROUBACEAE

AILANTHUS Desf., *nom. cons.*

Ailanthus excelsa Roxb., Pl. Coromandel 1: 24, t. 23. 1795; FBI 1: 518. 1874; FPM 1: 163. 1915; FAP 1: 180. 1997.

Local name: *Peddaman.*

Deciduous tree with large leaf-scars, to 15 m high; branches brown stellate hairy. Leaves imparipinnate, 30 cm long; petiole to 10 cm long; leaflets subopposite, obliquely ovate, 1.5–9 × 1–5 cm, pubescent on both sides, base oblique, truncate, margin irregularly toothed, apex acute, acuminate; petiolule to 2 cm long; lateral veins 9–17 pairs. Inflorescence axillary or terminal panicle, 30 cm long; peduncle to 15 cm long; bract lanceolate, to 2 mm long; pedicel 6 mm long. Flowers 8 mm across, polygamous. Calyx-tube short, pubescent; lobes ovate-triangular, *ca.* 2 mm long. Petals 5, greenish yellow, ovate-lanceolate, 3 – 4 × 2 mm, reflexed, concave, spreading. Stamens 10. Ovary sparsely hairy, 5-locular, free, 4 mm across. Fruit samara, 1-5 partite, each single-seeded with large membranous wing. Seeds pendulous, compressed, radicle small.

Very rare, in the study area. Fl. & Fr. Jan. – Mar. Way to Thimmamma Marrimanu (ATP), *KRKS & BSS* 41947.

OCHNACEAE

OCHNA L.

Ochna obtusata DC., Ann. Mus. Natl. Hist. Nat. 17: 411. 1811; var. **gamblei** (King ex Brandis) Kanis, Blumea 16: 34. 1968; FAP 1: 182.1997; *O. gamblei* King ex Brandis, Indian Trees 128. 1906; FPM 1: 166. 1915. *O. beddomei* Gamble, Fl. Madras 1:166.1915. & Bull. Misc, Inform. 1916: 34. 1916.

Local name: *Kuka moi.*

Deciduous small tree, 1–4 m high, stem thick brown; branchlets stout; bark thick, whitish. Leaves apically clustered, subsessile, variable, obovate-oblong, broadly elliptic, 9–16 × 2–6 cm, glabrous, white waxy glaucous, base attenuate-obtuse, margin distinctly crenate, apex acute-acuminate, petiole to 1 cm. Inflorescence many-flowered, compound raceme, peduncle 0.5–3 cm long. Sepals 5, ovate or ovate-oblong, 10–20 × 5 – 10 mm. Petals 5 – 8, yellow, 3–4.5 cm across. Stamens *ca.* 50; filaments 2 mm; anthers 6 mm. Ovary 10-lobed; ovule 1 per lobe. Drupe 10-lobed, 1.5 × 1 cm. Seeds many, brown, elliptic 1 × 0.8 cm.

Common, in dry deciduous forest. Fl. & Fr.: Mar. – Jun. Kalasamudram (ATP), *KRKS & BSS* 39559, 39826; Batrepalli (ATP), *KRKS* 39157.

BURSERACEAE

Key to Genera

1a. Bark geyish-brown, scaly; leaflets 6-15 pairs, oblong to lanceolate.... **Boswellia**

1b. Bark greenish yellow; leaflets < 3 pairs, ovate-suborbicular........ **Commiphora**

BOSWELLIA Roxb. ex Colebr.

Boswellia serrata Roxb. ex Colebr., Asiat. Res. 9: 379. t. 5. 1807; FBI 1: 528. 1875; FPM 1: 168. 1915; FAP 1: 184. 1997. *B. serrata* Roxb. ex Colebr. var. *glabra* (Roxb.) A.W. Bennett in Hook.f., Fl. Brit. India 1: 528. 1875. *B. glabra* Roxb., Pl. Coromandal 3: t. 207. 1811.

Local name: *Andaka, Anduga-pisunu, Tellaguggilamu.*

Large tree, to 15 m high, bark brown, scaly; branchlets with annual swelling at intervals. Leaves compound, 15–23 cm long, rachis pubescent, leaflets imparipinnate, 7–19, sessile, variable in size, shape and degree of pubescence, ovate-lanceolate, elliptic-oblong, 0.5–8 × 1–2 cm, coriaceous, whitish glaucous below, green above,

Plate 4

A. *Ochna obtusata* var. *gamblei*, B. *Gymnosporia heyneana*, C. *Ampelocissus tomenosa*, D. *Cardiospermum corindum*, E. *Moringa concanensis* ;, F. *Alysicarpus heyneanus*

base rounded or cuneate, margin entire or serrate, apex obtuse, acute. Panicle axillary, up to 21 cm long; bracts subulate, pubescent, caducous. Flowers pink, fragrant. Calyx tube broadly campanulate, 1.5–3 × 2.5–4 cm; lobes ovate-oblong or triangular. Petals ovate-oblong, 5–7 × 2.5–4 mm, shortly clawed, obtuse, pubescent outside. Stamens 2.5–5 mm long, glabrous. Capsules, 3-gonous, greenish yellow, ovoid 1.2 – 2.4 × 1 – 1.4 cm. Seed 1, cordate.

Common, in dry deciduous forest. Fl. & Fr.: Mar. – Aug. Kalasamudram (ATP), *KRKS & BSS* 39503.

COMMIPHORA Jacq. *nom. cons.*

Commiphora caudata (Wight & Arn.) Engl. in A. & C. DC., Monogr. Phan. 4: 27. 1883; FPM 1: 171. 1915; FAP 1: 185. 1997. *Protium caudatum* Wight & Arn., Prodr. Fl. Ind. Orient. 176. 1834. *P. caudatum* Wight & Arn. var. *roxburghianum* (Wight & Arn.) A.W. Bennett in Hook. f., Fl. Brit. India 1: 530. 1875. *P. roxburghum* Wight & Arn., Prodr. Fl. Ind. Orient. 176. 1834.

Local name: *Adavi raavi, Adavi mamidi.*

Deciduous unarmed aromatic tree, to 12 m high, bark green, smooth, turning brown or yellow which peels off in thin papery flakes. Leaves alternate, 8 cm, imparipinnate, 3–7-foliolate, leaflets sub-sessile, glabrous, ovate-elliptic or ovate, light fragrant, 2.7–10.5 × 1.3–5 cm, chartaceous, glossy above, base cuneate-attenuate, margin entire, apex long acuminate or caudate. Flowers unisexual or bisexual in axillary or terminal panicles, pale green to red or light yellow. Calyx campanulate or cupular, 4-lobed; lobes ovate or deltoid, 2–4 mm long, as long as broad, acute, persistent. Petals 4, linear-oblong or oblanceolate, apiculate at apex, 4–6 mm long. Disk cupular, intrastaminal. Stamens 8, free, alternately long and short, 1–3.5 mm long. Ovary ovoid-oblong, 2-locular. Drupes, ellipsoid, 0.7–1.7 × 0.8–1.5 cm, black. Seed solitary, brown, 3 mm long, with four radiating wings.

Occasional, in dry deciduous forests. Fl. & Fr.: Mar. – Jul. Batrepalli (ATP), *KRKS, BSS & SSR* 39514, 39096.

MELIACEAE

Key to Genera

1a. Fruit capsule; leaflets ≥ 7 cm long, margin entire .... **Soymida**

1b. Fruit drupe; leaflets ≤ 5 cm long, margin crenate .... (2)

2a. Bark longitudinally furrowed, rough; fruit ovoid, oblong-ellipsoid, pulpy .... **Azadirachta**

2b. Bark not furrowed, smooth; fruit globose-ovoid, dry .... **Melia**

AZADIRACHTA A. Juss.

Azadirachta indica A. Juss., Mem. Mus. Hist. Nat. 19: 221. t. 13. f. 5. 1830; FPM 1: 177. 1915; FAP 1: 188. 1997. *Melia azadirachta* L., Sp. Pl. 385. 1753; FBI 1: 544. 1875. *Melia indica* Brandis, Forest Fl. N.W. India 67. 1874.

Local name: *Vepa.*

Deciduous tree, to 40 m high, bark longitudinally furrowed, reddish-brown. Leaves alternate, compound, 20–25 cm long, leaflets imparipinnate, subopposite, 4–6 × 1–1.5 cm, glabrous, base oblique, attenuate, margin distinctly crenate, apex acuminate. Flowers white, 3 × 2 mm, in axillary panicles; pedicel 0.2 cm; peduncle 0.5 cm. Ovary subglobose, glabrous. Drupe ovoid, ellipsoid, 1.5 cm long, yellow, fleshy. Seed 1, whitish brown, elliptic oblong with coarse surface.

Common, on roadsides and villages, rare in forests. Fl. & Fr.: Feb. – Jul. Mobbuchintala palle (KDP), *KRKS, SSR, KRS & TP* 33658.

The plant is well-known for its medicinal value. Seed oil is antiseptic.

MELIA L.

Melia azedarach L., Sp. Pl. 384. 1753; FBI 1: 544. 1875; FPM 1: 176. 1915; FAP 1: 189. 1997.

Local name: *Thuraka vepa.*

Deciduous tree, to 10 m high, bark thick brown, smooth. Leaves bipinnate, 30–60 cm long; secondary pinna to 8 cm long, leaflets opposite, imparipinnate, glabrous, 3–3.5 × 1–1.5 cm, base attenuate, margin deeply crenate, apex acuminate. Flowers lilac, in axillary lax panicles. Pedicel 0.7 cm long. Ovary glabrous. Drupe, ellipsoid-subglobose, dry when ripe, 1.2 cm in diam., glabrous, yellow. Seed 1, rugose, globose, light brown.

Rare in dry deciduous forests. Cultivated in villages and agriculture lands for wood. Fl. & Fr.: Mar. – Jul. Batrepalli (ATP), *KRKS & BSS* 37981.

Seeds are edible. Wood valuable for domestic purpose.

SOYMIDA A. Juss.

Soymida febrifuga (Roxb.) A. Juss., Mém. Mus. Natl. Hist. Nat. 19: 251. 1832; FAP 1: 190. 1997; FBI 1: 567. 1875; FPM 1: 185. 1915.

Local Name: *Somi, Somida*

Lofty glabrous tree, up to 15 m high; branchlets with persistent leaf-scars. Leaves alternate, up to 35 cm long, even pinnate; leaflets 6 pairs, opposite, ovate-oblong, 7–9 × 3–5 cm, coriaceous, glabrous, base rounded, margin entire, apex obtuse; petiole to 6 cm long. Flowers white in terminal and axillary panicles. Sepals 5, ovate, 2 mm pubescent, imbricate. Petals 5, obovate, 5 mm, pubescent without. Disk flat. Staminal tube cupular, 10-lobed. Stamens 10. Ovary ovoid, 5-celled; ovules many. Capsule obovoid, to 7.5 cm long, woody, 5-celled, septifragally 5-valved; valves 2-lamellate, separating from the thick 5-winged axis. Seeds oblong, 4 × 1.5 cm, compressed, winged at both ends.

Planted along road sides. Fl. & Fr.: Feb. – Apr. Mabbuchintalapalle (KDP), *KRKS & BSS* 41804.

OPILIACEAE

1a. Petals connate, urceolate; flowers tetramerous in spikes; stigma on a short style, shallowly 4-lobed .... **Cansjera**

1b. Petals free; flowers pentamerous in racemose cymes; stigma sessile, simple .... **Opilia**

CANSJERA Juss. *nom. cons.*

Cansjera rheedii J.F. Gmel., Syst. Nat. 1: 280. 1791; FBI 1: 582. 1875; FPM 1: 193. 1915; FAP 1: 194. 1997.

Twining shrubs, up to 5 m long, branches sometimes spiny, younger ones as well as the inflorescence, petioles and tube of the flower pubescent. Leaves ovate or oblong-lanceolate, 3-5 × 2-3 cm, thin coriaceous, glabrous, base obtuse, sometimes oblique, margin entire, apex acuminate to caudate, petiole 5 mm. Flowers green or yellowish-green, 0.2-0.3 cm long, axillary, solitary or 2 or 3 in a spikes; corolla urceolate, greenish yellow, tube 2.5-3 mm long, lobes recurved, ca 0.5 mm; filaments *ca.* 2 mm long, anthers broadly ovate, reaching as far as throat of perianth tube; disk scales ovate, acute, irregularly toothed, slightly fleshy. Drupe ovoid or ellipsoidal, reddish-orange when ripe.

Common, in hilly areas. Fl. & Fr.: Nov. – Mar. Kalasamudram RF (ATP), *EC* 2982.

OPILIA Roxb.

Opilia amentacea Roxb., Pl. Coromandel 2 : 31 t. 158. 1802; FBI 1: 583. 1875; FPM 1: 192. 1915; FAP 1: 195. 1997

Scandent shrubs or low trees, up to 10 m high, branchlets pubescent. Leaves elliptic or oblong-lanceolate, 5-12 × 3-5 cm, inequilateral at times, thin coriaceous, glabrous, base obtuse, attenuate or subacute, margin entire, apex acuminate; petiole 8-9 mm long. Flowers greenish yellow, numerous in axillary racemose cymes, resembling cones in bud; bracts peltate, 2-3 mm wide, pedicels 1-2 mm long. Perianth 1-2 mm long, ovate, shortly pubescent outside. Stamens 5, antipetalous, filaments filiform, *ca.* 1.5 mm long, anthers ovoid; disk lobes club-shaped to ovoid; Ovary ellipsoid. Drupe ovoid-globose, 1 × 1 cm, apiculate, yellow-red when ripe; seed 1, ovoid, blackish.

Occasional, in dry deciduous forests. Fl. & Fr.: Apr. –Jul. Kalasamudram RF (ATP), *EC* 2982.

CELASTRACEAE

Key to Genera

1a. Trees; leaves ovate; flowers in cymes .... **Cassine**

1b. Shrubs; leaves obovate; flowers other than cymes .... (2)

2a. Inflorescence terminal panicle; fruit subglobose .... **Celastrus**

2b. Inflorescence axillary fascicle; fruit obovoid .... **Gymnosporia**

CASSINE L.

Cassine albens (Retz.) Kosterm., Gard. Bull. Sing. 39(2): 178. 1986. *Schrebera albens* Retz., Obsev. Bot. 6: 25. t. 3. 1791. *Mangifera glauca* (Vahl) Pers., Syn. 1: 241. 1806; FBI 1: 623. 1877 *p.p.*; FPM 1: 211. 1918. *Celastrus glaucus* Vahl, Symb. Bot.

2: 42. 1791. *Elaeodendron glaucum* M. Lawson in Hook.f., Fl. Brit. India 1: 623. 1875. *p.p.*; FPM 1: 211. 1918. *Cassine glauca* (Rottb.) Kuntze, Revis. Gen. Pl. 1: 114. 1891; FAP 1: 97. 1997. *Mangifera glauca* Rottb., Nye Saml. Kongel. Danski Vidensk. Skr. 2: 534. t. 4. f. 1. 1783.

Local name: *Bira, Kannilu, Nerasi.*

Tree, to 10 m high, branchlets reddish, subterete. Leaves opposite, ovate to elliptic-oblong, 2–7 × 1.2–4.2 cm, glabrous, base cuneate-obtuse, margin crenate, apex acute or obtuse; lateral veins 5–7 pairs; petiole to 3 cm long. Flowers 5-merous, in lax cymes at upper part of the branchlets; bracts ovate, acute, small. Calyx lobes unequal, orbicular, margins membranous. Petals greenish, oblong, margins membranous, 5 × 3 mm. Stamens on the lobes of the disk, shorter than petals, anthers 5. Ovary 1. Fruit drupaceous, ovoid, 1.2 cm long, apiculate. Seed 1, crustaceous.

Occasional, along the hill slopes of dry deciduous forests. Fl. & Fr.: April. – Jul. Thimmamma marrimanu (ATP), *KRKS* 37971.

CELASTRUS L.

Celastrus paniculatus Willd., Sp. Pl. 1: 1125. 1797; FBI 1: 617. 1875; FPM 1: 208. 1918; FAP 1: 197. 1997.

Local name: *Bavanji, Brukata.*

Scandent shrub or small tree, to 3 m high, stem dark brown, white lenticellate, branches weak, pendulous. Leaves alternate, obovate-orbicular, elliptic-ovate, 5–8 × 3–6 cm, glabrous, base cuneate, margin crenate, apex acute-acuminate; petiole to 2 cm long, lateral veins 5–8 pairs. Inflorescence terminal panicle, flowers unisexual. Male flowers: calyx 5-lobed; lobes semi-orbicular, ciliate, up to 2 mm long. Petals oblong to obovate-oblong, up to 3 mm long. Stamens 3 mm long. Female flowers: Sepals, petals and disk as in the male flowers. Stamens sterile. Ovary globose. Fruit capsular, subglobose, 0.5 cm diam., 3-locular, 1-seed per locule. Seeds 3-angular, ellipsoid, brownish, arillate, 3 × 4 mm.

Common, in dry deciduous forests. Fl. & Fr.: April. – Aug. Batrepalli (ATP), *KRKS & BSS* 39596.

Note: Seeds yield an oil which is used in certain stomach disorders.

GYMNOSPORIA (Wight & Arn.) Benth. & Hook.f.

Gymnosporia heyneana (Roth) M.A. Lawson Fl. Brit. India 1: 620. 1875. *Maytenus emarginata* (Willd.) Ding Hou in Fl. Males. I. 6(2): 241. 1962; sensu FAP 1: 199. 1997. *Celastrus emarginatus* Willd., Sp. Pl. 1.2: 1128. 1798. *Gymnosporia montana* (Roth) Benth., Fl. Austral. 1: 400. 1863; FPM 1: 209. 1918.

Local name: *Danti.*

Shrub up to 4 m high, spiny or unarmed; spines bear leaves and flowers; stem glabrous, pinkish. Leaves obovate, 2–5 × 1–4 cm, glabrous, coriaceous, base attenuate, margin crenate-serrate, apex acute-obtuse; petiole to 2 cm long, lateral veins 5–6 pairs. Flowers white in axillary fascicles or dichotomously branched cymes; bracts deltoid, fimbriate. Sepals 5-lobed, 1 mm long, minutely ciliate. Petals 5, oblong,

1 mm across. Stamens 5. Ovary immersed in the disk; stigma 3-fid. Capsule pyriform, 3-angular, 3-locular, obcordate, 0.8 cm, yellow, splitting into 3 valves. Seeds 1 per locule, curved, orange brown, half of the seed covered with fleshy white aril.

Common, in scrub forest of Mabbuchintalapalle and Bhumaiahgari palle. Fl. & Fr.: Sept.-Dec. Mabbuchintalapalle (KDP), *KRKS & SSR* 39026; *KRKS* 266 (HJCB-E).

Note: This species missapplied for *Gymnosporia emarginata* by many taxonomists. *G. emarginata* differs from this species with smaller green leaves, stem blackish brown, spines have no leaves, flowers clustered in axils, and functionally male and female, ovary red. *G. emarginata* specimen have been collected from Palamaneru forest and Sadasivakona Chittoor district, but not from Thummalapalle Mining Area.

RHAMNACEAE

Key to Genera

1a. Plants unarmed .... **Ventilago**

1b. Plants armed .... **Ziziphus**

VENTILAGO Gaertn.

Ventilago madraspatana Gaertner, Fruct. Sem. Pl. 1: 223. t. 47. f. 2. 1788; FBI 1: 631. 1875; FPM 1: 218. 1918; FAP 1: 206. 1997 var. **madraspatana**.

Climbing shrubs; branches terete, striate, sparingly puberulous or glabrous. Leaves elliptic-ovate or lanceolate, *ca.* 8 × 3 cm, thin- coriaceous, base acute, margin entire, apex obtuse or acute. Flowers pale green in terminal and axillary branched panicles. Calyx lobes spreading. Petals emarginated with a tooth in the middle. Stamens 1.5 mm long; disk glabrous. Ovary villous, style arms divergent or curved. Fruits of winged nuts, about 3 cm across, enclosed at base by adherent calyx tube near base, wings glabrous, 1- seeded. Seeds globose, brown.

Occasional in drier hilly forests. Fl. & Fr.: Dec.-Apr. Kalasamudram RF (ATP), *NY* 1201.

ZIZIPHUS P. Mill.

Key to Species

1a. Trees; leaves elliptic-orbicular; fruit villose, green, not pulpy, hard .... **Z. xylopyra**

1b. Shrubs; leaves glabrous above, ovate-elliptic; fruit glabous, black or dark red, pulpy .... (2)

2a. Stragglers; leaves ferruginous above with brown hairs; fruit black when ripe, sour taste .... **Z. oenoplia**

2b. Trees or shrubs; leaves glabrous above, rounded at apex; fruit orange red when ripe, sweet and sour taste .... **Z. jujuba**

Ziziphus jujuba Mill., Gard. Dict. ed. 8. 1. 1768; FBI 1: 632. 1875; FPM 1: 219. 1918.

Key to Varieties

1a. Tree; fruit more pulpy, >0.8 cm diam ....var. **jujuba**

1b. Shrub; fruits less pulpy, < 0.8 cm diam ....var. **fruticosa**

Ziziphus jujuba Mill. var. **fructicosa** Haines, For. Fl. Chota Nagpur 270. 1990; FBI 1: 632. 1875; FPM 1: 219. 1918. *Z. mauritiana* Lam. var. *fruticosa* (Haines) Sebest. & Balakr., Indian Forest. 89: 525. 1963; FAP: 208. 1997. *Z. mauritiana* Lam. var. *fructicosa* (Haines) Raizada & Saxena, Indian Forest. 92: 326. 1966. *nom. illeg.*

Local name: *Regu, Reni, Bogi pallu.*

Armed much branched shrub, to 3 m high, branches pubescent. Leaves alternate, variable, broadly elliptic or ovate, 2–3 × 1–1.5 cm, tomentose beneath, glabrous above, base subcordate, margin crenate, apex acute or retuse, basally 3-nerved, convergent venation. Flowers greenish in axillary fascicles, 5-merous. Sepals and petals 5, ovate-oblong. Ovary inserted in disk. Fruit glabrous, globose, orange, 0.6 - 1 cm in diam., apiculate tip, sweet taste; less pulpy. Kernel reticulately rugose, hard, globose. Seeds 2, compressed, 3 mm diam., dark brown, elliptic, diskoid.

Common, in scrub forests and agriculture lands. Fl. & Fr.: Sept.-Mar. Kalasamudram (ATP), *KRKS, BSS & TP* 39876.

Ziziphus jujuba Mill. var. **jujuba**

Local name: *Regu, Reni.*

Armed small tree, to 5 m high, stem brown, bark longitudinally peeling off. Leaves alternate, elliptic-orbicular, 2–3.5 cm across, nearly symmetrical, basal veins 3, convergent venation, glabrous above, tomentose below, base subcordate, margin crenate, apex truncate or obtuse, stipular thorns 2, curved one up to 1 cm long and straight one up to 2 cm long; petiole 2–3 mm long. Flowers greenish in axillary clusters, 5-merous. Fruit drupe, 1.5 cm in diam., more pulpy, globose, kernel hard, rugose. Seeds 2, compressed, 3 mm diam., dark brown.

Common, in scrub forest and margins of agriculture land. Sometimes cultivated. Fl. Fr.: Sept.-Mar. Kalasamudram (ATP), *KRKS, BSS & TP* 39808.

Ripe fruits are edible, eaten by local people.

Ziziphus oenoplia (L.) Mill., Gard. Dict. ed. 8: 3.1768; FBI 1: 634. 1875; FPM 1: 220. 1918; FAP 1: 209. 1997. *Rhamnus oenoplia* L., Sp. Pl. 194. 1753.

Local name: *Paranu, Pariki.*

Armed straggling shrub, to 3 m high, stem pinkish brown. Leaves with brown hairs, alternate, ovate-elliptic or elliptic-oblong, 2–3 × 0.8–2 cm, ferruginous hairy above, tomentose below, base subcordate, margin crenate, apex acute, basally 3–4-nerved, convergently reticulate. Thorns 2–3 mm long, axillary. Flowers green in axillary clustered fascicles; pedicel 3–4 mm long. Fruit 0.6 cm diam., glabrous, black when ripe, kernel hard, reticulate, rugose. Seeds 1–2, flattened, 2 mm across.

Common, in dry deciduous forests and scrub forests. Fl. & Fr.: Sept. – Dec. Batrepalli (ATP), *KRKS* 36951; Kalasamudram (ATP), *KRKS & BSS* 39818.

Ziziphus xylopyra (Retz.) Willd., Sp. Pl. 1: 1104. 1797; FBI 1: 634. 1875; FPM 1: 220. 1918; FAP 1: 210. 1997 (*'xylopyrus'*). *Rhamnus xylopyrus* Retz., Observ. Bot. 2: 11. 1781.

Local name: *Goti, Gotiki.*

Armed small tree, to 7 m high, branches rusty tomentose. Leaves alternate, elliptic-oblong, base cordate, margin crenate, apex mucronate or obtuse, yellowish hairy below, puberulose above, basal nerves 3, divergent; thorns 2, axillary, curved one 0.5 cm long and straight one 1 cm long. Flowers green in axillary, clustered cymes, pubescent, 5-merous. Fruit globose, villous, 0.5–1 cm diam., not pulpy, mucronate or apiculate at apex, kernel 0.8–0.9 cm diam., irregularly reticulate, hard. Seeds 3, compressed black, 1–2 × 1–1.5 mm.

Common, in hills of scrub forest. Fl. & Fr.: May – Sept. Kalasamudram (ATP), *KRKS & KRS* 36456.

VITACEAE

Key to Genera

1a. Flowers 5-merous; disk annular or 5- 10-lobed .... **Ampelocissus**

1b. Flowers 4-merous; disk 4-lobed .... (2)

2a. Flower buds flask-shaped; nectariferous disk of 4, free glands .... **Cyphostemma**

2b. Flower buds globose or oblong; nectariferous disk entire .... (3)

3a. Inflorescence axillary; berries 2–4-seeded .... **Cayratia**

3b. Inflorescence leaf-opposed; berries 1–2-seeded .... **Cissus**

AMPELOCISSUS Planch. *nom. cons.*

Ampelocissus tomentosa (Heyne ex Roth) Planch., J. Vigne Amer. 8: 375. 1884; FPM 1: 230. 1918; FAP 1: 211. 1997. *Vitis tomentosa* Heyne ex Roth, Nov. Pl. Sp. 157. 1821; FBI 1: 650. 1875.

Local name: *Atukula baddu, Nallagummadi theega.*

Climbing, stout shrub, up to 8 m long; young branches slender, angular, whitish woolly tomentose. Leaves simple, 5–7-lobed, 7–15 × 8–12 cm, tomentose above, woolly below, base cordate, margin serrate, apex acute-apiculate, basal veins 5; petiole to 10 cm long. Inflorescence dichotomous umbellate cyme, extra-axillary; peduncles *ca.* 10 cm long, stout; peduncle modified in to tendrils; tendrils branched; primary branch of inflorescence 1 – 2 cm long; cymes 2 – 4 cm long. Flowers *ca.* 1.5 mm long, red, subsessile. Calyx cupular, 5-lobed, villose. Petals oblong-ovate, 1.2 × 0.7 mm, incurved at apex. Stamens *ca.* 1 mm long. Disk annular, restricted to lower half of ovary. Ovary *ca.* 1 mm across. Berries globose, 8–10 mm across, black. Seeds

1–4, oblong, 4–5 mm acute at base, adaxial surface with an elliptical pit from which fissures radiate towards margin.

Occasional, in dry deciduous forests. Fl. & Fr.: Nov. – Feb. Idupulapaya RF (KDP), *KRKS & SSR* 39054.

CAYRATIA A. L. Juss., *nom. cons.*

Cayratia pedata (Lam.) Juss. ex Gagnepain, Notul. Syst. (Paris) 1: 346. 1911; FPM 1: 236. 1918; FAP 1: 212. 1997. *Cissus pedata* Lam., Encycl. 1: 31. 1783. *Vitis canarensis* Dalz., Hook. Kew J. Bot. 3: 123. 1851; FBI 1: 655. 1875.

Large, slender climbing shrub, to 7 m long; branches subterete, pubescent, tendrils branched, leaf-opposed, pubescent. Leaves 5–9-foliolate, pedate, pubescent to glabrate; petiole 7–13 cm long, pubescent, terminal leaflets lanceolate to elliptic, 10–13 × 2–3.5 cm, base subcordate, margin serrate, apex acuminate; lateral leaflets 7–10 × 2–3 cm, base oblique to acute, margin serrate, apex acuminate. Inflorescence dichotomous cyme, 5–15 cm long, peduncles 2–6 cm, pubescent. Flowers 2–3 mm long; pedicels 2–3 mm long. Calyx cupular, truncate, pubescent. Petals white, incurved at apex, 2–2.5 × 1–1.5 mm, pubescent to glabrous. Stamens *ca.* 2 mm long. Disk entire, cupular, 4-lobed, encloses the ovary. Ovary *ca.* 1.5 mm across. Fruit berry, sub-globose, flattened at apex, 1–1.5 cm across. Seeds 2–4, obellipsoid, circular pit in the centre of adaxial side, 6 × 5 mm.

Occasional, in dry deciduous forest in moist places. Fl. & Fr.: Jan. – Nov. Namalagundu (KDP), *KRKS & BSS* 39658.

CISSUS L.

Key to Species

1a. Erect subshrubs; leaves simple, up to 20 cm across........................ **C. arnottiana**

1b. Climbing shrubs or (often tree froms); leaves lobed or simple, up to 10 cm across..(2)

2a. Stem and branches 4-angular.-leaves simple (-lobed), veins obscure..**C. quadrangularis**

2b. Stem rounded...(3)

3a. Stem glabrous, leaf simple, ovate, crenate; climbing........................... **C. repens**

3b. Stem pubescent; leaf lobed, ovate-rhomboid, entire; climber or often tree form.. **C. vitiginea**

Cissus arnottiana Shetty & P. Singh, Kew Bull. 44: 473. 1989. *Vitis pallida* sensu M. Lawson in Hook.f., Fl. Brit. India 1: 647. 1875, *p.p.*, (*non* Wight & Arn., 1833). *Cissus pallida* sensu Planchon in DC., Monogr. Phan. 5: 477. 1887, *p.p.* (*non* salisb., 1796); FPM 1: 234. 1918; FAP 1: 241. 1997.

Local name: *Adavi gummadi, Buradoni, Nelagummadi, Tella budari.*

Erect sub shrub, to 1.5 m high, stem woody, reddish, glabrous, white glaucous, pubescent at younger leaves and stem. Tendrils leaf-opposite, simple. Leaves simple, suborbicular, 9–20 × 10–18 cm, glabrous, base truncate-subcordate, margin dentate-serrate, apex acute-apiculate, basal veins 5; petiole to 13 cm long, Inflorescence compound cyme, leaf-opposed, peduncle to 5 cm long; pedicels 2 – 4 mm long, pubescent. Flowers *ca.* 3 mm diam., white. Calyx cupular, 4-lobed; lobes 2 mm long, truncate. Petals white, 2.5 × 1.5 mm, glabrous. Stamens *ca.* 2 mm long. Disk 4-lobed, fleshy, white, covers ovary. Ovary *ca.* 1 mm across. Fruit berry, subglobose, glabrous, *ca.* 0.8 cm across, black, mucronate, edible. Seeds 1–2, subglobose-obovate, 0.5 cm across, dark brown.

Common, in scrub and dry deciduous forests. Fl. & Fr.: Jun. – Nov. K.K. Kottala (KDP), *KRKS & KRS* 37657; Kalasamudram (ATP), *KRKS* 39542.

Cissus quadrangularis L., Syst. Nat. ed. 12. 2: 124. 1767; FPM 1: 233. 1918; FAP 1: 215. 1997. *Vitis quadrangularis* (L.) Wall. ex Wight & Arn., Prodr. Fl. Ind. Orient. 125. 1834; FBI 1: 645. 1875.

Local name: *Nalleru, Nallerutiga.*

A perennial rambler, stem succulent, 4-angular, angles winged, tendrils simple. Leaves simple or 3–5-lobed; if simple ovate-reniform, 2–6 × 2–5 cm, if lobed lobes obovate, 3–5 × 2–3 cm, middle larger than laterals, glabrous, base truncate-cordate, margin crenate-dentate, apex obtuse-acute, basal nerves 3; petiole to 2 cm long. Inflorescence leaf-opposed, umbellate cyme; peduncle 1.5–3 cm long. Flowers white, 3.5 mm across. Calyx cupular, 4-lobed; lobes elliptic-ovate, 2 × 1.5 mm, reddish outside, glabrous. Corolla white, ovate-oblong, 1.5 mm long, hooded, glabrous. Stamens *ca.* 1.5 mm long. Disk 4-lobed, covering ovary. Ovary 1.5 mm across; style short, stout. Fruit berry, globose, 5 mm across, red-black, glabrous. Seeds 1-2, obovoid, 5 × 4 mm, smooth, thick brown, longitudinally encircling ridged raphe.

Common, in scrub forest. Fl. & Fr.: Jul. – Nov. Thummalapalle (KDP), *KRKS, KRS, SSR & TP* 33641; Kalasamudram (ATP), *KRKS & KRS* 36468.

Cissus repens Lam., Encycl. 1: 31. 1783; FPM 1: 233: 1918; FAP 1: 215. 1997. *Vitis repens* (Lam.) Wight & Arn. in Wight, Cat, 58. 1833; FBI 1: 646. 1875.

Local name: *Kuppikodda, Nelaboddu.*

Slender, climbing shrub, 4–7 m long; stem white-glaucous, glabrous, angular, tendrils branched, young stem and leaves pubescent. Leaves simple, ovate, 5–8 × 4–7 cm, glabrous above, sparsely hairy below when young, base cordate or truncate, margin serrate-crenate, apex acute-acuminate, basal veins 5; petiole 2–3.5 cm long; pubescent; stipule pubescent. Inflorescence corymbose, umbellate cyme, leaf-opposed, 5–10 cm long, bracteolate; peduncles 2.5–5 cm long. Flowers 2.5–3 mm long, contorted at apex; pedicels 3 – 6 mm long, glabrous. Calyx cupular, 4-lobed; lobes sub truncate, glabrous. Petals white, oblong-ovate, 2 × 1 mm. Disk 4-lobed, covering ovary. Ovary 1 mm across. Fruit berry, ellipsoid, 5 mm across, fleshy, black. Seed 1, 5 × 3 mm, longitudinally encircling ridged raphe.

Occasional, on stream sides of dry deciduous forests. Fl. & Fr.: Aug. – Nov. Kalasamudram (ATP), *KRKS & BSS* 39507; Batrepalli (ATP), *KRKS* 307 (HJCB-E).

Cissus vitiginea L., Sp. Pl. 117. 1753; FPM 1: 234. 1918; FAP 1: 216. 1997. *Vitis linnaei* Wight & Arn., Prodr. Fl. Ind. Orient. 126. 1834; FBI 1: 649. 1875.

Local name: *Godhuma theega, Gummadi, Nela gummadi, Varupaku.*

Climbing shrubs, often arborescent trees, 2–5 m high, young branches densely yellowish pubescent; tendrils when present stout, simple or branched, leaf-opposed. Leaves alternate, rotund-ovate, simple or 3–5-lobed, 5–9 × 4–8 cm, sparsely pubescent above, dense below, base truncate-cordate, margin irregular dentate, apex acute, basal nerves 3; petiole 1.5–2 cm long, whitish pubescent. Inflorescence leaf-opposed umbellate cyme or corymbs; peduncle to 2 cm long. Flowers *ca.* 2 mm long; pedicels 2 – 4 mm long, pubescent. Calyx cupular, 4-lobed, pubescent. Petals white, oblong-ovate, *ca.* 2 × 1 mm, hooded, pubescent. Stamens *ca.* 1 mm long. Disk 4-lobed, fleshy, covering ovary. Ovary *ca.* 1 mm across. Fruit berry, pyriform, apiculate, 5 – 7 mm across, glabrous, mucronate, 0.5 cm in diameter, purple with a glaucous bloom on surface, 1-seeded. Seeds 7–8 × 5–7 mm, tessellated on either side of raphe, with angular facets.

Common, in dry deciduous forests. Fl. & Fr.: Feb. – Nov. Kalasamudram (ATP), *KRKS & KRS* 37962; Batrepalli (ATP), *KRKS & BSS* 39004, 39134.

CYPHOSTEMMA (Planch.) Alst.

Cyphostemma setosum (Roxb.) Alston in Trimen, Handb. Fl. Ceylon 6 (Suppl.); 53. 1931; FAP 1: 216. 1997. *Cissus setosa* Roxb., Fl. Ind. 1: 428. 1820; FPM 1: 235. 1918. *Vitis setosa* (Roxb.) Wall. ex Wight & Arn., Prodr. Fl. Ind. Orient. 127. 1834; FBI 1: 654. 1875.

Local name: *Barrebachali, Pullabachali.*

Herbaceous straggling shrub with glandular bristly setose branches, succulent; tendrils branched, leaf-opposed. Leaves subsessile, 3-foliolate, sometimes the lower ones simple, glabrescent above, glandular-hispid below; upper leaflets elliptic to obovate-oblong, 5–12 × 3–5 cm, base acute to obtuse, margin irregularly serrate to laciniate, apex acute to obtuse; petiole to 3 mm long; stipules broadly ovate, acute; petiolules up to 1 cm long. Inflorescence dichotomous or trichotomous, lax, divaricate, 4–5 cm long cymes; peduncles 2.5–8.5 cm long, glandular-hispid. Flower buds flask-shaped. Flowers lageniform, *ca.* 2.5 mm long; pedicels 2–3 mm long, glandular-hispid. Calyx cupular, 4-lobed; lobes subtruncate, glabrous. Petals 4, oblong, hooded, *ca.* 2 × 1 mm, glabrous. Stamens *ca.* 2 mm long. Disk of 4 truncate glands. Ovary *ca.* 1.5 mm across. Fruit berry, ovoid, 6–7 mm across, red, glandular-hispid, 1-seeded. Seeds subglobose, 7 – 8 × 5–6 mm, crenate on margin.

Common, in dry deciduous forests. Fl. & Fr.: Jul. – Oct. Kalasamudram (ATP), *KRKS & TP* 39896; Batrepalli (ATP), *KRKS* 267 (HJCB-E).

SAPINDACEAE

Key to Genera

1a. Climbers with branched tendrils. -herbaceous.......................... **Cardiospermum**

1b. Shrubs and trees...(2)

2a. Leaves simple; fruit winged. -leaves shiny .... **Dodonaea**

2b. Leaves compound; fruit not winged .... (3)

3a. Fruit saponaceous; seed gobose, black .... **Sapindus**

3b. Fruit not saponaceous; seed compressed, brown .... **Schleichera**

CARDIOSPERMUM L.

Key to Species

1a. Capsules pyriform, winged at the angles; seeds with large heart-shaped hilum .... **C. halicacabum**

1b. Capsules subglobose, ovate when young, not winged; seeds with small, orbicular hilum .... **C. corindum**

Cardiospermum corindum L., Sp. Pl. 526. 1753. *C. canescens* Wallich, Pl. Asiat. Rar. 1: 14. 1829; FBI 1: 670. 1875; FPM 1: 245. 1918; FAP 1: 222. 1997.

Climber, to 6 m long; branches herbaceous, softly hairy, angular, tendrils branched, bearing flowers and fruits. Leaves biternate, leaflets 3–5, ovate, 1–2.5 × 0.6–1.2 cm, sparsely hairy above, densely pubescent below, lateral veins 5–6 pairs, base obtuse, margin dentate or irregularly toothed, apex acute, mucronulate; petiole up to 5 cm long; petiolules 0.1-1 mm long. Inflorescence umbellate cyme on axillary, coiled tendrils; peduncles up to 10 cm long, the lowest pair of pedicels transformed into spiral tendrils. Calyx campanulate; lobes ovate, to 3 mm long, pubescent. Petals white, oblong, obtuse at apex. Stamens yellow. Ovary 3 mm long. Capsule subglobose, 4–3.5 cm, sparsely hairy, wings absent. Seeds globose, smooth, black, with white orbicular hilum.

Common, in dry deciduous forest and field bunds. Fl. & Fr.: Throughout the year. Thummalapalle (KDP), KRKS, *KRS, SSR & TP* 33643.

Cardiospermum halicacabum L., Sp. Pl. 366. 1753; FBI 1: 670. 1875; FPM 1: 244. 1918; FAP 1: 222. 1997.

Local name: *Budaga theega.*

Annual or perennial climber, to 6 m long; tendrils bearing flowers; branches herbaceous, furrowed, pubescent. Leaves biternate, 3–8 cm long; leaflets deltoid or ovate, 1–2.5 × 0.6–1 cm, deeply incised, sparsely hairy above, densely hairy below, base obtuse-acuminate, margin irregularly dentate, apex acute-apiculate; petiole up to 5 cm; petiolules 0.1–1 mm long. Inflorescence umbellate cyme on axillary coiled tendrils; cymes resting on 2 opposite circinate tendrils; peduncle to 8 cm long. Flowers very small, white. Outer sepals rounded-obovate; inner rounded, larger than the outer, very thin. Petals rounded, spatulate. Capsule on short stalks, pyriform, 1–1.5 cm across, inflated, winged at the angles, trigonous, truncate at both ends. Seeds globose, black; hilum heart-shaped.

Common, in open forests near hedges and field bunds. Fl. & Fr.: Throughout the year. Idupulapaya (KDP), *KRKS & BSS* 39086.

DODONAEA L.

Dodonaea viscosa (L.) Jacq., Enum, Syst. Pl. 19. 1760; FBI 1: 697. 1875; FPM 1: 253. 1918; FAP 1: 223. 1997.

Local name: *Bandaraku.*

Shrubs to 4 m high; branches terete, often 4-angular, younger parts shiny and sticky or scurfy puberulous. Leaves simple, elliptic, oblanceolate-obovate, 1.2-5 × 0.3-1.2 cm, glabrous, viscid with shining yellowish resinous exudation, base abruptly tapering, margin entire, apex acute, lateral veins up to 35; petiole to 4 mm long. Inflorescence panicled cyme, to 7 cm long. Flowers greenish yellow, 1 mm across. Sepals oblong, to 2 mm long. Petals 0. Capsules membranous, compressed, wings 2 or 3, 1.5 cm long. Seeds 1–2, black, 2–3 mm across.

Common, in forests and scrub forest in much disturbed open gravelly and strong areas. Fl. & Fr.: Aug.– Jan. Thummalapalle (KDP), *KRKS & KRS* 36402.

Note: Twigs locally used for broom preparation.

SAPINDUS L.

Sapindus emarginatus Vahl, Symb. Bot. 3: 54. 1794; FPM 1: 250. 1918; FAP 1: 224. 1997. *S. trifoliatus* auct. non L. 1753; FBI 1 : 682. 1875.

Local name: *Kunkudukaya.*

Deciduous tree, 8–10 m high; branches terete, tomentose when young, bark rough, grey. Leaves alternate, 10–18 cm long; leaflets opposite, ovate-elliptic to oblong or obovate, 4–10 × 2.5–4 cm, glabrescent above, villous tomentose below, base acute, margin entire-undulate, apex emarginated-obtuse, lateral veins 10–12 pairs; petiole of basal leaf smaller than the apical leaf. Inflorescence terminal panicle, to 20 cm long. Flowers greenish white, 2–3 mm long. Sepals 5, hairy outer side. Petals 5, clawed, white, shining hairs. Stamens 8; anthers oblong. Ovary 3-lobed, densely hairy. Fruit drupe, ovoid, wrinked, brown, with saponaceous pulp. Seed 1 per lobe, hard, *ca.* 1 cm across, blackish red; cotyledons whitish yellow, edible.

Occasional, in dry deciduous forests. Cultivated in gardens and agriculture lands. Fl. & Fr.: Nov. – Mar. Chinnapalli (ATP), *KRKS & KRS* 37814.

Note: Fruits are used for headbath and also an important ingredient in shampoo and soap preparations.

SCHLEICHERA C.L. Willdenow *nom. cons.*

Schleichera oleosa (Lour.) Oken., Allg. Naturgesch 3(2): 1341. 1841; FAP 1: 225. 1997. *Pistacia oleasa* Lour., Fl. Cochinch. 2: 615. 1790. *Schleichera trijuga* Willd., Sp. Pl. 4: 1096. 1806; FBI 1: 681. 1875; FPM 1: 248. 1918.

Trees, to 13 m high, branches greyish, younger parts brown tomentose. Leaves alternate, to 30 cm long; leaflets opposite, paripinnate, 3 pairs, elliptic oblong, elliptic-ovate, 7–15 × 5–8 cm, brown, glabrous above, yellowish brown below, base obliquely obtuse, margin entire, apex acute-obtuse, lateral veins 10–19 pairs. Inflorescence axillary or terminal corymbose or panicles. Flowers polygamo-dioecious, greenish yellow. Calyx small, cupular; lobes 4–6, valvate or obscurely imbricate. Petals

'0'. Disk undulate, glabrous. Stamens 5–8, inserted within the disk. Ovary ovoid, narrowed, 3-celled, ovule 1 in each locule; stigma 3-lobed. Fruit dry, crustaceous drupe, ovoid, 2.5 × 1.5 cm, indehiscent, tipped with style like spin. Seeds 1-2, compressed, brown, enclosed in fleshy pulpy-aril.

Common, in dry deciduous forests. Fl. & Fr.: Jan. – May. Kalasamudram (ATP), *KRKS & BSS* 39194.

Wood is used for agricultural implements and carts.

ANACARDIACEAE

Key to Genera

1a. Shrubs; spines up to 3 cm long .... **Searsia**

1b. Trees; spines absent .... (2)

2a. Leaves compound; fruit subreniform, red .... **Lannea**

2b. Leaves simple; fruit not as above .... (3)

3a. Stamens usually twice the number of petals .... **Buchanania**

3b. Stamens as many as petals .... (4)

4a. Only one stamen fertile .... **Mangifera**

4b. All stamens fertile .... **Semecarpus**

BUCHANANIA Spreng.

Buchanania axillaris (Desr.) Ramam. in Saldanha & Nicolson, Fl. Hassan Dist. 374. 1976; FAP 1: 227. 1997. *Mangifera axillaris* Desr. in Lam., Encycl. 3: 697. 1792. *Buchanania angustifolia* Roxb., Pl. Coromandel t. 262. 1820; FBI 2 : 23. 1876; FPM 1: 259. 1918.

Vern. : *Sarapappu, Peddmorli.*

Trees, up to 8 m high, branchlets stout, glabrous, deeply fissured. Leaves simple, oblong-elliptic, 5-10 × 3-6 cm, base acute, margin undulate, apex emarginate coriaceous, lateral nerves about 15 pairs. Flowers white in axillary and terminal panicles. Calyx lobes 5, ovate, glabrous. Petals 5, oblong, reflexed. Stamens 10, filaments subulate, anthers oblong; disk 1.5 mm in diam. Ovary sub-oval, immersed in disk, style short, stigma truncate. Drupes compressed-globose. Seed 1, gibbose.

Occasional, in hilly forest areas. Fl. & Fr.: Jun.-Dec. Kalasamudram RF (ATP), *EC* 2918, 2996.

Nuts from the fruits are edible.

LANNEA A. Rich.

Lannea coromandelica (Houtt.) Merr., J. Arnold. Arbor. 19: 353. 1938; FAP 1: 228. 1935. *Dialium coromandelicum* Houtt., Nat. Hist. 2: 39. t. 5. f. 2. 1774. *Odina wodier* Roxb., Fl. Ind. 2: 293. 1832; FBI 2: 29. 1876; FPM 1: 263. 1918.

Local name: *Gumpin.*

Tree to 15 m high, bark dark brownish grey with longitudinal furrows; branchlets greyish black, leticellate, younger parts brown tomentose. Leaves compound, alternate, 15–35 cm long, petiole 7–13 cm long; leaflets 3–6 pairs, imparipinnate, elliptic-oblong, ovate-oblong, 3–6 × 4–5 cm, brown tomentose, base obliquely obtuse, margin entire, apex acuminate; petiolules to 2 mm long. Inflorescence terminal panicle to 15 cm long. Flowers unisexual yellowish-green, sub sessile. Male flowers 3 mm across. Sepals ovate, ciliate. Female flowers 2–2.5 mm across. Ovary superior, 1-loculed. Fruit drupe, subreniform, 1 × 0.7 cm, red. Seed 1, compressed, testa membranous, radicle thick, superior.

Occasional, in dry deciduous forests. Fl. & Fr.: Apr. – Jul. Batrepalli (ATP), *KRK & BSS* 39100; Kalasamudram (ATP), *KRKS* 41856.

MANGIFERA L.

Mangifera indica L., Sp. Pl. 200. 1753; FBI 2 : 13. 1875; FPM 1: 259. 1918; FAP 1: 229.1997.

Vern.: *Mamidi.*

Trees, up to 20 m high, branches glabrous, terete. Leaves simple, oblong or elliptic-lanceolate, dark green, 10-20 × 3-6 cm, alternate, shining, coriaceous, nerves 25 pairs, base cuneate-subacute, wavy along margins, apex acuminate. Flowers male and bisexual, yellowish-white in terminal tomentose panicles; pedicel 1.5-3 mm long; calyx 5-lobed, lobes tomentose; petals elliptic, with branched ridges, tips reflexed; stamens 5, 1 fertile, rest sterile, small, teeth like; disk distinctly 5-lobed, ovary obliquely ovoid, style 2 mm long, lateral. Drupes globose, obliquely gibbous on one side, stone compressed, fibrous; seeds oblong.

Common, in hill forests. Universally cultivated in gardens, avenues and 'topes' and run wild. Fl. & Fr.: Aug. – Jul. Kalasamudram RF (ATP), *NY* 397.

Fruits are edible. It is known as king of fruits.

SEARSIA F. A. Barkley

Searsia mysorensis (G. Don) Moffett, Bothalia 37 (2): 170. 2007. *Rhus mysorensis* G. Don, Gen. Hist. 2: 74. 1832. *R. sinuate* Thunb., Prodr. Pl. Cap. 52. 1794; FPM 1: 264. 1918; FAP 1: 230. 1997.

Shrub, to 4 m high, branches reddish, hirtellous, thorns about 2.5 cm long, bearing leaves and flowers. Leaves alternate, 3-foliolate, lateral leaflets smaller than middle leaflet, obovate, 0.5–1.5 × 0.3–0.7 cm, sparsely hairy above, pubescent below, base acute, margin basally entire, apically toothed, apex obtuse-acute. Inflorescence axillary and terminal panicles, 3–10 cm long. Flowers unisexual. Male flowers 1.5 mm across, pedicel to 5 mm long. Calyx lobes ovate, 0.3 mm long. Petals yellow, oblong, 1 mm long, ovate. Stamens 5, 1 mm long. Disk 1 mm across. Female flowers 2 mm across. Ovary 0.8 × 0.6 mm, glabrous; style 3-fid; stigma capitate. Fruit drupe, subglobose, 4–5 mm across, orange-red, pulpy. Seed 1 per locule, 2 mm across.

Common, in scrub forests. Fl. & Fr.: Oct. – Feb. Mabbuchintalapalle (KDP), *KRKS & SSR* 39018.

SEMECARPUS L. f.

Semecarpus anacardium L. f., Suppl. Pl.: 182. 1781; FBI 2: 30. 1876; FPM 1: 266. 1918; FAP 1: 230. 1997.

Local name: *Jiri.*

Deciduous tree, to 10 m high, young branches terete, tomentose, watery latex. Leaves alternate, oblong-obovate, 9–15 × 4–7 cm, adaxially puberulose, abaxially pubescent, base obtuse, margin entire, apex obtuse, petiole to 3 cm long, lateral veins 10–13 pairs. Inflorescence terminal panicle, 15–40 cm long. Flowers 0.3 cm across. Male flowers sessile. Sepals pubescent out side, oblong. Petals 5, greenish-white, 2 × 1 mm. Stamens 5, 2 mm long. Disk 1 mm in diam. Female flowers 3 mm in diameter. Ovary 1.5 mm; style densely hairy. Fruit drupe, obliquely ovoid, oblong, glabrous, black, hypocarp fleshy, turbinate, 1 – 2 × 1 – 1.5 cm, pubescent. Seed 1 per locule, pendulus, membranous.

Occasional, in dry deciduous forests. Fl. & Fr.: Jun. – Oct. Batrepalli (ATP), KRKS *& BSS* 39588.

MORINGACEAE

MORINGA Adans.

1a. Leaves tripinnate; leaflets apically rotund, nerves obscure; flowers white .... **M. pterigosperma**

1b. Leaves bipinnate; leaflets apically emarginated, nerves prominent; flowers white-purple .... **M. concanensis**

Moringa concanensis Nimmo ex Gibs. in Dalz. & Gibs., Bombay Fl. 311. 1861; FBI 2: 45. 1876; FPM 1: 270. 1918; FAP 1: 231. 1997.

Local name: *Advai munaga.*

Deciduous tree, up to 15 high; bark thick, grey, fissured; branchlets warty, tomentose. Leaves bipinnate to tripinnate, to 40 cm long, petiole to 8 cm long; leaflets 9 – 13, chartaceous, elliptic-obovate, 2.5–5 × 0.8–3 cm, glaucous beneath, base oblique, margin entire, apex emarginated or slightly apiculate; petiole to 3 mm. Panicles to 15 cm long, axillary or terminal, divaricate, pubescent; peduncle to 5 cm long; bracts and bracteoles linear, 1–2 mm long. Flowers creamy or purple-white, *ca.* 1.5 cm across. Calyx lobes 5, oblong, subequal, 0.8–1.2 × 0.4–0.6 cm, tomentose. Petals 5, oblong-obovate, 1.5 × 0.5 cm, unequal. Stamens 5. Ovary stipitate, pubescent. Capsule elongated, 15–40 cm long, pointed, sharply 3-angled. Seeds numerous, with thick membranous hyaline wings.

Occasional, along streams in dry deciduous forests. Fl. & Fr.: Mar. – Jul. Namalagundu (KDP), *BSS* 39623.

Note: Fuits not edible. Dried stem and branches are used as floating material by swimming learners.

Moringa pterigosperma Gaertn., Fruct. 2: 314. 1791; FBI 2: 45. 1876; FAP 1: 232. 1997. *Moringa oleifera auct. non*. Lam. 1: 398. 1785; FPM 1: 270. 1918.

Local name: *Munaga.*

Trees, up to 10 m high; bark grey-white with longitudinal wrinkles, corky. Leaves tri-pinnate, *ca.* 40 cm long, petiole to 6.5 cm long; leaflets small, ovate, elliptic-obovate, to 2.5 × 1.5 cm, base obtuse, margin entire, apex obtuse. Inflorescence axillary and terminal panicle to 25 cm long. Flowers strongly scented. Sepals green to 5 mm long. Petals white, oblong-obovate. Capsule elongated to 30 cm long, 9-ribbed, loculicidally dehiscent. Seeds 2–3 cm long, woody; rounded-trigonous, winged, brown.

Occasinally grown in agriculture fields and backyards. Fl. & Fr.: Jan. – Jul. Mabbuchintalapalle (KDP), *KRKS* 41803.

Note: Fruits and leaves are edble. Stem and branches are used as floating material. It is widely cultivated.

FABACEAE

Key to Genera

1a. Trees (except *Sesbania bispinosa*-shrub) .... (2)

1b. Herbs, shrubs or climbers .... (6)

2a. Flowers flame-coloured; leaves coriacious .... **Butea**

2b. Flowers other than flame-coloured; not coriaceous.. (3)

3a. Leaflets alternate .... (4)

3b. Leaflets opposite or sub-opposite.. (5)

4a. Pods orbicular, one-seeded, compressed with broad wing............ **Pterocarpus**

4b. Pods elongate, 2–3-seeded, not winged.. **Dalbergia**

5a. Leaflets imparipinnate; calyx cup-shaped; pod woody, flat, 5–6 × 3–4 cm, 1-seeded .... (6)

5b. Leaflets paripinnate; calyx lobed; pod not woody, cylindrical elongated or at least 30 cm long, ca. 40-seeded .... **Sesbania**

6a. Leaflets 5; pod blattened, elliptic-obliong, 6 cm long; seeds 1 or 2.... **Pongamia**

6b. Leaflets 11-13; pod sub cylindrical, elongate, 14 cm long; seeds 5-10 .... **Gliricidia**

7a. Leaves even-pinnate.. (8)

7b. Leaves odd-pinnate or digitate.. (10)

8a. Climbers; stamens 9.. **Abrus**

8b. Prostrate or erect herbs or shrubs; stamens 10 .... (9)

9a. Leaves gland-dotted below; fruit jointed and prickly; decumbent herbs; leaflets 2, stamens monadalphus .... **Zornia**

9b. Leaves not gland-dotted; fruit without joints and without prickles; erect shrubs; leaflets 8–10 pairs, stamens monadalphus .... **Mundulea**

10a. Plants erect or prostrate herbs or herbaceous climbers or shrubs .... (11)

10b. Plants stragglers or woody climbers .... (23)

11a. Leaves pinnately, 6–many foliolate .... (12)

11b. Leaves 1–3- or digitately 3–5-foliolate .... (13)

12a. Flowers in axillary lax recemes; pods sutures between seeds; plants aquatic .... **Aschynomene**

12b. Flowers in terminal receme; pod entire; plants terrestrial .... **Tephrosia** (*p.p*)

13a. Flowers yellow (except *Crotalaria verrucosa*- blue) .... (14)

13b. Flowers other than yellow .... (18)

14a. Pod turgid .... **Crotalaria**

14b. Pod not turgid .... (15)

15a. Climbers or twining herbs .... (16)

15b. Erect or decumbent herbs .... (17)

16a. Pods flattened; seeds compressed. - seeds 2–3 .... **Rhynchosia**

16b. Pod cylindrical; seeds ellipsoid. - fruit cylindrical; seeds 9–12 .... **Vigna**

17a. Erect under shrub, silky villose; root stock woody; pod hooked at apex .... **Stylosanthes**

17b. Prostrate herb, hirsute; root stock not woody; pod not hooked .... **Rothia**

18a. Venation striate .... **Tephrosia** (*p.p*)

18b. Venation reticulate .... (19)

19a. Pod breaking into articles .... (20)

19b. Pod not breaking as above .... (21)

20a. Calyx scarious, articles turgid .... **Alysicarpus**

20b. Calyx herbaceous, articles compressed .... **Desmodium**

21a. Anthers apiculate .... **Indigofera**

21b. Anthers not apiculate .... (22)

22a. Pod viscid, transverse lines absent between seeds .... **Pseudarthria**

22b. Pod not viscid, tranverse lines present between seeds .... **Cajanus**

23a. Stem woody (lianas), twiners or stragglers .... (24)

23b. Stem herbaceous, twiners only .... (28)

24a. Leaflets 5–7, elliptic-obovate .... (25)

24b. Leaflets 3, rhomboid .... (27)

25a. Fruit 1–3-jointed, prickly, warted .... **Ormocarpum**

25b. Fruit otherwise .... (26)

26a. Fruit stiff hairy, beaked, 5–8-seeded .... **Clitoria**

26b. Fruit glabrous, blunted or not beaked, 2–3 seeded .... **Derris**

27a. Flowers dark brown, up to 30; fruit 'S'-shaped with itching brown hairs .... **Mucuna**

27b. Flowers other than brown, up to 10; fruit straight, itching hairs absent .... **Canavalia**

28a. Flowers 1–3; petals white; seeds diskoid, light brown .... **Macrotyloma**

28b. Flowers 4–30; yellow or light purple .... (29)

29a. Pod with golden yellow spreading hairs, transversely depressions between seeds .... **Cajanus**

29b. Pod pubescent or sericeous, transvers depressions absent between seeds .... (30)

30a. Leaves rhomboid; flowers blue; fruit glabrous .... **Teramnus**

30b. Leaves elliptic-oblong; flowers pink; fruit pubescent .... **Galactia**

ABRUS Adans.

Abrus precatorius L., Syst. Nat. ed. 12. 2: 472. 1767; FBI 2: 175. 1876; FPM 1: 349. 1918; FAP 1: 236. 1997. *Glycine abrus* L., Sp. Pl. 753. 1753.

Local name: *Gurivinda, Guruginja.*

Climbing slender shrub, up to 6 m long, branchlets sericeous. Leaves alternate, paripinnate, 10–12 pairs of leaflets, stipule deciduous, stipels minute; leaflets oblong, 1.5–2.8 × 1–1.2 cm, thin, sparsely hairy, base obtuse, margin entire, apex obtuse-acute; petiole *ca.* 1 cm long; petiolel *ca.* 1 mm long. Inflorescence axillary raceme to 13 cm long; peduncle *ca.* 10 cm long. Flowers densely arranged at apex on 3 cm long raceme. Calyx narrowly campanulate, pubescent; lobes acuminate. Corolla rose pink; standard petal broadly ovate; wing petal narrow, oblong; keel petal curved. Stamens 9, monadelphous. Ovary subsessile, pubescent. Pod oblong, 2.5–3.5 × 0.8–1 cm, flattened, turgid, 5 to 7-seeded, mucronate at apex. Seeds hard, 5 mm across, shiny, red with a black terminal blotch around the hilum.

Common, in dry deciduous and scrub forests. Fl. & Fr.: Aug.–Mar. K. K. Kottala (KDP), *KRKS, KRS* 36436; Batrepalli (ATP), *KRKS* 36952.

Note: Seeds show variation in colours, some are white instead of red and black. The leaf and its roots are medicinal.

AESHCHYNOMENE L.

Key to Species

1a. Branchlets woody; racemes viscid; calyx glabrous; pod upper margin entire, inner margin indented .... **A. indica**

1b. Branchlets pith like; racemes hispid; calyx hispid; pod indented along both margins .... **A. aspera**

Aeschynomene aspera L., Sp. Pl. 713. 1753; FBI 2: 152. 1876; FPM 1: 331. 1918; FAP 1: 237. 1997.

Local name: *Neeti jiluga.*

Erect under shrub, up to 1.5 m high, branchlets glabrous, swollen. Stipules apically setaceous, lanceolate, to 1 cm long. Leaves imparipinnate, to 9 cm long; leaflets 30–35 pairs, alternate, oblong, 6–10 × 1–3 mm, glabrous, base obtuse, oblique, apex obtuse, margin entire; petiole to 8 mm. Raceme axillary, to 3 cm long, 2–4-flowered; bracts ovate, acute. Flowers 1.5 cm across; peduncle hispid, to 1.5 cm long; pedicel to 1 cm long. Calyx hispid, tube to 4 mm long; lobes rounded, 2 mm long, lower one 3.5 mm long. Corolla yellow; standard petal 1.2 × 1.5 cm, wing petal 8.5 × 4 mm; keel petal 1.8 × 0.8 cm. Stamens isodiadelphous (5+5). Ovary tomentose, 1 cm long; style 6 mm long. Pod nearly straight, 5.5 × 0.8 cm; articles 4–8, usually warted or echinate on the faces, indented along both margins. Seeds to 6 mm across, brownish.

Common, near water bodies. Fl. & Fr. Aug.–Oct. Somulavaripalli (ATP), *KRKS & KRS* 37817.

Aeschynomene indica L., Sp. Pl. 713. 1753; FBI 2: 151. 1876; FPM 1: 331. 1918; FAP 1: 237. 1997.

Local name: *Tiga jeluga.*

Annual suffruticose erect herb, to 70 cm high. Leaves pinnate, to 8 cm long; leaflets 17–25 pairs, linear-oblong, alternate, 2.5–6.5 × 1–1.5 mm, glabrous, obtuse at both ends, margin entire; petiole to 7 mm long. Racemes lax, to 2 cm long, 1–4-flowered; bract ovate, toothed, bracteoles 2, oblong-lanceolate; peduncle to 1.5 cm; pedicel to 5 mm long. Flowers 1–1.3 cm across. Calyx membranous, tube 3 mm long; lobes 4–5 mm long, acuminate, subequal. Corolla yellow with red nerves and hairy above; standard petal obovate, 1 × 0.7 cm; wing petal obtuse, keel petals obtuse-acute. Stamens isodiadelphous (5+5). Ovary linear, 4 × 1 mm. Legume linear, 4 × 0.5 cm long, flattened, 6–8-jointed, slightly falcate, upper suture indented, articles 4–10, smooth, glabrous, tuberculated. Seed shining, 2.5 mm across, brown dotted, compressed.

Common, in marshy areas and along the streams. Fl. & Fr.: Jul.–Dec. Chinnapalle (ATP), *KRKS & KRS* 37829.

ALYSICARPUS Neck. ex Desv. *nom. cons.*

Key to Species

1a. Leaves orbicular, truncate at base and apex; stem prostrate .... **A. hamosus**

1b. Leves other than orbicular, obtuse or acute at apex; stem erect or decumbent .... (2)

2a. Calyx cupular; lobes to 1 mm long, glabrous .... **A. bupleurifolius**

2b. Calyx imbricate or connate at base only; lobes 3 to 8 mm long, hirsute .... (3)

3a. Calyx lobes connate at base, slightly longer than the first joint of the pod; articles inflated, not wrinkled, hairy, reticulate venation. (4)

3b. Calyx lobes free, much longer than the first joint of the pod; articles not inflated, transversely wrinkled, glabous, reticulate venation absent but rugose .... (5)

4a. Fruit slightly moniliform; leaves only elliptic .... **A. vaginalis**

4b. Fruit moniliform; leaves oblong-lanceolate. .... **A. longifolius**

5a. Stem herbaceous, villose hairy; flowers distinctly arranged on receme; raceme up to 3 cm long......................... **A. roxburghianus**

5b. Stem woody; pubescent or distinctly hairy; flowers densely arranged on raceme; raceme more than 4 cm long .... (6)

6a. Calyx acuminate; fruit 2–3-articled .... **A. scariosus**

6b. Calyx acute; fruit 4–5-articled .... (7)

7a. Herbs to 50 cm high; raceme not branched... **A. rugosus**

7b. Sub-shrubs to 1.5 m high; racemes branched .... **A. heyneanus**

Alysicarpus bupleurifolius (L.) DC., Prodr. 2: 352. 1825; var. **gracilis** (Edgew.) Baker in Hook. f., Fl. Brit. India 2: 158. 1876; FPM 1: 338. 1918; FAP 1: 239. 1997; *p.p. Alysicarpus gracilis* Edgew., J. Linn. Soc. Bot. 9: 312. 1869.

Prostrate or ascending herbs, to 45 cm long. Leaves alternate, linear-lanceolate, glabrous, to 3 × 0.3 cm, base acute, margin entire, apex acute, stipules lanceolate, thin. Racemes, terminal or axillary, flowers distinctly arranged on peduncle, purple, to 5 mm across. Calyx longer than first joint of the pod, cupular, connate; lobes 5. Fruit slightly moniliform, inflated, glabrous, 1-3-jointed, apex mucronate. Seed brown, flattened, 1-2 mm diameter.

Common along the sandy soil of streams edges, wastelands and agriculture lands. Fl. & Fr.: Aug. – Jan. Thummalapalle (KDP), *KRKS & KRS* 3765; K. K. Kottala (KDP), *KRKS* & *KRS* 39030.

Alysicarpus hamosus Edgew., Journ. Asiat. Soc. Bengal 21: 171. 1853; FBI 2: 157. 1876; FPM 1: 338. 1918; FAP 1: 239. 1997.

Creeping herbs with long hairs; branchlets to 60 cm long. Leaves orbicular 1.5-2 × 2-2.5 cm, base truncate, margin entire, apex truncate. Racemes axillary, lax, less than 10-flowered. Flowers bluish purple, to 5 mm across. Calyx campanulate; lobes 5, smaller than first joint of the pod. Fruit 2 × 0.3 cm, not moniliform, reticulately veined, flattened, inflated, 3–5-jointed, hairy, apex mucronate. Seed 1 mm across, brown, flattened.

Common, in wastelands and agriculture lands. Fl. & Fr.: Jun.–Nov. Namalagundu (KDP), *KRKS & BSS* 39660; Thummalapalle (KDP), *KRKS* 41836.

Alysicarpus heyneanus Wight & Arn., Prodr. Fl. Ind. Orient. 234. 1834. *A. rugosus* (Willd.) DC., var. *heyneanus* Baker in Hoook. f., Fl. Brit. India 2: 159. 1876: FPM 1: 339. 1918; FAP 1: 242. 1997.

Small shrub, up to 1.5 m high, stem villous. Leaves obovate, 4–4.5 × 1–1.5 cm, pubescent on lower side, glabrous on upper side, base attenuate, margin entire, apex emarginate, petiole 0.8 cm; stipule ovate, glabrous, thin. Inflorescence axillary, raceme, to 50 cm long. Flowers paired; pedicel 1 cm. Calyx up to 3rd joint of pod or longer than 3rd joint of pod. Corolla thick pink. Fruit 3–5-articled, to 1 cm long, moniliform, tangential ribbed, hard, glabrous, apex beaked. Seed brown 1 × 2 mm, shining.

Very rare, in the agriculture lands and wastelands. Fl. & Fr.: Oct. –Jan. Obulareddypalli (ATP), *KRKS* & *BSS* 41926.

Alysicarpus longifolius Wight & Arn., Prodr. Fl. Ind. Orient. 233. 1834; FBI 2: 159. 1876; FPM 1: 338. 1918; FAP 1: 240. 1997.

Erect herb, to 50 cm high; stem pubescent. Leaves oblong-lanceolate, 5-7 × 1-3 cm, glabrous, base truncate-round, margin entire, apex acute to round; petiole 1.5 cm. Racemes terminal; bracts ciliate at margin, white papery, covering the stem. Calyx pubescent, longer than the first joint of the fruit. Fruit 5-7-articled, reticulate, pubescent, inflated. Seed flattened, brown.

Common, in wastelands Fl. & Fr.: Oct.-Nov. Batrepalli (ATP), *KRKS* & *BSS* 40353; Namaladundu (KDP), *KRKS & BSS* 40172.

Alysicarpus roxburghianus Thoth. & Pramanik, Bull. Bot. Surv. India 21: 189. 1981; FAP 1: 242. 1997. *Hedysarum bupleurifolium* Roxb., Pl. Coromandel t. 194. 1805. non L. 1763. nec. Roxb. Fl. Ind. 3: 346-347. 1832.

Decumbent herb or undershrub, up to 50 cm; branchlets villous hairy. Leaves linear-lanceolate, 2.8 – 10 × 0.3 – 0.6 cm, base truncate-rounded, margin entire, apex acute or obtuse, pubescent lower side; petiole 0.5 cm. Raceme axillary to 15 cm long. Flowers small, to 6 mm across, purple. Calyx longer than the first joint of the pod; lobes minute. Fruit monilform, transversely ribbed, hard, apex beaked. Seed brown, flattened, 3 mm across.

Common, weed in Agriculture land. Fl. & Fr.: May-Oct. Chinnapalle (ATP), *KRKS* & *BSS* 37806.

Alysicarpus rugosus (Willd.) DC., Prodr. 2: 353. 1825; FPM 1: 338. 1918. *Hedysarum rugosum* Willd., Sp. Pl. 3: 1172. 1802; FAP 1: 242. 1997.

Diffuse or decumbent herb, to 45 cm long; branchlets purple, pubescent, hard. Leaves elliptic ovate to oblong, 1.5–3.4 × 0.8–1 cm, base acute, margin entire, apex round. Inflorescence, axillary, up to 7 cm long. Flowers pinkish, to 5 mm long. Calyx 0.5 cm long, boat-shaped, thin, longer than the first joint of the pod. Pod moniliform, 2–3-articled, apex mucronate. Seed brown, shiny, ovate.

Rare in dry deciduous forest. Fl. & Fr.: Oct.-Dec. Rachakuntapalle (KDP), *KRKS, SSR* & *TP* 39036; Vemula (KDP), *KRKS* & *BSS* 41964.

Alysicarpus scariosus (Rottl. ex Spreng.) Grah. ex Thw. Ennum. Pl. Zeyl. 88. 1858. *Hedysarum scariosum* Rottl. ex Spreng. Syst. Veg. 3: 319. 1826. *Alysicarpus rugosus* (Willd.) DC. var. *styracifolius* Baker in Hook. f., Fl. Brit. India 2: 159. 1876; FPM 1: 336. 1918; FAP 1: 243. 1997.

Key to Varieties

1a. Stem glabrescent; raceme filiform; calyx more than 1 cm long.... var. **scariosus**

1b. Stem finely pubescent, receme terete; calyx less than 1 cm .... var. **pilifer**

Alysicarpus scariosus (Rottl. ex Spreng.) Grah. var. **scariosus**

Dwarf diffuse herb, to 20 cm long, with spreading hairs and white glaucous. Leaves elliptic-ovate, oblong-lanceolate, 3–5 × 0.3–0.5 cm, white glaucous hairy, base obtuse, margin entire, apex obtuse; petiole to 5 mm long. Racemes to 3 cm long; peduncle to 5 mm long, filiform. Flowers dense on raceme. Calyx 6 mm long with spreading hairs, boat-shaped, margin ciliate, longer or equal to 3rd joint of the pod. Corolla purple or pinkish-red, with yellow dot. Pod 3-jointed, to 8 mm long; articles transversely ribbed, apically beaked. Seeds compressed, 2 mm across, brown.

Common, in moist palces of agriculture lands and scrub forests. Fl. & Fr.: May – Aug. Kalasamudram (ATP), *KRKS & KRS* 39153.

Alysicarpus scariosus (Rottl. ex Spreng.) Grah. var. **pilifer** (Prain) Pramanik & Thoth., Bull. Bot. Surv. India 24: 113. 1983. *Alysicarpus rugosus* (Willd.) DC. var. *pilifer* Prain, J. Asiat. Soc. Beng. 66: 384. 1897.

Erect herb, up to 75 cm high, branchlets terete, striate, sparsely hirsute. Stipules 1 cm long. Leaves oblong, ovate or lanceolate, 3–5 × 0.4–0.8 cm, pubescent on midrib, base and apex obtuse, margin entire, nerves prominent. Racemes 5–8 cm long. Flowers 3 mm long. Calyx 6 mm long, up to 3rd joint of pod, ovate, glumaceous, ciliate. Corolla pink; petals glabrous. Ovary glabrous, style incurved. Pod moniliform, 2 × 1 mm, articles 3–5, transversely ribbed, hard, blackish when mature and dry, beaked at apex. Seeds orbicular, brown, 3 × 2 mm.

Rare in dry deciduous forests. Fl. & Fr.: Aug. – Nov. Kalasamudram (ATP), *KRKS* 40075; Namalagundu (KDP), *KRKS & BSS* 40186.

Alysicarpus vaginalis (L.) DC., Prodr 2: 253. 1835. var. **vaginalis**; FBI 2: 158. 1876; FPM 1: 338. 1918; FAP 1: 244. 1979. *Hedysarum vaginalis* L., Sp. Pl. 746. 1753.

Prostrate, perennial herb, to 40 cm long, rooting at nodes; branches glabous. Leaves variable elliptic-oblong, 2–4.5 × 1–2 cm, glabrous, base truncate-retuse, margin entire, apex obtuse; petiole to 1 cm. Raceme axillary, to 8 cm long. Flowers purple, paired at each bract; bracts deciduous, scarious. Calyx 5-lobed; lobes to 3 mm long, slightly longer than the first joint of the pod. Corolla purplish pink. Pod terete-subterete, 1.5 cm long, not constricted between the joints, pubescent, reticulate venation, inflated, apex mucronate. Seeds 3 – 8, 0.5 mm across, elliptic, brown, shiny, flattened.

Very common weed in agriculature fields and along road sides. Fl. & Fr.: Jun. – Nov. Bhumaiahgari palle (KDP), *KRKS & KRS* 37617; Somulavaripalli (ATP), *KRKS* & *KRS* 40189.

BUTEA J. Koen. ex Roxb. *nom. cons.*

Butea monosperma (Lam.) Taub., in Engler & Prantl, Pflanzenfam. 3(3): 366. 1894; FAP 1: 246. 1997. *Erythrina monosperma* Lam., Encycl. 1: 391. 1785. *Butea frondosa* Koen. ex Roxb., Asiat. Res. 3: 369. 1792; FBI 2: 194. 1876; FPM 1: 357. 1918.

Local name: *Moduga.*

Deciduous tree, up to 12 m high; bark grey, trunk crooked and irregular; young shoots tawny tomentose. Leaves trifoliolate; leaflets coriaceous, laterals broadly ovate, 16.5 × 15 cm, glabrous above, sericeous below, base truncate-cuneate, margin undulate, apex emarginated; obliquely rounded at base in lateral leaflets; petiole to 14 cm long; petiolule to 7 mm long. Racemes 30–35 cm long. Flowers 4.5 cm across; pedicel 2 cm long. Calyx dark brown, velvety, tube 1 cm long; lobes 5 mm long, triangular; lower one deltoid, 4 mm long. Corolla scarlet red or flame-coloured; standard petal lanceolate, 3.7 × 2.2 cm, silky pubescent without; wing petals falcate, adnate to keels, 4.2 × 1.2 cm; keel petals incurved, 4.3 × 1.3 cm, Staminal sheath 3.5 cm long; stamens diadelphous (9+1). Ovary tomentose, 2 cm long. Pod reticulately nerved, 18 × 5 cm, sutures thickened, grey tomentose. Seeds 1 or 2, obovate.

Common, in dry deciduous forests and scurb forests. Fl.: May – Jul. & Fr.: Sept. – Nov. Kalasamudram (ATP), *KRKS &BSS* 40019.

The tree is leafless when in flower. Presenting a beautiful sight. It is known as Flame of the forest. Gum which is extracted from bark is called as 'Butea gum or kino gum'.

CAJANUS A.P. DC. *nom. cons.*

Canjanus scarabaeoides (L.) Thouars., Dict. Sci. Nat. 6: 617. 1817. *Dolichos scarabaeoides* L., Sp. Pl. 726. 1753. *Atylosia scarabaeoides* Benth. in Miq., Pl. Jungh 242. 1852; FBI 2: 215. 1876; FPM 1: 369. 1918.

Local name: *Adavi kandi.*

Slender, twining herb, to 3 m long, grey villous. Stipule 1–1.5 mm long. Leaves tri-foliolate, middle leaflet elliptic-oblong, lateral leaflets obliquely elliptic-oblong, 2–3 × 1–1.9 cm, densely pubescent, base obtuse, margin entire, apex rounded; stipule 1–1.5 cm, stipels 1 mm long. Flowers 2–6 in axillary clusters. Calyx tube 1.5 mm long, densely pubescent; lobes lanceolate; upper ones 2.5 mm long; lower one 3.5 mm long. Corolla yellow; standard petal 7 × 4 mm; wing petals 6.5 mm long; keel petals 7 × 2 mm. Staminal sheath 7 mm long; stamens diadelphous (9+1). Ovary woolly, 3.5 × 1 mm. Pod 2–3-seeded, with golden yellow hairs, flattened, 1–1.5 × 0.7 cm. Seeds elliptic, shiny, 4 × 5 mm, brown.

Common in dry deciduous forests. Fl. & Fr.: Oct.–Jan. Rachakuntapalle (KDP), *KRKS & SSR* 39032.

CANAVALIA A. P. DC. *nom.cons*

Canavalia gladiata (Jacq.) DC., Prodr. 2: 404. 1825; FAP 1: 248. 1997. *Dolichos gladiatus* Jacq., Coll. Bot. 2: 276. 1788. *Canavalia ensiformis* sensu Baker in Hook. f. Fl. Brit. India 2: 195. 1876; FPM 1: 359. 1918.

Local name: *Thammakaya.*

Perennial, stout, woody climber, to 10 m long; branches glabrous, terete. Stipules subulate, deciduous. Leaves 3-foliolate; leaflets ovate-elliptic or rhomboid, 7–11 × 5–8 cm, glabrous, base obtuse, margin entire, apex acute; petiole to 10 cm long; petiolule 0.8 cm, pubescent. Racemes axillary, to 30 cm long, 12–16-flowered. Bracts minute, ovate, caducous. Flowers purple, 3 cm diameter. Calyx campanulate, tube 1.5 × 1 cm; lower one 3-lobed. Corolla white or pink; standard petal 2.8 × 2.5 cm, wing petals oblong-spatulate; keel petals oblong, obtuse. Staminal sheath 2.3 × 0.8; stamens monadelphous (10). Ovary pubescent, 2 × 0.3 cm. Pod flattened, to 12 × 2.5 cm, linear-oblong, beaked, 8–12 seeded. Seeds ellipsoid, compressed 3 × 2 cm, red, hilum 1.5 cm long.

Common, in dry deciduous forests. Fl. & Fr.: Mar.– Jul. Rachakuntapalle (KDP), *SSR* & *TP* 33617.

Usually cultivated for its edible fruits as vegetable.

CLITORIA L.

Clitoria ternatea L., Sp. Pl. 753. 1753; FBI 2: 208. 1876; FPM 1: 365. 1918; FAP 1: 249. 1997. *Clitoria ternatea* L. var. *albiflora* (Voigt) Fantz., Moseasoa 6: 164. 1990.

Local name: *Shanku-puspi.*

Perennial twiner, to 5 m long, branches somewhat woody when mature, tomentose when young. Stipules to 2.5 mm long, lanceolate. Leaves compound, alternate, 5.6 cm long, 5–7- foliolate; leaflets odd-pinnate, elliptic-ovate or obovate, 3–5 × 2–3 cm, base rounded, margin entire, apex obtuse-retuse, pubescent; petiole 1.5 cm long; petiolule 1–2 mm long; stipules 1 mm long. Flowers blue or white, axillary, solitary or paired. Bract to 3 mm long, lanceolate; bracteoles foliaceous, 6 × 8 mm. Calyx tube 1.2 cm long; lobes ovate, acuminate, pubescent, 8 × 5 mm. Corolla lip-shaped; standard petal spatulate, 4.5 × 3.2 cm, pubescent without; wing petals spatulate, blue, 2.3 cm long, keel petals blue, 2.5 cm long. Staminal sheath 1.5 cm long; stamens diadelphous (9+1). Ovary 1.7 cm long, white villous. Pod flattened, up to 10 cm long, white pubescent. Seeds 11–13, thick brown, flattened, shiny, 0.4 cm in diameter.

Common, weed in agriculture fields and foot hills of dry deciduous forests. Fl. & Fr.: Throughout the year. Mabbuchintalapalle (KDP), *KRKS & KRS* 33664; Vempalle (KDP), *KRKS & BSS* 40308; Nuthna kaluva (KDP), KRKS 41949

Note: Plants with two different coloured – blue and white have been collected from the study area. As per literature white flowered variety is *Clitoria* *ternatea* L. var. *albiflora* (Voigt) Fantz. Cultivated in gardens for its showy flowers.

CROTALARIA L.

Key to Species

1a. Leaves digitately 3-foliolate .... (2)

1b. Leaves simple .... (4)

2a. Pods subglobose; seeds two .... **C. medicaginea**

2b. Pods oblong or cylindrical; seeds many .... (3)

3a. Diffuse herbs........ **C. orixensis**

3b. Erect shrubs .... **C. laburnifolia**

4a. Plants with closely set numerous leaves on branches; flowers solitary in most of the leaf-axils........ **C. hebecarpa**

4b. Plants and flowers not as above .... (5)

5a. Racemes with few, often 1-flowered; mostly prostrate trailing herbs .... (6)

5b. Racemes many-flowered; erect herbs, undershrubs or shrubs .... (8)

6a. Stipule present........ **C. hirsuta**

6b. Stipule absent .... (7)

7a. Leaves ovate-oblong or suborbicular, 1-2 (-3) cm long; seeds 18-20 .... **C. angulata**

7b. Leaves orbicular, 1.0–1.5 cm long; seeds 4 .... **C. globosa**

8a. Pods hairy; flowers in racemes or panicles........ (9)

8b. Pods glabrous (except *C. pusilla*); always in racemes .... (11)

9a. Inflorescence raceme; no glands on bracts........ **C. verrucosa**

9b. Inflorescence panicle; bracts with yellow oil glands, foetid .... (10)

10a. Plant dwarf, umbrella shaped; branches many, densely silky hairy herbs; seed brownish-black........ **C. ramosissima**

10b. Plant large, cone-shaped; branches less, silky pubescent shrub; seeds tan colour........ **C. paniculata**

11a. Pods twice as long as or longer than calyx .... **C. retusa**

11b. Pod shorter or slightly exceeding calyx........ (12)

12a. Small herb, to 15 cm high; pods 5–6 mm long .... **C. pusilla**

12b. Tall herb 0.5 to 1.5 m high; pods more than 6 mm long........ (13)

13a. Leaves linear-lanceolate, glabrous above; calyx with densely brown silky hairs, plant with short appressed hairs .... **C. calycina**

13b. Leaves linear, villous with long hairs on both; calyx densely sericeo-pilose, plant with long hairs .... **C. mysorensis**

Crotalaria angulata Mill., Gard. Dict. ed. 8. no. 9. 1768; FAP 1: 253. 1997. *C. biflora* (L.) L., Mant. Pl. 273. 1771; FBI 2: 66. 1876; FPM 1: 292. 1918. *Astragalus biflorus* L., Mant. Pl. 273. 1771.

Annual prostrate herbs, to 30 cm long, villose hairy, dichotomously branched. Leaves simple, ovate or suborbicular, alternate 1–1.5 × 0.6–1 cm, villose on both surfaces, base obliquely truncate or cordate, margin entire, apex acute; petiole 1 mm long. Raceme lax, leaf-opposed, 1.3–2 cm long. Flowers 1–2; peduncle to 3 cm long; pedicels 2–4 mm long; bracts 2, setaceous or filiform. Calyx tube 2 mm long; lobes *ca.* 5 mm long, linear, densely silky hairy. Corolla yellow, scarcely exserted; standard petal ovate-oblong, *ca.* 6 × 4 mm; wing petals oblong, *ca.* 5 × 2 mm; keel petals gibbous with short twisted beak, *ca.* 5 × 3 mm. Staminal sheath 2 mm long; stamens monadelphous (10), filaments 3 mm long. Ovary shortly stipitate, 2.5 mm, style pubescent, 5 mm long. Pod shortly stalked, subglobose, 1 × 1 cm, hirsute, pericarp membranous, brown. Seeds 7–10, reniform, smooth, brown, 2 mm in diam., compressed.

Occasional, in dry deciduous forest near moist places. Fl. & Fr.: Jul. – Dec. Batrepalli (ATP), *KRKS & SSR* 37872.

Crotalaria calycina Schrank, Pl. Rar. Hort. Monac. t. 12. 1. 1817; FBI 2: 72. 1876; FPM 1: 295. 1918.

Annual erect herbs, to 80 cm high; branches terete, densely coarsely pilose with apressed hairs. Leaves simple, subsessile, oblong-linear to linear-lanceolate, 3–8 × 0.5cm, abaxially (lower) densely brownish pilose, adaxially (upper) pilose on mid-vein, base narrow, margin entire, apex acute-apiculate. Stipules filiform, *ca.* 1 mm, persistent or caducous. Flowers in terminal lax racemes, usually condensed or headlike, 3–12-flowered; bracts lanceolate, slightly ovate-falcate, 1–2 cm. Pedicel 2–4 mm long, thick; bract and bracteole short. Calyx 5-lobed; lobes 1.5–2 cm long, densely brown silky hairs, lanceolate. Corolla yellow, included in calyx; standard obovate-suborbicular, 1.5–2 cm, base with 2 appendages, apically pubescent; wings narrowly elliptic, *ca.* as long as standard; keel with a long straight shortly twisted beak. Pod glabrous, 2 – 2.5 cm long, 0.6 – 0.8 cm broad, black when mature, subcylindric. Seeds 20 – 25, reniform, glabrous.

Occasional, in foothills of dry deciduous forests. Fl.& Fr.: Jun.–Dec. Somulavaripalli (ATP), *KRKS* & *BSS* 41903.

Crotalaria globosa Wight in Wight & Arn., Prodr. Fl. Ind. Orient. 190. 1834; FBI 2: 66. 1876; FPM 1: 292. 1918; FAP 1: 256. 1997.

Prostrate, much branched, silky hairy herbs; branches terete. Leaves simple, orbicular, 0.6–1.5 × 0.5–1 cm, densely hairy beneath and less above, base subcordate or rounded, margin entire, apex acute or mucronulate, somewhat obtuse; petiole *ca.* 1 mm long. Raceme leaf-opposed, *ca.* 2 cm long. Flowers 1–3, *ca.* 6 mm long; pedicels 2–2.5 mm long; bracts setaceous, 1–1.5 mm long; bracteoles very small. Calyx lobes connate at base, lanceolate or linear-lanceolate, *ca.* 5 mm long, acuminate, densely hairy. Corolla yellow; standard petal ovate-oblong, *ca.* 5.5 × 4 mm, prominently

Plate 5

A. *Crotalaria angulata*, B. *C. laburnifolia*, C. *Galactia tenuiflora*, D. *Indigofera oblongifolia*, E. *Ormocarpum sennoides*, F. *Pterocarpus santalinus*

veined with short, curved claw; wing petals oblong, *ca.* 5 × 3 mm; keel petals slightly ventricose with short twisted beak. Staminal sheath 3 mm long; stamens monadelphous (10); ovary ovate-oblong, 4–5 mm long. Pod globose, 8 × 5 mm, brown, pericarp hard, coriaceous and covered with scattered short stiff hairs. Seeds *ca.* 4, reniform, *ca.* 2 × 2 mm, yellowish-brown.

Occasional, in foot hills of forests and agricultural fields. Fl. & Fr.: Jul.–Oct. Kalasamudram (ATP), *KRKS* 39170.

Crotalaria hebecarpa (DC.) Rudd, Phytologia 54: 28. 1983. *Goniogyna hebecarpa* DC., Ann. Sci. Nat. Bot. 4: 92. 1825. *G. hirta* (Willd.) Ali, Taxon 16. 463. 1967; FAP 1: 286. 1997. *Hallia hirta* Willd., Sp. Pl. 3. 1169. 1802. *Heylandia latebrosa auct. non* (L.) DC.; FBI 2: 65. 1876; FPM 1: 280. 1918.

Prostrate, annual herbs, to 40 cm long, hirsute; branches dichotomously branched. Leaves simple, numerous, close-set on branches, subsessile, obliquely-ovate or cordate-ovate, 0.7–1.4 × 0.4–0.7 cm, hirsute, base obliquely cordate or truncate, margin entire, apex acute; petiole to 1 mm long. Flowers axillary, solitary, yellow; 4 mm across; pedicel 2 mm long. Calyx campanulate, turbinate; lobes lanceolate, acute, the two upper lobes slightly connate, the three lower lobes cleft to the base. Corolla yellow, exserted; standard petal obovate-oblong or orbicular, 5.5 × 6 mm; wing petals oblong-obovate, 4 × 2 mm; keel petals connate, 4 × 2.2 mm long. Staminal sheath 2 × 2.5 mm, filaments 2.5 mm long. Ovary sessile, 1.5 mm long. Pod flattened, 4 × 2.5 mm, silky hairy, surrounded at the base by the persistent calyx and tipped by the style. Seeds 1–2, reniform or cordiform, *ca.* 2 mm across, brown.

Common, in scrub forest and grass lands. Fl. & Fr.: Aug.-Dec. Thummalapalle (KDP), *KRKS & KRS* 36912.

Crotalaria hirsuta Willd., Sp. Pl. 3(2): 978. 1802; FBI 2: 68. 1876; FPM 1: 293. 1918; FAP 1: 256. 1997.

Erect, sometimes diffuse herbs, to 70 cm high; branches strigosely hairy. Stipules 2, ovate, 0.5 cm long. Leaves simple, membranous, ovate, elliptic, 3–4 × 1.9–2.3 cm, pubescent on both sides, base acute, margin entire, apex acute-apiculate; petioles 1–2 mm long. Raceme lax, leaf-opposed or terminal; peduncles 3–7 cm long, stout, recurved. Flowers 3–5, yellow,1.7 cm long; pedicels 3–6 mm long, slender; bracts cordate or lanceolate, 2–4 × 1–2 mm; bracteoles 2, minute, acuminate. Calyx 5-lobed, tube very short; lobes linear-lanceolate, 9–15 × 1.5–2 mm, acuminate at apex, sparingly pubescent. Corolla yellow, slightly exserted; standard petal obovate-rhomboid, 2.5 × 0.8 cm, glabrous; wing petals oblong or obovate-oblong, 1–1.5 × 0.2–0.4 cm; keel petals 1 × 0.7 cm, beak twisted. Staminal sheath 6 mm long; stamens monadelphous (10). Ovary sessile or subsessile, oblong, 5–7 mm long, softly hirsute. Pods subsessile, oblong, 2.5 × 0.7 cm, hirsute with bulbous based brown hairs, apex beaked with persistent style. Seeds 8-12, reniform, 2.5 × 2 mm, brown, shiny.

Occasional in open exposed hilly slopes and near rocks in moist places. Fl. & Fr.: Aug.-Dec. Kalasamudram (ATP), Idupulapaya (KDP), *KRKS, KRS* & *BSS* 37689, 40322.

Crotalaria laburnifolia L., Sp. Pl. 715. 1753; FBI 2: 84. 1876; FPM 1: 301. 1918; FAP 1: 257. 1997.

Local name: *Dedda-giligicha.*

Robust shrub, to 2.5 m high, glabrous, white glaucous. Leaves 3-foliolate, petiole *ca.* 9 cm long, estipulate; leaflets ovate-elliptic or oblong, 2.5–4 × 1.5–2.5 cm, membranous, glabrous, base obtuse-subacute or cuneate, margin entire, apex subacute, apiculate; petiolule 0.5 cm long. Raceme terminal, to 20 cm long. Flowers yellow, 3 × 2 cm; pedicel 1–1.5 cm long; bracts linear, setaceous, *ca.* 3 mm long; bracteoles linear, *ca.* 1 mm long. Calyx campanulate, *ca.* 4 mm long, lobes subequal, 4–5 mm long, ovate or lanceolate, turbinate, glabrous. Corolla yellow, *ca.* 2.5 cm long; standard petal ovate, 2 × 2.5 cm; wing petals obovate-oblong, *ca.* 1.8 × 0.7 cm, apex obtuse; keel petals *ca.* 2 × 1.5 cm, beak straight. Staminal sheath 1.5 cm long; stamens monadelphous, 1 cm long. Ovary oblong, *ca.* 1.5 cm long with elongate stipe. Pod 3–3.5 cm long, faintly nerved, glabrous; stalk 1–2 cm long, slender; persistent stigma 3 cm long. Seeds 15–30, reniform, *ca.* 3 mm long, yellow.

Rare in dry deciduous forest and weed in agriculture fields. Fl. & Fr.: June–Dec. Batrepalli (ATP), *KRKS & BSS* 39617.

Crotalaria medicaginea Lam., Encycl. 2: 201. 1790; FBI 2: 81. 1876; FPM 1: 299. 1918; var. **neglecta** (Wight & Arn.) Baker in Hook. f., Fl. Brit. India 2: 81. 1876; FPM 1: 300. 1918; FAP 1: 259. 1997. *Crotalaria neglecta* Wight & Arn., Prodr. Fl. Ind. Orient. 192. 1834.

Diffuse herb, to 40 cm long, dichotomously branched, slightly pubescent; root stock yellow. Leaves 3-foliolate; leaflets oblong, obovate, 1–2 × 0.5–0.8 cm, pubescent, base acute, margin entire, apex emarginate or truncate; petiole 1.5 cm long, petiolule 2 mm long. Racemes terminal and leaf-opposed, 6–9-flowered. Flowers yellow, 0.5 cm long. Calyx campanulate, tube 1.5 mm long; lobes, 2 mm long, pubescent. Corolla yellow; standard petal obovate, pubescent on the back; wing petals oblong, 4–5 mm long, obtuse at apex; keel petals ovate oblong, 6 mm long. Staminal sheath 1.5–2 mm long. Ovary sessile, ellipsoid, 1.5–2 mm long. Pod globose, pubescent, persistent stigma 0.3 cm long at apex, curved. Seeds 2, reniform, shiny, greenish-brown.

Common, in agriculture lands and dry deciduous forests. Fl. & Fr.: Sept. - Jan. Rachakuntapalle (KDP), *KRKS* 36982; Kalasamudram (ATP), KRKS *& BSS* 39845; Gandi (KDP), *KRKS & BSS* 41987.

Crotalaria mysorensis Roth, Nov. Pl. Sp. 338. 1821; FBI 2: 70. 1876; FPM 1: 295. 1918; FAP 1: 259. 1997.

Erect undershrub with brown silky hairs, to 1.5 m high; branchlets densely hairy. Stipule persistent, to 6 mm long, conspicuous, linear-subulate. Leaves simple, membranous, subsessile; oblong-lanceolate, lanceolate, 3–8 × 0.4–1.5 cm, sericeo-pilose on both sides, base acute-obtuse, margin entire-ciliate, apex acute-apiculate. Raceme leaf-opposed or terminal, to 18 cm long, pilose. Flowers yellow, crowded, 1–1.8 cm long; pedicels 0.4–1.3 cm long, slender; bracts linear-lanceolate, 1.2–1.8 cm, acuminate at apex; bracteoles 2, beneath the calyx, linear or lanceolate, 5–11 mm long. Calyx tube 1.5–2 mm long; lobes 5, densely sericeo-pilose; upper 2-lipped; lower 3-lipped, acuminate at apex. Corolla yellow, sometimes with red venation, equal to calyx; standard petal obovate-elliptic or oblong, 10–14 × 5.5–6 mm, shortly acuminate, pilose on the back at apex; wing petals oblong, *ca.* 12 × 2.5 mm, transverse ridges up to half of its breadth; keel petals 9–12 × 4–5 mm, margins ciliate, beak 4–5 mm long, twisted. Ovary oblong, *ca.* 6 × 1 mm, glabrous, stalk ca 0.5 mm long. Pods oblong, 2.5 × 0.8 cm, beaked, dark brown or yellowish-brown, glabrous, stalk 2.5 cm long. Seeds 2–13, reniform-elongate, *ca.* 2 × 1.7 mm dark brown or black, smooth.

Rare in dry deciduous forests. Fl. & Fr.: Nov.–Jan. Btrepalli (ATP), *KRKS* 42004.

Crotalaria orixensis Rottl. ex Willd., Ges. Naturf. Freunde Berlin Neue Schrif. 4: 217. 1803; FBI 2: 83. 1876; FPM 1: 300. 1918; FAP 1: 260. 1997.

Diffuse perenniall herb, much branched; branches terete, slender, pubescent, 30–40 cm long. Leaves 3-foliolate; petioles 1.5-2.5 cm long, hairy; stipules linear-lanceolate, foliaceous, 2.5–3 mm long; leaflets obovate-elliptic, 1.5–2.5 × 0.8–1.6 cm, pubescent, base cuneate, margin entire, apex obtuse; petiolules 1 mm long. Racemes leaf-opposed, up to 15 cm long; peduncles up to 3 cm long. Flowers yellow distinctly arranged on axis, to 1 cm across; bract ovate, to 4 mm long, deflexed; bracteoles linear, *ca.* 2 mm long; pedicel 1.5 cm. Calyx tube to 2 mm long; lobes 5, deeply divided into 2 lips, 0.5 cm, lanceolate, acute, silky outside. Corolla yellow, 0.7 cm long, slightly exserted; standard petal obovate, *ca.* 8 × 6 mm, pubescent outside; wing petals *ca.* 5 × 2 mm; keel petals up to 6 mm long, beak straight, twisted. Staminal sheath short, to 2 mm long; filamensts 5 mm long. Ovary 1.5 mm long, glabrous. Pods long-stipitate, oblong, 1–1.2 × 0.8–0.9 cm, glabrous, cylindrical, brown. Seeds 10–12, brown, reniform, shiny.

Common, in wastelands and agriculture fields. Fl. & Fr.: Sept.–Feb. Batrepalli (ATP), *KRKS & BSS* 40159; Obulareddypalli (ATP), *KRKS & BSS* 41927; way to Gollalaguduru (KDP), KRKS 41970.

Crotalaria paniculata Willd., Sp. Pl. 3: 980. 1802; FBI 2: 81. 1876; FPM 1: 299. 1918; FAP 1: 261. 1997.

Stiff, erect, sticky, foetid undershrub, silky pubescent, much branched, up to 1.5 m high; stem tomentose. Leaves simple, oblong-lanceolate or oblanceolate, 5–8 × 1.5–3 cm, pubescent on both sides, base cuneate or obtuse, margin entire-ciliate, apex acute-obtuse; petioles 1–2 mm long; petiolule 2 mm long; stipules curved, falcate with reflexed margins, glabrous, minute. Panicles terminal, elongated with numerous curved small bracts, sticky with yellow foetid glands, margin reflexed; bracteoles 2, smaller than bracts, situated at the base of the calyx-tube. Flowers numerous, 1.5–2.5 cm long; pedicels to 0.5–1 cm long. Calyx campanulate, tube 2–3 mm long; lobes 5, lanceolate or oblong, 1–8 mm long, densely brown-silky, upper and lower lobes equal, margins recurved, viscous. Corolla yellow, exserted; standard petal reddish-brown at base, obovate or orbicular, 1.5–1.7 cm long, silky on the back side; wing petals yellow, obovate or oblong, keel petals connate, ovate-oblong, much incurved. Staminal sheath *ca.* 5 mm long; free filaments 4–8 mm long. Ovary sessile, adpressed-pubescent; style long. Pods sessile, ellipsoid or ovoid, 10–12 × 6–7 mm, sparsely pubescent with recurved, persistent style. Seed 1, *ca.* 4 × 3 mm, reniform, black.

Common in plains of dry deciduous forests. Fl. & Fr.: Dec.–Feb. Batrepalli (ATP), *KRKS & SSR* 37869.

Crotolaria pusilla Heyne ex Roth, Nov. Pl. Sp. 335. 1821; FBI 2: 70. 1876; FPM 1: 296. 1918; FAP 1: 262. 1997.

Suberect-erect, annual, villous herb, to 15 cm high; branches many, villose hairy. Leaves subsessile, simple, linear-lanceolate, 0.5–2 × 0.2–0.3 cm, silky hairy below, pubescent above, base cuneate, margin entire-ciliate, involute, apex acute; stipules absent. Racemes terminal; peduncles 4 cm long, pubescent. Flowers yellow, 3–6, crowded, to 1.5 cm long; bracts very minute. Calyx deeply 5-lobed, tube *ca.* 0.5 mm long, lobes linear-lanceolate, 2.5 × 0.5 mm, acuminate at apex, densely villous. Corolla 0.3 cm long; standard petal obovate-oblong, 3–3.5 × 1–1.5 mm, pilose on the back at apex; wing petals obovate-oblong, 2.5 × 1 mm; keel petals obovate, 2.5–3 × 0.5–1 mm, beak *ca.* 1 mm long, twisted, claw *ca.* 0.5 mm long. Staminal sheath *ca.* 1 mm long; free filaments 1–2 mm long. Ovary sessile, oblong or ovate-oblong, 1.5–2 mm, softly hirsute. Pod sessile, oblong, villous, 0.4–0.5 × 0.2 cm. Seeds 3–6, reniform, brownish-black.

Occasional, in foot hills of dry deciduous forests. Fl. & Fr.: Jul.–Dec. Batrepalli (ATP), *KRKS & SSR* 37870; Kalasamudram (ATP), *KRKS & BSS* 40083.

Crotalaria ramosissima Roxb., Fl. Ind. 3: 268. 1832; FBI 2: 81. 1876; FPM 1: 299. 1918; FAP 1: 263. 1997. var. **ramosissima.**

Erect, dwarf, bush-like herb, to 40 cm high, white or light yellow villous hairy, sticky, branches very closely arranged on main stem. Leaves simple, sub sessile, obovate-oblong or lanceolate, 1–2.8 × 0.2–0.6 cm, silky hairy on both sides, base acute, margin entire-ciliate, apex acute or apiculate. Panicles short, close, terminal. Flowers 4–8, reddish yellow, sticky glandular; bracts and bracteoles to 4 mm long, alternate, cordate, ovate, recurved, blackish when dry, margin recurved. Calyx reddish, yellowish, oily glandular, sticky, 0.6 cm long, villose. Corolla yellow, to 1 cm long; standard petal obovate, 1 × 1.5 cm, pubescent, red outside; wing petals 7 × 2.5 mm, yellow; keel petals 10 × 4 mm, yellow, beak 3 mm long, twisted. Staminal sheath 4 mm long, filaments 7 mm. Ovary pubescent, 1.2 × 0.3 cm. silky. Pods sessile, ovoid, 7.5 × 0.5 mm, as long as calyx, silky, flattened. Seed 1, brownish-black, reniform, shiny.

Common, weed in agriculture fields and wastelands. Fl. & Fr.: Non.–Feb. Thummalapalle (KDP), *KRKS & SSR* 39013.

Crotolaria retusa L., Sp. Pl. 715. 1753; FBI 2: 75. 1876; FPM 1: 293. 1918; FAP 1: 263. 1997.

Local name: *Potti-gilligichaka.*

Erect, pubescent under shrub, to 1 m high; branches herbaceous, pubescent. Leaves simple, sub-sessile, obovate-lanceolate or oblanceolate, 1.5–4 × 0.5–1 cm, glabrous above, pubescent below, base attenuate, margin entire, apex obtuse, retuse. Racemes terminal, lax, up to 25 cm long; peduncles up to 3 cm long. Flowers 10–13, light yellow, pedicel to 1 cm long; bracts subulate, 3–5 mm long; bracteoles 2, scarious, *ca.* 1 mm long, appressed to calyx. Calyx campanulate, tube *ca.* 5 mm long, glabrescent, lobes 2-lipped, 5–8 mm long, upper lip 2-lobed, lower lip 3-lobed, acuminate at apex. Corolla to 1.5 cm long; standard petal *ca.* 1.8 × 1.6 cm, glabrous, wing petals obovate, *ca.* 1.6 × 0.8 cm; keel petals ovate, *ca.* 1.8 × 1 cm, beak spirally twisted. Staminal sheath *ca.* 8 mm long. Ovary subsessile, oblong, *ca.* 1 cm long, glabrous, stigma to 2.5 cm. Pod glabrous, oblong-terete, shortly stalked, 2.9–3.8 × 0.8–1 cm, beak curved, brown, much exceeding the persistent calyx. Seeds 12-16, reniform, thick brown, 0.2 cm across.

Occasional, near water bodies, fields bunds and in open forest patches. Fl. & Fr.: Jul.–Nov. Sanevaripalli (ATP), *KRKS* 40181.

Crotalaria verrucosa L., Sp. Pl. 715. 1753; FBI 2: 77. 1876; FPM 1: 297. 1918; FAP 1: 265. 1997.

Erect herb, to 60 cm high, branches 4-angled, woody, pubescent. Leaves simple, ovate, deltoid or rhomboid, 4–7 × 3–5.5 cm, pubescent, base attenuate, margin irregularly undulate or entire, apex obtuse-acute; petiole 4–5 mm; stipule foliaceous, semilunate, 0.5 × 1 cm. Raceme close, terminal or lateral; peduncles 8–20 cm long. Flowers 8–12, bluish white or light purple, 2 × 2 cm; pedicels 5–7 mm long; bracts subulate, *ca.* 4 mm long, glabrous; bracteoles at the base of pedicels or a little above, linear-filiform, 2–3 mm long, reflexed, glabrous. Calyx campanulate to 1 cm long; tube *ca.* 4 mm long; lobes 4–5, membranous, lanceolate, upper *ca.* 7 mm long, lower *ca.* 6.5 mm long, faintly pubescent. Corolla 1.5 cm long, standard petal broad-ovate, 2–2.5 × 1–1.5 cm, obtuse-retuse at apex, pubescent on the upper surface; wing petal obovate, *ca.* 1.5 × 0.7 cm; keel petals *ca.* 1.5 × 0.6 cm, rounded about the middle, beak short, spirally twisted. Staminal sheath *ca.* 6 mm long. Ovary stipitate, oblong, *ca.* 7 mm long; style *ca.* 1 cm long. Pods shortly stalked, oblong or oblong-terete, 3–4 × 1–1.5 cm, minutely and sparsely pubescent, yellowish-brown, nerves prominent, much exceeding the calyx. Seeds 15–18, reniform, shiny, 0.2 cm across.

Occasional, in wastelands and borders of agriculture fields. Fl. & Fr.: Sept.–Jan. Somulavaripalli, *KRKS & KRS* 37816; Malakavemula (ATP), *KRKS* 41993.

DALBERGIA L. f. *nom. cons.*

Key to Species

1a. Leaflets 3–7, 5–8 × 3–6 cm .... (2)

1b. Leaflets above 9, 2–4 × 1–2 cm .... **D. lanceolaria**

2a. Apex of leaflets cuspidate, acuminate, leaflets broadly elliptic, leaf rachis zig zag.. **D. sissoo**

2b. Apex of leaflets retuse or obtuse, leaflets orbicular-elliptic, leaf rachis straight.. **D. latifolia**

Dalbergia lanceolaria L.f., Suppl. Pl. 316. 1782. subsp. **paniculata** Thoth., Bull. Bot. Surv. India 25: 171. 1985. *D. paniculata*Roxb., Pl. Coromandel. t.114. 1799; FBI 2: 236. 1876; FPM 1: 383. 1918; FAP 1: 267. 1997.

Local name: *Sopera, Porla-parchari, Patsaru.*

Large deciduous tree, to 25 m high; trunk irregularly fluted, bark smooth, grey, wood greyish white, rather soft; young branches tomentose, terete, black when dry. Leaves pinnately compound, to 11 cm long; petiole to 2 mm long; petiolule to 3 mm long; leaflets 9–13, ovate-oblong, terminal one obovate-oblong, 1.3–2.6 × 1–1.5 cm, glabrous above, tomentose below base round-obtuse, margin entire, apex obtuse or emarginated, sometimes truncate. Panicles axillary, to 15 cm long, on young branches; peduncles 2 cm long; pedicles 3 mm long, rusty tomentose. Flowers white, 0.5 cm across. Calyx campanulate, rusty tomentose, tube 4 mm long; lobes 2 mm long, rusty tomentose, ciliate at margin. Corolla white; standard petal ovate-oblong 1 × 0.5 cm; wing petals 7.5 × 3 mm; keel petals 7 × 3 mm; staminal sheath 5 mm long, filaments 3 mm long; stamens isodiadelphous (5+5). Ovary glabrous, 3 mm long, style 2 mm long. Pod pubescent, glabrous towards apex, 4–6.5 × 1.2 cm, flattened, elliptic-oblong, tapering at both ends, black when dry. Seed 1, rarely 2, 0.8 × 0.5 cm, black, kidney-shaped, flattened.

Common, in dry deciduous forest. Fl. & Fr.: Sept.–Jan. K.K. Kottala (KDP), *KRKS, KRS*& *TP* 39009.

Note: Twigs black when dry. The timber is useful for furniture.

Dalbergia latifolia Roxb., Pl. Coromandel t. 113. 1799; FBI 2: 235. 1876; FPM 1: 383. 1918; FAP 1: 267. 1997.

Local name: *Jittegi, Zitregi.*

Large deciduous tree, to 20 m high; bark light grey, rough; wood dark purple with black streaks; branchlets terete, glabrous. Leaves imparipinnate, to 23 cm long; petiole to 6 cm long; leaflets 5–7, elliptic-obovate, 3–5 × 3–5 cm, glabrous, brown when dry, base attenuate-acute, margin entire, apex emarginate; petiolule 0.7 cm long, glabrous. Panicles axillary, to 9 cm long, much branched; peduncle 1 cm long; pedicel 1 mm. Flowers white with pale yellow tinge, 4 mm across. Calyx tube 3 mm long; lobes subequal, to 1.5 mm long, pubescent, acute. Corolla white, with pale yellow tinge; petals clawed; standard petal 7 × 4.5 mm, obtuse; wing petals 7 × 2.5 mm, apex obtuse; keel petals 6.5 × 3.5 mm; stamens monadalphous (9), staminal sheath 5 mm long, filaments 2.5 mm long. Ovary glabrous, 4 mm long, style 2 mm long. Pod oblong-lanceolate, 4.8 × 1.5–2 cm, abruptly narrowed to a stipe, flat, glabrous. Seeds 1–3, flat, brown, kidney-shaped.

Common, in hilly areas of dry deciduous forest. Fl. & Fr.: Jul.–Feb. Batrepalli (ATP), *KRKS & BSS* 39568.

Note: Wood is very valuble for furniture.

Dalbergia sissoo Roxb., Fl. Ind. 3: 223. 1832; FBI 2: 231. 1876. FAP 1: 267. 1997. *Dalbergia sissoides* Graham ex Wight & Arn., Prodr. Fl. Ind. Orient. 265. 1834; FPM 1: 383. 1918.

Local name: *Sissoo*.

Deciduous tree, up to 15 m high, densely foliaceous; bark thick grey, exfoliating in narrow longitudinal strips; branchlets pilose. Leaves imparipinnate, rachis to 13 cm long; petiole to 3 cm long; leaflets rhomboid, obovate, 2–6 × 2–6 cm, glabrous, base cuneate, margin entire, apex acute or abruptly acuminate, shining above. Panicles 7 cm long. Flowers white, 5 mm across; peduncles 2 cm long; pedicels 1 mm. Calyx tube 3.5 mm; lobes 2 mm. Corolla yellowish white; stamens monadelphous (9). Pod strap-shaped, 5–7.5 cm long, pale brown, glabrous, indehiscent, cuneate at base. Seeds 1-4, kidney-shaped, flat, light brown.

Planted on roadsides. Fl. & Fr.: Mar.–Aug. Rekkamanu (ATP), *KRKS & KRS* 37929.

DERRIS Lour. *nom. cons.*

Derris scandens (Roxb.) Benth., J. Linn. Soc. Bot. 4 (Suppl.) 103. 1860; FBI 2: 240. 1876; FPM 1: 387. 1918; var. **scandens** Thoth., Fasc. Fl. Ind. 8: 28. 1982; FAP 1: 269. 1997. *Dalbergia scandens* Roxb., Pl. Coromandel t. 192. 1805.

Local name: *Nalla tiga*.

Straggling shrub, to 8 m high; branches rusty tomentose. Leaves imparipinnate, 7–9-foliolate; leaflets elliptic-oblong or obovate, 1–5 × 2–2.5 cm, glabrous, base truncate-cuneate, margin entire, apex acute, emarginated; petiole 4.5 cm long; petiolule 3 mm long. Panicles terminal and axillary, to 15 cm long; peduncle to 3 cm long; pedicel slender, to 5 mm long. Flowers pale pink, 7 mm across; pedicel 1.5 cm long. Calyx campanulate, tube 4 mm long; lobes *ca.* 1 mm, glabrous. Corolla creamy, exserted; standard petal obovate, broad, 8 × 6.5 mm; wing petals obliquely oblong, 7 × 2.5 mm; keel petals obtuse, incurved, 6.5 × 2.5 cm; staminal sheath 8 mm long; stamens monadelphous (10). Ovary to 3 mm long, glabrous. Pod flat, reticulately veined, shortly apiculate, glabrous, 4–6 cm long; stalk 0.5 cm long. Seeds 1–4, flattened, brown.

Common, in dry deciduous forests. Fl. & Fr.: Jul.–Aug. Rachakuntapalle (KDP), *KRKS & KRS* 37922.

DESMODIUM Desv. *nom. cons.*

Key to Species

1a. Leaves 3-foliolate; prostrate herb .... (2)

1b. Leaves simple; sub-erect herb or stragglers .... (3)

2a. Rooting at nodes present; branchlets terete, filiform; leaflets 0.5–0.7 cm long; flowers 3–5 clusters, in axillary fascicles .... **D. triflorum**

2b. Rooting at nodes absent; branchlets angulate, strong; leaflets 1.8–8 cm long; flowers in terminal racemes or racemose panicles .... **D. dichotomum**

3a. Stipules more than 1 cm long, apex setaceous; calyx-lobe twice longer than the tube .... **D. gangeticum**

3b. Stipules less than 5 mm long, apex acuminate; calyx-lobes equal to the tube .... **D. velutinum**

Desmodium dichotomum (Willd.) DC., Prodr. 2: 336. 1825; FAP 1: 271. 1997. *Hedysarum dichotomum* Willd. Sp. Pl. 3: 1180. 1802.

Herbaceous-suffrutescent with procumbent, diffuse branches; branches angular, grooved with tawny to brownish, densely puberulent with long hairs. Leaves 3-foliolate; stipules obliquely ovate-acuminate, often becoming reflexed, 5–10 × 2–5 mm, strongly striate, puberulent without, ciliate, persistent; stipels lanceolate to ovate, attenuate, 1.5–9 × 0.5–3 mm, glabrous, ciliate along margin, persistent; petioles angular, ridged and grooved, moderately uncinulate-pubescent and sparsely pilose with long tapering hairs, 1–7 cm. long; leaflets ovate to elliptic or orbicular, base obtuse, margin entire, apex acute, abundantly appressed-pilose above, velutinous beneath with long hairs along the veins, terminal leaflet 1.8–8 × 2–6 cm, lateral leaflets similar, smaller, 1.2–7 × 1–4.8 cm. Inflorescence racemose or racemose-panicle; rhachis densely uncinulate-pubescent; primary bracts narrowly ovate, rather shortly acuminate, 2.5–4.5 × 0.5–1.5 mm, puberulent, each subtending a fascicle of 2–3 pedicels; secondary bracts each subtending a single pedicel, lanceolate-ovate, similar, 1.5–3 × 0.5–1 mm; pedicels minutely puberulent, usually more densely so on the upper half, mostly somewhat spreading, 3.5–5 mm long. Flowers pink or blue, small. Calyx pilose throughout; upper lobe deeply bifid, 2–3.2 mm long, lower lobe ± equal, 2–3 mm long. Corolla slightly exceeding calyx; standard obovate, entire or retuse, narrowed to cuneate base, 3.3–3.5 × 2.5–3 mm; wings *ca.* hatchet-shaped, slightly auricled, shortly clawed, 2.5–3 × 1–1.3 mm; keel-petals obscurely scythe-shaped, scarcely auricled, shortly clawed, 3–4.3 × 1 mm. Fruit shortly stipitate (stipe 0.5–1.5 mm. long), 4–6-articled; articles quadrate at maturity, nearly evenly indented on both sutures, 2–3 × 2.5–3 mm, uncinulate-puberulent throughout. Seed oval-reniform, 2 × 1.5 mm, brown.

Rare in deciduous forests. Fl. & Fr.: Nov. – Feb. Batrepalli (ATP), *KRKS* 41969.

Desmodium gangeticum (L.) DC., Prodr. 2. 327. 1825. *Desmodium gangeticum* var. *maculatum*; FBI 2: 168. 1876; FPM 1: 345. 1918; FAP 1: 272. 1997. *Hedysarum gangeticum* L., Sp. Pl. 746. 1753.

Local name: *Yitanarum.*

Under shrub, stem woody, up to 1 m high; branchlets sericeous, somewhat angular. Stipules to 1 cm long, narrowly ovate, apex setaceous. Petiole to 5 cm long. Leaves simple, oblong-obovate or ovate, 4–8 × 2–4.5 cm, chartaceous, base obtuse-subacute, margin entire, apex acute, sometimes acuminate, glabrescent above, appressed tomentose below. Racemes terminal, to18 cm long; peduncle to 5 cm long; pedicel to 3.5 mm long; bracts linear, to 3 mm long. Flowers violet, to 5 mm across. Calyx tube 1 mm long; lobes 4, triangular, pubescent, acuminate. Corolla violet, or white; standard petal broadly obovate, 5 mm across; wing petals 3 mm long; keel petals 4 mm long. Staminal sheath 3.5 mm long; stamens diadelphous (9+1). Ovary sessile, 3 mm long. Pod 2 cm long, moniliform, 2 × 0.3 cm, lower margin deeply undulate, upper margin straight, hooked at apex, pubescent, articles 6–8. Seeds 2 mm across, compressed.

Occasional, in dry deciduous forests as under growth. Fl. & Fr.: Sept.–Nov. Batrepalli (ATP), *KRKS* & *BSS* 40354.

Desmodium triflorum (L.) DC., Prodr. 2: 334. 1825; FBI 2: 173. 1876; FPM 1: 347. 1918; FAP 1: 275. 1997. *Hedysarum triflorum* L., Sp. Pl. 749. 1753.

Local name: *Muntamandu.*

Prostrate herb, to 50 cm long, branches often rooting at nodes, sparsely hirsute. Leaves tri-foliolate, to 1.5 cm long; leaflets obovate-circular or elliptic; terminal leaflet bigger than lateral leaflets, 0.5–0.7 × 0.4–0.7 cm, sparsely hirsute, base obtuse, margin entire, apex obtuse or emarginate. Petiole to 0.5 cm long; petiolule 0.3 cm long. Stipules obliquely-ovate, acute at apex, 3.5 mm long; stipels minute. Flowers 3 mm across, 3–5 clusters, in axillary fascicles; primary bracts ovate, 2 mm long, acuminate; secondary bracts to 2 mm long, setaceous; pedicel to 6 mm long. Calyx tube 1 mm long, pubescent; lobes 5, upper lobes 1 mm long, lower lobes 2 mm long. Corolla pink to violet; standard petal obovate, 3 mm long; wing petals 2 mm long; keel petals 4 mm long. Staminal sheath 4 mm long; stamens diadelphous (9+1). Ovary 4 mm long. Pod 1.5 × 0.3 cm, pubescent, articles 4–6, hooked, lower margin deeply indented, upper margin entire. Seeds broadly oblong, 1.2 mm, compressed.

Common, in wastelands and near the water bodies. Fl. & Fr.: Nov.–Feb. Somulavaripalle (ATP), *KRKS* & *KRS* 37828.

Desmodium velutinum (Willd.) DC., Prodr. 2: 328. 1825; FAP 1: 276. 1997. *Hedysarum velutinum* Willd., Sp. Pl. 3: 1174. 1802. *Desmodium latifolium* (Roxb. ex Ker.) DC., Prodr. 2: 328. 1825; FBI 2: 168. 1876; FPM 1: 346. 1918. *Hedysarum latifolium* Roxb. ex Ker., Bot. Reg.5: t. 355. 1819.

Erect under shrub, to 1.5 m high; branchlets terete, rough, densely brown tomentose. Leaves simple, broadly ovate, 7–11 × 5–8 cm, tomentose on both sides, lateral veins 6–7 pairs, base slightly truncate-obtuse, margin undulate, apex obtuse-acute; petiole 1–2 cm; petiolule to 3 mm long, densely scabrous-hirsute; stipule linear-cordate or ovate, 0.4 cm long; stipels 1.5 mm long. Panicles axillary and terminal, branched, to 15 cm long; peduncle to 2.5 cm long; pedicel to 1.5 mm long. Flowers 3–5 in a cluster, purple, 3 mm across. Calyx tube 0.5 mm, 4-lobed, pubescent; upper lobes connate, 0.5 mm long; lower one 1 mm long. Corolla rose-purple; standard petal orbicular-obovate, 6 × 4 mm; wing petals oblong, 6 × 2 mm; keel petals slightly incurved, 6 × 2 mm. Staminal sheath 4 mm long; stamens diadelphous (9+1). Ovary sessile, 3 mm long. Pod pale yellow, 2 × 1.7 cm, pubescent, indented on both margins, articles 4–6, oblong, hispid.

Common, in dry deciduous forests along streams. Fl. & Fr.: Aug.–Nov. Batrepalli (ATP), *KRKS & BSS* 40356; Batrepalli (ATP), *KRKS* 311 (HJCB-E).

ELEIOTIS A.P. DC.

Eleiotis monophylla (Burm. f.) DC., Legum. 7: 350. 1825; FAP 1: 278. 1997. *Glycine monophylla* Burm. f., Fl. Ind. 161 t. 50 f. 2. 1768. *Eleiotis sororia* DC., Prodr. 2: 348. 1825; FBI 2: 153. 1876; FPM 1: 333. 1918.

Local name: *Nelanapa.*

Annual, procumbent, slender herb, to 4 m long, branchlets glabrous, triquetrous. Leaves 1-foliolate, orbicular or broadly elliptic, 1–2.5 × 1–2.8 cm, glabrous above, sparsely hairy below, base cordate, margin entire, apex emarginate; petiole to 2 cm long; stipules ovate, linear, 0.4 cm long. Racemes lax, axillary, to 8 cm long; pedicel 0.5 cm; bracts broadly ovate, 2.5 mm long. Flowers single or paired, 5 mm across, distantly arranged on raceme; pedicels to 5 mm long. Calyx tube 0.5 mm long, campanulate, puberulous; lobes 5, 1 mm long. Corolla reddish, 0.5 cm across; standard petal obovate, emarginate, 3.8 × 4 mm, wing petals oblong, 1.2 × 2 mm, adnate to keels; keel petals obtuse, 3.5 × 1.5 mm. Staminal sheath 2.5 mm long, stamens diadelphous (9 + 1), 0.2 cm long. Ovary sessile, sparsely pubescent, 1.5 mm long, glabrous. Pod boat-shaped, curved, 0.4 cm long, glabrous, reticulate, compressed, dorsal margin straight, prominently nerved, mucronate. Seeds 1–2, flattened, 1 mm across, dotted.

Occasional, weed in groundnut fields. Fl. & Fr.: Jul.–Oct. Kalasamudram (ATP), *KRKS & BSS* 40133.

GALACTIA P. Br.

Galactia tenuiflora (Klein ex Willd.) Wight & Arn., Prodr. Fl. Ind. Orient. 206. 1834; FBI 2: 196. 1876; FPM 1: 356. 1935; FAP 1: 284. 1997.

Key to Varieties

1a. Leaflets 2.5–3 cm long, lateral veins 6 pairs, base subcordate; inflorescence to 1.5 cm, pod 6-seeded .... var. **minor**

1b. Leaflets 3–6 cm long, lateral veins 8 pairs, base obtuse; inflorescence to 8 cm, pod 7–8-seeded .... var. **tenuiflora**

Galactia tenuiflora (Klein ex Willd.) Wight & Arn. var. **minor** Baker in Hook. f. Fl. Brit. India 2: 192. 1876.

Annual climbing or twining herbs, to 2 m long; branchlets slender, appressed pubescent. Leaves 3-foliolate, to 10.5 cm long; leaflets oblong-elliptic, 5–5.5 × 1.2–1.5 cm, glabrescent above, sparsely pubescent below, base obtuse, margin entire, apex obtuse-emarginate, lateral veins 7–9 pairs; petiole to 5 cm long; stipules 2 mm long, setaceous, stipels 1 mm long. Raceme axillary, up to 2 cm long. Flowers 4 mm across, twin on slender rachis; peduncle 1.7 cm; pedicel 4 mm long; bracts 1 mm long, deciduous. Calyx campanulate, tube 1.5 mm long; lobes 5, 3 mm long, lanceolate, acute, sparsely pubescent. Corolla purple, slightly exserted; standard petal oblong, 6 × 4 mm; wing petals narrowed, 6 × 1.5 mm; keel petals 6 × 2 mm. Staminal sheath 4 × 1.5 mm; stamens diadelphous (9+1). Ovary pubescent, flat, 3.5 × 0.7 mm. Pod straight 4 × 0.6 cm, faintly puberulous, obliquely constricted between the seeds, margin thick, apex spiny. Seeds 6, circular or elliptic, glabrous, 1.5 mm diameter, compressed.

Common, in dry deciduous forests. Fl. & Fr.: Jul.–Dec. Kalasamudram (ATP), *KRKS & BSS* 39505.

Galactia tenuiflora (Klein ex Willd.) Wight & Arn. var. **tenuiflora**

Slender twining herb, appressed pubescent, to 7 m long. Leaves 3-foliolate, 7 cm long; leaflets oblong-ovate, 3–4.5 × 0.6–1 cm, sparsely silky pubescent above, densely silky pubescent below, base rounded, margin entire, apex acute-rounded, somewhat apiculate, lateral veins 8–10 pairs; petiole to 4 cm long; stipules setaceous, minute. Raceme axillary, to 13 cm long, white pilose. Flowers bluish-purple, 1 cm across, distinctly arranged; pedicel 0.3 cm; bracts ovate, 1.5 mm long. Calyx campanulate, tube 2 mm long; lobes 4, 4 mm long, hairy, acute. Corolla 0.7 cm long, purple-white; standard petal 1.1 × 0.7 cm; wing petals 1.1 × 0.3 cm; keel petals 1 × 0.4 cm. Staminal sheath 5 × 1.5 mm; stamens diadelphous (9+1). Ovary flat, pubescent, 3 mm long. Pod flattened, curved, 2–2.5 × 0.5-0.6 cm, silky hairy, mucronate at apex. Seeds 5–8, flattened, dark blackish-brown, 2 mm diameter.

Occasional, in dry deciduous forest. Fl. & Fr.: Aug.-Dec. Chinnapalle forest (ATP), *KRKS, BSS* & *TP* 40377.

GLIRICIDIA Kunth

Gliricidia sepium (Jacq.) Kunth ex Walp., Repert. 1: 679. 1842; FAP 1: 285. 1997. *Robinia sepium* Jacq., Enum. Pl. Carib. 28. 1760.

Lacal name: *Seema kanuga.*

Small deciduous tree, to 9 m high; bark greyish; branchlets lenticellate, white dotted, pubescent. Leaves imparipinnate, to 22 cm long; pinna whitish; petiole to 4 cm long; petiolule to 6 mm; leaflets, 7–9 pairs, 3.5–5 × 1–2 cm, ovate-oblong, glabrous base attenuate, margin entire, apex acute-acuminate. Raceme axillary, 10–13 cm long. Flowers pinkish white, paired on axis; bracts 0.5 mm; pedicels solitary or paired. Calyx undivided, cupular, 0.5 cm across; lobes obscure. Corolla exserted, rose pink; petals clawed; standard petal orbicular, 2 × 1.5 cm, with 2 callosiries above the claw; wing petals oblong, to 2 × 0.6 cm, auricled; keel petals obovate, 2.2 × 0.8 cm, incurved, apex obtuse. Staminal sheath 1.6 cm; stamens diadelphous (9+1); stamens free. Ovary sessile, 1.5 cm; style incurved, glabrous. Pod 10–14 × 2 cm, flattened, glabrous. Seeds 5–10, compressed, 0.6 cm in diameter.

Planted on roadsides. Fl.: Jan.–Feb. Fr.: May–Jul. Kalasamudram (ATP), *KRKS* 37964.

Note: Flowering comes when leaves are shed. Very beautiful at the time of flowering.

INDIGOFERA L.

Key to Species

1a. Leaves simple .... (2)

1b. Leaves compound .... (3)

2a. Prostrate; leaves cordate, villose; pods 2-seeded; flowers pink ...... **I. cordifolia**

2b. Decumbent; leaves linear, pubescent; pods 1-seeded; flowers red .... **I. linifolia**

3a. Leaves tri-foliolate .... (4)

3b. Leaves multi-foliolate .... (6)

4a. Pods 2-seeded, 5 mm long; inflorescence subcapitate; leaf gland-dotted .... **I. glandulosa**

4b. Pods more than 2-seeded; inflorescence elongate; leaf not gland-dotted .... (5)

5a. Erect herbs or sub-shrubs; branchlets and leaves silvery pubescent; pod 3 cm long, beaked. .... **I. trita**

5b. Prostrate or decumbent herbs; branchlets and leaves greenish pubescent; pod 1.5 cm long .... **I. prostrata**

6a. Shrubs or undershrubs .... (7)

6b. Herbs .... (10)

7a. Pods straight or slightly curved; leaflets 2–5 pairs .... (8)

7b. Pods curved; leaflets more than 5 pairs .... (9)

8a. Plant to 90 cm high, hispid-villose, hairs 2 mm long; leaflets 3–4 pairs; pod *ca.* 4-angled, hispid, not beaked, straight .... **I. astragalina**

8b. Plant to 1.5 m high, white pubescent; leaflets 1- or rarely 2 pairs; pod terete, pubescent, beaked at apex and slightly curved upwards .... **I. oblongifolia**

9a. Pods to 1 cm long, pubescent, 3–4-seeded; branchlets silvery, leaves obovate .... **I. caerulea**

9b. Pods to 2.5 cm long; torulose, *ca.* 10-seeded; branchlets appressed pubescent. -pod slightly curved; shrubs; leaflets less than 20............ **I. tinctoria**

10a. Prostrate herbs .... **I. linnaei**

10b. Erect herbs .... (11)

11a. Pods glabrous, subcylindric; flowers red; hairs simple (plant wiry) .... **I. glabra**

11b. Pods brown tomentose, cylindric; flowers pink-rose; hairs glandular-viscid .... **I. colutea**

Indigofera astragalina DC., Prodr. 2: 228. 1825; FAP 1: 288. 1997. *I. hirsuta* sensu Baker in Hook.f. FBI 2: 98. 1876 *p.p.* non L. 1753; FPM 1: 312. 1918.

Annual erect herb or subshrub, to 1 m high, branches dense white pilose. Leaves 5.5–13.5 cm long, imparipinnate; petiole 2–3.5 cm long; stipules linear, 0.7 cm long; leaflets 3–5 pairs, oblong-obovate or elliptic-oblanceolate, 0.6–1.5 × 0.4–1 cm, villous, base attenuate, margin entire, apex obtuse-apiculate; petiolule 2 mm long. Raceme axillary, to 20 cm long; peduncle 1.5 cm long, pedicel 3 mm long. Flowers 0.5 cm across, pink or brick red, as long as calyx. Calyx 5-lobed, villous, lanceolate, pilose with brown hairs. Corolla pink; standard petal broadly obovate, 5.5 mm; wing petals 5 mm long; keel petals 6 mm long. Staminal sheath to 4 mm long; stamens diadelphous (9+1). Ovary sessile, 2 mm long, villous, stigma curved. Pod straight, 4-gonous, 1.5 × 0.4 cm, tomentose. Seeds 4–6, angular, brown, 2 mm across.

Common, in dry deciduous forest in moist places. Fl. & Fr.: Sept.–Feb. K. K. Kottala (KDP), *KRKS* & *KRS* 37655; Batrepalli (ATP), *KRKS* 37874, 279 (HJCB-E).

Indigofera caerulea Roxb., Fl. Ind. 377. 1832. var. **caerulea** Ali, Bot. Not. 111: 564. 1958; FAP 1: 289. 1997. *I. argentea* L. var. *caerulea* (Roxb.) Baker in Hook. f., Fl. Brit. India 2: 99. 1876. *I. articulata* sensu Gamble, Fl. Madras 1: 312. 1918.

Much branched, woody, small shrub, to 1.2 m high; branches white canescent. Leaves imparipinnate, to 10 cm long; leaflets obovate-elliptic, 1–2.5 × 0.8–1.3 cm, glabrous above, pubescent below, base acute, margin entire, apex obtuse; stipules 1 mm long, setaceous. Raceme simple, axillary, 5–6 cm long; bracts 2 mm long. Flowers 4 mm across, pink, closly arranged on axis. Calyx pubescent, 1.2 mm, deltoid; lobes 5, shorter than the tube. Corolla pink, 4 mm long; standard petal ovate, sessile, 4 mm long, tomentose; wing petals 3.5 mm long, keel petals 4 mm long with short spur. Staminal sheath 2 mm long; stamens diadelphous (9+1). Ovary 0.5 cm long, subsessile. Pods pubescent, 0.5–1 cm long, slightly curved, 1–4-seeded, mucronate at apex, slight constricted between seeds, dark brown. Seeds 1–4, brown, ellipsoid, rounded at both ends, 1 × 2 mm, shiny.

Common, in wastelands and occasional, in dry deciduous forest and scrub forests. Fl. & Fr.: Jun.–Oct. Mabbuchintalapalle (KDP), *KRKS & KRS* 37681; Kalasamudram (ATP), *KRKS* 39555.

Indigofera colutea (Burm. f.) Merr., Philipp. J. Sci. 19: 355. 1921; FAP 1: 290. 1997. *Galega colutea* Burm. f., Fl. Ind. 172. 1768. *Indigofera viscosa* Lam., Encycl. 3: 247. 1789; FBI 2: 95. 1876; FPM 1: 311. 1918.

Local name: *Banka thuttara.*

Erect herb, to 70 cm high, much branched; branches viscous glandular hairy. Leaves imparipinnate, to 9 cm long, viscous; stipules linear, to 1.5 cm long, hairy. Leaflets 5–7 pairs, elliptic oblong, 0.9–1.2 × 0.3–0.5 cm, hirsute above and below, base obtuse, margin entire-ciliate with gland tipped hairs, apex obtuse -apiculate. Panicles axillary, 3–5 cm long, glandular hairy; bracts linear, 4 mm long. Flowers pinkish-red; peduncle to 2 cm long; pedicels 2 mm long. Calyx 5-lobed, glandular hairy, tube 1 mm long; lobes 5, 1.5 mm long. Corolla 4 mm long; exserted, slightly pubescent; standard petal oblong-elliptic, 4 × 3 mm, base narrow; wing petals oblong, 4 × 1.2 mm; keel petals obovate, 4.7 × 1.7 mm with spur. Staminal sheath 3 mm long; stamens diadelphous (9+1). Ovary 4 mm long, hispid. Pod oblong, straight, 2–2.2 × 0.2 cm, glandular hairy, beaked at apex. Seeds 7–14, ovoid–cuboid or ellipsoid, 1 × 0.5 mm, blunt ends, pitted, dark brown or red.

Common, weed in agriculture lands and wastelands. Fl. & Fr.: Jun.–Aug. Kalasamudram (ATP), *KRKS* 39552.

Indigofera cordifolia Heyne ex Roth, Nov. Pl. Sp. 357. 1821; FBI 2: 9. 1876; FPM 1: 309. 1918; FAP 1: 290. 1997.

Local name: *Bhudidaku.*

Prostrate herb, to 40 cm long, much branched, branches densely grey tomentose. Leaves simple, subsessile, ovate-cordate, 0.6–1.5 × 0.5–1.5 cm, white or grey tomentose, base cordate-truncate, margin entire, apex acute-mucronate; petiole 1 mm long; stipule linear, 2 mm long. Raceme axillary, 3–5-flowered, condensed; bracts 2 mm long. Flowers pinkish-red, 3 mm across; pedicel 2 mm long. Calyx tube 2 mm long, hispid, lobes 5, to 2.5 mm long, ovate. Corolla exserted; standard petal spatulate, densely tomentose without, base gradually tapering; wing petals oblong, 3 × 1 mm; keel petals oblong, 2.5 × 1 mm. Staminal sheath 2.5 × 1 mm; stamens diadelphous (9+1). Ovary tomentose, 1 mm long. Pod oblong, 4 × 1.5 mm, sub-cylindrical, straight, beaked at apex. Seeds 2, globose, 2 mm across, pitted.

Common in rocky drier regions of scrub forest and Agriculture land. Fl. & Fr.: Nov.–Feb. Thummalapalle Mining site (KDP), *KRKS & KRS* 36903.

Indigofera glabra L., Sp. Pl. 751. 1753; FPM 1: 311. 1918; FAP 1: 290. 1997. *I. pentaphylla* Murr., Syst. veg. ed. 13. 564. 1774; FBI 2: 95. 1876.

Local name: *Pulchintaku.*

Annual, erect or suberect herb, to 50 cm high; branches weak, terete, villose, sometimes glabrous. Leaves imparipinnate, 2–4 cm long; leaflets 3–7-foliolate, obovate-elliptic, 0.5–1.4 × 0.4–0.8 cm, pubescent, base rounded, margin entire, apex obtuse; petiole 1 cm long; petiolule 1 mm long; stipule 0.5 cm, lanceolate. Raceme axillary, distinctly pubescent, 3–7-flowered, arranged distantly. Flowers reddish-pink; peduncle 1 cm long; pedicel 0.5 cm long. Calyx 5-lobed, ovate-oblong, sparsely hairy. Corolla 0.4 cm long; standard petal orbicular, to 4 mm long; wing petals spatulate, 3 × 2 mm; keel petals 3 mm across, beak straight. Staminal sheath 3 mm long; stamens diadelphous (9+1). Ovary pubescent, to 2 mm long. Pod linear, to 2 cm long, reddish brown, glabrescent, cylindric. Seeds 8–12, brown, subcuboid, 1 mm in diameter, truncate ends.

Common, in wastelands and scrub forest. Fl. & Fr.: Aug.–Dec. Kalasamudram (ATP), *KRKS & BSS* 39190; Somulavari palle (ATP), *KRKS & KRS* 37818; Batrepalli (ATP) *KRKS* 37878.

Indigofera glandulosa Roxb. ex Willd., Sp. Pl. 3: 1327. 1803; FBI 2: 94. 1876; FPM 1: 309. 1918; FAP 1: 290. 1997.

Erect or decumbent subshrub, to 70 cm high, branches downy hairy, yellowish. Leaves 3-foliolate, 1 cm long; leaflets obovate–oblong, 1.5–2 × 0.5–0.8 cm, pubescent above, dotted below with glands and downy, base cuneate, margin entire, apex obtuse; petiolule 2 mm long. Raceme axillary, capitate, 1 cm long. Flowers pinkish-red, 4 mm across, sessile; pedicel to 2 mm long. Calyx tube 1 mm long, lobes 5, pubescent, linear, 2 mm long, acuminate. Corolla red, 2.5 mm long; standard petal oblong-obovate, 2.5 mm long; wing petals obovate, to 1.5 mm long; keel petals 2 mm long with 1.5 mm spur. Staminal sheath 2 mm long; stamens diadelphous (9+1). Ovary subsessile, hirsute, 2 mm long. Pod oblong, 6 × 2.5 mm, winged, pubescent, gland dotted, brown, 0.5 × 0.3 cm, mucronate at apex, longitudinally yellow lined, tuberculate at margin, pubescent, 2-seeded. Seeds 1.5 mm in diameter, dark brown, glabrous.

Rare in the forest area and agricultural fields. Fl. & Fr.: Sept.–Dec. Obulareddypalle (ATP), *KRKS* & *BSS* 41929; Mabbuchintalapalle (KDP), *KRKS* 270 (HJCB-E).

Indigofera linifolia (L. f.) Retz., Observ. Bot. 4: 29. 1786; FBI 2: 92. 1876; FPM 1: 309. 1918; var. **linifolia**; FAP 1: 292. 1997. *Hedysarum linifolium* L. f., Suppl. Pl. 331. 1781.

Prostrate, annual herb, to 25 cm long; branches silvery-canescent. Leaves simple, linear-oblong or lanceolate, 2–2.5 × 0.3–0.5 cm, white pubescent, base cuneate, margin entire, apex acuminate to mucronate; petiole 1–1.5 mm. Raceme axillary, 1 cm long; bracts 5 mm long, linear. Flowers red, 8–12; peduncle reduced; pedicel to 1 mm. Calyx ovate, campanulate, white pubescent, 0.3 cm long; lobes 5, acuminate. Corolla reddish, 3 mm long; standard petal ovate, 3.5 × 2.5 mm, pubescent; wing petals oblong, 3 × 1.2 mm; keel petals 4 × 1 mm. Staminal sheath 1 mm long; stamens diadelphous (9+1). Pod globose, pubescent, 2 mm diam. Seed 1, kidney-shaped, brown, 1 mm diam.

Common along crevices of rocky boulders of scrub forests and wastelands. Fl. & Fr.: Oct.–Jan. Bhumaiahgari palle (KDP), *KRKS* & *KRS* 37611.

Indigofera linnaei Ali, Bot. Not. 111: 549. 1958; FAP 1: 292. 1997. *Hedysarum prostratum* L., Mant. Pl. 102. 1767. *Indigofera enneaphylla* L., Mant. Pl. 2: 272. 1771, *nom. illeg.* FBI 2: 94. 1876; FPM 1: 309. 1918.

Local name: *Yerra palleru, Chalapacchi, Cheragaddam.*

Prostate spreading herb, to 30 cm long; branches slender, white hirsute; stipules ovate–lanceolate, hirsute, to 3 mm long. Leaves imparipinnate, 1.5 cm long; leaflets obovate, 0.4–0.6 × 0.2–0.3 cm, hirsute on both sides, base attenuate, margin entire, apex retuse obtuse; petiole 1 mm long. Raceme axillary,1 cm long, 10–12-flowered. Flowers red, 3 mm across; pedicels 1 mm long. Calyx tube 1 mm across, pubescent, lobes 5, ovate, 2 mm long, acute. Corolla red, 0.2–0.3 cm diameter; standard petal obovate, 5 × 4 mm; wing petals 5.5 × 1.8 mm, keel petals 5 × 1.2 mm with 1 mm spur. Staminal sheath 2.5 mm long; stamens diadelphous (9+1). Ovary stipitate, 2 mm long. Pod ellipsoid, 4 × 2 mm, hirsute. Seeds 2, globose, 3 mm across.

Very Common, in wastelands and roadsides. Fl. & Fr.: Jun.–Jan. K. K. Kottala (KDP), *KRKS & KRS* 36420.

Indigofera oblongifolia Forrsk., Fl. Aegypt-Arab. 137. 1775; FPM 1: 310. 1918; FAP 1: 293. 1997. *I. paucifolia* Del., Descr. Egypt. Hist. Nat. 251. 1812; FBI 2: 97. 1876.

Erect, woody undershrub, up to 1.5 m high; branchlets silvery grey pubescent; stipules lanceolate, acuminate. Leaves 1–5-foliolate, to 3 cm long; leaflets elliptic-oblong, 1.5–2.5 × 0.5–1.3 cm, alternate, white pubescent, base acute, margin entire, apex acute-apiculate; petiole 5 mm long; petiolule 1 mm long. Racemes axillary, to 7 cm long, 20–50-flowered; peduncle to 1 cm long; pedicel to 2 mm long. Calyx silvery outside, 2 mm long; lobes as long as the tube; lobes acute, to 2 mm long. Corolla red, to 3 mm long; standard petal densely hairy on the back, obovate, to 3 mm long; wing petals obovate or spatulate, to 2 × 1 mm long; keel petals 2 × 1.5 mm, beak 1 mm long. Staminal sheath to 1.5 mm long; stamens diadelphous (9+1). Ovary 2 mm long, pubescent. Pod to 2 cm long, brown, pubescent, 4–8-seeded, beaked at apex, slightly constricted between seeds. Seeds ellipsoid, brown-red, 2 × 1 mm, blunt ends.

Occasional, near water bodies and scrub forests. Fl.& Fr.: Dec.–Aug. Mabbuchintalapalle (KDP), *KRKS & KRS* 36938.

Indigofera prostrata Willd., Sp. Pl. 3: 1226. 1802; FPM 1: 310. 1918; FAP 1: 294. 1997.

Prostrate herb, to 35 cm long; branches appressed pubescent, brown. Leaves 3-foliolate; stipules subulate; petiole to 1 cm long, pubescent; leaflets obovate, 0.5–1 × 0.3–0.4 cm, pubescent on both sides, base attenuate, margin entire, apex subobtuse. Racemes axillary, 0.2–0.5 cm long, 5–10-flowered. Flowers pink, to 3 mm across. Calyx 5-lobed, pubescent, ovate, 0.2 cm long. Corolla 0.4 cm long, pink; standard petal obovate, to 3 mm long; wing petals to 2 mm long; keel petals 2 × 2 mm. Staminal sheath 2 mm long; stamens diadelphous (9+1). Ovary to 1.5 mm long. Pods straight, pendulous from stalk, 1–1.5 cm long, pubescent with scabrid hairs, mucronate, slightly flattened, brown. Seeds 3–6, elliptic, blunt ends, 0.1 cm long.

Common, in wastelands and scrub forest. Fl. & Fr.: Aug.–Jan. K. K. Kottala (KDP), *KRKS* 37654.

Indigofera tinctoria L., Sp. Pl. 751. 1753; FBI 2: 96. 1876; FPM 1: 319. 1918; FAP 1: 295. 1997.

Woody undershrub, up to 2 m high, branchlets terete, faintly angular when young, appressed pubescent. Leaves imparipinnate, to 7 cm long; leaflets opposite, 7–11, ovate-oblong, 1–3 × 0.5–1.5 cm, glabrous above, puberulous below, base acute, margin entire, apex apiculate to acute; stipules 2 mm long; petioles 1.5 cm long. Racemes axillary, spike like; bracts setaceous, 1 mm long. Flowers 5 mm across;

peduncle very slender; pedicel 2 mm. Calyx hairy, lobes more than twice as long as tube; lobes lanceolate. Corolla pink; standard petal sub-orbicular; wing petals obovate; keel petals oblong-obovate, twisted, beaked, staminal sheath 3.2 mm long; stamens diadelphous (9+1). Ovary hairy, 3.5 mm long. Pod straight or slightly curved, 2.8 × 0.2 cm, cylindric, slightly falcate towards tips, torulose, thick. Seeds *ca.* 10, globose.

Common, in wastelands and near ponds. Fl. & Fr.: Sept.–Dec. Kalasamudram (ATP), *KRKS* 39512.

Indigofra trita L. f., Suppl. Pl. 335. 1781.

Key to Varieties

1a. Leaflets orbicular-obovate or elliptic, base attenuate-rounded, apex acute-rounded .... var. **subulata**

1b. Leaflets oblanceolate-lanceolate, base attenuate, apex apiculate.......... var. **trita**

Indigofera trita L. f., Suppl. Pl. 335. 1781; FPM 1: 310. 1935; var. **subulata** (Vahl ex Poir.) Ali, Bot. Not. 111: 558. 1958; FAP 1: 296. 1997. *I. subulata* Vahl ex Poir. in Lam., Encycl. Suppl. 3: 150. 1813; FBI 2: 96. 1876; FPM 1: 311. 1918.

Undershrub, to 50 cm high; branches brown pubescent. Leaves 3-foliolate; petiole 2 cm long; petiolule 2 mm long; leaflets orbicular-obovate or elliptic, 1.5–3 × 1–3 cm, pubescent on both sides, base attenuate-rounded, margin entire, apex acute-rounded, apical leaflet bigger than lateral leaflets. Racemes axillary, 3–3.5 cm long, 15–25 flowers densely arranged; peduncle to 2 cm long; pedicel to 1 mm long. Flowers pink, 0.4 cm long; bract to 2 mm long, acuminate. Calyx tube to 1 mm long; lobes 5, linear, 0.3 cm long, acuminate. Corolla 0.4 cm long, pink; standard petal obovate-oblong, 3 mm long; wing petals spatulate 3 mm long; keel petals 3 × 3 mm, tube 2 mm long. Staminal sheath to 2 mm long; stamens diadelphous (9+1). Ovary 2 mm long, pubescent. Pod 4-anglar, slightly curved upwards, pubescent, 2.5 cm long, 0.2 cm wide, apex sharply beaked. Seeds 6–10, 2 mm long, ellipsoid, brown.

Common, along the forest path, in open hilly slopes and also along the moist field bunds. Fl. & Fr.: Mar.–Oct. Idupulapaya (KDP), *KRKS & BSS* 39079; Kalasamudram (ATP), *KRKS & BSS* 39171.

Note: According to Gamble and Pullaiah (Flora of A.P.) this variety *subulata* has silvery pubescence; leaf base cuneate, apex emarginated, raceme 20 cm long (Gamble) and 20–40-flowered, 5 cm long, where as in our specimen leaves are orbicular, base rounded, apex rounded, raceme to 5 cm long, pod 2.5 cm long.

I. trita subsp. *subulata* var. *subulata* (Vahl ex Poir.) Ali differs from the proper species in having large leaflets and petioles, more than 12-flowered raceme, drooping pods which are not curved at the base. In our specimen pod slightly curved.

Indigofera trita L. f. var. **trita**; FAP 1: 296. 1997.

Local name: *Nakanaru.*

Grey pubescent under shrub, to 1 m high; branches whitish pubescent, hard. Leaves 3- foliolate, leaflets oblanceolate-lanceolate, apical leaflet bigger than the lateral leaflets, 0.5–2 × 0.3–0.5 cm, grey pubescent, base attenuate, margin entire, apex apiculate; petiole 0.5 cm, petiolule 0.2–0.4 cm. Racemes axillary, to 5 cm long, 20–25-flowered. Flowers reddish, to 3 mm across. Calyx reduced, to 1 mm, cupular, pubescent. Corolla reddish, 0.3 cm long, glabrous. Pod 1.5–2 cm long, brown grey pubescent, furrowed upwards, hard, apex mucronate. Seeds brown, ellipsoid, blunt ends, 1 mm diameter.

Common, in wastelands and agriculture lands, near water bodies. Fl. & Fr.: Aug.–Nov. Kalasmaudramu (ATP), *TP & EC* 2839: Idupalapaya (KDP), *KRKS & BSS* 39079.

MACROTYLOMA (Wight & Arn.) Verdc.

Macrotyloma uniflorum (Lam.) Verdc., Kew Bull. 24: 322. 1970; FAP 1: 298. 1997. *Dolichos uniflorus* Lam., Encycl. 2: 299. 1786. *D. biflorus* auct. non L.; FBI 2: 210. 1876; FPM 1: 367. 1918.

Local name: *Wulavalu.*

Twining or climbing herb, to 1.5 m long; branches villous, terete. Leaves 3-foliolate; leaflets rhomboid, 2–4 × 1.5–3 cm, chartaceous, villous, base oblique or rounded, margin entire, apex obtuse, apiculate; petiole 2.5 cm long; petiolule 0.2 cm. Raceme axillary, 2–3-flowered; peduncle to to 3 mm long; pedicels 1.5 mm long; bracts lanceolate, straight, 7 mm long; bracteoles linear, 3 mm long. Flowers whitish-yellow, to 1cm long. Calyx tube 3 mm long, pubescent; lobes 5, linear, 4 mm long, acute, pubescent. Corolla whitish-yellow; standard petal obovate, 4 mm long; wing petals oblong, to 3 mm long; keel petals 3 × 3 mm. Staminal sheath 5 mm long; stamens diadelphous (9+1). Ovary 5 mm long, pubescent. Pod flattened, up to 5 cm long, 0.5 cm width, pubescent, shortly beaked at apex. Seeds 6–8, suborbicular-elliptic, shiny, hard, brownish dotted, 0.3 cm diameter.

Cultivated crop and sometimes escape from fields and grows naturally in wastelands. Fl. & Fr.: Nov.–Jan. Nuthnakaluva (ATP), *KRKS & BSS* 41934.

MUCUNA Adans. *nom. cons.*

Mucuna pruriens (L.) DC., Prodr. 2: 405. 1825; FBI 2 : 189. 1876; FAP 1: 302. 1997. *Dolichos pruriens* L., Herb. Amb. 23. 1754. *Mucuna prurita* Hook., Bot. Misc 2: 348. 1831; FPM 1: 356. 1918.

Local name: *Teetikoila, Piliadagu kai, Adavi chintala.*

Slender, climbing, annual herb, to 8 m long; branches terete, downy pubescent. Leaves 3-foliolate; petiole 20–25 cm, pubescent; petiolule to 1 cm long; leaflets oblique–rhomboid, 5–12 × 5–8 cm, silvery white below, pubescent above, base subtruncate or subacute, margin entire, apex apiculate. Raceme axillary, to 50 cm long; pendulous to 13 cm long; pedicels 0.5 cm long, yellowish tomentose. Flowers

20–30, dark violet, 3.5 – 4 cm long. Calyx brown, itching hairy, tube 2.5 mm long; lobes 4, triangular, acute at apex. Corolla purple, standard petal 2 × 1.6 cm; wing petals 3.3 × 1.2 cm; keel petals 3.7 × 0.5 cm. Staminal sheath 3.5 cm long; stamens diadelphous (9+1). Fruit curved, 'S'-shaped, brown pubescent, hairs itching. Seeds to 6, 6 × 4 mm, brownish-black, shining.

Occasional, climbing on shrubs in outskirts of forest and in agriculture fields. Fl. & Fr.: Nov.–Mar. Nuthna Kaluva (ATP), Thalupula. *KRKS & BSS* 41924.

MUNDULEA (DC.) Benth.

Mundulea sericea (Willd.) A. Chev., Compt. Rend. 1925, 180. 1521; FAP 1: 302. 1997. *Cytisus sericeus* Willd. Sp. Pl. 2: 1121. 1802. *Mundulea suberosa* (DC.) Benth. in Miq. Pl. Jungh. 248. 1852; FBI 2: 110. 1876; FPM 1: 314. 1918.

Local Name: *Verribillu, Neelamurri*

Virgately branched shrub or a small tree, up to 5 m high, bark when young smooth, greenish yellow-brown and when mature very rough, longitudinally fissured, very corky, yellowish brown. Young branchlets densely pilose-velvety. Leaves alternate, numerous, imparipinnate, unusual pale silvery bluish-green that gleams in the sunshine; rhachis 3-17 cm. long, slightly to densely silky- pubescent; stipules small, linear; leaflets 9–19, the lower ones often alternate, narrowly ovate-lanceolate, subsessile, 1–3 × 0.8–2 cm, base obtuse, margin entire, apex apiculate, more or less glabrous and bright green above, nearly glabrous to densely silky and silvery beneath. Inflorescence a rather dense terminal raceme. Flowers large on pilose 1–3 cm long pedicels. Calyx silky, deeply campanulate, segments short, broadly triangular, acute, the upper two connate. Corolla lilac, occasionally white, up to 2.5 cm. long; standard silky, broadly oval or obovate, with a long claw; wing-petals falcate, oblong, coherent with the keel near the base; keel obtuse, the apex incurved; upper stamen monadelphous, free and geniculate at the base, united upwards with the rest in a closed tube, filaments flattened upwards, anthers uniform. Ovary multiovulate; style incurved, subterete, glabrous, the apex inflexed; stigma capitate. Fruit a pod, 7.5–9.5 × 0.5–1 cm, linear, flat, densely velvety with short golden brown hairs, sutures rather thick, tardly dehiscent. Seeds 6–10, about 4 mm broad, yellowish brown.

Occasional, clumps, on fully sun-exposed slopes and foot hills of deciduous forests. Fl. & Fr.: Sep. – Dec. Talupula (ATP), *KRKS* & *BSS* 41921.

Note: Fruits and seeds are used as fish poision.

ORMOCARPUM P. Beauv. *nom. cons.*

Ormocarpum sennoides (Willd.) DC., Prodr. 2: 375. 1825; FBI 2: 152. 1876; FPM 1: 332. 1918. *Heydysarum sennoides* Willd., Sp. Pl. 3: 1207. 1807. *Ormocarpum cochinchinense auct non* (Lour.) Merr., sensu FAP 1: 303. 1997.

Shrub, up to 2 m high; branchlets glabrescent. Leaves odd-pinnate, leaflets 6 pairs, sub-opposite-alternate, obovate, 1–1.5 × 0.5–1 cm, chartaceous, base cuneate, margin entire, apex retuse or emarginate, petiole minutely prickled. Inflorescence axillary, 3–5-flowered lax raceme; bracts scarious, 2 mm long; bracteoles 2 mm long, inserted below calyx tube. Flowers yellow. Stamens 5 + 5. Ovary stipitate, 8 mm long, glandular-pubescent. Legume 2–4 distant jointed, prickly or echinate. Seeds oblong, 4 mm across.

Very rare, in dry deciduous forest. Fl. & Fr.: Aug. – Nov. Idupulapaya (KDP), *KRKS* & *SSR* 39067.

Note: *Ormocarpum sennoides* has prickles on pod while *O. cochinchinensis* has no prickles on pod. After comparison with the type specimen from Missouri Botanical Garden, we identified our specimen as *O. sennoides*.

PONGAMIA Vent. *nom. cons.*

Pongamia pinnata (L.) Pierre, Fl. For. Cochinch. t. 385. 1899; FAP 1: 304. 1997. *Cytisus pinnatus* L., Sp. Pl. 741. 1753. *Pongamia glabra* Vent., Jard. Mal. t. 28. 203; FBI 2: 240. 1876; FPM 1: 385. 1918.

Local name: *Kanuga.*

Evergreen tree, up to 15 m high; branches glabrescent. Leaves odd-pinnate; petiole 4 cm long; petiolule 1 cm long; leaflets 5, ovate, 5–7 × 3–3.5 cm, glabrous, base cuneate, margin entire, apex acuminate. Racemes axillary, to 10 cm long. Flowers whitish pink. Calyx cupular, 0.8 cm long. Corolla purplish white. Pod flattened, indehiscent, obliquely oblong, woody. Seed 1, flattened, thick brown.

Common, along roadsides and cultivated for bio-diesel. Fl. & Fr.: Feb.–Jul. Batrepalli (ATP), *KRKS & BSS* 39094.

PSEUDARTHRIA Wight & Arn.

Pseudarthria viscida (L.) Wight & Arn., Prodr. Fl. Ind. Orient. 209. 1834; FBI 2: 154. 1876; FPM 1: 334. 1918; FAP 1: 305. 1997. *Hedysarum viscidum* L., Sp. Pl. 747. 1753.

Under shrub, to 1 m high, branchlets viscid pubescent, sticky. Leaves 3-foliolate; leaflets ovate-rhomboidal, lateral leaflets obliquely ovate, 5–8 × 3.5–7 cm, base cuneate, margin repand-sinuate, apex acute, middle leaflet rhomboid, 5–9 × 3.5–8 cm, base rounded, margin entire, apex acute. Receme terminal, to 15 cm long, viscid hairy. Flowers paired on the rachis, 4 mm long, pink, branches villous; peduncle to 2 cm long. Calyx campanulate, tube 1.5 mm. Corolla rose, exserted, petals clawed; standard petal to 3 mm long; wing petals 2.5 mm long; keel petals 2.5 mm. Pod linear-oblong, flattened, articulated, jointed between seeds, viscid hairy. Seeds 4–8, compressed, reniform, dark brown.

Common, in dry deciduous forests. Fl. & Fr.: Sept.–Dec. Kalasamudram (ATP), *KRKS* 37687; Batrepalli (ATP), *KRKS & BSS* 40355.

PTEROCARPUS L. *nom. cons.*

Key to Species

1a. Leaves 3-foliolate, greyish pubescent below, lateral veins not very conspicuous; flowers in short raceme; calyx greyish tomentose, deeply lobed, apex acuminate; pod with narrow wings................ **P. santalinus**

1b. Leaves 5–7-foliolate, glabrous on both sides, lateral veins very conspicuous; flowers in long panicles; calyx pubescent, shortly lobed, apex acute; pod with broad wings .... **P. marsupium**

Pterocarpus marsupium Roxb., Pl. Coromandel t. 116. 1799; FBI 2: 239. 1876; FPM 1: 385. 1918; FAP 1: 306. 1997.

Local Name: *Yegisa, Yegi.*

A large deciduous tree, to 17 m high; bark thick grey, longitudinally fissured, exfoliating in irregular scales; wood dark red and juice exudes by incisions. Leaves imparipinnate, 5–7-foliolate, petiole 3 cm long, petiolule 0.3 cm long; leaflets elliptic-oblong, ovate, 6.5-9.5 × 4.5-6 cm, coriaceaous, shiny, glabrous above, sparsely hairy below only on veins, base truncate or rounded, margin entire, apex emarginate. Panicles axillary and terminal, 17 cm long. Flowers 2 cm across, yellow; pedicel to 3 mm long. Calyx dark brown, pubescent; tube 6 mm long. Corolla golden yellow; standard petal 1.2 × 1 cm; wing petals 1.1 × 0.6 cm; keel petals 9 × 6 mm. Staminal sheath 4 × 6 mm; stamens monadelphous (10), becoming diadelphous (9+1). Ovary tomentose, 7 mm. Pod stipitate, flat, 3.5 cm across, orbicular, stipitate, dark reddish brown; convex, pubescent, winged. Seeds 1 or 2, flat, 8 mm across.

Occasional, in dry deciduous forests. Fl. & Fr.: May–Feb. Batrepalli (ATP), *KRKS & BSS* 39587.

Note: Wood is used for making furniture. The bark is used as an astringent and cure for toothache.

Pterocarpus santalinus L. f., Suppl. Pl. 318. 1781; FBI 2: 239.1876; FPM 1: 384. 1918; FAP 1: 306. 1997.

Local name: *Yerra chandanam, Rakta chandanam.*

Moderate sized trees, to 15 m high; bark blackish-brown, deeply divided into rectangular plates by vertical and horizontal cracks, exuding deep red juice when cut; wood dark red. Leaves 3-foliolate, broadly ovate-elliptic, 4–7.5 × 3–6 cm, glabrous above, pubescent below, base truncate-rounded, margin entire, apex emarginated; petiole 4 cm; petiolule 0.5 cm long; secondary veins 25 –35 pairs. Racemes axillary and terminal, *ca.* 8 cm long; peduncle to 1.5 cm; pedicel to 4 mm. Flowers yellow, to 2 cm across. Calyx tomentose, tube 5 mm long; lobes ovate. Corolla yellow; standard petal ovate, 2.1 × 1.5 cm; wing petals 1.8 × 0.8 cm; keel petals 1.4 × 0.3 cm. Staminal sheath 4 mm long; stamens monadelphous (10). Ovary stipitate, 7 mm long. Pod gradually narrowed into stipe, 3–4 cm across, brown, broad winged. Seed 1, reddish brown.

Common, in dry deciduous forests. Fl. & Fr.: Jul.–Oct. Idupulapaya (KDP), *KRKS & BSS* 40313.

Note: Endemic to Kadapa, Nellore, Chittoor and Prakasam districts and Common in Idupulapaya, also cultivated for valuble timber in all southern states of India. Wood is very valuable and being smuggled. The tree growth is extremely slow and takes decades to gain thickness. It is in demand in countries like China and Japan where it it used in making musical instruments, sculpture, furniture and medicines.

RHYNCHOSIA Lour. *nom. cons.*

Key to Species

1a. Erect herbs or shrubs .... (2)

1b. Trailing herbs or climbers .... (5)

2a. Calyx accrescent, conspicuous .... (3)

2b. Calyx persistent, not conspicuous .... (4)

3a. Branches week, some times climbing, plant green pubescent, pod one-seeded .... **R. rufescens**

3b. Branches strong, not climbing, plants silvery pubescent, pod two-seeded .... **R. ravii**

4a. Bracts 1.5 mm long; peduncle to 1.5 cm long; pedicel articulated; pod glabrous, not septate between the seeds .... **R. cana**

4b. Bracts 3 mm long; peduncle to 3 cm long; pedicel not articulated; pod puberulous, septate between seeds, -leaves reduced towards apex .... **R. suaveolens**

5a. Racemes capitate .... (6)

5b. Racemes elongated .... (7)

6a. Leaves, calyx and pods gland-dotted; pod ovoid, transversely wrinkled .... **R. capitata**

6b. Leaves, calyx and pods not gland-dotted; pod oblong, not wrinkled .... **R. densiflora**

7a. Leaflets obtuse, pod < 1.6 cm long; pods curved .... **R. minima**

7b. Leaflets acute or acuminate; pods > 2.2 cm long, straight .... **R. bracteata**

Rhynchosia bracteata Benth. in Hook. f., Fl. Brit. India 2: 225. 1878; FPM 1: 376. 1918; FAP 1: 309. 1997.

Suffruticose climber, to 3 m long; branchlets striate, woody, clothed with brown velvety pubescence. Leaves 3-foliolate; petiole up to 5 cm long, pubescent; petiolule 2 mm long; apical leaflet rhomboid, lateral leaflets obliquely ovate, 3–4 × 4–6 cm, thinly downy above, strongly below, base cuneate, margin entire, apex acute. Raceme axillary, many-flowered, to 25 cm long; peduncle 7 cm long; pedicel 1 mm long, pubescent; bracts to 6 mm, ovate, acuminate, softly pubescent. Flowers 4 mm across. Calyx campanulate, tube 4 mm; lobes to 1 cm long, pubescent. Corolla yellow *ca.* 1.5 cm long; standard petal auricled, pubescent on the back, 1.5 × 1.2 cm; wing petals 1.2 × 0.2 cm; keel petals 1.1 × 0.5 cm. Staminal sheath 9 × 3 mm; stamens diadelphous (9+1). Ovary sparsely pubescent, 7 × 2 mm. Pods curved, brown, pubescent, compressed, mucronate, 2.5 × 0.4 cm, glandular, prominent style to 7 mm long. Seeds 2, brown, flattened, ellipsoid.

Rare in dry deciduous forest. Fl. & Fr.: Mar.–Jul. Batrepalli (ATP), *KRKS & BSS* 39599.

Rhynchosia cana (Willd.) DC., Prodr. 2: 386. 1825; FBI 2: 222. 1876; FPM 1: 374. 1918; FAP 1: 309. 1997. *Glycine cana* Willd., Sp. Pl. 3: 1063. 1802.

Under shrub, up to 1 m high, branchlets viscid, glandular pubescent. Leaves 3-foliolate, petiole 4–5 cm long; petiolule 2 mm long, pubescent. Leaflets apical one elliptic-rhomboid, lateral leaflets obliquely rhomboid, 3–4.5 × 2–3.5 cm, pubescent above, villous below, base rounded, cuneate, margin entire, apex acute-apiculate. Racemes axillary; peduncle to 1.5 cm long; pedicel nodding, to 6 mm long. Flowers yellow, solitary or in pairs, 0.6 cm across. Calyx pubescent, tube 2 mm long; lobes lanceolate-acuminate, 0.3 cm long. Corolla 1 cm across, yellow; standard petal orbicular, 7 × 6 mm; wing petals 6.5 × 2 mm; keel petals 6 × 3. Staminal sheath 4 × 1.5 mm; stamens diadelphous (9+1). Ovary sessile, 1.5 mm long. Pod 1 cm long, compressed, brown. Seeds 2, reniform, reddish.

Common, in dry deciduous forests at rocky and sandy atone areas of plains. Fl. & Fr.: Nov.–Apr. Rachakuntapalle (KDP), *KRKS & KRS* 37634; Pulivendula Ghat (KDP), *KRKS* 274 (HJCB-E).

Rhynchosia capitata DC., Prodr. 2: 386. 1825; FPM 1: 374. 1918; FAP 1: 309. 1997. *R. aurea* sensu Baker in Hook. f., Fl. Brit. India 2: 221. 1876.

Annual, trailing or twining herb, to 2 m long, slightly villous. Leaves 3-foliolate; petiole 5-10 cm long, villous; petiolule 2 mm long, villous; leaflets rhomboid, 3–5 × 3–6 cm, dotted and pubescent below, glabrescent above, base rounded, margin entire, apex obtuse. Raceme capitate, axillary, 10–15-flowered; peduncle to 12 cm long, hirsute; pedicels to 1 cm long. Flowers yellow, to 1.5 cm across. Calyx tube 2 mm long, lobes 2 mm long, pubescent. Corolla yellow, 1.5 cm long; standard petal rounded-obovate, 6 × 4 mm; wing petals obovate; keel petals 3 × 3 mm. Staminal sheath 4 mm long; stamens diadelphous (9+1). Ovary sericeous, 3 mm long. Pod flattened, ellipsoid, brown, villous, transversely ridged. Seeds 2, flattened, dark brown.

Common, in agriculture land and wastelands. Fl. & Fr.: Oct.–Dec. Thummalapalle (KDP), *KRKS & KRS* 36909, 36971, 39041; Kalasamudram (ATP), *KRKS & KRS* 39540.

Rhynchosia densiflora (Roth.) DC., Prodr. 2: 386. 1825; FBI 2: 226. 1876; FPM 1: 376. 1918; FAP 1: 310. 1997. *Glycine densiflora* Roth, Nov. Pl. Sp. 348. 1821.

Slender twiner, soft villous, to 3 m long. Leaves 3-foliolate; petiole 3–5 cm long, pubescent, apical leaflet rhomboid; lateral leaflets obliquely ovate, 4–5 × 2.5–4 cm, thin, pubescent on both sides, base cuneate, margin entire, apex acute. Raceme axillary, to 2 cm long; bract 2, 0.5 cm long, pubescent. Flowers capitate, yellow, to 4 mm long. Calyx tube 1 mm long; lobes 5, 0.5 cm, pubescent. Corolla yellow, 0.8 cm long; standard petal obovate, to 4 mm long; wing petals oblong, to 3 mm long; keel petals 3 × 3 mm. Staminal sheath to 3 mm long; stamens diadelphous (9+1).

Ovary to 2 mm long, pubescent. Pod oblong, up to 1 cm long, flattened, glandular villous. Seeds 2, flattened, slightly elongated, reddish.

Occasional, in dry deciduous forest and field bunds. Fl. & Fr.: Oct.–Jan. Chinnapalli (ATP), *KRKS & KRS* 37700; Batrepalli (ATP), *KRKS* 312 (HJCB-E).

Rhynchosia minima (L.) DC., Prodr. 2: 385. 1852; FBI 2: 223. 1876; FPM 1: 375. 1918; FAP 1: 311. 1997. *Dolichos minimus* L., Sp. Pl. 726. 1753.

Twining herb, to 2 m long, pubescent. Leaves 3-foliolate; petiole 0.8–2.5 cm; leaflets rhomboid–obovate, 1–2 × 1–2 cm, pubescent on both sides, apical and lateral leaflets approximetly same, base cuneate, margin entire, apex retuse. Raceme lax, axillary, 2–5-flowered; peduncle to 7 cm long; pedicel minute. Flowers yellow, 5 mm across. Calyx tube 2 mm long; lobes linear, 3 mm long, pubescent. Corolla yellow, 0.5 cm long, red striations on standard petal. Staminal sheath 2 mm long; stamens diadelphous (9+1). Ovary pubescent, to 2 mm long. Pod brown, pubescent, flattened, mucronate, 0.5–1 cm long, curved. Seeds 1–2, flattened, 0.3 cm diameter, circular.

Common, in agriculture lands and wastelands. Fl. & Fr.: May–Sept. Thummalapalle (KDP), *KRKS & KRS* 33681; Kalasamudram (ATP), *KRKS & BSS* 39539; Pulivendula Ghat (KDP), *KRKS & BSS* 39663.

Rhynchosia ravii K. Prasad & A. Narayana Swamy, Phytotaxa 175 (3): 156. 2014.

An erect shrub, 1–2 m high, with glandular hairs, greyish-white, branches terete. Leaves 6–10 cm long, alternate, 3-foliolate; petiole 2–3.5 cm long; stipule ovate-lanceolate, 3.5–5 × 1–1.5 mm, acuminate, caducous; leaflets dimorphic, with greyish-white tomentose, veins impressed above, strongly raised beneath; terminal leaflet ovate-oblong 3.5–5 × 1.5–2 cm, base obtuse, margin entire, apex acute; petiolule 8–10 mm long; lateral leaflets obliquely oblong-ovate, 2–4 × 1.5–1.8 cm, base obtuse, margin entire, apex acute; petiolule 1.5–2 mm long; stipels 2 per leaflet, subulate, *ca.* 1 × 0.4 mm, caducous. Inflorescence an axillary raceme, 1–3 racemes per axil, filiform, pubescent; peduncle terete, 1–3.5 cm long; racemes 5–6 cm long, laxly 6–8-flowered. Bracts ovate, *ca.* 5 × 2.6 mm, pubescent. Pedicels short, filiform, *ca.* 1.2 mm long, pubescent. Flowers yellow *ca.* 1 cm long. Calyx pale green with red tinge, *ca.* 1 cm long, campanulate, tube 2–4 mm long, with dense, pubescent; lobes unequal, obliquely oblong-ovate, 6–8 × 4–4.5 mm, acute, 1–3-veined. Corolla standard yellow, orbicular, 7–8 × 7–7.5 mm, emarginate at apex, with two inflexed auricles at the base, claw 1.6–2 mm long; wings oblong-obovate, 7.5–8.2 × 2–2.5 mm, rounded at apex, claw *ca.* 2 mm long; keels light yellow, incurved, connate half, falcately obovate, 7–7.5 × 2.6–3 mm, rounded at apex, claw *ca.* 2 mm long. Stamens 9+1, white, unequal, bundle 6–8 × 1.2–1.8 mm, filament *ca.* 7.5 mm long; anthers yellow. Ovary sessile, oblong-elliptic, 4 × 1.5 mm, hairy, style *ca.* 6 mm long, glabrous, stigma pale yellow, capitate, *ca.* 0.2 mm long. Pod compressed, longer than the calyx, septate between the seeds, oblong, *ca.* 1.4 × 0.7 cm, incurved beak, tomentose. Seeds 2, subreniform, *ca.* 4 × 3 mm, strophiolate.

Occasional, in the type locality, open deciduous forest along with grass. Fl. & Fr.: Nov. – Apr. Nigidi forest, Batrepalli (ATP), *KRKS* 277, 431 (HJCB-E).

Rhynchosia rufescens (Willd.) DC., Prodr. 2: 387. 1825; FBI 2: 220. 1876; FPM 1: 373. 1918; FAP 1: 311. 1997. *Glycine rufescens* Willd., Ges. Naturf. Freunde Berlin Neue Schriften 4: 222. 1803.

Erect shrub with trailing branches, to 3 m long, viscid pubescent. Leaves 3-foliolate; petiole 4–5 cm long, viscid pubescent, apical leaflets smaller than the lateral leaflets; leaflets elliptic–oblong, ovate, 2–4 × 1–2.5 cm, viscid on both sides, base rounded-cuneate, margin entire, apex acute-acuminate. Racemes or panicles clustered, axillary; flowers paired and distinct on peduncles; peduncle 4–6 cm long; pedicel 0.5 cm, bract 0.2 cm long, ovate. Flowers yellow, to 1 cm long. Calyx 5-lobed; lobes 0.8 cm, elliptic-oblong, accrescent, conspicuous, covering the corolla and pod, pubescent, margins thin. Corolla yellow, 0.7 cm long; standard petal orbicular, 8 ×6 mm; wing 6.5 × 1.5 mm; keel petals 6 × 3 mm. Staminal sheath 5 × 1.7 mm; stamens diadelphous (9+1). Ovary stipitate, 2 mm long. Pod ovoid, apically beaked, brown. Seed 1, flattened, ovoid, strophiolate.

Common, in dry deciduous forest of Rachakuntapalle. Fl. & Fr.: Dec.–Mar. and Jul.–Oct. Rachakuntapalle (KDP), *KRKS & BSS* 41813.

Rhynchosia suaveolens (L.f.) DC., Prodr. 2: 387. 1825; FBI 2: 221. 1876; FPM 1: 374. 1918; FAP 1: 312. 1997. *Glycine suaveolens* L. f., Suppl. Pl. 326. 1781.

Erect undershrub, up to 2 m high; branchlets yellowish viscid-pubescent. Leaves 3-foliolate; petiole 3–5 cm; leaflets rhomboid–ovate, 3–4 × 2–3 cm, pubescent on both sides, base cuneate, margin entire, apex acuminate. Racemes lax, axillary 1-2-flowered; peduncle to 3 cm long; pedicels 2 mm long. Flowers yellow, to 0.8 cm across. Calyx tube 2 mm long; lobes 3 mm long, ovate, viscid hairy. Corolla yellow; standard petal 7 mm long; wing petals 5 mm long; keel petals 5 × 4 mm. Staminal sheath to 6 mm long; stamens diadelphous (9+1). Ovary to 4 mm long, pubescent. Pod oblong, pubescent, exceeding calyx, compressed. Seeds 2, strophiolate, flattened.

Common, in dry deciduous forests. Fl. & Fr.: Nov.–March. Batrepalli (ATP), *KRKS & BSS* 39598; Pulivendula Ghat (KDP), *KRKS* 274 (HJCB-E).

ROTHIA Pers. *nom. cons.*

Rothia indica (L.) Druce, Bot. Exch. Club. Soc. Brit. Isles 3: 423. 1914; FAP 1: 312. 1997. *Trigonella indica* L., Sp. Pl. 778. 1753. *Rothia trifoliata* Roth, Pers. Syn. 2: 302. 1807; FBI 2: 63. 1876; FPM 1: 279. 1918.

Annual, decumbent herb, up to 25 cm, branchlets silvery hirsute. Leaves 3-foliolate; petiole 0.6–1 cm long; leaflets oblanceolate, unequal, 1–1.3 × 0.3–0.4 cm, pubescent, base cuneate, margin entire, apex acute; stipules ovate-oblong, 0.5 cm. Raceme axillary, 2-flowered. Flowers yellow; stipules ovate-oblong, 0.5 × 0.2 cm 1 cm long. Calyx pubescent, tube 2.5 mm long; lobes 5, acuminate. Corolla yellow, pubescent, slightly longer than the calyx; standard petal elliptic-ovate, 4.5 mm long; wing petals spatulate, to 4 mm long; keel petals 3 × 4 mm. Staminal sheath 4 mm

Plate 6

A. *Rhynchosia rufescens*, B. *R. ravii*, C. *Zornia diphylla*, D. *Senna holosericea*, E. *Albizia thompsonii*, F. *Neptunia triquetra*

long; stamens monadelphous. Ovary densely villous, 4 mm long. Pod flattened, appressed villous, 5 × 0.2 cm, shortly beaked at apex. Seeds 30–35, 1 mm across, brownish, shiny.

Occasional, in Agriculture lands and wastelands. Fl. & Fr.: Through the year. Batrepalli (ATP), *KRKS & SSR* 37873; *KRKS* 3963.

SESBANIA Scop. *nom. cons.*

Key to Species

1a. Shrubs; stem and branches prickly; flowers to 1.2 cm long, yellowish with brown dots; pods 25 cm long.................................. **S. bispinosa**

1b. Trees; stem and branches not prickly; flowers to 10 cm long, whitish-yellow, light green striations; pod 48 cm long................. **S. grandiflora**

Sesbania bispinosa (Jacq.) W. F. Wight, U. S. D. A. Bur. Pl. Industr. Bull. 137: 15. 1909; FAP 1: 313. 1997. *Aeschynomene bispinosa* Jacq., Icon. Pl. Rar. 3: t. 564. 1792. *Sesbania aculeata* (Willd.) Poir. in Lam., Encycl. 7: 128. 1806; FBI 2: 114. 1876; FPM 1: 323. 1918. *Coronilla aculeata* Willd., Sp. Pl. 3: 1147. 1802.

Local name: *Tentua, Avisa, Suiminta.*

Shrub, annual, to 2.5 m high; branchlets smooth, prickly, white glossy. Leaves 25–30 cm long; stipule linear, to 0.8 cm long; stipels 0.5 mm long; petioles 0.8 cm long; petiolule 1 mm long; leaflets paripinnate, up to 47 pairs, linear-oblong, 1 × 0.2 cm, glabrous, base rounded, margin entire, apex obtuse-apiculate. Raceme lax, axillary, 6–8-flowered. Flowers yellowish; peduncle to 2 cm long, filiform; pedicel to 1 cm long, filiform. Calyx campanulate, cupular, 0.5 cm across, lobes minute, glabrous. Corolla yellowish; standard petal yellow with brownish-purple or pink dotted, 8 mm long; wing petals 7 mm long; keel petals 7 × 6.5 mm. Staminal sheath 6.5 mm long; stamens diadelphous (9+1). Ovary 1.5 cm long. Pod subterete, torulose, linear, 25–30 cm long, glabrous, straight, beaked, beak up to 1 cm. Seeds ellipsoid, 0.4 × 0.2 cm, brown.

Common, in agriculture lands and along the streams. Fl. & Fr.: Oct.–Feb. Somulavaripalli (ATP), *KRKS & KRS* 37817; Thummalapalle (KDP), *KRKS* 370 (HJCB-E).

Sesbania grandiflora (L.) Poir. in Lam., Encycl. 7: 127. 1806; FBI 2: 115. 1876; FPM 1: 323. 1918; FAP 1: 313. *Robinia grandiflora* L., Sp. Pl. 722. 1753.

Local Name: *Pedda Avisa.*

Tree, to 15 m high; bark greyish-brown, corky; wood soft; branchlets pubescent when young, glabrous when mature, green. Leaves even pinnate, up to 30 cm long; leaflets elliptic-oblong, 1.5–3 × 0.7–1 cm, glabrous above, pubescent below, base cuneate, margin entire, apex obtuse. Raceme axillary, lax, 3–5-flowered; peduncle to 5 cm long; pedicels 2 cm long. Flowers white, with green lines, 11 cm long. Calyx campanulate, 5-lobed, 3 cm long. Corolla white, up to 10 cm long; standard 9 × 6 cm, reflexed; wing petals 10 × 2.5 cm; keel petals 11 × 3.5 mm. Staminal sheath 10 cm long; stamens diadelphous (9+1). Ovary to 10 cm long. Pods up to 48 × 0.5 cm, curved, compressed between seeds, 4 cm long beak at apex. Seeds up to 30, ellipsoid, 8 × 4 mm.

Cultivated on borders of Agriculture lands and roadsides. Fl. & Fr.: Oct.–Mar. Chinnapalle (ATP), *KRKS & KRS* 37697.

Note: Leafy juice is medicinal for headaches and mumps.

STYLOSANTHES O. Swartz.

Stylosanthes fruticosa (Retz.) Alston, in Trimen, Handb. Fl. Ceylon 6: 77. 1931; FAP 1: 316. 1997. *Arachis fruticosa* Retz., Observ. Bot. 5: 26. 1789. *Stylosanthes mucronata* Willd., Sp. Pl. 3. 1166. 1802, *nom. illeg.*; FBI 2: 148. 1876; FPM 1: 326. 1918.

Local Name: *Saillee kumpa.*

Stiff herb to 50 cm high, often decumbent; branches villous, green. Leaves 3-foliolate; petiole 1 cm, stipules large, adnate to the petiole, 1.5 cm, chartaceous; leaflets linear-ovate, 1.5–2.5 × 0.3–0.5 cm, lateral ones smaller than the apical one, base cuneate, margin entire, apex mucronate. Bracts scarious, linear, lanceolate, 6 mm long. Flowers in axillary or terminal heads. Calyx connate 0.3 cm, tube being relatively long, slender; lobes subconnate. Corolla yellow; standard orbicular, 5 × 5 mm; wing petals oblong-obovate, 4 × 2 mm; keel petals incurved, 4 × 2 mm. Stamens monadelphous (10). Pod oblong, 0.2 cm, 1-2-jointed, tipped with the falcate indurated lower part of the style.

Common, in all dry areas, wastelands and roadsides. Fl. & Fr.: Nov.–Feb. Kalasamudram (ATP), *KRKS & KRS* 36450.

Note: Cultivated for fodder.

TEPHROSIA Pers. *nom. cons.*

Key to Species

1a. Leaves simple. .... **T. strigosa**

1b. Leaves compound .... (2)

2a. Prostrate herb .... (3)

2b. Erect herbs or under shrubs .... (4)

3a. Flowers white; pod hairy .... **T. pumila**

3b. Flowers purple; pod glabrous .... **T. maxima**

4a. Leaves glabrescent; fruits glabrous; flowers purple .... **T. purpurea**

4b. Leaves densely pubescent; fruits villous; flowers white .... **T. villosa**

Tephrosia maxima (L.) Pers., Syn. 2: 329. 1807; FPM 1: 319. 1918; FAP 1: 318. 1997. *Galega maxima* L., Syst. Nat. ed. 10: 1172. 1759. *Tephrosia purpurea* (L.) Pers. var. *maxima* (Pers.) Baker in Hook. f., Fl. Brit. India 2: 113. 1876.

Prostrate herb, to 70 cm long; young branches sparsely hairy, mature branches hard. Leaves imparipinnate, 3–5 cm long; leaflets 9–13, obcordate, 0.8-1 × 0.3-0.5 cm, glabrous, base cuneate, margin entire, apex obcordate or emarginate; petiole to 1.5 cm long; petiolule to 2 mm long; stipules ovate to 3 mm long. Raceme lax, axillary or terminal, to 20 cm long; peduncle to 10 cm long; pedicels to 5 mm long; bracts lanceolate, acuminate, to 3 mm long. Flowers, 3–8, to 1 cm long. Calyx connate 0.3 cm across; lobes 5, irregularly spreading, pubescent. Corolla purple; standard petal orbicular to 5 mm long; wing petals obovate, 3 mm long; keel petals 3 × 4 mm. Staminal sheath 3 mm long; stamens diadelphous (9+1). Pod flattened, oblong, 3–6 × 0.4 cm. Seeds 8-13, ellipsoid, brown, 3 × 2 mm.

Occasional, in dry open areas with sandy soils of the scrub forests and foot hills of dry deciduous forests. Fl. & Fr.: Nov.–Dec. Kalasamudram (ATP), *KRKS & TP* 37693.

Tephrosia pumila (Lam.) Pers., Syn 2: 330. 1807; FAP 1: 319. 1997. *Galega pumila* Lam., Encycl. 2: 599. 1788. *Tephrosia purpurea* (L.) Pers. var. *pumila* (Lam.) Baker in Hook. f., Fl. Brit. India 2: 113. 1876. *T. procumbens* (Buch.-Ham.) Benth., Gen. Index to Trans. Linn. Soc. 101. 1866; FPM 1: 320. 1918. *Galega procumbens* Bunch.-Ham., Trans. Linn. Soc. 13: 546. 1822.

Prostrate herb, branches villous, whitish, to 40 cm long. Leaves up to 3.5 cm long, villous; leaflets 7–13, obovate, 0.5–1.5 × 0.3–0.7 cm, pubescent, base cuneate, margin entire, apex truncate–obtuse, all leaflets bent downwards on rachis; stipules setaceous, 4 mm long. Inflorescence lax pseudoraceme, leaf-opposed; bracts 4 mm long; peduncle to 3 cm long; pedicel 1.5 mm. Flowers 3–7. Calyx connate, cupular, 0.3 cm across, pubescent. Corolla white, 0.8 cm long; standard petal suborbicular, 6 × 5.5 mm, tomentose outside; wing petals oblong, 6.5 × 2 mm; keels 5.5 × 3 mm. Staminal seath 8.5 mm long; stamens diadelphous (9+1). Ovary densely tomentose, 4 mm. Pod flattened, pubescent, 3–4 × 0.3 cm, mucronate. Seeds 8–12, ellipsoid, brown, 0.3 diameter, compressed.

Occasional, in wastelands and agriculture lands. Fl. & Fr.: May–July. Thummalapalle (KDP), *KRKS & KRS* 33682.

Tephrosia purpurea (L.) Pers., Syn 2: 329. 1807; FBI 2: 112. 1876; FPM 1: 320. 1918. *Cracca purpurea* L., Sp. Pl. 752. 1753.

Local name: *Vempali.*

Glabrescent, erect under shrubs, to 80 cm high. Leaves imparipinnate to 7 cm long, leaflets 7–13, obovate, 1–2 × 0.3–0.6 cm, glabrous, base cuneate, margin entire, apex rounded. Racemes terminal, up to 25 cm long; peduncle to 3 cm long; pedicel to 6 mm long. Flowers purple, 1 cm long; bracts 3 mm long. Calyx connate, cupular, 0.3 cm across, lobes 5, pubescent, acuminate. Corolla purple, 1.2 cm across; standard petal obovate-orbicular, 1 cm long; wing petals 9 mm long; keel petals 9 × 7 mm. Staminal sheath to 8 mm long; stamens diadelphous (9+1). Ovary to 9 mm long. Pod flattened, to 5 cm long, pubescent, apex mucronate. Seeds 7–10, ellipsoid, brown, 0.3 cm diameter.

Very common, in wastelands and scrub forest or even at foot hills of deciduous forests. Fl. & Fr.: Through the year. Thummalapalle (KDP), *SSR & TP* 33638.

Uses: Used as green manure.

Tephrosia strigosa (Dalz.) Santapau & Maheshw., J. Bombay Nat. Hist. Soc. 54: 805. 1857; FAP 1: 320. 1997. *Macronyx strigosus* Dalz., Hooker's J. Bot. Kew Gard. Misc. 2: 35. 1850. *Tephrosia tenuis* Wall. ex Dalz. & Gibs., Bombay Fl. 61. 1861; FBI 2: 111. 1876; FPM 1: 318. 1918.

Tufted, grass like, decumbent herb, to 20 cm long; branches appressed pubescent. Leaves simple, pubescent, linear, 1.5–4 × 0.2–0.6 cm, base acute, margin entire, apex acute-apiculate; stipules small, 2 mm long, deltoid. Flowers axillary, solitary, pink-blue; pedicel to 1.5 cm long, filiform; bracts minute, to 1 mm long. Calyx connate; tube 1.5 mm long, lobes 5, pubescent, 2 mm. Corolla 3 mm long; standard petal to 2.5 mm long; wing petals 2 mm long; keel petals 2 × 2 mm long. Staminal sheath to 2 mm long; stamens diadelphous (9+1). Ovary to 1.5 mm long, pubescent. Pod flattened, pubescent, linear, thin, 2 × 0.3 cm, apex apiculate, curved. Seeds 3–5, elliptic, 2 mm across, compressed.

Common in dry deciduous forests and scrub forests, grows along with grasses. Fl. & Fr.: Jul.–Feb. Bhumaiahgaripalle (KDP), *KRKS & KRS* 36999; Mabbuchintalapalle (KDP), *KRKS* 282 (HJCB-E).

Tephrosia villosa (L.) Pers., Syn. 2: 329. 1807; FBI 1: 113. 1876; FAP 1: 321. 1997. *Galega villosa* L., Syst. Nat. ed. 10. 2: 1172. 1758. *Tephrosia hirta* (Bunch.-Ham.) Benth., Gen. Index to Trans. Linn. Soc. Lond. 101. 1866; FPM 1: 318. 1918. *Galega hirta* Buch.-Ham., Trans. Linn. Soc. London 13: 546. 1822.

Local Name: *Tella vempali.*

Erect, silky villous under shrub, to 60 cm high. Leaves imparipinnate, to 7 cm long; leaflets 7–13, obovate, 1–1.5 × 0.3–0.5 cm, pubescent, base cuneate, margin entire, apex obtuse; stipule to 3 mm long, acuminate. Racemes terminal, to 15 cm long; peduncle to 4 cm long, pedicels to 1 mm long, villous. Flowers white, 1–3 at a point on rachis, 1.5 cm across. Calyx villous, 5-lobed, tube to 2 mm long; lobes acuminate, 3 mm. Corolla white, 0.5 cm; standard petal orbicular, to 8 mm long, pubescent; wing petals to 7 mm long; keel petals 7 × 6 mm. Staminal sheath to 6 mm long; stamens diadelphous (9+1). Ovary villous, to 4 mm long. Pod villous, linear, 3–4 × 0.3 cm, curved upwards, apex beaked. Seeds 5–8, elliptic, compressed, brown, 0.3 cm diameter.

Common along roadsides and wastelands, weed in cultivated lands. Fl. & Fr.: May –Aug. K.K. Kottala (KDP), *KRKS & KRS* 36453.

TERAMNUS P. Br.

Teramnus labialis (L.f.) Spreng., Syst. Veg. 3: 235. 1826; FBI 2: 184. 1876; FPM 1: 352. 1918; FAP 1: 321. 1997. *Glycine labialis* L. f., Suppl. Pl. 325. 1781.

Slender twining herb; branchlets appressed tomentose. Leaves 3-foliolate, petiole tomentose, 3.5 cm long, petiolule 0.3 cm, tomentose, terminal leaflet bigger than the lateral leflets, ovate or obliquely rhomboidal, 4–6 × 2–3 cm, base cuneate, margin entire, apex acute-mucronulate. Panicles axillary, to 20 cm long, branches densely pubescent; peduncle to 3 cm long; pedicels to minute. Flowers 3 at a point on peduncle; bracts 1 mm long, pubescent. Calyx cupular, tube to 1 mm long, lobes 5, ovate-lanceolate, pubescent, 4 mm long. Corolla white-pink; standard 5 mm long, pubescent; wing petals 4 mm long, spatulate; keel petals 4 × 3 mm. Staminal sheath to 3 mm long; stamens monadelphous (10). Ovary to 4 mm long, pubescent. Pod flattened, pubescent 3.5–5 × 0.3 cm, brown, apex beaked, curved. Seeds 5–11, flattened, elliptic, 0.3 cm across, compressed.

Common, in dry deciduous forest and scrub forests on the plains as well as hill tops. Fl. & Fr.: Oct. – Jan. Bhumaiahgaripalle (KDP), *KRKS* 37613; Rachakuntapalle (KDP), *KRKS & KRS* 37921; Idupulapaya (KDP), *KRKS & BSS* 40302.

VIGNA Savi *nom. cons.*

Key to Species

1a. Leaflets distinctly lobed .... (2)

1b. Leaflets entire, obscurely lobed .... (3)

2a. Stipules lanceolate; leaflets 3 - 5-lobed; peduncle to 15 cm long .... **V. aconitifolia**

2b. Stipules ovate-oblong; leaflets 3-lobed; peduncle more than 15 cm .... **V. trilobata**

3a. Pods 6 cm long, long hairy; seeds green.. **V. radiata**

3b. Pods 10 cm long, appressed hairy; seeds white or brown.......... **V. unguiculata**

Vigna aconitifolia (Jacq.) Marechal, Bull. Jard. Bot. Nat. Belge. 39: 160. 1969; FAP 1: 324. 1997. *Phaseolus aconitifolius* Jacq., Obs. 3: 2. t. 52. 1768; FBI 2: 202. 1876; FPM 1: 363. 1918.

Local name: *Pilli pesara.*

Slender, trailing herb, branchlets villous or hispid, up to 70 cm long; stipules lanceolate. Leaves 3-foliolate, each 3–5-lobed; lobes linear, villous; leaflets ovate, 2–3 × 2–3 cm, base cuneate, margin entire, apex acute. Racemes lax, axillary, clustered on peduncle at apex; peduncle to 15 cm long; bract ovate, 0.3 cm long. Flowers yellow, to 1 cm across. Calyx connate, campanulate; tube to 2 mm long; lobes 0.1–0.2 mm long. Corolla yellow to purple; standard petal orbicular, 4 × 7 mm; wing petals obliquely obovate, 5 × 3 mm; keel petals deeply falcate, 5.5 × 1.5 mm. Staminal sheath 4.5 × 1.5 mm; stamens diadelphous (9+1). Ovary terete, to 4 mm. Pod cylindrical or subcylindrical, 2–4.5 × 0.3 cm pubescent, beaked. Seeds ellipsoid, blunted ends, 5–8, 0.3 × 0.2 cm, green.

Occasional, in moist shady hilly areas. Fl. & Fr.: Aug.–Nov. Thummalapalle (KDP), *KRKS & KRS* 36913, 36992; Chinnapalli (ATP), *KRKS & KRS* 37802.

Note: This species is used as fodder.

Vigna radiata (L.) Wilczek, Fl. Congo Belge 6: 386. 1954. var **radiata**; FAP 1: 325. 1997. *Phaseolus radiatus* L., Sp. Pl. 725. 1753; FPM 1: 363. 1918. *P. mungo* L. var. *radiatus* (L.) Baker in Hook. f., Fl. Brit. India 2: 203. 1876.

Local name: *Pesalu, Pesara.*

Erect or shortly twining herb, up to 1 m long; branches some times trailing and spreading, hispid hairy. Leaves 3-foliolate, terminal leaflet broader than the lateral leaflets, ovate-obliquely ovate, 5–8 × 3–5 cm; hispid on both sides, base cuneate, margin entire, apex acute; petiole 12–16 cm, hispid; petiolule 0.4 cm; stipule ovate, 1 cm long, stipel 0.5 cm long, linear. Racemes axillary, capitate; peduncle 20-25 cm long. Flowers yellow, to *ca.* 1 cm long; bracts peltate, 8 mm long; bracteoles linear. Calyx 5-lobed; lobes 0.2 cm long, hispid. Corolla yellow–purple; standard petal orbicular to obovate, 1 cm long; wing petals obovate to spatulate, 6 mm long; keel petals 6 × 5 mm, beaked at apex with pocket of staminal sheath, staminal sheath 5 mm long; stamens diadelphous (9+1). Ovary to 5 mm long, pubescent. Pod 3–5 × 0.2 cm, cylindrical, pubescent, dehiscent. Seeds ellipsoid, 4 × 3 mm, edges blunted, hyline swelling, dark greenish brown.

Cultivated in dry fields. Fl. & Fr.: Jul. – Jan. Batrepalli (ATP), *KRKS & BSS* 39644; Nuthnakaluva (ATP), *KRKS & BSS* 41933.

Note: Used for fodder.

Vigna trilobata (L.) Verdc., Taxon 17: 172. 1968; FAP 1: 326.1997. *Dolichos trilobatus* L., Mant. Pl. 1: 101. 1767; FBI 2: 262. 1876. *Phaseolus trilobatus* (L.) Schreb., Nov. Actorum Acad. Caes. Leop Carol Nat. Cur. 4: 132. 1770; FPM 1: 362. 1918.

Local name: *Pillipesara.*

Trailing herb, branchlets villous, up to 60 cm long. Leaves 3-foliolate; stipules ovate, 0.5 × 0.3 cm, pubescent, acute at apex; petiole, up to 5 cm long, villous; petiolule 0.3 cm long, villous; stipel elliptic, acuminate; leaflets, 2–2.5 × 2–2.5 cm, pubescent on both surfaces, 3-lobed, each lobe obovate, base cuneate, margin entire, apex acute. Raceme axillary; peduncle 10–15 cm long, pubescent. Flowers clustered at apex of peduncle, 8–10, yellow, 0.8 cm long; bracts ovate, 0.3 cm long. Calyx cupular, 5-lobed, 4 mm across. Corolla 0.8 cm across; standard petal orbicular, 8 × 7 mm; wing petals spatulate, 6 mm long; keel petals 6 mm long, curved at apex with staminal sheath. Staminal sheath 5 mm long; stamens diadelphous (9+1). Ovary to 6 mm long, pubescent. Pod linear, subcylindrical, pubescent, up to 4 cm long, greenish brown. Seeds 6–8, ellipsoid, brown with black dots; hilum elliptic, slightly protruding; aril slightly developed.

Occasional, in agriculture lands and wastelands. Fl. & Fr.: Nov.–Jan. Nuthana kaluva (ATP), *KRKS & BSS* 41932.

Uses: Used as a fodder.

Vigna unguiculata (L.) Walp., Report 1: 779. 1842. subsp. **cylindrica** (L.) Estt. in Hendrick, Vegetables New York 1: 11. 1931; FAP 1: 326. 1997. *Phaseolus cylindricus* L., Herb. Amb. 23. 1754. *Dolichos unguiculatus* L., Sp. Pl. 725. 1753. *Vigna catjang* (Burm. f.) Walp., Linnaea 13: 533. 1839; FPM 1: 365. 1918. *Dolichos catjang* Burm. f., Fl. Ind. 161. 1768.

Local name: *Alasanda.*

Suberect herb, branchlets sometimes twining, glabrous. Leaves 3-foliolate, leaflets deltoid, obliquely ovate, 7–10 × 5–6 cm, base cuneate, margin entire, apex acute-apiculate; petiole 7 cm long; petiolule 0.4 cm long; stipule ovate, to 1.2 × 0.8 cm; stipel 0.2 × 0.1 cm. Raceme axillary, lax, 3–5-flowered. Flowers 1.5 cm long. Calyx cupular, 0.5 cm diam. Corolla whitish purple. Pod 10–15 cm long, linear, pubescent, subcylindrical, apex beaked. Seeds 15–20, ellipsoid, brown or whitish, 0.5 × 0.3 cm.

Cultivated in dry fields, sometimes as an escape. Fl. & Fr.: Nov.–Jan. Nuthnakaluva (ATP), *KRKS & BSS* 41931.

Note: All subspecies of *V. unguiculata* are cultivated.

ZORNIA J. F. Gmel.

Key to Species

1b. Branchlets pubescent; leaflets lanceolate, dotted below; bracts mucronate and gland dotted; calyx pubescent; pod 4–5 articulate; prickles pubescent along and hairy at apex .... **Z. gibbosa**

1a. Branchlets glabrous; leaflets ovate, not dotted below; bracts acuminate and not dotted; calyx glabrous; pod 3–4 articulate; prickles glabrous, apex only pubescent .... **Z. diphylla**

Zornia diphylla (L.) Pers., Syn 2: 318. 1807; FAP 1: 327. 1997; FBI 2: 147. 1876; FPM 1: 325. 1918. *Hedysarum diphyllum* L., Sp. Pl. 747. 1753. *Zornia diphylla* (L.) Pers. var. *zeylonensis* (Pers.) Baker in Hook. f., Fl. Brit. India 2: 148. 1876. *Z. zeylonensis* Pers., Syn. 2: 318. 1807; FPM 1: 325. 1918.

Decumbent herb; branchlets glabrous. Leaves 2-foliolate; petiole 1.8 cm; stipule lanceolate, perfoliate, subulate, glabrous, apex acute. Leaflets ovate, 1.2–1.5 × 0.8–1.5 cm, glabrous, base rounded-oblique, margin entire, apex apiculate. Raceme 7 - 10 cm long, flowers distinctly arranged; bracts ovate, 8 × 4 mm, pubescent. Calyx 5-lobed, pubescent; lobes 3 mm long. Corolla yellow, 0.8 cm long; standard petal cordate obovate, 6 × 6 mm; wing petals 6 × 3 mm; keel petals 5 × 3 mm. staminal sheath 4 mm; stamens monadelphous. Ovary densely pubescent, 3.5 mm long. Lomentum 1.5 cm long, linear, 3–4 articled, prickled, 5 × 3 mm, apically straight, mucronate, flattened, articles moniliform, glabrous, hairy at apical region. Seeds 1 per article, flattened.

Rare in dry deciduous forests at higher altitudes. Fl. & Fr.: Aug.–Dec. Idupulapya (KDP), *KRKS & BSS* 40312.

Zornia gibbosa Span., Linnaea 15: 192. 1841; FAP 1: 327. 1997. *Z. diphylla* auct. non (L.) Pers. 1807; FBI 2: 147. 1876. *p.p.*; FPM 1: 325. 1918.

Decumbent, much branched herb, branchlets pubescent, dotted. Leaves 2-foliolate; petiole 1.5 cm long, glabrous, dotted; stipule subulate, glabrous, dotted. Leaflets lanceolate, 2.5 - 3 × 0.5 - 0.6 cm, glabrescent, marginally pubescent, base obliquely rounded-truncate, margin entire, apex mucronate. Raceme axillary, 5 - 7 cm long; bracts foliaceous 6 × 3 mm, gland dotted, mucronate. Flowers distinctly arranged on raceme. Calyx 5-lobed, glabrous, tube 2 mm long; lobes 1.5 mm long. Corolla yellow, 0.5 cm long; standard petal broadly ovate, 5 mm long; wing petals 5 mm long, spatulate; keel petals oblong 4 × 4 mm. Staminal sheath 3 mm long; stamens monadelphous. Ovary 3 mm long, pubescent. Lomentum 1.5 cm long, articles 4 – 5, flattened, moniliform, prickled, 3.5 × 2 mm, apically curved, mucronate, prickles pubescent, apically scabrid hairy. Seeds 1 per article, flattened.

Occasional, in open dry deciduous forests and scrub forests along the open hilly slopes. Fl. & Fr.: Jun.–Nov. Batrepalli (KDP), *KRKS & BSS* 39577.

CAESALPINIACEAE

1a. Leaves of 2 leaflets .... (2)
1b. Leaves of sevaral leaflets .... (3)
2a. Petals present; leaflets adnate at their length up to middle or above the middle .... **Bauhinia**
2b. Petals absent; leaflets free .... **Hardwickia**
3a. Leaflets reduced to scales .... **Parkinsonia**
3b. Leaflets not reduced .... (4)
4a. Stamens 3. -fruit is pulpy, seeds hard .... **Tamarindus**
4b. Stamens 5 – 10 .... (5)
5a. Branches prickly .... **Pterolobium**
5b. Braches not prickly .... (6)
6a. Flowers to 7 cm across .... **Delonix**
6b. Flowers to 4 cm across .... (7)
7a. Pod one to two-seeded, prickly .... **Caesalpinia**
7b. Pods more then two-seeded, terete or flattened .... (8)
8a. Extra floral glands absent; pod elongate, cylindric .... **Cassia**
8b. Extra floral glands present; pods various, flattened or sub cylindrical .... (9)
9a. Bracteoles 2; androecium actinomorphic; pods hairy and flattened, dehiscent at upper suture .... **Chamaecrista**

9b. Bracteoles absent; androecium zygomorphic; pods glabrous (except *S. uniflora* and *S. holosericea*) indehiscent or inertly dehiscent through one or both sutures .... **Senna**

BAUHINIA L.

Bauhinia racemosa Lam., Encycl. 1: 390. 1785; FBI 3: 276. 1878; FPM 1: 406. 1919; FAP 1: 330. 1997.

Local name: *Are manu.*

Deciduous tree, up to 12 m high; bark dark grey or whitish, rough; branchlets tomentose. Leaves widely depressed ovate, ovate-orbicular, 1.5–3.5 × 3–3.5 cm, thin-coriaceous, leaflets connate to two-thirds of their length, 7–8-nerved, pubescent below, glabrous above, base cordate or truncate, margin entire, apex deeply emarginated; petiole to 1 cm. Racemes terminal or axillary to 12 cm long. Flowers to 1.3 cm across; bracts lanceolate, to 2 mm long; bracteoles subulate, to 1 mm; pedicel to 2 mm long. Calyx tube campanulate, 2 mm, puberulous without; lobes spathaceous, 5 × 2 mm. Petals 5, creamish-yellow, 5 × 1 mm, narrow at base. Stamens 10; filaments 4–7 mm long, pilose at base. Ovary compressed, to 8 mm long, pubescent. Pod flattened, oblong, 12–16 × 1.5–2 cm, woody, straight or curved to 220° angle, dark-brown or blackish, apex apiculate, indehiscent. Seeds *ca.* 8, flattened, ovoid, brown, smooth.

Common along the hilly slopes and forest paths of dry deciduous and scrub forests. Fl. & Fr.: Mar. – Nov. Thummalapalle (KDP), *KRKS & KRS* 36925; Rachakunta palle (KDP), *KRKS & BSS* 40167.

CAESALPINIA L.

Caesalpinia bonduc (L.) Roxb., Fl. Ind. 2: 362. 1832; FBI 2: 255. 1878; FAP 1: 332. 1997. *Guilandina bonduc* L., Sp. Pl. 381. 1753. *Caesalpinia bonducella* (L.) Fleming, Asiat. Res. 11: 159. 1810. *nom. illeg.*; FBI 2: 254. 1878. *C. crista* L., Sp. Pl. 381. 1753; FPM 1: 393. 1919.

Local name: *Gatchakai.*

Large, straggling, very thorny perennial shrub, 7–8 m high; branchlets downy pubescent, prickles recurved. Leaves bipinnate, 20–40 cm long; pinnae 10–15 cm long, 4–6 pairs, prickled; leaflets 6–9 pairs, ovate-elliptic 1.5–3.5 × 1–2.5 cm, glabrescent above, pubescent below, base truncate-rounded, margin entire, apex obtuse, apiculate; stipules foliaceous. Racemes terminal and supra axillary, to 5 cm long, prickled, bracts, linear-lanceolate, deciduous, 2 cm long; pedicel to 1.5 cm. Flowers pale yellow, 1.2 cm across. Calyx tube campanulate, to 5 mm across; lobes 3 mm long. Sepals 5, unequal, oblong-obovate, tomentose. Stamens 10, declinate, subequal; filaments attenuate, base glandular-villous, 4.5–7 mm. Ovary stipitate, globose, with ring of hairs around. Pods oblong-obovoid, somewhat flattened, 5–7 × 2.5–4.5 cm, brick red, densely clothed with sharp prickles, apex beaked to 1 cm long, dehiscent. Seeds 1 or 2, subglobose, hard, 0.8 × 1 cm, smooth, dark grey or lead coloured.

Common, in scrub forest and borders of agriculture lands. Fl. & Fr.: Dec.–Apr. Kalasamudram (ATP), *KRKS & BSS* 39983.

CASSIA L.

Cassia fistula L., Sp. Pl. 377. 1753; FBI 2: 261. 1878; FPM 1: 400. 1919; FAP 1: 336. 1997; V. Singh, Monogr. Cassiinae 25. 2001.

Local name: *Rela manu.*

Deciduous tree, about 10 m high; bark pale yellow, smooth; branchlets glabrous. Leaves 15–30 cm long; leaflets 4–6 pairs, sub-opposite, oblong-broadly ovate, 6–13 × 4–5 cm, glabrous, base cuneate, margin entire-slightly undulate, apex gradually tapering, emarginate; petiole 3–5 cm long; petiolule to 0.6 cm, dark brown. Racemes terminal, 15-40 cm long, pendulous; peduncle 2.5 cm; pedicel elongated, 3–7 cm. Flowers yellow, 2 cm across. Calyx lobes 5, ovate 1 × 0.5 cm. Corolla lobes 5, obovate, 6–9.5 × 3.5–5.5 mm. Stamens 10 [3+3+4]. Ovary appressed pubescent, 8 mm long. Pod shortly stipitate, oblong, terete, 40–60 cm long, cylindrical, dark brown, apex beaked, hard. Seeds *ca.* 50, circular, grey, flattened.

Common on hilly slopes of dry deciduous forests, also planted as ornamental as well as avenue tree. Fl. & Fr.: Mar. – Dec. Chinnapalle (ATP), *KRKS & KRS* 37808.

CHAMAECRISTA (L.) Moench.

1a. Leaflets 6, ovate or obovate; raceme 20-flowered; fruit sub-cylindric .... **C. absus**

1b. Leaflets more than 6, oblong; raceme 3-flowered; fruit flattened .... **C. pumila**

Chamaecrista absus (L.) Irwin & Barneby, Mem. New York Bot. Gard. 35 (2): 664. 1982; V.Singh, Monogr. Cassiinae 55. 2001. *Cassia absus* L., Sp. Pl. 376. 1853; FBI 2: 265. 1878; FPM 1: 403. 1919; FAP 1: 335. 1997.

Local name: *Chanupala vittulu.*

Erect annual herb, up to 50 cm high; branchlets viscid-pubescent. Leaves 5 cm long; leaflets broadly ovate or obovate, 2.5–3.5 × 1–2 cm, chartaceous, viscid-hairy, base cuneate, margin entire, apex acute, obtuse-rotund; petiole to 4 cm long, petiolule to 2 mm long, rachis with subsessile glands between leaflets. Stipules linear, subulate, to 6 mm. Raceme terminal or axillary, 2–10 cm long, 3–7-flowered; bracts *ca.* 2 × 1.5 mm, acuminate. Flowers 0.8 cm across. Calyx 5-lobed, viscid hairy, 3–4 mm long. Sepals 5-lobed, unequal, reddish yellow or flame-coloured. Anthers 5, unequal. Ovary sessile, densely bristly. Pod flattened, stellate hairy, oblong, 3–4.5 × 0.5–0.6 cm, oblique. Seeds 4–6, ovoid, black, shining.

Common, in dry deciduous forests and wastelands. Fl. & Fr.: Jul. – Nov. Rachakuntapalle (KDP), *KRKS & KRS* 37628.

Chamaecrista pumila (Lam.) V. Singh, J. Econ. Taxon. Bot. 16 (3): 600. 1992 & Monogr. Cassiinae 55. 2001. *Cassia pumila* Lam., Encycl. Meth. Bot. 1 (2): 651. 1785; FBI 2: 266. 1878; FPM 1: 403. 1919; FAP 1: 339. 1979.

Prostrate or decumbent herb, to 40 cm long, branchlets downy-villous. Leaves 1.5–5 cm long; leaflets 16–25 pairs, sessile, linear-oblong, 0.2–0.6 × 0.1 cm, chartaceous, glabrescent above, sparsely villous below, base obliquely cuneate, margin ciliate, apex acute; petiole to 3 mm long, with basal stipitate gland; stipules lanceolate, 3–4 mm long. Flowers 1 to 3 in supra axillary, pale to bright yellow, 0.5 cm across. Calyx lobes 5, lanceolate, pubescent, apex aristate to acuminate. Petals 5, ovate-elliptic, as long as calyx. Stamens 5, all equal and fertile, antheriferous, erect. Ovary villous. Pod flattened, compressed, 2–3.5 × 0.4 cm, strap-shaped, obtuse, pubescent, longitudinally dehiscent, valves coiling. Seeds 7–12, dark brown, rhomboid or obovate-oblong, beaked, flattened, 0.3 × 0.2 cm.

Common on moist hilly slopes of dry deciduous forest growing along with grasses. Fl. & Fr.: Sept. – Jan. Thummalapalle (KDP), *KRKS & KRS* 36988; Batrepalli (ATP), *KRKS & SSR* 37868; Rachakuntapalle (KDP), *KRKS & KRS* 37916; K. K. Kottala (KDP), *KRKS & KRS* 37662; Idupulapaya (KDP), *KRKS & BSS* 40311.

DELONIX Rafin.

Key to Species

1a. Bark grey; wood soft, white; pods 30–40 cm long, very strong .... **D. regia**

1b. Bark black; wood strong, yellow-white; pods 8–15 cm long, not much strong .... **D. elata**

Delonix elata (L.) Gamble, Fl. Madras 396. 1919; FAP 1: 342. 1997. *Poinciana elata* L., Cent. Pl. 2: 16. 1756; FBI 2: 260. 1878.

Local name: *Thella sunkesula.*

Deciduous tree, up to 12 m high; wood strong, yellow-white; branchlets warty puberulent. Leaves alternate, 20–25 cm long; pinnae 6–8 pairs, opposite, 5–7 cm long; leaflets opposite, 17–20 pairs, oblong, 0.3 × 0.9 mm, glabrous, base oblique, margin entire, apex obtuse; petiole to 4 cm long; stipules caducous. Raceme terminal, corymbose, to 10 cm long; peducle to 1.5 cm long; pedicel to 2.5 cm long. Flowers white-yellow, turning orange, 5 cm diam. Calyx ovate oblong, up to 3 cm long, outer green, inner creamish, pubescent. Corolla white-yellowish, obovate, suborbicular, shortly clawed, margins curled and fimbriate. Stamens 10, up to 8 cm long. Ovary villose, 2 cm long. Pod 8–15 cm long, pale to dark brown, linear, glabrous-reticulate. Seeds *ca.* 10, transverse, oblong.

Planted as avenue tree. Fl. & Fr.: Jan. – Mar. Rachakuntapalle (KDP), *KRKS & KRS* 37647.

Delonix regia (Boj. ex Hook.) Rafin., Fl. Tellur. 2: 1837; FPM 1: 396. 1919. *Poincina regia* Boj. ex Hook., Bot. Mag. t. 2884. 1829.

Local name : *Errasunkesula.*

Deciduous tree, up to 20 m high; wood soft, thick, white; branchlets glabrous. Leaves alternate, 40–50 cm long; pinnae opposite, 18–20 pairs; leaflets 14–20 pairs, glabrous, 0.2 cm long, base oblique-truncate, margin entire, apex obtuse; stipules foliaceous, to 2 cm long. Racemes terminal, corymbose, branched. Flowers 5 cm diam., red. Posterior one red, with white lines. Calyx obovate, green outer side, red lower side. Corolla red. Stamens 5, red. Ovary 2 cm long. Pod up to 40 cm long, 5 cm wide, very hard, dark brown. Seeds up to 30, oblong-subcylindric, 3 – 5 cm long, greyish or brownish.

Planted along road sides. Fl. & Fr.: Jun–Aug. K. K. Kottala (KDP), *KRKS* & *KRS* 36435.

HARDWICKIA Roxb.

Hardwickia binata Roxb., Pl. Coromandel t. 209. 1811; FAP 1: 342. 1997; FBI 2: 270. 1878; FPM 1: 412. 1919.

Local name: *Yepi, Narepi.*

Deciduous tree, up to 15 m high; wood hard and heavy; branchlets whitish brown. Leaves 2-foliolate; leaflets ovate-trapezoid, 3–4 × 1.6–3 cm, glabrous, coriaceous, base acute, margin entire, apex obtuse; rachis with apical deciduous bristle. Panicle axillary and terminal, lax, 8 cm long; peduncle to 3 cm; bracts minute; pedicel to 3 mm. Flowers pale yellow, 6 mm across. Calyx lobes 5, petaloid, greenish-cream, oblong-ovate. Petals 0. Stamens 10; filaments to 4 mm long. Ovary 2 mm long, sessile. Pods strap-shaped, 7.5 × 1.6 cm, flattened, winged, glabrous, brown, dehiscent at apex. Seed 1, greyish.

Common, in dry deciduous forests and scrub forests of some regions. Fl. & Fr.: Aug. – Jan. Batrepalli (ATP), *KRKS* & *BSS* 39612.

Note: Endemic to South India.

PARKINSONIA L.

Parkinsonia aculeata L., Sp. Pl. 375. 1753; FAP 1: 343. 1997; FBI 2: 260. 1878; FPM 1: 397. 1919.

Local name: *Simathumma.*

Armed shrub, up to 3.5 m high; branchlets tomentose; spines erect, hard, to 1 cm long. brown. Leaves 2-pinnate, up to 30 cm long; main rachis short, spiny; pinnae 2–3 pairs, to 25 cm long, flattened, green, bearing small leaflets; Leaflets sessile, elliptic, subopposite, distinctly arranged, 1–2 × 1 mm, glabrous, base acute, margin entire, apex obtuse. Racemes terminal, to 20 cm long. Flowers yellow, 1.5 cm across. Calyx cupular, green, lobes 5, boat-shaped. Corolla yellow, thin, with dark veins, obovate, 6 × 5.6 mm. Stamens 5. Ovary 3 mm long, glabrous. Pod glabrous, up to 20 cm long, torulose, turgid above, finely nerved, attenuate at apex, dehiscent, edible. Seeds light brown, glabrous, shiny.

Occasional, in outskirts of villages. Fl. & Fr.: Jul. – Sept. Kurumavili (ATP), *KRKS* & *KRS* 37942.

PTEROLOBIUM R. Br. ex Wight & Arn. *nom. cons.*

Pterolobium hexapetalum (Roth) Santapau & Wagh, Bull. Bot. Surv. India 5: 108. 1963; FAP 1: 344. 1997. *Reichardia hexapetala* Roth, Nov. Pl. Sp. 210. 1821.

Pterolobium indicum A. Rich., Tent. Fl. Abyss. 1: 247. 1847; FBI 2: 259. 1878; FPM 1: 395. 1919.

Local name: *Korintha, Pariki, Seekkampa.*

Straggling shrub, armed with recurved thorns; branchlets dark pink, pubescent. Leaves 2-pinnate, 12–18 cm long; pinnae prickled, 6–9 pairs, 4–5 cm long; leaflets 8–10 pairs, ovate-oblong, 1–1.5 × 0.5–1 cm, glabrous, base cuneate-truncate, margin entire, apex obtuse. Panicles axillary or terminal, to 15 cm long, small prickled. Flowers white, small, 0.5 cm across. Pedicel to 1 cm long, filiform; bracts subulate, caducous. Calyx cupular, to 3 mm across, pubescent. Corolla white, with red tinge, unequal, spatulate, clawed to 1 mm long. Stamens 10, free, filaments villose at base. Ovary sessile, pubescent, to 2.5 mm long. Pod indehiscent, samaroid, 1-seeded, apically winged. Seed solitary at base, flattened, 3 mm across.

Common along the stream sides on hilly slopes of dry deciduous forests. Fl. & Fr.: Jul. – Dec. Kalasamudram (ATP), *KRKS* & *KRS* 36498.

SENNA P. Mill.

Key to Species

1a. Foliar glands conspicuous, present on the petiole or leaf-rachis .... (2)

1b. Foliar glands absent on the petiole and or rachis .... (7)

2a. Foliar glands borne somewhere on the petiole .... (3)

2b. Foliar glands borne on the leaf-rachis between the pairs of leaflets .... (4)

3a. Leaflets 5–7-paired; pods subcylindric; seeds 30–40 .... **S. sophera**

3b. Leaflets 3–4 paired; pods flattened; seeds 15–20 .... **S. occidentalis**

4a. Small tree or large shrub; leaf rachis to 20 cm long; leaflets to 7 cm long .... **S. surattensis**

4b. Shrubs or herbs; leaf rachis 6–9 cm long; leaflets to 4 cm long .... (5)

5a. Stipules broadly reniform, persistent; pods flat, more than 1 cm wide .... **S. auriculata**

5b. Stipules linear, subpersistent or caducous; pods terete or subcompressed, striated or 4-gonous, sometimes torulose .... (6)

6a. All anthers rounded at the apex; seed-areole 1.3–2 mm wide, about as wide as seed, extending towards hilum .... **S. tora**

6b. Three large anthers narrowed at the apex like neck of the bottle, rest anthers rounded at the apex; seed-areole slit-like, up to 1 mm wide, not extending towards hilum .... **S. uniflora**

7a. Fertile stamens 10; staminodes absent .... **S. siamea**

7b. Fertile stamens 7; staminodes 3 .... (8)

8a. Pods slightly curved, to 5 cm long, pubescent or glabrous .... (9)

8b. Pods straight, to 14 cm long; glabrous, shiny .... **S. montana**

9a. Prostate herbs; ovary glabrous; pod reticulately veined........................ **S. italica**

9b. Erect herbs or shrubs; ovary pubescent or villous; pod not reticulately veined .... (10)

10a. Raceme not exceeding the leaf rachis; pod subreniform, not veined, villous .... **S. holosericea**

10b. Raceme exceeding the leaf rachis; pod straight or upwardly slightly falcate, shallowly elevated over the seeds, transversely septate .... **S. alexandrina**

Senna alexandrina Gars. ex Mill., Gard. Dict. ed. 8. no. 1. 1768; V. Singh, Monogr. Cassiinae 104. 2001. *Cassia senna* L., Sp. Pl. 377. 1753; FAP 1: 340. 29. 1790; FBI 2: 264. 1878; FPM 1: 404. 1919.

Local name: *Senna.*

Subshrubs, to 1 m high, branchlets yellowish, glabrescent. Leaves 5–7 cm long; rachis eglandular; leaflets 4–8 pairs, opposite, ovate-oblong, lanceolate, 1.1–2.2 × 0.4–0.7 cm, glabrous above, pubescent below, base cuneate, rounded, margin entire, apex mucronate; petiole 1–1.8 cm long, pubescent; petiolule 1–2 mm. Raceme terminal, to 30 cm long, pedicels to 1 cm long. Flowers yellow, 1 cm across. Calyx-lobes 5, ovate, 7–9 mm long, glabrous. Corolla lobes 5, ovate, oblong. Stamens 10; fertile 7, (2+1+4), of which two anthers *ca.* 1 cm long, one *ca.* 4.5 mm long and four 3–4 mm long; staminodes three. Pod short stipitate, oblong-obovate, compressed, flat, 5 × 2 cm, greenish black, shortly beaked at apex, with persistent style, indehiscent. Seeds *ca.* 8, ovoid, membranous, 7 × 4 mm.

Common, in foot hills of the forests and scrub forests. Fl. & Fr.: Sept. – Dec. Velpula (KDP), *KRKS & KRS* 37672.

Senna auriculata (L.) Roxb., Fl. Ind. 2: 349. 1832; V. Singh, Monogr. Cassiinae 119. 2001. *Cassia auriculata* L., Sp. Pl. 379 1753; FBI 2: 263. 1878; FPM 1: 402. 1919; FAP 1: 336. 1919.

Local name: *Thangedu.*

Shrub, 2–3 m high, bark dark brown, branchlets finely pubescent. Leaves pinnately compound, 5–9 cm long, paripinnate, leaflets 6–8 pairs, elliptic-oblong, oblong-ovate, 1–15 × 0.5–0.8 cm, sparsely pubescent below, glabrous above, base rounded, margin entire, apex obtuse, apiculate; petiole 1.5 cm, rachis with linear, erect, stipitate glands, between all pairs of leafletes, stipules reniform, foliaceous, 0.7–1 cm, persistent. Corymbs 3–6-flowered, 10–15 cm long, peduncle 4–5 cm long; pedicel 2–3 cm; bracts ovate-lanceolate, 0.5 × 0.3 cm. Flower bud obovate. Flowers yellow, to 7 mm across. Calyx lobes 5, outer two larger, to 8 mm long, inner three smaller, to 6 mm long, ovate-orbicular. Petals with orange veins. Stamens 10, fertile 7, staminodes 3. Pod flattened, 7–12 × 2 cm, linear, torulose, apex obtuse, beaked, puberulous. Seeds 6–12, ovoid, laterally compressed.

Common, in wastelands and scrub forests in much disturbed forest and in waste places. Fl. & Fr.: Almost through the year. Rachakuntapalle (ATP), *SSR* 33602.

Note: It is very adaptable and grows well as pioneer even on barren gravelly areas and eroded soils.

Senna holosericea (Fresen.) Greuter, Willdenowia 15: 429. 1986; V. Singh, Monogr. Cassiinae 142: 2001. *Cassia holosericea* Fresen., Flora 22 (1): 54. 1839.

Suffruticose shrubs or under shrubs, 70–90 cm high, branches subterete, velvety when young. Leaves 8–15 cm long, rachis densely pubescent, eglandular, single appendage at apex or rachis in between the terminal pair of leaflets on the lower side to 1 mm long; petioles 1.2–2 cm long, stipules 5–7 mm long, linear-lanceolate, very acute with a small auricle at the base on the side remote from the petiole, rigid, densely pubescent. Leaflets 5–8 pairs, 1.7–2.5 × 0.7–0.9 cm, ovate, oblong, obtuse or retuse, clothed on both surfaces with soft velvety pubescence, base obliquely rounded or subacute, margin entire, apex obtuse, mucronate; petiolule very short, to 1 mm long; thick brown, velvety. Flowers light yellow in narrow axillary racemes, shorter than the leaves, to 9 cm long, peduncle velvety; pedicels to 0.4 cm long, densely pubescent; calyx 0.4–0.5 × 0.2–0.25 cm, divided to the base, segments oblong, obtuse, membranous, veined, base cuneate, apex obtuse, only outer two sepals pubescent; petals 0.4–0.5 × 0.3–0.35 cm, obovate-oblong or ovate-oblong, base cuneate, apex obtuse, shortly clawed, yellow, conspicuously reticulate with darker veins; stamens 10, of which the 3 upper are reduced to small staminodes, to 1.5 mm long (including filament), the remaining in 7 perfect, of which the 2 lower are much larger, to 4 mm long, rest are small, to 3.5 mm long; ovary densely pubescent, to 0.65 cm (0.4 ± 0.25 cm). Pods 3.5–4 × 1.2–1.8 cm, flat, thin and papery, recurved, rounded at both the ends, velvety - pubescent, slightly turgid and with the persistent base of the style on the shorter side of the pod, some what remote from its apex. Seeds 4–8, obovoid, 5–7 × 2–3 mm, retuse at apex and cuneate at base, funicle 0.5 cm long, flattened, glabrous, yellowish brown.

Rare in dry scrub forests. Fl. & Fr.: Sept. – Jan. Rachakuntapalle (KDP), *KRK* 39043; Namalagundu (KDP), *KRK & BSS* 40183.

Note: Very closely allied to *Senna italica* Mill. which is common, in Southern Peninsular India particularly identical shape of the pod. But it can be distinguished from *Senna italica* by its erect habit, height (*ca.* 1 m high), lanceolate stipules, mucronate leaf tip and velvety pod without longitudinal crest on the valves above the seeds, absence of reticulate venation.

Senna italica Mill., Gard. Dict. ed. 8. No. 2. 1768. subsp. **micrantha** (Brenan) Lock, Kew Bull. 43 (2): 339. 1988; V. Singh, Monogr. Cassiinae 151. 2001. *Cassia italica* (Mill.) Lam. ex Andr., Fl. Pl. Anglo Egypt. Sudan 2: 117. 1952; FAP 1: 337. 1997. *Cassia obovata* Callad., Hist. Cass. 92. t. 15 A. 1816; FBI 2: 264. 1878. *C. obtusa* Roxb., Fl. Ind. 2: 344. 1824; FPM 1: 403. 1919.

Local name: *Nela thangedu, Nelaponna, Sunamukhi.*

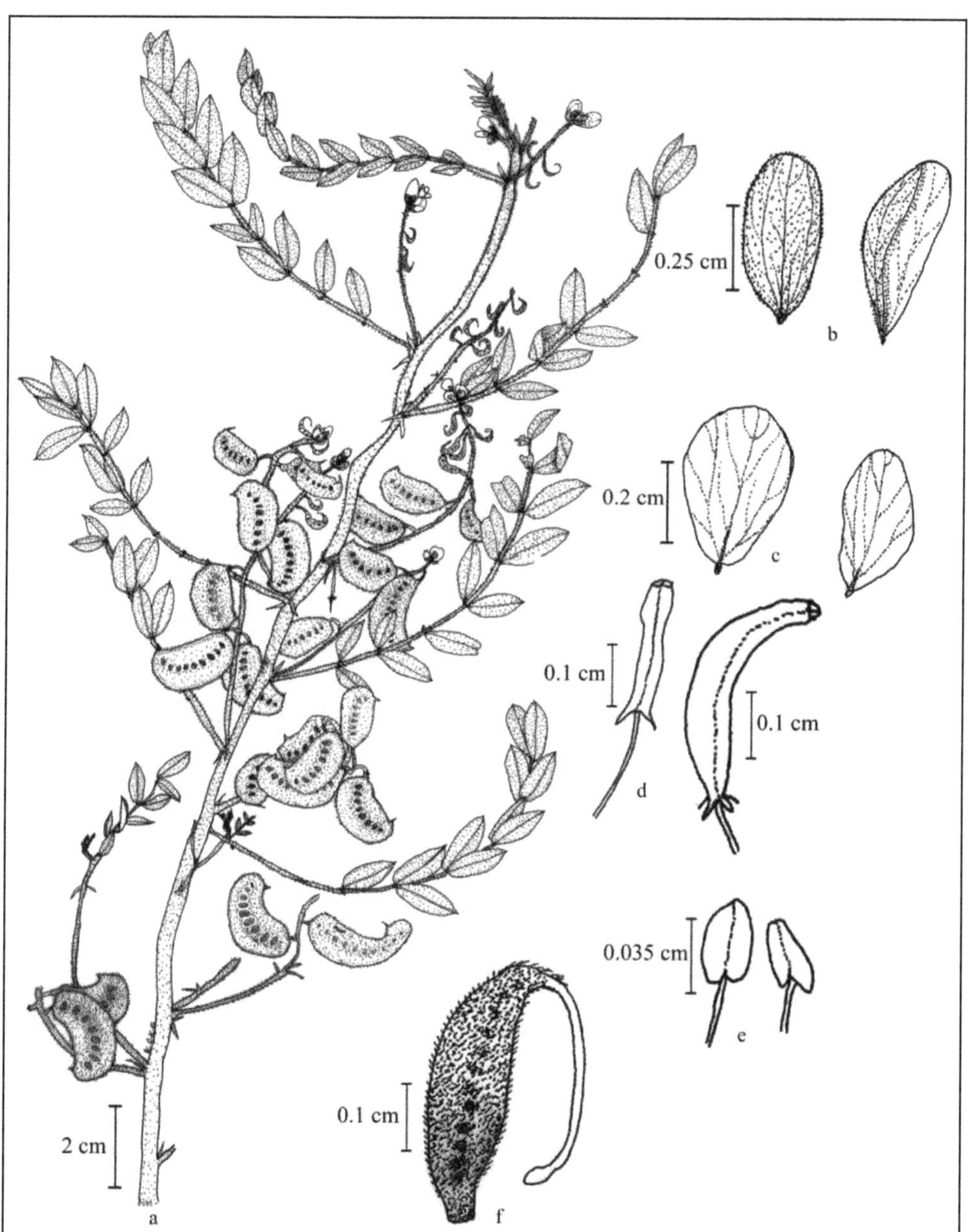

Figure 6: ***Senna holosericea*** **(Fresen.) Greuter.**

a. Flowering and fruiting twig; b. Calyx lobes; c. Corolla lobes; d. Fertile stamens; e. Staminodes; f. Pistil.

Prostrate, perennial herb, 50–60 cm long; branchlets spreading from root stock, glabrescent or clothed with appressed pubescence. Leaves 4–6 cm long, paripinnate; leaflets 4–6 pairs, obovate-oblong, 1.5–2 × 0.6–0.8 cm, glabrous above, pubescent below, base cuneate-rounded, margin entire, apex obtuse-apiculate; rachis eglandular; petiole 1.3–1.5 cm long; petiolule 0.2 cm; stipule ovate, persistent. Racemes axillary or terminal, 3–8 cm long, 5–8-flowered. Flowers light yellow, 0.8 cm across. Calyx lobes 5, oblong-obovate. Petals 5, ovate, subequal. Stamens 10, unequal, staminodes 2. Ovary to 1 cm long, glabrous. Pod short stipulate, glabrous, 2.5–4 × 1.5 cm, flat, valves thin, reticulately veined. Seeds 6–10, ovoid, flat, brown.

Occasional, in dry deciduous forests, scrub forests and wastelands. Fl. & Fr.: Jul. – Dec. Thummalapalle (KDP), *KRKS & KRS* 36922.

Senna montana (Heyne ex Roth) V. Singh, J. Econ. Taxon. Bot. 16 (3): 600. 1992 & Monogr. Cassiinae 154. 2001. *Cassia montana* Heyne ex Roth, Nov. Pl. Sp. 214. 1821; FBI 2: 264. 1878; FPM 1: 402. 1919; FAP 1: 338. 1997.

Local name: *Pagadi tangedu, Konda tangedu.*

Shrub, up to 3 m high; branchlets glabrous, somewhat lenticellate. Leaves 10–15 cm long, paripinnate; petiole to 2 cm long; petiolule to 3 mm long; leaflets 9–13 pairs, opposite, elliptic-oblong, obovate, 2–3 × 0.7–1 cm, glabrous, base rounded, margin entire, apex obtuse, mucro to 2 mm long; rachis eglandular. Corymbs terminal and axillary, to 20 cm long; pedicel to 2.5 cm long. Flower bud *ca.* 1 cm across. Flowers yellow, to 2 cm across. Calyx lobes 5, ovate, boat-shaped. Petals 5, ovate, 7–9 × 2.5–6 mm. Stamens 10; fertile 7; staminodes 3. Pod short stipitate, flat, straight, glabrous, 8–9 × 1.3–1.5 cm, brown, thin, shiny. Seeds 14–18, elliptic-oblong, brown, *ca.* 0.5 × 0.3 cm.

Common, in dry deciduous forests and scrub forests. Fl. & Fr.: May – Dec. Somulavaripalli (ATP), *KRKS & KRS* 37820; Kalasamudram (ATP), *KRKS & KRS* 36469.

Senna occidentalis (L.) Link, Handbuch 2 : 140. 1831; V. Singh, Monogr. Cassiinae 170. 2001. *Cassia occidentalis* L., Sp. Pl. 377. 1753; FBI 2: 262. 1878; FPM 1: 401. 1919; FAP 1: 339. 1997.

Local name: *Kashindha, Kasturi thulasi.*

Under shrub, to 1.5 m high; branchlets glabrous. Leaves 10–15 cm long, rachis ribbed, pubescent in the groove; gland solitary, on petioles, *ca.* 1.5 mm long, ovoid-conical, sessile; leaflets 4–5 pairs, ovate, oblong-lanceolate, 3.5–5 × 2–3 cm, base rounded or cuneate, margin entire, apex acuminate; petiolule 3 mm. Raceme axillary and terminal, to 15 cm long; peduncle to 2 cm; bracts linear-lanceolate. Flowers yellow, to 1 cm across. Calyx lobes 5, ovate. Petals 5, yellow-golden yellow, 1–1.5 cm long. Stamens 10; fertile 7 (3+4); staminodes 3. Ovary white pubescent, hooded. Pod short stipitate, flat, linear-oblong, acute at apex, brown with yellow margins, 10–12 × 0.8 cm. Seeds 15–20, ovoid, smooth, glabrous.

Common, in wastelands. Fl. & Fr.: Throughout the year. K. K. Kottala (KDP), *KRKS & KRS* 36433.

Senna siamea (Lam.) Irwin & Barneby, Mem. New York Bot. Gard. 35 (1): 98. 1982; V. Singh, Monogr. Cassiinae 193. 2001. *Cassia siamea* Lam., Encycl. Meth. Bot. 1 (2): 648. 1785; FBI 2: 264. 1878; FPM 1: 402. 1919; FAP 1: 340. 1997.

Local name: *Chettu tangedu, Seema tangedu, Kurumbi.*

Evergreen trees, to 20 m high; branchlets spreading, ribbed, pubescent. Leaves 10–20 cm long; rachis eglandular; leaflets 4–15 pairs, elliptic-oblong or obovate-oblong, 2–7 × 0.8–2.5 cm, glaucous above, pubescent below, subcoriaceous, base rounded, margin entire, apex emarginate; petiole 2–4 cm long. Corymbose panicles 20–50 cm long, terminal; pedicels 1–3 cm long. Bracts subtending the pedicels, spatulate. Flowers yellow, to 2 cm across. Sepals 5–7 mm long, unequal, ovate-orbicular, obtuse. Petals yellow, 1.2–2 cm long, orbicular to obovate, sessile. Stamens 10, all fertile. Ovary somewhat 4-gonous, pubescent. Pods 10–30 cm long, strap-shaped, flat, thickened on sutures, dehiscent, depressed between the seeds. Seeds 20–30, oblong, flat, light brown.

Common, road side and planted. Fl.: Dec. – Jan.; Fr.: Mar. – Jun. Batrepalli (ATP), *KRKS & SSR* 37897.

Senna sophera (L.) Roxb., Fl. Ind, 2: 347. 1832. var. **purpurea** (Roxb. ex Lindl.) V. Singh, J. Econ. Taxon. Bot. 16 (3): 600. 1992. *Cassia purpurea* Roxb. ex Lindl., Bot. Reg. t. 856. 1824. *C. sophera* L. var. *purpurea* (Roxb. ex Lindl.) Benth., Trans. Linn. Soc. London 27: 533. 1871; FBI 2: 263. 1878. *C. sophera* L., Sp. Pl. 379. 1753; FPM 1: 401. 1919; FAP 1: 340. 1997.

Local name: *Pyditangedu.*

Subshrubs, up to 2 m high; branchlets glabrous often tinged with purple. Leaves 7–15 cm long; gland on rachis, usually below the middle of the petiole, solitary, 0.5–1 mm in diam., globose to somewhat oblong or dome-shaped; leaflets 5–7 pairs, oblong, lanceolate, 2–5 × 0.8–2 cm, glaucous, base cuneate, margin entire, apex acute-mucronate; petiole 2-2.5 cm; petiolule 1-2.5 mm, glabrous. Racemes axillary and terminal, few-flowered; pedicel to 2 cm long; bracts to 1.5 mm long, green, ovate; bracts to 1 mm long, green, ovate. Flowers yellow, to 2 cm across. Calyx lobes 5, obtuse, green, 5 – 7 mm long. Petals 5, subequal, ovate, obtuse, 1–1.2 cm long. Stamens 10, fertile 7; staminodes 3. Ovary stipitate; 5 × 1 mm, glabrous, incurved. Pod curved, 7–12 × 0.6–0.8 cm, linear-oblong, sub-cylindrical, greenish brown. Seeds 30–40, broadly ovoid, flat, acute at apex, dark brown.

Occasional, in wastelands, outskirts of villages. Fl. & Fr.: Dec. – May. Kalasamudram (ATP), *KRKS & TP* 37688; Thimmamma marrimanu (ATP), *KRKS & KRS* 37934; Batrepalli (ATP), *KRKS* 278 (HJCB-E).

Note: Seed arrangement in this species is vertical to sutures, which is the key character for easy identification.

Senna surattensis (Burm. f.) Irwin & Barneby, Mem. New York Bot. Gard. 35 (1). 1982; V. Singh, Monogr. Cassiinae 215. 2001. *Cassia surattensis* Burm. f., Fl. Indica 97. 1768; FAP 1: 341. 1997. *C. glauca* Lam., Encyl. 1: 647. 1785; FBI 2: 265. 1878; FPM 1: 403. 1919.

Local name: *Metta tangedu, Konda tangedu.*

Shrub or small tree, to 5 m high, branchlets sparsely pubescent, ribbed. Leaves 10–20 cm long, rachis with clavate glands between leaflets; leaflets 3–5 pairs, opposite, elliptic-oblong, 3–8 × 1.5–2.8 cm, glabrescent above, sparsely pubescent below, base cuneate, margin entire, apex obtuse-acute; petiole to 6 cm long; petiolule to 4 mm long, pubescent; stipule to 2 cm long, curved. Raceme axillary, some times terminal, corymbose, 5–20-flowered; peduncle to 12 cm long; pedicel to 4 cm long; bract reflexed. Flowers yellow, 2 cm across. Sepals 5, ovate, unequal, yellow, glabrous; outer 2 half the length of inner ones, inner ones 5–8 mm long. Petals 5, yellow, ovate, clawed, broadly oblong, obtuse, 0.8–2.5 cm long. Stamens 10, subequal, all fertile, differs in length (2+1+4+3). Ovary stipitate, silky hairy. Pod flat, 12 × 1.6 cm, compressed, strongly nerved, yellowish brown. Seeds 20–30, oblong, ellipsoid, beaked, compressed, smooth, dark brown, shining, 5–7 × 2.5–4 mm, oblong, narrowed at hilum end, blackish, shiny.

Rare in forest, planted on field bunds. Fl. & Fr.: Aug. – Dec. Vempalli (KDP), *KRKS & SSR* 39045.

Senna tora (L.) Roxb., Fl. Ind. 2: 340. 1832; V. Singh, Monogr. Cassiinae 222. 2001. *Cassia tora* L., Sp. Pl. 376. 1753; FBI 2: 263. 1878; FPM 1: 401. 1919; FAP 1: 341. 1997.

Local name: *Tantipu, Tagirise, China kashinda, Tella kasinda.*

Annual herb, to 1 m high, branchlets sparsely pubescent. Leaves to 7 cm long, leaflets 3 pairs, opposite, obovate, oblong, 2–3.5 × 1.5–2 cm, glabrescent above, pubescent below, base cuneate-rounded, margin entire, apex obtuse-mucronate or cuspidate; petiole 3–3.5 cm long; petiolule 2 mm long, pubescent; rachis with 2 glands; glands linear-cylindric, between two lowest pairs of leaflets; stipules 5 – 10 mm long, linear. Raceme axillary; peduncle to 1 cm; pedicel to 1.5 cm. Flowers yellow, 1.5 cm across. Sepals 5, pubescent, ovate, 5–6 mm long, clawed, apex obtuse. Petals 5, obovate, subequal, clawed, to 1 cm long. Stamens 10; fertile 7; staminodes 3. Ovary pubescent, to 5 mm long. Pod long, to 20 cm long, beaked at apex, puberulous. Seeds 20–25, oblong, longitudinally tetrangular, brown.

Common weed of road sides and on foot hills and wastelands. Fl. & Fl.: Sept. – Nov. Batrepalli (ATP), *KRKS & KRS* 36963; K. K. Kottala (KDP), *KRKS & KRS* 37650.

Senna uniflora (Mill.) Irwin & Barneby, Mem. New York Bot. Gard. 35 (1): 258. 1982; V. Singh, Monogr. Cassiinae 228. 2001. *Cassia uniflora* Mill., Gard. Dict. ed. 8. n.2. 1768; C.S. Reddy *et al.*, Suppl. Fl. Andhra Pradesh 58. 2008.

Annual, erect herb, to 1 m high; branchlets villose. Leaves 4–6 cm long; leaflets 3 pairs, obovate-oblong, 2–3 × 1.2–2 cm, tomentose on both sides, base cuneate-rounded, margin entire, ciliate, apex obtuse, mucronate; glands between leaflets on rachis, slender, stalked. Petiole 2–2.5 cm long; petiolule 2 mm long, tomentose. Racemes axillary, 2 cm long, 5–8-flowered, compactly arranged; peduncle to 2 cm long, villous. Flowers 0.5 cm across; pedicel to 0.8 cm; bracts lanceolate. Calyx 5-lobed, claw 0.3–0.5 cm long, ovate, villous. Corolla 5-lobed, yellow, claw to 0.5–0.8

cm. Stamens 10; fertile 7; stminodes 3. Ovary 4 mm long, sessile. Pod linear, oblong, tomentose, 3–4 × 0.3 cm, transversely ribbed, apex beaked. Seeds 5–10, elliptic oblong, truncate at both ends, sub-cylindric, brown, shiny, 0.4 × 0.3 cm.

Very rare, in wastelands and along the roads. Fl. & Fr.: Aug. – Dec. Batrepalli (ATP), *KRKS & BSS* 40335.

Note: It is generally very common on road sides, but in the study area it has very restricted occurrence.

TAMARINDUS L.

Tamarindus indica L., Sp. Pl. 34. 1753; FAP 1: 344. 1997; FBI 2: 273. 1878; FPM 1: 409. 1919.

Local name: *Chinta.*

Large deciduous tree, 15–30 m high with wide trunk, branchlets glabrous; bark fissured, grey-brown. Leaves even pinnate, 6–8 cm long; leaflets 10–15 pairs, glabrous, oblong, 1.5 × 0.5 cm, base oblique (or) rounded, margin entire, apex obtuse. Raceme terminal or axillary, to 7 cm long. Flowers yellow, to 1 cm long. Calyx turbinate, lined by disk; lobes 4, lanceolate, membranous, imbricate; corolla yellow, with red nerves, 3 bigger to 7 mm long and 2 smaller, to 5 mm long, imbricate. Stamens 3, monadelphous, sheath to 6 mm long. Ovary stipitate, stalk adnate to the disk, many-ovuled. Pod hard, linear-oblong, incurved, yellowish, peeling off, somewhat compressed, mesocarp pulpy and leathery, septate endocarp. Seeds obovate-orbicular, compressed, testa hard, dark brown, shiny, annulated margin.

Common, in waste lands, road sides and in forests, also planted. Cultivated near villages. Fl.: Jun. – Aug. Fr.: Sept. – Feb. K. K. Kottala (KDP), *KRKS* & *KRS* 36432.

Note: Pulp of ripe fruit known as Tamarind is a good sources of income for local people. It is used in several dishes as condiment and the wood as fuel.

MIMOSACEAE

Key to Genera

1a. Herbs (prostrate), with angular stem. -sensitiv to touch .... **Neptunia**

1b. Trees, shrubs, stragglers (with hooks) ....(2)

2a. Plants unarmed ....(7)

2b. Plants armed (except *Acacia auriculiformis*) ....(3)

3a. Inflorescence bicoloured; branchlets spine-tipped; pods indehiscent .... **Dicrostachys**

3b. Inflorescence one coloured; brachlets with either internodal prickles or stipular spines at nodes; pods dehiscent ....(4)

4a. Flowers 4-merous; stamens 4 or 8; pods bristly or spinous on margins ....**Mimosa**

4b. Flowers 5-merous (exept *A. horrida*); stamens 10 or many; pods unarmed ...(5)

5a. Pinnae pairs 3 or more; pods compressed...**Acacia**

5b. Pinnae pairs 2; pods inflated ....(6)

6a. Stamens many, connate; pods spirally twisted, without pulp; leaflets 2 ....**Pithecellobium**

6b. Stamens 10, free; pods straight, pulpy; leaflets many ....**Prosopis**

7a. Flower heads less than 1.5 cm across .... **Leucaena**

7b. Flowers heads more than 3 cm across ....**Albizia**

ACACIA P. Mill.

Key to Species

1a. Unarmed trees, phyllodes present .... **A. auriculiformis**

1b. Armed trees, shrubs or stragglers, phyllodes absent ....(2)

2a. Thorns non-stipular, scattered on internodes; stragglers-flowers in heads ....(3)

2b. Thorns stipular, inserted at nodes; trees ....(4)

3a. Leaflets 18–35 pairs, overlapping, thorns <2 mm long; branches dense hairy .... **A. torta**

3b. Leaflets10–15 pairs, not overlapping, thorns >2 mm long, branches glabrous ....**A. caesia**

4a. Flowers in cylindrical spikes ....(5)

4b. Flowers in globose heads ....(7)

5a. Pinnae 10–25 pairs ....**A. catechu**

5b. Pinnae 2–8 pairs ....(6)

6a. Thorns hollow, white; leaf-rachis eglandular ....**A. horrida**

6b. Thorns not hollow; gland on the uppermost pinnae of leaf-rachis .... **A. ferruginea**

7a. Pinnae < 1 cm long; pod to 2 cm width; spines shorter to 0.5 cm .... **A. planifrons**

7b. Pinnae > 1 cm long; pod to 1 cm width; spines longer to 4 cm ....(8)

8a. Bark greenish when young, blackish when mature; flowers yellow in heads in axillary, 1 or 3-clustered; pod moniliform, grey pubescent, pulpy when young...**A. nilotica**

8b. Bark whitish green when young, light brown when mature; flowers white in a terminal paniculate racems; pod not moniliform, flat, brown pubescent, not pulpy.. **A. leucophloea**

Acacia auriculiformis A. Cunn. ex Benth. in Hook. J. Bot. Kew Gard. Misc. 1: 377. 1842.

Local Name: *Australian tumma.*

Unarmed, medium sized tree, up to 10 m high, branchlets glabrous, slightly angular. Phyllodes alternate, elliptic, narrowly oblong, falcate, 8–12 × 1.6–2.3 cm, coriaceous, with 3–4-nerved, base attenuate, margin entire, apex tapering, obtuse; petiole to 2 cm long; stipules to 1.5 mm. Leaves bipinnate deciduous. Spikes axillary, crowded, 4 – 9 cm long. Flowers yellow, small, 2.5 mm across. Calyx lobes 5, campanulate, 2 mm across. Corolla lobes 5, yellow, lanceolate, 2 mm long. Ovary subsessile, to 1 mm long, pubescent. Pod coiled, 3–6 cm long, outer margin wrinkled, inner margin entire. Seeds 5–7, brown with yellow aril.

Commonly planted along roads and in gardens, also forest areas. Fl.: Aug. – Dec. & Fr.: Jan. – Mar. Pulivendula ghat (KDP), *KRKS* & *BSS* 39628.

Acacia caesia (L.) Willd., Sp. Pl. 4: 1090. 1806; FPM 1: 428. 1919; FAP 1: 347. 1997. *Mimosa caesia* L. Sp. Pl. 522. 1753. *Acacia intsia* Willd. var. *caesia* (L.) Baker in Hook. f., Fl. Brit. India 2: 297. 1878.

Local name: *Korintha, Yerracheeki.*

Large straggler, to 6 m long; branchlets glabrous, internodal thorns hooked, to 3 mm long, yellowish. Leaves alternate, bipinnate, 10–15 cm long; pinnae *ca.* 6 pairs, 3–8 cm long; leaflets 10–15 pairs, oblong, 0.5–0.8 × 0.4 cm, puberulous, base truncate, margin entire, apex obtuse, apiculate; petiole to 6 cm long; prickled on the underside with a convex, elongate gland at base; rachis prickled, with glands opposite to lowermost and 2 uppermost pairs of pinnae. Panicles terminal, to 18 cm long, pubescent. Flower heads globose, 2 or 3 in a cluster, 1 cm across; peduncle to 4 cm long; rachis prickled; bracts lanceolate, apex cuspidate, 0.5–1 cm long; bracteoles to 1 mm long. Flowers 3 mm across. Calyx lobes 5, to 2 mm long, pubescent. Corolla creamish-white, lobes 5, ovate, 4 mm long. Stamens many, basally connate. Ovary stipitate, oblong, to 2 mm long, pubescent. Pod stipitate, 7–15 × 1–1.5 cm, thin, flat, glabrous, base and apex obtuse, horned. Seeds *ca.* 10, greyish-brown, flattened, 5 mm across.

Common, in edges of ponds, foot hills of dry deciduous forests. Fl.: Nov. – Dec. & Fr.: Jan. – Mar. Kalasamudram (ATP), *KRKS* 42007.

Acacia catechu (L. f.) Willd., Sp. Pl. 4: 1079. 1806; FBI 2: 295. 1878; FPM 1: 427. 1919; FAP 1: 348. 1997. *Mimosa catechu* L. f., Suppl. Pl. 439. 1781. *Acacia chundra* (Roxb. ex Rottl.) Willd., Sp. Pl. 4: 1078. 1806. *Mimosa chundra* Roxb. ex Rottl., Ges. Naturf. Freunde Berlin. Neue Schriften 4: 27. 1803.

Local name: *Sandra.*

Moderate-sized deciduous tree, to 5 m high, bushy; branches with stipular thorns, pinkish, glabrous; spines 2 at node, leaf-opposed, hooked; bark whitish grey, transversely peeled. Leaves bipinnate, to 9 cm long, minute spines distinctly present on lower side of the rachis and pinnae; pinnae 1–1.5 cm long; leaflets up to 11 pairs, 3 × 1 mm, glabrous, oblong-elliptic, base oblique-obtuse, margin entire, apex obtuse; petiolule minute. Cylindric spike axillary, 8–12 cm long. Flowers yellowish white, 4 mm across. Calyx lobes 5, minute, 1.5 mm across, campanulate, glabrous. Corolla lobes 5, glabrous, spatulate, 3 mm long. Stamens numerous, anther elliptic, yellow. Ovary 1 mm long, pinkish, glabrous. Pod reddish pink, flat, glabrous, subacute at base, beaked at apex, wavy at margin, constricted between seeds. Seeds 3–6, compressed, circular.

Common, in scrub forest and foot hills of dry deciduous forest. Fl.: May – Jun. & Fr.: Jul. – Oct. Kalasamudram (ATP), *KRKS* & *KRS* 36495; K.K. Kottala (KDP), *KRKS* 37661.

Acacia ferruginea DC., Prodr. 2: 458. 1876; FBI 2: 295. 1878; FAP 1: 349. 1997; Chakrabarthy & Gangopadhyay, J. Econ. Taxon. Bot. 20: 611. 1996.

Local name: *Ansandra.*

Deciduous thorny tree, to 10 m high; bark grey, cracking into pieces, branchlets glabrous, stipular spines hooked. Leaves 6–8 cm long, bipinnate; prickles present on main rachis; a gland in the middle of the petiole and another gland between the terminal pair of pinnae, glabrous; pinnae 3–4 pairs, to 5 cm long; leaflets opposite, 13–17 pairs, 0.6–0.8 × 0.2 cm, oblong, glaucous, glabrous, base obliquely-subcordate, margin entire, apex obtuse-emarginate. Spike axillary, 8 × 1 cm, solitary or pairs. Flower light yellow, to 3 mm across. Calyx 5-lobed, to 1 mm, campanulate, puberulous without. Corolla 5-lobed, yellow, to 1 mm. Stamens numerous, 4 mm long. Ovary 1 mm long, stipitate. Pod 7–16 × 2 cm, grey, lanceolate, glabrous, apex obtuse, horned, nerved. Seeds 4–6, grey.

Occasional, in dry deciduous forest. Fl.: Mar. – Apr. & Fr.: Aug. – Nov. Kalasamudram (ATP), *KRKS* & *BSS* 39618.

Acacia horrida (L. f.) Willd., Sp. Pl. 4: 1082. 1806; FAP 1: 349. 1997. *Mimosa horrida* L. f., Suppl. Pl. 438. 1781. *Acacia latronum* (L.f.) Willd., Sp. Pl. 4: 1077. 1806; FBI 2: 296. 1878; FPM 1: 427. 1919. *Mimosa latronum* L.f., Suppl. Pl. 438. 1781.

Local name: *Paki thumma.*

Umbrella-shaped, armed, small tree, to 5 m high; bark dark reddish-brown; branchlets dark brownish warty, striate, glabrous; spines straight, divergent, up to 7 cm long, white, hollow, inhabited by ants. Leaves bipinnate, 3–4 cm long; glands present on petiole, below the first pair of pinnae; pinnae 3–4 pairs, to 1.5 cm; leaflets 5–9 pairs, glabrous, 2–4 × 1 mm, oblong-elliptic, base oblique-truncate, margin entire, apex obtuse. Spike cylindric, 2–3, axillary. Flowers white, 6 mm across. Calyx 4-lobed, glabrous, campanulate, 1 mm long. Petals 4-lobed, glabrous, tube 2 mm long, lobes 1 mm long, triangular. Stamens numerous, filaments 4 mm, basally connate. Ovary sub-sessile, 0.75 mm. Pod flat, curved, 4 × 1.5 cm, glabrous, dark brown, mucronate at apex. Seed 4-5, grey, 0.5 × 0.4 cm, some what compressed, obovate, centre of the seed darker than periphery, horned.

Common, in scrub forests and foot hills of dry deciduous forest. Fl.: Mar. – May & Fr.: May – Aug. Thummalapalle (KDP), *KRKS, KRS, SSR* & *TP* 33645.

Acacia leucophloea (Roxb.) Willd., Sp. Pl. 4: 1083; FBI 2: 294. 1878; FPM 1: 427. 1919; FAP 1: 350. 1997. *Mimosa leucophloea* Roxb., Pl. Coromandel t. 150. 1800.

Local name: *Tella thumma.*

Moderate to large size tree, 12–15 m high; bark whitish-grey; branches puberulous when young; spines unequal and different sizes from 0.2 to 2 cm long, puberulous. Leaves 2–5 cm long; main rachis puberulous, disk-shaped glands 4, present on main rachis between first pair and apical 3 pairs of pinnae; pinnae 3–9 pairs, 2–2.5 cm long; leaflets up to 25, puberulent, 4 × 1 mm, oblong, base truncate, margin entire, apex acute. Panicles terminal, tomentose, about 20 cm long. Flower heads numerous, globose; peduncle of head 0.6–1 cm long; bracts ovate, minute. Flowers to 1 mm across, pale yellow. Calyx 5-lobed, pubescent, teeth triangular, apex ciliate. Sepals 5-lobed, 1.5 mm long, lobes triangular. Stamens numerous, to 2.6 mm long. Ovary stipitate, to 0.5 mm long. Pod sessile, flat, 10–15 × 0.5–1 cm long, thin, straight or slightly curved, turgid; rusty tomentose. Seeds 10–15, angular, ovoid, to 6 × 4 mm, pale brown.

Occasional, in edges of dry deciduous forests and scrub forests. Fl.: Jun. – Aug. & Fr.: Aug. – Nov. Rachakuntapalle (KDP), *KRKS* 36927.

Acacia nilotica (L.) Willd. ex Del. subsp. **indica** (Benth.) Brenan, Kew Bull. 12: 84. 1957; FAP 1: 350. 1997. *Mimosa nilotica* L., Sp. Pl. 521. 1753. *Acacia arabica* (Lam.) Willd. var. *indica* Benth., London J. Bot. 1: 500. 1842. *A. arabica* auct. non (Lam.) Willd. 1806; FBI 2: 293.1878; FPM 1: 425. 1919. *Mimosa arabica* Lam., Encycl. 1: 19. 1783.

Local name: *Thumma.*

Tree, up to 15 m high, crown spreading; bark blackish-brown, longitudinally cracked; branchlets greyish, pubescent. Stipular spines paired, white, up to 4.5 cm long, slender, unequal. Leaves bipinnate, main rachis 5.5–9 cm long, pubescent; glands between all pinnae, disk-shaped; pinnae 3–4 pairs, pubescent 2–3 cm long; leaflets 14–16, glabrous or sparsely hairy when young, 0.3–0.5 × 0.1 cm, oblong, apical pair obovate-oblong, base obtuse, margin entire, apex obtuse. Flower heads yellow, globose, axillary, 2–6 in a cluster; peduncle to 2 cm long, pubescent; pedicel to 1 cm long; bracts 2, ovate, to 3 mm long; bracteoles, ovate, 0.1 cm. Flowers 20–30 in a head. Calyx 5-toothed, campanulate, 1 × 2 mm, glabrous without. Corolla 5-toothed, bright yellow, 2 × 3 mm, glabrous. Stamens numerous, to 4 mm long. Ovary stipitate, 1 mm long. Pod stipitate, 10–18 × 1–1.5 cm, glaucous, moniliform, greyish-white, pubescent, base acute, apex acuminate, beaked. Seeds 8–12, ovoid, dark brown, compressed, shining, blunt edges, light brown in periphery, dark in center, horned, 5 mm across.

Common, in scrub forests and wastelands. Fl.: Jun. – Aug. & Fr.: Aug. – Oct. K. K. Kottla (KDP), *KRKS* & *KRS* 36437; Kalasamudram (ATP), *KRKS* & *KRS* 36454.

Note: The timber is used for agricultural implements and the bark for tannin. The gum is used for various purposes and also in medicines.

Acacia planifrons Koening ex Wight & Arn., Prodr. Fl. Ind. Orient. 276. 1836; FBI 2: 293. 1873; FPM 1: 426. 1919; FAP 1: 351. 1997.

Local name: *Godugu thumma.*

Deciduous armed tree, up to 10 m high; crown spreading; bark dark brown; branchlets warty tomentose. Leaves bipinnate, 3–5 cm long, pubescent; petiole 1 cm long; one gland near middle of petiole and another one between upper most pinnae; pinnae 3–6 pairs, 0.5–1.5 mm long, closely placed near end of the rachis. Stipular spines two types, curved one 4 mm long; straight one divergent, 2–3 cm long, white with a brown apex. Leaflets 6–13 pairs, linear-oblong, 0.2–0.4 cm long, glabrous, base oblique, margin entire, apex rounded. Flower heads yellow, to 1 cm across, in axillary fascicles; peduncle to 1 cm; bracteoles erect. Flowers to 2 mm across. Calyx 5-lobed, 1 mm, campanulate, teeth very short, acute. Corolla 5-toothed, 2 mm long; lobes short, acute, villous. Stamens numerous; filaments 4 mm long, exserted, free; anthers small. Ovary sessile, 1 mm; style filiform. Pod to 10 cm long, acute, turgid, slightly moniliform, pubescent and pulpy when young, blackish when dry. Seeds 3-8, elliptic, compressed, dark blackish-brown.

Very rare, near dry deciduous forests. Fl.: Nov. – Jan. & Fr.: Jan. – Mar. Thimmamma Marimanu (ATP), *KRKS* & *KRS* 37931.

Acacia torta (Roxb.) Craib, Bull. Misc. Inform. 1915. 410. 1915; FPM 1: 428. 1919; FAP 1: 353. 1997. *Mimosa torta* Roxb., Fl. Ind. 2: 566. 1832. *Acacia pennata* sensu Baker in Fl. Brit. India 2: 297. 1878.

Large prickly shrub or straggler, to 6 m long; branchlets pubescent, dark brown; internodal thorns 1 – 2 mm, distinctly arranged, curved. Leaves alternate, main rachis 15–20 cm long, densely pubescent, with hooked spines on lower side; disk-shaped gland present at the middle of the petiole and 4 glands between pinnae (one in each) towards apex; pinnae 7–13 pairs, 4–7 cm long; leaflets 18–35 pairs, closely arranged or ovarlapping, linear-oblong, 4–7 × 2 mm, glabrous on both sides, but hairy on margins, base truncate, margin entire, apex acute, mid nerve towards distal margin. Panicles terminal and axillary, to 15 cm long. Flower heads creamy-white, 1 cm across, 1–3 in a cluster. Flowers 2 mm across. Calyx tube 5-lobed, 1 mm long, oblong-obovate, pubescent. Corolla 5-lobed, glabrous, oblong, to 2 mm long. Stamens numerous (44), basally connate. Ovary pubescent, to 3 mm long, stalk 1 mm. Pod *ca.* 14 × 2 cm, pale to dark – brown, flat, linear, apiculate. Seeds 5-7, compressed.

Rare in dry deciduous forests. Fl.: Sept. – Nov. & Fr.: Dec. – Feb. Batrepalli (ATP), *KRKS* 40188.

ALBIZIA Duraz.

Key to Species

1a. Pods with thick mesocarp, indehiscent; areoles absent in pods .... **A. saman**

1b. Pods without mesocarp, dehiscent; areoles present in pods .... (2)

2a. Pinnae 6–13 pairs; leaflets 1–1.5 mm wide .... (3)

2b. Pinnae 3–6 pairs; leaflets 1–2 cm wide (except *A. thompsonii*) .... (4)

3a. Stipules foliaceous, leaflets acute at apex .... **A. chinensis**

3b. Stipules minute; leaflets rounded or obtuse at apex .... **A. amara**

4a. Leaf rachis glabrous .... **A. lebbeck**
4b. Leaf rachis pubescent .... (5)
5a. Diskoid gland below middle of petiole; leaflets to 1 cm long .... **A. odoratissima**
5b. Diskoid gland above middle of petiole; leaflets to 0.8 cm long .... **A. thompsonii**

Albizia amara (Roxb.) B. Boivin, Encycl. 19: 34. 1838; FBI 2: 301. 1878; FPM 1: 432. 1919; FAP 1: 354. 1997. *Mimosa amara* Roxb., Pl. Coromondel t. 122. 1799.

Local name: *Nallrange.*

Unarmed deciduous tree, up to 10 cm high; bark dark brown, rugose, younger branches yellowish tomentose. Leaves bipinnate, 5–12 cm long; rachis pubescent with disk-shaped gland at middle of the petiole and between apical pinnae; pinnae 6–10 pairs, 1.5–5 cm long, rachis pubescent; leaflets glabrescent above, puberulous below, oblong, 3–4 × 1 mm, base obtuse, margin ciliate, apex obtuse. Flower heads solitary or 2–3 fascicled in upper axils, creamy-white to pale yellow, 2 cm across; bracts to 5 mm long, caducous; pedicel to 1 mm. Flowers yellow, to 2 mm across. Calyx 5-lobed, to 2 mm, acute, ciliate at apex. Petals 5-lobed, creamish, to 3 mm long, pubescent without. Stamens many; filaments to 9 mm long. Ovary stipitate, to 2 mm. Pods flat, compressed, greyish-brown, 8–20 × 2–3 cm, strongly nerved, straight, base and apex rotund, horned, indehiscent. Seeds 6–8, ovoid, 1 × 0.5 cm, brick red.

Common along forest out skirts and path of dry deciduous forests. Fl.: May – Jul. & Fr.: Jul. – Oct. Thummalapalle (KDP), *KRKS, KRS, TP* & *SSR* 36401.

Albizia chinensis (Osbeck) Merr., Amer. J. Bot. 3: 575. 1916; FAP 1: 354. 1997; Chakrabarthy & Gangopadhyay, J. Econ. Taxon. Bot. 20: 584. 1996. *Mimosa chinensis* Osbeck, Dagbok. Ostind. Resa 233. 1757. *Albizia stipulata* (Roxb.) Boivin, Encycl. 19: 33. 1838; FBI2: 300. 1878. *Mimosa stipulata* Roxb., Fl. Ind. 2 : 549. 1832; *"stipulacea". Albizia marginata* (Lam.) Merr., Philipp. J. Sci 5: 23. 1910; FPM 1: 433. 1919.

Lacal name: *Bandichinduruge, Bandichoragu.*

Large deciduous trees, up to 20 m high, bark grey, horizontally furrowed; branchlets densely yellowish pubescent. Leaves 18–22 cm long, petiole gland below the middle of petiole; rachis with gland between base of lower pinna; pinnae 10–13 pairs, 4.5–10 cm, stipules very large, 2 × 2 cm, ovate, deciduous, leaflets 15–25 pairs, sessile, narrow-elliptic, 5–7 × 1.5–2 mm, pubescent, base very oblique, truncate, margin ciliate, apex acute, midnerve very close to the upper margin. Flower heads 1–few, 2 cm across, in axillary or terminal racemes; peduncle to 1 cm; bracts stipular, persistent, pedicel to 1 mm. Calyx 5-lobed, to 3 mm, campanulate, pubescent half way down, lanceolate. Stamens many, filaments to 8 mm, basally connate. Ovary stipitate, to 4 mm, style 1 cm. Pods flat, compressed, 15–25 × 3–4 cm, thin, brown, straight or wavy margin, often wrinkled over seeds. Seeds 8-12, ovate, much compressed, dark brown, smooth.

Occasional, in hill regions of the deciduous forests. Fl. & Fr.: April-December. Kalasamudram RF (ATP), *NY* 399.

Note: Wood is used for building purposes.

Albizia lebbeck (L.) Benth., Hook. Lond. J. Bot. 3: 87. 1844; FBI 2: 298. 1878; FPM 1: 432. 1919; FAP 1: 355. 1997. *Mimosa lebbeck* L., Sp. Pl. 516. 1753.

Local name: *Dirisana.*

Deciduous tree, to 20 m high, bark brown-grey, streaked; branchlets yellowish, glabrous. Leaves bipinnate, rachis to 18 cm long; petiole 5–8 cm long, glabrescent; glands near middle of the petiole, between pinnae and between leaflets; pinnae 2–4 pairs, 5–12 cm long, leaflets 4–8 pairs, oblong, terminal pair of leaflets obovate, 2–4.8 × 0.8–2.2 cm, minute, base obliquely obtuse, margin entire, apex rounded; midvein towards distal margin; secondary veins prominent. Flowers greenish-yellow, fragrant, heads in axillary umbellate clusters. Calyx funnel-shaped, 3–4.5 × 1.5–3 mm, puberulous. Corolla funnel-shaped, 7–10 mm long, lobes ovate (or) oblong to lanceolate 2.5–5 × 1–2.5 mm long, puberulous towards apex. Stamens 2.5–4 cm long. Ovary 2–2.5 mm long, glabrous. Pods strap-shaped, 11–35 × 2.5–5 cm, flat, narrowed and rounded or rostrate at apex, sometimes shortly sinuate margin, glabrous, yellow with obscure to faint venation, attenuate to acute at base. Seeds 4–15, elliptic to suborbicular, flattened, 9–10 × 6–7 mm; areole 5–8 mm across.

Common, along road sides and rarely outskirts of forests. Fl.: Jan. – Apr. & Fr.: May – Aug. Mabbuchintalapalle (DKP), *KRKS* & *KRS* 36943.

Albizia odoratissima (L. f.) Benth., London. J. Bot. 3: 88. 1844; FBI 2: 299. 1878; FPM 1: 431. 1919; FAP 1: 355. 1997. *Mimosa odoratissima* L. f., Suppl. Pl. 437. 1781.

Large, unarmed, evergreen trees, to 10 m high; bark grey with irregular cracks; wood dark brown; branchlets appressed pubescent. Leaves bipinnate, 15–20 cm long, main rachis downy pubescent; petiole to 4.5 cm long, glabrous, with a diskoid gland below middle; rachis slightly pubescent, with a gland between uppermost pair of pinnae; pinnae 3–5 pairs, 6–13 cm long; leaflets 6–15 pairs, oblong-elliptic, terminal one obovate, 1–2 × 0.5–1 cm, glabrescent, base oblique, truncate, margin entire, apex obtuse. Panicles to 12 cm, umbellate-corymbs. Flower heads white, globose, 1–1.5 cm across, sessile; bracts subulate, to 4 mm, sessile. Flowers 2–3 mm across. Calyx 5-lobed, to 1 mm long, pubescent without, margin ciliate. Corolla toothed, to 4.5 mm long, pubescent without, acute, ciliate. Stamens many, filaments to 9 mm. Ovary stipitate, 1.5 mm long. Pods subsessile, compressed, flat, 13–15 × 2–2.5 cm, straight, smooth, reticulate above seeds, indehiscent, brown. Seeds 8–12, ovoid, 1.5 × 0.7 cm, yellow or brown.

Very rare, in dry deciduous forests. Fl.: May – Jun. & Fr.: Nov. – Feb. Kalasamudram (ATP), *KRKS & BSS* 39197.

Albizia saman (Jacq.) F.Muell., Select. Pl. ed. 2. 12. 1876. *Samanea saman* (Jacq.) Merr., J. Wash. Acad. Sci. 6: 47: 1916; FAP 1: 365. 1997. *Mimosa saman* Jacq., Fragm. Bot., 15. t. 19. 1800. *Enterlobium saman* (Jacq.) Prain ex King, J. Asiat. Soc. Bengal 66: 252. 1898; FPM 1: 435. 1919.

Large tree, 15–20 m high; bark whitish, rough, branchlets yellowish-tomentose, spreading. Leaves alternate, bipinnate, 10–16 cm long. Pinnae 2–4 pairs, 6–8 cm long; leaflets 6–7 pairs, obovate, inequilateral, 1–3.5 × 0.7–2.5 cm, puberulous or glabrous above, pubescent below, base truncate, cuneate, margin entire, apex obtuse, apiculate; petiole 3.5–4.5 cm; rachis tomentose, with glands opposite to almost all pinnae. Petiole reduced. Corymbose panicles terminal or axillary. Flower heads pinkish-brown, 2 cm across; peduncle 4–6 cm; bracts lanceolate, 0.6–1 cm; bracteoles linear, to 5 mm; pedicel to 1.5 mm. Flowers 20–25 per head, each 2.5–3.5 × 0.5 cm. Calyx 5-lobed, to 6 mm, pubescent, ovate, acute, triangular, connate from above the middle. Corolla 5-toothed, to 6 mm, connate from above the middle, densely pubescent, stamens many, monadelphous, very longer than corolla, to 2.6 cm long, apex pinkish, base white. Ovary stipitate, to 6 mm, glabrous. Pods linear-oblong, 10–13 × 1–1.5 cm, slightly curved, woody, septate, indehiscent, brownish-black, beaked at apex. Seeds 8–10, ovoid.

Common, on road sides. Fl.: Mar. – Apr. & Fr.: Jul. – Dec. Rekkamanu (ATP), *KRKS* & *KRS* 37926.

Albizia thompsonii Brandis, Indian. Forest. 25: 284. 1899; FPM 1: 432. 1919; FAP 1: 356. 1997.

Local name: *Kondala*

Large deciduous tree, to 15 m high; bark greyish, longitudinally streaked, internally dark red when cut; young shoots ochraceous to entellous; branchlets brownish, pubescent, terete, initially sparsely puberulous, finally glabrous. Leaves bipinnate, rachis 10–17 cm long, densely ochraceous pilose; petiole 3–8 cm long; gland near base of petioles, 2–3 mm diameter; pinnae 3–6 pairs, 5–10 cm long, glands at the junction of distal pair of pinnae; leaflets sessile, 10–25 pairs, asymmetric, oblong, 10 × 5–7 mm, unequal at base, margin entire, acute to obtuse and mucronate, tomentellous when young, glabrous above with age, black when dry, main vein slightly towards distal margin; lateral nerves obscure to faint. Flower heads pale yellow in axillary fascicles, peduncle 2–5 cm; pedicel 2–3 mm. Calyx narrowly campanulate, 2–2.5 × 1.2–1.5 mm, puberulous, acute. Corolla 4–4.5 × 2.5–3 mm; lobes oblong-lanceolate, 1.5–2 × 0.7–1 mm, puberulous towards apex. Stamens 8–12 mm long. Ovary 1.5 mm long, hirsute, short stipitate. Pod strap-shaped, 12–18 × 2.5-3 cm, flat, thin, rounded to apiculate at apex, brown, stipe 1 cm long. Seeds 6–10, elliptic, flat, 5–6 × 3.5–4.5 mm.

Occasional, in dry deciduous forests. Fl.: Feb. – Apr. & Fr.: Jun. – Nov. Kalasamudram (ATP), *KRKS* 41887.

DICHROSTACHYS (DC.) Wight & Arn. *nom. cons.*

Dichrostachys cinerea (L.) Wight & Arn., Prodr. Fl. Ind. Orient. 271. 1834; FBI 2: 288. 1878; FPM 1: 419. 1919. subsp. **cinerea** var.**cinerea**; Brenan & Brummit,

Bot. Soc. Brot. (Ser. 2) 39. 110. 1965; FAP 1: 357. 1997. *Mimosa cinerea* L., Sp. Pl. 520. 1853, non L. 517. 1753.

Local name: *Velthuru kampa*.

Armed small tree, up to 5 m high; bark grey, longitudinally cracked; branchlets spiny tipped, pubescent. Leaves alternate or 2–3 in a cluster, bipinnate, to 6 cm long; petiole to 1 cm, rachis softly pubescent, with small tipped glands between each pair of pinnae, up to 1 mm long; pinnae 4–10 pairs; leaflets 10–25 pairs, linear-oblong, to 3 × 1 mm, closely arranged, glabrous above, pubescent beneath, apex subacute, margin entire, base oblique, sessile. Spikes axillary, cylindrical, to 6 cm long. Flowers numerous, yellow, dense at upper half, bisexual, the lower pink, sterile; peduncle to 1.5 cm long; bracts oblong, 1 mm long. Pod linear, flat, 5–7 cm long, irregularly twisted, dark brown, glabrous. Seeds 6–10, ovoid, compressed.

Planted on field borders and wastelands frequent in open hilly slopes and forest paths. Fl.: May – Aug. & Fr.: Oct. – Dec. Kalasamudram (ATP), *KRKS* & *KRS* 36465.

LEUCAENA Benth.

Leucaena latisiliqua (L.) Gillis, Taxon 23: 190. 1974; FAP 1: 358. 1997. *Mimosa latisiliqua* L., Sp. Pl. 519. 1753. *Leucaena glauca* auct. non Benth.; FBI 2: 290. 1878; FPM 1: 419. 1919.

Local name: *Banidichiguru*.

Large unarmed deciduous trees, 6 – 20 m high; bark rough; wood delicate, branchlets glabrescent. Leaves bipinnate, to 15 cm long, main rachis pubescent, ending with appendage. Flowers white in dense globose heads in axils of leaves. Pods 5 – 12 × 1 – 1.5 cm, straight, flat, dehiscent; valves twisted. Seeds 20 – 30, dark brown, elliptic, compressed.

Planted on field bunds and wastelands. Fl.: Jun. – Aug. & Fr.: Oct. – Nov. Ambakampalli (ATP), *KRKS & BSS* 40490.

MIMOSA L.

Key to Species

1a. Leaf main rachis to 15 cm long; leaflets distinctly arranged on pinnae .... **M. polyancistra**

1b. Leaf main rachis below 10 cm long; leaflets compactly arranged on pinnae .... (2)

2a. Internodal spine 5 mm long and 5 mm wide at base; leaf rachis to 8 cm long .... **M. prainiana**

2b. Internodal spine 3 mm long and 1 mm wide at base; leaf rachis to 4 mm long .... **M. hamata**

Mimosa hamata Willd., Sp. Pl. 4: 1033. 1806; FBI 2: 291. 1878; FPM 1: 421. 1919; FAP 1: 358. 1997.

Much branched, prickly shrub, up to 2 m high; branchlets pubescent, greyish, with numerous straight prickles. Leaves bipinnate, 3–6 cm long, main rachis pubescent, some times prickly; stipules minute, subulate, pinnae 4–6 pairs, 1.5–2 cm long; shortly stalked; leaflets 6–10 pairs, oblong, apical one ovate, 3–4 × 2–2.5 mm, hairy beneath, nearly glabrous above, base oblique, petiolules very short, margin entire, hairy, apex acute, mucronate. Flower heads pink, globose, axillary, solitary, 1 cm across; peduncle to 3 cm long. Flowers many in a head, 2 mm across. calyx 5-toothed, to 0.5 mm long, short, campanulate. Corolla 4-toothed, pink, to 3 mm long, divided nearly half way down, valvate, connate at base, glabrous without. Stamens 8, much exserted, to 4 mm long, connate at base; anther minute, 0.5 mm. Ovary subsessile, globose, pubescent. Pod flat, strongly falcate, 3–6 × 1–1.5 cm, reddish brown, pubescent, prickly along both sutures, somewhat curved, slightly moniliform. Seeds 4–6, light red, ellipsoidal.

Common, in wastelands and road sides of village outskirts. Fl. & Fr.: Jun. – Nov. Bhumayyagaripalle (KDP), *KRKS* & *KRS* 37614.

Mimosa polyancistra Benth., Trans. Linn. Soc. London 30: 422. 1875; FPM 1: 421. 1919; FAP 1: 359. 1997.

Local name: *Errasandra, Pariki.*

Straggling shrub, up to 3 m high, branchlets prickly, terete, glabrous; prickles recurved, strong. Leaves up to 15 cm long; rachis glabrescent, prickly, stipule minute, linear, pinnae 5–6 pairs, to 4 cm long, gland stipitate, minute at base of pinnae; leaflets 5–7 pairs, distantly arranged, elliptic-oblong, ovate, 4–6 × 3–4 mm, glabrous above, faintly puberulous below, base cordate, margin entire, apex obtuse. Flower heads globose, pink, 2–3-clusters, axillary, 1 cm across; bracts linear, to 2 mm glabrous; bracteoles minute; peduncle to 3 cm long. Flowers pink 1 mm across. Calyx 4-toothed, to 0.7 mm, connate, glabrous. Corolla 4-lobed, to 2 mm long, sometimes two petals connate, glabrous, apically pubescent. Stamens 8, to 0.5 cm long. Ovary stipitate, 0.5 cm long, pubescent, stigma 0.3 cm. Pod flattened, *ca.* 8 × 1 cm, curved, dark brown-pinkish, prickly on periphery, apiculate at apex with 5 mm long filiform spines, glabrous. Seeds 5–7, flattened, orbicular.

Rare in open hilly slopes of dry deciduous forests in foot hills. Fl. & Fr.: Jun. – Nov. Idupulapaya (KDP), *KRKS* & *BSS* 40319.

Mimosa prainiana Gamble, Fl. Madras 1: 421. 1919 & Kew Bull. 1920: 5. 1920; FAP 1: 360. 1997.

Local name: *Sandra.*

Straggling shrub to 4 m high; branchlets pubescent with curved prickles. Leaves alternate, bipinnate, 5–8 cm long; rachis slender, pubescent, prickled; stipules to 1.5 mm long, linear-lanceolate, pubescent; pinnae 4–5 pairs, 1.5–2.5 cm long, shortly stalked; rachis sometime prickled, basally gland-tipped; leaflets 3–5pairs, elliptic-oblong, 3–4 × 2–3 mm, pubescent below, glabrescent above, base oblique,

Plate 7

A. *Mimosa polyancistra*, B. *Terminalia coriacea*, C. *Ammannia octandra*, D. *Rotala densiflora*, E. *R. fimbriata*, F. *R. illecebroides*,

subcordate, margin entire, apex subobtuse. Flower heads pink, axillary, solitary or 3 clustered, 1 cm across; peduncle to 2.5 cm, prickles at middle, pubescent; bracts minute, lanceolate, pubescent; bracteoles minute. Flowers 2 mm across, pink. Calyx 4-lobed, pubescent, 0.5 mm, connate at base. Corolla 4-lobed, connate at base, 2 mm long, glabrous without. Stamens 8, filaments filiform. Ovary stipitate, slightly pubescent at base. Pod 6–8 × 1.5 cm, 5–7-jointed, obtuse, dark brick red in colour, minutely pubescent, slightly curved. Seeds 5–8, small, ovoid.

Occasional, in forest foot hills and scrub jungles. Fl.: Jun. – Aug. & Fr.: Sept. – Dec. Kalasamudram (ATP), *KRKS* & *KRS* 36496; *KRKS* & *BSS* 39548.

NEPTUNIA Lour.

Neptunia triquetra (Vahl) Benth., Hook. J. Bot. 4: 355. 1842; FBI 2: 286. 1878; FPM 1: 416. 1919; FAP 1: 360. 1997. *Mimosa triquetra* Vahl, Symb. Bot. 3: 102. 1794. *Desmanthus triquetrus* Willd., Sp. Pl. 4: 1045. 1806.

Local name: *Athipathi.*

Perennial, prostrate glabrous herb, slightly woody at base; branches abruptly slender, compressed, angled. Leaves bipinnate, 2–6 cm long, sensitive; pinnae 2–3 pairs, 5 cm long; a small gland between the lower pair of pinnae; petiole to 2 cm; stipules obliquely ovate, 3 × 2 mm, acuminate, 3-nerved, glabrous; leaflets 10–15 pairs, subsessile, oblong, 5 × 2 mm, glabrous, base obliquely subcordate, margin entire, apex obtuse, somewhat mucronate. Flower heads light yellow, axillary, globose, solitary; peduncle to 8 cm, glabrous; bract ovate, 3 × 2 mm, acuminate, 3-nerved, glabrous; bracteoles 2, ovate, 3 × 3 mm, acuminate, 5-nerved, glabrous. Flowers to 10 in a head, 2 mm across, yellow. Calyx 5-lobed, minute, campanulate. Corolla 5-lobed, acute at apex, connate at base, to 3 mm. Stamens 10, much exserted; filaments slender, to 4 mm, subequal; anthers small, a gland crested. Ovary stalked, glabrous, style to 0.8 mm long; stalk 1.5 mm long. Pod oblong, 2–3 × 1–1.5 cm, glabrous, thick brown, transverse nerves reticulate, base stalked, to 0.5 cm, margin slightly wavy, apex apiculate. Seeds 5–9, oblong, compressed, brown, glabrous, 0.5 × 0.4 cm.

Occasional, in agriculture lands especially in black soils. Fl. & Fr.: Nov. – Jan. Gollala guduru, near Pulivendula (KDP) *KRKS* & *BSS* 41972.

PITHECELLOBIUM C. F. P. Mart. *nom. cons.*

Pithecellobium dulce (Roxb.) Benth., London J. Bot. 3: 199. 1844; FBI 2: 302. 1878; FPM 1: 434. 1919; FAP 1: 361. 1997. *Mimosa dulcis* Roxb., Pl. Coromondel. 1: 67, t. 99. 1798.

Local name: *Sima chinta.*

Large tree, to 15 m high; branches with prickles, stipular thorns straight; bark dark grey; wood reddish; branchlets densely tomentose. Leaves bipinnate; petiole 1.5 cm, with a solitary apical concave gland; pinnae 2 pairs; leaflets paired, obliquely oblong, 2–3 × 0.5–1 cm, chartaceous, glabrous, base and apex obtuse, margin entire; petiole to 1.5 cm long; petiolule to 5 mm long. Panicles terminal and axillary, 9–12 cm long; peduncle 1-1.5 cm. Flower heads white. Flowers to 1 mm across. Calyx 5-lobed, campanulate, 1 mm long, pubescent. Corolla 5-toothed, creamish, to 2.5 mm long, connate; lobes valvate, densely tomentose without. Stamens many, monadelphous; filaments 4–5 mm. Ovary sessile, to 1 mm glabrous. Pods strap-shaped, 13–15 × 0.5–1 cm, torulose, spirally twisted, submoniliform, turgid, dehiscent. Seeds black, 5–7, broadly ovate, covered by reddish-white aril, aril edible.

Common, in village outskirts and wastelands. Fl.: Nov. – Jan. & Fr.: Feb. – May. Idupulapaya (KDP), *KRKS* & *BSS* 40307.

A native of Tropical America; it is natulaised. Plants usually cultivated for its edible flesh aril.

PROSOPIS L.

Key to Species

1a. Large shrub; spines straight; fruit pulpy, not moniliform, yellow when dry .... **P. juliflora**

1b. Large tree; spines hooked; fruit not pulpy, moniliform, brown when dry .... **P. cineraria**

Prosopis cineraria (L.) Druce, Rep. Bot. Soc. Exch. Club. Soc. Brit. Isles. 3: 422. 1914; FAP 1: 362. 1997. *Mimosa cineraria* L., Sp. Pl. 517. 1753, (*cinerea*). *Prosopis spicigera* L., Mant. Pl. 68. 1767; FBI 2: 288. 1878; FPM 1: 421. 1919.

Local name: *Jammi.*

Middle-sized, glabrous, thorny tree, 8–10 m high, bark grey, rough with deep cracks, branchlets grey, dotted, armed with scattered broad-based conical prickles, 0.2–0.8 cm. Leaves bipinnate, alternate, oblong or ovate-lanceolate, 5–15 × 3–7.5 cm, coriaceous, glabrescent, nerves 6 pairs, base subacute-obtuse, oblique, margin entire, apex acuminate, ovoid glands 2 on the rachis. Spikes axillary, clustered, to 7 cm long. Flowers cream-coloured; bracts small, membranous. Calyx 5-lobed, 2 × 3 mm, campanulate. Corolla 5-toothed, 2–3 mm, yellow. Stamens 10, free, exserted. Ovary sessile, to 2.5 mm, glabrous; style filiform. Pod straight, some times curved, 5–10 × 0.5 cm, moniliform, slender, cylindric, torulose, glabrous, pendent. Seeds 10–15, dull brown, oblong, ellipsoid.

Common, in scrub forest and wastelands. Fl.: Mar. – May & Fr.: May – Sept. Chinnakudala (KDP) near Pulivendula, *KRKS* & *BSS* 40465.

Note: The plant is worshiped by Hindus during Dasara festival.

Prosopis juliflora (Swartz.) DC., Prod. 2: 447. 1825. *Mimosa juliflora* Swartz., Prodr. 85. 1788. *Prosopis chilensis* sensu FAP 1: 362. 1997.

Local name: *Sarcarthumma.*

Armed shrub or small tree, 8–10 m high; branchlets glabrous; spines axillary, 1–1.5 cm long, straight, divaricate. Leaves alternate, bipinnate; pinnae 1–2 pairs, 4–6 cm long, leaflets 15–17 pairs, oblong, 0.7–1.5 × 0.2–0.3 cm, glabrous, base and apex obtuse, margin entire. Spikes axillary in 2–4 clusters, 2–8 cm long; bracteole linear to 1.5 mm; pedicel to 0.5 mm. Flowers yellow, 2 mm across. Calyx 5-lobed, campanulate, *ca.* 1 × 0.7 mm, triangular. Corolla 5-toothed, 1.5 × 1 mm. Stamens to 10, free, filament 4 mm long. Ovary stipitate, to 2 mm long. Pod drupaceous, 8–15 × 0.8–0.9 cm, straight or slightly curved, compressed, indehiscent, pulpy, yellow when dry. Seeds 15–19, ovoid, shiny, grey.

Very Common, in all wastelands and road sides. Fl.: Oct. – Mar. & Feb. – May. Kalasamudram (ATP), *KRKS* & *BSS* 40403.

Note: The species, a native of Mexico and Central America.

DROSERACEAE

DROSERA L.

Key to Species

1a. Acaulescent; leaves spatulate or spoon-shaped in basal rosette; stipulate .... **D. burmanni**

1b. Caulescent; leaves linear-lanceolate; estipulate .... **D. indica**

Drosera burmanni Vahl, Symb. Bot. 3:50. 1994; FBI 2: 424. 1878; FPM 1: 452. 1919; FAP 1: 368. 1997.

Local name: *Kanmara mogga.*

Acaulescent herb. Leaves in basal rosette, reddish to green, obovate-orbicular or spoon-shaped, 1–2 × 0.5–1 cm, base alternate with glandular tentacles; stipules 3-partite, scarious. Scape terminal, circinate, 5–15-flowered; peduncles 4–12 cm long; bracts linear, to 1.5 mm long; pedicel 2–5 mm long. Flowers 8 mm across. Calyx-lobes 5, oblong, to 3.5 mm. Corolla –lobes 5, pinkish, oblong-oblanceolate. Stamens 5, filaments 3.5 mm. Ovary ovoid, 1.5 mm, styles 5, apically shortly lobed, to 2 mm. Capsule *ca.* 1.5 mm long. Seeds many, black, finely scrobiculate, shiny, 0.2 mm acroos.

Common, in open wet areas associated with grasses near streams. Fl. & Fr.: Aug. – Jan. Kalasamudram (ATP), *KRKS* 40099; Batrepalli (ATP), *KRKS* 310 (HJCB-E).

Drosera indica L., Sp. Pl. 282. 1753; FBI 2: 424. 1878; FPM 1: 452. 1919; FAP 1: 368. 1997.

Slender, caulescent herb, to 10 cm high, straggling, stem with glandular hairs. Leaves estipulate; lower leaves deflexed, to 2.5 cm, upper ones alternate, narrow, linear, 1–3.5 cm long; shortly petioled, glandular tentacled above, glabrous below. Racemes, to 1.5 cm, 3-20-flowered. Flowers whitish, to 3 mm across; peduncle to 2 cm long; pedicel to 1 mm long; bracts linear, to 1.5 mm long. Calyx-lobes 5, elliptic-lanceolate, 2 mm long, glandular pubescent. Corolla lobes 5, rose, to 4.5 mm long. Stamens 5; filaments 3 mm long. Ovary ovoid, 2 mm; styles 3, bifid to the base, to

2 mm. Capsule broadly oblong. Seeds apiculate, scrobiculate, black, shiny, 0.2 mm across.

Occasional, in wet areas associated with cyperoids and grasses near tanks, on moist pockets. Fl. & Fr.: Aug. – Mar. Kalasamudram (ATP), *KRKS* 40131.

COMBRETACEAE

Key to Genera

1a. Stragglers or lianes .... **Combretum**

1b. Trees .... (2)

2a. Flowers in globose heads; fruits to 3 mm across with apical beak .... **Anogeissus**

2b. Flowers in branched spikes or panicles; fruits more than 1 cm across, without apical beak .... **Terminalia**

ANOGEISSUS (DC.) Guill. & Perr.

Anogeissus latifolia (Roxb. ex DC.) Wall. ex Guill. & Perr., Fl. Seneg. Tent 7: 280. 1832; FBI 2: 450. 1878; FPM 1: 466. 1919; FAP 1: 374. 1997. *Conocarpus latifolia* Roxb. ex DC., Prodr. 3 : 16. 1828.

Local name : *Tirman, Chirumanu, Velama.*

Large deciduous tree, to 20 m high; bark greyish, scaly, peeling off; branchlets grey-pubescent, glabrous later. Leaves sub-opposite, chartaceous, variable, elliptic-obovate or suborbicular, 3–6 × 3–5 cm, glabrescent above, pubescent below, base rounded or obliquely cuneate, margin entire, apex obtuse-rounded, sometimes emarginate; petiole 0.6–1 cm long. Flowers yellow, axillary, in pedunculate cymose-heads, 4 mm across, sessile, numerous; bracts 1 – 2 pairs, subfoliaceous on peduncles of cymes; bracteoles minute, spatulate. Calyx 5-toothed, tube 3 × 1.5 mm, persistent. Petals absent. Stamens 10 in 2-rows; filaments to 1.5 mm long. Ovary to 2 mm, 1-celled, densely tomentose, ovules 2, pendulous. Fruits clustered into a globose head, brownish yellow, 0.3–0.5 cm long, orbicular, compressed, narrowly 2-winged, crowned by persistent calyx, apically beaked, to 3 mm long. Seeds obovate, solitary, pale brown, ellipsoid.

Common, in dry deciduous forests. Fl.: Aug. – Oct. & Fr.: Oct. – Dec. Kalasamudram (ATP), *KRKS* & *BSS* 39563; Idupulapaya (KDP), *KRKS* & *BSS* 40316.

COMBRETUM Loefl. *nom. cons.*

Combretum albidum G. Don, Trans. Linn. Soc. London 15: 429. 1827; FAP 1 : 375. 1997. *C. ovalifolium* Roxb., Fl. Ind. 2: 226. 1832; FBI 2: 458. 1878; FPM 1: 469. 1919.

Local name: *Yadatheega, Putangi, Bondikota, Bandi kattu tiga.*

Large climbing shrub or liana, to *ca.* 10 m long; branchlets greyish brown. Leaves sub-coriaceous, opposite, orbicular to elliptic-lanceolate, 3.7–11.5 × 3.8–7 cm, base cuneate, margin entire, apex acute, pale brown when mature; petiole 0.7–2.2 cm, glabrous. Panicled spikes terminal, to 15 cm long. Flowers dense, greenish yellow; bracteoles minute, to 1 mm. Calyx tube glabrous, 4 mm, lobes 4, 1 mm long, reflexed. Petals 4, oblong, 1.5 × 0.5 mm. Stamens 8, biseriate, incurved; filaments subulate, 2.5 mm; anthers 0.2 mm. Ovary 1-celled, 2 mm, ovules 2–5. Fruits shiny, pale yellowish-brown, papery, 4-winged, 1.8–2.2 × 1.5–2 cm. Seed 1, testa membranous.

Common along the streams in hilly areas in dry deciduous forests near foot hills. Fl.: Feb. – Mar. & Fr.: May – Jul. Kalasamudram (ATP), *KRKS* 39945.

TERMINALIA L. *nom. cons.*

Key to Species

1a. Fruit winged .... (2)

1b. Fruit not winged .... (3)

2a. Bark smooth; branchlets brown tomentose; leaves linear oblong, serrate .... **T. arjuna**

2b. Bark deeply cracked; branchlets whitish pubescent; leaves ovate-lanceolate .... **T. coriacea**

3a. Petiole glands present; petiole to 2 cm long; fruit ellipsoid, glabrous .... **T. chebula**

3b. Petiole glands absent; petiole to 8 cm long; fruit globose, villous .... **T. bellirica**

Terminalia arjuna (Roxb. ex DC.) Wight & Arn., Prodr.Fl. Ind. Orient. 314. 1834; FBI 2: 447. 1878; FPM 1: 465. 1919; FAP 1: 377. 1997. *Pentaptera arjuna* Roxb. ex DC., Prodr. 3: 14. 1828.

Local name: *Tella maddi, Neerumaddi, Tittimaddi.*

Evergreen tree, up to 30 m high; outer bark stiff, flaking-off in pieces; inner bark whitish, smooth, young branches brown tomentose. Leaves thick-coriaceous, sub-opposite, oblong, 8–17 × 3–5.5 cm; glabrescent above except mid vein, villous or pubescent below, base obliquely cuneate, margin ciliate, serrate towards apex, apex acute-obtuse; petiole to 0.5 cm, tomentose; glands two on the petiole, closer to the base of lamina. Panicled spikes axillary, to 10 cm long. Flowers dull yellow, 5 mm across; bracts 2 mm long; peduncle to 5 cm long. Calyx 3 mm long, lobes 5, light yellow, triangular, hairy within. Stamens 10 in 2 series, 1–2 mm long. Ovary ovoid, 1.5 mm, 1-celled; ovules 2 or 3; style 3 mm long. Fruit dark brown, 2 × 4 cm, obovoid-oblong; woody, fibrous, apex notched; wings 5, equal, projecting striated with numerous curved lines, dark brown to blackish when dry.

Occasional, in dry deciduous forests, especially near streams. Fl.: Apr. – Jul. & Fr.: Nov. – Jan. Batrepalli (ATP), *KRKS* & *BSS* 39604.

Terminalia bellirica (Gaertn.) Roxb., Pl. Coromondel t. 198. 1805 *"bellerica"*; FBI 2: 445. 1878; FPM 1: 463. 1919; FAP 1: 378. 1997. *Myrobalanus bellirica* Gaertn., Fruct. 2: 90. t. 97. 1791.

Local name: *Tadi, Thandra, Bhutavasamu, Tani, Vibhitakamu.*

Large deciduous tree, up to 30 m high; crown extensive, bark brown-grey with vertical cracks; branchlets warty, whitish-pubescent with persistent leaf scars. Leaves crowded at the ends of branchlets, broadly elliptic-obovate, 9–14 × 7–13 cm, glabrous, minutely puberulous below when young; petiole to 7 cm long, glands abscent. Panicled spikes axillary, 7-15 cm long, slender. Flowers pale yellow, to 5 mm across; peduncle 2 cm long. Calyx-lobes 5, 2 mmlong, cupular; tube 1 mm long, densely villous, triangular. Stamens 10 in 2 series; filaments 3 mm long. Ovary 2 mm long. Drupe globular, 2.5–3 × 2.3 – 3 cm, grey tomentose, 5-angled when dry, hard, not winged. Seed 1, 2.5 cm across, grey, 5-angled, horned.

Common, in dry deciduous forest especially near water. Fl.: Apr. – Jun & Fr.: Aug. – Dec. Batrepalli (ATP), *KRKS* & *SSR* 36964; *KRKS* & *BSS* 39081.

Note: The fruits are one of the impotant constituents of the Triphala of Ayurveda.

Terminalia chebula Retz., Observ. Bot. 5: 31. 1789; FBI 2: 446. 1878; FPM 1: 464. 1919; FAP 1: 379. 1997.

Local name: *Karakkai, Nalla Karaka.*

Moderate size deciduous tree, 5–10 m high, much branched; branches glabrous; bark brown, shallowly fissured. Leaves alternate, thin-coriaceous, elliptic-oblong or obovate, 5–12 × 3–7.5 cm, glabrous, base rounded or oblique-acute, margin entire, apex obtuse-apiculate; petiole 1.5–2.8 cm long; sulcate above; glands 2, just below the leaf blade, rounded, dark brown. Spikes simple, axillary, to 7 cm long. Flowers dull white, to 6 mm across. Calyx campanulate, 3 mm long, glabrous out side, villous within; lobes 5. Stamens 10, biseriate; filaments 2 mm long. Ovary inferior, 2 mm long; style 3.2 mm long. Fruit yellowish-green, 2.5–4 × 1.3–1.3 cm, ovoid or obovoid, faintly 5-angled, glabrous. Seed solitary, greyish.

Occasional, in dry deciduous forest. Fl.: Aug. – Nov. & Fr.: Dec. – Apr. Kalasamudram (ATP), *KRKS* & *BSS* 39180, *KRKS* 41849.

Note: Fruits of this species are one of the constituents of Triphala churnam.

Terminalia coriacea (Roxb.) Wight & Arn., Prodr. Fl. Ind. Orient. 315. 1834; FPM 1: 465, 1919; FAP 1: 379. 1997. *Pentaptera coriacea* Roxb., Fl. Ind. 2: 438. 1832. *Terminalia tomentosa* (Roxb. ex DC.) Wight & Arn. var. *coriacea* (Roxb.) C. B. Clarke in Hook.f., Fl. Brit. India 2: 448. 1878.

Local name: *Jani, Nall maddi, Tani.*

Modarate sized deciduous tree, up to 11 m high; branchlets pubescent; bark black, deeply cracked transversely (crocodile skin type). Leaves alternate, coriaceous, elliptic-ovate, 3–7 × 8.5–14 cm, brown puberulous above, greyish white, densely tomentose below, base obliquely rounded or cuneate, margin entire, apex obtuse-acute or apiculate; petiole 1–1.5 cm long; glands 2, sub-globose, sessile, at the base of midrib beneath. Panicles terminal, to 14 cm long. Flowers dull yellow, to 5 mm across. Calyx 5-lobed, brown, pubescent, 2 mm long. Stamens 10, 2 seriate. Ovary 2.5 mm long. Fruit yellowish white, 4.5 × 3 cm, wings 5, equal, velvety, thin lines on wings, apex notched. Seed solitary, oblong, brownish.

Very rare, in dry deciduous forest. Fl.: Jun. – Aug. & Fr.: Nov. – Dec. Kalasamudram (ATP), *KRKS* 39560.

MYRTACEAE

SYZYGIUM Gaertn. *nom. cons.*

Syzygium cumini (L.) Skeels, U.S.D.A. Bur. Pl. Industr. Bull. 248. 2. 1912; FAP 1: 383. 1997. *Myrtus cumini* L., Sp. Pl. 471. 1753. *Eugenia jambolana* Lam., Encycl. 3: 198. 1789; FBI 2: 499. 1878. *Syzygium jambolanum* (Lam.) DC., Prodr. 3: 259. 1828; FPM 1: 481. 1919.

Local name: *Neredu, Jambu.*

Large tree to 35 m high; branchlets glabrous; bark greyish, smooth; wood reddish-grey. Leaves decussate, elliptic or ovate-lanceolate, 5.5–11.5 × 2.5–6.5 cm, coriaceous, glabrous, glossy above, base cuneate-obtuse, margin entire, apex acuminate; petiole to 2 cm long. Panicles axillary, to 10 cm long. Flower whitish cream, 1 cm across. Calyx turbinate, 2 mm long, glabrous; lobes 4. Petals 4, cream, orbicular, 2.5 mm long, fugacious. Stamens many; filaments red. Ovary 2 mm, 2-celled. Berry globose or oblong-ovoid, 1–2.5 × 1–1.5 cm, deep purple, glabrous, glossy. Seed solitary with fleshy cotyledons.

Common, in forest, also planted in agriculture lands and as avenue tree. Fl.: Apr –Jun. & Fr.: Jul – Aug. Batrepalli (ATP), *KRKS* & *BSS* 39092; *KRKS* 39640.

Note: Fruits are edible.

LECYTHIDACEAE

CAREYA Roxb.

Careya arborea Roxb., Pl. Coromondel t. 218. 1811 & Fl. Ind. 2: 638. 1832; FBI 2: 511. 1879; FPM 1: 488. 1919; FAP 1: 385. 1997.

Local name: *Kumbhi, Budda budwa, Budadhermi, Gadava.*

Deciduous tree, to 30 m high; bark dark brown, deeply vertically cracked; branchlets pubescent with persistent leaf scars. Leaves apically clustered, coriaceous, 18–32 × 6–14.4 cm, glabrous, base attenuate, margin crenulate, apex shortly acuminate; petiole to 1 cm long, whitish glaucous. Spikes terminal. Flowers yellowish-white, showy, 4–18 pairs on woody peduncles; bracts lanceolate, 8 × 4 mm. Calyx tube campanulate, funnel-shaped, 2 × 1.5 cm, glabrous, limb 4-lobed, lobes imbricate, 7 × 9 mm, persistent; petals 4, 3.8 × 2 cm, oblanceolate. Stamens many in several rows; staminal column 6 mm long. Ovary 4–5-celled, 6.2 cm long. Berry globular, 5–7 cm diam., yellowish, rind thick, crowned with the limb of calyx and remains of style. Seeds numerous embedded in pulp.

Occasional, in dry decidous forest. Fl.: Mar. – Apr. & Fr.: Jun. – Jul. Batrepalli (ATP), *KRKS* & *BSS* 39606.

MELASTOMACEAE

OSBECKIA L.

Osbeckia zeylanica L. f., Suppl. Pl. 215. 1781; FBI 2: 516. 1879; FPM 1: 494. 1919; FAP 1: 390. 1997.

Erect herb, up to 30 cm high, bulbous based hairs throughout, branches quadrangular, hirtellous along angles. Leaves oblong or lanceolate, 2–2.5 × 0.5–1 cm, membranous, 3-nerved, hirsute, base subacute-cuneate, margin entire-serrulate, apex acute; petiole to 3.5 mm long. Flowers in terminal or axillary, 3–5-flowered sub-umbellate; bracteoles lanceolate-linear, to 3.5 mm long; pedicel to 3 mm long. Calyx tube urceolate, 4 mm long, ribbed, stellate-bristly, appendages stipitate, to 5 mm, lobes 4, linear. Petals 4, pink to light purple, broadly obovate, 8 × 3 mm, ciliate at apex. Stamens 8, filaments to 4 mm long; spur to 1 mm long. Ovary 2.5 mm long; style to 1 cm long. Capsule, 6.5 × 3.5 mm. Seeds many, to 0.5 mm across, black.

Common, in dry deciduous forest near moist localities and open areas. Fl. & Fr.: Jul. – Nov. Rathnachalam Anjaneya swamy temple, Pulivendula ghat (KDP), *KRKS* & *BSS* 40058; Kalasamudram (ATP), *KRKS* & *BSS* 40080.

LYTHRACEAE

Key to Genera

1a. Herbs; branchlets shiny; flowers in axillary cymes or solitary; capsule to 0.2 cm .... (2)

1b. Trees or shrubs; branchlets not shiny; flowers in terminal corymbs; capsule 0.8–3 cm .... (3)

2a. Stem light pinkish; leaves lanceolate, 1–1.5 cm long; flowers solitary .... **Rotala**

2b. Stem green or dark brown; leaves obovate-oblanceolate or lanceolate, 3–4 cm long; flowers in axillary cymes .... **Ammannia**

3a. Flowers tetramerous; stamens 8; capsule irregularly dehiscent; seeds pyramidal, wingless .... **Lawsonia**

3b. Flowers hexamerous; stamens numerous; capsule regularly dehiscent; seeds flat, winged .... **Lagerstroemia**

AMMANNIA L.

Key to Species

1a. Erect herbs; axillary cymes multi-flowered .... (2)

1b. Decumbent herbs; axillary cymes 3-flowered .... **A. octandra**

2a. Leaves sessile, auriculate-cordate at base; flowers pedicillate, dichotomously arranged; petals conspicuous, pinkish; stamens 4; capsule reddish, not covered by calyx lobe .... **A. multiflora**

2b. Leaves petiolate, not auriculate at base; flowers subsessile, cymose or clusters; petals inconspicuous, reddish, stamens 8; capsule dark red, covered by calyx lobes .... **A. baccifera**

Ammannia baccifera L., Sp. Pl. 120. 1753.

Key to Subspecies

1a. Stem herbaceous; leaf base cuneate; flowers pedicellate ....subsp. **baccifera**

1b. Stem woody; leaf base sub-cordate; flowers sub sessile ....subsp. **aegyptiaca**

Ammannia baccifera L. subsp. **aegyptica** (Willd.) Koehne, Bot. Jahrb. Syst. 1: 259 1880; FBI 2: 569, 1879; FPM 1: 510. 1919; FAP 1: 391. 1997. *Ammannia aegyptiaca* Willd. Hort. Berol. t. 6 1809.

Erect herb, to 50 cm high, stem woody, usually much branched, the branches 4-angled, glabrous, dark brown. Leaves simple, opposite, linear-lanceolate or oblanceolate, 4–6.5 × 0.5–1.2 cm, glabrous, mostly longer than the internodes, becoming smaller towards the apex of stem, base truncate or sub-cordate, margin entire, apex sub-obtuse; petiole to 1 mm or absent. Cymes axillary, compactly arranged. Flowers globose, sessile; pedicels at centre rarely 1 mm, at periphery 3 mm, in cluster. Calyx 4-lobed, tube broadly bell-shaped, 1–2.5 mm long, lobes dentate, minute, 2 appendages present innerside of the calyx tube. Petals 4, reddish, minute, rounded at apex. Stamens 4, included, filaments red. Capsule globose, 1–1.5 mm in diameter, bright red to purple. Seeds globose; bright red, winged, ovate.

Common, in moist places like stream bed soils and near stagnant water tanks. Fl.: & Fr.: Aug. – Dec. Kalasamudram (ATP), *KRKS* & *BSS* 40140.

Ammannia baccifera L., subsp. **baccifera**; FBI 2: 569. 1879; FPM 1: 510. 1919; FAP 1: 391. 1997.

Erect herb, to 40 cm high; branches glabrous, 4-angled near the main axis. Leaves linear-lanceolate, 1–5 × 0.2–0.4 cm, glabrous, base acute, margin entire, apex acute. Cymes axillary in clusters, subsessile; pedicel 2 mm long of peripheral flowers, sessile of centre flowers. Flowers red, to 3 mm across. Calyx lobes 4, minute, cupular, tube 0.2 mm broad. Corolla lobes 4, reddish. Stamens 4, opposite. Capsules depressed-globose, red. Seeds minute, black.

Occasional, in moist localities. Fl.: & Fr.: Oct. – Dec. Batrepalli (ATP), *KRKS* & *SSR* 37883.

Ammannia multiflora Roxb. Fl. Ind. 1: 447 1820; FAP 1: 392. 1997; FBI 2: 570. 1879; FPM 1: 509. 1919.

Annual erect herb, to 30 cm high, glabrous. Leaves linear-lanceolate, 0.5–3 × 0.2–0.3 cm, base auriculate, cordate, margin entire, apex acute. Flowers lax, in axillary dichasial cymes, middle flowers sessile, lateral flowers pedicillate; peduncle filiform, to 1.5 cm long; pedicels 3 mm long. Bracts 2, minute, lanceolate, 0.5 mm long, glabrous. Calyx cupular, 1.5 mm across, glabrous, 4-lobed, lobes 3-angular. Corolla light pink, slightly larger than calyx, 3 mm long, 1 mm broad, obovate, margin irregular. Stamens 4, filaments 1.5 mm long, anthers yellow. Ovary globose, style filiform, stigma capitate. Capsule 2 mm across, red, thin-layered, irregularly open. Seeds many, minute, light brown-yellow, 3-angular.

Occasional, in moist places of open deciduous forests in plains. Fl. & Fr.: Nov. – Jan. Batrepalli, Nigidi forest (ATP), *KRKS* 42013, 432 (HJCB).

Note: It grows among grasses and sedges and difficult to find.

Ammannia octandra L. f., Suppl. Pl. 127. 1781; FBI 2: 571. 1879; FPM 1: 510. 1919; FAP 1: 392. 1997.

Prostrate or decumbent herbs, up to 30 cm long; branchlets 4-angular, glabrous. Leaves sessile, cordate, ovate-oblong, 1.5 – 2.5 × 1 – 1.2 cm, base cordate, margin entire, apex obtuse; petiole to 1 mm long. Cymes axillary, 3-flowered. flowers red, 3 mm across. Calyx 4-lobed; tube 2 mm long; lobes triangular, 2 mm long, covering capsule at mature stage. Sepals 4, pinkish, spatulate, 3 mm long. Stamens 4, opposite calyx-lobes. Ovary obpyramidal, 2 mm long. Capsule obpyramidal, apex truncate, covered by persistent calyx. Seeds many, dark brown, trigonous, 1 mm across.

Occasional, in moist locations of dry deciduous forests. Fl. & Fr.: Oct. – Feb. Batrepalli (ATP), *KRKS* 42039.

LAGERSTROEMIA L.

Lagerstroemia parviflora Roxb., Pl. Coromandel t. 66. 1796; FBI 2: 575. 1879; FPM 1: 513. 1919.

Local name: *Chennangi, Gullakaraka, Chinnagi.*

Deciduous trees, up to 20 m high, bark ash-coloured; branchlets puberulous. Leaves opposite decussate, lanceolate or elliptic, coriaceous, glabrous and shining above, pale or greyish, some times hoary–tomentose beneath, base acute or cuneate, subacute, margin entire, apex obtuse- retuse, secondary nerves 6-8 pairs, arcuate, prominent beneath, petiole to 4 mm. Flowers white in terminal and axillary, 8-12 cm long panicles, fragrant. Calyx tube 7 mm, with a ring inside; lobes 6, to 3 mm. Petals 6, white, ovate–suborbicular, to 6 mm. Stamens many, filaments to 1 cm. Ovary 2.5 mm, style to 1 cm. Capsule ellipsoid, brown, 1-3 × 1-1.5 cm, dehiscent by 3 or 4 valves. Seeds many, flattened, winged at tips.

Common, in deciduous forests. Fl. & Fr.: May-Nov. Kalasamudram (ATP), *KRKS* 41846.

Note: Wood hard and used for agricultural and building purpose.

LAWSONIA L.

Lawsonia inermis L., Sp. Pl. 349. 1753; FPM 1: 514. 1919; FAP 1: 393. 1997. *L. alba* Lam., Encycl. 3: 106. 1789; FBI 2: 573. 1879.

Local name: *Gorintaku.*

Bushy shrub, up to 3 m high; branches spine-tipped, with greyish-white, rough bark. Leaves elliptic-ovate, 2–2.5 × 0.6–2.5 cm, glabrous, base acute, margin entire, apex acute; petiole 0.2 mm long. Panicles terminal, to 15 cm long. Flowers white, 4 mm across. Calyx 4-lobed, cupular; lobes 3 mm long; tube 2 mm long. Corolla 4-lobed, white. Stamens 8 in 4-pairs, alternating to corolla lobes. Capsule globose, with persistent calyx, 0.8–1 cm across, 4-locular. Seeds pyramidal, 1 mm across, obovate, 4-angular, brown.

Common, in villages and occasional in forests. Fl.: Apr. – Jun. & Fr.: May – Aug. Mabbuchintalapalle (KDP), *KRKS* & *BSS* 40064.

Note: Leaves are made into paste and used for dyeing nails and palms and also hair by some people. The leaf paste is used for curing skin diseases.

ROTALA L.

Key to Species

1a. Leaves in whorls .... **R. verticillaris**

1b. Leaves opposite decussate .... (2)

2a. Stem filiform; leaves sessile, minute, 3 angular, acute, 3 mm long; flowers < 1 mm across, sepals and petals lanceolate .... **R. illecebroides**

2b. Stem not filiform; leaves not minute more than 3 mm long; flowers >1 mm across, sepals and petals not as above .... (3)

3a. Internodes less than 5 mm long; flowers 3 mm across; corolla larger than calyx .... (4)

3b. Internodes more than 5 mm long; flowers 1.5 mm across; corolla as long as calyx .... **R. mexicana**

4a. Leaves ovate-oblong, lanceolate; petal bi-lobed at apex; anthers shorter than petals (inserted) .... **R. densiflora**

4b. Leaves linear-lanceolate; petals multifid (dissected); anthers as long as petals .... **R. fimbriata**

Rotala densiflora (Roth ex Roem. & Schult.) Koehne in Engler, Bot. Jahrb. Syst. 1: 164. 1880; FPM 1: 508. 1919. *Ammannia densiflora* Roth ex Roem. & Schult. in L., Syst. Veg. 3 : 304. 1818.

Erect, simple or branched, slender, flaccid, glabrous herbs, 10–40 cm high, sometimes creeping below, stems 4-angled, winged. Leaves decussate, ovate-oblong, lanceolate, 0.5–2 × 0.1–0.4 cm, base subcordate-obtuse, margin entire, apex subacute or obtuse. Flowers pink, solitary, axillary; floral bracts almost dimorphic - on main stem and lower branches like foliar, on upper branches bracts like and secondary exceeding the flowers; bracteoles longer than sepals with distinct – midribs. Calyx-tube campanulate, 1 mm long, lobes 5 or rarely 4, triangular to shallowly triangular, 0.5 mm long, sepal appendages linear or capillary, as long as or up to 2 times longer than sepal lobes or very rarely rudimentary. Petals 5 or rarely 4, bright pink or rarely white, 0.5–1 mm long, at least as long as the sepal lobes. Stamens 5 or rarely less, inverted on the lower half of the sepal tube; anthers inserted. Capsule globose, as long as calyx tube, dehiscent by three valves. Seeds numerous, reddish-pale brown, glabrous.

Common, in moist places. Fl. & Fr.: Sep.-Oct. Kalasamudram RF (ATP), *NY* 1177.

Rotala fimbriata Wight, Icon. Pl. Ind. Orient. t. 217. 1840; FPM 1: 508. 1919; FAP 1: 395. 1997. *Ammannia pentandra* Roxb. var. *fimbriata* (Wight) C. B. Clarke in Hook. f., Fl. Brit. India 2: 569. 1879.

Erect herb, up to 40 cm high, weakly 4-angled, pink, glabrous. Leaves decussate, linear-lanceolate to oblong, 0.5–1.5 × 0.2–0.3 cm, glabrous, sessile, base cordate, margin entire, apex acute-obtuse. Flowers axillary, solitary, to 3 mm across, light pink. Calyx campanulate, 2–2.5 mm long; lobes 5, or 6, dentate, 0.5 mm long. Petals 5 or 6, bright pink, pinnately divided into linear segments, 2.3 mm long. Stamens 5, inserted, 2 mm long; anthers exserted, brown. Capsules elongated, ellipsoid, 3 mm long, 3-valved, brown. Seeds minute, brown, trigonous, 0.2 mm across.

Endemic to Peninsular India. Very rare, found in small ditches of rocks and stagnant water pits. Fl. & Fr.: Nov. – Jan. Batrepalli (ATP), *KRKS* Kalasamudram (ATP), *KRKS* 42034.

Rotala illecebroides (Arn. ex C.B. Clarke) Koehne, Bot. Jahrb. Syst. 1: 161. 1881; FAP 1: 395. 1997; FPM 1: 508. 1919. *Ammannia pentandra* Roxb. var. *illecebroides* Arn. ex C.B. Clarke in FBI 2: 569. 1879.

Annual, very slender, erect herb, to 20 cm high, stem filiform. Leaves sessile, ovate-deltoid, glabrous, 2 – 3 mm, base truncate, margin entire, apex acuminate. Flowers axillary, solitary, 1 mm across, light rose. Bract lanceolate, glabrous, bracteoles same as bracts, lesser in size. Calyx lobes acuminate-lanceolate, base wider, 2 mm long. Corolla similar to calyx lobes, smaller in size. Capsule globose, 4-valved, dehiscence regular. Seeds minute, brown, pitted.

Occasional, in moist places of plains of deciduous forests, grows alongwith grass, very difficult to find. Fl. & Fr.: Nov. – Jan. Kalasamudramu (ATP), *KRKS* 41876; Batrapalli (ATP), *KRKS* HJCB 0436.

Note: To find the species in field one has to sit and search in moist sandy places.

Rotala mexicana Schltdl. & Cham., Linnaea 5: 67. 1830. *R. rosea* (Poir.) C.D.K. Cook, Boissiera 29: 86. 1979; FAP 1: 396. 1997. *Ammania rosea* Poir. in Lam., Encycl. (Suppl.) 329. 1810. *A. pentandra* Roxb., Fl. Ind. 1: 427. 1832; FBI 2: 568. 1879.

Plate 8

A. *Rotala rosea*, B. *Cucumis prophetarum*, C. *Trianthema triquetra*, D. *Mollugo disticha*, E. *Dentella repens*, F. *Gardenia latifolia*,

Erect herb, to 30 cm high; stem 4-angular. Leaves decussate, linear-lanceolate to ovate, 1–1.5 cm long, glabrous, dark green, base cuneate or obtuse, margin entire, apex obtuse or acute; floral bracts like foliage leaves and exceeding the flowers, bracteoles shorter than sepal tube. Flowers axillary, solitary, monomorphic, sessile 1.5 mm across. Calyx-tube campanulate, 5-lobed, greenish or pinkish, shallowly triangular, 0.25 mm long, appendages subulate, longer, rarely shorter than the calyx-lobes or sometimes rudimentary. Petals 5 or rarely 4, or rudimentary, pink, distinctly longer than wide, 0.25 mm long, as long as or shorter than the sepal lobes. Stamens 5 or less; anthers included within the sepal tube. Capsules about 2 times as long as the sepal tube, 3-valved. Seeds numerous, yellow to straw-coloured, 0.1 mm.

Occasional, near moist locations, associated with grass. Fl. & Fr.: Oct. – Jan. Kalasamudram (ATP), *KRKS* 39867; Batrepalli (ATP), *KRKS* 42035.

Rotala verticillaris L., Mant. Pl. 2: 175. 1771; FAP 1: 397. 1997; FPM 1: 508. 1919. *Ammannia rotala* sensu Clarke in Hook.f. FBI 2: 567. 1879 non F. Muell.

Erect or shortly creeping herb, up to 10 cm high, rooting at the node; stems reddish. Leaves in whorls of 4–10, linear-lanceolate, or subulate, one veined, 3–10 × 1 mm, base attanuate, unequal, margin entire, apex acute. Flowers sessile, rose coloured, axillary, solitary. Bracts subulate. Calyx campanulate, small, scarious with 5 deltoid, acute short lobes. Petals caducous. Stamens 3–5, arising from the base of the calyx-tube, exserted. Ovary three-celled, thin septate; stigma subsessile. Capsule ellipsoid, longer than broad, dehiscing by 3 valves from apex, exceeding calyx-tube. Seeds *ca.* 45, ovoid, compressed.

Occasional, in moist places and plains and foot hills of deciduous forest. Fl. & Fr.: Jan.- Mar. Kalasamudram (ATP), *NY* 673.

ONAGRACEAE

LUDWIGIA L.

Key to Species

1a. Herbs, to 50 cm high; branches glabrous; leaves to 4 cm long; fruit subterete, to 1.5 cm long; lateral roots spongy, grow erect from soil for respiration .... **L. perennis**

1b. Shrubs, to 2 m high; branches white tomentose; leaves to 7 cm long; fruit terete to 4 cm long; roots not as above.................. **L. octovalvis**

Ludwigia octovalvis (Jacq.) Raven, Kew Bull. 15: 476. 1962; FAP 1: 400. 1997. *Oenothera octovalvis* Jacq., Enum. Syst. Pl. 19. 1760. *Jussiaea suffruticosa* L., Sp. Pl. 388. 1753; FBI 3: 587. 1879; FPM 1: 516. 1919.

Shrub or sub-shrub, *ca.* 2 m high; branches densly tomentose, angular, old stem 1 cm diameter. Leaves lanceolate, oblong-ovate, 2.5–4 × 0.5–2 cm, chartaceous, base attenuate, margin entire, apex attenuate or acute; petiole 0.5 cm long. Flowers yellow, 2.5 cm across, axillary, solitary; pedicel 1.5 cm long; bracteoles subulate, at the base of calyx-tube; bracts sessile, ovate. Calyx 4-lobed; tube 2 cm long, pubescent, lanceolate, 8 × 4 mm. Corolla lobes 4, yellow, obovate, 1 × 0.8 cm, base cuneate, apex rounded. Stamens 8; filaments to 3 mm long; short ones epipetalous; anthers oblong. Ovary 2 cm long; 4-celled. Capsule terete, 4 × 0.6 cm, thin-valved, 8-ribbed, pubescent. Seeds multiseriate, numerous with a prominent raphe, brown.

Occasional, near water and wet places on rocky crevices of the rocks. Fl. & Fr.: Oct. – Jan. Batrepalli (ATP), *KRKS* & *SSR* 37885, *KRKS & BSS* 40404.

Ludwigia perennis L., Sp. Pl. 119. 1753; FAP 1: 401. 1997. *L. parviflora* Roxb., Fl. Ind. 1: 419. 1832; FBI 2: 588. 1879; FPM 1: 517. 1919.

Erect herb, to 30 cm high; branches glabrous, glossy. Lateral roots spongy and grow erect from ground up to water surface for respiration. Leaves linear, elliptic-lanceolate, 2.5–4 × 0.5–0.8 cm, glabrous, base attenuate, margin entire, apex acuminate; petiole 0.8 cm long. Flowers yellow, axillary, solitary; bracteoles subulate, 1 mm long; pedicel 2 mm long. Calyx-tube 1 cm long; lobes 4, lanceolate, 4 mm long. Corolla lobes 4, yellow, elliptic, to 4 mm across, base cuneate, apex obtuse. Stamens 4; filaments 2 mm long. Ovary 1 cm long, 4-celled, inferior. Capsule sub-terete, 0.5–1.5 cm, apically broader than base, thin-valved, 4-ribbed, glabrous. Seeds numerous, multiseriate, free, ellipsoid-ovoid, smooth with narrow raphe, brown.

Common, in wet areas, edges of ditches and paddy fields. Fl. & Fr.: Aug. – Nov. Batrepalli (ATP), *KRKS* 36965.

PASSIFLORACEAE

PASSIFLORA L.

Passiflora foetida L., Sp. Pl. 959. 1753; FBI 2: 599. 1879; FPM 1: 524. 1919; FAP 1: 404. 1997.

Local name: *Tellasumiki, Adavi motala, Guje teega.*

Viscid climbing herb, to 10 cm long; branches sticky hairy with salty taste, glandular hairs. Tendrils axillary, simple; stipules sub-reniform, 0.5–1 cm across, deeply cleft into glandular processes. Leaves 3–5-lobed, 5–10 × 4–7 cm, chartaceous, glandular pubescent, base cordate, margin ciliate, apex acute; petiole glandular hairy, 2–5 cm long. Flowers white, axillary, solitary, 4 cm across; peduncle 3–4 cm; bracts and bracteoles 2–3 cm across, deeply pinnatisect, glandular pubescent. Calyx tube short; lobes ovate-lanceolate, 1.5 cm long, apically spurred. Corolla slightly shorter than calyx, outer one hairy, inner one glabrous, to 2 mm long. Disk prominent, annular. Androgynophore 4–6 mm long, filaments to 5 mm long, flat; anthers to 3.5 mm long. Ovary globose, base to 5 mm long; stigmas 3, capitate, glandular. Berry subglobose, 4 × 3 cm, oily and sticky, pulpy. Seeds ellipsoid, 5 × 2 mm, gelly coated, deeply pitted, black.

Common, on edges of water tanks, wastelands and roadsides. Fl. & Fr.: Jun. – Nov. Kalasamudram (ATP), *KRKS* & *KRS* 36902.

Note: Fruits edible.

CUCURBITACEAE

Key to Genera

1a. Tendrils branched .... (2)

1b. Tendrils simple .... (4)

2a. Flowers < 2.5 cm across, male flowers solitary; leaves deeply lobed .... (3)

2b. Flowers > 4 cm across, male flowers in raceme; leaves usually angular or shallowly lobed .... **Luffa**

3a. Male flowers clustered, coaxillary with female; fruit < 1.5 cm across .... **Diplocyclos**

3b. Male flowers solitary, not coaxillary with female; fruit > 4 cm across .... **Citrullus**

4a. Bracts ciliate; stamens free; anther cells straight .... **Ctenolepis**

4b. Bracts glabrous; stamens connate; anther cells triplicate (S-shaped) .... (5)

5a. Calyx tube closed at base by scales .... **Momordica**

5b. Calyx tube without scales .... (6)

6a. Male flowers solitary; flowers white .... **Coccinia**

6b. Male flowers in clusters or racemes; corolla and calyx campanulate .... (7)

7a. Connectives apically crested; trailing herbs .... **Cucumis**

7b. Connectives apically uncrested .... (8)

8a. Ovary rostrate at apex .... (9)

8b. Ovary not rostrate .... **Mukia**

9a. Stamens inserted at the base of calyx tube; dioecious .... **Solena**

9b. Stamens inserted in the mouth of calyx tube; monoecious .... (10)

10a. Leaves angular, strongly aromatic; connectives 2-partite; fruit indehiscent .... **Kedrostis**

10b. Leaves deeply lobed, not aromatic; connectives faintly 2-fid; fruit dehiscing by operculum .... **Corallocarpus**

CITRULLUS Schrad. *nom. cons.*

Key to Species

1a. Fruit bigger, up to 40 cm across.-pulp sweet, cultivated (sometimes escape) .... **C. vulgaris**

1b. Fruit not more than 10 cm across.-pulp bitter, grows in wastelands .... **C. colocynthis**

Citrullus colocynthis (L.) Schrad., Linnaea 12. 414. 1838; FBI 2: 620. 1879; FPM 1: 536. 1919; FAP 1: 406. 1997. *Cucumis colocynthis* L., Sp. Pl. 1011. 1753.

Local name: *Papara kaya.*

Trailing, scabrid, herbaceous climber; tendrils slender, short, sparsely villose-hirsute hairy, simple or 2–3-fid; stem angular; hairs with tuberous base. Leaves ovate or narrowly triangular, deeply 3–7-lobed, lobes pinnatifid-sinuous, 6–12 × 4–6 cm, rigid, densely villous hirsute below; lobes obtuse or acute, margin crisped; terminal lobes larger than lateral lobes. Flowers monoecious, yellow, both male and female flowers solitary, axillary. Male flowers 1 cm across; pedicel to 1.5 cm long. Calyx-tube 2 mm; lobes linear, 2 mm, hirsute without. Corolla yellow, hirsute; petals ovate, 8 × 4 mm. Stamens 3; filaments 1.5 mm long; anthers to 5 mm across. Female flowers: ovary ovoid, 8 × 4 mm; style to 1 mm. Fruits globose, slightly depressed, 5–7 × 4.6–6 cm, striped green and white when young, yellow when ripe, fruit wall hard. Seeds blackish brown, ovate or ovate-oblong, 5 × 3 mm, hard.

Common, in wastelands and village outskirts. Fl. & Fr.: Aug. – Dec. Gandi River (KDP), *KRKS* & *BSS* 40327.

Citrullus vulgaris Schrad., Enum. Pl. Afric. Austral 279. 1836. *C. lanatus* (Thunb.) Matsum. & Nakai, Cat. Sem. Spor. 30, no. 854. 1916.

Spreading herb to 10 m long. Leaves to 15 cm long, dissected, lobes rounded at apex. Male flowers axillary, solitary or in cymes. Female flowers axillary, solitary. Ovary villous, stigma 5-lobed. Fruit up to 40 cm across, green and white bands alternate, pulpy. Seeds ovate, flattened, dark brown.

Cultivated as commercial crop, some times escape from the cultivation. Fl. & Fr.: Dec. – Feb. Kalasamudram (ATP), *KRKS* 40405 (wild).

COCCINIA Wight & Arn.

Coccinia grandis (L.) Voigt. Hort. Suburb. Calc. 59. 1845; FAP 1: 406. 1997. *Bryonia grandis* L., Mant. Pl. 126. 1767. *Coccinia indica* Wight & Arn., Prodr. Fl. Ind. Orient. 1: 347. 1834; FPM 1: 537. 1919. *Cephalandra indica* (Wight & Arn.) Naud., Ann. Sci. Nat. Bot. (Ser. 5) 5: 16. 1866; FBI 2: 621. 1879. *Coccinia cordifolia* auct. non (L.) Cogn. A. & C. DC. Monogr. Phan. 3: 529. 1881.

Local name: *Thonda kaya.*

Climbing herb, stem glabrous, smooth, angular, tendrils simple, glabrous. Root tuberous. Leaves alternate, 5-lobed, chartaceous, glabrous, punctuate above, glandular below; lobes divided, to 4 cm long, margin entire-dentate, apex rounded to apiculate, leaf base sinus subrotund-cordate; petiole to 5 cm long. Flowers white, solitary, axillary. Male flowers 3 cm across; pedicel 5–7 cm long. Calyx 5-lobed; tube to 3 mm long; lobes triangular-lanceolate. Corolla 1.5 × 1 cm; stamens inserted at the base of the calyx tube; anther connate. Female flower 2.5 cm wide. Calyx 1–2 × 1 cm. Ovary oblong, glabrous. Fruit oblong-ovoid, blood red when ripe, 5 × 2.5 cm, reticulately veined, fibrous. Seeds ovoid, compressed, granular, white.

Common, on shrubs of scrub forest and agriculture field hedges. Fl. & Fr.: Jan. – Aug. Thummalapalle (KDP), *KRKS* & *KRS* 36408.

CORALLOCARPUS Welwit.

Corallocarpus epigaeus (Rottl. & Willd.) C. B. Clarke in Hook. f., Fl. Brit. India 2: 628. 1879; FPM 1: 541. 1919; FAP 1: 407. 1997. *Bryonia epigaeus* Rottl. & Willd., Ges. Naturf. Freunde Berlin Neue Schriften 4: 228. 1803.

Monoecious climber; branchlets glabrous. Leaves broadly suborbicular, deeply 3–5-lobed, 2.5–4.5 × 5–4 cm; lobes obovate, chartaceous, pubescent, lobulate, base cordate, margin irregularly dentate, apex obtuse, apiculate; petiole 1 – 3.5 cm long. Tendrils simple. Raceme of male flowers to 15 cm long; peduncle to 4 cm long. Flowers 2 mm wide, yellow. Calyx tube campanulate, pubescent without, 3 mm long; lobes 5, ovate, 1.5 mm long. Corolla campanulate, greenish-yellow, to 1 mm long. Stamens free, inserted into the middle of calyx tube. Female flowers 10 × 2 mm. Corolla yellow. Ovary oblong-ovoid, beaked. Fruits ellipsoid, 1.5 × 0.7 cm, glabrous, rostrate, operculate. Seeds ovoid, smooth, whitish.

Common, in scrub forest and dry deciduous forest. Fl. & Fr.: Jun. – Oct. Thummalapalle (KDP), *KRKS* 36984; Rachakuntapalle (KDP), *KRKS* & *KRS* 37623; Batrepalli (ATP), *KRKS* 39642.

CTENOLEPIS Hook.

Ctenolepis garcinii (Burm. f.) C.B. Clarke, Fl. Brit. India 2: 629. 1879; FAP 1: 407. 1997. *Sicyos garcinii* Burm. f., Fl. Indica 211. 1768. *Blastania garcinii* (Burm. f.) Cogn. in A. & C. DC. Monogr. Phan. 3: 629. 1881; FPM 540. 1919.

Climber vine, branches scabrid. Leaves palmately 5-lobed, 3–7 × 5–7 cm; lobes obovate, membranous, scabrid, with small tuberculous hairs, base cordate, margin crenulate, denticulate, apex obtuse, mucronate; tendrils simple; petiole 5–10 cm long. Flowers monoecious; bracts stipuliform, ovate-rounded, 1 cm across, ciliate. Male flower raceme axillary, to 1 cm long. Flower 4 mm across, yellow. Calyx campanulate, 1 mm across, linear. Corolla greenish-yellow. Anthers inserted in calyx tube, connate. Female flowers solitary, axillary, 1.5 mm across; peduncle 3 mm. Bracts stipuliform, ovate-rotund, ciliate, 1 cm. Fruit subreniform, 4.5 × 8 mm, glabrous, orange-red. Seeds 1 or 2, oblong, smooth, sulcate on one face, convex on other face.

Common, in foot hills of dry deciduous forest and scrub forest. Fl. & Fr.: Nov. – Jan. Thummalapalle (KDP), *KRKS* & *KRS* 36990; Thurakapalli (ATP), *BSS* & *KRKS* 41998; Mabbuchintalapalle (KDP), *KRKS* 281 (HJCB-E).

CUCUMIS L.

Key to Species

1a. Leaves simple .... **C. melo**

1b. Leaves 5–7-lobed .... (2)

2a. Fruit softly echinate; leaves 7-lobed .... **C. prophetarum**

2b. Fruit smooth or not echinate; leaves 5-lobed .... **C. callosus**

Cucumis callosus (Rottl.) Cogn. in Engler, Pflanzenr. 4. 275. 2: 129. 1924; FAP 1: 408. 1997. *Bryonia callosa* Rottl., Ges. Naturf. Freunde Berlin 4: 210. 1803. *Cucumis trigonus* Roxb., Fl. Ind. 2: 722. 1824; FBI 2: 619. 1879; FPM 1: 535. 1919.

Local name: *Budama.*

Weak climbing herb; stem scabrid with tuberculate hairs, angular. Leaves 5-lobed, 5–7 × 4–7 cm; lobes obovate, base cordate, margin crenulate, apex rounded; petiole to 4 cm long; tendrils simple. Male flowers in axillary clusters, 4 mm across, yellow. Calyx 5-lobed, pubescent. Corolla 5-lobed, thin. Anthers connate in calyx tube. Female flower inferior. Calyx lobes 5, ovate. Corolla 5-lobed, yellow. Ovary ovoid-oblong. Fruit green when young, pubescent, whitish striation, yellow when mature, ovoid, 3.5 × 2 cm. Seeds white, smooth, ovoid, 4 × 2 mm.

Common, in foot hills of dry deciduous forest, scrub forest and agriculture lands. Fl. & Fr.: Mar. – May. Thummalapalle (KDP), *KRKS* & *KRS* 33689; Idupulapaya (KDP), *KRKS* & *SSR* 39060.

Cucumis melo L., Sp. Pl. 1011. 1753; subsp. **agrestis** (Naud.) Pang. La Turgu. Agri. 534. 1933. *C. melo* L. var. *agrestis* Naud., Ann. Sci. Pl. 4: 614. 1805; FPM 1: 535. 1919.

Climbing herb, branches scabrid, angular. Leaves rounded-cordate 3.5–7 × 4–7.2 cm, rough hairy on both surfaces, base cordate, margin dentate-crenate, apex obtuse; petiole to 10 cm long; tendrils simple. Male flowers yellow in axillary cluster. Male flowers 0.5 cm across. Calyx pubescent. Corolla yellow. Anthers connate, curved. Female flowers solitary, axillary. Corolla 1 cm across. Ovary white pubescent, silky shiny; peduncle to 3 cm long. Fruit oblong, or ellipsoid-ovoid, 4 × 2 cm. Seeds white, shiny, smooth ellipsoid, 5 × 3 mm.

Occasional, in foot hills the forest and agriculture lands. Fl. & Fr.: Oct. – Jan. Idupulapaya (KDP), *KRKS* & *SSR* 39052; Kalasamudram (ATP), *KRKS* 40117.

Cucumis prophetarum L., Cent. Pl. 1. 32. 1755; FBI 2: 619. 1879; FPM 1: 535. 1919.

Local name: *Mullapapara.*

Climbing perennial herb, stem scabrid, hirsute. Leaves 5–7-lobed, ovate, 3–5 × 3–5 cm, tuberculate rough hairs on both sides; base cuneate, margin denticulate or crenulate, apex obtuse; petiole to 3 cm long; tendrils simple, filiform. Male flowers yellow, axillary, solitary. Calyx pubescent, triangular, 2 mm long. Corolla 3 mm across, 5-lobed; pedicel 0.5 mm. Female flower axillary, solitary. Calyx 5-lobed, triangular, pubescent. Corolla yellow, 5-lobed. Ovary glabrous, spinulate, pedicel 3 cm long. Fruit spinulate, soft echinate, ovoid, 3.5 × 2.5 cm, glabrous. Seeds white, shiny, glabrous, ellipsoid, 3 × 2 mm.

Common, in wastelands. Fl. & Fr.: Mar.–Aug. Kalasamudram (ATP), *KRKS* & *KRS* 36449.

DIPLOCYCLOS (Endlic.) Post & Kuntze

Diplocyclos palmatus (L.) Jeffrey, Kew Bull. 15: 132. 1962; FAP 1: 409. 1997. *Bryonia palmatus* L., Sp. Pl. 1012. 1753. *B. laciniosa* L., Sp. Pl. 1013. 1753 *p.p.*, excl. typus; FBI 2: 622. 1879. *Bryonopsis laciniosa* sensu Naud., Ann. Sci. Nat. Bot. Ser. 4. 12. 141. 1859; FPM 1: 534. 1919.

Local name: *Magalingam.*

Annual slender climbing herb with root stock, branchlets glabrous, angled. Leaves deeply palmately 5-lobed, 7 × 9.5 cm, membranous, scabrid above, glabrous below; lobes elliptic-oblong, base sub-cordate, margin sub-crenulate, apex acuminate; tendrils 2-fid, glabrous; petiole to 6 cm long. Flowers unisexual, male and female in clusters. Male flowers yellow, co-axillary, 1.5 cm wide. Calyx-tube broadly campanulate, 3 mm long; lobes 5, linear, 0.5 mm. Corolla greenish-yellow, broadly campanulate; petals 5, ovate, to 4 mm long. Stamens 3; filaments free, 1 mm inserted in the middle of calyx-tube. Female flower yellow, axillary 3–5 in a cluster. Ovary globose, 0.5 cm across, glabrous, smooth, shiny. Fruit globose, 1.5 cm across, white and green irregular vertical lines. Seeds ovoid, 5 × 3 mm, turgid.

Common in forest outskirts, wastelands on shrubs. Fl. & Fr.: Oct. – Jan. Nuthna Kaluva (ATP), *KRKS* & *BSS* 41925.

KEDROSTIS Medik.

Kedrostis foetidissima (Jacq.) Cogn. in DC., Monog. Phan. 3: 634. 1881; FAP 1: 411. 1997. *Trichosanthes foetidissima* Jacq., Collect. 2: 341. 1789. *Kedrostis rostrata* (Rottl.) Cogn. in DC., Monog. Phan. 3: 636. 1881; FPM 1: 540. 1919. *Bryonia rostrata* Rottl., Ges. Naturf. Freude Berlin Neue Schriften Ser. 4: 213. 1803. *Rhynchocarpa foetida* sensu Clarke in Hook.f., Fl. Brit. India 2: 627. 1879. *p.p.* non Schrader 1838.

Local name: *Muru thonda, Pamudonda.*

Climbing herb, branches glandular-pubescent. Leaves broadly ovate-cordate, faintly 3-5-angled, 4–5 × 4–6 cm, chartaceous, softly pubescent on both sides, margin irregular dentate, apex acute, petiole to 4 cm long; tendrils simple. Flowers unisexual, yellow, axillary. Male flowers in raceme, 8 mm wide; peduncle to 1.5 cm; pedicel 8 mm, Calyx tube campanulate, 2 mm wide, pubescent. Corolla yellow, rotate, 5 mm wide, glandular, pubescent. Stamens 3, inserted in the middle of calyx tube. Female flowers axillary, solitary, 8 mm wide; pedicel 2 mm. Ovary oblong-fusiform, glandular pubescent, beaked. Fruit flask-shaped, 2.5 cm, rostrate, orange when ripe. Seeds 4, obovate, 3 mm, blackish, granular.

Rare on shrubs in scrub forest and borders of agriculture lands. Fl. & Fr.: Sept. – Nov. Thummalapalle (KDP), *KRKS* & *KRS* 36984; Rachakuntapalli (KDP), *KRKS* & *KRS* 37622; N. S. P. Kottala (ATP), *KRKS* & *BSS* 40480.

Note: Plant semisucculent and possesses strong foetid smell.

LUFFA Mill.

Luffa acutangula (L.) Roxb., Fl. Ind. 3: 713. 1832 incl. var. **amara** C.B.Clarke in FBI 2: 615. 1879; FAP 1: 411. 1997; FPM 1: 533. 1919. *Cucumis acutangula* L., Sp. Pl. 1011. 1753.

Local name: *Adavibeera kaya.*

Climbing shrub, sparingly scabrous, stem angular, tendrils bifid. Leaves 5–7-lobed, 5–7 × 6–9 cm, scabrid on both surfaces, base cordate, margin denticulate, apex mucronate; petiole 5–7 cm long. Flowers yellow, axillary. Male flowers in raceme to 12 cm long; pedicel 1.5 cm; bract ovate, 2 × 1 mm. Calyx 5-lobed, pubescent. Corolla 5-lobed. Anthers 3, connate. Female flowers solitary, axillary, peduncle to 8 cm long. Calyx 5-lobed, strong. Corolla 5-lobed. Ovary oblong-obovate, pubescent. Fruit 10–11-sharp-ribbed, oblong-obovate, glabrous, fibrous. Seeds ellipsoid, black, tuberculate shiny.

Common, in scrub and foot hills of dry deciduous forest and roadsides in Mining Area. Fl. & Fr.: Mar. – Jul. Thummalapalle (KDP), *KRKS* & *SSR* 33662.

MOMORDICA L.

Momordica cymbalaria Fenzl ex Naudin, Ann. Sci. Nat., Bot., Sér. 4, 12: 134. 1859. *Luffa tuberosa* Roxb., Fl. Ind. 3: 717. 1832; FAP 1: 412. 1997. *Momordica tuberosa* (Roxb.) Cogn., Monogr. Phan. 3: 454. 188; FPM 1: 532. 1919.

Local name: *Kasara.*

Perennial climbing or prostrate herb, branches spreading, to 1 m long; root stock tuberous. Tuber sub-globose, 7 × 4 cm, brown. Leaves cordate, 5–7-lobed, scabrid 5–7 × 5–6 cm, base cordate, margin denticulate, apex obtuse; petiole to 5 cm long; tendrils simple. Flowers yellow, axillary. Male flowers 5–7 on 3 cm long peduncle. Flowers to 3 mm across; pedicel to 0.5 cm, bract ovate. Calyx 5-lobed, pubescent. Corolla yellow, 5-lobed, obovate. Anthers 3, connate. Female flower solitary, axillary. Calyx 5-lobed, triangular. Corolla 5-lobed, yellow, lobes ovate, to 5 mm long. Ovary oblong-ellipsoid, angular; angles 5–7, scabrid. Fruit ovoid, 2 × 1 cm, tapering at both ends, 5–7-angled, green, splitting irregularly. Seeds ovoid, 0.4 × 0.3 cm, black or dark brown, hard, shiny.

Common, in black soils in agriculture lands. Fl. & Fr.: July–Oct. Chinna kudala (KDP), *KRKS* & *BSS* 40462.

Note: Fruits are collected and used as vegetable.

MUKIA Arnott

Mukia maderaspatana (L.) M. Roem., Fam. Nat. Syn. Monogr. 2. 47. 1846; FAP 1: 415. 1997. *Cucumis maderaspatana* L., Sp. Pl. 1012. 1753. *Mukia scabrella* (L. f.) Arn., Hook. J. Bot. 3: 276. 1841; FBI 2: 623. 1879. *Melothria maderaspatana* (L.) Cogn. in A. & DC. Monogr. Phan. 3: 623. 1881; FPM 1: 539. 1919.

Local name: *Budda budama.*

Climbing herb, branches scabrous-hispid. Leaves deltoid, cordate, 3–5-lobed, 4–6.5 × 4–6 cm, chartaceous, scabrid above, shortly hispid below, base cordate, margin denticulate, apex mucronate; petiole to 3 cm long; tendrils simple. Flowers axillary. Male flowers 3–6 in a cluster, 6 mm across. Calyx-tube 3 mm, villose, lobes linear. Corolla yellow, ovate, 3 mm. Stamens 3; anthers oblong, 2 mm long. Female flowers 2, axillary. Ovary globose, 3 mm, villose. Fruit sub-globose sub-sessile, 1 × 0.8 cm, glabrescent, yellowish, turing red at maturity. Seeds 4, white, ovate, compressed, 4 × 2 mm, pitted.

Common, in scrub and dry deciduous forest. Fl. & Fr.: Aug. – Dec. Idupulapaya (KDP), *KRKS* & *SSR* 39057; Somulavaripalli (ATP), *KRKS* & *BSS* 40178.

SOLENA Lour.

Solena amplexicaulis (Lam.) Gandhi in Saldanha & Nicolson, Fl. Hassan 179. 1976; FAP 1: 415. 1999. *Bryonia amplexicaulis* Lam., Encycl. 1: 496. 1785. *Melothria amplexicaulis* (Lam.) Cogn. in DC., Monogr. Phan. 3: 621. 1881; FPM 1: 539. 1919.

Climbing herb, branches glabrous, angular, tendrillar. Leaves polymorphous, cordate-ovate, 7–9 × 5–8 cm, chartaceous, scabrous, base cordate, margin dentate, apex mucronate-apiculate; tendrils simple. Flowers axillary, yellow. Male flowers in clusters of 6–9 flowers. Calyx-tube 2 mm across, 5-lobed. Corolla 5-lobed, yellow. Stamens 3, connate. Female flower axillary, solitary. Calyx 5-lobed, 3 mm long. Corolla yellow, elliptic-ovate. Ovary oblong-ellipsoid, to 1 cm long; peduncle 0.5 cm. Fruit sub globose-ellipsoid, green with white lines, glabrous. Seeds white, ovate, pitted 3 × 2 mm.

Common, in dry deciduous forest. Fl. & Fr.: Sept. – Dec. Batrepalli (ATP), *KRKS* & *BSS* 39615; Kalasamudram (ATP), *KRKS* & *BSS* 39506.

CACTACEAE

OPUNTIA L.

Opuntia stricta (Haw.) Haw., Syn. Pl. Succ. 191 1812. var. **dillenii** (Ker-Gawl.) Benson, Cact. Succ. J. America 41: 126.1969; FAP 1: 422. 1997. *Cactus strictus* Haw., Misc. Nat. 188. 1803. *Opuntia dillenii* (Ker–Gawl.) Haw., Suppl. Pl. Succ. 79. 1819; FBI 2: 657. 1879; FPM 1: 548. 1919. *Cactus dillenii* Ker-Gawl., Bot. Reg. 3: t. 255. 1818.

Local name: *Papachi, Brahma jamudu.*

Erect, branching, bushy, xerophytic shrub, up to 4 m high, branches drooping, large, flat limbs, lower often longer and corky, median 28–38 cm long and 6–8 mm thick, mostly widely obovate; areoles 7–10, articulations often with a ring of glochidiae; spines 5–7 per areole, 1 to 5 cm long, yellow, rough surfaced, small spines 20–40 per areole, 1–2 mm long, yellow. Young leaves subulate, fleshy, 4 mm, sub-spinulose, swollen, deciduous. Perianth many, outer tepals to 1.5 cm, inner ones obovate, 3 × 2 cm, mucronate. Petals many, yellow orbicular. Stamens many; filaments to 2 cm long. Ovary 2.5 cm long, glochidiate; style apically branched. Berry obovate, fleshy, areolar and glochidiate, yellow spines surrounding the areole, one star-shaped spine present at apex; pulp light red, sour in taste. Seeds brown, many, curved, 0.4 cm across.

Common, in scrub forest and wastelands. Fl. & Fr.: Jun. – Dec. Kalasamudram (ATP), *KRKS& BSS* 40406.

Note: It is a native of Tropical America and frequently seen near village outskirts. Fruits are edible and used for preparation of jams. Medicine for Gynic problems.

MOLLUGINACEAE

Key to Genera

1a. Leaves radical or radical and cauline; flowers in dichasial to polychasial cymes .... **Mollugo**

1b. Leaves all cauline; flowers in clusters, cymes or solitary .... (2)

2a. Flowers pink, in cymes; staminodes many, petaloid .... **Corbichonia**

2b. Flowers greenish-white, in clusters; staminodes 0 .... (3)

3a. Fruit syncarpous, loculicidal; seeds strophiolate, many per cell .... **Glinus**

3b. Fruit apocarpous, indehiscent; seeds estrophiolate, 1 per cell .... **Gisekia**

CORBICHONIA Scop.

Corbichonia decumbens (Forssk.) Exell, J. Bot. 73. 80. 1935; FAP 1: 423. 1997. *Orygia decumbens* Forssk., Fl. Aegypt.-Arab. 103. 1775; FBI 2: 661. 1879; FPM 1: 551. 1919.

Procumbent herb, up to 30 cm long, glabrous, branchlets whitish. Leaves obovate-elliptic 2–3.5 × 1.5–2.5 cm, glabrous, fleshy, base cuneate, margin entire, apex obtuse, mucronate, chartaceous. Inflorescence leaf-opposed or terminal, cymes or lax recemes, to 7 cm long; peduncle to 4 cm long; pedicel to 4 mm long; bracts lanceolate, to 3 mm long. Flowers 7 mm across, pink. Calyx lobes 5, ovate, green, apiculate, 4 mm long. Corolla 5-lobed, pink, dissected to 15–20 segments. Stamens many, staminodes 7, petaloid. Ovary ovoid, to 3 mm long. Capsule 5-locular, ovoid, 0.5 cm across, loculicidal. Seeds many per locule, black, kidney-shaped with concentric rings, black, 0.5 mm across.

Common, in scrub forests and dry deciduous forests. Fl. & Fr.: Aug. – Jan. Bhumaiah garipalle (KDP), *KRKS* 37619.

GISEKIA L.

Gisekia pharnaceiodes L., Mant. Pl. 562. 1771; FBI 2: 661. 1879; FPM 1: 553. 1919; FAP 1: 423. 1997.

Local name: *Isaka chenchalaku.*

Prostrate annual herb, branches shiny, glabrous. Leaves unequal, opposite, elliptic-oblong, 2–4 in whorl, 0.5–1.5 × 0.3–1 cm, base obtuse-subacute, margin entire, apex obtuse, glabrous on both sides, slightly tuberculed with epidermal growths on lower side; petiole to 5 mm long. Inflorescence of lax cyme, axillary, 5–10 in a cluster. Flowers pinkish-green, 1 mm across. Sepals 5, green, thin, hooded, rough with white crystals, ovate. Petals 0. Stamens 5; filament *ca.* 1 mm. Ovary 5-locular; pedicel 2–3 cm long. Capsule 5-locular, 2 mm across, indehiscent. Seeds 5, black, 0.3 mm, base acute, pitted, shiny.

Common, weed in groundnut crop in sandy soils. Fl. & Fr.: June–Aug. Kalasamudram (ATP), *KRKS* &*BSS* 39169; Chinnapalli (ATP), *KRKS* 37995.

Note: Used as leafy vegetable and medicine for stomacache.

GLINUS L.

Key to Species

1a. White woolly hairy herb; leaves spatulate; pedicels to 2 mm long .... **G. lotoides**

1b. Pubescent herb; leaves oblong; pedicels to 1.5 cm long .... **G. oppositifolius**

Glinus lotoides L., Sp. Pl. 463. 1753; FAP 1: 424. 1997. *Mollugo lotoides* (L.) Kuntze, Revis. Gen. Pl. 214. 1891; FPM 1 : 652. 1919. *M. hirta* Thunb., Prodr. Fl. Cap. 24. 1794; FBI 2 : 562. 1879.

Decumbent hirsute or woolly herb, to 30 cm long. Leaves spatulate, 3–5 in a whorl, obovate, 0.3–1.5 × 0.3–0.8 cm, hirsute, base narrowed, margin entire, apex obtuse-acute; petiole to 0.5 cm long. Inflorescence axillary fascides. Flowers white, 3-4, 5 mm across. Sepals 5, ovate, pubescent, acute, 0.3–0.5 × 0.2–0.3 cm, greenish-white, margin scarious. Stamens 5; staminodes usually 5. Ovary 3-locular, to 3 mm long; styles 3. Capsule ovoid, 0.5 cm across, enclosed in persistent sepals. Seeds black, kidney-shaped, 0.3 mm across, many per cell, papillate.

Common, in wastelands and edges of ponds. Fl. & Fr.: Jun. – Aug. Kalasamudram (ATP), *KRKS* & *KRS* 36476; *KRKS* & *BSS* 39112; Batrepalli (ATP), *KRKS* 39093.

Glinus oppositfolius (L.) A. DC., Bull. Herb. Boissier (Ser.2) 1: 552. 1901; FAP 1: 424. 1997. *Mollugo oppositifolius* L., Sp. Pl. 89. 1753.

Prostrate-ascending, much spreading herb, branches pubescent. Leaves 3–5 in a whorl, oblong-oblanceolate or elliptic-obovate, unequal, 0.5–2 × 0.2–0.4 cm, chartaceous, base cuneate, margin entire, apex acute-mucronate; petiole to 3 mm long. Inflorescence axillary in a cluster. Flowers 4 mm across, 5–7 in a cluster; pedicel to 0.5–1.5 cm. Sepals 5, green-white, oblong-lanceolate, 5 mm long, glabrous, margin scarious, white. Stamens 5; filaments 2.5 mm; staminodes bifid, 2.5 mm. Ovary 3-locular, 2.5 mm long; styles 3. Capsule oblong, 4 × 2 mm. Seeds many, subreniform, 0.5 mm long, blackish, granular-papillate, kidney-shaped.

Common, near water bodies and wastelands. Fl. & Fr.: May – Aug. Kalasamudram (ATP), *KRKS*& *KRS* 36476; *KRKS* & *BSS* 39112; Batrepalli (ATP), *KRKS* 39093.

MOLLUGO L.

Key to Species

1a. Leaves all radical; stem reduced .... **M. nudicaulis**

1b. Leaves radical and cauline .... (2)

2a. Cymes in umbels or whorled cymes; leaves narrow-linear or aciculate .... **M. cerviana**

2b. Cymes in elongate racemes; leaves other than linear or aciculate.................. (3)

3a. Branchlets pubescent, inflorescence rachis zigzag, pedicels straight, seeds obscurely granular... **M. disticha**

3b. Branchlets glabrous, inflorescence rachis straight, pedicels curved, prominantly granular .... **M. pentaphylla**

Mollugo cerviana (L.) Ser. in DC., Prodr. 1: 392. 1824; FBI 2: 663. 1879; FPM 1: 553. 1919; FAP 1: 425. 1997; var. **spathulifolia** Fenzl, Ann. Wiever Mus. Naturgesh. 1: 379. 1836. *Pharnaceum cerviana* L., Sp. Pl. 272. 1753.

Erect herb, to 10 cm high, branchlets glaucous, filiform, glabrous. Leaves in whorls, radical ones 5–7, elliptic, *ca.* 8 mm across, above the root stock; cauline ones 3–4, narrowly linear or acicular, 1–1.5 cm long, chartaceous, glaucous below, base attenuate, margin entire, apex obtuse, apiculate; petiole 1 mm. Cymes terminal or axillary, multichasial cyme; peduncle short, to 1.5 cm, filiform; bract subulate, 0.5 mm long; pedicel to 5 mm long, filiform. Flowers 2.5 mm across, light pink-green. Sepals 5, subequal, elliptic, 1.5–2 mm obtuse, margin scarious. Petals 0. Stamens 5–8; filaments 1 mm long. Ovary 3-lobed, 3-celled; styles 3. Capsule ellipsoid, to 2 mm across. Seed ovoid-subreniform, minute, glandular or tuberculate, black.

Occasional on sandy soils in scrub forests and agriculture lands. Fl. & Fr.: May – Aug. Kalasamudram (ATP), *KRKS* & *BSS* 39119.

Mollugo disticha Ser. Prodr. 1: 392 1824. FAP 1: 425. 1997; FBI 2: 663. 1879; FPM 1: 553. 1919.

Erect annual herb to 30 cm high. Branches pubescent, stem angular. Leaves 3 or 4 at a node, subsessile, linear oblong or elliptic-oblong, 0.5 – 2 × 0.2 – 0.5 cm, glabrous. Inflorescence axillary compound cyme; rachis zigzag. Flowers white on straight pedicels to 3 mm across. Sepals elliptic, thin, margins transparant, 3 × 2 mm. Petals obovate. Capsule ovoid, 5-locular. Seeds 3–4 in a locule, curved ovate, light red, obscurely granulate, 0.5 mm across.

Rare in deciduous forests. Fl. & Fr.: Aug. – Nov. Gandi river, opposite to Hanuma temple in scrub jungle (KDP), *KRKS* 40326.

Mollugo nudicaulis Lam., Encycl. 4: 234. 1797; FBI 2: 664. 1879; FPM 1: 553. 1919; FAP 1: 425. 1997.

Erect herb, up to 10 cm high. Leaves radical, spatulate, to 4 × 2 cm, membranous, glabrous, base attenuate, margin entire, apex obtuse-retuse; petiole 1 cm long.

Cymes terminal, corymbose, multichasial; peduncle erect-patent, 7 cm long; bracts 1 mm long; pedicel to 4 mm. Flowers 8 mm across, white. Sepals 5, elliptic, 3 mm long. Stamens 3–5; filaments 1.5–2 mm. Ovary oblong, 3-loculed; styles 3. Capsule ellipsoid, 2 mm long. Seeds granular, minute, kidney-shaped, black, 0.2 mm acorss.

Very Common, in scrub forests and foot hills of dry deciduous forests along hill paths in rocky crevices and agriculture lands. Fl. & Fr.: May – Dec.K. K. Kottala (KDP), *KRKS* & *KRS* 36414.

Mollugo pentaphylla L., Sp. Pl. 89. 1753; FPM 1: 553. 1919; FAP 1: 426. 1997. *M. stricta* L., Sp. Pl. (ed. 2) 131. 1762; FBI 2: 663. 1879.

Ascending herb, up to 20 cm high; branchlets glabrous. Radical leaves oblanceolate-spatulate; cauline ones 3–5 in a whorl, obovate or elliptic, 1–4.5 × 0.5–1.5 cm, chartaceous, base cuneate-attenuate, margin entire, apex obtuse-subacute, apiculate; petioles 2 mm long. Cymes leaf-opposed and terminal, multichasial, 5 cm long; peduncle 8 cm long; bracts 1 mm; pedicels curved, 5–8 mm long. Flowers white, 2 mm across. Sepals 5, ovate-oblong, 2 mm; stamens 3. Ovary ellipsoid, 1.8 mm, 3-celled; styles 3. Capsule oblong, 2 mm across. Seeds dark brown-black, prominently granular, kidney-shaped, 0.3 mm.

Common, in foot hills and exposed slopes of forests. Fl. & Fr.: Mar – Dec. Rachakuntapalli (KDP), *SSR* 33619; Kalasamudram (ATP), *KRKS* & *KRS* 36480.

AIZOACEAE

Key to Genera

1a. Flowers in sub-umbellate clusters; branches pubescent; sepals free, with thin margin; ovary 3–5- celled, perianth tube free from the petiole .... **Zaleya**

1b. Flowers solitary; branches glandular hairy; sepals connate, not as above; ovary 1-2-celled. perianth tube adnate to the basal part of the petiole .... **Trianthema**

TRIANTHEMA L.

Key to Species

1a. Leaves to 5 mm long, tapering at both ends; branches filiform; sepals 2 mm long; capsule to 2 mm across .... **T. triquetra**

1b. Leaves to 2 cm long, rounded at base, truncate at apex; branches not filiform; sepals 4 mm long; capsule to 4 mm across .... **T. portulacastrum**

Trianthema portulacastrum L., Sp. Pl. 223. 1753; FPM 1: 550. 1919; FAP 1: 427. 1997. *T. monogyna* L., Mant. Pl. 69. 1767; FBI 2: 660. 1879.

Local name: *Ganjaraku, Galijeru.*

Prostrate or ascending herb, up to 8 cm long; branchlets glandular or glabrescent, thickened at the nodes, shiny. Stipules triangular, to 2.5 mm, membranous. Leaves subopposite, unequal, succulent, broadly obovate to elliptic, 1.2–4 × 0.8–3 cm, base cuneate, margin entire, apex obtuse-truncate; petiole to 2.5 cm, base winged, membranous. Flowers white, axillary, solitary in a pouch or between forks of branches; bracteoles lanceolate, to 3.5 mm. Calyx tube 1.5 mm long; lobes 5, oblong-lanceolate, to 5 mm. Corolla lobes 5, white, 3 mm, glabrous. Stamens *ca.* 20; filaments to 3 mm long. Ovary 3 mm long, glabrous; style 1, to 3 mm long. Capsules turbinate, circumscissile, glabrous, partly concealed in the petiolar hood, 4 mm long, obconical. Seeds dispersed by operculum; operculum cone-shaped. Seeds black, 0.5 mm, glabrous, smooth.

Common along road sides, in wastelands and weed in agriculture fields. Fl. & Fr.: Through year. Thummalapalle (KDP), *KRKS, KRS, SSR* & *TP* 36407; Kalasamudram (ATP), *KRKS* & *BSS*39145.

Note: Plants with whitish green flowers and reddish-pink flowers can be found in this species at the same place.

Leaves are made into paste and decoction to control liver and kidney troubles.

Trianthema triquetra Rottl. ex Willd., Ges. Naturf. Freunde Berlin Neue Schriften 4: 181. 1803; var. **triquetra**; FPM 1: 551. 1919; FAP 1: 427. 1997. *T. crystallina* auct. non Vahl. 1790. FBI 2: 660. 1879.

Prostrate herb, up to 30 cm long; branches glandular hairy, filiform, radiating from a root stock, reddish, studded with minute papillae. Stipules triangular. Leaves linear-lanceolate to elliptic-lanceolate, succulent, 0.5–1 × 0.2–0.3 cm, shiny, base subacute, margin entire, apex obtuse, apiculate; petiole 4 mm long, dilated at the base and sheathing. Flowers pink, axillary, solitary or some times clustered. Sepals 5, tube 1.5 mm; lobes 0.5 mm long, triangular. Petals 5, oblong. Stamens 5, not exceeding calyx-lobes. Ovary obconical, style 0.5 mm. Capsule 3 mm diameter, opeculum cone-shaped. Seeds covered with thin layer. Seeds 1–2, black, reniform with concentric rings.

Common, in foot hills of dry deciduous forest and scrub forest near moist places in saline soils. Fl. & Fr.: Jun. – Sept. Kalasamudram (ATP), *KRKS* 39528.

ZALEYA Burm.f.

Zaleya decandra (L.) Burm. f., Fl. Ind. 110. t. 31. f. 3. 1768; FAP 1: 427. 1997. *Trianthema decandra* L., Mant. Pl. 70. 1767; FBI 2: 661. 1879; FPM 1: 551. 1919.

Decumbent herb, up to 50 cm long; branchlets pubescent, rough. Leaves uneqal, elliptic-oblong, oblong-lanceolate, 2–3 × 1–1.5 cm, sub-succulent, papillose, base acute, margin entire, apex obtuse or rounded; petiole to 7 mm long, dilated, amplexicaul. Flowers pink, axillary, 6–8 in sub-umbellate cluster; bracteoles linear, 1.5 mm; pedicel to 2 mm long. Calyx tube 2 mm, 5- lobed, 2 mm long, margins thin. Corolla lobes 5, 4.5 mm long, pink. Stamens 10, unequal; filaments to 2 mm. Ovary terete, 3 mm long; styles 2, ciliate, to 2 mm long. Capsule truncate at apex, separating into 2 parts, each with 2 seeds. Seeds orbicular or reniform, rugulose, 4-angular, black, 2 mm long.

Common, in dry lands and road sides. Fl. & Fr.: Jun. – Nov. Mabbuchintalapalle (KDP), *KRKS, KRS, SSR* & *TP* 33657.

Note: Leaves and roots are taken internally for curing Hepatitis 'B'. (as per the information of local practioner)

ALANGIACEAE

ALANGIUM Lam.

Alangium salvifolium (L. f.) Wanger in Engl., Pflanzen. IV. 220 b. 41: 9. 1910. subsp. **salvifolium** Turill & Milne. Redh., Fl. Trop. E. Africa. Alangiaceae 3. 1958; FPM 1: 572. 1919; FAP 1: 436. 1997. *Grewia salvifolia* L. f., Suppl. Pl. 409. 1781. *Alangium lamarckii* Thw., Enum. Pl. Zeyl. 133. 1859; FBI 2: 741. 1879.

Local name: *Uduga, Ankolamu, Urga.*

Deciduous small tree, up to 5 m high; trunk often with numerous holes; branchlets appressed tomentose. Leaves chartaceous, ovate-oblong or elliptic-lanceolate, 6–15 × 1.8–4.3 cm, glabrous above, glabrescent or pubescent below, base obliquely obtuse, margin entire, apex subacute to sub-acuminate, 3-nerved from base; nerves irregular; petiole to 1 cm long; tomentose. Bracts ovate, 1 mm, deciduous. Flowering starts when complete leaf fall. Flowers in axillary fascicles, white, to 2.5 × 1.5 cm. Calyx tube cupular, 2.5 cm across, tomentose, whitish, lobes ovate, 0.5 mm long. Corolla lobes 10, white, 2.5 × 3.2 cm, linear oblong; reflexed. Stamens 20, oblong, to 1 cm long. Ovary turbinate, to 2 mm across. Berry ellipsoid, 2 × 1 cm, red or blackish-purple, pubescent, pulpy, crowned by calyx lobes at apex. Seed solitary, ovoid, 1.2 × 0.8 cm across, brown, glabrous, hard; cotyledons cuboid, 1 × 0.6 cm, one side surface brown, smooth, other side notched, rough brown, acute at apex, base obtuse.

Common, on forest outskirts near streams and ponds. Fl. & Fr.: Jul. – Aug. Kalasamudram (ATP), *KRKS* & *BSS* 39123; Batrepalli (ATP), *KRKS* & *BSS* 37979.

Note: Fruits edible.

RUBIACEAE

Key to Genera

1a. Trees or large shrubs .... (2)

1b. Herbs or undershrubs .... (12)

2a. Ovules more than 2, usually many in each cell .... (3)

2b. Ovules single in each cell .... (8)

3a. Flowers in globose heads .... (4)

3b. Flowers not in globose heads .... (5)

4a. Fruit a syncarpium; flowers to 10; apical bud covered with ovate stipule .... **Morinda**

4b. Fruit aggregated capsule; flowers to 100; apical bud covered with foliaceous stipule .... **Mitragyna**

5a. Armed shrub .... (6)

5b. Unarmed trees or shrubs .... (7)

6a. Berry <1 cm across; seeds few, not embedded in pulp; pedicel to 20 mm long .... **Benkara**

6b. Berry >2 cm across; seeds many, embedded in pulp; pedicel to 5 mm long .... **Catunaregam**

7a. Petiole to 0.5 cm long; lateral veins 10–20 pairs; ovary one-celled .... **Gardenia**

7b. Petiole to 2 cm long; lateral veins to 7 pairs; ovary two-celled .... **Deccania**

8a. Corolla lobes valvate .... (9)

8b. Corolla lobes contorted .... (10)

9a. Leaves 8 cm long; stipular spines absent .... **Psydrox**

9b. Leaves 2 cm long; stipular spines to 2 cm long .... **Canthium**

10a. Stipules interpetiolar .... **Ixora**

10b. Stipules intrapetiolar .... (11)

11a. Style glabrous; flowers four-merous .... **Pavetta**

11b. Style hairy; flowers five-merous .... **Tarenna**

12a. Leaves to 4 mm long; flowers 5-merous .... **Dentella**

12b. Leaves more than 10 mm long; flowers 4-merous .... (13)

13a. Ovules many in each cell .... (14)

13b. Ovules single in each cell .... (15)

14a. Leaves whorled; flowers in terminal raceme .... **Kohautia**

14b. Leaves opposite, flowers in axillary and terminal cymes .... **Oldenlandia**

15a. Flowers dimorphic in terminal corymbos .... **Knoxia**

15b. Flowers monomorphic in axillary fascicles .... (16)

16a. Capsule dehiscence septicidal at apex; nerves on leaf not prominent .... **Spermacoce**

16b. Capsule dehiscence circumcissile; nerves on leaf prominent .... **Mitracarpus**

BENKARA Tirveng.

Benkara malabarica (Lam.) Tirveng., Taxon 32: 440. 1983; FAP 2: 468. 1997. *Randia malabarica* Lam., Encycl. 3: 25. 1789; FBI 3: 111. 1880; FPM 2: 616. 1921.

Armed much branched shrub, up to 3 m high; branches woody, greyish-brown. Leaves coriaceous, decussate, clustered, obovate-oblong, 3–4.5 × 1.3–2.3 cm, glabrous, base cuneate, margin entire, apex obtusely apiculate; petiole to 0.7 cm long; stipule inter petiolar, connate. Inflorescence axillary, umbel like, corymbose cyme. Flowers whitish, to 5 mm across; bracts and bracteoles free, sometimes connate. Calyx turbinate, 4 mm across, lobes 5, triangular 1 mm, acute, deciduous.

Corolla salver-form, 2 cm long; tube cylindric with a ring of hairs within; lobes 5, oblong, 3 mm long, recurved, obtusely acute. Stamens 5, between the lobes, exserted; filaments 0.5 mm long; anthers linear, 3 mm long. Ovary 1 mm long; style 0.2 mm long. Berry globose, 0.8 cm across, orange, two-locular. Seeds 2 per locule, dark brown, 0.3 cm across.

Common in thorny- scrub foresta and outskirts of dry deciduous forest. Fl.: Jul. – Sept. & Fr.: Nov. – Jan. Rachakuntapalle (KDP), *KRKS* & *SSR* 37923.

CANTHIUM Lam.

Canthium coromandelicum (Burm. f.) Alston, in Trimen, Hand Book Fl. Ceylon 6. Suppl., 152. 1931. *Gmelina coromandelina* Burm. f., Fl. Ind. 132. 1768. *Canthium parviflorum* Lam., Encycl. 1: 602. 1785; FBI 3: 136. 1880; FAP 2: 470. 1997. *Plectronia parviflora* (Lam.) Bedd., Fl. Sylv. S. India 124. 1872; FPM 2: 625. 1921.

Local name: *Balusa, Balija.*

Armed shrub, up to 4 m high; stem hard, spines supra-axillary, horizontal up to 2 cm long; branches 4-angled, greyish brown, glabrous. Leaves elliptic-ovate, 2–3 × 1–2 cm, glabrous, shiny above, dull green below, base acute to attenuate, margin entire, apex acute-obtuse; petiole 0.4 cm long; stipules ovate-linear 0.2 cm long, interpetiolar. Cymes axillary, lax, decussate. Flowers greenish-white to 3 mm across. Calyx cupular, 4-lobed; lobes 2.5 × 1 mm, glabrous. Corolla 5 mm across; tube 1.5 cm long, villous within; lobes 4, obovate, 2.5 mm long, acute, spreading or reflexed. Stamens 4, sub-sessile; anthers 0.7 mm long, pointed at apex. Ovary 1 × 0.7 mm. Drupe globose-subglobose, 1 cm across, glabrous, green shiny when young, yellow when mature, brown pulpy when ripe. Seeds 1–2, brown, curved, 0.8 × 0.5 cm, rugose.

Very Common in scrub forest. Fl.: Mar. – May & Fr.: Jun. – Sept. Rachakuntapalle (KDP), *KRKS* 37919.

Note: Fruits sweet, pulp with granules and edible.

CATUNAREGAM Wolf

Key to Species

1a. Leaves silvery whitish tomentose, acute at both ends; stipules triangular .... **C. tomentosa**

1b. Leaves pubescent, acute at base; stipules acuminate .... **C. spinosa**

Catunaregam spinosa (Thunb.) Tirveng., Bull. Mus. Hist. Nat. Paris 3, Ser. Bot. 35. 1978 & Taxon 27: 515. 1978; FAP 2:470. 1997. *Gardenia spinosa* Thunb., Diss. Gard. No. 7. 1780. *Randia dumetorum* (Retz.) Poir. var. *floribunda* (DC.) Gamble, Fl. Madras 2: 616. 1921. *Randia brandisii* Gamble, Fl. Madras 2: 616. 1921.

Local name: *Manga.*

Armed, large shrub, up to 4 m high; branches 4-angled when young, grey with rough surface when mature. Spines decussate, hard, 1.5–2 cm long, supra-axillary. Leaves clustered on very short branchlets, elliptic-obovate, 1.5–2.8 × 1–2 cm, glabrescent above, pubescent below, base attenuate, margin entire, apex obtuse; petiole to 2 mm long; stipule ovate, 3 × 2 mm, acuminate. Flowers axillary, solitary or in subsessile cymes on pair of straight axillary spines, whitish-yellow. Calyx-tube to 4 mm long, pubescent; lobes 5, ovate, 3 mm long, acute, pubescent without. Corolla 5-lobed, pubescent without; tube 0.5 cm long; lobes elliptic-ovate, acute-obtuse at apex. Stamens 5, alternate to corolla lobes at mouth of corolla-tube; anther yellowish-white, lanceolate, 2 mm long. Ovary inferior in corolla tube, obovate. Berry globose, glabrous, 3 × 2 cm, ribbed or smooth, claw at apex with persistent calyx. Seeds many, in pulp, white, 2 × 2 mm, curved.

Common, in scrub forests, foot hills of dry deciduous forests and wastelands. Fl.: May – Jul. & Fr.: Sept. – Nov. Thummalapalle (KDP), *KRKS, SSR &TP* 36404.

Catunaregam tomentosa (Blume ex DC.) Tirveng., Taxon 27: 515. 1978; FAP 2: 471. 1997. *Gardenia tomentosa* Blume ex DC., Prodr. 4: 379. 1830. *Randia tomentosa* (Blume. ex DC.) Hook. f., Fl. Brit. India 3: 110. 1880.

Local name: *Kotuvagoch.*

Armed tree, up to 6 m high; bark grey, rough; branches greyish tomentose; spines straight, decussate, supra-axillary, pubescent ¾ from base, remaining glabrous, 1.5–2.5 cm long. Leaves elliptic-oblong, 1.6–3.5 × 1.2–2.2 cm, villose on both sides with silvery hairs, not much as above, base acute-attenuate, margin entire, apex acute; lateral veins 5 pairs; petiole to 3 mm long, pubescent; stipule triangular, 4 mm wide, 6 mm long, acuminate. Flowers white-yellow, axillary, solitary. Calyx cupular, pubescent; tube 5 mm long; lobes 5, ovate, 5 × 3 mm, pubescent without. Corolla 5-lobed, tube 6 mm long, pubescent without and mouth; lobes pubescent on both sides. Stamens 5, alternate to the corolla lobes, lanceolate, 3 mm long, pubescent. Ovary globose, pubescent, 2 mm diam. Berry globose-subcylindrical, 8–10-ribbed, pubescent, 3 × 2 cm, pulpy, green when young, yellow when mature, rim at apex with persistent calyx. Seeds many in pulp, hard, white, rugose, curved, 2 × 3 mm.

Very rare, in scrub forest and dry deciduous forest. Fl.: Apr. – Jul. & Fr.: Sept. – Nov. K. K. Kottala (KDP), *KRKS & BSS* 40067.

DECCANIA Tirveng.

Deccania pubescens (Roth) Tirveng., Nordic J. Bot. 3: 456. 1983 var. **candolleana** (Wight & Arn.) Tirveng. *l.c.*; FAP 2: 472. 1997. *Randia candolleana* Wight & Arn., Prodr. Fl. Ind. Orient. 1: 399. 1834; FBI 3: 113. 1880; FPM 2: 616. 1921. *Gardenia pubescens* Roth, Nov. Pl. Sp. 151. 1821.

Bushy tree, up to 6 m high; bark greyish, rough; branches with elevated leaf scars, densely pubescent. Leaves chartaceous, clustered apically on branchlets, orbiculate or ovate to obovate, 9–12 × 6–10 cm, pubescent below, puberulent above, blackish when dry, lateral veins 7–9 pairs, base attenuate, margin entire, apex obtuse; petiole to 2 cm long, pubescent; stipules inter petiolar, triangular, 3 × 3 cm, acute at apex, pubescent. Inflorescence short, axillary or sub-terminal cymes.

Flowers white; pedicel to 1 cm long. Calyx lobes 5, cuspidate, 3 mm long. Corolla 1.2 cm in diam.; lobes broadly oblong, glabrous within, much longer than the tube; tube silky without. Stamens 5, at mouth of corolla tube. Ovary globose, glabrous. Berry globose, 1 cm diam., green when young, black when ripe, pulpy. Seeds many, brown, smooth.

Occasional, in dry exposed hilly areas of dry deciduous forests. Fl.: May – Jul. & Fr.: Jul. – Oct. Kalasamudram (ATP), *KRKS* & *BSS* 39553.

DENTELLA J.R.Forst. & G. Forst.

Dentella repens (L.) J.R.Forst. & G. Forst., Char. Gen. Pl. 26. t. 13. 1776; FBI 3: 42. 1880; FPM 2: 590. 1921; FAP 2: 473. 1997. *Oldenlandia repens* L., Mant. Pl. 40. 1767.

Small, prostrate herb, branches glabrous, semisucculent, to 12 cm long, dichotomously branched. Leaves ovate, 2 × 1 mm, glabrous, base acute, margin entire, apex acute; petiole 1 mm; stipules connate, scarious, pubescent. Flowers white, minute, sessile, solitary, axillary and in the forks of the dichotomous branches. Calyx tube globose, limb tubular, 5-lobed, persistent. Corolla funnel-shaped, hairy within; lobes 5, 2–3-toothed. Stamens 5, in the middle of the corolla-tube; filament short; anthers short. Ovary 2-celled; ovules many; stigma filiform. Fruit spiny, globose, not pulpy, indehiscent, 2-celled. Seeds many, minute, angular, testa puncticulate, shiny.

Common, near water bodies, like ponds and streams, grows when dry. Fl. & Fr.: Mar. – Jun. Chinnapalli (ATP), *KRKS* & *BSS* 37996; Kalasamudram (ATP), *KRKS* 39148.

GARDENIA Ellis *nom. cons.*

Key to Species

1a. Calyx-lobes triangular, acute; leaves < 7 cm long; nerves irregular .... **G. gummifera**

1b. Calyx-lobes lanceolate, acuminate; leaves > 9 cm long; nerves regular .... (2)

2a. Berry globose; placentae 3; calyx and corolla-lobes 7; calyx-lobes >2 cm long .... **G. latifolia**

2b. Berry ellipsoid; placentae 2; calyx and corolla-lobes 5; calyx-lobes<1.5 cm .... **G. resinifera**

Gardenia gummifera L. f., Suppl. Pl. 164. 1781; FBI 3: 116. 1880; FPM 2: 618. 1921; FAP 2: 474. 1997.

Local name: *Bikki, Manchi Bikki, Chinna mitta.*

Deciduous small tree, to 5 m high; bark greyish-brown, smooth, brown pieces peeling off when mature; woody branches twisted. Leaves obovate-oblong, 3–7 × 1.5–4.5 cm, sessile, rough hairy above, glabrescent below, base acute to obtuse, margin entire, apex obtuse-acute, 15–18 pairs of nerves; stipule connate surrounding the stem like ring, apical bud gummy. Flowers yellowish-white, 1–3 in axils. Calyx truncate, connate, tube 3 mm long; lobes 5, 3-angular, 1.5 mm long, puberulous without, acute. Corolla 4.5 cm across, tube 1.8 cm; lobes 5, obovate, puberulous without, obtuse. Stamens 5, alternate to corolla lobes at mouth. Berry ellipsoid or oblong with numerous longitudinal lines or smooth, 4 × 3 cm, beak stout, crown with calyx. Seeds many, rugose.

Common, in dry deciduous forests. Fl.: Jan. – Apr., Fr.: May – Aug. Pulivendula ghat (ATP), *KRKS* & *BSS* 39629; Batrepalli (ATP), *KRKS* & *BSS* 39650.

Note: Fruits edible, sweet like Guava and local people selling as Bikkipallu. The growing tip exudes a yellow resin.

Gardenia latifolia Ait., Hort. Kew. 1: 294. 1789; FBI 3:116. 1880; FPM 2: 619. 1921; FAP 2: 475. 1997.

Local name: *Pedda bikki, Adavi bikki, Tharpi chettu, Pedda manga.*

Deciduous tree, up to 10 m high; bark glaucous-grey, exfoliating; wood smooth. Leaves broadly elliptic-oblong, 6–30 × 2.5–18.5 cm, glabrescent above, pubescent below, base obtuse-acute, margin entire, apex obtuse to sub-acute; petiole to 3 mm; stipule surrounding the stem like a ring, 0.5 cm across. Flowers white, in axillary cyme or solitary. Calyx campanulate; tube 1.4 cm; lobes 5–7, lanceolate, 1.8 cm long, acuminate. Corolla-tube *ca.* 6 cm long, thin-fleshy; lobes 5 – 7, to 3 cm long, oblong-lanceolate. Stamens 5–9, subexerted; anther 1.9 cm long. Ovary oblong, 0.9 cm long, stigma 2-lamellate. Berry globose, 4 × 3.5 cm, with a crown of calyx and 4–5 placentae, pubescent. Seeds many, rugose, 0.2 cm across.

Common, in dry deciduous forests. Fl.: Feb. – Apr. & Fr.: May – Nov. Batrepalli (ATP), *KRKS* 37948, *KRKS* & *BSS* 39099.

Gardenia resinifera Roth, Nov. Pl. Spec. 150. 1821; FAP 2: 475. 1997. *G. lucida* Roxb., Fl. Ind. 2: 553. 1824; FBI 3: 115. 1880; FPM 2: 618. 1921.

Local name: *Karing, Chinna karinga, Verri bikki.*

Deciduous tree, up to 10 m high; bark greenish-grey, flaking off; wood yellowish-white. Leaves thin-coriaceous, elliptic, 5–8 × 2.3–6.5 cm, puberulent, base acute-obtuse, margin entire, apex sub-acute or obtuse; petiole to 3 mm; lateral veins 8–18 pairs; stipule connate surrounding the stem. Flower yellowish, axillary, solitary. Calyx-tube 0.7 cm long; lobes 5, triangular oblong, 0.5 cm long, lanceolate, puberulous without, acuminate. Corolla-tube 8 cm long; lobes 5, obovate, thin fleshy. Stamens 5, exerted, 9 mm long. Ovary ovoid, 0.5 × 1 cm. Berry ellipsoid, 2 × 1.5 cm, with a crown of lanceolate calyx lobes. Seeds rugose, whitish, 3 mm across, many in pulp, hard.

Common, in dry deciduous forest. Fl.: Apr. – May & Fr.: Jul. – Sep. Kalasamudram (ATP), *KRKS* & *BSS* 39562; 39109.

Note: The growing tip exudes a yellow resin.

IXORA L.

Ixora pavetta Andr., Bot. Repos 2: t. 78. 1799. *I. arborea* Roxb. ex Smith in Rees. Cyclop. 19: 5. 1811; FAP 2: 485. 1997. *I. parviflora* Vahl, Smb. Bot. 3: 11. t. 52: 1794; FBI 3: 142. 1880; FPM 2: 631. 1921. *nom. illegit.*

Local name: *Kori, Gorivi.*

Small tree, to 5 m high; branches glabrous, 4-angular, greyish; bark brown, hard. Leaves oblong-obovate or elliptic-obovate, 10–17 × 3.5–6 cm, glabrous, base sub-cordate, margin entire, apex obtuse or acute; petiole to 0.8 cm; stipules inter-petiolar, 3-angular, acute, 3 × 4 mm long. Inflorescence terminal, dichotomously branched cymes; peduncle up to 4 cm, puberulent; bracts ovate, 2 mm long; pedicels 3 mm long; bracteoles minute, 0.5 mm. Flowers white, 4 mm across. Calyx truncate, glabrous; tube 1 mm long; lobes 4, triangular, glabrous. Corolla tube 7 mm long, puberulous without, lobes 4, obtuse. Stamens 4, exserted. Ovary 1 mm diam. Berry sub-globose, didymous, 0.6 × 0.5 cm, reddish. Seeds 2, plano-convex, whitish brown, striate, 6 mm across.

Common, in dry deciduous forests. Fl.: Jan. – Mar. & Fr.: Apr. – Jun. Batrepalli (ATP), *KRKS* & *BSS* 39082.

KNOXIA L.

Knoxia wightiana Wall. ex Wight & Arn., Prodr. Fl. Ind. Orient. 440. 1834; FBI 3: 120. 1880; FPM 2: 622. 1921; FAP 2: 487. 1992.

Erect woody sub-shrub, to 1.5 m high; branches dichotomous, 4-angular, white villose, blackish when dry. Leaves ovate–lanceolate, 3–8 × 0.5–1.8 cm, pubescent, dark green above, villose pale greenish below, base obtuse-acute, margin entire, apex acute, nerves 8 pairs; petiole to 0.8 cm long; stipules 6 at node, lanceolate, pubescent, 0.4 cm long. Inflorescence terminal, branched raceme; peduncles to 4 cm; bract ovate, 2 mm; pedicel to 1 mm; bracteole minute. Flowers whitish blue. Calyx cupular; tube 2 mm across; lobes 4, 0.5 mm acute, triangular. Corolla-tube 7 mm long; lobes 4, glabrous without, 2 × 4 mm. Stamens 4, included. Ovary 1 mm long. Capsule 2.5 × 1 mm, didymous, truncate, mericarps very small, separating from the base. Seeds 2, brown, oblong 2 × 1 mm.

Common in dry deciduous forests, especially near rocky crevices and slopes. Fl. & Fr.: May – Dec. Pulivendula ghat (KDP), *KRKS* & *SSR* 39008.

KOHAUTIA Cham. & Schltdl.

Kohautia aspera (B. Heyne ex Roth) Bremek., Verh., Kon. Ned. Akad. Wetensch., Afd. Natuurk., Sect. 2. 48(2): 113. 1952. *Hedyotis aspera* B. Heyne ex Roth, Nov. Pl. Sp. 94. 1821; FAP 2: 478. 1997. *Oldenlandia aspera* (Roth) DC., Prodr. 4: 428. 1830; FBI 3: 68. 1880; FPM 2: 602. 1921.

Annual stout, erect herb, to 25 cm high; branches hirsute, 4-angular. Leaves whorled, linear, 2.5–4 × 0.2–0.5 cm, pubescent, base attenuate, margin entire, apex acute, sub-sessile; stipules lanceolate, to 5 mm long. Inflorescence terminal, compound cyme, to 10 cm long; peduncle to 6 cm long; pedicel to 2 mm long. Flowers white, to 2 mm across. Calyx-tube 2 mm diam.; lobes 4, horn-shaped, to 2

Plate 9

A. *Knoxia wightiana*, B. *Morinda pubescens* var. *stenophylla*, C. *Amberboa ramosa*, D. *Cynthillium albicans*, E. *Epaltes divaricata*, F. *Blepharispermum subsessile*

mm long. Corolla-tube 4 mm long; lobes 4, white, 2.5 mm long. Anthers 4 at corolla mouth, < 1 mm long. Ovary inferior in calyx tube. Capsule 4 mm across, 4-celled, brown. Seeds black, 0.2 mm, different shapes, rugose.

Very common, in wastelands and agricultural fields. Fl. & Fr.: Aug. – Oct. Thummalapalle, near mining site (KDP), *KRKS* & *KRS* 36989.

MITRACARPUS Zuccari

Mitracarpus hirtus (L.) DC. Prodr. 4: 572. 1830. *M. villosus* (Sw.) DC. Prodr. 4: 572. 1830; FAP 2: 488. 1997. *M. verticillatus* (Schum. & Thonn.) Vatke, Linnaea 40: 196. 1876.

Erect or often decumbent pubescent herbs, 20–50 cm high, stem quadrangular, light brown, strigose. Leaves ovate or ovate-lanceolate, 2–4 × 1.5–2 cm, membranous, sessile, base obtuse, margin entire, apex acute, nerves prominent below; stipule pectinate, teeth of lobes lanceolate, filiform. Flowers white, verticillate in axillary fascicles, 4-merous. Bracteoles filiform. Calyx-tube small, unequally 4-lobed, larger 2 lobes *ca.* 1 mm long. Corolla white *ca.* 1.5 mm long, funnel-shaped; lobes 4, ovate, to 3 mm long. Stamens 4, attached at throat. Ovary 2-locular, ovule 1-per locule. Capsule straw-coloured *ca.* 2 mm long, ovoid, dehiscence circumcissile. Seeds *ca.* 1, minute, pale yellow-brown, compressed ellipsoid-rectangular, ventrally separated into 4 distinct areas by 4 impressed lines radiating from the hilum, rugulose and reticulate.

Common, in plains of deciduous forests. Fl. & Fr.: Oct. – Jan. Batrepalli (ATP), *KRKS* & *SSR* 37881.

MITRAGYNA Korth. *nom. cons.*

Mitragyna parvifolia (Roxb.) Korth., Observ. Naud. Indic. 19. 1839; FPM 2: 585. 1921; FAP 2: 488. 1997. *Nauclea parvifolia* Roxb., Pl. Coromandel t. 52. 1796 & Fl. Ind. 1: 513. 1832. *Stephegyna parvifolia* Korth., Verh. Nat. Gesch. Ned. Bezitt. Bot. 161. 1840; FBI 3: 25. 1880.

Local name: *Batta genupu.*

Deciduous tree, up to 12 m high; branches 4-angular, lenticellate, grey. Leaves oblong-obovate, 5–11 × 3.5–9 cm, glabrous, base acuminate, margin entire, apex obtuse-acute; petiole to 1 cm; stipules inter-petiolar, foliaceous, membranous, oblong-elliptic, caducons, 1 × 1 cm. Inflorescence axillary, globose head, to 3 cm across; peduncle 5 mm long. Flowers white, to 150 in a head. Calyx truncate; lobes 5, triangular; 1 mm across. Corolla salver-form; tube 5 mm long; lobes 5, calyculate, 1 mm long. Stamens 5, attached near the throat, exserted. Ovary 2-celled; stigma oblong, 2 mm long. Capsule aggregated into a globose head, 1 cm across; follicles 2 in each capsule. Seeds many, flat, 1 mm, winged.

Occasional, in hilly slopes of deciduous forests near streams. Fl. & Fr.: Mar. – Aug. Kalasamudram, near Muniswara temple (ATP), *KRKS* & *BSS* 39102; Kalasamudram (ATP), *KRKS* 37961.

Note: Wood is used for building, furniture and cots.

MORINDA L.

Key to Species

1a. Leaves 6-14 cm long; cymes many-flowered; corolla tube 3 cm long .... **M. pubescens**

1b. Leaves 15-22 cm long; cymes 2-3-flowered; corolla tube 1.5 cm long .... **M. angustifolia**

Morinda angustifolia Roxb., Pl. Coromandel 3: 32. 1815; FBI 3: 156. 1880; FPM 2: 651. 1921; FAP 2: 489. 1997.

Small tree, up to 4 m high, young branches black when dry, pubescent, brown when mature, bark flaking off; wood hard, yellow. Leaves coriaceous, large, elliptic or obovate-lanceolate, 15–22 × 3–10 cm, base caudate-acuminate, margin undulate, apex shortly acuminate or acute, glabrous above, scabrid below; petiole to 2 cm long; stipules inter-petiolar, ovate, 3 × 2 mm. Cymes axillary, 2–3-flowered; peduncle to 2.5 cm long. Flowers white, to 3 cm long. Calyx merged in the head of cyme, with ovary and forms fruit. Corolla tube 1.5 cm, lobes 5, oblong. Stamens 5, alternate to the corolla lobes. Ovary included in globose head. Fruit formed by syncarpous ovaries in cymose flowers, 1–2 cm across, turbinate, black when ripe. Seeds many, inserted in pulp of fruit, black.

Common, in foot hills of the dry deciduous forests. Fl. & Fr.: May – Sept. Idupulapaya (KDP), *KRKS* & *BSS* 39076.

Morinda pubescens J.E. Smith, in Rees. Cyclop. 24: n. 3. 1813.

Key to Varieties

1a. Leaves narrowly linear-oblong .... var. **stenophylla**

1b. Leaves broadly obovate-elliptic .... var. **pubescens**

Morinda pubescens J. E. Smith var. **pubescens** Verdc., Kew Bull. 37: 543. 1983; FAP 2: 490. 1997. *M. tinctoria* Roxb., Fl. Ind. 2: 197. 1824, non Noronha 1790; FBI 3: 156. 1880; FPM 2: 652. 1921.

Local name: *Togara, Bandamaddi.*

Small tree, up to 6 m high; branches pubescent, greenish brown when dry; bark corky, longitudinally cracked, greyish brown; wood red. Leaves ovate, elliptic-oblong, 7–12 × 4–7 cm, pubescent, prominently nerved, base acute, margin undulate-entire, apex acuminate; stipules inter-petiolar, 3-angular, 2-lobed, pubescent, 6 × 6 mm; petiole to 1 cm long, pubescent. Flowers white in globose heads, solitary or 2-nate, leaf-opposed. Calyx connate merged in head. Corolla tube 3 cm long, pubescent, lobes 5, ovate, glabrous within. Stamens 5, alternate to corolla lobes. Ovary inserted in syncarpous, false berries. Fruit syncarpium, 2 cm diam., irregularly globose, pubescent, notched. Seeds oblong-ovoid, black, 3 × 2 mm.

Common, in foot hills and wastelands of all types of forests. Fl. & Fr.: May – Oct. Thummalapalle (KDP), *KRKS* & *KRS* 33700.

Morinda pubescens J. E. Smith var. **stenophylla** (Speng.) Kumari in Henry *et al.*, Fl. Tamilnadu 2: 14. 1987; FAP 2: 490. 1997. *M. stenophylla* Spreng., Syst. Veg. 1: 749. 1824. *M. tinctoria* Roxb. var. *stenophylla* (Spreng.) Gamble, Fl. Madras 2: 652. 1921.

Shrub, up to 3 m high; stem villose, reddish. Leaves linear or oblong, 6–14 × 1–2 cm, pubescent above, villose below, base acute, margin undulate, apex acute, lateral veins 24 pairs; petiole to 1.5 cm; stipules broad, inter petiolar, surrounding the stem, 0.7 × 0.8 cm. Cymes axillary. Flowers white. Fruit 0.5 × 0.4 cm.

Very rare, in scrub forest and foot hills of the dry deciduous forest. Fl. & Fr.: May – Nov. K. K. Kottala (KDP), *KRKS* & *BSS* 40068; Batrepalli ((ATP), *KRKS* 275 (HJCB-E).

OLDENLANDIA L.

Key to Species

1a. Flowers white; peduncles 1–4-flowered.(-plants erect)....................................(2)

1b. Flowers other than white; peduncles 1–many-flowered..................................(3)

2a. Apex of capsules rounded, protruded...**O. herbacea**

2b. Apex of capsules flat, not protruded .... **O. corymbosa**

3a. Flowers in axillary capitate clusters; plants mostly prostrate, scabrid hairy ....**O. umbellata**

3b. Flowers in axillary and terminal lax cymes; plant erect, glabrous......**O. affinis**

Oldenlandia affinis (Roem. & Schult.) DC., Prodr. 4: 428. 1830. *Hedyotis affinis* Roem. & Schult., Syst. 3: 194. 1818; FAP 2: 478. 1997. *Oldenlandia dichotoma* (Koen. ex Roth) Hook. f., Fl. Brit. India 3: 67. 1880; FPM 2: 601. 1921. *Hedyotis dichotoma* Koen. ex Roth, Nov. Pl. Sp. 93. 1821.

Annual erect, slender paniculately branched herb, up to 40 cm high, branches glabrous, 1 mm dia. Leaves heteromorphic, basal ones one-whorled, spatulate, 0.8 × 0.5 cm, base attenuate, apex obtuse, apical ones on branches linear- lanceolate, 1–6 × 0.3–0.5 cm, glabrous above, sparsely pubescent below, base acute, margin entire-revolute, apex acuminate; petiole minute or sessile; stipules ovate, paired, minute, < 1 mm long. Cymes lax, axillary and terminal, helicoid; peduncles filiform, to 10 cm long; pedicels filiform, to 2–3 cm long; bract minute, < 1 mm long. Flower light blue, 2–3, distinctly with long pedicel. Calyx tube 2 mm diam., lobes 4, minute, 1 mm long. Corolla-tube to 1.5 mm long, slender; lobes 4, 2 mm across. Stamens 4, to 0.2 mm long. Ovary inferior in calyx tube, 1 mm across. Capsule 2 mm across, loculicidally dehiscent as 4 cells. Seeds many, to *ca.* 0.1 mm, brown, ovate, hexagonal marks surrouding pit.

Common, in moist areas near rocks and rocky shades in dry deciduous forests. Fl. & Fr.: Aug. – Nov. K. K. Kottala (KDP), *KRKS* 37669; Kalasamudram (ATP), *KRKS* & *BSS* 40137; Batrepalli (ATP), *KRKS* & *BSS* 40352.

Oldenlandia corymbosa L., Sp. Pl. 119. 1753; FBI 3: 64. 1880; FPM 2: 600. 1921. *Hedyotis corymbosa* (L.) Lam., Encycl. 1: 272. 1792; FAP 2: 480. 1997.

Erect or procumbent herb, to 30 cm high; branches sympodial, 4-angular, glabrous; black when dry, glossy. Leaves sessile, linear 2–3.5 × 0.2–0.3 cm, glabrous above, scabrid below, base obtuse, margin entire or revolute, apex acute, stipules ovate, to 1 mm long. Flowers white, axillary, generally 3 (solitary); pedicel to 1 cm long, filiform. Calyx-tube 1 mm long; lobes 4, opposite to corolla lobes, to 0.5 mm long. Corolla-tube 5 mm long; lobes 4, 0.5 mm long. Anthers 4 at the mouth of the tube. Ovary 0.5 mm, inferior. Capsule globose with persistent calyx, 4-celled. Seeds to 0.2 mm, ovate, blackish, rugose, dotted.

Common, in moist, shady and sandy soils in all types of forests. Fl. & Fr.: Jun. – Oct. Kalasamudram (ATP), *KRKS* & *BSS* 39513.

Oldenlandia herbacea (L.) Roxb., Fl. Ind. 1: 445. 1820; FPM 2: 601. 1921. *Hedyotis herbacea* L., Sp. Pl. 102. 1753; FAP 2: 422. 1997. *Oldenlandia heynii* R. Br. ex G. Don, Gen. Syst. 3: 531. 1834; FBI 3: 65. 1880.

Erect herb, to 30 cm high, branching dichotomous; branches 4-angular, glabrous. Leaves sessile, linear, 10–25 × 1–1.5 mm, glabrous above, distinctly scabrid below, base obtuse, margin entire or revolute, apex acuminate; stipule 1 × 2 mm long, partially 2-lobed. Flowers white, axillary, solitary (2 or 4 per node), pedicel to 2 cm long, filiform. Calyx-tube 1 mm diam.; lobes minute, ovate, 0.5 mm long. Corolla-tube 3 mm long; lobes 4, ovate, acute. Anthers 4, at tube mouth, 0.5 mm long. Ovary sub-globose. Capsule hairy, loculicidal, 2.5 × 2 mm, crown protruded, 4-celled. Seeds minute, blackish, 0.2 mm, very widely ovate, pitted.

Very Common, in all wastelands and weed of agriculture fields. Fl. & Fr.: Jul. – Sept. (sometimes Jan. – March). Thimmamma marrimanu (ATP), *KRKS* & *KRS* 37937.

Oldenlandia umbellata L., Sp. Pl. 119. 1753; FBI 3: 66. 1880; FPM 2: 601. 1921. *Hedyotis puberula* (G. Don) R. Br. ex Arn., Pugill. Pl. Ind. Or. 24. 1836; FAP 2: 483. 1997. *Oldenlandia puberula* G. Don, Gen. Syst. 3: 530. 1834. *O. wighitii* Hook. f., Fl. Brit. India 3: 66. 1880; FPM 2: 601. 1921.

Local name: *Cherivelu.*

Annual or perennial herb, prostrate much branched with woody base, up to 15 cm high; branches crowded at base, scabrid, whitish, 4-angled. Leaves linear, whorled 1.5–2 × 0.1–0.3 cm, scabrid, base acute, margin entire or revolute, apex acute; petiole 2 mm long; stipule linear, to 2 mm long. Umbels axillary, 3–5-flowered; peduncle to 1.5 cm long. Flowers bluish-white; pedicel 4 mm long; bract lanceolate, to 2 mm long; bracteole to 1 mm long. Calyx-tube 1 mm diam.; lobes 4, ovate, 1 mm long, scabrid. Corolla white, 4-lobed, 4 mm long. Stamens 4 at tube mouth, exserted, oblong. Ovary 2 mm, hairy. Capsule 4-celled, loculicidal, sub-globose, 2 × 2 mm, crown protruded. Seeds brown, 0.3 mm, hexagonal, pitted, very widely ovate.

Common, on road sides and wastelands. Fl. & Fr.: Aug. – Jan. Thummalapalle (KDP), *SSR* & *TP* 33628.

PAVETTA L.

Key to Species

1a. Leaves tomentose, elliptic-oblong; flowers in terminal cymes ..... **P. tomentosa**

1b. Leaves glabrous, obovate-oblong; flowers in terminal and axillary cymes .... **P. indica**

Pavetta indica L., Sp. Pl. 110. 1753; FBI 3: 150. 1880; FPM 2: 633. 1921; FAP 2: 495. 1997.

Bushy, large shrub, up to 3 m high; branches 4-angular, glabrous, light brown when young, black when dry. Leaves chartaceous, obovate-oblong or oblanceolate, 6–9 × 2.5–5 cm, glabrous, base acute, margin entire, apex acute or obtuse; petiole 1–2 cm long; stipules inter petiolar, connate, 0.6 × 0.4 cm, ovate. Cymes terminal and axillary, trichotomously branched corymbs; peduncle to 2 cm; pedicel to 0.5 cm. Flowers white, to 1.5 cm long. Calyx-tube 2 mm long; lobes 4, minute. Corolla-tube 1 cm long; lobes 4, obovate, recurved, pubescent within, 4 × 1.5 mm. Stamens 4, alternate to the corolla lobes; anthers extruded, 4 mm long. Ovary inferior, 1 mm across; style 2 cm long; stigma 4 mm long, bifid, swollen, oblong. Berry fleshy, globose, glabrous, 0.6 × 0.4 cm, crown 1.5 mm across. Seeds 4 in 2 locules, 4 × 2 mm, brown.

Common, in scrub and dry deciduous forests. Fl.: Jul. – Sept. & Fr.: Oct. – Jan. B. Kottapalli (ATP), *KRKS* 41822; Batrepalli (ATP), *KRKS* 41899.

Pavetta tomentosa Roxb. ex Smith in Rees, Cycl. 26: 2. 1813; FAP 2: 495. 1997; FPM 2: 1979. *P. indica* L. var. *tomentosa* (Smith) Hook.f. FBI 3: 150. 1880; FPM 2: 683. 1921.

Small tree, up to 3 m high. Leaves elliptic-oblong, 8–12 × 4.5–7 cm, base acute-rounded, margin entire, apex acute, softly pubescent above, tomentose below; petiole 1.5 cm; stipules connate, to 1 cm, prominant. Flowers white in terminal, trichotomously branched corymbose cymes. Calyx tomentose, teeth minute. Corolla lobes 4, oblong, tube 1–1.3 cm long, greenish. Style far exserted. Fruit a drupaceous berry, globose or subglobose, didymous.

Common, in foot hills of forests. Fl. & Fr.: May – Feb. Nigidi forest, Kalasamudram (ATP), *KRKS* 40407.

PSYDRAX Gaertn.

Psydrax umbellata (Wight) Bridson, Kew Bull. 48 (4): 762. 1993. *Canthium umbellatum* Wight, Icon. Pl. Ind. Orient. 3: t. 1034. 1845;FBI 3: 132. 1880. *Canthium dicoccum* var. *umbellatum* (Wight) Santapau & Merchant, Bull. Bot. Surv. India 3: 107. 1962; FAP 2: 469. 1997. *Plectronia didyma* (Gaertn.) Kurz. var. *umbellata* (Wight) Gamble, Fl. Madras 2: 624. 1921.

Local name: *Nalla manthanam, Alli, Balusu, Nalla balusu, Nakkina, Nakkeri.*

Evergreen tree, up to 10 m high; wood pale yellow; young branches 4-angled, glabrous, green; stipules lanceolate-ovate 0.4 × 0.5 cm. Leaves coriaceous, elliptic-ovate, obovate, 5–13 × 4–7 cm, glabrous, shiny green above, dull green below, base acute, attenuate, margin entire, apex acuminate; lateral veins 7 pairs; petiole 1–1.5 cm, glabrous. Inflorescence axillary, pedunculate umbels to 5 cm across. Flowers whitish, to 3 mm across. Calyx cupular, tube 0.3 cm across; lobes 5, 1 mm long, glabrous. Corolla-tube 3 mm long, villous within; lobes 4, oblong, 4 mm long, acute. Stamens 4, attached at apex of tube, 2 mm long. Ovary 2 mm long, inserted in calyx tube, 2-celled. Drupe obovoid, 0.6 × 0.7 cm, brown, two-seeded. Seeds obovate, granular, 4 × 3 mm.

Common, in dry deciduous forests. Fl.: Feb. – Apr. & Fr.: Jun. – Sept. Batrepalli (ATP), *KRKS* 37943.

SPERMACOCE L.

Key to Species

1a. Leaves oblong-obovate, margin entire; stipules pectinate................................(2)

1b. Leaves linear-lanceolate, margin revolute; stipules not pectinate...... **S. pusilla**

2a. Stem terete with densely pilose hairs; leaves highly flexuous; corolla-tube equalling the lobes.. **S. hispida**

2b. Stem 4-angled, with minutely appressed hairs; leaves rarely flexuous; corolla-tube 3-4 times longer than lobes, occasionally equal...**S. articularis**

Spermacoce articularis L.f., Suppl. Pl. 119. 1782; Sivarajan & Vasu, Taxon 35: 366. 1986. *Borreria articularis* (L.f.) F.N. Williams, Bull. Herb Boissier, Ser. 2. 5 : 956. 1905.

Local name: *Madanak.*

An annual, erect herb, up to 25 cm high, stem 4-angled, glabrous. Leaves opposite, decussate, elliptic-obovate, 1.5-4 × 1-2 cm, base acute-attenuate, margin entire, apex acute, subsessile; stipules connate with the base of leaves into a sheath, pectinate. Flowers in axillary and terminal fascicles, pinkish white. Calyx-tube up to 0.3 cm long, lobes 4. Corolla funnel-shaped, lobes 4. Stamens 4, exserted. Style filiform, up to 0.75 cm long, stigma capitate. Capsule globose, hispid, 0.5 × 0.2 cm; seeds narrow, oblong, testa reticulate.

Common, grows as weed and in rocky crevices of the forests. Fl. & Fr.: Jun – Feb. Kalasamudram RF (ATP), *NY* 651.

Spermacoce hispida L., Sp. Pl. 102. 1753; FBI 3: 200. 1880; FAP 2: 498. 1997. *Borreria hispida* (L.) K. Schum. in Engl. & Prantl., Pflanzen. Fam. 4(4): 144. 1891; FPM 2: 654. 1921.

Annual, erect, hispid herb, sometimes it grows decumbent, up to 30 cm high; branches obscurely 4-angled. Leaves ovate-obovate, 2.3–4 × 1–1.8 cm, base acute-attenuate, margin entire-undulate, apex acute; subsessile or petiolate to 2 mm; stipules connate, interpetiolar, ciliate with sheath, pectinate, 0.5 × 0.5 cm. Inflorescence axillary fascicle, few-flowered. Flowers whitish-pink, to 17 in a cluster. Calyx-tube to 1 mm long, hispid without; lobes 4, equal, obovate, acute-apiculate. Corolla-tube 2.5 mm long, with a ring, hairs below within; lobes 4, oblong, acute. Stamens 4, exserted. Capsule hispid, 0.5 × 0.3 mm, sub-globose. Seeds broadly ellipsoid, coarse, closely pitted, black.

Common, in forests and wastelands. Fl. & Fr.: Throughout the year. Batrepalli (ATP), *KRKS* & *SSR* 36961; Kalasamudram (ATP), *KRKS* & *KRS* 36482.

Spermacoce pusilla Wall. in Roxb., Fl. Ind. 1: 379. 1820; FAP 2: 499. 1997. *Borreria stricta* (L. f.) K. Schum., in Engl. & Prantl. Pflanzen. fam. 4 (4): 143. 1891; FPM 2: 654. 1921. *Spermacoce stricta* L. f., Suppl. 120. 1781; FBI 3: 200. 1880.

Erect herb, up to 30 cm high; branches 4-angular, hispid-scabrid. Leaves sessile, lanceolate, hispid-scabrid, base truncate, margin entire-revolute, apex acute; stipules broad, truncate with marginal bristles. Inflorescence axillary fascicle. Flowers pink, to 3 mm long. Calyx truncate, villose without; lobes 4, linear, acute. Corolla tubular; tube narrowly cylindric, 1.5 mm long; lobes 4, oblanceolate, acute. Stamens 4, longer than corolla lobes. Capsule oblong, 3.5 mm. Seeds 2 mm long, longitudinally striated.

Common, in foot hills of all forests and wastelands. Fl. & Fr.: Aug. – Nov. Batrepalli (ATP), *KRKS* & *BSS* 40359.

TARENNA Gaertn.

Tarenna asiatica (L.) Kuntze ex K. Schum., Bot. Tidsskr. 24: 332. 1902. *Chomelia asiatica* (L.) Kuntze, Revis. Gen. Pl. 1: 278. 1891; FPM 2: 613. 1921; FAP 2: 472. 1997. *Rondeletia asiatica* L., Sp. Pl. 172. 1753. *Webera corymbosa* Willd., Sp. Pl. 1: 1224. 1798; FBI 3: 102. 1800.

Local name: *Komari, Papidi, Konda papidi, Komi.*

Evergreen shrub to small tree; branches 4-angled, shiny and gummy when young, greyish brown when mature, apical bud gummy. Leaves glossy above, turn black when dry, coriaceous, oblong or elliptic-oblong, 4–15 × 2–5 cm, glabrous, base acute, margin entire, slightly revolute, apex acute-acuminate; petiole to 1.5 cm long; stipule inter petiolar with broad persistent triangular base, cuspidate blade. Inflorescence terminal corymbose cyme, to 11 cm long. Flowers white, fragrant, 1.5 cm across. Calyx campanulate; lobes 5, 1 mm long, puberulous without. Corolla salver-form; tube 5 mm long, villous within; lobes 5, oblong, 3.5 mm long, flexed. Stamens 5, between the corolla lobes, exerted, filament 0.5 mm long; anther linear, 3 mm long. Ovary bilocular, 1 mm diam. Berry green when young, black when mature, globose, 6 mm diam., with 2 mm long crown of calyx lobes at apex. Seeds 6–8, black, 2 mm diam., rugose.

Common, in foot hills of dry deciduous forests and scrub forests. Fl.: Jul. – Sept. & Fr.: Oct. – Jan. Somulavaripalli (ATP), *KRKS* 37944; Batrepalli (ATP), *KRKS* 37823.

ASTERACEAE

Key to Genera

1a. Flowers all ligulate; sap milky; pappus fine capillary or silky .... (2)

1b. Flowers all tubular or the outer only ligulate; sap watery; pappus other than silky .... (3)

2a. Achenes slightly or perfectly beaked and narrowed at apex .... **Launaea**

2b. Achenes not beaked, truncate and narrowed at both ends .... **Sonchus**

3a. Involucral bracts 1-seriate .... (4)

3b. Involucral bracts more than 1-seriate .... (5)

4a. Flowers homogamous, not rayed; leaves dimorphic .... **Emilia**

4b. Flowers heterogamous, rayed; leaves not dimorphic .... **Senecio**

5a. Heads homogamous, not rayed; pappus strictly present .... (6)

5b. Heads heterogamous, rayed or not (except *Lagascea*); pappus present or absent .... (12)

6a. Leaves opposite, rarely upper ones alternate .... **Ageratum**

6b. Leaves alternate .... (7)

7a. Involucral bracts not spinescent; style branches elongated .... (8)

7b. Involucral bracts spinescent; style branches not elongated .... (9)

8a. Involucral bracts many *ca*. 20–40; heads compact, many-flowered *ca*. 30–45 .... **Cyanthillium**

8b. Involucral bracts not many *ca*. 10; heads loosley arranged, *ca*. 6-flowered .... **Blepharispermum**

9a. Style branches thickened towards base .... (10)

9b. Style branches not thickened towards base .... **Dicoma**

10a. Heads compound; leaves spinescent, deeply lobed .... **Echinops**

10b. Heads solitary; leaves simple, lyrate, lobed .... (11)

11a. Leaves spinescent; flowers purple; branches decumbent .... **Amberboa**

11b. Leaves not spinescent; flowers lilac; branches erect .... **Tricholepis**

12a. Receptacle epaleaceous .... (13)

12b. Receptacle paleaceous .... (19)

13a. Anther base obtuse .... **Grangea**

13b. Anther base sagittate or tailed .... (14)

14a. Heads aggregated into dense globose or ovoid clusters .... **Sphaeranthus**

14b. Heads not aggregated into clusters .... (15)

15a. Stems winged .... **Epaltes**

15b. Stem not winged .... (16)

16a. Heads rayed .... (17)

16b. Heads not rayed .... (18)

17a. Pappus 2-seriate, outer row of scales forming cup, inner of setaceous free hairs .... **Pulicaria**

17b. Pappus 1-seriate, of setaceous free hairs .... **Pentanema**

18a. Involucral bracts broad, scarious .... **Gnaphalium**

18b. Involucral bracts other than above .... **Blumea**

19a. Leaves opposite, all cauline .... (20)

19b. Leaves alternate, radical and cauline .... (26)

20a. Inner involucral bracts partly or completely enclosing the achenes of ray florets; pappus of ray and disk florets absent; bisexual florets sterile, ray florets fertile .... **Acanthospermum**

20b. Inner involucral bracts do not enclose the achenes of ray florets, almost free from achenes; pappus of either ray and disk florets present or rarely absent; bisexual florets fertile, ray florests usually sterile .... (21)

21a. Heads one-flowered, clusterd together or aggregated in compound fashion, subtended by leafy bracts, ligules absent .... **Lagascea**

21b. Heads separate, not aggregated in compound fashion, not subtended by leafy bracts; ligules present .... (22)

22a. Pappus of awns .... (23)

22b. Pappus of scales, bristles or absent .... (24)

23a. Pappus awns plumose, achenes densely silky .... **Tridax**

23b. Pappus awns retrorsely barbellate, achenes hispid or glabrous .... **Bidens**

24a. Receptacle conical .... **Acmella**

24b. Raceptacle flat, or slightly convex .... (25)

25a. Bisexual flowers yellow, palea broad, dentate at apex .... **Blainvillea**

25b. Bisexual flowers white, palea narrow, acute to gland-tipped .... **Eclipta**

26a. Prostrate, decumbent or scapigerous low herbs .... **Glossocardia**

26b. Erect herbs to sub-shrubs .... (27)

27a. Heads unisexual, female heads eligulate, 2-flowered, covered by hooked spines, anemophilous .... **Xanthium**

27b. Heads bisexual, female florets ligulate, at periphery, not spiny.... **Parthenium**

ACANTHOSPERMUM Schrank

Acanthospermum hispidum DC., Prodr. 5: 522. 1836; FPM 2: 704. 1921; FAP 2: 509. 1997.

Local name: *Kukkamullu.*

Erect herb–under shrub, to 40 cm high, branches sparsely pubescent, dichotomously branched. Leaves opposite, subsessile, pubescent, elliptic obovate, 3–8 × 1–3 cm, base attenuate, margin entire-dentate towards apex, apex apiculate. Heads solitary, axillary in forks, 1 cm across. Florets heterogamous, yellow; rayed, subsessile. Involucral bracts 2-seriate; outer 5, elliptic to oblong, 4–6 mm long, basally connate, pubescent above, ciliate along margin; inner connate, 4–6 mm long, enveloping the ovary of ray florets, compressed, prickly, apically 2-awned, awns divaricate, apex curved, 3–5 mm long; receptacle slightly convex, 2–3 mm long. Ray florets 8, tube minute, glabrous; ligule elliptic, 1.5 mm across, 3-veined, glandular on dorsal side; style 1.5 mm long, recurved, papillate. Bisexual florets 7; tube 1 mm, cylindrical, pubescent; limb 1 mm, 5-lobed, lobes ovate, elliptic, 0.8–1 mm long, pubescent, subacute; stamens 5, exserted to the level of lobes; style not exserted, papillate. Pappus absent. Achene of ray obovate, 5 mm long, glabrescent, spiny, enclosed with echinate inner achenes of bisexual florets, 1 mm long, glandular, sterile, apex truncate.

Common, in wastelands. Fl. & Fr.: Throughout the year. K. K. Kottala (KDP), *KRKS* & *SSR* 36443.

Note: It is a native of South America.

ACMELLA Pers.

Acmella paniculata (Wall. ex DC.) R.K.Jansen, Syst. Bot. Monogr. 8: 67. 1985. *Spilanthes paniculata* Wall. ex DC., Prodr. 5: 624. 1836; FAP 2: 538. 1997. var. *paniculata*; FBI 3: 307. 1881.

Decumbent or prostrate herb, to 1 m long, rooting at lower side; stem striated, fistular, glabrous in lower parts, sparsely pilose when young, branches trichotomous, rarely dichotomous. Leaves simple, opposite, ovate-lanceolate, 2–3.5 × 1–2.5 cm, pilose, base attenuate, margin distantly serrate, apex acute; petiole 0.5–1.5 cm long, pilose. Heads 2 or 3, terminal and axillary, cone-shaped, yellow, 0.8 × 0.5 cm, homogamous, not rayed; peduncle to 9.5 cm long. Involucral bracts 10, 2-seriate, ovate, 4 mm long, 3-veined with distinct midvein, sparsely hairy on dorsal side, margin ciliate, inner ones scarious on margins, apex dentate towards apex; palea boat-shaped, 5 mm long, vertically folded, receptacle oblong, white, rugose. Florets around 80, 2 mm long, glabrous; corolla 4–5-lobed, lobes ovate, papillate, subacute. Stamens 4-5, anthers oblong, 1 mm, hood ovate, obtuse, base sagittately auricled; style 1.5 mm, branches linear. Pappus of two unequal bristles, half to that of floret in length. Achenes dimorphic, marginal trigonous, inner ones laterally compressed, black, hairy at margin, ciliate, apex slightly obcordate.

Rare near moist places and streams in dry deciduous forest. Fl. & Fr.: Jul. – Nov. Namalagundu (KDP), *KRKS* & *BSS* 39655.

AGERATUM L.

Ageratum conyzoides L., Sp. Pl. 938. 1753; FBI 3 : 243. 1881; FPM 2 : 672. 1921; FAP 2 : 510. 1997. var. **conyzoides.**

Erect, aromatic herb-undershrub, 30–50 cm high; stem terete, whitish pilose. Leaves opposite, upper ones alternate, rhomboid ovate, 4–8 × 3–5 cm, distinctly sericeous, base truncate to acute, margin dentate, apex acute; petiole 2–4 cm long. Heads terminal, white, 4–6 in corymbs, 3–4 mm across, homogamous, pedicels 1 cm. Involucral bracts 2–3-seriate, 18–24; outer oblong-lanceolate, scarious, acute; inner ones broadly lanceolate, scarious, ciliate to serrate, short acuminate; receptacle convex, 1–2 mm across, glabrous. Florets 50–60 in a head; corolla whitish, tube glandular, 1.5-2 mm long, 5-lobed, acute; stamens 5, anthers hooded; style 2–2.5 mm long, exserted; stigmas 2, papillate. Pappus of 5 scales, lanceolate, 2 mm, plumose, along the margins. Achenes 2 mm long, black, 5-angled, glabrous or scabrid.

Common, in wastelands and agricultural lands especially near water. Fl. & Fr.: Feb. – July. Batrepalli (ATP), *KRKS* 37954.

AMBERBOA (Pers.) Less.

Amberboa ramosa (Roxb.) Jafri, Scientist 3: 29. 1959. *Carduus ramosus* Roxb., Fl. Ind. 3: 407.1832. *Volutarella ramosa* (Roxb.) Sant., Pl. Saurashtra 22.1953. *V. divaricata* (DC.) Benth. & Hook.f., Gen. Pl. 2: 416. 1873; FBI 3: 383. 1881;FPM 2: 727. 1921. *Microlonchus divaricatus* DC., Prodr. 6: 562. 1838. *Tricholepis procumbens* Wight, Ic. t. 1139. 1846. *Oligochaeta ramosa* (Roxb.) Wagenitz, Verroffent. Geobot. Inst. 37: 323. 1962; FAP 2: 532. 1997.

Annual decumbent herbs, dichotomously branched; branches 25–50 cm long, angled, smooth or scabrid. Leaves usually clasping to the stem, oblong or obovate, entire, toothed or pinnatifid, 2.5–5 × 0.6–1.5 cm; lobes mucronate, often undulate or crisped. Heads *ca.* 2 cm across. Involucral bracts 8–9 seriate, recurved spinescent awns, spines *ca.* 6–8 mm long; outer ones ovate-lanceolate, 3–7 mm long, pubescent, margin thin, scarious, whitish pubescent, apex spinescent; middle ones linear-lanceolate, spinescent; inner most oblong to oblong-lanceolate, to 16 mm long, margins scarious, spinescent, acuminate, white tomentose. Receptacular bristles short. Corolla *ca.* 1.2 cm long, straight, pale purple. Achenes narrow *ca.* 5 mm long, 4–5-angled, grooved and punctate between angles. Pappus silvery brown, *ca.* 1.2 cm long.

Rare in foot hills of the forests and scrub jungles in black soil. Plant somewhat hard to touch. Mabbuchintalapalle (KDP), *KRKS* 42041.

BIDENS L.

1. Ligule white; outer involucral bracts narrowly spatulate .... **B. bipinnata**
1. Ligule yellow; outer involucral brcts linear or linear-lanceolate.... **B. biternata**

Bidens bipinnata L., Sp. Pl. 832. 1753; FAP 2: 512. 1997. *Bidens pilosa* L. var. *bipinnata* (L.) Hook. f., Fl. Brit. Ind. 3: 309. 1881.

Erect herb, to 1.2 m high; stems quadrangular, sericeous at internodes, pilose at nodes. Leaves opposite, petiolate, to 6 cm, sparsely pilose, lower ones bipinnate compound, upper most once pinnate, to 12 cm long, lateral segments 1–2 pairs, ovate-lanceolate, 2–3 × 1–5 cm, base acute to attenuate, margin dentate, apex acuminate. Heads solitary, terminal on divaricate branches, to 1.6 cm across, heterogamous, rayed, peduncle to 10 cm long. Involucral bracts around 21, sub-equal or outer slightly shorter than inner; outer ones 13, narrowly oblanceolate, 5 mm, 3-veined, glandular, acute at apex; inner ones 8, ovate-elliptic, brown, sparsely pilose, glandular, obtuse at apex. Receptacle 4–5 mm across, flat-slightly concave, paleaceous. Ray florets 5; corolla white, tube 1.5 mm long, glabrous; ligule 5 mm long, broadly elliptic, 5-nerved, retuse or minutely 3-lobed. Bisexual florets around 40; corolla-tube 1.2 mm long, cylindrical, glabrous, limb 3 mm across, veined, sparsely hairy along veins, 5-lobed; lobes 0.4 mm long, triangular; stamens 5, anthers 1.5 mm long, hood ovate, triangular; style 5 mm, branches 1 mm. Pappus 2–3, retrorsely barbellate, awns, 3 mm long, divergent. Achene of ray 5 mm long, linear, sterile, hispid, 4-ribbed. Achenes of bisexual florets linearly elliptic, 10–12 mm long, narrowed at both ends, brown in lower half, glabrous, 4-ribbed, awns 2, 2.5 mm long.

Common, in scrub forests and weed in ground nut crop. Fl. & Fr.: Sept. – Dec. Rachakuntapalle (KDP), *KRKS* 37621.

Bidens biternata (Lour.) Merr. & Sherff, Bot. Gaz. 88: 293. 1929. *Coreopsis biternata* Lour., Fl. Cochinch. 508. 1790. *Bidens pilosa* auct. non L. FBI 3: 309. 1881

Erect herb, to 1 m, stems quadrangular, striated, internodes to 30 cm, glabrous. Leaves opposite, petiolate, sparsely pilose, lower one bipinnately compound, uppermost one pinnate, to 12 cm long, bipinnate one with 2-3 lateral secondaries, lateral segments 1-2 pairs, terminal one 2-3-lobed, segments one-veined from the base, glabrous or sparsely pilose on both surfaces. Heads solitary, terminal on divaricate branches, yellow, heterogamous, rayed. Involucral bracts around 21, subequal. Receptacle 4-5 mm across, flat to slightly concave, paleaceous. Ray florets 5. Corolla 6.5 mm, tube 1.5 mm, glabrous, ligule 5 mm, strongly 5-nerved, sessile, glandular and sparsely hairy along veins on dorsal side, retuse or minutely 3-lobed. Bisexual florets around 40. Corolla 4.2 mm, tube 1.2 mm, cylindrical, glabrous, 5-lobed, lobes triangular, acute. Stamens 5, black, anthers 1.5 mm, hood ovate, acute, base sagittate, auricles free, obtuse; style 5 mm, branches 1mm, elongate, hairy on dorsal side, acute. Pappus barbellate, awns 3 mm. Achenes of ray 5 mm, linear, sterile, sparsely hispid at apex, 4-ribbed, colorless.

Open forests near and around streams along hilly slopes. Fl. & Fr.: Nov. – Jan. Kalasamudram RF (ATP), *TP* & *EC* 2847.

BLAINVILLEA Cass.

Blainvillea acmella (L.) Philipson, Blumea 6: 350. 1950; FAP 2: 514. 1997. *Verbesina acmella* L., Sp. Pl. 901. 1753. *Blainvillea latifolia* (L. f.) DC. in Wight, Contrib. Bot.

Indica 17. 1834. & Prodr. 5. 492. 1836; FBI 3: 305. 1881. *Eclipta dichotoma* L. f., Suppl. 378. 1781. *Blainvillea rhomboidea* Cass., Dict. Sci. Nat. 29: 494. 1823; FPM 2: 706. 1921.

Erect herb, to 1 m high; stem terete, striated, hispid. Leaves simple, opposite-alternate, elliptic or ovate-lanceolate, 5–7 × 2–5 cm, hispid, rough hairy on both sides, base acute, margin dentate or serrate, apex acuminate, 3-veined at base; petiole to 2 cm long, hispid. Heads solitary, terminal in the forks of dichotomies or in upper axils, white 6 mm across, heterogamous, rayed. Peduncles to 1.5 cm long. Involucral bracts 8, 2-seriate; outer longer than inner ones, ovate, broadly ovate-elliptic of lanceolate, 3–6 mm long, hispid, apex sharply acute, 2-lobed; inner ones scarious, 3–5-veined, pilose, acute, 2–3-lobed. Receptacle 2–2.5 mm across, convex, slightly conical, paleaceous, palea obovate, 5 mm long, scarious, folded, 3-veined, setose hairy, apex dentate or 3-lobed. Ray florets 5; corolla white, tube 1.2 mm, glabrous, ligule 0.8 mm long, 2–3-lobed; lobes obtuse; style 2 mm long, stigma awns broad, subacute. Bisexual florets 12; corolla white, tube 1 mm, glabrous, limb 1 mm long, 5-lobed; lobes ovate; stamens 5, black, anthers oblong-elliptic, hood broad, obtuse; style 1.5 mm long, stigma awns 0.3–0.4 mm, papillate. Pappus unequal, short bristles, hispid. Achenes obovate, 3–5 mm long, dimorphic, ray triquetrous, disk compressed, black, whitish pubescent, wrinkled, hispid at apex, truncate.

Common, weed in ground nut crop and field bunds. Fl. & Fr.: Jun. – Nov. Kalasamudram (ATP), *KRKS* & *BSS* 39550, 39830.

BLEPHARISPERMUM Wight

Blepharispermum subsessile DC. in Wight, Contr. Bot. India 12. 1834; FBI 3: 276. 1881; Pant in Hajra *et al.*, Flora of India 13: 43. 1995.

Local Name: *Kukundara, Gangaapatri*

Erect glabrous herb, to 90 cm high; stems clustered from the common root stock. Leaves simple, alternate, glaucous beneath, glabrous, elliptic oblong, tapering, 3–6.5 ×1–3 cm, petiole to 2–3 mm long, midvein prominent, lateral veins 6–7 pairs, base acute, margin entire, apex acute. Heads aggregated into globose cluster. Bracts of common receptacles leaf-like, 10 in two whorls alternating, narrowly oblong, 1–2 × 0.5–1 cm, one-veined, glabrous, acuminate; bracts of individual heads 1, glabrous, elliptic, *ca.* 5 mm long, one-veined, apex obtuse, axial ones 5, lanceolate, *ca.* 9 mm long one-veined. Female florets 2, tubular, 3 mm long, glandular, 3 – 4-lobed, lobes ovate, obtuse; style 3.5 mm, exserted, branches broad, recurved, glabrous. Bisexual florets 4, 4.5 mm long, glandular above the middle, 5-lobed, lobes lanceolate, 1.8 mm, subacute. Stamens 5, anthers 2 mm, linear-oblong, hood obtuse, sagittately auricled. Style 5 mm, branches 1 mm, recurved, glabrous. Pappus of unequal bristles. Achenes obovoid, 3.5–4 mm, black, trigonous, densely ciliate along the margins, slightly obcordate at apex.

Rare in dry deciduous forest. Fl. & Fr.: Nov. – Jan. Batrepalli (ATP), *KRKS* 276 (HJCB-E).

Note: Only two individuals have been found on slopes of dry deciduous forests of Nigidi. It has anti-inflammatory properties; used internally and externally for rheumatic pans.

BLUMEA DC. *nom. cons.*

Key to Species

1a. Heads solitary, axillary; pappus yellowish white................................**B. obliqua**

1b. Heads clustered or paniculate, terminal; pappus white. ....(2)

2a. Heads sessile in interrupted or compact, globose clusters in spicate panicles; receptacle pubescent, slightly concave ....**B. fistulosa**

2b. Heads peduncled, in lax panicles; receptacle minutely pilose......... **B. laciniata**

Blumea fistulosa (Roxb.) Kurz, J. Asiat. Soc. Bengal 46: 187. 1877; FAP 2: 515. 1997. *Conyza fistulosa* Roxb., Fl. Ind. 3: 429. 1832. *Blumea glomerata* DC. in Wight, Contrib Bot. Ind. 15. 1834; FBI 3: 262. 1881; FPM 2: 686. 1921. *B. nodiflora* Hook. f., Fl. Brit. India 3: 262. 1881.

Erect herb, to 1 m high; stem terete, brown, villous when young, pubescent when mature. Leaves alternate, simple, sessile, elliptic-oblanceolate to oblanceolate, 1.5–5.5 × 0.5–2.3 cm, base attenuate, clasped to stem, margin distantly dentate, apex acute; one-veined. Heads terminal in compact, globose clusters on spicate panicles, 4–5 mm acorss, yellow, heterogamous, not rayed, sessile; panicle 5 cm long. Involucral bracts linear-lanceolate, to 5 mm long, purplish tinged, slightly longer than florets, pubescent, glandular on dorsal side. Receptacle flat to slightly concave, 1.5–2 mm across, pubescent. Female florets filiform, 3.5 mm long, glabrous, corolla 2–3-lobed; style 3.5 mm long, stigma awns linear. Bisexual florets 3.5–4 mm, glabrous, corolla 5-lobed; lobes ovate-triangular, pubescent, papillate and acute at apex; stamens 5, anthers 1.5 mm, hood ovate, acute, base sagittately auricled; style 3.5 mm long, stigma awns linear, acute. Pappus 3.5 mm long, uniseriate, white, slender, setose. Achene oblong, to 1 mm, brown, puberulous, 4-angled.

Common, in wastelands and edges of streams. Fl. & Fr.: Jul. – Oct. Batrepalli (ATP), *KRKS* & *BSS* 39631.

Blumea laciniata (Wall. ex Roxb.) DC., Prodr. 5: 436. 1836; FBI 3: 264. 1881; FPM 2: 688. 1921; FAP 2: 516. 1997. *Conyza laciniata* Wall. ex Roxb., Fl. Ind. 3 427. 1832.

Annual, erect herb, up to 1.2 m high; stem brown, stout, covered with glandular hairs. Leaves 5–12 × 1–10 cm, lower leaves lyrately lobed, petioled; upper ones obovate, sessile, clasped to stem, tapering towards base, margin entire to coarsely dentate, apex acute. Heads yellow, in lax, terminal panicles, 5–7 mm across; involucral bracts linear-lanceolate, white-hairy with prominent midrib. Receptacle minutely pilose. Corolla of female florets 3–4.5 mm long, 2–3-lobed, glabrous; of bisexual florets 4–5 mm long, 5-lobed; lobes papillate, pubescent. Achenes *ca.* 1 mm long, brown, ribbed, sparsely pubescent.

Rare in edges of streams and rocky crevices. Fl. & Fr.: Dec. – Apr. Chinnapalli, (ATP), *KRKS* & *BSS* 37993.

Blumea obliqua (L.) Druce Rep. Bot. Soc. Exch. Club Brit. Isles 4 (2): 609. 1917; FAP 2: 517. 1997. *Erigeron obliquus* L., Mant. Pl. Altera 573. 1771. *Blumea amplectens* DC. in Wight Contr. Bot. India 13. 1834; FBI 3: 260. 1881; FPM 2: 681. 1921.

Annual erect herbs, to 70 cm high; dichotomously branched; stems branched, yellowish, puberulous. Leaves ovate-oblong to oblanceolate or elliptic oblong to lanceolate, 0.5–5 × 0.2–2 cm, base half amplexicaul to obtuse, margins serrate-dentate, apex apiculate, scabrid to velutinous on both surfaces. Heads terminal and axillary, solitary or two, pinkish, heterogamous, 6–10 mm in diam., long pedunculate. Involucral bracts linear-lanceolate, 1–8 mm long, hairy on dorsal surface, reflexed at maturity; outer ones apiculate; inner acuminate, often tinged with purple, margins scarious. Receptacle slightly convex, areolate, glabrous. Corolla of bisexual florets yellow, tubular, 4–5 mm long, 5-lobed; lobes acute, papillate, pubescent; corolla of female florets filiform, 3-4 mm long, 2–3-lobed, glabrous. Achenes dark brown, oblong, not ribbed, pubescent. Pappus yellowish white, 3–4 mm long.

Common, wastelands and roadsides and weed in agricultural fields. Fl. & Fr.: Jul. – Mar. Kalasamudram (ATP), *EC* 1164.

CYANTHILLIUM Blume

Key to Species

1a. Leaves silky; heads to 1 cm across; grows only in forest .... **C. albicans**

1b. Leaves green pubescent; heads to 0.5 cm across; grows in all places .... **C. cinerea**

Cyanthillium albicans (DC.) H. Rob., Proc. Biol. Soc. Washington 112(1): 229. 1999. *Vernonia albicans* DC., in Wight, Contrib. Bot. Ind. 6. 1834; FPM 2. 473. 1921; 1995; FAP 2: 540. 1997.

Erect herb to 1.5 m high; stem terete, striated, solid, tomentose and sessile glandular, branched. Leaves simple, alternate, petiole 2–5 mm long, ovate or elliptic-lanceolate, 2–5 × 0.4–4 cm, sparsely pilose, yellowish pilose above, whitish, tomentose below, base cuneate, margin irregular dentate, apex acute, venation prominent below. Heads in corymbs, terminal, to 1 cm across, pinkish-purple, bracteate, homogamous, not rayed, peduncle to 1.5 cm long, densely white tomentose and glandular. Involucral bracts around 40, 3–4-seriate; outer ones short, 1.5–3 mm long, linear-lanceolate, rigid, one-veined, pubescent, yellow glands on dorsal side; inner ones broad, elliptic-lanceolate, 5 mm long, scarious, 3-veined, hairy and sessile glandular on dorsal side, ciliate along margins, apiculate to sub-acuminate. Receptacle 2.5 mm across, slightly convex and flat, alveolate, glabrous. Florets around 45, pinkish-purple; corolla 6–7 mm long, tube 3 mm long, glabrous, limb 4 mm long, sessile glandular, deeply lobed; lobes 5, 2 mm long, lanceolate, unicellular hairy and sessile glandular on dorsal side, papillate and acute at apex; stamens 5, anthers 2 mm long, inserted, yellow, hood triangular, ovate, acute, base sagittately auricled, auricles acute; style to 7 mm long, stigma awns 1.5–2 mm long, acute. Pappus 2-seriate, outer of short persistent scales, united at base, free and ciliate at apex, inner ones 32, free, slender, white, deciduous, setaceous. Achenes oblong or obovate, 1–1.2 mm long, brown, ribbed, densely white hairy along ribs and glandular between the ribs and at base.

Occasional, in dry deciduous forest at higher altitudes. Fl. & Fr.: Oct. Jan. Idupulapaya (KDP), *KRKS* & *SSR* 39064; Chinnapalli (ATP), *KRKS* & *BSS* 39508, 40374.

Cyanthillium cinereum (L.) H.Rob., Proc. Biol. Soc. Washington 103: 252. 1990. *Conyza cinerea* L., Sp. Pl. 862. 1753. *Vernonia cinerea* (L.) Less., Linnaea 4: 291. 1829; FBI 3: 233. 1881; FPM 2: 676. 1921; FAP 2: 541. 1997.

Local name: *Sahadevachettu.*

Erect herb, to 60 cm high; stem striated, puberulous and glandular. Leaves simple, alternate, subsessile, broadly elliptic, 1.5–5 × 0.4–5 cm, sparsely puberulous on both surfaces, base cuneate, attenuate, margin entire, undulate, apex obtuse, apiculate, petiolate, to 1.5 cm long. Heads in terminal, corymbose, divaricate cymes, 5 mm across, bluish-purple, bracteate, homogamous, not rayed. Pedicel to 1.5 cm long, puberulous. Involucral bracts 24, 4-seriate, unequal; outer ones 1.5–3 mm; inner ones broad, rigid, elliptic-lanceolate, 3.5 mm long, pubescent, yellow glands on dorsal side, ciliate at margin, acuminate at apex. Receptacle 1 mm across, convex slightly flat, naked. Florets around 30; corolla bluish-purple; tube 2 mm long, glabrous, shortly-lobed, to 0.5 mm long, ovate, glandular on dorsal side, minutely papillate and acute at apex; stamens 5, anthers 0.5 mm, hood ovate; style 3 mm long, branches 0.5 mm long, acute. Pappus 2-seriate; outer of short, free scales; inner ones around 30, free, 3 mm long, white, deciduous, setaceous slender hairs. Achenes obovate, 1 mm long, densely hairy, black.

Common, in wastelands and open forests. Fl. & Fr.: Aug. – Apr. Idupulapaya (KDP), *KRKS* & *BSS* 40196.

DICOMA Cass.

Dicoma tomentosa Cass., Bull. Sci. Soc. Philom. Paris 1818: 12. 1818; FBI 3: 387. 1881; FPM 2: 728. 1921; FAP 2: 522. 1997.

Erect herb, to 30 cm high; stem white tomentose, dichotomously branched. Leaves sessile, simple, alternate, linear, 1–5 × 0.3–0.7 cm, tomentose with woolly hairs, base attenuate and clasped to stem, margin serrate-denticulate, apex acute. Heads axillary, solitary, homogamous, 1.5 cm across; peduncles to 0.5 cm long. Involucral bracts 3–5 seriate, 24–34, outer short and inner longer than the florets, to 1.3 cm long, base obtuse, one-veined, margin serrate-denticulate, apex spinescent, yellowish. Receptacle 3–4 mm across, pitted, margins of pits elevated and fimbriate. Florets 9–15, 10 mm long, corolla white, tube 2.5 cm long, cylindrical, pubescent with colleters towards apex, limb 4.5 mm long, deeply lobed; lobes 5, oblong-lanceolate, 3 mm, pubescent with colleters on dorsal side; stamens 5, anthers oblong, 4 mm, hood elongate, subacute, base sagittately tailed; style 7 mm long, cylindrical, stigma apically lobed, sparsely hairy. Pappus 3-seriate, united at base to form a ring, equal to florets, barbellate hairy; inner most 10, paleaceous scales lanceolate. Achene obovate, brown, 3 mm long, 10-ribbed, densely clothed with silky hairs, developed from base.

Common, in foot hills of dry deciduous forests, scrub forests and wastelands. Fl. & Fr.: Jul. – Jan. Batrepalli (ATP), *KRKS* 37957, 301-HJCB-E; Kalasamudram (ATP), *KRKS* & *BSS* 40409.

ECHINOPS L.

Echinops echinatus Roxb., Fl. Ind. 3: 447. 1832; FBI 3: 358. 1881; FPM 2: 724. 1921; FAP 2: 522. 1997.

Erect herb to under shrub, to 1 m high, spiny; stem terete, striated, white tomentose in younger part. Leaves simple, alternate, sessile, oblong, elliptic-oblong, 3–5 × 10–15 cm, hispidulous above, white tomentose below, spinescent, base amplexicaul, margin lobed, spinescent, apex acuminate by spine, strongly veined. Heads many, collected into globose clusters, compound head solitary, globose, terminal on branches with long spines, to 5 cm across; peduncle 2–3 cm, white tomentose. Common involucral bracts 8, reflexed completely oblanceolate, 7 mm long, glabrous. Common receptacle globose, 4 mm long, brown, marked with spherical scars of fallen partial heads. Simple head one-flowered, to 2 cm long, homogamous, not rayed. Involucral bracts many-seriate, unequal; outer most of numerous, whitish, barbellate hairs, 1.4 cm long, followed by 7–12 bracts, linear, oblanceolate, elliptic 0.7–2 cm long, glabrous, ciliate margin, spinescent in one bract; inner most 5, united to form a sheath, spinescent 1.4 cm long. Receptacle narrow, glabrous. Florets 1.2 cm long; corolla white, tube 6 mm long, glabrous, 5-lobed; lobes deeply divided, lanceolate, 6 mm long, acute; stamens 5, anthers oblong, 4 mm, hood 1 mm long, oblong, obtuse, base tailed; tails branched; style 14 mm long, stigma awns 2.5 mm long, flat, glabrous, obtuse. Pappus 1–1.5 mm long, uniseriate, short scales, connate at base, persistent. Achene obovate, 5 mm long, ribbed, densely villose whitish.

Common, in wastelands. Fl. & Fr.: Aug. – Dec. Kalasamudram (ATP), *KRKS* 41890; *KRKS* & *BSS* 40410.

ECLIPTA L.

Eclipta prostrata (L.) L., Mant. Pl. 2: 286. 1771; FAP 2: 523. 1997. *Verbesina prostrata* L., Sp. Pl. 902. 1753. *Eclipta alba* (L.) Hassk., Pl. Jav. Rar. 528. 1848; FBI 3: 304. 1881; FPM 2: 705. 1921. *Vebesina alba* L., Sp. Pl. 902. 1753.

Local name: *Gunta galagara, Brungarajam.*

Erect to decumbent herb, sometimes prostrate, to 40 cm high, rooting at nodes, stem terete, hirsute. Leaves opposite, simple, elliptic or elliptic-lanceolate, 2–4 × 0.5–1 cm, scabrid on both sides, base acute-obtuse, margin entire to distantly serrulate, apex acute, 3-veined from base; petiole to 3 mm long. Heads 3–6-clustered, axillary or terminal, white, 5 × 8 mm, heterogamous, rayed; peduncle 2 mm long. Involucral bracts 10–12, 2–3-seriate; outer most ovate-elliptic, 5 mm long, longer than the inner, scabrid on dorsal, margin ciliate, acute; inner scabrid on dorsal, margin ciliate, acute; inner most 3 to 4, oblong to obovate, 2 mm long, scarious, margin ciliate, apex obtusely apiculate to acute. Receptacle 3–4 mm across, convex, paleaceous. Palea linear-oblanceolate, 2 mm, ciliate along the margin. Ray florets *ca.* 60, in 2–3 marginal series; corolla white, tube 0.8 mm long, glabrous; ligule of outer florets oblanceolate,

1.2–1.7 mm long, 2-lobed; lobes obtuse; inner linear, apex obtuse, pubescent; style 1.5 mm long, stigma awns short, broad, obtuse. Bisexual florets around 25; corolla white; tube 0.4–0.5 mm long, glabrous, limb 1 mm long, minutely hispid, 5-lobed; stamens 5, black, hood absent or very short; style 1.2 mm long, stigma papillate. Pappus of 2 very short scales, connate, the rim dentate. Achene obovate to oblong 2 mm long, triquetrous, tuberculate or wrinkled, margined, hairy at apex.

Common, in moist places. Fl. & Fr.: Throughout year. Velpula (KDP), *KRKS* & *KRS* 37677.

EMILIA Cass.

Emilia sonchifolia (L.) DC. in Wight, Contrib. 24.1834; FBI 3: 336. 1881; FPM 2: 716. 1921; FAP 2: 524. 1997. *Cacalia sonchifolia* L., Sp. Pl. 835. 1753.

A slender erect or diffuse herbs sometimes decumbent, glabrous or pilose. Leaves radical and cauline, crowded at the base; radical leaves petiolate 4–10 × 1–3 cm, lyrate-pinnatifid with large terminal lobes, semi-orbicular; cauline leaves alternate, obovate to spatulate more or less amplexicaul and auricled at base. Heads homogamous, up to 12 mm long, not rayed, 1–3 on long slender peduncles. Involucral bracts cylindric, uniseriate, 6–8 mm long, oblong-lanceolate, coherent, almost equalling the corollas. Receptacle flat, naked. Florets bisexual. Corolla pink, 3–3.5 mm long, wider to the mouth, lobes 5, tube slender. Stamens 5; anthers 1–1.5 mm long. Ovary *ca.* 1.5 mm long; style-arms acute, exserted, slightly ecurved. Achenes 1–2 mm long, 5-ribbed, hispid. Pappus white silky, capillary, 5–6 mm long.

Common, in wastelands and moist places of open deciduous forests. Fl. & Fr.: Jul. – Mar. Batrepalli (ATP), *KRKS* 438 (HJCB); Kalasamudram (ATP), *EC* 1183.

EPALTES Cass.

Key to Species

1a. Erect herbs; stem 2-winged to 1 mm wide, distinctly denticulate; heads 0.7 cm across; leaves thin.. **E. pygmaea**

1b. Decumbent herb; stem 3-winged to 3 mm wide, entire, irregularly or folded; heads 1.3 cm across; leaves thick.................. **E. divaricata**

Epaltes divaricata (L.) Cass., Bull. Soc. Philom. Paris 139. 1818; FBI 3: 274. 1881; FPM 2: 691. 1921; FAP 2: 524. 1997.

Decumbent herb, to 25 cm high, branches thick, glabrous, 3-winged, wings entire towards base, denticulate towards apex. Leaves alternate, simple, sessile, linear-oblong, 1.5–6 × 0.3–1.3 cm, glabrous, thick, base decurrent on stem, margin distantly denticulate, apex acute. Heads solitary or 1–3 together, terminal on divaricate branches, pink, 1.3 cm across, heterogamous, not rayed; peduncle to 1 cm long, winged. Involucral bracts multiseriate, unequal; outer most lanceolate, 2 mm long, shorter than inner ones, greenish in middle, scarious margin, acuminate; middle ones ovate-lanceolate, 4 mm long, broadly scarious, pubescent, acuminate; inner most oblong, 4 mm long, scarious, acuminate at apex. Receptacle flat to concave, 5 mm across, alveolate, glabrous. Female florets around 200; corolla pinkish-purple, 3 mm long, glabrous, 3–4-lobed; lobes obtuse; style 2 mm, exserted, stigma awns linear, slender. Bisexual florets 24; corolla pinkish-purple, 3.2 mm long; tube 2.5 mm, glabrous, limb 1 mm long, 5-lobed; lobes ovate-triangular, acute; stamens 5, anthers linear-oblong; hood truncate; style 3.2 mm long, exserted, spatulate, stigma awns linear. Pappus absent. Achene obovoid, 0.5–0.8 mm long, brown, minutely tuberculate, striated.

Occasional, near streams and ponds mixed with grass. Fl. & Fr.: Oct. – Jan. Bartepalli (ATP), *KRKS* 42038.

Epaltes pygmaea DC., Prodr. 5: 461. 1836; FBI 3: 274. 1881; FPM 2: 691. 1921.

Erect herb, to 30 cm high; branches thin, two-winged; wing to 1 mm wide, distinctly denticulate. Leaves alternate, sessile, linear-elliptic, oblanceolate, 1.5–4 × 0.2–0.5 cm, glabrous, membranous, base attenuate, margin distinctly denticulate, apex acute. Heads violet, solitary or few, terminal, 0.3–0.7 cm across; peduncle to 1 cm long, winged. Involucral bracts multiseriate, broad; outer oblong, 1 mm long, acuminate, midvein green, pubescent; inner ones oblong-ovate, 2–5 mm long, pubescent, acuminate, margin scarious; inner most thin, 5 mm long, acuminate, equal to florets. Receptacle naked, flat, 5 mm across. Female florets in periphery on receptacle, fertile; corolla-tube 2 mm long; lobes minute; style 2 mm long. Bisexual florets in centre on receptacle, tubular, sterile; corolla 3–5-fid. Pappus absent. Achenes minute, to 0.6 mm long, ellipsoid, ribbed, smooth, brown.

Rare near moist places and streams of dry deciduous forests. Fl. & Fr. Dec. – Feb. Kalasamudram (ATP), *KRKS* 41881; near Thalupula (ATP), *KRKS* & *BSS* 41920.

GLOSSOCARDIA Cass.

Glossocardia bosvallia (L. f.) DC. in Wight, Contrib. Bot. Ind. 19. 1834; FPM 2: 709. 1921; FAP 2: 526. 1997. *Verbesina bosvallia* L. f., Suppl. Pl. 1781. *Glossocardia linearifolia* Cass., Dict. Sci. Nat. 19: 62. 1821; FBI 3: 308. 1881.

Local name: *Adavi kotthimera, Thurukoni meesalu.*

Diffuse herb, to 20 cm long, branches many from base, ribbed, glarbous or sparsely puberulous. Leaves bipinnate, in rosette at base and alternate on stem, basal leaves petiolate, decurrent on stem, 3 × 1 mm, apices apiculate. Heads solitary, terminal or leaf-opposed, yellow, 1 cm long, heterogamous, rayed; peduncle to 1.5 cm long, puberulous. Involucral bracts 7; outer most short, ovate, 2.5–3 mm, glabrous, broadly scarious margin, sparsely ciliate, sharply acute; inner 4, imbricate, obovate, to 6 mm long, brown-striated, slightly folded, glabrous, margins broadly scarious, obtuse and ciliate at apex. Receptacle narrow, to 0.8 mm long, flat, glabrous in centre, paleaceous towards margins, palea 2, elliptic-lanceolate, slightly folded, glabrous, very broadly scarious, obtuse to sub-acute. Ray florets only one in number; corolla yellow, tube 1.6 mm long, sparsely hairy; ligule 1.4 mm, 2–3-lobed; lobes oblong, to 0.8 mm long, obtuse; style 2 mm long, stigma awns short, blackish. Bisexual florets 5; corolla yellow, 2.5 mm long; tube 1 mm long, glabrous to sparsely hairy, limbs 1 mm long, 3–4-lobed; lobes elliptic-ovate, 0.6 mm long, glabrous, obtuse; stamens

3–4, anthers linear-oblong, hood short, obtuse to retuse, obtuse at base; style to 2.5 mm long, exserted, stigma awns to 1 mm long, black, distinctly papillate. Pappus 2–3, 2 mm long, black, glabrous. Achenes linear-elliptic, to 8 mm long, black, densely yellow villose, two-spinulate at apex, spines divaricate, 2 mm long.

Common, in scrub forests and wastelands. Fl. & Fr.: Aug. – Nov. Velpula (KDP), *KRKS*& *KRS* 37675.

GNAPHALIUM L.

Gnaphalium polycaulon Pers., Syn. Pl. 2. 421. 1807; FAP 2: 527. 1997. *G. indicum* auct. Pl. (non. L. 1753): Hook. f., Fl. Brit. Ind. 3: 289. 1881; FPM 2: 698. 1921.

Erect or decumbent-ascending herb, 15–20 cm high; stems appressed white woolly, simple or branched from base. Leaves simple, alternate, sessile, oblanceolate to spatulate, 2–6 × 0.3–1 cm, white woolly, base narrow, margin entire, apex acute, one-veined from base. Heads 2–3 mm across, yellowish white, in spikes, terminal and axillary; spikes to 5 cm long, florets heterogamous, not rayed, in sub-sessile heads. Involucral bracts 16, 2-seriate, oblong or oblong-elliptic, 2.2–2.5 mm long, broadly scarious, obtuse, glandular, woolly on the outer ones; lower one glabrous. receptacle flat, 1 mm, glabrous. Female florets around 80; corolla 1.5 mm, tube glabrous, minutely lobed and densely pubescent; style 1 mm, truncate and papillate at apex. Bisexual florets 5; corolla 1.5 mm, tube glabrous, 5-lobed, lobes ovate, acute pubescent, with colletors on dorsal side; stamens 5, anthers linear, 1 mm long, hood ovate, obtuse, base sagittately tailed; tails shorter than thickened part; style 1mm long, stigma branches linear, short, truncate, papillate at apex. Pappus 10, uniseriate, free, slender, white hairy, exceeding the floret. Achene oblong, 0.3–0.6 mm long, papillate, black.

Common, near ponds in high altitudes, grows with grass. Fl. & Fr.: Nov. – Feb. Batrepalli (ATP), *KRKS* 41879.

GRANGEA Adans.

Grangea maderaspatana (L.) Poir., Encycl. Suppl. 2: 825. 1912; FBI 3: 247. 1881; FPM 2: 680. 1921; FAP 2: 528. 1997. *Artemisia maderaspatana* L., Sp. Pl. 849. 1753.

Local name: *Adavimacha pathri.*

Prostrate herb, to 25 cm, stem terete, striated, pubescent and glandular. Leaves radical and cauline, alternate, pinnatifid, sub-sessile, oblanceolate or oblong in outline, 7 × 2.5 cm, whitish pubescent, segments 0.5–1.5 × 0.2–1 cm, lobed, terminal segments 3–many-lobed; lobes acute. Heads yellow, solitary, sometimes 2, terminal, 8 mm across, heterogamous, not rayed. Peduncle to 1.5 cm, pubescent. Involucral bracts 3-seriate; outer 2 series sub-equal; outer and middle ones narrowly elliptic-oblong, pubescent, scarious at apex; innermost few, oblanceolate, 3 mm, scarious, glabrous, obtusely lobed or dentate. Receptacle 2.5–3 mm across, convex, fleshy, glabrous. Female florets around 120, marginal, 2–3-seriate; corolla 1.5 mm long; tube 0.6 mm long, glandular, limb 0.5–1 mm long, along with style, 5-lobed; lobes ovate. Bisexual florets *ca.* 70; corolla 1–5 mm long; tube 0.5 mm long, glandular, limb to 1 mm, 5-lobed; lobes ovate, 0.5 mm long, sparsely glandular on dorsal side, acute;

stamens 5, anthers linear-oblong, 0.5 mm, hood ovate, acute, base obtuse; style to 1 mm exserted, branches very short, acute. Pappus a laciniate cup. Achene obovate, 1.5 mm, green, ribbed, glandular and hairy, stalked.

Common, in wastelands and sandy soils near water bodies of dry deciduous forests. Fl. & Fr.: Aug. – Nov. Kalasamudram (ATP), *KRKS* & *SSR* 36472; Batrepalli (ATP), *KRKS* 36960.

LAGASCEA Cav.

Lagascea mollis Cav., Anal. Cienc. Nat. 6: 332. 1803; FBI 3: 302. 1881; FPM 2: 703. 1921; FAP 2: 530. 1997.

Local name: *Metthanallaku.*

Erect herb, to 50 cm high; stem pale, slender, striate, terete, pubescent, grey. Leaves simple, opposite, ovate, 1–4 × 0.6–2.5 cm, pubescent above, white silky below, base obtuse, margin coarsely serrate, apex acute; petiole to 3 cm long, densely pubescent. Heads one-flowered, aggregated into terminal compound capitula; compound head subtended by leafy bracts, 1–1.5 cm across, white, bisexual. Involucral bracts 5, connate at base, densely pubescent. Corolla-tube glabrous, 5 mm long, limb 5-lobed; lobes ovate, sparsely hairy on dorsal side; stamens 5, brown, anthers 2 mm, hood ovate, subacute, base sagittate, auricles acute; style 4 mm, exserted, black, branches slender, acute, hairy. Pappus absent. Achene obovate, 2 mm long, ribbed, glabrous except apex (hairy).

Common, on road sides and as weed in cultivated fields. Fl. & Fr.: May. – Sept. Thummalapalle (KDP), *KRKS*, *KRS* & *SSR* 33679.

Note: It is native of Central America, naturalized.

LAUNAEA Cass.

Launaea intybacea (Jacq.) Beauverd, Bull. Soc. Bot. Geneve ser. 2, 2: 114. 1910. *Lactuca intybacea* Jacq., Pl. 162. 1784. *L. runcinata* DC., in Wight, Contrib. Bot. Ind. 26. 1834; FBI 3: 403. 1881; FPM 2: 731. 1921; FAP 2; 529. 1997.

Erect herb, to 1 m high, stem terete, striated, glabrous, divaricately branched above or simple. Leaves simple, chiefly radical and cauline, sessile, ovate-triangular; cauline ones linear to lanceolate, 2–13 × 1–4 cm, glabrous; radical ones lyrately lobed; lobes triangular, terminally hastate, base clasped to stem, spinulosely ciliated, apex obtuse to acute. Heads solitary or fascicled, on racemose branches, 1 cm long, homogamous, rayed; peduncle to 3 cm long, glabrous, bracteate. Involucral bracts *ca.* 16, two-seriate; outer most minute, to 1 mm long; inner ones 8, oblong-lanceolate, 10 mm long, scarious at margins, obtuse. Receptacle slightly concave, 2.5–3 mm across, glabrous. Florets 24; corolla pinkish; tube 6 mm, glabrous, ligule oblong, 6 mm long, hairy at base, 5-lobed; lobes ovate, 0.4 mm long, obtuse; stamens 5, anthers linear, 0.2 mm long, hood ovate, obtuse, base sagittately auricled, auricles acute; style 8 mm, branches 1–1.5 mm long, slender, setose hairs on dorsal side, often recurved, acute. Pappus 5 mm long, copious, white, silky, persistent. Achene 3 mm, obovate to obovate-cuneate, black, ribbed, tuberculated; beak 1 mm long, colourless, dialated to form a cup at apex.

Common, on road sides and wastelands. Fl. & Fr.: Aug. – Dec. Batrepalli (ATP), *KRKS* & *SSR* 37887; Kalasamudram (ATP), *KRKS* 40136.

PARTHENIUM L.

Parthenium hysterophorus L., Sp. Pl. 988. 1753; FAP 2: 533. 1997.

Local name: *Vayyari bhama, Congres pulu.*

Annual erect herb, to 1.5 m high, stem angular, pubescent. Leaves pinnatified, 2–15 × 0.5–7 cm, pinnules opposite, oblong-lanceolate, base decurrent, apex acute, sparsely to densely scaberulose and gland-dotted on both surface. Heads in terminal panicles, heterogamous, white. Female florests in outer rim, 5–6, corolla limbs reniform, orbicular to oblong. Disk florests 12–30. Achenes dorsally compresed, obovoid, 1.5–2 mm long, pappus like enations erect, deltate to ovate, 0.5–1 mm.

Very Common, in all wastelands and weed in cultivated fields. Fl. & Fr.: Throughout year. Thummalapalle (KDP), *KRKS, KRS, SSR* & *TP* 33676.

PENTANEMA Cass.

Pentanema indicum (L.) Ling, Acta Phyt. Sin. 10: 179. 1965; FAP 2: 533. 1997. var. **indicum**. *Inula indica* L., Sp. Pl. ed. 2. 2: 1236. 1763. *Vicoa auriculata* Cass., Ann. Sci. Nat. Ser. 1.17: 418. 1829; FBI 3: 247. 1881. *V. indica* (L.) DC., in Wight, Contrib. Bot. Ind. 10. 1834; FPM 2: 701. 1921.

Erect herb, to 1.5 m high; stem terete, reddish black, sparsely puberulous, branches alternate. Leaves simple, alternate, sessile, scabrid on upper, densely pilose on lower surface, lanceolate, linear, oblong-lanceolate, 1.5–6 × 0.5 cm, base cordate, amplexicaul, auricled, margin serrulate, revolute, apex acute. Heads few, paniculate, terminal, yellow, 1 cm across, heterogamous, rayed; peduncles to 4 cm long, puberulous. Involucral bracts 3-seriate, unequal; outer ones linear-lanceolate, 2 mm long, green, pilose, glandular on dorsal, acuminate; inner most oblong-lanceolate, 4.5 mm long, scarious, glandular on dorsal, serrate, acute; receptacle convex, 3–3.5 mm across, fimbriate. Ray florets around 27; corolla-tube cylindrical, 3 mm long, glabrous, ligule yellow, 2-lipped, ventral one very small, recurved, dorsal one spreading, oblong, 8 mm long, yellowish, sparsely glandular on dorsal side, 3-lobed, lobes obtuse; style 4.5 mm long, stigma awns linear, 1 mm long, glabrous, obtuse. Bisexual disk florets around 150; corolla tube glabrous, 5-lobed, lobes ovate, 0.4–0.5 mm long, glandular on dorsal side, acute; stamens 5, anthers linear, 2–2.3 mm long, hood ovate, obtuse, auricled; style to 4 mm long, stigma awns linear, 1 mm, glabrous, obtuse. Pappus absent in ray, uniseriate in bisexual florets, white, setaceous hairy. Achene to 1 mm, glabrous in ray florets, hairy and striated in bisexual florets.

Common, in scrub forest and foot hills of the dry deciduous forests. Fl. & Fr.: Sept. – Dec. K. K. Kottala (KDP), *KRKS* 37656.

PULICARIA J. Gaert.

Pulicaria wightiana (DC.) C.B.Clarke, Comp. Ind. 178. 1876; FBI 3: 298. 1881; FPM 2: 702. 1921; FAP 2: 535. 1997. *Poloa wightiana* DC., Guill. Arch. Bot. 2: 515. 1833.

Local name: *Adavi chamanthi.*

Erect herb, to 50 cm high; stem striated, puberulous. Leaves simple, sessile, puberulous, oblong 2–4 × 0.3–0.5 cm, base sub-auricled, sessile, margin serrate to entire, apex acute or obtuse. Head solitary, terminal, yellow, 2.5 cm across, heterogamous, rayed; peduncle to 3 cm long, puberulous, fistular, widened at apex below the head. Involucral bracts 3-seriate, sub equal; outer spatulate, 7.5 mm long, herbaceous, slightly exceed the inner one, veined, pilose and sessile glandular on dorsal side, margin ciliate, apex obtuse; middle ones 6 mm long, scarious below the middle and foliaceous above, pilose and glandular, obtuse; inner ones linear-lanceolate, 5.5–6 mm; scarious, one-veined, glandular towards apex, ciliate along margins, apex acute. Receptacle slightly convex, 5 mm across, alveloate, glabrous. Ray female florets around 30, 1.2 cm long, exceed the head, to 5-6 mm long; tube 3 mm long, pubescent, ligule oblong, 9 mm long; style 4 mm long, bifid, stigma awns linear, 1 mm long, narrowly obtuse. Bisexual disk florets around 300, corolla tube to 5 mm long, equal in length to involucral bracts, 5-lobed; lobes ovate-triangular, each lobe 2–5-dentate, usually acute, rarely obtuse; stamens 5, anthers 2.5 mm long, exserted to the level of tube, oblong, hood ovate, acute, base sagittately tailed; style 4 mm long, bifid, stigma awns linear. Pappus two seriate, outer short scales; inner 4 mm long, white, deciduous. Achenes oblong, 1–1.2 mm, striate, brown, truncate at apex, pubescent.

Common, in wastelands and weed in Agriculture fields. Fl. & Fr.: Jul. – Nov. Thummalapalle (KDP), *KRKS* 36915.

SENECIO L.

Senecio tenuifolius Burm. f., Fl. Ind. 181. t. 60. f. 4. 1768; FBI 3: 345. 1881; FPM 2: 722. 1921; FAP 2: 536. 1997.

Erect herb, to 25 cm high; stem striated, glabrous, branches from above the middle. Leaves pinnatified, upper ones deeply lobed to base, alternate, sessile, glabrous, 1–1.25 × 0.1–0.3 cm; lobes subacute to acute. Heads solitary or paired, terminal, yellow, 8 mm across, heterogamous, rayed, bracteolate; peduncle 4 cm long, glabrous. Involucral bracts 12–13, 1-seriate, elliptic-oblanceolate, 4 mm long, entirely reflexed at maturity, glabrous, greenish in the center, broadly scarious, ciliate, acute. Receptacle 1.5 mm across, glabrous. Ray female florets 6; corolla yellow, 4 mm long, tube short, hairy, elliptic, 3–4-veined, glabrous, 3-lobed; lobes acute; bisexual disk florets 20–32; corolla 2.5 mm long, 5-lobed, lobes ovate, 0.5 mm long, glabrous, acute; stamens 5, anthers 1.2 mm long, hood ovate, auricles obtuse; style 2 mm long, stigma awns 0.5 mm, truncate and papillate at apex. Pappus 1–2 mm long, around 30, scabrid, uniseriate, yellowish free hairs, basally form a ring. Achene obovate, 2 mm, ribbed, hairy between ribs.

Occasional, in dry deciduous forests. Fl. & Fr.: Aug. – Dec. Kalasamudram (ATP), *KRKS & BSS* 39033.

SONCHUS L.

Sonchus oleraceus L., Sp. Pl. 794. 1753; FBI 3: 414. 1881; FPM 2 : 732. 1921; FAP 2: 537. 1997.

Local name: *Palakusalu.*

Erect herb, to 50 cm, stem fistular, striated, glabrous below, glandular when young. Leaves radical and cauline, sessile, clasped to stem, alternate, glabrous, cauline ones subsessile, lanceolate; radical ones lobed, lateral lobe 7–14 × 1.2–7.5 cm, base of lower leaves amplexicaul, obtuse, upper ones acutely sagittate, margin sharply denticulate, apex acuminate. Heads cymose in upper umbel, bracteate, yellow, to 1.5 cm across, whitish tomentose at base when young, homogamous, all ligulate; peduncle to 5 cm, glandular hairy. Involucral bracts 28, 3–4 seriate, outer lanceolate, 4–5 mm long, minutely setose hairy on dorsal side, narrowly scarious on margins, apex obtuse and scarious; middle ones oblong-lanceolate, 7–10 mm long, scarious towards margins, setose on dorsal surface, apex obtuse and ciliate; inner most elliptic or oblong-lanceolate, 7–9 mm long, scarious, acute. Receptacle convex, 7 mm across, glabrous. Florets 130-150; corolla-tube 5 mm, pubescent; ligule 4 mm, elliptic, hairy below the middle, 5-lobed, lobes obtuse; stamens 5, 1–1.5 mm long, completely exserted, hood blackish, obtuse; style 9–10 mm long, exserted, to 4–5 mm long, stigma awns 1–1.5 mm long, linear, setose hairy on dorsal side, obtuse; pappus around 6.5 mm long, heterogenous; outer white, silky, slender hairs forming a ring at base; inner of 17–20 setaceous hairs. Achene towards periphery sterile, towards center fertile, obovoid, 3 mm long, flat, 4-ribbed, muricate on ribs.

Occasional, in wastelands and roadsides. Fl. & Fr.: Oct. – Feb. Thalupula (ATP), *KRKS* & *BSS* 41941.

SPHAERANTHUS L.

1a. Heads spherical .... **S. indicus**

1a. Heads ovoid .... **S. africanus**

Sphaeranthus africanus L., Sp. Pl. ed. 2. 1314. 1763; FBI 3: 285. 1881; FPM 2: 692. 1921.

Annual erect, aromatic herb, with spreading branches, stem winged, wings dentate. Leaves spatulate or oblanceolate, 2-4 × 0.5-1 cm, base narrowed, margin denticulate, apex acute to apiculate. Heads many, aggregated to form a compound head, solitary, terminal, purple, ovoid. Achenes oblong or elliptic, 3-ribbed.

Occasional, on the edges of streams, ditches, ponds and wastelands. Fl. & Fr.: October – May. Kalasamudramu (ATP), *GO* 2458.

Sphaeranthus indicus L., Sp. Pl. 927. 1753; FBI 3: 275. 1881; FPM 2: 692. 1921; FAP 2: 538. 1997.

Suberect or erect aromatic herb, 10–30 cm high; stem glandular, spreading, somewhat strong, whitish pilose, continuously winged; wings sharply denticulate; divaricately branched. Leaves simple, alternate, sessile, glandular and pilose, obovate or spatulate, 1–2.5 × 0.5–1 cm, base subamplexicaul and decurrent on stem, sharply dentate at margin, apex mucronate. Heads many, aggregated to form a compound head; compound heads solitary, terminal on upper branches, pink-purple, spherical, 1.2 cm across; peduncle to 3 cm long, winged, glandular and pubescent with spreading hairs. Common, receptacle 4 mm across, globose, whitish in the centre. Subtending bracts of simple heads lanceolate, 3.5 mm long, equaling the involucral bracts, glandular at apex, ciliate at margin. Involucral bracts 11, lanceolate, oblong-lanceolate or sub-spatulate, 3.5 mm long, glabrous, ciliate along margins, acute. Receptacle very small, glabrous. Female florets 9–10; corolla 2 mm long; basal half swollen, thick and wide; upper half thin, tubular, sparsely glandular, 3-lobed; lobes glabrous, subacute; style 2 mm, slightly exserted. Bisexual florets 2; corolla 3 mm long, glandular at base, lower half swollen, globose, upper half thin, spreading, 5-lobed, lobes ovate, glandular on dorsal side, obtuse; stamens 5, anthers 1.5 mm long, hood ovate, obtuse, base sagittately auricled; style 3 mm long, exserted to 0.5 mm long, stigma awns linear, obtuse, papillate. Pappus absent. Achenes of female florets elliptic-oblong, 1 mm long, hairy, often 3-angled. Achenes of bisexual florets oblong, 1 mm long, cylindrical, glabrous, rarely glandular.

Common, near paddy fields and water bodies. Fl. & Fr.: Jun. – Apr. Kalasamudram (ATP), *KRKS* & *KRS* 36458.

TRICHOLEPIS DC.

Tricholepis radicans (Roxb.) DC., Prodr. 6: 584. 1838; FBI 3: 381. 1881. FPM 2: 726. 1921. *Carduus radicans* Roxb., Fl. Ind. 3: 408. 1832.

Annual erect herb, 30–40 cm high, branches striated, puberulous and gland dotted. Leaves simple or lyrate, sessile, gland dotted on both sides, 2–10 × 0.4–4 cm, base obtuse, amplexicaul, margin remotely irregularly toothed or pinnatifid, apex acute-apiculate, one-veined from base, vein distinct on lower surface. Heads solitary, terminal on branches, 2 cm across, lilac, homogamous, not rayed; peduncle to 5 cm, glabrous, gland dotted. Involucral bracts 7 seriate; outer ones ovate to ovate-elliptic, 3–7 mm long, pubescent, margin serrate, scarious, apex spinescent, spines deflexed; middle ones elliptic-oblong; inner most oblong to oblong-lanceolate, to 16 mm long, margins scarious, acuminate, white tomentose, serrate. Receptacle flat, 3–4 mm long, paleaceous; palea linear, to 6 mm long, glabrous, entire, acute. Florets around 25; corolla 15 mm long, lanceolate, subacute; stamens 5, anthers to 6 mm, hood elongate, 2 mm long, acute, base sagittately auricled; auricle connate on dorsal side, equal to the thickened part, filaments free, 2 mm, papillose; style to 15 mm, branches spreading or recurved, papillate, obtuse, swollen and hairy at the base of forking. Pappus many seriate in different heights, setaceous, connate at base to form a ring. Achene 2–4 mm long, angled, glabrous, basal areola oblique.

Rare in foot hills of dry deciduous forests, near streams and wastelands. Fl. & Fr.: Sept. – Dec. Kalasamudram (ATP), *KRKS* 40108; Way to Gollalaguduru (KDP), *KRKS* & *BSS* 41967.

TRIDAX L.

Tridax procumbens L., Sp. Pl. 900. 1753; FBI 3. 311. 1881; FPM 2: 711. 1921; FAP 2: 539. 1997.

Local name: *Nallaku, Palakalaku, Gayamaku.*

Procumbent herb, to 30 cm long, stem terete, scabrid. Leaves simple, opposite; petiole to 1 cm long, scabrid and gland dotted, ovate-lanceolate, 1.5–3 × 0.8–2 cm, base cuneate, incised dentate or lobed; lobes ovate, acute, apex acute. Heads solitary, terminal on branches, yellow, heterogamous, rayed; peduncle to 16 cm long, scabrid. Involucral bracts 10, 2-seriate, herbaceous, hirsute, ovate, boat-shaped, oblong, cuspidate at apex. Ray florets 5; corolla light yellow, 3-lobed, 9 mm long, pubescent. Bisexual florets around 40; corolla 8 mm long, pubescent, 5-lobed; stamens 5, anthers 3 mm long, exserted, hood triangular-ovate, acuminate, plumose, persistent. Pappus many, barbellate. Achenes obconical, 2 mm, truncate at apex, ribbed, densely white silky.

Very common, in wastelands. Fl. & Fr.: Through the year. Thummalapalle (KDP), *SSR* 33639.

XANTHIUM L.

Xanthium strumarium L., Sp. Pl. 987. 1753; FBI 3: 303. 1881; FPM 2: 703. 1921; FAP 2: 542. 1997. *X. indicum* Koenig, in Roxb., Fl. Ind. 3: 601. 1832.

Local name: *Marula mathangi.*

Sub shrubs, to 2 m high; stem green-brown, hispidulous, rough, solid. Leaves simple, alternate, broadly ovate, 3–10 × 1.5–10 cm, lobed, base cordate, margin dentate, apex acute, prominently 3-nerved; petiole to 15 cm long, hispidulous. Heads monoecious, cymes in axils or clustered at nodes on a recemose axis; each cyme has a male head at apex and female flower at base. Male heads 5–6 mm across, white-yellow; peduncle to 5 mm long, hispidulous and glandular. Involucral bracts 10, 2-seriate; outer lanceolate, 3 mm long, scarious, 3-veined, pubescent on dorsal side, margin ciliate, apex acuminate; inner elliptic, 3 mm long. Receptacle cylindrical, 3 mm long, elongate, paleaceous; palea oblanceolate to sub-spatulate, 2–2.2 mm long, one-veined, scarious, thick at apex. Florets around 60; corolla-tube 2 mm long, 5-lobed; lobes ovate, glandular; stamens 5, black, exserted, anthers oblong, 1 mm long; hood incurved, ovate; filaments form a cylindrical tube. Female heads solitary, ovoid, to 1 cm long, 2-flowered. Involucral bracts 2-seriate; outer 8, free, linear to linear-lanceolate, 5 mm long, 3-veined, pubescent; inner connate, covered with uncinate spines, spines 5 mm long, spreading, hispidulous at base, hooked at apex with two stout rostra, rostra 5 mm long, slightly curved, hooked; corolla absent. Ovary ovoid, 5 mm, brown, glabrous; style 8 mm, running through rostra terminally, elongated, acute. Pappus absent. Achene elliptic-ovoid, to 2 × 0.5 cm, greyish, hooked spiny, tapering at base, fine point at apex, forked with 0.5 cm long spines. Matured burr narrowly ovoid, 2.2 × 1.5 cm including rostra and spines, brown, hispidulous and glandular.

Common, in field bunds and moist localities. Fl. & Fr.: Sept. – Mar. Kalasamudram (ATP), *KRKS* & *BSS* 40412.

LOBELIACEAE

LOBELIA L.

Lobelia alsinoides Lam., Encycl. 3: 588. 1792. *L. trigona* Roxb., Fl. Ind. 2: 111. 1824; FBI 3: 423. 1881; FPM 2: 736. 1921.

Annual decumbent herb 5-15 cm high, rooting from lower nodes, stem 3-winged. Leaves 1.5–2 × 0.7–1.5 cm, ovate or ovate-orbicular, base cordate, margin crenate, apex acute or obtuse; petiole 3 mm broad, upper leaves sessile, elliptic, acute, smaller in size. Stipules 2, narrowed, lanceolate, to 3 mm long. Flowers blue forming lax, terminal, leafy racemes. Sepals 5 mm long, lanceolate; corolla blue,10 mm long, upper 3 lobes obovate, lower 2 lobes narrower. Stamens 5, penicillate at the tip. Capsule 6 mm long, glabrous. Seeds many, trigonous in top view, ellipsoid in side view, grooved, brown.

Common, in marshy areas of deciduous forests and plains. Fl. & Fr.: Aug. – Jan. Pulivendula Ghat road near Anjaneyaswamy temple (KDP), *BSS* & *KRKS* 40057; Near Thalupula (ATP), *KRKS* & *BSS* 41910; Batrepalli (ATP), *KRKS* 42023.

PLUMBAGINACEAE

PLUMBAGO L.

Plumbago zeylanica L., Sp. Pl. 151. 1753; FBI 3: 480. 1882; FPM 2: 744. 1921; FAP 2: 545. 1997.

Perennial under shrub, to 1 m high, branches scandent, striate, stem cylindrical, longitudinally grooved, glabrous, green. Leaves alternate, thin, elliptic-ovate, 3.5–7.5 × 2.5–4.5 cm, glabrous, base acuminate, margin entire, apex acuminate, 7–9-nerved; petiole to 1 cm long, auriculate at base; stipules auriculate. Raceme terminal, 15 cm long, rachis glandular; bract ovate 2 × 3 mm, glabrous, acuminate at apex, pedicel 2 mm, glandular hairy. Flowers white, to 2 cm long. Calyx tubular, to 1 cm long, ribs 10, with glandular hairs, 5-toothed. Corolla salver-form, tube narrow, cylindrical, glabrous, lobes 5, obovate, apiculate at apex. Stamens 5, free; filaments to 2 cm long; anthers white, exserted, parallel. Ovary 3 mm long; locules 5, style filiform, stigma 5-fid. Capsule membranous, enclosed in persistent calyx, 5-lobed; lobes deeply canaliculate. Seed one per locule, dark brown.

Common along stream sides of scrub forests, wastelands and edges of agriculture fields. Fl. & Fr.: Sept. – Jan. Rachakuntapalle (KDP), *KRKS* & *KRS* 37636, 37917.

SAPOTACEAE

MADHUCA J. F. Gmel.

Madhuca longifolia (J. Koening) Macbr. Contr. Gray Herb. (n. s) (53). 17. 1918; FAP 2: 550. 1997. *Bassia longifolia* J. Koenig ex L., Mant. Pl. 563. 1771; FBI 3: 544. 1882; FPM 2: 763. 1921.

Key to Varieties

1a. Leaves linear-lanceolate; anther connectives 3-toothed; berry ellipsoid, glabrous, 1-seeded .... var. **longifolia**

1b. Leaves elliptic-ovate; anthers connectives apiculate; berry ovoid, tawny tomentose, 2–4-seeded .... var. **latifolia**

Madhuca longifolia (J.Koenig) Macbr. var. **latifolia** A. Chev., Rev. Int. Bot. Appl. Agric. Trop. 23: 149. 1943. *Madhuca indica* J. F. Gmel. Syst. Nat. 2: 799. 1791; FAP 2 : 549. 1997. *Bassia latifolia* Roxb., Pl. Coromandel t. 19. 1795 & Fl. Ind. 2 : 526. 1832; FBI 3 : 544. 1882; FPM 2 : 763. 1921.

Local name : *Ippachettu.*

Deciduous tree, up to 18 m high, bark thick, dark with vertical cracks, milky latex, wood reddish-brown, shallowly inter-locked grained; branchlets glabrous, spreading, young branches brown villose. Leaves clustered near the ends of branches, chartaceous, thick, elliptic or elliptic-oblong (obovate-oblong), 8–20 × 4–10 cm, glabrous above, brown pubescent below with loosely attached, scaly peeling off, base acute, margin entire, apex acute or shortly acuminate, nerves 10–12 pairs; petiole to 3.5 cm long, rounded, brown. Flowers creamy white, in axillary fascicles, drooping, pedicels densely tomentose. Calyx-lobes 5, elliptic-oblong, brown, 1 cm long. Corolla overlapping, creamy-white, glabrous, oblong. Stamens more than 12, equal, clustered. Ovary sub-globose, brown, 0.5 cm across. Style 0.5 cm long, cylindrical; stigma capitate, light yellow. Berry ovoid-oblong, 4 × 2 cm, fleshy, greenish, brown when mature, mucronate with persistent style and stigma. Seeds 2–4, brown, 1–2 cm long, glossy, slightly broad, testa hard, base obtuse, apex acute.

Occasional, in dry deciduous forests. Fl.: Feb. – Apr. & Fr.: Jun. – Aug. Kalasamudram (ATP), *KRKS* 41896.

Madhuca longifolia (J. Koening) Macbr. var. **longifolia**

Local name: *Pedda ippa, Natu ippa.*

Deciduous tree, 15–20 m high; bark dark grey to brown, scaly; wood reddish, close grained; branchlets brownish-yellow, glabrous. Leaves chartaceous, oblanceolate, 5.5–8 × 2–3 cm, glabrous, green above, pale beneath, base narrowly acute, margin entire, apex acute; secondary nerves 10–12 pairs, raised above; petiole to 2 cm long, glabrous. Umbels axillary, long pedicellate. Flowers creamish-white, 1 cm across. Calyx lobes 5, brown, pilose, elliptic oblong. Corolla elliptic-oblong, 0.8 cm long. Stamens clustered; clusters to 5 mm long. Style filiform, yellowish; stigma diskoid. Berry ellipsoid, glabrous, brown scaly, 4 × 2 cm. Seeds 1, compressed, light brown, smooth, glossy with a fairly thick and hard testa.

Occasional, in dry deciduous forest. Fl.: Mar. – Apr. & Fr.: May – Jul. Batrepalli (ATP), *KRKS* 41857.

EBENACEAE

DIOSPYROS L.

Key to Species

1a. Leaves to 30 cm long, thick, coriaceous; fruit 3–4.5 cm across, calyx not enlarged and supported by thick lobes.-bark longitudinally deeply furrowed .... **D. melanoxylon**

1b. Leaves to 8 cm long, thin, not coriaceous; fruit up to 2 cm across; calyx enlarged and not supported by lobes .... (2)

2a. Leaves glabrous; calyx lobes not attached to the fruit at base, flat foliacious, spreading; bark smooth.................................... **D. sylvatica**

2b. Leaves hairy; calyx lobes attached to the fruit at base, not foliacious; bark deeply cracked by scaly structures (peeling off rectangular scales) .... **D. chloroxylon**

Diospyros chloroxylon Roxb., Pl. Coromandel t. 4: 1795; FBI 3: 560. 1882; FPM 2: 776. 1921; FAP 2: 553. 1997.

Local name: *Nela ulimiri, Ullinda, Ellinda, Thorika.*

Evergreen shrub to tree, up to 4 m high; bark dark brown, deeply cracked, peeling off in rectangular small scales, branchlets rusty tomentose, brown-grey. Leaves thinly-coriaceous, elliptic or elliptic-oblong, 1.6– 6 × 0.8–3 cm, slightly pubescent above, fulvous below, base acute-obtuse, margin entire, apex obtuse or subacute, lateral veins 10 pairs; petiole to 1 cm long, greyish brown. Inflorescence axillary, fascicled clusters or cymes, 4–6 flowers in a cluster, white. Flowers unisexual. Male flowers in small fascicles, axillary. Calyx cupular; lobes 4, ovate, concave, 2.5 mm long, sericeous without, obtuse. Corolla tubular, campanulate; lobes 4, ovate, 5 mm long, thinly hairy on the back, acute. Stamens 16, biseriate, inner short; filaments to 1 mm long; anthers lanceolate, 1 mm long, tip acute. Female solitary, globose, axillary. Calyx lobes 4, to 3.5 mm across, orbicular, densely tomentose without. Corolla white; lobes 4, with a tuft of hairs at the back of each corolla-lobe. Stamens 16, glabrous, staminodes 8. Berry brown, globose, 1 cm across, smooth, shining, fruiting calyx enlarged, sub-membranous, rusty hairy, flat. Seeds 2–3.

Rare, open hilly areas in dry deciduous forests. Fl.: Mar. – May & Fr.: Jun. – Sept. Kalasamudram (ATP), *KRKS* & *BSS* 39523.

Diospyros melanoxylon Roxb., Pl. Coromandel t. 46. 1795; FBI 3: 564. 1882; FPM 2: 772. 1921; FAP 2: 554. 1997.

Local name: *Beedi aku, Tuniki, Thimiki, Manchi tumiki, Tellagada.*

Deciduous tree, up to 8 m high, bark greyish brown or black, with longitudinal-irregular cracks; wood reddish-brown with an irregular black striations, hard wood; young branches tawny tomentose. Leaves coriaceous, broadly elliptic-oblong, 6–30 × 5–15 cm, pubescent above, villose below, base acute-obtuse, margin entire or irregular, apex subacute or obtuse, lateral veins 9-12 pairs. Petiole to 1 cm long, dark brown, pubescent. Inflorescence umbellate clustered cymes, axillary. Flowers unisexual. Male flowers 3–12, tawny-tomentose, 2.5 cm long. Calyx lobes 5, ovate, to 9 mm, thick, tomentose. Corolla white, 1 mm across, tubular, thin-fleshy, tube 1 cm long, appressed-hairy whithout; lobes 5, ovate, reflexed. Stamens 10–12, filaments 4 mm long. Female flowers solitary, pedicel stout. Calyx lobes 5, ovate, margin revolute. Ovary globose, 4 mm across, tomentose; style 3 mm long; staminodes 5. Berry globose, 3–4.5 cm across, yellow when ripe, smooth, supported by the thick calyx. Seeds 4–6, compressed, oblong, brown, 5 mm long, shining with a wrinkled testa.

Common in dry deciduous forests and occasional in scrub forests. Fl.: Mar. – Jun. & Fr.: Jul. – Oct. Batrepalli (ATP), *KRKS* & *BSS* 39091.

Note: The leaves are used in rolling local cigarettes called beedies.

Diospyros sylvatica Roxb., Pl. Coromandel 1: 38. 1795; FBI 3: 559. 1882; FPM 2: 773. 1923; FAP 2: 555. 1997.

Local Name: *Tella gada, Gadaluti, Gada, Pasara gatha, Tella gatha, Gatha.*

Dioecious trees, to 18 m high, bark thick, dull grey to black, smooth; blaze yellow; branchlets puberulous. Leaves simple, alternate, estipulate, lamina 5–15 × 2–5 cm, elliptic, ovate, elliptic-oblong, often elliptic-obovate, base acute or cuneate, margin entire, apex acute, obtusely acuminate, glabrous, shining, membranous; lateral nerves 5–8 pairs, pinnate, obscure, intercostae reticulate, obscure; petiole 5–10 mm long, slender, glabrous. Flowers unisexual, white. Male flowers: 6–10 in axillary cymes, 6 mm across; peduncle 5 mm long, puberulous; calyx 3 mm long, campanulate; lobes 4, ovate, acute, pubescent outside; corolla 6 mm long, campanulate, lobes 4, triangular, obtuse; stamens about 24; filaments in unequal pairs, ciliate at apex. Female flowers; solitary or 3 together, 8–10 mm across; calyx much longer than in male, lobes ovate-oblong, acute; corolla campanulate; lobes triangular, obtuse; staminodes 4, inserted on the corolla and alternate with its lobes; ovary superior, globose, pubescent, 6–8-celled, 1-ovule in each cell; style 3–4, free, glabrous, bilobed with fringed margins. Fruit a berry, 1.6–1.8 cm across, globose, dark green, fruiting calyx enlarged, flat, foliaceous, spreading. Seeds 2–8, oblong, rugose, shining, albumen ruminate.

Rare, in deciduous forests. Fl. & Fr.: Jan. – Apr. Batrepalli (ATP), *KRKS* 39572, 39594.

OLEACEAE

Key to Genera

1a. Climbing or erect shrubs; corolla lobes 4–10; fruit globose berry ..... **Jasminum**

1b. Trees; corolla lobes 5–8; fruit obovoid, capsule .... **Schrebera**

JASMINUM L.

Key to Species

1a. Climbing shrubs; leaves chartaceous .... (2)

1b. Erect shrubs; leaves coriaceous; calyx glabrous or nearly so ...... **J. cuspidatum**

2a. Leaves 3-foliate; calyx lobes glabrous, 2 mm long .... **J. auriculatum**

2b. Leaves simple; calyx lobes pubescent, minute, triangular. .... **J. arborescens**

Jasminum arborescens Roxb., Fl. Ind. 1: 94. 1820. *J. roxburghianum* Wall. ex C. B. Clarke in Hook f., Fl. Brit. India 3: 595. 1882; FPM 2: 789. 1923; FAP 2: 561. 1997.

Local name: *Garuda malli, Erri malli.*

Climbing shrub, branches softly pubescent. Leaves opposite, elliptic, ovate-oblong, 2–3.5 × 1–2 cm, sparsely pubescent above, villous below on veins, base acute-obtuse, margin entire, apex apiculate, lateral veins 3 pairs; petiole 3 mm long, pubescent. Inflorescence terminal and axillary, lax, corymbose cymes. Flowers white, to 1.5 cm long. Calyx connate, 3 mm across, pubescent, 5-lobed; lobes minute. Corolla tube to 1 cm, glabrous; lobes 5–7, glabrous, ovate, 0.5 cm long. Stamens 2; anthers oblong, attached at the mouth of corolla tube, ovate, 1 mm long. Ovary 1 mm across in calyx tube. Berry globose, glossy, black when mature, 0.5 cm across. Seed 1, brown, testa thin.

Occasional, in dry deciduous forests especially in foot hills. Fl. & Fr.: Jun. – Sept. Kalasamudram (ATP), *KRKS* & *KRS* 36457.

Jasminum auriculatum Vahl, Symb. Bot. 3: 1 1794; FBI 3: 600. 1882; FPM 2: 790. 1923; FAP 2: 559. 1997.

Climbing scandent shrubs, pubescent. Lower leaves 3-foliolate; lateral leaflets small, 1 × 0.5 cm, ovate, middle leaflet large, 4 × 3 cm, chartaceous, densely pubescent on the nerves beneath when young, base rounded, margin entire, apex mucronate; petiole to 2 mm long. Flowers white in panicled cymes, axillary and terminal; bracts 0.5 cm, ovate, bracteoles 2, linear. Calyx 5-lobed, 2 mm. Corolla lobes 7 or 8, 8 mm, twisted, oblong-lanceolate, shortly mucronate. Stamens 2. Ovary 1 mm, oblong; style 4.5 mm, stigma bifid. Berry globose, 7 mm across, black when ripe.

Occasional, along the outskirts in deciduous forests. Fl. & Fr.: Jul. – Dec. Kalasamudram (ATP), *EC* 2860.

Jasminum cuspidatum Rottl., Ges. Naturf. Freunde Berlin Neue Schriften 4: 192. 1803; FAP 2: 559. 1997. *J. rigidum* Zenker Pl. Ind. 5. t. 6. 1835; FBI 3: 598. 1882; FPM 2: 790. 1923.

Local name: *Adavi malli, Naga malli, Pedda malli.*

Shrub, to 3 m high, rarely scandent, stems glabrous, young parts pubescent. Leaves opposite, elliptic-oblong or obovate, 1.8–5.8 × 0.6–1.5 cm, coriaceous, shining, nerves to 6 pairs, base acute, margin entire, apex acute, petiole to 4 mm long. Cymes terminal and axillary, corymbose,1–5-flowered. Flowers white, fragrant. Calyx 5-lobed, unequal, glabrous or near to pubescent, ovate, connate at base, acute at apex. Corolla 3.5 cm across; tube 2.5 cm long; lobes 5–7, obovate, 1.7 cm long, acute. Ovary oblong, 1.5 mm long; style 2.5 cm long; stigma forked. Berry oblong-globose, bilobed, shining, black when ripe. Seed 1 per carpel, testa thin.

Very rare, in dry deciduous forests. Fl. & Fr.: Jan. – May. Rathnachalam Anjaneya swamy temple (KDP), *KRKS* & *BSS* 40380.

SCHREBERA Roxb. *nom. cons.*

Schrebera swietenioides Roxb., Pl. Coromadel 2: 1. t. 101. 1789 & Fl. Ind. 1: 109. 1832; FBI 3: 604. 1882; FPM 2: 792. 1923.

Local name: *Magalinga, Pasarai.*

Deciduous tree, up to 10 m high; bark pale grey, nearly smooth, lenticellate; branches glabrous. Leaves imparipinnate, to 25 cm long; petiole 7 cm long; 7–9-foliolate; leaflets coriaceous, ovate-lanceolate, 2.5–8.5 × 1–3.5 cm, glabrous above, villous beneath, petiolules to 5 cm long, 2 cm long to the terminal leaflet, base obliquely attenuate, margin entire, apex acute, acuminate; lateral veins 4–6 pairs. Inflorescence trichotomously branched cymes, teminal and axillary, to 14 cm long, bracts caducous; pedicel to 5 mm long. Flowers yellowish-brown, to 8 cm long. Calyx truncate, unequally 4–7-toothed, 3.5 mm long, puberulous without, ciliate, acute. Corolla 1.3 cm across, salver-form; tube 1.5 cm long, white, pubescent at base; lobes 5–8, 0.5 cm long, pilose within, obtuse, undulate. Stamens 2, partly exserted; filaments 1.2 cm long, hairy, adnate to tube; anthers oblong. Ovary depressed, 2-lobed, capitate. Capsule obovoid, 5.5 × 3 cm, pustulate, loculicidal, 2-valved, woody, light brown, pendulous, dotted with scabrous specks, swollen base. Seeds 20–30, flat, winged, 2–3.7 cm, brown.

Very rare, in dry deciduous forests. Fl.: May – Jul. & Fr.: Aug. – Nov. Chinnapalli (ATP), *KRKS* & *BSS* 140370.

NYCTANTHACEAE

NYCTANTHES L.

Nyctanthes arbor-tristis L., Sp. Pl. 6. 1753; FBI 3: 603. 1882; FPM 2: 792. 1923; FAP 2: 565. 1997.

Local name: *Parijatham.*

Bushy shrub or small tree, to 4 m high, rough all over with stiff white hairs; young branches sharply quadrangular, hairy, scabrid. Leaves opposite, ovate to obovate, 3–7.5 × 1.8–4 cm, coriaceous, rough and scabrous, base rounded, margin undulate, irregular, apex acuminate; lateral veins 7 pairs, conspicuous; petiole 1 cm long. Inflorescence, trichotomously branched, axillary and terminal cyme, to 12 cm long; peduncle 4-angled, hairy, 6 cm long; bracts broadly ovate. Flowers yellowish white with rose tinge. Calyx tubular; tube 0.5 cm long. Corolla tube 1 cm long, lobes 6–7, obovate, glabrous; each lobe divided into 2 sublobes at apex. Anthers 2, at mouth of corolla tube. Capsule 1–2 × 0.8–1.8 cm, compressed, separating into 2, flat, 1-seeded carpels, prominently reticulate.

Common, in villages near temples, rare in forest. Fl.: & Fr.: May – Oct. Bhumaiahgari palle (KDP), *KRKS* & *KRS* 37620.

APOCYNACEAE

Key to Genera

1a. Armed shrubs or small trees; fruit berry .... **Carissa**

1b. Unarmed shrubs, trees, lianas (except *Catharanthus pusillus*-herb) ....(2)

2a. Corolla lobes overlapping to right ....(3)

2b. Corolla lobes overlapping to left ....(4)

3a. Small trees; cyme to 10 cm long, villous; flowers to 2 cm long........**Holarrhena**

3b Woody climber; corymbs to 2 cm long, pubescent; flowers to 5 mm long ....**Ichnocarpus**

4a. Corolla-throat with corona scales .... **Wrightia**

4b. Corolla-throat without corona scales ....(5)

5a. Corolla salver- shaped ....**Catharanthus**

5b. Corolla funnel-shaped..(6)

6a. Corolla yellow; fruit a drupe.. **Cascabela**

6b. Corolla other than yellow; fruit follicular mericarp ....(7)

7a. Xerophytic tree; corolla thick, variously coloured ....**Plumeria**

7b. Mesophytic liana; corolla thin, white.. **Vallaris**

CARISSA L. *nom. cons.*

Key to Species

1a. Leaves ovate-elliptic, apex acute-mucronate, ovary 2- ovuled...... **C. spinarum**

1b. Leaves obovate-oblanceolate, apex obtuse-emarginate, ovary 4-ovuled ....**C. carandas**

Carissa carandas L., Mant. Pl. 52. 1767; FBI 3: 630. 1882; FPM 2: 804. 1923; FAP 2: 570. 1997.

Local name: *Kalivi, Pedda vakakaya.*

Large, evergreen, thorny shrub, to 3 m high, with stout, sharp, horizontal spines at nodes, spines to 3 cm long, glabrous; branches terete, alternate, sometimes dichotomously branched. Leaves chartaceous, elliptic-oblong obovate-oblong, 2–6.5 × 1.5–3.5, glabrous, base obtuse-acute, margin entire, apex obtuse or emarginate; lateral veins 7–9 pairs, prominent below; petiole to 0.3 cm long. Inflorescence axillary, corymbose cymes; peduncle up to 1.1 cm long. Flowers white, to 2 cm across. Calyx-lobes 5, unequal, ovate, glabrous, 0.3 cm long. Corolla-tube up to 1.3 cm long, glabrous; lobes 5, elliptic-oblong, 1 cm long, mucronate. Stamens 5, attached from above the middle of tube; anthers 2.5 mm long, ovate. Ovary 2 mm long; style 8 mm long. Berry ellipsoid, 2.5 cm across, green when young, reddish pink when mature, black when ripen, glabrous, sweet with light sour. Seeds 2, 1 per cell, brown, 0.5 cm across, mucronate at apex.

Occasional, in scurb jungles. Fl. & Fr.: May – Aug. Idupulapaya (KDP), *KRKS* 41867.

Note: Fruits are edible and sold by local people as Pedda kali kayalu. Ripe fruits are cooling and acidic in taste and is much relished.

Plate 10

A. *Carissa carandas*, B. *Brachystelma ciliatum*, C. *Caralluma adscendens* var. *attenuata*, D. *C. adscendens* var. *fimbriata*, E. *Ceropegia bulbosa* var. *lushii*, F. *Ceropegia pullaiahii*

Carissa spinarum L., Mant. Pl. 559. 1771; FBI 3: 631. 1882; FPM 2: 805. 1923; FAP 2: 571. 1997. *C. paucinervia* A. DC., Prodr. 8: 333. 1844; FBI 3: 631. 1882; FPM 2: 806. 1923; FAP 2: 571. 1997.

Local name: *Kalikaya, Chinna kalivi, Konda kalivi.*

Shrub to 2 m high, sometimes spreading; spines divaricate, pubescent, to 2 cm long; branchlets pubescent. Leaves elliptic-ovate, lower pairs orbicular, 1.5–3 × 0.5–2 cm, pubescent on both surfaces, base acute, margin entire, apex mucronate; nerves 5–7 pairs, prominent on both sides; petiole to 1 mm long, pubescent. Inflorescence axillary, dichotomous cyme; peduncle to 1 cm long, villose, bracts 1 mm long, lanceolate, villose. Flowers white 1.5 cm long. Calyx lobes 5, ovate, villose, 1.5 mm long. Corolla-tube 0.6 cm long, distinctly pubescent; lobes 5, ovate-lanceolate, margin ciliate, acuminate. Anthers 5, attached to corolla tube. Ovary 1 mm across, covered by calyx lobes. Berry 0.6 cm across, pubescent, green at young, light rose at mature, black when ripe. Seeds in blackish pulp, brown, 0.3 cm across.

Very Common, in scrub forests. Fl. & Fr.: Jul. – Oct. Thummalapalle (KDP), *KRKS* & *SSR* 33644; Idupulapaya (KDP), *KRKS* & *BSS* 40328.

Note: Ripe fruits are edible and are commonly sold by local people in the markets.

CASCABELA Rafin.

Cascabela thevetia (L.) Lippold, Feddes Repert. 91: 52. 1980; FAP 2: 572. 1997. *Cerbera thevetia* L., Sp. Pl. 209. 1753. *Thevetia nerifolia* Juss. ex Steud., Nom. Bot. (ed. 2.) 2: 680. 1841 (*'nereifolia'*); FPM 2: 821. 1923.

Local name: *Pachaganneru, Adavi ganneru.*

Large evergreen shrub to small tree, up to 3 m high, latex milky; bark greyish, lenticellate. Leaves alternate, crowded, coriaceous, narrowly linear-oblong, 7–10 × 0.6–1 cm, glossy above, glabrous, base cuneate, margin entire, slightly revolute, apex acute; petiole to 0.5 cm long; stipules 3–5-fid, light brown, each ovate, glabrous. Cymes axillary, over topped by axillary branches; peduncle to 2.5 cm long; pedicel to 2 cm long; bracts ovate, 0.5 cm long, glabrous. Flowers yellow, to 5 cm across. Calyx-lobes 5, unequal, shortly united at base, lanceolate, to 1 cm long, acute. Corolla yellow (some light rose), 5 cm across, funnel-shaped; tube 4 cm long, narrowed; lobes 5, ovate, overlapping or twisted, 2 cm across, acute, glabrous. Stamens 5, at the throat of narrow tube, included; anthers oblong-cordate, 2 mm long, mucronate. Ovary depressed-conic, 2 mm across, 2-locular; style 1.5 mm long; stigma conic with basal ring. Drupe broadly turbinate, 3 cm across, slightly compressed laterally, glabrous, 4- angled. Seed solitary, hard, light brown, glabrous, 1.5 cm long, 1 cm wide.

Common, in wastelands and near water bodies. Fl. & Fr.: Sept. – Jan. Velpula (KDP), *KRKS* & *KRS* 37671.

CATHARANTHUS G. Don

Key to Species

1a. Erect herb, to 30 cm high; leaves ovate-lanceolate; flowers 1.5 cm long; fruit 2 cm long .... **C. pusillus**

1b. Erect shrub, to 75 cm high; leaves obovate-oblong; flowers 3 cm long; fruit 4 cm long .... **C. roseus**

Catharanthus pusillus (Murr.) G. Don, Gen. Syst. 4: 95. 1837; FAP 2: 572. 1997. *Vinca pusilla* Murr., Nov. Comment. Soc. Regiae Sci. Gott. 3: 66. 1733; FBI 3. 640. 1882. *Lochnera pusilla* K. Schum. in Engler & Prantl, Pflanzenfam. 4 (2): 145. 1895; FPM 2: 809. 1923.

Local name: *Palaganneru, Nelaganneru, Chinnabilla ganneru, Erimirapa, Katuka kalugu.*

Erect herb, to 30 cm high; branchlets terete below, glabrous. Leaves ovate-lanceolate, 2–5 × 0.8–1.7 cm, glabrous, base obtuse-acute, margin entire, apex acute; petiole to 0.5 cm long; stipule ovate, 0.2 cm long, glabrous. Flowers white, axillary or terminal, solitary or 3-flowered corymbs; pedicel to 4 mm long. Calyx-tube 2 mm long; lobes 5, lanceolate. Corolla salver-form; tube up to 7 mm long; lobes white, glabrous, 2.5 mm long, acuminate. Stamens 5, anthers 1 mm long at apex of corolla tube. Ovary of two carpels, 1.2 mm long; style 6 mm long, stigma columnar. Follicles 1.5–2.5 cm long, glabrous, slender. Seeds 3–8, oblong-elliptic, black, 3 × 1.2 mm, ribbed, ribs 12, longitudinal, tuberculate dotted.

Common, in agriculture fields. Fl. & Fr.: May – Sept. Thummalapalle (KDP), *KRKS* & *KRS* 33688.

Catharanthus roseus (L.) G. Don, Gen. Syst. 4: 95. 1837; FAP 2: 573. 1997. *Vinca rosea* L., Syst. ed. 10. 944. 1759; FBI 3: 640. 1882. *Lochnera rosea* (L.) Reich. Consp. Regini Veg. 134. 1828; FPM 2: 809. 1923.

Local name: *Billa ganneru, Nitya kalyani, Nitya sumangali.*

Erect shrub, to 75 cm high; branches pubescent, sub-quandrangular, green. Leaves oblong-obovate (or) elliptic-oblong, 4–6 × 1.2–2 cm, pubescent on both sides, base acute, margin entire, apex obtuse-apiculate; petiole to 0.5 cm long; stipules dissected, fimbriate. Flowers white or pink in axillary pairs, shortly pedicellate, to 3 mm long; stipules lanceolate, to 2 mm long. Calyx-tube 1.5 mm long; lobes 5, pubescent, ovate-lanceolate, 5 mm long. Corolla salver-form; tube cylindrical, pubescent, to 1.2 cm long; lobes 5, glabrous, obovate, overlapping, mucronate at apex. Anthers white, at mouth of the corolla tube. Ovary of 2 carpels; style filiform; stigma columnar. Follicles paired, pubescent, up to 4 cm long. Seeds 3–8, black, tuberculate.

Common, in gardens, near temples and field bunds. Fl. & Fr.: Throughout the year. Rekkamanu cross (ATP), *KRKS* & *KRS* 37928.

Note: Plant latex used for Bee bites and Scorpion stings. Medicinally used for treating cancers.

HOLARRHENA R. Br.

Holarrhena pubescens (Buch.-Ham.) Wall. ex G. Don, Gen. Hist. 4: 78. 1837; FAP 2: 573. 1997. *Echites pubescens* Buch.-Ham., Trans. Linn. Soc. London 13: 521. 1821. *Holarrhena antidysenterica* Wall. ex DC., Prodr. 8: 413. 1844; FBI 3: 644. 1882; FPM 2: 811. 1923.

Local name: *Kolamukki, Palavareni, Tedlapala, Palankudu, Kodaga, Kodisipala.*

Deciduous, small tree, to 4 m high; branches pubescent. Leaves membranous, pubescent, dense below, ovate-oblong, sometims elliptic, 7–18 × 5–10 cm, base obtuse-acute, margin entire, apex acute-acuminate; lateral veins 14 pairs; petiole to 0.6 cm long, pubescent. Inflorescence extra axillary, corymbose cyme; pedicels to 5 cm long, pubescent; peduncles to 1 cm long, pubescent; bracts ovate, pubescent, 0.3 cm long. Flowers white, 2 cm across. Calyx-tube 2 mm long; lobes 5, ovate-lanceolate, 1 mm long. Corolla-tube 1.3 cm long, pubescent, salver-form; lobes 5, white, glabrous, elliptic-oblong, obtuse, 1.2 cm long. Anthers 5, at the mouth of the tube, lanceolate. Ovary 0.2 cm across. Follicles distinct, linear, oblong, smooth, terete, slender, curved, 20–30 × 0.5 cm. Seeds oblong, 2 cm long; coma 4.5 cm long.

Very rare, in dry deciduous forest. Fl.: Jun. – Aug. Fr.: Sept. – Nov. Kalasamudram (ATP), *KRKS* & *BSS* 40414.

ICHNOCARPUS R.Br.

Ichnocarpus frutescens (L.) R. Br., Mem. Wern. Nat. Hist. Soc. 1: 62. 1811; FBI 3: 669. 1882; FPM 2: 820. 1923; FAP 2: 574. 1997. *Quirivelia frutescens* (L.) M. R. Almeida & S. M. Almeida, J. Bombay Nat. Hist. Soc. 90: 427. 1994. *Apocynum frutescens* L., Sp. Pl. 213. 1753.

Local name: *Palatheega.*

Climbing shrub, to 10 m long; branches slender, brown pubescent. Leaves elliptic-oblong, 3–7× 0.7–2.7 cm, glabrous above, puberulent below, more on veins; petiole to 1 cm long, pubescent; stipule minute. Inflorescence paniculate cyme, axillary and terminal, pubescent; peduncle to 8 cm at terminal, 2 cm at axillary; pedicel to 0.5 cm. Flowers small, to 4 mm long. Calyx cupular, puberulent, lobes 5, sub-equal, ovate, acute at apex. Corolla creamish-yellow, 6 mm across, salver-form; lobes 5, 2 mm wide, overlapping to right, villous within, acuminate apex. Stamens 5, attached below the middle; anthers oblong, sagittate, 1 mm long. Ovary two-carpellry. Follicles paired, cylindrical, 8–12 × 0.3 cm. Seeds linear, 1.7 × 0.2 cm, basally furrowed, coma apical, to 1 cm long.

Occasional, on road sides, borders of agriculture fields and dry deciduous forests. Fl. & Fr.: Aug. – Dec. Idupulapaya (KDP), *KRKS* & *BSS* 40306.

PLUMERIA L.

Plumeria rubra L., Sp. Pl. 209. 1753; FAP 2: 575. 1997. *P. acutifolia* Poir., in Lam. Encycl. Suppl. 2: 667. 1812; FBI 3: 641. 1882; FPM 2: 821. 1923.

Local name: *Devaganneru.*

Small tree, to 6 m high, fleshy branches with milky latex. Leaves spirally arranged, oblong-lanceolate or obovate-oblanceolate, 20–25 × 7–8 cm, glabrous to sparsely pubescent below, base acute, margin entire, apex cuspidate; lateral veins to 30 pairs. Inflorescence axillary, umbellately divided cymes with long peduncle. Flowers whitish yellow or whitish pink, fragrant. Follicles paired, to 30 cm long.

Ornamental plant in gardens and near temples. Fl. & Fr.: Jan. – Aug. Near Muniswara temple in Kalasamudram forest (ATP), *KRKS* & *BSS* 40415.

VALLARIS N. L. Burm.

Vallaris solanacea (Roth) Kuntze, Revis. Gen. Pl. 2: 447. 1891; FPM 2: 815. 1923; FAP 2: 577. 1997. *Peltanthera solanacea* Roth, Nov. Pl. Sp. 132. 1821. *Vallaris heynei* Spreng., in L., Syst. Veg. 1: 635. 1824; FBI 3: 650. 1882.

Large twining shrub, glabrous, stem dotted with lenticels. Leaves minutely dotted above, ovate to elliptic-oblong, 2–6 × 1–2.8 cm, glabrous, base acute, margin entire, apex acuminate, veins 10–14 pairs; petiole to 1 cm long; stipules dissected to 5–8 lobes, minute. Cymes extra axillary, dichotomously branched; peduncle 4 cm long; pedicel to 1.5 cm long. Flowers white, many. Calyx cupular, lobes 5, unequal, lanceolate, to 5 mm long, alternating with minute glands, hispid, acute. Corolla 2.5 cm across, salver-form; tube 5 mm long, cylindric; lobes 5, overlapping, 1.5 cm long, obtuse. Stamens 5, at throat of the tube; anthers lanceolate, 2 mm long. Ovary 2-locular, free. Follicles connate, slowly splitting later, cylindric, gradually tapering, 20 × 2.5 cm. Seeds oblong, 1 cm long, grey; coma silky.

Rare in foot hills of dry deciduous forest. Fl. & Fr.: Jun. – Sept. Batrepalli (ATP), *KRKS* & *BSS* 39581.

WRIGHTIA R. Br.

Key to Species

1a. Branchlets tomentose; flowers greenish or cream-coloured; follicles 2, connate through out .... **W. arborea**

1b. Branchlets glabrous; flowers white; follicle 2, connate at apex only .... **W. tinctoria**

Wrightia arborea (Dennst.) Mabb., Taxon 26: 533. 1977; FAP 2: 578. 1997. *Periploca arborea* Dennst., Schluess. Hort. Malab. 13. 23 & 25. 1818. *Wrightia tomentosa* Roem. & Schult. in L., Syst. Veg. 4: 414. 1819; FBI 3: 653. 1882; FPM 2: 816. 1923

Local name: *Tella pala, Tedla pala, Pala adavi-venkudu, Pala bariki.*

Small deciduous tree, up to 10 m high; bark light grey, corky; branchlets softly white tomentose. Leaves elliptic-obovate, 2–6 × 2–3 cm, tomentose both surfaces, base obtuse, margin entire, apex acuminate; nerves 7–10 pairs; petiole to 3 mm long, villous. Inflorescence terminal corymbose cyme; peduncles to 7 cm long, tomentose; pedicels to 2 cm long; bracts minute, ovate. Flowers white, to 3 cm across. Calyx-tube 1 mm long; lobes 5, pubescent, broadly ovate, apex acute. Corolla creamish-white; tube 0.2 cm long, spurred; lobes 5, oblong basally ciliate, overlapping. Anthers lanceolate, apically ciliate, 5 mm long. Ovary 2 mm long, 2- loculed. Follicles 2, connate throughout, 15–20 × 1.5–2 cm, pustulate, apically beaked, with longitudinal sutures on either side. Seeds slender, attenuate at apex; coma apically 2 cm long.

Occasional, in exposed hilly slopes of dry deciduous forest. Fl.: May – Jul. & Fr.: Aug. – Nov. Thummalapalle (KDP), *KRKS* & *KRS* 36403.

Wrightia tinctoria (Roxb.) R. Br., Mem. Wern. Nat. Hist. Soc. 1: 774. 1811; FBI 3: 653. 1882; FPM 2: 815. 1923; FAP 2: 578. 1997. *Nerium tinctorium* Roxb., Orient. Repert. 1: 39. cum tab. 1791.

Local name: *Palavara, Akupala, Doddipala, Teddapala, Ankudu, Palakordsha, Reppala, Chittiankudu, Jeddepala.*

Deciduous tree, to 10 m high, with milky latex; bark pale, scaly, smooth, branchlets glabrous. Leaves membranous, elliptic-oblong, 4–10 × 1.5–4 cm, glabrous, base acute, margin entire, apex cuspidate, nerves 8–10 pairs; petiole to 3 mm long. Inflorescence terminal, lax, trichotomous cyme, to 20 cm long. Flowers white, fragrant, to 2.5 cm across. Calyx-tube up to 0.2 cm long, glabrous; lobes 5, divided more than half way down, oblong, obtuse. Corolla white, up to 2.3 cm across; tube up to 0.3 cm; lobes 5, oblong, obtuse. Corona scales of two different lengths; shorter ones up to 0.2 cm long; longer ones up to 0.5 cm long; stamens adnate to the throat of corolla tube; anthers up to 0.7 cm long, pubescent on margins. Ovary up to 0.1 cm long; style to 0.6 cm long. Follicles 2, distinct, apically jointed, cylindrical, curved, 20 × 0.6 cm, green, glabrous, pointed. Seeds pointed at the apex with coma, often more than 2 cm long.

Common, in foot hills of dry deciduous forest and scrub forest. Fl.: Jan. – Mar. Fr.: Aug. – Oct. Kalasamudram (ATP), *KRKS* 39948.

ASCLEPIADACEAE

Key to Genera

1a. Pollen masses 2 in each anther locule .... **Secamone**

1b. Pollen masses 1 in each anther locule .... (2)

2a. Pollen masses pellucid on margin or at apex; anthers without membranous appendage .... (3)

2b. Pollen masses without pellucid layer; anthers with membranous appendage .... (7)

3a. Corona double, outer corolline and inner staminal .... **Leptadenia**

3b. Corona double, staminal .... (4)

4a. Leafless herbs with fleshy 4-angled stems and branches .... (5)

4b. Leafy herbs with terete stems .... (6)

5a. Inflorescence raceme; leaf rudiments 1–5 mm long, lanceolate, basally accompanied by pair of stipular glands .... **Caralluma**

5b. Inflorescence umbel; leaf rudiments broadly cordate, without stipular glands .... **Boucerosia**

6a. Corolla tubular, more than 1 cm long; lobes connate at apex and form a beak .... **Ceropegia**

6b. Corolla campanulate or minutely tubular, to 1 cm long; lobes free or connate at apex and not form any structure .... **Brachystelma**

7a. Erect shrubs .... **Calotropis**

7b. Twiners, stem herbaceous or woody .... (8)

8a. Corona single .... (9)

8b. Corona double .... (13)

9a. Corolla divided half way down, rotate; flowers large, to 1.5 cm across, lateral, in lax racemes or cymes .... (10)

9b. Corolla divided near to the base, rotate or campanulate; flowers small, axillary in umbellate cymes .... (11)

10a. Corolla lobes thin; corona processes dilated at base; leaves linear-lanceolate .... **Oxystelma**

10b. Corolla lobes thick; corona processes annular, fleshy; leaves broadly ovate .... **Holostemma**

11a. Corona segments laterally compressed, produced into spur at base .... **Pentatropis**

11b. Corona segments not compressed, spur absent .... (12)

12a. Straggling or twining shrubs; usually leafless; follicles smooth ... **Cynanchum**

12b. Twining shrubs; leafy, follicles echinate .... **Pergularia**

13a. Corolla rotate .... (14)

13b. Corolla urceolate, campanulate .... **Gymnema**

14a. Corolla green; fuit hard .... **Dregea**

14b. Corolla not green; fruit soft .... **Tylophora**

BOUCEROSIA Wight & Arn.

Key to Species

1a. Peduncle sparsely branched; flowers less than 50 in number per umbel .... **B. umbellata**

1b. Peduncle extensively branched; flowers more than 50 in number per umbel .... **B. lasiantha**

Boucerosia lasiantha Wight, Icon. Pl. Ind. Orient. 4(2): 1.t. 1286. 1848; FBI 4: 78. 1883. *Caralluma lasiantha* (Wight) N. E. Br., Gard. Chron. 12: 369. 1892; FPM 2: 862. 1923; FAP 2: 583. 1997.

Perennial, fleshy, branched herb, up to 50 cm high, bushy; branches erect, acute angled, not tapering, quadrangular. Leaves reduced, opposite, sessile, deltoid, glabrous, leaf scars spine-like. Inflorescence terminal, umbellate cyme; peduncle branched; bracts 2–5 mm long, linear, pedicels terete, 1–2 cm long. Flowers dark purple to reddish pink. Calyx 5-lobed, divided up to base; lobes linear, 5 mm long, apex acute, glabrous, glandular within. Corolla 5-lobed, rotate, *ca.* 1.5 cm long, apex acute or apiculate, dark brown, velvety within, long ciliate along the margin from upper half. Corona staminal, biseriate; outer 5-lobed, truncate; inner 5-lobed, apex acute. Pollinia 5; pollen masses paired, attached to carpusculum, yellow, waxy; gynostegium 2.6 mm long. Follicles paired, 10–15 cm long, linear-lanceolate, tapering, glabrous. Seeds oblong, glabrous, margin scarious, base rounded, dark brown; coma at apex loosely attached, 3–4 cm long.

Common, in foot hills of dry deciduous forest and scrub forest near rocky crevices in dry places, associated with spiny shrubs. Fl. & Fr.: Jun. – Aug. Batrepalli (ATP), *KRKS* & *BSS* 39609.

Boucerosia umbellata (Haw.) Wight & Arn., Contrib. Bot. Ind.: 34. 1834; FBI 4: 77. 1883. *Caralluma umbellata* Haw., Syn. Pl. Succ. 47, 1812; FPM 2: 861. 1923; FAP 2: 584. 1997.

Perennial, fleshy, branched herb, up to 60 cm high, bushy, branches erect, acute angled, not tapering, quadrangular. Leaves reduced, opposite, sessile, deltoid, glabrous, leaf scars spine-like. Inflorescence terminal umbellate cyme; bracts 2–8 mm long, linear; pedicels terete 1–2 cm long. Flowers dark brown to reddish pink. Calyx 5-lobed, divided up to base; lobes linear, 5 mm long, apex acute, margin minutely hairy, glabrous, glandular within. Corolla campanulate, 1.5 cm long; lobes 5, apex acute, purplish transverse bands, glabrous. Corona staminal, biseriate; outer 5-lobed, dark purple, truncate, 9 mm across; inner 5-lobed, apex acute. Pollinia 5, pollen masses paired, attached with corpusculum, yellow, waxy. Gynostegium 2.6 mm long. Follicles paired, 12–16 cm long, linear, tapering, glabrous. Seeds 4–8 mm, ovoid; coma 2.5 cm long.

Occasional, in foot hills of dry deciduous forest. Fl. & Fr.: May – Sept. Batrepalli (ATP), *KRKS* 37947.

BRACHYSTELMA R. Br. *nom. cons.*

Brachystelma ciliatum Arekal & Ramakrishna, Curr. Sci. 50 (3): 145. 1981; Albers & Meve, Ill. Handb. Succ. Pl. (Asclepiadaceae) 2: 30. 2002.

Perennial tuberous herb, tuber brown, subglobose-globose, 4 cm across and dorsiventrally compressed, 5.6 × 5.6 × 1.5 cm. Shoot erect, 30–35 cm high, 1.5–2 mm in diam., rarely branched, purplish, pubescent, buried part of stem 3–4 cm, white, glabrous. Leaf scales opposite, 3–4 pairs from the base, ovate, *ca.* 3 × 2 mm, brown puberulent, sessile. Leaves opposite in middle of the shoot, 4–5 pairs, linear, linear-lanceolate, 5–9 × 0.2–0.3 cm; nerves 9–12 pairs, semi-succulent, brown, midrib pubescent beneath, margins entire, pubescent, acute at base, acuminate at apex, pubescent, sessile with 2–3 ovate glands at base, green-pink; epidermal hairs unicellular. Flowers axillary, in 4-flowered umbels, pendulous; pedicels filiform, to 3.8 cm long, pubescent, stout and erect in fruits; bracts 4 at each node, each *ca.* 1 mm, ovate, pubescent. Calyx 5-lobed, minute, *ca* 0.5 mm, connate at base, ovate, pubescent. Corolla 5-lobed, connate at base, lobes oblong to ovate, *ca.* 5 × 2 mm, glabrous, narrowing at apex with 2 mm long, shiny cilia, greenish–yellow, dark pink at apex and middle; tube *ca.* 1 mm long, lobes reflexed when mature. Corona 2-whorled; outer whorl 5-lobed, *ca.* 2.5 mm across, cupular, thinly pubescent, dark pink dotted, 10-lobed at margin; inner whorl 5-lobed, alternate with stamens, ovate-oblong, extended to staminal column; staminal column 5-lobed, yellow, elliptic, winged. Pollinarium yellow, 60 μm across, erect; corpusculum brown. Ovary two carpellary, to 1 mm long; stigma light greenish-yellow, pentagonal, truncate, minute, sticky. Follicles 1–2, linear-lanceolate, 7–8 × 0.2–0.3 cm, light reddish brown, glabrous, unequal, curved at apex when young, acuminate and stronger when mature. Seeds 8–12, brown, ovate, 7–8 × 2–25 mm, thin; coma white, 2–2.5 cm long.

Very rare, in dry deciduous forests, especially near moist places and sandy soils. Fl. & Fr.: Nov. – Jan. Batrepalli (ATP), *KRKS* 41870.

Note: It grows among grasses and could not be located easily. The plant can be recognized easily when in flower and in fruit particularly in dehiscence.

CALOTROPIS R. Br.

Key to Species

1a. Flower bud cone-shaped, white-pink; corolla lobes recurved to back; corona rounded; large shrub to 2.5 m .... **C. gigantea**

1b. Flower bud sub-globose to depressed, white-purple; corolla lobes erect; corona bifid; small shrub to 1.5 m......................... **C. procera**

Calotropis gigantea (L.) R. Br., in Ait. Hort. Kew. ed. 2. 2: 78. 1811; FBI 4: 17. 1883; FPM 2: 832. 1923; FAP 2: 582. 1997. *Asclepias gigantea,* L., Sp. Pl. 214. 1753.

Local name: *Tella jilledu, Racha jilledu.*

Large bushy shrub to small tree, 2.5–5 m high; stem woody, terete, soft white woolly, gooved, latex milky. Leaves coriaceous, elliptic to oblong, woolly on both surfaces, 6–11 × 3–9 cm, base cordate with auricles, margin entire, apex acute, nerves prominent in lower side; petiole to 1 cm long. Inflorescence umbellate panicle, axillary. Flowers pink-white; flower bud cone-shaped, to 1.5 cm long. Calyx 5-lobed, 4–5 mm long, saucer-shaped, glandular within. Corolla rotate, white-pink, 5-lobed, recurved, glabrous. Corona staminal, fleshy, completely adnate and shorter than the staminal column, glabrous with a recurved spur at the base. Pollinia 5, yellow, waxy. Gynostegium 1.2–2.3 cm long. Follicles paired, horizontal, 3.8–10 cm long, oblong, glaucous, slightly curved. Seeds ovate to oblong, plano-convex, 0.9 × 0.6 cm; coma at apex, long silky.

Very common, in all dry areas and road sides, very dominant near mining site and surrounding places in scrub forest. Fl. & Fr.: Throughout the year. Thummalapalle (KDP), *KRKS, KRS, SSR* & *TP* 33651 (pink flowers); Thummalapalle (KDP), *KRKS, KRS, SSR* &*TP* 33650 (white flowers).

Calotropis procera (Ait.) R. Br., in Ait. Hort. Kew. ed. 2. 2: 78. 1811; FBI 4: 18. 1883; FPM 2: 832. 1923; FAP 2: 582. 1997.

Local name: *Jilledu.*

Bushy shrub, up to 1.5 m high, stem terete, latex milky. Leaves coriaceous, 5–15 × 2–10 cm, white cottony pubescent on both sides, base cordate, auricled, margin entire, apex acute; nerves 4–8 pairs, prominent; petiole 2–5 mm long. Inflorescence umbellate cyme, axillary; peduncle fleshy. Flowers white-purple, to 1.5 cm across; flower buds subglobose, depressed. Calyx 5-lobed, 4 mm long, margin papery, glabrous. Corolla rotate, pink, lobes 5, glabrous. Corona staminal, five, fleshy, completely adnate to staminal column and equaling or exceeding gynostegium, 5–6 mm long. Pollinia 5; corpusculum pink, adnate at apex; pollen sac yellow. Gynostegium 2 mm long, with two carpels. Follicles paired, horizontal, turgid, apex blunt, rounded. Seeds ovate, brown; coma at apex, long silky.

Very common, in all dry places and road sides, very dominant near mining site and surrounding places in scrub forest. Fl. & Fr.: Throughout the year. Thummalapalle (KDP), *KRKS* & *KRS* 36920; Somulavari palle (ATP), *KRKS* & *BSS* 40179.

CARALLUMA R. Br.

Caralluma adscendens (Roxb.) R. Br., Mem. Wern. Soc. 1: 14. 1810; FBI 4: 76. 1883; FPM 2: 861. 1923. *Stapelia adscendens* Roxb., Pl. Coromondel. 1: 28-29. 1795.

Key to Varieties

1a. Distal portion of the stem clearly attenuate; flowers only on the attenuated portion, singly or double on the axis .... var. **attenuata**

1b. Distal portion of the stem not attenuate; flowers on cylindrical axis near the tip .... var. **fimbriata**

Caralluma adscendens (Roxb.) R. Br.var. **attenuata** (Wight) Grav. & Mayur., Bull. Madr. Govt. Mus. n.s. N. H. 4(1): 13. 1931; FAP 2: 583. 1997. *Caralluma attenuata* Wight, Icon. Pl. Ind. Orient. 4 (1): 15. t. 1268. 1848; FBI 4: 76. 1883; FPM 2: 861. 1823.

Local name: *Kundeti kommulu.*

Fleshy branched herb, up to 1 m high; stem quadrangular, angles blunt, 1 cm across at base, apically tapering and cylindrical, glabrous; rooting at base. Leaves oblong-ovate, 2–3 mm long, glabrous, succulent, caducous. Flowers in terminal portion, axillary, solitary or paired, dark purple; pedicel terete, 3–5 mm long. Calyx 5-lobed, longer than the corolla tube, lanceolate, glabrous, *ca.* 0.5 cm long. Corolla lobes 5, ovate-lanceolate, margin pubescent with pendulous, purple hairs, dark pinkish-red from middle to apex, base broad with transverse purple lines, apex apiculate. Corona 2-seriate; outer annular, two filiform filaments, 1.5 mm long, alternating with anther lobes, glabrous; inner 5-lobed, flask-like, overlapping anther-lobes, *ca.* 1 × 0.4 mm, apex truncate or crenate. Stamens 5, 1.6 mm long;

pollinia 5, pollen masses solitary in each anther cell, yellow, waxy; caudicles light brown; pollen mass attached to corpusculum. Ovary with two carpels, to 1.5 mm long. Follicles usually solitary, 2.3–5.5 × 3.5–4.5 mm, beaked, glabrous. Seeds many, brown, ovate, 9.5 × 4.5 mm, base rounded, margin scarious; coma white at apex.

Very Common, in scrub forest and foot hills of dry deciduous forest. Fl. & Fr.: Through the year. Kalasamudram (ATP), *KRKS* & *BSS* 39174, *KRKS* & *KRS* 36460, *KRKS* & *BSS* 39173 B; Batrepalli (ATP), *KRKS* 37946; Rachakuntapalle (KDP), *KRKS* 37911; Thummalapalle (KDP), *KRKS* & *KRS* 36908.

Caralluma adscendens (Roxb.) R. Br. var. **fimbriata** (Wall.) Grav. & Mayur., Bull. Madr. Govt. Mus. (n.s.) N. H. 4(1): 12. 1931. *Caralluma fimbriata* Wall., Pl. Asiat. Rar. 1: 7. t. 8. 1830; FBI 4: 77. 1883; FAP 5: 42. 2008; FPM 2: 861. 1923.

Local name: *Kundeti kommulu*

Succulent, stem quandrangular, up to 40 cm high, not tapering, 1 cm across at base. Leaves succulent, 0.3 cm, ovate, acute. Flowers dark purplish-pink in axillary, 1–2 or 3; pedicel pendulous. Calyx-lobes ovate, 1.5 mm long, shorter than the corolla tube, glabrous. Corolla-lobes ovate, oblong-lanceolate, ciliate margin from middle and apex apiculate, base transversly striated with dark purple-pink lines. Corona 2-seriate. Follicle 8–10 cm long, tapering, glabrous.

Common, in scrub forest, foot hills of dry deciduous forest associated with spiny shrubs and wastelands. Fl. & Fr.: May – Aug. Sometimes through the year. Thummalapalle, (KDP), KRKS 36908; Kalasamudram (ATP) *KRKS* & *BSS* 39173 A; Pulivendula ghat near Anjaneyaswamy temple (KDP) *KRKS*& *BSS* 39671 & 39673.

CEROPEGIA L.

Key to Species

1a. Leaves 0.5–1 cm long; stem succulent; follicle to 15 × 1.5 cm............... **C. juncea**

1b. Leaves more than 1 cm long; stem not succulent; follicles to 12 × 0.8 cm...... (2)

2a. Leaves succulent, ovate-lanceolate; flowers in axillary umbels; corolla lobes ovate .... **C. bulbosa**

2b. Leaves not succulent, linear lanceolate; flowers in axillary cymes; corolla lobes linear-lanceolate.. (3)

3a. Plant up to 20 cm high; stem not twining; corolla lobes twining ...... **C. spiralis**

3b. Plant up to 2 m high; stem twining; corolla lobes not twining...... **C. pullaiahii**

Ceropegia bulbosa Roxb., Pl. Coromandel 1: 11. t. 7. 1795 & Fl. India 2: 28. 1824; FBI 4: 67. 1883; FPM 2: 857. 1923; FAP 2: 584. 1997.

Key to Varieties

1a. Leaves ovate-oblong or orbicular; stem greenish .... var. **bulbosa**

1b. Leaves linear-lanceolate; stem whitish green .... var. **lushii**

Ceropegia bulbosa Roxb. var. **bulbosa.**

Local name: *Nimmatai, Manchi manda.*

Twining herb, glabrous, slender perennial, fleshy, up to 2 m long, stem slightly furrowed, brown striated; root-tubers rounded or sub-globose, brown, 4 × 4 × 2 cm. Leaves chartaceous, glabrous, fleshy, orbicular, ovate-oblong 2.5–4.5 × 1.5–2.6 cm, base acute, margin entire to slightly lyrate, apex acute-apiculate; petiole to 0.5 cm, glabrous, brown-lined. Inflorescence umbellate cyme, axillary, 3–7 flowers in umbel, pedunculate with minute pedicel and bracts. Flowers light greyish, to 2.5 cm long. Calyx 5-lobed, linear lanceolate, glabrous, 2 mm long. Corolla greyish purple, 2.5 cm long, glabrous; tube oblong, inflated at base, hairy within; lobes 5, hairy within, connate at apex, up to 0.5 cm long. Corona two seriate; outer minute, saucer-shaped; inner slender, linear or sickle-shaped. Staminal column 1 mm long. Pollinea 5, in between inner corona and staminal column, yellow sticky, 0.5 mm across. Gynostegium 1 mm long. Follicles 2, slender, 11–15 cm long, glabrous, tapering towards apex, horizontal. Seeds ovate, brown, 3 mm long; coma white at apex.

Occasional, in foot hills of dry deciduous and scrub forests near marshy areas. Fl. & Fr.: Jul. – Nov. Nigidi village (ATP), *KRKS* 41852; Kalasamudram (ATP) *KRKS* 41897.

Note: Leaves and tubers are edible. Leaves slightly sour taste.

Ceropegia bulbosa Roxb., Pl. Cormondel. 1: 11. t. 7. 1795 & Fl. India 2: 28. 1824; FBI 4: 67. 1883; FPM 2: 857. 1923; FAP 2: 584. 1997. var. **lushii** (Grah.) Hook. f., Fl. Brit. India 4: 68. 1883; FPM 2: 857. 1923. *C. lushii* Grah., Bot. Mag. t. 3300. 1834.

Local name: *Colanimmata.*

Fleshy twiner, to 50 cm long, whitish green. Leaves very narrowly linear-lanceolate 3–5 × 0.5–1 cm, glabrous, succulent, base acute, margin entire, apex acuminate, midvein prominent; petiole to 3 mm long. Cymes axillary, leaves reduced towards apex, flowers present in apical axillaries only. Flowers greenish-white, to 1.5 cm long. Calyx lobes 5, acuminate, to 2 mm long. Corolla-tube 2 cm long, base inflated, 0.5 cm across; cylindrical tube to 1.5 cm long; lobes ovate-oblong, dark purple, apically connate, ciliate. Corona two series, to 2 mm diam. Pollinia five, pollen masses yellow, waxy with pellucid layer, attached by light-brown caudicle, dark brown corpuscle. Follicle to 5 cm long.

Very rare, in dry deciduous forests. Fl. & Fr.: Aug. – Dec. Batrepalli (ATP), *KRKS* 42002.

Ceropegia juncea Roxb., Pl. Coromandel. t. 10. 1795; FBI 4: 68. 1883; FPM 2: 856. 1923; FAP 2: 585. 1997.

Local name: *Ganupu nimmata.*

Fleshy twining perennial herb, glabrous, root tuberous. Leaves scale like, lanceolate, 1 × 0.2 cm, subsessile, fleshy, deciduous. Inflorescence axillary cyme or 3–5-flowered umbel, greenish; pedunculate, to 1 cm long; bracts linear, 3 mm long. Flowers greenish, pinkish dotted and purple, to 4 cm long. Calyx-lobes 5, linear-lanceolate, 4 mm long. Corolla brownish, purple, tube 2 cm long, curved, base slightly inflated; lobes 5, linear-lanceolate, 2 cm long, membranous, acute, with long, purple deciduous hairs near apex, pilose within, tube hairy near the mouth. Corona two series; outer corona of 5 bidentate-deltoid, ciliate lobes; inner linear, erect, hooked at apex. Pollinia 5, oblong, 0.6 mm, yellowish. Follicle narrow, 12.5–15 cm long, tapering horizontally arranged to each, glabrous. Seeds brown, ovate, thin, 0.5 × 0.3 cm; coma at apex.

Common, in scrub forest and foot hills of dry deciduous forest. Pulivendula ghat, near Rathnachalam Anjaneya swamy temple (KDP) *KRKS* & *BSS* 40033.

Ceropegia pullaiahii Kullayisw. *et al.*, Nordic J. Bot. 31: 166. 2013.

Local name: *Bodanimmatagadda.*

Twining, perennial herbs, to 2 m long, glabrous or sparsely hairy. Rootstock tuberous; tubers 3.8 × 3.6 × 2.7 cm, depressed globose; roots 3.2–4.5 cm, white, smooth. Leaves subsessile, linear 10– 14.5 × 0.2–0.4 cm, distinctly hairy above, scabrous below, puberulent at margin and apex, base acute, margin entire, apex acuminate; midrib puberulous beneath; petiole to 1–3 mm, distinctly puberulous. Inflorescence axillary, 3-flowered cyme; peduncle 3.5–5.0 mm, glabrous. Flowers bracteate; pedicel 6.5–6.6 mm long; bracts lanceolate, 2 × 0.3 mm, ovate, acute. Calyx lobes 5, 8–10 × 1.5 mm, linear, acuminate, glabrous. Corolla tubular, 5.5–8.5 cm long, erect; base of tube inflated, globose, 1.0–1.7 × 1. cm, green with purple striations; corolla tube narrow cylindrical, widened above or narrowly conical, 2.4–3.2 cm long, 0.5 cm wide at base, 1.3 cm wide at mouth, mouth dark purple above, yellowish white with purple blotching, densely hairy within; lobes linear, 2.0–3.6 cm long, erect, slightly overlapping, jointed at the apex, forming a narrow beak, with recurved margins; outer corona cupular, 0.7 cm in diameter, with lobes 2.5 × 3.5 mm, bifid, glabrous, purple with interspersed yellow striae; inner corona lobes linear, 4.5 mm long, erect, glabrous, basally swollen, yellowish-pink, dome-shaped. Staminal column, 2.3 mm wide; stigma, 1.2 mm wide; gynostegium sessile, 2.8–3.0 mm in length, apically narrow, with acute tip, purple. Pollinarium 1.0 × 1.3 mm; pollen sacs erect with short caudicle. Pollen key winged; corpuscle pink, rounded at apex, gradually narrowed downwards. Gynostegium 2 mm long. Follicles divergent, 10.3 × 0.3 cm, glabrous, brownish purple. Seed ovoid, flattened, 7 × 3 mm, dark brown, glabrous; coma 2 cm long, silky white.

Rare in dry deciduous forests. Fl. & Fr.: Oct. – Dec. Nigidi forest, Batrepalli (ATP), *KRKS* 41845, 41850.

Ceropegia spiralis Wight., Icon. Pl. Ind. Orient. t. 1267. 1848; FBI 4: 66. 1883; FPM 2: 856. 1923; FAP 2: 586. 1997.

Local name: *Teega nimmatai, Teega ganapa.*

Slender erect herb, up to 20 cm long, glabrous; root tubers sub-globose, brown, 3 × 3 × 2 cm. Leaves chartaceous, linear, 7–10 × 0.2–0.3 cm, glabrous, base acute,

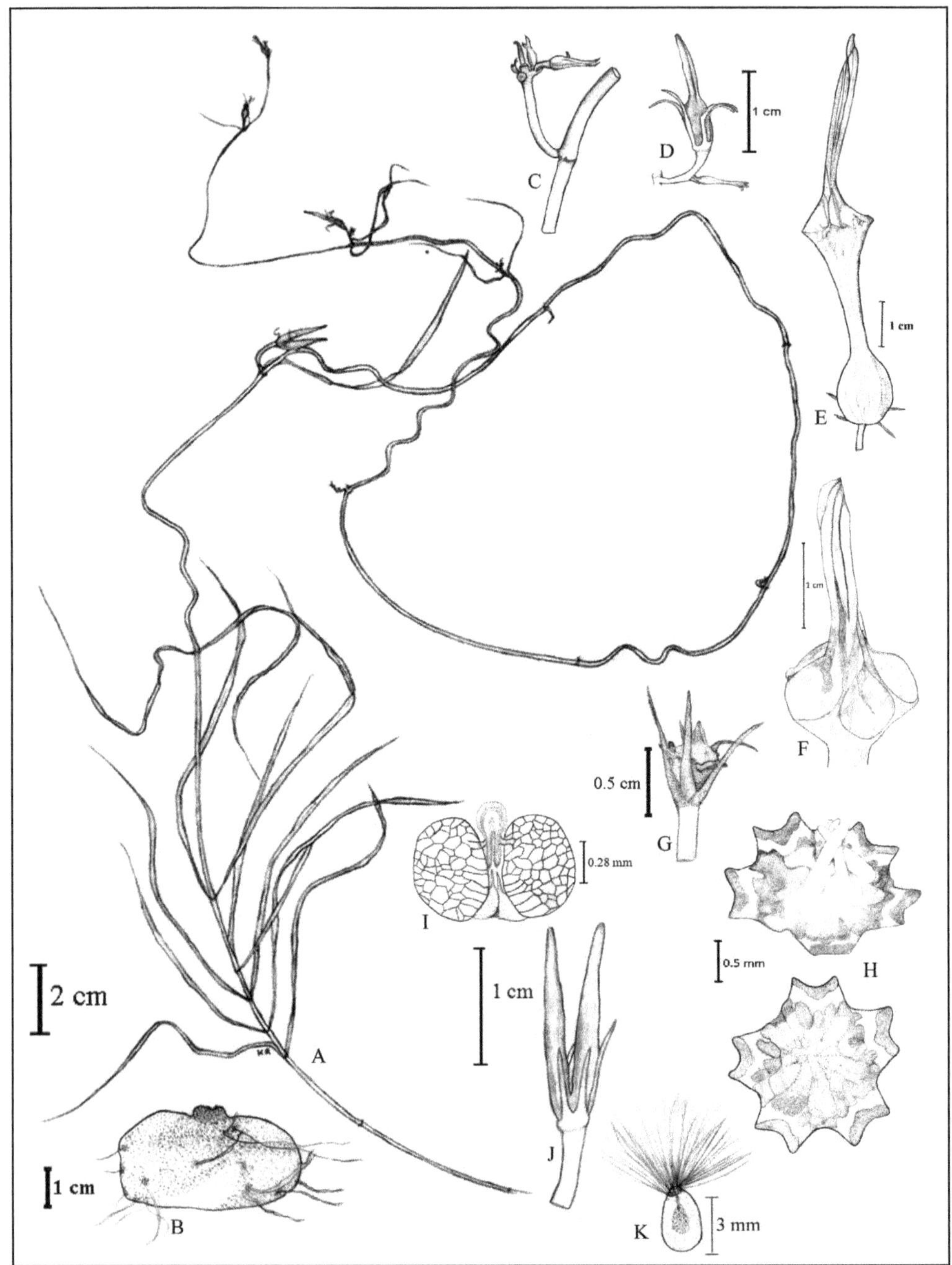

Figure 7: ***Ceropegia pullaiahii*** **Kullayis.** ***et al.***

A. Twig, B. Tuber, C. Cyme, D. Flower bud, E. Flower, F. Closer view of corolla tube mouth and lobes, G. Corona with calyx, H. Closer view of corona, I. Pollinarium, J. Young fruit, K. Seed.

Plate 11

A. *Ceropegia spiralis*, B. *Cynachum sarcomedium*, C. *Cryptolepis buchanani*, D. *Decalepis hamiltonii*, E. *Tylophora fasciculata*, F. *Hoppea dichotoma*

margin entire, apex acute; petiole to 2 mm long. Cymes axillary, 1–3-flowered. Flowers greenish, to 4 cm long. Calyx lanceolate, filiform, to 0.5 cm long. Corolla greenish-yellow, light purple, 4–5 cm long, stripes below, tube subcylindric, hairy within, above base, lobes 5, twisted, glandular hairy, linear lanceolate, apically connate. Corona biseriate; outer corona of bifid with deltoid lobes; inner corona linear, extended to staminal column. Pollinia oblong, 0.5 mm. Gynostegium 2 mm long. Follicles very slender, about 12.5 cm long, 0.2 cm diam. Seeds ovate, brown.

Very rare, in dry deciduous forest at high altitudes. Fl. & Fr.: Oct. – Dec. Chinnapalli (ATP), *KRKS* & *BSS* 40373.

CYNANCHUM L.

Key to Species

1a. Style apex rounded; follicles glabrous; seed entire; stem green .... **C. viminale**

1b. Style apex acuminate; follicles villous; seed toothed; stem whitish .... **C. sarcomedium**

Cynanchum sarcomedium Meve & Liede, Kew Bull. 67: 753. 2012. *Sarcostemma intermedium* Decne., in DC. Prodr. 8: 538. 1844; FBI 4: 27. 1883; FPM 2: 838. 1923; FAP 5: 44. 2008.

Local name: *Kondapala teega.*

Much branched, leafless, twining shrub; stem whitish, glabrous, latex milky. Inflorescence terminal umbel; pedicels erect, 6–9 mm long, pubescent. Flowers white, to 4 mm across. Calyx lobes 5, connate at base, free above, ovate, glabrous 1 × 1 mm, margin scarious, apex obtuse. Corolla white, rotate, lobes 5, free, oblong, glabrous. Corona staminal, biseriate; outer annular ring 10-lobed, membranous; inner five-lobed, adnate at the base of anthers. Stamens 5. Pollinia five; pollen masses solitary in each anther cell, yellow, waxy; caudicles dark-brown. Ovary 2 mm long, pubescent. Follicle usually single, 3.7–8 × 1 cm, linear, apex acute, glabrescent. Seeds oblong, base rounded, margin scarious, toothed; coma silky white.

Rare in dry deciduous forest, associated with rocks. Fl. & Fr.: May – Aug. Kalasamudram (ATP), *KRKS* & *BSS* 39122, *KRKS* 39185; *KRKS* 40077.

Cynanchum viminale (L.) L., Mant. Pl. 2: 392. 1771. subsp. **viminale.** Meve & Liede, Kew Bull. 67: 753. 2012. *Euphorbia viminale* L., Sp. Pl. 1: 452. 1753. *Sarcostemma viminale* (L.) R. Br., Mem. Wern. Soc. 1: 51. 1809, Printed in 1811 et Prodr. Nov. Holl. 464. 1810; subsp. *viminale*; Ali Fl. Pakistan 150: 30. 1983. L., Sp. Pl. 452. 1753. *Sarcostemma acidum* (Roxb.) Voigt, Hort. Sub. Cal. 542. 1845; FAP 2: 595. 1997. *Asclepias acida* Roxb., Fl. Ind. 2: 31. 1832.

Local name: *Konda pala, Somalatha, Chittipala, Dudhi, Pandiri jamudu, Akuleni jamudu, Palmakashtam.*

Much branched, leafless, climbing, succulent shrub; stem glabrous, cylindric, green, latex milky-white; internodes 10–15 cm long. Inflorescence terminal umbel;

bracts 2 mm long, lanceolate; pedicel to 1.3 cm long, glabrous, filiform. Flowers white, 0.7 cm across. Calyx 5-lobed, imbricate, ovate, pubescent, 1 mm long. Corolla whitish, overlapping; lobes 5, lanceolate, 0.6 × 0.25 mm, glabrous, apex acute. Corona staminal, cone-shaped, creamish, two-seriate; outer 5-lobed, cupular, adnate to back of staminal column, 1.5 mm high; inner erect, 5-lobed, adnate below on the back of staminal column, slightly curved, acute at apex, fleshy. Stamens 5. Pollinia 5; pollen masses solitary in each anther cell, yellow, waxy; caudicle brown. Gynostegium 2.5–3 mm long. Follicles paired, 4.5–7 cm long, 3–5 mm across, lanceolate, tapering towards both ends. Seeds brown, ovate, base rounded, margin thin; coma white to 2 cm long.

Common, in scrub forest and foot hills of dry deciduous forest. Fl. & Fr.: Mar. – Jul. Thummalapalle (KDP), *SSR* 33609, *KRKS* & *KRS* 36985.

DREGEA E. Meyer

Dregea volubilis (L. f.) Benth. ex Hook. f., FBI 4: 46. 1883. *Asclepias volubilis* L. f. Suppl. Pl. 170. 1781. *Wattakaka volubilis* (L. f.) Stapf, Curtis Bot. Mag. Subt. 8876. 1922; FAP 2: 599. 1997. *Marsdenia volubilis* (L. f.) T. Cooke., Fl. Bombay 2: 166. 1904; FPM 2: 846. 1923.

Local name: *Bandi guruginja, Bandi-gurga, Pedda-gurga.*

Extensive perennial woody, stout, large, hoary twiner with watery latex; branches lenticellate; bark brown, rough, some what corky. Leaves broadly ovate, 8–15 × 1.5–7 cm, glabrous, base rounded-truncate, margin entire, apex acuminate; veins 4–6 pairs, lowest 1 or 2 pairs arising from base; petiole 2–4 cm long, 2-glands near lamina. Inflorescence drooping umbellate, axillary cyme, 20–25-flowered; peduncle 2.5 cm long; pedicel to 2 cm long; bracts concave, orbicular, 0.7 mm across. Flowers creamy-yellowish, to 5 mm across, glabrous. Calyx-lobes 5, lanceolate, imbricate, 3 mm across, coriaceous, puberulous without, ciliate, subacute, glandular. Corolla greenish, 1.5 cm across, rotate, lobes 5, ovate, overlapping to right in bud, 7 × 4 mm, thick-herbaceous, ciliate, subacute, margin revolute. Corona staminal, uniseriate, 5-lobed, adnate at the base of staminal column. Pollinial bags 0.5 mm across, sub-sessile; receptacles 0.2 mm long. Ovary glabrous, globose, 1.5 mm long; style 1.5 mm long. Follicles solitary or paired and divaricate, oblong, 9 × 2.5–3 cm, gradually narrowed and apically blunted, fugaciously tomentose, hard. Seeds many, ovate, yellowish-pink, margin thin; coma soft, dull white.

Occassional in foot hills of dry deciduous and scrub forests, near agriculture lands and field bunds. Fl. Aug. – Oct. Fr.: Nov. – Jan. Kalasamudram (ATP), *KRKS* 39923.

GYMNEMA R. Br.

Gymnema sylvestre (Rotz.) R. Br. ex Schult., in Roem. & Schult., Syst. Veg. 6: 57. 1820; FAP 2: 589. 1997. *Marsdenia sylvestris* (Retz.) P.I. Forst., Austral. Syst. Bot. 8: 694. 1995. *Periploca sylvestris* Retz., Observ. Bot. 2: 15. 1781; FBI 4: 29. 1883; FPM 2.839. 1923.

Local name: *Podapatri, Chinnapalateega, Puttapodara, Puttabhadra*

Large, woody, much branched climber, young branches pubescent, latex milky-white. Leaves ovate or elliptic-obong or ovate-elliptic, 1.8–5 × 1.1–3.2 cm, pubescent along veins beneath, base obtuse or rounded, margin entire, apex acute, lateral veins 3–4 pairs; petiole to 1.2 cm long. Cymes extra-axillary; peduncles to 1 cm long; pedicel to 8 mm long. Flowers greenish yellow, to 2 mm across. Calyx 5-lobed, 1.5 mm long. Corolla greenish-yellow, campanulate, 3 mm long, fleshy. Corona corolline, uniseriate, 5-lobed, ridged. Stamens 5, 1.5 mm long. Follicles paired, lanceolate, to 7 cm long, apex beaked. Seeds with a thin, broad, marginal wing, brown, glabrous.

Common, in dry deciduous forest and road sides. Fl. & Fr.: Oct. – Feb. Rachakuntapalle (KDP), *SSR* 33604.

HOLOSTEMMA R. Br.

Holostemma ada-kodien Schult., Syst. Veg., 6: 95. 1820; FAP 2: 591. 1997. *H. annulare* (Roxb.) K. Schum., in Engl. & Prantl. Pflanzenfam. 4 (2): 250. 1895. *Asclepias annularis* Roxb., Fl. Ind. 2: 37. 1832. *Holostemma rheedei* Wall., Pl. Asiat. Rar. 2: 51. 1831; FBI 4: 21. 1883; FPM 2: 834. 1923.

Local name: *Dudipalatiga, Polagurugu, Palatura, Palajilledu, Teega jilledu.*

Perennial climber, to 8 m long; stem shining, glabrous, latex milky. Leaves chartaceous, ovate, 4.5–13 × 2.9–9 cm, puberulous below, base deeply cordate, lobes rounded, margin entire, apex apiculate or acuminate; nerves 4–6 pairs; petiole 1–2.5 cm long; glands 5–9 at base of the midrib above. Inflorescence umbellate, few-flowered cyme, axillary; peduncle to 1 cm long; pedicels to 5 mm long. Flowers pinkish-red or white, to 1.5 cm across; flower-buds cone-shaped. Calyx-lobes 5–7 mm long, glabrous, ovate. Corolla rotate, 1.5 cm across; lobes 5, leathery with scarious margin, ovate. Corona staminal, cone-shaped, uniseriate, annular ring arising from the base. Stamens 5. Pollinia pendulous, pollen sac oblong; caudicle brownish. Ovary sub-globose, 1.5 mm long; style 4.5 mm long; stigma conic, 5 angular. Follicles 1–2, turgid, 5–7 × 3.5 cm, boat-shaped, glabrous. Seeds ovate, oblong, brown, coma white, margin thin.

Occasional, in dry deciduous forest and scrub forest. Fl. & Fr.: Oct. – Jan. K. K. Kottala (KDP), *KRKS* & *KRS* 37659.

Note: Corona fleshy, sweet, edible because of having more nectar.

LEPTADENIA R. Br.

Leptadenia reticulata (Retz.) Wight & Arn., in Wight, Contrib. Bot. Ind. 47. 1834; FBI 4: 6. 1883; FPM 2: 850. 1923; FAP 2: 592. 1997.

Local name: *Thummipalaku, Mukkutummudu, Pala teega, Guttipala teega.*

Twining, perennial, pubescent shrub, to 7 m long, latex watery, viscous. Leaves ovate- lanceoate, chartaceous, 3–8 × 1.5–5 cm, pubescent above, densely pubescent beneath, base truncate-acute, margin entire, apex acute; nerves 5–6 pairs; petiole to 1.5 cm long. Inflorescence axillary, umbellate cyme; peduncle to 1 cm long; pedicel 0.5 cm long. Flowers green, to 3 mm across. Calyx cupular; lobes ovate, 1.5 × 1 mm, pubescent. Corolla yellowish-green, 6 mm across, lobes 5, 2 mm long, pubescent without. Corona two seriate; outer corona corolline; inner staminal, of 5 shallow annular lobes. Pollinia 0.5 mm across. Ovary half-inferior, 1 mm long; stigma conic, 2-fid. Follicles paired, cylindric, bluntly acute, 6 × 0.5 cm, acute both ends. Seeds oblong, winged, 3.5 × 1 mm, margin scarious; coma brownish white.

Occasional, in foot hills of dry deciduous forests, scrub forests and borders of agriculture lands. Fl. & Fr.: May – Aug. Thummalapalle (KDP), *KRKS* & *KRS* 36405.

Note: Plant extract is used as a tonic and stimulant.

OXYSTELMA R. Br.

Oxystelma esculentum (L.f.) Sm., Cycl. 25: 1813. *Periploca esculentum* L.f., Suppl. Pl. 168. 781. *Oxystelma esculentum* R.Br. ex Schult. in L. Syst. Veg. 6: 89. 1820; FBI 4: 17. 1883; FPM 2: 833. 1923.

Local name: *Dudipala.*

Very pretty, graceful, twining herb; stem terete, glabrous, latex milky. Leaves narrow linear-lanceolate, sub-coriaceous, 3–10 × 0.3–0.5 cm, glabrous, base rounded-subacute, margin entire, apex finely acuminate; veins 8–12 pairs; petiole to 2 cm, glabrous. Inflorescence axillary, corymbose cymes; peduncle terete, to 8 cm long; pedicel filiform. Flowers white or rose, marked with purple veins, to 1 cm across. Calyx lobes 5, 3.7 mm long, glabrous, glandular within. Corolla sub-campanulate or saucer-shaped, white or pink with purple nerves, villose; lobes 5, connate at base, erect, 1.5 × 1 cm. Corona two seriate, cone-shaped; outer corolline; inner staminal. Follicles glabrous, 3.5–7 × 1–1.8 cm, ovate-lanceolate, slightly curved, tapering at the apex. Seeds brown, ovate, 0.5 cm across; coma white.

Occasional, near water bodies associated with *Cyperus* sp. Fl. & Fr.: Aug. – Nov. Naganupula (ATP), *KRKS* & *BSS* 40494.

PENTATROPIS Wight & Arn.

Pentatropis capensis (L. f.) Bull., Kew Bull. 9: 284. 1955; FAP 2: 594. 1997. *Cynanchum capense* L. f., Suppl. Pl. 169. 1781. *Pentatropis microphylla* (Roth) Wight, Contrib. Bot. Ind. 52. 1834; FBI 4: 19. 1883; FPM 2: 835. 1923. *Asclepias microphylla* Heyne ex Roth, Nov. Pl. Spec. 177. 1821.

Local name: *Chekurutiva, Eddupulla teega, Chitapataku, Kakanasika, Poolapala.*

Slender, glabrous, twining shrub; stem terete, latex watery. Leaves chartaceous, succulent, ovate-oblong or elliptic-oblong, 1–2 × 0.5–1 cm, glabrous above, glabrescent below, base truncate, obtuse, margin entire, apex apiculate; veins 4–5 pairs, glands 1–2 at the base of midrib above; petiole to 0.4 cm long. Inflorescence axillary, 3–4-flowered, umbellate cyme, 1 cm across on slender filiform pedicels. Flowers purplish pink or green, to 3 mm across. Calyx 5-lobed; lobes lanceolate, 2 × 1 mm, glabrous, glandular within, margin scarious. Corolla rotate, 8–10 mm long; tube 2–3 mm long; lobes 5, 6–8 × 2.5 mm, lanceolate, apex acute, pubescent within. Corona staminal, uniseriate, lobes 5, connate at base, adnate to the back of the stamens. Stamens 5, 3 mm long; pollinia 5; pollen masses solitary in each anther cell, sac yellow, waxy, attached by caudicles. Ovary 1 mm long. Follicles usually single, 5 × 1 cm, lanceolate, acuminate, minutely winged, glabrous. Seeds broadly ovate, 5 × 3 mm, rounded at base, brown, margin thin, to 0.5 mm wide; coma silky white.

Common, in scrub forests and road sides. Fl. & Fr.: Jul. – Nov. Mabbuchintalapalle (KDP), *KRKS* & *KRS* 36944.

PERGULARIA L.

Pergularia daemia (Forssk.) Chiov., Result. Sci. Miss. Stefan. Paoli Somal. Ital. 1: 115. 1916; FAP 2: 594. 1997. *Asclepias daemia* Forssk., Fl. Aegypt-Arab. 51. 1775. *Daemia extensa* (Jacq.) R. Br., Mem. Wern. Soc: 1: 50. 1809; FBI 4: 20. 1883. *Cynanchum extensum* Jacq., Misc. 2: 353. 1781. *Pergularia extensa* (Jacq.) N. E. Br. in Harvey & Sonder, Fl. Cap. 4 (1): 758. 1908; FPM 2: 836. 1923.

Local name: *Dushtapu teega, Juttapaku.*

Slender, climbing shrub, with foetid smell, latex milky-white, hispid with sticky hairs. Leaves cordate, broadly ovate, 3–8 × 3–6 cm, hispid, base deeply cordate with incision up to 1 cm deep; lobes rounded, margin entire, apex acuminate or cuspidate; veins 4–6 pairs; petiole to 5 cm long, pubescent. Inflorescence axillary, racemose cyme; peduncle terete; bracts lanceolate. Flowers white-light pink, to 1.5 cm across. Calyx 5-lobed, free, imbricate, 2.5 × 1 mm, lanceolate, hairy without, glandular within, 2.5 × 1 mm. Corolla campanulate, 1 cm long, tube 3 mm long, pubescent within; lobes 5, 10 × 4 mm, lanceolate, apex acute, margin ciliate. Corona biseriate, 5-lobed; outer flap-like, membranous, 1 × 1 mm, adnate to the base of staminal coloumn, forked to below; inner 5-lobed, laterally compressed, 6 mm long. Stamens 5, 8 mm long. Pollinia 5; pollen masses yellow, waxy, caudicles brown, erect. Ovary 8 mm long. Follicles paired, 4–9 × 0.5–2 cm, pubescent, softly echinate. Seeds 7 × 4 mm, ovate, base rounded, brown; coma silky white.

Common, on the hedges along forest outskirts in scrub forests and road sides. Fl. & Fr.: Oct. – Jan. Rachakuntapalle (KDP), *SSR* 33611.

Note: Leafy juice used for catarrhal affections, infantile diarrhea and in rheumatic swellings.

SECAMONE R. Br.

Secamone emetica (Retz.) R. Br., ex Schult. Syst. Veg. 6: 124. 1820; FBI 4: 13. 1883; FPM 2: 829. 1923; FAP 2: 596. 1997. *Periploca emetica* Retz., Observ. Bot. 2: 14. 1781.

Local name: *Vanthula teega, Pala teega.*

Climbing wiry much branched shrub; stem green, glabrous when young, brown, thick, corky and deeply-cleft when mature. Leaves narrowly linear-lanceolate 3–5 × 0.3–0.6 cm, glabrescent above, pubescent on mid-vein above, punctate below, pubescent on mid vein below, base acute, margin revolute, apex acute, pointed, lateral veins 4–7 pairs; petiole 1–3 mm long, pubescent. Inflorescence axillary, dichotomous cyme; peduncles 8 mm, puberulous; bracts small, ovate, 1 × 0.5 mm, pubescent without; pedicels terete, 2–3 mm long, puberulous. Flowers yellowish, to 3 mm across. Calyx 5-lobed; lobes free, broadly ovate, 1 × 1 mm, glandular within. Corolla campanulate, 4 mm long; tube 1 mm long; lobes 5, 3 × 1.5 mm, ovate, glabrous. Corona staminal, uniseriate, 5-lobed, adnate to stamens. Stamens 5. Pollinia 5; pollen masses 2 in each anther cell, making total 20 pollen masses, yellow, waxy; caudicles brown. Gynostegium *ca.* 2 mm long. Follicles paired, divaricate, 5–6 × 0.5-1 cm, linear-lanceolate, apex beaked, minutely winged. Seeds ovate-oblong, base rounded, brown; coma silky-white.

Common, in dry deciduous forest along the rocky crevices. Fl. & Fr.: Jul. – Mar. Kalasamudram (ATP), *KRKS* & *BSS* 39565; 37972.

TYLOPHORA R. Br.

Key to Species

1a. Flowers greenish-yellow; follicles two, horizontal, 7 × 1.2 cm, tapering at apex .... **T. indica**

1b. Flowers reddish-brown; follicle usually one, 7 × 3 cm, spindle-shaped .... **T. fasciculata**

Tylophora fasciculata Buch.-Ham. ex Wight, Contr. Bot. India 50. 1834; FBI 4: 40. 1883; FPM 2: 842. 1923; FAP 2: 597. 1997.

Slender twining shrub, to 2 m long; stem pubescent, twining, trailing or climbing; roots as fasciculate tubers; latex watery. Leaves coriaceous, ovate-lanceolate or elliptic-lanceolate, 1.5–6 × 1–2.7 cm, glabrescent above, pubescent beneath, base rounded to truncate or acute, margin entire, apex acute; veins 3–6 pairs, arising from base; petiole 1–3 cm, pubescent. Inflorescence extra-axillary or axillary, branched umbels; peduncles filiform, 1–1.5 cm long, pubescent; pedicels 0.5 cm long; bracts lanceolate, 1.5 mm long, pubescent. Flowers chocolate-red or brown, to 1 cm across. Calyx 5-lobed, ovate-lanceolate, pubescent above. Corolla rotate, light brown or chocolate red, 6 mm across; lobes 5, 2.5 × 1.5 mm, elliptic-oblong, pubescent within, with silky hairs. Corona staminal, uniseriate, 5-lobed, completely adnate to staminal column, orbicular, apex acute. Pollen masses solitary in each anther cell, yellow, waxy; caudicles dark-brown. Gynostegium long, 2 mm long. Follicles usually solitary, ovoid, 5–7 × 2–3 cm, glabrous, spindle-shaped. Seeds many, 7 × 3 mm, broadly ovate, base rounded, brown, 0.5 mm wide; coma silky-white.

Rare in dry deciduous forests at high altitude near rocky areas. Fl. & Fr.: Apr. – Dec. Kalasamudram (ATP), *KRKS* 39558; Batrepalli (ATP), *KRKS* 280 (HJCB-E).

Tylophora indica (Burm. f.) Merr., Philipp. J. Sci. Bot. 19: 373. 1921; FAP 2: 598. 1997. var. **indica;** Maheshwari, Fl. Delhi 220. 1963. *Cynanchum indicum* Burm. f., Fl. Ind. 70: 1768. *Tylophora asthmatica* (L. f.) Wight & Arn., in Wight Contrib. Bot. Ind. 51. 1834; FBI 4: 44. 1883; FPM 2; 843. 1923. *Asclepias asthmatica* L. f., Supl. 171. 1781.

Local name: *Meka meyani aku, Verripaala, Kaka Paala, Kukka pala, Poda pacchali.*

Much branched climbing shrub; stem cylindric, tomentose, with long fleshy roots, latex watery. Leaves subcoriaceous, ovate-oblong or elliptic-oblong, 1–5.5 × 0.5–4 cm, pubescent along veins above, densely hairy beneath, base rounded or truncate, rarely cordate, margin ciliate-entire, apex acute to apiculate; veins 5–8 pairs; petiole 1–1.5 cm. Inflorescence axillary branched umbel, 20–30-flowered; peduncles terete, 1.5–2.5 cm long, pubescent; pedicels 1.5–2 cm, filiform; bracts ovate-lanceolate, pubescent, 1.5 mm. Flowers greenish-yellow, 0.8 cm across, glabrous. Calyx-tube minute; lobes linear-lanceolate, margin ciliate, apex acute, glandular within. Corolla rotate, 7 mm across; tube up to 1 mm long; lobes 5, 3 × 2.5 mm long, ovate, apex obtuse, pubescent within. Corona staminal, uniseriate; lobes 5, adnate below to staminal column, with free points above, 1.5–2 × 1–1.5 mm, ovate, fleshy. Stamens 5. Pollinia 5; pollen masses solitary in each anther cell, yellow, waxy; caudicles light-brown. Gynostegium 2 mm long. Follicles paired, divaricate, 5.4–7 × 0.7–1.2 cm, lanceolate, tapering, puberulous. Seeds ovate-oblong, 7.5 × 4.5 mm, base rounded, brown; coma silky-white.

Occasional, in scrub forests and foot hills of dry deciduous forests, associated with spiny shrubs. Fl. & Fr.: May – Aug. Kalasamudram (ATP), *KRKS* & *BSS* 39137; 39564.

PERIPLOCACEAE

Key to Species

1a. Corona processes connate with the stamens .... **Decalepis**

1b. Corona processes connate with the corolla .... (2)

2a. Corolla tube as long as or longer than than the lobes, velvety ..... **Cryptostegia**

2b. Corolla tube shorter than the lobes .... (3)

3a. Corolla lobes valvate .... **Hemidesmus**

3b. Corolla lobes imbricate, overlapping to the right .... **Cryptolepis**

CRYPTOLEPIS R. Br.

Cryptolepis buchanani R. Br. ex Roem.& Shult., Syst. Veg. 4: 409. 1819; FBI 4: 5. 1883; FPM 2: 826. 1923; FAP 2: 587. 1997.

Twining shrub, latex milky, stem light brown dotted, lenticellate. Leaves opposite, elliptic-ovate, 5–8 × 3–6 cm, glabrous, base obtuse, margin entire, apex apiculate; nerves 5–9 pairs, upwards and connected in a intramarginal nerve; petiole 1–2.5 cm long, glabrous; stipule minute. Inflorescence axillary cyme; peduncle 6 cm; secondary peduncle 2 cm; bract 2 mm long, ovate; pedicel 2 cm long. Flowers yellowish-white, to 2.3 cm long. Calyx imbricate, glabrous, 5-lobed; lobe ovate, 3 mm long. Corolla-tube to 8 mm long; lobes overlapping, 1.5 cm long; oblong-lanceolate, 1 cm long, acute. Corona corolline, 5-lobed, arising from the base of corolla, 1 mm long. Ovaries 1.5 mm. Follicles paired, stout, 7 × 1.5 cm, basally inflated, pointed above. Seeds obovoid, flat, 0.7 × 0.3 cm, apically narrowed.

Occassional at foot hills in shady regions of dry deciduous forest. Fl. & Fr.: Nov. – Feb. Rachakuntapalle (KDP), *KRKS* & *KRS* 37913.

CRYPTOSTEGIA R.Br.

Cryptostegia grandiflora R. Br., Bot. Reg. 5: t. 435. 1819; FBI 4: 6. 1883; FPM 2: 862. 1923; FAP 2: 587. 1997.

Local name: *Rabbaru pala teega, Rabbaru teega.*

Woody climbing shrub or liana, latex milky; stem glabrous when young, dotted with lenticels and woody when mature, hard. Leaves elliptic-oblong, 7–10 × 3–5 cm, sub-coriaceous, glabrous, glossy, base acute or truncate, margin entire, apex sub-acute; nerves 8–14 pairs, closely arranged, nearly parallel; petiole to 1 cm long. Cyme axillary, 3–5-flowered; peduncle 4 cm long; pedicel to 2 cm long. Flowers whitish-pink, to 6 cm long. Calyx-tube 7 mm long; lobes 5, ovate-lanceoalte, imbricate, 1.5 × 0.4 cm, thick, acute at apex. Corolla purple-pink, 5 cm across, funnel-shaped; tube 3.5 cm; lobes 5, broadly obovate-ovate, overlapping, twisted in bud. Corona of 5 lobes, fleshy, scales alternating with corolla, inserted near the base of tube, 2-fid. Ovary oblong; style 3 mm long; stigma conical; pollinia granular, 4 × 1.5 mm, sub-spatulate. Follicles paired, 8–12 × 1.8–4 cm, triangular, horizontal, tip blunt, glabrous, furrowed, hard, green. Seeds many in a cluster, 5–8 × 1.5–2.5 mm, ovate, winged, glabrous; coma silky-white.

Common, near water bodies, wastelands and road sides. Fl. & Fr.: Through out the year. Gollala guduru (KDP), *KRKS* & *BSS* 41963.

DECALEPIS Wight & Arn.

Decalepis hamiltonii Wight & Arn., Contrib. Bot. Ind. 64. 1834; FBI 4: 11. 1883; FPM 2: 828. 1923; FAP 2: 589. 1997.

Local name: *Nannari gadda, Maredu kommulu, Madina kommulu.*

Large glabrous climbing shrub; branches terete, glabrous, swollen and slightly winged at nodes. Leaves ovate, 2.5–7 × 1.5–5 cm, sub-coriaceous, base attenuate to truncate, margin entire to undulate, apex sub-acute to obtuse; petiole 1–2 cm long. Cymes 3-chotomous; peduncle to 2 cm long; pedicel to 1 cm long; bracts and bracteoles lanceolate, 1.5 mm long. Flowers creamish-yellow, to 3 mm across. Calyx-lobes oblong, tinged with brown, 2 mm long, chartaceous, valvate, acute at apex. Corolla creamish, 3 mm across, campanulate; lobes spreading, valvate, villous within, acute, recurved. Stamens connivent; pollinia horizontal; pollinial bags closely adherent, flat, 0.6 mm across; caudicle indistinct; receptacle minute. Corona two seriate, staminal; outer scale-like, truncate, 1 mm across; inner flat, adhering to gynostegium, 2 mm long. Glands 5 at the base of corolla, alternating with stamens, 1 mm, 2-fid. Ovary sub-globose, bicarpellary, 1 mm long; style 0.7 mm long; stigma obtuse or flat. Follicle oblong to lanceolate; cylindric, 5.5 × 3 cm; epicarp thick, wrinkled. Seeds ovate, 6 × 4 mm, chartaceous long tipped; coma white, silky.

Common, in dry deciduous forests. Fl. & Fr.: Jun. – Sept. Kalasamudram (ATP), *KRKS* 39520.

HEMIDESMUS R. Br.

Hemidesmus indicus (L.) R. Br., in Ait., Hort. Kew. 2.2: 75. 1811; FBI 4: 5. 1883; FPM 2: 825. 1923; FAP 2: 590. 1997. *Periploca indica* L., Sp. Pl. 211. 1753.

Key to Varieties

1a. Plant glabrous; leaves linear-lanceolate .... var. **indicus**

1b. Plant pubescent; leaves obovate-elliptic, oblong ....var. **pubescens**

Hemidesmus indicus (L.) R. Br. var. **indicus**

Local name: *Sugandhipala, Sarasaparilla, Nannari.*

Slender, twining shrub, glabrous, root stock aromatic. Leaves narrowly oblong, lanceolate, 3–8 × 0.5–1.5 cm, glabrous above, some what glaucous below, green or brown when dry, base rounded to cordate, margin sometimes revolute-entire, apex obtuse or broadly acute, tip mucronate; petiole to 0.5 cm. Cymes in extra axillary umbels. Flowers dark brown, crowded in subsessile cymes, to 2 mm across. Calyx lobes 5, broadly ovate, 2 × 1.8 mm, rounded to truncate, margin thin, apex acute. Corolla chocolate red, 6 mm across; lobes 5, ovate, 3 × 2 mm, acute. Corona corolline, uniseriate, scales small, fleshy; stamens 5; pollinia 0.5 mm across. Ovary 2 mm long. Follicles 8–17 cm long, slender, tapering to a point at the apex, glabrous. Seeds flattened, black, glabrous; coma silvery-white.

Occasional, in scrub forest and dry deciduous forest. Kalasamudram (ATP), *KRKS* 39556.

Note: Root decoction is used for curing high ferver and skin diseases. It is well known drug plant in Ayurveda and other stystems, used for purifying blood.

Hemidesmus indicus (L.) R. Br. var. **pubescens** (Wight & Arn.) Hook. f., Fl. Brit. India 4: 5. 1883; FPM 2: 825. 1923; FAP 2. 591. 1997. *H. pubescens* Wight & Arn., in Wight, Contr. Bot. Ind. 63. 1834.

Local name: *Sugandhi pala.*

Twining shrub; root stock aromatic; branchlets pubescent. Leaves elliptic to broadly oblong, orbicular, 3.5–7.5 × 2.2–5.4 cm, softly pubescent above, velvety tomentose below, base broadly acute to rounded, margin entire, apex rounded, apiculate; veins 4–5 pairs; petiole 3 mm long, pubescent. Inflorescence axillary cyme. Flower purplish brown, 6 mm across, up to 9 flowers in a cluster of cyme. Calyx lobes 5, ovate to lanceolate, 2.5 × 1 mm, villous without, base rounded, margin ciliate, apex shortly acuminate. Corolla dark purple, 0.8–1 cm across; lobes 5, ovate to obovate, 4.5 × 2.5 mm, acute. Corona corolline, 5-lobed, finger like processes, adnate to corolla-tube, alternating with lobes; stamens 5, 1.5 mm long; pollinia 0.6 mm across. Ovary 2 mm long. Follicles 14 × 0.5 cm, apex incurved, divaricate. Seeds many, linear-oblong, dark brown.

Common, in wastelands and dry deciduous forests. Fl. & Fr.: Oct. – Dec. K. K. Kottala (KDP), *KRKS & KRS* 37660; Kalasamudram (ATP), *KRKS & BSS* 40417.

LOGANIACEAE

STRYCHNOS L.

Key to Species

1a. Trees, to 15 m high; leaves elliptic, basally 5-veined; petiole to 1 cm long; fruit yellow, hard, to 5 cm across; bark not furrowed........ **S. nux-vomica**

1b. Trees, to 10 m high; leaves oblong, basally 3-veined; petiole to 3 mm long; fruit black, not hard, to 1.5 cm across; bark furrowed........ **S. potatorum**

Strychnos nux-vomica L., Sp. Pl. 189. 1753; FBI 4: 90. 1882; FPM 2: 868. 1923; FAP: 602. 1997.

Local name: *Mushti, Mushini, Visha mushti, Pedda mushti.*

Deciduous tree, up to 15 m high, bark thin, yellowish grey, smooth, covered with tubercles; branches with axillary spines; spines short, sharp, strong. Leaves elliptic-orbicular, coriaceous, 3.5–7 × 3–6 cm, base rounded margin entire, apex rounded; veins 5, arise from base of lamina, lateral veins irregular, inconspicuous, convergent venation; petiole to 1 cm long. Inforescence terminal compound cyme; pedicel 3 mm; bracts 5 mm long; bracteoles 1.5 mm long. Flowers greenish-white, fragrant, 4 mm long. Calyx 2 mm long; lobes 5, pubescent. Corolla greenish, 8 mm across; lobes 5, to 2.5 mm long. Stamens 5; filaments 0.1 mm long. Ovary 1.5 mm long, pubescent; style 9 mm. Berry to 5 cm across, thick-shelled, orange-yellow, when ripe, shiny. Seeds 4, compressed, 2 cm across, diskoid, ashy-grey, with silky hair radiating from the centre, concave on one side and convex on the other; testa horny, embeded in pulp.

Common in core zone, near Rachakuntapalle, foot hills of dry deciduous forests. Fl. & Fr.: Mar. – Apr.; Fr.: Jun. – Aug. Rachakuntapalle (KDP), *KRKS* & *KRS* 37629.

Note: It is very useful medicinal plant. Strycnine is extracted from seeds.

Strychnos potatorum L. f., Suppl. Pl. 148. 1781; FBI 4: 90. 1882; FPM 2: 868. 1923; FAP 2: 603. 1997.

Local name: *Chilla ginja, Chilla, Induga, Indiba, Chilaka mushti, Katakamu.*

Deciduous tree, to 10 m high; bark dark black or brown, deeply vertically cracked, branches fluted, dichotomously branched. Leaves ovate or elliptic-oblong, 4–7 × 2.5–4 cm, glabrous, shining, bright green, base rounded-acute, margin entire, apex acute; veins 3–4 pairs, raised from mid vein, convergent venation; petiole 2–3 mm long; stipule ovate. Inflorescence axillary cyme; bracts 5 mm long; bracteoles 1.5 mm long; pedicel 2 mm long. Flowers whitish, 2.5 mm long. Calyx cupular; lobes 5, 1.5 mm across, imbricate, puberulous. Corolla whitish, 4 mm across; lobes 5, reflexed, 2.5 mm long. Stamens 5; anthers 1 mm long. Ovary 1 mm long, 2-celled, ovules few. Berry globose, 1.3 cm diam., black when ripe, glabrous, thin-shelled, tuberculate, shiny. Seeds usually 1(2), sub-orbicular, greyish, testa horny, bluntly lenticular, shining, with short adpressed, silky hairs.

Common, along the hilly slopes, some times on foot hills of dry deciduous forests. Fl. & Fr.: Feb. – May; Fr.: Oct. – Mar. Rachakuntapalle (KDP), *KRKS* & *SSR* 39020; Kalasamudram (ATP) *KRKS* & *BSS* 39120.

Note: Seeds are used to clear turbid water.

GENTIANACEAE

Key to Genera

1a. Ovary 2-celled; stamens connivent around style, exserted, dehiscence porous .... **Exacum**

1b. Ovary 1-celled; stamens away from style, included, dehiscence longitudinal .... (2)

2a. Corolla regular; flowers sessile or subsessile; corolla white.......... **Enicostemma**

2b. Corolla irregular; flowers petiolate; corolla coloured .... (3)

3a. Flowers yellow; stigma subentire.......... **Hoppea**

3b. Flowers pink or white; stigma deeply 2-lobed .... **Canscora**

CONSCORA Lam.

Key to Species

1a. Flowers sessile; filaments of the upper stamen not broadened below the anther .... **C. heteroclita**

1b. Flowers pedicillate; filaments of the upper stamen broadened below the anther .... **C. alata**

Canscora alata (Roth) Wallich, Numer. List. 4363. 1831. *Exacum alatum* Roth in Roemer & Schultes Syst. Veg., ed. 15 bis. 3: 159. 1818. *Canscora decussata* (Roxb.) Schult. & Schult.f., Mant. 3: 229. 1827; FBI 4: 104. 1883; FPM 2: 879. 1923; FAP 2: 604. 1997.

Annual erect herbs, glabrous, 7–60 cm high. Stems 4-angular, decussately branched basally and apically, wings of branches *ca.* 2 mm broad. Leaves sessile, ovate-lanceolate, 8–40 × 4–30 mm, 3-nerved, base attenuate, margin entire, apex acute. Cymes terminal or axillary, 1–7 flowered; pedicels winged, 2–25 mm long; bracts green, linear-lanceolate, 3–5 × 1–3 mm. Flowers 4-merous. Calyx green, acuminate, with 8 prominent wings, lobes-4, ovate, persistent, 8–10 × 2–5 mm. Corolla white, tubular, zygomorphic; tube 8–10 mm long; lobes-4, obovate, 2 larger lobes, 4 × 2 mm, 2 smaller lobes fused together, 3 × 2 mm. Stamens 4, anisomorphic, oblong, sagittate at base. Ovary green, globose, 4–6 × 1–2 mm; style, 4 mm long; stigma bilobed. Capsule brown, oblong-ovate, 6 × 2 mm. Seeds rectangular with shallowly sunken sides, *ca.* 0.3 × 0.2 mm; testa brown, reticulate.

Occasional, along the edges of streams and moist rocky crevices in deciduous forests. Fl. & Fr.: Sep. – Dec. Nigidi forest, Batrepalli (ATP), *KRKS* 40082.

Canscora heteroclita (L.) Gilg., Nat. Pflanzenfam. 4 (2): 76. 1895; FAP 2: 605. 1997. *Gentiana heteroclita* L., Mant. Altera 2: 560. 1770. *Canscora sessiliflora* Roem. & Schult. f., Mant. Syst. Veg. 3: 230. 1827; FBI 4: 104. 1883; FPM 2: 879. 1923.

Annual, erect herb, to 30 cm high, dichotomosly branched; stem 4-winged, glabrous. Leaves basally 3–4 pairs, broadly ovate, 0.7–1.5 × 0.4–0.6 cm, base rounded to obtuse, margin entire, apex acutely apiculate, petiole minute, upper leaves linear-lanceolate, 0.5 × 0.2 cm. Cymes dichasial, becoming monochasial, axillaty; peduncle to 5 cm long; central flowers sessile; lateral ones pedicellate; pedicel winged; bract linear to 0.5 cm long. Flowers pink-rose, to 5 mm long. Calyx lobes 4, 6 mm across, acuminate; tube 1 cm long, glabrous. Corolla rose to bright pink, 2 mm across; tube 8 mm long; lobes 4, obovate, to 4 mm long, obtuse. Stamens 4; anthers 1 mm long. Ovary cylindrical, 3.5 mm long, 2-lobed, one-celled. Capsule oblong-compressed, 0.7 × 0.2 mm, brownish. Seeds rounded or rhomboid, 0.2 mm across.

Rare near stream sides and moist places of dry deciduous forests. Fl. & Fr.: Sept.– Nov. Kalasamudram (ATP), *KRKS* 40082, 42022; Batrepalli (ATP), *KRKS* 286 (HJCB-E).

ENICOSTEMA Blume *nom. cons.*

Enicostema axillare (Lam.) A. Raynal, Adansonia 9: 75. 1969. subsp. **axillare**; FAP 2: 606. 1997. *Gentiana axillaris* Lam., Tabl. Encycl. 1 (2): 487. 1793. non Raf. 1828. nec Reichb. 1823. *Enicostema littorale* auct. non Blume; FBI 4: 101. 1883; FPM 2: 875. 1923.

Local name: *Vellaragu, Gulvidi, Nela-guli, Nela gulimidi, Guramadaku*

Perennial, erect herb, 7–15 cm high; stem 4-angular, tender parts white glaucous. Leaves decussate, lanceolate-oblong, densely on stem, 3–6 × 0.3–0.5 cm, base cuneate, margin entire, apex acute; petiole to 0.2 mm long. Cymes axillary, sessile clusters, fascicled, ebracteate. Flowers white, to 4 mm across, 5-merous. Calyx cupular, 3 mm long; lobes 5 (6), acute, glabrous. Corolla white, 2 mm across, 6 mm long; tube 4 mm long; lobes 5 (6), 2 mm long, glabrous. Stamens 4–5 (6). Anthers 1 mm long. Ovary 1.5 × 1 mm, glabrous, ovules many; style 1.2 mm long; stigma capitate. Capsule 3.5 × 3 mm, septicidally 2-valved. Seeds circular, minute, reticulate.

Occasional, in scrub forests, foot hills of dry deciduous forests and wastelands. Fl. & Fr.: Aug. – Nov. Velpula village (KDP), *KRKS* & *KRS* 37676.

EXACUM L.

Exacum pedunculatum L., Sp. Pl. 112. 1753; FBI 4: 97. 1883; FPM 2: 874. 1923; FAP 2: 608. 1997.

Annual erect herb, glabrous, up to 30 cm high. Stem 4-angular, pale green. Leaves opposite, elliptic-lanceolate, 1–2 × 0.2–1 cm, glabrous, base acute, margin entire, apex acute, basally 3–5-nerved; petiole minute, to 2 mm long. Cymes terminal, much branched dichasial, to 3 cm long; pedicel to 1 cm long. Flowers light blue, to

1.5 cm across. Calyx-lobes 4, dorsally winged, 0.5 cm, acuminate. Corolla lilac, 1.5 cm across; lobes 4, 0.8 cm long, obtuse, acute. Stamens 4, yellow; filaments 1.5 mm long. Ovary 2 mm long, 2-celled; style curved; stigma capitate. Capsule 0.5 × 0.4 cm glabrous. Seeds circular, minute.

Common, near water bodies and ponds and associated with *Cyperus* spp., especially in plains. Fl. & Fr.: Oct. – Jan. Kadiri-Rayachoty road side (ATP), *KRKS* & *SSR* 39070.

HOPPEA Willd.

Hoppea dichotoma Willd., Ges. Naturf. Freunde Berlin Neue Schriften 3: 435. 1801; FBI 4: 100. 1883; FPM 2: 877. 1923; FAP 2: 608. 1997.

Small herb, to 12 cm high, divaricately branched, stem and branches 4-angular, glabrous. Leaves elliptic-ovate, 6–7 × 2.5–4 mm, base cuneate, margin entire, apex apiculate, sessile, glabrous. Cymes lax, terminal; peduncle 0.7–1 cm; pedicel to 0.3 cm. Flowers pale yellow, 2 mm across. Calyx campanulate; lobes 4, elliptic-lanceolate, 3 mm long, shortly acuminate. Corolla lobes 4, oblong-ovate, 2 mm long, ridged within, acute. Stamens 4; anthers cordate. Ovary oblong-globose, 1.3 × 1 mm, style 1 mm. Capsule globose, glabrous, enclosed in persistent sepals. Seeds numerous, minute, pale to dark green, brown, muricate, elliptic.

Common, in moist places, near streams in dry deciduous forests. Kalasamudram (ATP), *KRKS* 40100; Batrepalli (ATP), *KRKS* 285 (HJCB-E).

BORAGINACEAE

Key to Genera

1a. Leaves crenate-serrate, oblique; style divided....................................... **Coldenia**

1b. Leaves entire or undulate, symmetric; style undivided.................................. (2)

2a. Flowers axillary or terminal, solitary; ovary 4-lobed; fruit of 4 nutlets attached to a carpophore .... **Trichodesma**

2b. Flowers in terminal helicoid or scorpioid spikes or racemes; ovary not lobed; fruit a drupe of 4 pyrenes, not attached to the carpophores .... **Heliotropium**

COLDENIA L.

Coldenia procumbens L., Sp. Pl. 125. 1753; FBI 4: 144: 1883; FPM 2: 892. 1923; FAP 2: 611. 1997.

Local name: *Hamsa padu, Chepputattaku.*

Annual, prostrate, scabrid herb, branches spreading, to 15 cm long, densely pilose; closely appressed to substratum. Leaves simple, obovate-oblong, 1–1.2 × 0.8 cm, pilose above, densely so below, base cuneate, margin crenate-dentate, apex obtuse, emarginate, white-gladular pubescent along the nerves; petiole to 1 cm long. Flowers white, minute, to 2 mm across, axillary, solitary, subsessile; stalk to 1 mm long, pubescent; bracts absent. Calyx-lobes 4, ovate, 1.5 mm long, thick, scabrous without, truncate at base, acute at apex. Corolla 2 mm across, tubular; tube 1.5

mm; lobes 4, ovate, imbricate, 0.5 mm across, acute. Stamens 4, included, attached at the middle of tube; filaments 0.4 mm; anthers ovate, 0.4 mm long. Ovary conic, 4-lobed, 1 mm long; ovules 1 per cell; style arms 2, capitate. Fruit drupe of 4 connate pyrenes, pubescent, minute protuberant like spines at ventral side, spiny at apex. Seed solitary in each pyrene.

Common, in all waste, moist places and near ponds. Fl. & Fr.: Jun. – Nov. Kalasamudram (ATP), *KRKS* 36474.

HELIOTROPIUM L.

Key to Species

1a. Erect herbs or sub-shrubs .... (2)

1b. Spreading or procumbent herb .... (5)

2a. Inflorescens a raceme; stamens apically connivent; leaves linear-lanceolate .... (3)

2b. Inflorescens a spike; stamens free; leaves other than above .... (4)

3a. Racemes to 10 cm long; pedicel to 1.5 mm long; bracts linear; flowers dense .... **H. bracteatum**

3b. Racemes to 3 cm long; pedicel to 1 mm long; bracts ovate; flowers lax .... **H. strigosum**

4a. Nutlets connate in pairs, angled, apically beaked; leaves to 8 cm long .... **H. indicum**

4b. Nutlets free, rounded or tuberculate, obtuse; leaves to 5 cm long .... **H. subulatum**

5a. Leaves glabrous, succulent .... **H. curassavicum**

5b. Leaves hairy, not succulent .... (6)

6a. Leaves white villous, spatulate; spike elongate to 7 cm long..... **H. ovalifolium**

6b. Leaves scabrous hairy, ovate-lanceolate; spike crowded in upper leaves < 1 cm .... **H. scabrum**

Heliotropium bracteatum R. Br., Prodr. 493. 1810; FBI 4: 151. 1883; FPM 2: 897. 1923; FAP 2: 612. 1997.

Local name: *Guttakandi.*

Annual, erect herb, up to 50 cm high; stems and branchlets softly sericeous. Leaves simple, linear-lanceolate, 2–4 × 0.3–0.5 cm, strigose-hirsute on both sides, base cuneate, margin revolute, 1-nerved; lateral nerves obscure; petiole up to 0.3 cm long. Racemes terminal; peduncle sparingly branched, up to 9 cm long; bracts large, lanceolate, 2 mm long; pedicels up to 1 mm long. Flowers white, to 3 mm across, densely arranged on peduncle. Calyx lobes 5, ovate, 2.5 mm, scabrous without, acute, ciliate margin. Corolla white, salver-form, 2.5 mm across; tube 2 mm, sparsely strigose or almost glabrous; lobes 5, broadly ovate, acute, mid-nerve yellowish. Stamens 5; anthers ovate, apically connivent. Ovary globose, 4-locular;

style diskiform above; stigma pointed, cone-shaped. Nutlets 4, rounded, scabrid, 1 mm diam., loosely connate. Seed brown, 1 mm diameter.

Common, in dry deciduous and scrub forests in plains and slopes associated with grass. Fl. & Fr.: Sept. – Nov. Thummalapalle (KDP), *KRKS* 36921.

Heliotropium curassavicum L., Sp. Pl. 130. 1753; FPM 2: 896. 1923; FAP 2: 612. 1997.

Annual, ascending or prostrate, often sub-erect herb, to 50 cm long; branchlets glabrous. Leaves glabrous, fleshy, aromatic, lanceolate-spatulate, 2–3 × 0.2–0.8 cm, base cuneate to decurrent, margin entire, apex subacute; veins obscure; petiole to 0.2 cm long. Spikes terminal or sub-terminal, to 5 cm long; peduncle to 5 cm long; pedicels minute or flowers sub-sessile. Flowers white, to 3 mm across on helicoid spike. Calyx-lobes 5, oblong, 2 mm, glabrous, acute. Corolla 2.5 mm across, funnel-shaped, campanulate; tube 2 mm; lobes ovate, 1 mm, subacute. Stamens 5; anthers oblong, 0.7 mm long. Ovary sub-globose, 1 mm across; stigma conical, shortly 2-fid. Fruit pyramidal with 4 nutlets, each 1 mm across, glabrous, margins thickened, covered with calyx lobes. Seeds dark brown, 1 mm across.

Occasional, in waste and moist places especially in ponds and saline soils. Fl. & Fr.: Aug. – Oct. Dorigallu (ATP), *KRKS* 41820.

Heliotropium indicum L., Sp. Pl. 130. 1753; FBI 4: 152. 1883; FPM 2: 896. 1923; FAP 2: 613. 1997.

Local name: *Nagadanti*

Annual, erect, hispid herb or sub-shrub, up to 80 cm high. Leaves simple, alternate, elliptic-ovate, 5–9 × 3–5 cm, pubescent, base attenuate, margin undulate-dentate, apex acute-obtuse; veins reticulate; petiole up to 6 cm long. Spike terminal or leaf-opposed, helicoid, apically spherical, to 15 cm long. Flowers pale violet or white, to 4 mm across. Calyx-lobes 5, linear-lanceolate, 2.5 mm long, sparsely scabrous without, acute. Corolla white, salver-form, 2 mm across; tube 2 mm long; lobes 5, ovate, obtuse. Stamens 5; anther lanceolate, tips protruded. Ovary globose, 4-locular; style diskiform; stigma obtuse. Fruit separating into 2-seeded, 2-pointed pyrenes, nutlets connate in pairs, angled, 2.5 mm diam., apically beaked, brown, longitudinally 4-ridged out side. Seed brown, 1 mm across.

Common, in dry deciduous forests and plains, near water bodies and ponds. Fl. & Fr.: Sept. – May. Kalasamudram (ATP), *KRKS* & *BSS* 39101.

Heliotropium ovalifolium Forssk., Fl. Aegypt.-Arab. 38. 1775; FBI 4: 150. 1883; FPM 2: 896. 1923; FAP 2: 613. 1997.

Local name: *Nugu danti.*

Annual ascending or sub-erect herb, to 30 cm high, white villous; branchlets softly white sericeous, arises from thick root-stock. Leaves simple, obovate-spatulate, 1–2 × 0.5–0.8 cm, white villous above, more below, margin entire, apically obtusely apiculate; veins obscure; petiole to 0.5 cm. Spike terminal, helicoids to 5 cm long.

Flowers white, to 3 mm across. Calyx-lobes 5, obovate-lanceolate, unequal, 1.5 mm long, strigose on both sides, acute, persistent. Corolla white, campanulate, 2 mm across; tube 1.5 mm long, strigose without; lobes 5, ovate, acute. Stamens 5; anthers linear-lanceolate, free, bluntly apiculate. Ovary globose; stigma acute, minutely pubescent. Nutlets 4, single-seeded, pubescent, connate, apically beaked 2 mm across. Seed brown, 0.5 mm across.

Common, in moist places of agriculture fields, wastelands and near water bodies. Fl. & Fr.: Jun. – Dec. Kalasamudram (ATP), *KRKS* & *KRS* 36466.

Heliotropium scabrum Retz., Observ. Bot. 2: 8. 1781; FBI 4: 152. 1883; FPM 2: 897. 1923; FAP 2. 614. 1997. *H. marifolium* Retz., Observ. Bot. 2: 8. 1781; FBI 4: 152. 1883.

Local name: *Nela danti.*

Annual or biennial, stout, much branched, procumbent or spreading, woody herb, up to 15 cm long, growing in tufts; branchlets strigose hairy, closely apressed to substratum. Leaves simple, ovate-lanceolate or linear-lanceolate, 1–1.5 × 0.2–0.4 cm, base attenuate, margin entire, apex acute, flat, scabrous; veins obscure; petiole up to 2 mm long. Racemes 3–4, sub-terminal, crowded among the upper leaves; bracts conspicuous, lanceolate; pedicels up to 2 mm long. Flowers white, with yellow tinge in centre, to 5 mm across. Calyx lobes 5, obovate to lanceolate, strigose on both sides, acute. Corolla white, salver-form, 5 mm across; tube 2 mm, lobes 5, broadly ovate, strigose-hirsute without. Stamens 5, anthers ovate, connivent. Ovary globose, style diskiform above, stigma conical, pointed. Fruit brown, 4 nutlets, shortly bristly, each sub-globose with one seed. Seed brown, 1 mm across.

Common, in wastelands, plains and foot hills of dry deciduous and scrub forests. Fl. & Fr.: Aug. – Dec. Rachakuntapalle (KDP), *KRKS* & *KRS* 36924; Kalasamudram (ATP), *KRKS* & *KRS* 36473.

Heliotropium strigosum Willd., Sp. Pl. 1. 743. 1798; FBI 4: 151. 1883; FAP 2: 614. 1997. *H. brevifolium* Wall. in Roxb., Fl. Ind. 2: 2. 1824; FPM 2: 897. 1923.

Erect, small, much branched strigose-hirsute herb, up to 20 cm high; branches white-strigose, hispid. Leaves simple, linear or linear-lanceolate, 1.5–2 × 0.2–0.3 cm, white strigose, base acute, margin revolute, apex acute. Racemes terminal or axillary; bracts narrow-lanceolate, to 3 mm long, midvein conspicuous (leaves reduced to bracts towards apex). Flowers white, to 4 mm across. Calyx up to 2 mm long, strigose; lobes 5, unequal, one lobe brodly lanceolate, others narrow-lanceolate. Corolla white, funnel–shaped, strigose without, tube 3 mm long; lobes 5, not clearly segmented, short, plicate in the sinuses. Stamens 5; anthers minute, within the tube. Ovary subglobose; style minute, to 0.5 mm long; stigma conical-shaped, 0.5 mm across. Fruit broad, 2 mm across, depressed in middle, connate with 4-nutlets, each nutlet sub-globose, strigose. Seed one in each nutlet, brown, 1 mm across.

Common, in plains and foot hills of open dry deciduous forests, near rocks. Fl. & Fr.: Aug. – Nov. K. K. Kottala (KDP), *KRKS* & *SSR* 39011.

Heliotropium subulatum (Hochst. ex DC.) Vatke, Oesterr. Bot. Zeitschr. 25. 166. 1875; FPM 2: 896. 1923; FAP 5: 46. 2008. *Tournefortia subulata* Hochst. ex DC. Prodr. 9: 528. 1845. *Heliotropium zeylanicum* auct. non. Lam.; FBI 4: 148. 1883.

Local name: *Natteraku.*

Annual or perennial, erect, scabrous hispid sub-shrub, up to 80 cm high; branches with tuberculate trichomes, stinging. Leaves simple, lanceolate, 2.5–5 × 0.5–1.5 cm, base cuneate to attenuate, margin entire or slightly undulate, apex acute, densely strigose; petiole to 1 cm long. Racemes terminal to 18 cm long; peduncles branched, up to 20 cm long; pedicels minute or sessile. Flowers whitish-yellow, distinctly arranged on axis. Calyx-lobes 5, broadly ovate, strigose, 1 × 1 mm, acute at apex. Corolla whitish yellow; tube 4 mm long, hirsute without; lobes 5, 3 mm across, incurved when young, patent later, caudate-acuminate. Stamens 5, inserted in inflated part of the corolla-tube at apex, surrounding the style; anthers sessile, ovate-oblong, bi-fid at the apex. Ovary sub-globose, hirsute; style glabrous, stigma short, oblong, densely hairy, included in the corolla tube. Fruit with 4 nutlets, 2 mm across, brown, rounded or tuberculate on the back, separating first in pairs, later free with each.

Common, weed in agriculture lands, field bunds and wastelands. Fl. & Fr.: Sept. – Jan. (Throughout the year based on water avalability). Batrepalli (ATP), *KRKS* 41900.

Note: Leaves are used as medicine for wounds; leaf extract red, when fresh leaves are crushed. Hairs stinging.

TRICHODESMA R. Br. *nom. cons.*

Key to Species

1a. Leaf base cordate; corolla light violet; glands of the throat 4 .... **T. sedgewickianum**

1b. Leaf base auriculate; corolla bluish or pinkish-white; glands of the throat 10 .... **T. indicum**

Trichodesma indicum (L.) R. Br., Prodr. 149. 1810; FBI 4: 153. 1883; FPM 2: 899. 1923; FAP 2: 615. 1997. *Borago indica* L., Sp. Pl. 137. 1753.

Local name: *Guvvagutti.*

Annual, erect much branched herb, up to 70 cm high, grey-villous; branches hispid with tuberculate hairs with itching sense. Leaves simple, ovate-oblong, 3–5 × 1–1.5 cm, white dotted above, pubescent below with normal hairs, with tuberculate hairs above, base auriculate, margin entire, apex obtuse or sub-acute; veins conspicuous below, 3 pairs; sub-sessile. Flowers bluish or pinkish white with brown ring in centre, axillary, solitary; pedicels to 1 cm long, pubescent. Calyx 0.8 cm long, accrescent in fruit; lobes 5, lanceolate, hastate at base, margins strongly ribbed, villous. Corolla blue-white with brown ring at centre, 1 cm across; tube 4 mm long, throat with yellow glandular spots, lobes 5, rounded, apiculate and

Plate 12

A. *Heliotropium subulatum*, B. *Trichodesma sedgewickianum*, C. *Cordia monoica*, D. *Argyreia kleiniana*, E. *Rivea ornata*, F. *Xenostegia tridentata* subsp. *hastata*

recurved in flower. Stamens 5, exserted, filaments 2–3 mm long; anthers 1.2 cm, pilose, twisted. Ovary 1 mm, style 7.5 mm, sickle-shaped. Fruit pyramidal, 4-ribbed, nutlets ovoid, smooth. Seeds brown, ovate 3 × 2 mm.

Common, in wastelands and agricultural weed. Fl. & Fr.: May – Nov. Thummalapalle (KDP), *KRKS* & *KRS* 36409.

Trichodesma sedgewickianum S. P. Banerjee, Bull. Bot. Soc. Bengal 16: 10. 1962; FAP 2: 616. 1997. *T. amplexicaule* auct. *non*. Roth; FBI 4: 153. 1883; FPM 2: 899. 1923.

Annual, erect, hispid herb, up to 75 cm high; branchlets with tuberculate hairs. Leaves simple, broadly oblong, 4–9 × 2–4 cm, prominently hispid, tuberculate hairy, base cordate, margin entire, apex obtuse-acute, coriaceous; sub-sessile. Flowers light violet, white at the centre, axillary, solitary; pedicels up to 2.5 cm long; hispid. Calyx lobes 5, hastate or çordate at base, scabrous along nerves without, sparsely hirsute within, acute at apex. Corolla light violet, 0.7 cm across, salver-form, tube inflated over stamens, throat with 4 unequal spots towards one side, lobes 5, broadly ovate, obtuse, apiculate, not recurved. Stamens 5, included, anthers 3 mm, sericeous on the back. Ovary 1.5 mm, style 4 mm; stigma capitate. Fruit spinous, hairy, nutlets smooth, covered with persistent calyx, 0.5 cm long. Seeds thick brown, oblong-ovate.

Very rare, in dry deciduous forest at slopes, near stream sides. Fl. & Fr.: Sept. – Jan. Vemula (KDP), *KRKS* & *BSS* 41958.

CORDIACEAE

Key to Genera

1a. Calyx-lobes connate from above the middle, accrescent in fruit; style twice forked; arms 4; stigma simple; drupe pulpy .... **Cordia**

1b. Calyx-lobes divided to below the middle, not accrescent in fruit; style once forked; arms 2; stigma capitellate .... **Ehretia**

CORDIA L.

Key to Species

1a. Leavs fulvous-tomentose or scabrous below .... **C. macleodii**

1b. Leaves glabrescent or thinly pubescent below .... (2)

2a. Leaves rough above (like sand paper), spotted with whitish epidermal bulbous hairs; corolla < 0.5 cm across; filaments glabrous at base .... **C. monoica**

2b. Leaves glabrescent above, not spotted; corolla > 0.7 cm across; filaments hairy at base .... **C. dichotoma**

Cordia dichotoma Forst. f., Prodr. 18. 1786; FAP 2: 617. 1997. *C. obliqua* Willd., Phytogr. 4. t. 4. 1794; FBI 4: 137. 1883; FPM 2: 887. 1923.

Local name: *Bankiriki, Iriki, Nakkeri.*

Moderate-sized deciduous tree, up to 10 m high; branches drooping, hairy when young; bark thick, grey or brown with shallow furrows and wrinkles. Leaves simple, variable in size and shape, but usually elliptic-oblong or orbicular, 4–11 × 3–8 cm, glabrescent on both sides, base obtuse or cuneate, margin repand-crenate, apex acute; hairy at base of lateral nerves; petiole up to 4 cm long. Corymbose panicles terminal, to 20 cm long; peduncle up to 5 cm long; pedicel to 5 mm long. Flowers white to cream-coloured, fragrent, to 6 mm across. Calyx cupular, 6 mm across; lobes 5, irregular-dentate. Corolla white or cream, 8 mm across; lobes 5, recurved. Stamens 5, villous at base; filaments 4 mm long; anthers 1.5 mm long. Ovary globose, 2 mm long; style 2 mm long, 4-lobed. Drupe ovoid, cream-coloured, mucilaginous, base covered with accrescent calyx, apex mucronate. Seeds solitary, light brown, testa hard, 0.5 cm across.

Occasional, on road sides, wastelands and agriculture fields. Fl.: Dec. – Jan. & Fr.: Apr. – Jul. Idupulapaya (KDP), *KRKS* & *BSS* 39089.

Cordia macleodii Hook. f. & Thomson, J. Proc. Linn. Soc. Bot. 2: 128. 1858; FBI 4: 139. 1883; FPM 2: 888. 1923; FAP 2: 619. 1997.

Local name: *Pedda botava, Iriki.*

Small deciduous tree, up to 10 m high; branchlets brown-tomentose; bark thick, grey, corky. Leaves simple, alternate, ovate, 10–15 × 8–10 cm, rugose above, fulvous tomentose below, base truncate or apiculate, margin distantly denticulate, apex apiculate-acute, sometimes obtuse; basally 3-nerved; lateral veins 3–4 pairs; petiole to 4 cm long, pubescent. Corymbs large, terminal or axillary, to 7 cm long; peduncles to 5 cm long; pedicels 0.5 cm long, brown villose. Flowers white, to 0.6 cm long. Calyx-tubular, 3 mm long; lobes 5, brown villose, oblong-lanceolate, 4 mm long, acute at apex. Corolla white, campanulate; lobes 5, broadly ovate, acute at apex. Stamens alternating the corolla lobes; anthers oblong, white. Ovary globose, villose, 2 mm across; style arms 4; stigma capitate. Drupes ovoid, up to 2 × 1 cm, glabrous, acute, yellowish, 1–2-seeded, tomentose, fruiting calyx ribbed. Seeds 1–2, brown, testa hard, 0.4 cm across.

Rare in dry deciduous forests. Fl.: Nov. – Jan. & Fr.: Mar. – Jul. Idupulapaya (KDP), *KRKS* & *SSR* 39059.

Cordia monoica Roxb., Pl. Coromandel 1: 43. t. 58. 1796; FBI 4: 137. 1883; FPM 2: 888. 1923; FAP 2: 619. 1997.

Local name: *Bankeera.*

Small tree or shrub, up to 3 m high, stunted growth; branches brown fulvous-tomentose at younger parts. Leaves elliptic-oblong or elliptic-ovate, variable in tomentum, 2.5–3 × 2.1–3 cm, tomentose below, strigose above, bulbous hairy, base obtuse, margin entire, apex obtuse; lateral veins 3–4 pairs; petiole to 2 cm long, tomentose. Corymbose cymes terminal, 3 cm across; peduncle to 2 cm, tomentose; pedicels 0.5 cm. Flowers white, fragrant, 5 mm across. Calyx cupular; tube 5 mm long, brown tomentose; lobes 5, minute, ovate, acute, 0.2 mm long. Corolla white,

4 mm across; tube 5 mm long; lobes 5, acute. Stamens 5; filaments glabrous at base, anthers 1 mm long. Ovary ovoid, 1.5 mm across. Drupes ovoid, bright orange-red, gummy, fruiting calyx pubescent, lobes erect.

Occasional, in plains and foot hills of the dry deciduous forests. Fl.: Jul. – Sept. & Fr.: Oct. – Dec. Near Ratnachalam Anjaneyaswamy temple, Pulivendula Ghat (KDP), *KRKS* & *BSS* 39669.

EHRETIA R. Br.

Ehretia laevis Roxb., Pl. Coromandel 1: 42. 1796. *E. pubescens* Benth., in Royle, Ill. Bot. Himal. 306. 1836; FPM 2: 891. 1923; FAP 2: 621. 1997. *E. laevis* Roxb. var. *pubescens* (Benth.) C. B. Clarke, in Hook. f., FBI 4: 142. 1883.

Small tree up to 4 m high, with stout pale brown branchlets, bark blackish brown. Leaves simple spatulate, obovate-oblong or elliptic-oblong, 4–9 × 3–5 cm, tuberculate hirsute above, tomentose below, base acute or attenuate, margin entire, apex obtuse-acute or apiculate; lateral veins 4–5 pairs; petiole to 1.5 cm long, white tomentose. Corymbose cymes terminal, dichotomousely branched; peduncle to 3 cm long; pedicels 0.5 cm long. Flowers white, 3 mm across. Calyx-lobes 5, oblong, pilose without, apex obtuse, 2 mm long; tube 1 mm long. Corolla white, 5 mm across; tube 2 mm; lobes 5, oblong, 2.5 mm, acute. Stamens 5; filaments 1.5 mm long, basally hairy; anthers 1 mm long, ovate oblong. Ovary 0.5 mm long, globose; style 2-fid, 2 mm long. Drupes globose; red when ripe, with persistent calyx, 4-lobed, mucronate, 0.5 cm across. Seeds 4, brown, longitudinally striated, 3 mm across.

Very rare, in dry deciduous forests near streams. Fl.: Dec. – Mar. & Fr.: Apr. – Jul. Idupulapaya (KDP), *KRKS* & *SSR* 39061; Chinnapalli (ATP), *KRKS* & *BSS* 37994.

CONVOLVULACEAE

Key to Genera

1a. Styles 2, free; petiole to 2 mm long (prostrate; flowers blue) .... **Evolvulus**

1b. Style 1; petiole more > 2 mm long ....(2)

2a. Stigma filiform to elliptic, linear or oblong ....(3)

2b. Stigma bi-lobed, globular ....(5)

3a. Flower nocturnal; corolla white, salver-form; tube up to 9 cm long; stigma linear-oblong; seeds enveloped with a mealy pulp .... **Rivea**

3b. Flowers diurnal; corolla white with colour tinge, funnel-shaped; tube < 4 cm long; stigma filiform to elliptic or oblong; seeds without pulpy envelope ....(4)

4a. Stem and leaves stellate hairy; stigma bifid .... **Jacquemontia**

4b. Stem and leaves without stellate hairs; stigma linear, papillate .... **Convolvulus**

5a. Capsule circumscissile, upper part of the epicarp separate from the lower part; stem terete or often winged .... **Operculina**

5b. Capsule valvular or irregularly dehiscent; stem terete .... (6)

6a. Corolla with hairy on mid-petalline bands; fruits fleshy or leathery, dehiscent by operculum .... **Argyreia**

6b. Corolla without hairy mid-petalline bands; fruits dehiscent loculicidally or irregularly .... (7)

7a. Pollen grains spinulose; corolla variously coloured, often purple or reddish purple .... **Ipomoea**

7b. Pollen grains smooth; corolla nearly always yellow or white .... (8)

8a. Anthers mostly spiralled when dehisced; seeds rounded to truncate apically; stigma undulate or lobed, the papillae clavate or cylindrical .... **Merremia**

8b. Anthers straight when dehisced; seeds obtuse to rounded apically; stigma smooth, lobed to undulate, the papillate .... **Xenostegia**

ARGYREIA Lour.

Key to Species

1a. Leaves whitish or golden tomentose below .... **A. kleiniana**

1b. Leaves pubescent, not whitish or golden tomentose below .... **A. cymosa**

Argyreia cymosa (Roxb.) Sweet, Hort. Brit. 289. 1826; FBI 4: 190. 1883; FPM 2: 908. 1923; FAP 2: 624. 1997. *Lettsomia cymosa* Roxb., Fl. Ind. 2: 82. 1824.

Local name: *Maddi tige, Gantapala.*

Climbing shrubs, stem woody, herbaceous towards the tip, twining, pubescent. Leaves simple, deltoid to cordiform, 6–11 × 4–9 cm, base truncate to cordate, margin entire, apex acute to shortly acuminate, pubescent on both sides; midrib and lateral veins raised beneath; petiole up to 4.5 m long, pubescent like stem. Flowers in axillary, several-flowered cymes; peduncle up to 10 cm long, pubescent, bracts leafy, 2–4 in number, orbicular, 0.5–1.5 × 0.2–1.5 cm, apex acute or obtuse, pubescent on both sides, persistent; pedicels short, up to 1.5 cm long, pubescent. Bracts and bracteoles foliacious, to 2 × 1 cm, ovate, midrib prominent, petiolate (bracteoles smaller in size). Sepals unequal, outer 3 large, ovate to orbicular, apex obtuse, inner 2 small, oblong, apex acute, pubescent outside, glabrous inside. Corolla pale purple, darker near the centre, funnel-shaped, corolla-tube up to 4.5 cm, mouth 5-lobed, 3–3.2 cm across, densely silky on the mid-petaline bands. Stamens inserted, anthers up to 5 mm long. Ovary globular, glabrous, style single; stigma biglobose. Berry globose, glabrous. Seeds black, 2, 3 or 4, ovate to elliptic, glabrous to pubescent.

Occasional, in bushes in forests, but common, in open forests. Fl. & Fr.: Sep. – Jan. Kalasamudram RF (ATP), *NY* 645, *KRKS* 42021; Chinnapalli (ATP), *KRKS* 37801.

Argyreia kleiniana (Roem. & Schult.) Raiz., Indian For. 92: 302. 1966; FAP 2: 625. 1997. *Ipomoea kleiniana* Roem. & Schult., in L., Syst. Veg. 4: 789. 1819. *Argyreia bracteata* Choisy, Convolv. Or. 99. 1834; FBI 4: 188. 1883; FPM 2: 907. 1923.

Climbing, perennial shrub; stem robust, shortly tomentose. Leaves simple, broadly ovate, 8–10 × 6–8.5 cm, pilose or pulverulent, dotted above, velutinose below, base truncate to sub-cordate, margin entire, apex obtuse to acute; veins 8–10 pairs, prominently pubescent; petiole up to 7 cm long. Inflorescence axillary, 1–5-flowered, sub-capitate cymes; peduncle up to 13 cm long, terete, tomentose; bracts ovate-elliptic to elliptic-oblong, 2–2.5 × 0.8–1.5 cm, apically obtuse-acute, silky villous outside, glabrescent lower side. Flowers 4.5 cm across. Calyx-lobes unequel, ovate, 5–6 × 2.5–3 mm, apically acute, scarious, outer-lobes 2, slightly ciliate outside in the middle portion only, glabrous within. Corolla light purple, funnel-shaped; tube up to 4.5 cm long; mouth slightly 5-lobed, 6 cm across, mid petaline bands pilose out side. Stamens 5, inserted; anthers up to 4 mm long, straight; filaments attached above to the corolla, hyaline, hairy at base of filament. Ovary 1.5 mm; disk 2 mm long, glabrous, style single, glabrous; stigma biglobose, papillate. Fruit globose, yellowish, 1.5 cm across, mucronate, enclosed with calyx. Seeds 4, blackish, 5 mm diameter.

Occasional, in plains and near the streams of open dry deciduous forests. Fl.: Jul. – Sept. & Fr.: Oct. – Dec. Nigidi forest, near Mallalamma temple (ATP), *KRKS* 39643.

CONVOLVULUS L.

Convolvulus rottlerianus Choisy, Convolv. Orient. 95. 1834; FBI 4: 219. 1883; FPM 2: 925. 1923; FAP 2: 628. 1997.

Perennial undershrubs, to 90 cm high; stem erect, terete, silky villous. Leaves simple, linear, 2–3 × 0.1–0.4 cm, silky-villous on both sides, base attenuate, margin entire, apex acute to acuminate; midrib raised beneath; petiole absent. Inflorescence axillary, 1–3-flowered cymes; peduncle up to 3 cm long, filiform, terete, pilose; bracts small, 4–4.5 × 1 mm, hirsute; pedicels shorter than peduncle, up to 2.5 mm long. Flowers 5 mm across. Calyx-lobes 5, subequal, outer 2 slightly narrow, ovate-elliptic, 3.8–4.2 × 2–2.3 mm, silky villous out side, glabrous inside, inner 3 slightly broad, ovate-elliptic, 4–4.3 × 3–3.8 mm. Corolla white, campanulate, mouth 5-lobed; tube 3 mm long, 8 mm across, petaline bands hairy without. Stamens inserted; anthers 1 mm long. Ovary conical, glabrous, disk small; style glabrous; stigma linear, papillate. Capsule depressed globose, fruiting calyx nearly enclosed. Seeds 4, black, minutely pubescent.

Very rare, in foot hills of dry deciduous forests. Fl. & Fr.: Oct. – Mar. Idupulapaya (KDP), *KRKS* & *BSS* 39077.

EVOLVULUS L.

Evolvulus alsinoides (L.) L., Sp. Pl. 2: 392. 1762; FBI 4: 220. 1883; FPM 2: 923. 1923; FAP 2: 629. 1997. *Convolvulus alsinoides* L., Sp. Pl. 157. 1753.

Local name: *Vishnukrantham.*

Annual or perennial decumbent or erect herb, to 40 cm high; stem slender, terete, pale green, thinly or densely silky. Leaves simple, elliptic to linear-oblong or spatulate, 0.4–1.5 × 1–0.5 cm, silky pilose on both surfaces, base acute, margin entire, apex acute to acuminate; lateral veins not conspicuous; petiole to 2 mm long. Inflorescence axillary, 1–3-flowered cyme; peduncle up to 4 cm long, slender; bracts 2, linear-lanceolate, 2 mm long; pedicels up to 8 mm long. Flowers blue, to 1 cm across. Calyx subequal; outer 3 large, lanceolate or ovate-lanceolate, 3 × 0.5 mm; inner 2 short, 2 × 0.4 mm, glabrous within. Corolla funnel-shaped, sub-rotate; tube up to 1.5 mm long, 5–8 mm wide across, mid petaline bands pale blue and pilose. Stamens exserted; anthers small, 1 mm long, white. Ovary orbicular, 1 × 1 mm, glabrous; disk very small; styles 2, exserted, up to 7 mm long, united at the base, glabrous; stigmas 4, white, terete. Capsule globose, 3–4 × 3–5 mm, glabrous. Seeds usually 4, sometimes 2-3, ovoid, 1–2 × 1–1.5 mm, apically apiculate, glabrous, pale brown to black.

Common, in wastelands, agriculture lands, scrub forests and foot hills of dry deciduous forests. Fl. & Fr.: Oct. – Dec. Batrepalli (ATP), *KRKS* 40343.

IPOMOEA L.

Key to Species

1a. Corolla scarlet red; few to several-flowered, dichasial cyme; corolla salver-shaped; stamens and style exserted .... **I. hederifolia**

1b. Corolla other than red; solitary or compound or paniculate or capitate cymes, corolla funnel-shaped; stamens and style inserted .... (2)

2a. Sepals verrucose or cristate on the back, pubescent; pedicel filiform .... **I. obscura**

2b. Sepals smooth, glabrous or hirsute; pedicel not filiform .... (3)

3a. Plants glabrous or rarely pubescent; sepals mostly obtuse or orbicular, convex; seeds tomentose to villous .... (4)

3b. Plants pubescent to hispid; sepals oblong, ovate, lanceolate or linear .... (5)

4a. Shrubs; corolla finely tomentose outside; style minutely pubescent .... **I. carnea**

4b. Herbaceous or woody vines; corolla glabrous; style glabrous....... **I. staphylina**

5a. Plants pilose; flowers in subsessile capitate cymes, sepals oblong-spatulate, basally not hastate, hispid or pilose except inner side .... **I. eriocarpa**

5b. Plants pubescent or hispid; flowers in pedunculate cymes; sepals lanceolate or linear-lanceolate, hirsute .... (6)

6a. Flowers in dense capitate, pedunculate heads; bracts large, involucrate....... (7)

6b. Flowers not in capitate heads; bracts small, lanceolate or linear .... (8)

7a. Leaves pubescent to strigose beneath; outer bracts more than 5 mm wide; flowers white; seeds greyish pubescent .... **I. pes-tigridis**

7b. Leaves cottony-tomentose beneath; outer bracts less than 5 mm wide; flowers magenta or mauve; seeds glabrous or pubescent.......... **I. wightii**

8a. Leaves up to 5 cm long, entire; corolla salver-shaped, midpetaline bands pilose.......... **I. barlerioides**

8b. Leaves more than 5 cm long, entire or shallowly to deeply-lobed; corolla campanulate, midpetaline bands glabrous.......... (9)

9a. Corolla blue; calyx lobes 2.2 cm long, villous hairy.......... **I. nil**

9b. Corolla purple; calyx lobes 1.2 cm long, pubescent.......... **I. indica**

Ipomoea barleriodes (Choisy) Benth. ex C. B. Clarke, in Hook. f., FBI 4: 201. 1883; FPM 2: 915. 1923; FAP 2: 633. 1997. *Anisea barlerioides* Choisy, Mem. Soc. Phy. Geneve 6: 484. 1834.

Annual trailing herb; stem twisted or prostrate, slender, terete, yellowish, hirsute. Leaves simple, variable in shape and size, ovate to oblong-lanceolate, 1–5 × 0.4–2.8 cm, hirsute on both sides, base truncate to cordate, margin entire, apex obtuse to acute; petiole to 4 mm long, hirsute. Inflorescence axillary, solitary to few-flowered cymes; peduncle up to 4 cm long, pubescent; bracts 2, unequal, hirsute; pedicel up to 4 mm long. Flowers pink, to 4 cm across, salver-shaped. Calyx 5-lobed; lobes sub-equal; outer lanceolate; inner linear, acuminate, hirsute. Corolla pink, salver-shaped, mouth 5-lobed; tube up to 4 cm long; lobes up to 4 cm across, sparsely pilose without. Stamens inserted; anthers up to 3 mm long; filaments equal or subequal. Ovary conical, glabrous; disk small, annular; style inserted, up to 4.5 cm long, glabrous; stigma biglobose. Capsule mucronulate by persistent style base, 4-valved, fruiting calyx persistent and enlarged. Seeds 4, black, glabrous.

Occasional, on the hill slopes in scrub forests and road sides. Fl. & Fr.: Jul. – Jan. Batrepalli (ATP), *KRKS* 440 (HJCB).

Ipomoea carnea Jacq., Syst. Pl. 13. 1760; FPM 2: 919. 1923. subsp. **fistulosa** (Choisy) D. Austin, Taxon 26: 237. 1977; FAP 2: 633. 1997. *I. fistulosa* Mart. ex Choisy, in DC. Prodr. 9: 349. 1845.

Local name: *Rabbaru chettu.*

Perennial, erect, large, diffuse shrub, to 3 m high, latex milky; branches hollow, glabrous, rooting at lower portion of branches. Leaves ovate-cordate or often triangular, 6–15 × 4–9 cm, glabrescent above, sparsely pubescent below, base truncate or slightly cordate, margin entire, apex acuminate; lateral veins 10–13 pairs. Inflorescence axillary corymbose cymes; peduncle to 10 cm; pedicels 1–3 cm long; bract ovate, 0.5 × 0.3 cm, glabrous, margin thin, apex acute. Flowers whitish-pink, to 10 cm across. Calyx-lobes 5, ovate, 6 × 4 mm, sparsely pubescent without; outer 2 larger than inner three. Corolla pink, glabrous; tube 5–7 cm long, mouth of corolla slightly 5-lobed. Stamens inserted; anthers 8 mm long; filaments unequal. Ovary conical, glabrous, disk small; style inserted; stigma bi-globose, papillate. Capsule

4-valved, glabrous, ovate, 2 × 1.5 cm. Seeds 4, ovate, 1 × 0.6 cm, covered with long woolly, greyish brown hairs.

Common, in moist areas and ponds. Fl.: Jul. – Feb. & Fr.: Sept. – Mar. Thimmamma marrimanu (ATP), *KRKS* & *KRS* 37940.

Ipomoea eriocarpa R. Br., Prodr. 484. 1810; FBI 4: 204. 1883; FAP 2: 634. 1997. *I. hispida* (Vahl) Roem. & Schult., in L. Syst. Veg. 4: 238. 1819; FPM 2: 915. 1923. *Convolvuls hispidus* Vahl, Symb. Bot. 3. 29. 1794.

Local name: *Puriti teega.*

Twining, annual herb; stem pubescent or hispid with 1.5 mm long trichomes. Leaves simple, oblong, linear-oblong, hastate, 4–8 × 1.5–3 cm, glabrescent above, pubescent below, base truncate-cordate, margin entire, apex apiculate; midrib and lateral veins raised on both sides; petiole up to 5 cm long, pubescent. Inflorescence axillary, many-flowered, clustered cyme; peduncle short, to 7 mm long, hirsute; bracts 2, lanceolate; pedicel short, to 7 mm long. Flowers light pink, 2 cm across. Calyx-lobes 5, subequal; outer 2 long; inner 1 medium in size; inner most 2 small, pubescent, recurved. Corolla pink, funnel-shaped, mouth very shallowly lobed; tube up to 5 mm long, up to 1 cm across, mid petaline bands hirsute. Stamens sub-exserted; anthers up to 1 mm long; filaments sub-equal, 3 long, 2 short. Ovary 2 mm long, oblong, sericeous; style 2.5 mm, sub-exserted; stigma biglobose. Fruit capsule, hairy, globose, covered with persistent calyx, 0.4 cm across. Fruiting calyx accrescent, 0.6 × 0.3 cm. Seeds 4, grey, 3 × 2 mm, curved.

Common, in wastelands and agriculture lands. Fl. & Fr.: Nov. – Jan. Idupulapaya (KDP), *KRKS* 39058; Rachakuntapalle (KDP), *KRKS* & *KRS* 37625; Bakkannagaripalle (KDP), *KRKS* & *BSS* 41989.

Ipomoea hederifolia L., Syst. Nat. (ed. 10) 925. 1759; FAP 2: 635. 1997. *Quamoclit phoenicea* (Roxb.) Choisy, Mem. Soc. Phys. Geneve 6: 433. 1834; FPM 2: 919. 1923. *Ipomoea phoenicea* Roxb., Fl. Ind. ed. Carey & Wall. 2: 92. 1824.

Local name: *Kasiratnam.*

Annual twining herb; stem glabrous, woody at base; branchlets hollow. Leaves simple, ovate-cordate, 3–10 × 2.5–8 cm, glabrous, base cordate, margin entire or obscurely lobed apex acuminate; midrib and lateral veins raised on both sides; petiole up to 10 cm long. Inflorescence axillary, few to many-flowered cymes or dichotomously branched; peduncle up to 10 cm long; bracts small, triangular; pedicels angular, glabrous. Flowers scarlet-red, to 4 cm long. Calyx 5, sub-equal; outer 2 small, 5 mm long; inner 3 large, 8 mm long, glabrous. Corolla scarlet-red, salver-shaped; tube up to 3.5 cm long; mouth 5-lobed, to 2.5 cm across. Stamens exserted; anthers up to 2 mm long; filaments subequal. Ovary conical, glabrous, disk small; style exserted; stigma biglobose. Capsule globose, glabrous, 4-valved, septa thin, persistent. Seeds 4, black, pubescent, curved, 4 × 3 mm.

Rare in foot hills of dry deciduous forests and scrub forests. Occasional, in village outskirts and agriculture lands. Fl. & Fr.: Sept. – Mar. Mabbuchintalapalle (KDP), *KRKS* 41805.

Ipomoea indica (Burm.f.) Merr., Interpr. Rumph. Herb. Amboin. 445. 1917. *I. leari* Paxt., Mag. Bot. 6. t. 267. 1839; FPM 2: 919. 1923.

Twining, much branched, perennials, herbaceous to somewhat woody near the base, stems appressed pubescent to glabrate. Leaves alternate, rounded-ovate, commonly 3-lobed, also entire, 5–9 cm long, base cordate, margin entire, apex acuminate, densely pubescent below, to glabrate or glabrous above. Flowers purple 1-to few-flowered cymes, the cymes usually silky white-pubescent. Pedicels mostly 1 cm long. Calyx lobes lanceolate to ovate or broadly-ovate, 0.8–1.2 cm long, long acuminate to acuminate, appressed pubescent to glabrate. Corolla purple, rarely white, 5–7 cm long, limb 6–8 cm broad. Stamens included. Fruits capsular, to 1 cm broad, globose or somewhat flattened apically, glabrous. Seeds usually 4, dark brown, glabrous.

Occasional, in forests, often cultivated. Fl. & Fr.: Sept. – Dec. Idupulapaya (KDP), *KRKS* & *SSR* 39053.

Iponoea nil (L.) Roth, Cat. Bot. 1: 36. 1797; FAP 2: 236. 1997. *Convolvulus nil* L., Sp. Pl. ed. 2. 1: 219. 1762. *Ipomoea hederacea* auct. non (L.) Jacq.; FBI 4: 199. 1883; FPM 2: 917. 1923.

Local name: *Jirika kolli, Kollivittulu.*

Hirsute annual twiners. Leaves 3-lobed, rarely entire, 5–12 × 4.5–11 cm, ovate to suborbicular, base cordate, margin entire, apex acute to acuminate, pubescent, midrib and lateral veins raised beneath, prominently hirsute; petiole up to 3–8 cm long. Flowers blue, solitary or in cymes; peduncle 2–7 cm, sparsely or densely villose, elongated in fruiting stage; pedicel to 1 cm long. Calyx lobes 5, subequal, 1.5–2.2 × 0.2–0.4 cm, linear-lanceolate, acuminate, densely long-hirsute without. Corolla blue, throat white, 3–5 cm long, 4–5 cm across, funnel-shaped; filaments to 2.5 cm long. Capsule 0.8–1.2 cm long, subglobose, surrounded by the sepals. Seeds 5–6, broadly ovate, 6 × 5 mm, black, puberulent.

Common, in foot hills of deciduous forests and wastelands. Fl. & Fr.: Sep. – Dec. Rachakuntapalle (KDP), *KRKS* & *KRS* 37637.

Ipomoea obscura (L.) Ker.-Gawl., Bot. Reg. 3: t. 239. 1817; FBI 4: 207. 1883; FPM 2: 916. 1923; FAP 2: 632. 1997. *Convolvulus obscurus* L., Sp. Pl. ed. 2. 220. 1762.

Local name: *Nallakokkita.*

Annual twining or trailing herb; stem prostrate, herbaceous, terete, glabrous-hairy, latex milky. Leaves simple, ovate to broadly ovate or cordate, (in some forms 3-lobed) 2–6 × 2–5 cm, glabrous sparsely pilose on both sides, base cordate, margin entire, apex acuminate-apiculate; midrib and lateral veins raised beneath; petiole 2–3.5 cm. Inflorescence axillary, solitary or 3-flowered cymes; peduncle 2–5 cm, filiform; bracts 2, lanceolate, 2–3 mm long, pubescent; pedicel up to 1.5 cm, pubescent. Flowers whitish-yellow, funnel-shaped, to 3 cm across. Calyx-lobes 5, sub-equal, ovate-elliptic, 0.5 × 0.3 cm, margin thin, apex acute, pubescent. Corolla white or light-yellow, funnel-shaped; tube 2 cm long; mouth slightly 5-lobed. Stamens sub-exserted; anthers up to 2 mm long; filaments unequal, 2 large, to 1 cm long, 3 short, to 3 mm long, ciliate at base. Ovary 0.5 mm long, conical, glabrous, disk small; style 7 mm long; stigma biglobose. Capsule globose, glabrous, apically beaked, persistent calyx at the base. Seeds 4, ovoid, black, pubescent, 3 × 2 mm.

Common, on road sides, wastelands and scrub forests. Fl. & Fr.: May – Nov. Thummalapalle (KDP), *KRKS, KRS, SSR* & *TP* 33692; Bhumaiahgari palle (KDP), *KRKS* & *KRS* 36994.

Ipomoea pes-tigridis L., Sp. Pl. 162. 1753, FBI 4: 204. 1883; FPM 2: 918. 1923; FAP 2: 637. 1997.

Local name: *Mekamadugu, Chikunuvvu.*

Annual hispid herb; stem twining, terete, hollow, covered with bristly trichomes. Leaves palmatly 5–7-lobed; lobes obovate, lanceolate or elliptic, 3–5 × 1.5–2.5 cm, base obtuse, margin entire, apex acuminate, leaf base cordate, sericeous on both surfaces; lateral veins raised beneath; petiole up to 6 cm long, pubescent. Inflorescence axillary clustered cymes, enclosed with foliaceous bracts; outer bracts 2 large, 1.5 cm long, oblong; inner 4 small, oblong-lanceolate; peduncle 3–14 cm long; pedicels to 2 mm long. Flowers white, to 3 cm across. Calyx herbaceous, subequal; outer 2 lanceolate, 0.8–1 × 0.2–0.3 cm; inner 3 small, linear-lanceolate, 0.6–0.7 × 0.1–0.2 cm. Corolla white, funnel-shaped; tube up to 3 cm long, mid petaline bands pilose without; mouth slightly lobed to 10 divisions. Stamens inserted; anthers up to 1 mm long; filaments attached 2 mm above the corolla base, sub-equal, 5–6 mm long, ciliate at base. Ovary small, conical, 1 × 1 mm, glabrous; disk small, slightly lobed; style up to 1 cm long; stigma biglobose, papillate. Fruit capsule, ovoid, 7–8 × 6–7 mm, straw-coloured, enclosed with enlarged bracts and calyx. Seeds 4, black, 3 × 2.5 mm, greyish pubescent.

Common, in wastelands, scrub forests, dry deciduous forests and agriculture lands. Fl. & Fr.: Apr. – Nov. Kalasamudram (ATP), *KRKS* 39862.

Ipomoea staphylina Roem. & Schult., Syst. 4: 249. 1819; FBI 4: 210. 1883; FPM 2: 917. 1923; FAP 2: 639. 1997.

Local name: *Theendratheega.*

Perennial, large, straggling or climbing shrub; stem twining, woody at base, terete, glabrous; latex milky. Leaves simple, ovate or ovate-elliptic, 3–15 × 2.5–10 cm, base cordate, margin entire, apex acute or acuminate, glabrous, often with galls; petiole up to 9 cm long, canaliculated, glabrous. Inflorescence axillary, many-flowered, paniculate cymes; peduncle up to 15 cm long; bracts small, caducous; pedicels up to 1.3 cm long. Flowers whitish-purple, to 4 cm across. Calyx-lobes 5, sub-equal; outer 2 small, broadly ovate, 4 × 3 mm; inner 3 large, broadly ovate, 5 × 4 mm; glabrous, apex obtuse. Corolla white, purple in the throat; mouth slightly 5-lobed, 3 cm long; tube up to 1.8 cm across. Stamens inserted; anthers 2–3 mm long; filaments more or less equal, ciliate at base. Ovary conical, glabrous, style inserted; stigma biglobose. Capsule ovoid, 1.2–1.6 × 0.7–0.9 cm, valvate, mucronate with persistent style base. Seeds 4, elliptic-oblong, 5–5.5 × 2.5–3.5 mm, light brown, white comose, attached only at the apex, 7–8 mm long.

Common, in dry deciduous forests and foot hills, near water bodies. Fl. & Fr.: Jan. – Oct.Thimmamma marrimanu (ATP), *KRKS* & *KRS* 37938.

Ipomoea wightii (Wall.) Choisy, Mem. Soc. Phys. Geneve 6: 470. 1834; FBI 4: 203. 1883; FPM 2: 917. 1923. *Convolvulus wightii* Wall., Pl. Asiat. Rar. 2: 55. t. 171. 1831.

Climbing undershrubs, stem twining or prostrate, terete, yellowish, strigose, latex milky. Leaves simple, ovate, 5-8 × 6-7 cm, base cordate to truncate, margin entire to obscurely lobed, apex acute to acuminate, pilose above, white cottony tomentose below, coriaceous; midrib and lateral veins not conspicuous on both sides; petiole up to 5 cm long; pubescent as stem. Flowers axillary, few to many (up to 4)-flowered, dense to lax head-like cymose clusters; peduncle up to 7 cm long, strigose; bracts lanceolate, pilose; pedicels short or absent, up to 3 mm long. Sepals 5, nearly equal, linear-lanceolate, 10-12 × 2.5–4 mm, fulvous-hirsute. Corolla pink-purple, funnel-shaped, mouth slightly 5-lobed, tube up to 2 mm across, midpetaline bands pilose. Stamens inserted, anthers up to 2 mm long, filaments subequal, 2 long, 3 short, ciliate at base. Ovary conical, glabrous, disk small, style single, glabrous, stigma biglobose. Capsule globose, slightly pubescent. Seeds ovoid, black, minutely puberulous.

Common, on the hills at Idupulapaya. Fl. & Fr.: Oct. – Mar. Idupulapaya (KDP), *KRKS* 39056.

JACQUEMONTIA Choisy

Jacquemontia paniculata (Burm. f.) Hallier. f., Bot. Jahrb. Syst. 16: 541. 1893; FPM 2: 926. 1923; FAP 2: 640. 1997. *Ipomoea paniculata* Burm. f., Fl. Ind. 50. t. 21. f. 3. 1768. *Convolvulus parviflorus* Vahl., Symb. Bot. 3: 29. 1794; FBI 4: 220. 1883.

Annual climbing herbs, stem herbaceous, twining, terete, pubescent, latex milky. Leaves simple, ovate, oblong, 3–5 × 2–4 cm, glabrescent above, pubescent below, base truncate or cordate, margin entire, apex apiculate; lateral veins 7–9 pairs, raised from midvein; petiole up to 3 cm long, pubescent. Inflorescence axillary, clustered cymes; peduncle up to 5 cm long, pubescent; bracts small, subulate; pedicels up to 5 cm long, pubescent; bracts small, subulate; pedicels up to 1 cm long, pubescent. Flowers white, to 8 mm across. Calyx 5-lobed, ovate, 6 × 4 mm, pubescent without, acuminate at apex, enclose the capsule. Corolla white or pink, campanulate; mouth slightly 5-lobed, 6 mm across; tube up to 6 mm long. Stamens sub-exserted; anthers up to 2.2 mm long; filaments sub-equal, hairy at base. Ovary 1 mm long, ovoid, glabrous; style 6 mm long; stigma bifid. Capsule globose, 8-valved, valves ovate. Seeds 4, slightly winged, brownish or blackish, 3 × 2 mm.

Occasional, in foot hills of dry deciduous forests and weed in agriculture crops. Fl. & Fr.: Oct. – Dec. Idupulapaya (KDP), *KRKS* & *SSR* 39055.

MERREMIA Dennst. ex Endl. *nom. cons.*

Key to Species

1a. Leaves simple, reniform; flowers yellow .... **M. emarginata**

1b. Leaves palmately compound, lobes oblong; flowers white .... (2)

2a. Leaf margins entire, densely bulbous based hairy; seeds glabrous, brown.. **M. aegyptia**

2b. Leaf magins dentate, pubescent or glandular hairy; seeds pubescent or stellately hairy, black .... (3)

3a. Leaf palmately compound with 5 leaflets, sub-sessile, margin distinctly serrate; flowers axillary, solitary or 2–4-flowered cyme; peduncle to 4 cm long; pedicillar gland absent .... **M. cissoides**

3b. Leaf palmately compound with 5–7 leaflets, sessile, margin dentate; flowers axillary, 2–6-flowered cyme; peduncle to 20 cm long; pedicillar glands 4.. **M. dissecta**

Merremia aegyptia (L.) Urb., Symb. Antill. 4: 505. 1910; FPM 2: 928. 1923; FAP 2: 641. 1997. *Ipomoea aegyptia* L., Sp. Pl. 162. 1753. *I. pentaphylla* Jacq., Collectanea 2: 297. 1788; FBI 4: 202. 1883.

Local name: *Elika jemudu.*

Annual twining herbs; stem twining, herbaceous, terete, green or light red, densely hirsute, strigose, hairs up to 4 mm long, latex viscous, milky white, itching when touch to the skin. Leaves palmately compound, leaflets 5, lanceolate, 3–10 × 1–5 mm, hirsute on both sides, base acute, margin entire, apex acuminate; lateral veins raised beneath, 5–7 pairs; petiolule 0.3 cm long; petiole up to 7 cm long, hirsute. Inflorescence axillary, 1–6-flowered cymes; peduncle up to 3.5 cm long, hirsute, bracts lanceolate, caducous; pedicels up to 2.5 long, hirsute. Flowers white, to 6 cm long, funnel-shaped. Calyx 5-lobed, lobes unequal; outer most 2 larger, ovate, lanceolate, 1.5–2.5 cm long, apically acute or acuminate, hirsute; inner most pair small, 1–1.5 cm long, glabrous. Corolla waxy, white; tube 1.8–2 cm long; mouth slightly 5-lobed, 1.8–2 cm across. Stamens inserted; anthers 2–4 mm long, white, spirally twisted after dehiscence; filaments attached 3–5 mm above the corolla base, subequal, 3 large, 2 small, white, ciliate at dilated base. Ovary ovoid, 1.5–1.8 × 1.8–2 mm, glabrous; nectary disk annular; style inserted, up to 12 mm long, glabrous, slightly dilated at base; stigma white, biglobose, papillate. Capsule sub-globose, 1.1–1.3 × 1.2–1.4 cm, epicarp septifragally split, endocarp loculicidally split; transparent, septate between seeds; closed with persistent calyx. Seeds 4, vary in shape, ovate or orbicular, 3–5 × 4–5 mm, glabrous, shiny, light brown, dorsal surface with a longitudinal groove.

Common, on road sides and borders of field bunds on bushes. Fl. & Fr.: Oct. – Jan. Vempalli (KDP) *KRKS* & *SSR* 39049.

Merremia cissoides (Lam.) Hallier. f., in Engl. Bot. Jahrb. 16: 552. 1893. *Convolvulus cissoides* Lam., Tabl. Encycl. 1: 462. 1791. *Ipomoea cissoides* (Lam.) Griseb., Fl. Br. W. Ind. Island 473. 1861.

Annual twining, slender herbs; stem terete, green, hirsute, trichomes up to 3 mm long, latex gummy white. Leaves palmately compound, leaflets 5, sub-sessile, ovate-ellitpic, terminal one large, 3 × 0.7 cm, lateral pair middle sized, 2.5 × 0.5 cm, lower pair small size, 2 × 0.4 cm, base acute to acuminate, margin distinctly serrate, apex acuminate to apiculate, midrib raised beneath; lateral veins 6–10 pairs; petiolule 2 mm long; petiole 4.5–5 cm long, hirsute. Inflorescence axillary, solitary or in 2–4-flowered cyme; peduncle up to 3.5 cm long, terete, glandular hairy; bracts linear, 1.5–5 × 2–3 mm, apically acuminate, pubescent; pedicels up to 5 mm long. Flowers white, to 2.5 cm across, funnel-shaped. Calyx unequal; outer most 2 large, broadly rhomboid to ovate, 1.8–2 × 0.5–0.7 cm, apically long acuminate, margin wavy, hirsute; middle ones glabrous, smaller than outer ones, third one rhomboid to ovate, 1.6–1.7 × 0.3–0.4 cm, apically long acuminate. Corolla white; tube 8–16 mm long; mouth slightly 5-lobed, 2–2.5 cm across. Stamens sub-exserted, anthers whitish yellow, up to 3 mm long, tip curved after dehiscence; filaments white, sub-equal, 3 large, 2 small, white ciliate at dilated base. Ovary ovoid, up to 1 × 1 mm, whitish yellow, glabrous, 4-celled; disk annular; style subexserted, up to 12 mm; stigma white, biglobose, 0.5–1 mm across. Capsule compressed, globose, 4–5 × 8 mm, dehiscent, light brown, fruiting calyx slightly enlarged, accrescent. Seeds 4, subrotund, 4 × 3 mm, apically rounded, base obtuse, light brown, hairy.

Very rare, on road sides and wastelands. Fl. & Fr.: Aug. – Dec. Thimmamma marrimanu (ATP), *KRKS* 41943.

Merremia dissecta (Jacq.) Hallier f., Bot. Jahrb. Syst. 16: 552. 1893; FPM 2: 628. 1923; FAP 2: 641. 1997. *Convolvulus dissectus* Jacq., Obs. 2: 4. 1767. *Ipomoea sinuata* Ort., Hort. Matr. Dec. 7: 84. 1798; FBI 4. 214. 1883.

Annual twining herb; stem herbaceous, terete, younger parts green, older parts violet green, densely hirsute, latex milky white, gummy with itching sense. Leaves palmately 5–7-fid, divided to below the middle; lobes lanceolate, 5–8 × 2–3 cm, base shortly decurrent, margin dentate, apex acuminate, glabrous, midrib raised beneath; petiole up to 10 cm long, reddish purple, hirsute. Inflorescence axillary, 2–6-flowered cymes; peduncle up to 20 cm long, hirsute; bracts linear, caducous; pedicels up to 2.5 cm long, dilated towards the calyx, dilation 3–4 mm in diameter; pedicellar nectarines 4, on extreme top of the dilated pedicels. Flowers white, to 4 cm across, funnel-shaped. Calyx sub-equal, ovate-lanceolate, 2–2.3 × 1–1.4 cm, apically acute or acuminate or apiculate, purplish. Corolla creamy white, purple throat, funnel-shapped; tube 1 cm long; mouth slightly 5-lobed, 3.5 cm across. Stamens inserted; anthers whitish yellow, 6 mm long; filaments white, attached 7 mm above the corolla base, white ciliate at base. Ovary conical 1 × 2.5 mm, glabrous; disk annular, glabrous, 1 mm long, style inserted, up to 1.7 cm long, glabrous; stigma biglobose, papillate. Capsule depressed globose, 1–1.5 × 1.5–1.7 cm, dehiscent, light brownish, pedicels recurved, fruiting calyx persistent, accrescent, 3–3.5 × 1.5 cm, recurved. Seeds 4, widely ovate, 5–7 × 6–8 mm, glabrous, dull black.

Occasional, in wastelands and road sides, near agriculture fields. Fl. & Fr.: Nov. – Feb. Vempalli (KDP), *KRKS* 42052.

Merremia emarginata (Burm. f.) Hallier f., in Engl. Bot. Jahrb. 16: 552. 1893; FPM 2: 928. 1923. *Evolvulus emarginatus* Burm. f., Fl. Ind. 77.t.30.f. 1768. *Merremia gangetica* (L.) Cuf., Bull. Jard. Bot. Etat. 31 (Suppl.): 743. 1961; FAP 2: 641. 1997. *Convolulus gangeticus* L., Cent. Pl. 2: 9. 1756.). *Ipomoea reniformis* (Roxb.) Choisy, Mem. Soc. Phys. Geneve 6: 446. 1834; FBI 4: 206. 1883. *Convolvulus reniformis* Roxb., Fl. India ed. Carey & Wall. 2: 67. 1824.

Local name: *Eluka cheviaku, Eluka chevvaku.*

Annual or perennial herb; stem creeping; rooting at nodes, slender, terete, sparsely pubescent, latex white. Leaves simple, reniform to broadly ovate, 0.4–3 × 0.6–3.2 cm, glabrescent above, sparsely pilose below on the veins, base broadly cordate, margins coarsely crenate, apex obtuse to retuse or emarginated; lateral veins 3–5 pairs, raised beneath; petiole up to 3.8 cm long, violet green, sulcate above, shortly hairy. Inflorescence axillary, solitary or in 2–4-flowered cincinnus cyme; peduncle to 1.5 cm long. Flowers yellowish-white, to 2 cm across, funnel-shaped. Calyx subequal; outer 2 small, obovate to orbicular, ovate to elliptic, 3 × 1–2 mm, apically obtuse; inner single medium sized; inner most 2 larger, obovate-orbicular, 4.5–5 × 3.5–4 mm, apically emarginated, ciliate margin. Corolla yellowish-white; tube 4–8 mm long, veins red; mouth 5-lobed, up to 2 cm across, each bifid. Stamens subexserted; anthers 1 mm long; filaments attached 2 mm above the corolla base, 4–5 mm long, ciliate at base. Ovary conical 1 × 1 mm; disk very small; style subexserted, up to 3.5 mm, glabrous; stigma biglobose, papillate. Capsule subglobose, 4 × 6 mm in diam., longitudinally sulcate, glabrous, crowned by the style base, accrescent. Seeds 2, 3 × 4 mm in diam., glabrous, greyish brown.

Common, in wastelands and weed in cultivated fields, especially black soils. Fl. & Fr.: Dec. – Feb. Idupulapaya (KDP), *KRKS* & *BSS* 39078; Kalasamudram (ATP), *KRKS* & *BSS* 39526.

OPERCULINA S. Manso

Operculina turpethum (L.) S. Manso, Enum. Subst. Braz. 16. 1836; FPM 2: 929. 1923. *Convolvulus turpethum* L., Sp. Pl. 155. 1753. *Ipomoea turpethum* (L.) R. Br., Prodr. 485. 1810; FBI 4: 212. 1883. *Merremia turpethum* (L.) Shaw & Bhat in J. Bombay Nat. Hist. Soc. 74: 567. 1978; FAP 2: 643. 1997.

Local name: *Tella tegada, Nalla tegada, Erra tegada.*

Perennial, stout, twining shrubs; stem robust, herbaceous, usually winged, 3–5-angled, glabrous or sparsely pilose, latex gummy, white. Leaves simple, ovate-cordate, 6–12 × 7–12 cm, glabrous or pilose above, pubescent beneath, base cordate, margin entire, apex acute-apiculate; midvein and lateral veins prominent below; petiole up to 6 cm long, pubescent. Inflorescence axillary, 1–few-flowered cymes; peduncle up to 9 cm long; bracts elliptic, oblong, concave, 1.5–2.2 × 0.3–0.5 cm, apically mucronate, pubescent; pedicels up to 2 cm long, 4-angled. Flowers white, to 3 cm across, funnel-shaped. Calyx unequal; outer 3 large, broadly ovate,

2.3–2.5 × 1–1.5 cm, apically acute, pubescent; inner 2 small, ovate, 1.8–2 × 1–1.3 cm. Corolla white to creamy white, broadly funnel-shaped; tube up to 3.2 cm long; mouth slightly 5-lobed, 4–4.3 cm across. Stamens inserted; anthers up to 6 mm long, yellowish; filaments attached 1 cm above the corolla base, equal, up to 1.3 cm long, base papillose. Ovary globular, 1–1.3 × 1.8–2.1 mm, glabrous, disk slightly lobed, yellowish white, 1.5–2 mm long; style single, inserted, up to 1.8 cm long; stigma white, biglobose, papillate, 1–1.3 × 2–2.2 mm. Capsule depressed globose, 1–1.8 × 1.8–2 cm, grey when dry, splits around the middle of the capsule, crowned with a persistent style base. Seeds 4, orbicular, 5–7 × 4–6 mm glabrous, black.

Very rare, on field bunds. Fl. & Fr.: Nov. – Jan. Vempalli (KDP), *KRKS* & *BSS* 40305.

RIVEA Choisy

Key to Species

1a. Climbing shrub; leaves to 5 cm, pubescent below.-lower elevations, agriculture lands and wastelands.................... **R. hypocrateriformis**

1b. Erect shrubs; leaves to 15 cm, silvery white villous below.-higher elevations, near rocky crevices.. **R. ornata**

Rivea hypocrateriformis (Desr.) Choisy, Mem. Soc. Phys. Geneve 6: 408. 1834; FBI 4: 184. 1883; FPM 2: 903. 1923; FAP 2: 645. 1997. *Convolvulus hypocrateriformis* Desr., in Lam. Encycl. 3: 561. 1792.

Local name: *Boddikura, Niru boddi.*

Perennial, large, climbing shrub; stem twining, woody, terete, white tomentose, latex milky white. Leaves simple, broadly ovate-cordate, 3–5 × 4–6 cm, glabrous above, white tomentose below, base truncate-cordate, margin entire, apex cuspidate-apiculate; midrib and lateral veins raised beneath, 5–6 pairs; petiole 2–3 cm long, rounded, tomentose, gland is present below the leaf base on petiole, leaf extract jelly when crushed. Inflorescence axillary, 1–few-flowered cyme; peduncle up to 1.8 cm long, pubescent; bracts 2, linear-lanceolate, pubescent outside, 1 × 0.2 cm, acute at apex, peduncles up to 4 mm long. Flowers white, to 5 cm across, salver-shaped. Calyx-lobes 5, variable in size; outer 3 large, 0.8 × 0.5 cm; inner 2 small, 0.6 × 0.3 cm, ovate, oblong-ovate, pubescent, apex obtuse. Corolla white, salver-shaped, glabrous; tube up to 4 cm long; mouth 5-lobed, 5 cm across, mid-petaline bands sparsely pubescent without. Stamens inserted, sub-exserted; anthers up to 6 mm long; filaments subequal, 3 cm long, 2 short. Ovary conical, glabrous; style sub-exserted; stigma bifid, linear, oblong, papillose. Fruit capsular, laterally compressed, globose, 1–1.5 × 0.8–1.2 cm, brown, opening from the middle like a cup, inside white pulpy when young, articulate at the junction of peduncle. Seeds 1–2, ovate, 6 × 4 mm, glabrous to pubescent, brown, covered with white dry pulp.

Common, in scrub forests, agriculture lands and road sides. Fl. & Fr.: May – Jan. Thummalapalle (KDP), *KRKS* & *KRS* 33699.

Rivea ornata (Roxb.) Choisy, Convolv. Orient. 27. 1834; FBI 4: 183. 1883; FPM 2: 903. 1923; FAP 2: 645. 1997. *Lettsomia ornata* Roxb., Fl. Ind. 2: 85. 1824.

Local name: *Chandrapala, Samudrapala.*

Perennial, climbing shrub; latex milky; stem woody; branchlets tomentose, whitish. Leaves simple, cordate, 10–20 × 9–20 cm, glabrescent above, densely white tomentose below with pits at the base of lamina, shining, base cordate, margin entire, apex obtuse or acute; veins 10–14 pairs; petiole up to 12 cm long, tomentose. Inflorescence axillary, 1–few-flowered, subcapitate cymes; peduncle up to 10 cm long, pubescent; bracts 3–5, large; outer 1 longer, broadly ovate-concave, 5–5.5 × 3–4.2 cm; inner small, ovate-elliptic, 1.5–3 × 0.8–1.4 cm, acuminate, softly pilose; pedicels 1.2 cm long. Flowers 6 cm across. Calyx sub-equal; outer 2 slightly larger; inner 3 equal, elliptic ovate or broadly elliptic, 1–1.9 × 0.8–1.4 cm, apically obtuse, tomentose. Corolla whitish, funnel-shaped, throat darker; tube up to 1.4 cm long; mouth 5-lobed. Stamens inserted. Ovary 4-celled, style 4 cm, stigmas 2. Berry globose, yellow.

Common, in foot hills of dry deciduous forests, plains of deep forests. Fl.: & Fr.: Nov. – Feb. Chinnapalli (ATP), *KRKS* & *BSS* 40381.

XENOSTEGIA Austin & Stapl.

Xenostegia tridentata(L.) Austin & Stapl., Brittonia 32 (4): 533. 1980

Key to Subspecies

1a. Prostrate herb; pedicels tetra- or penta-angled; outer sepals mostly obtuse to emargiante, inner one attenuate to acuminate with an acute apex, fruiting sepal not exceeding the capsule .... subsp. **tridentata**

1b. Twining herb; pedicels terete, not angular; all sepals long acuminate, fruiting sepal exceeding the capsule and directed downwards after drying .... subsp. **hastata**

Xenostegia tridentata (L.) Austin & Stapl. subsp. **tridentata**. *Convolvulus tridentatus* L., Sp. Pl. 157. 1753. *Ipomoea tridentata* (L.) Roth, in Roem. Arch. Bot. (Leipzip) 1, 2: 38. 1798; FBI 4: 205. 1883. *Merremia tridentata* (L.) Hallier f., Bot. Jahrb. Syst. 16: 552. 1893; FPM 2: 928. 1923; FAP 2: 643. 1997.

Perennial, spreading herb, with thick root stock; branches to 1.5 m long, angular, green, glabrous; latex milky. Leaves simple, sessile, variable, linear, spatulate, linear-hastate or oblong-obovate, 1–3 × 0.2–1 cm, glabrous, base truncate, margin basally dentate, apically entire, apex acute or obtuse or emarginated, mucronulate; lateral veins slightly raised beneath; petiole to 1 mm long. Inflorescence axillary, solitary or 2–3-flowered cyme; peduncle up to 3.5 cm, terete or slightly angular, glabrous, bracts 2, minute, subulate; pedicels up to 1 cm long, 4-angular. Flowers yellowish-white, to 0.8 cm across. Calyx 5 mm long, lobes 5, subequal; outer 2 small, ovate; inner 3 large, lanceolate. Corolla pale yellow or white, with a purple throat, funnel-shaped; mouth stightly 5-lobed, up to 1.1 cm across; tube up to 7 mm long.

Stamens inserted; anthers up to 1 mm long; filaments equal, sparsely hairy at the base. Ovary 1.5 mm long, 4-celled, glabrous, disk small; style 5.5 mm long, inserted; stigma biglobose. Capsule ovoid, glabrous, 4-valved. Seeds 4, globose, dull black, ovate, 6 × 4 mm, glabrous to pubescent, brown, covered with white dry pulp.

Common, in dry deciduous forests and scrub forests near moist localities. Fl. & Fr.: Sept. – Jan. Rachakuntapalle (KDP), *SSR* & *TP* 33627.

Xenostegia tridentata (L.) Austin & Stapl. subsp. **hastata** (Desr.) Pullaiah *et al.*, Fl. Eastern Ghats 4: 260. 2011. *Merremia tridentata* (L.) Hallier f. subsp. *hastata* (Desr.) Oostr., Blumea 3: 317. 1939; FAP 2: 643. 1997. *Convolvulus hastatus* Desr., in Lam., Encycl. 3: 542. 1789; non Forssk. 1775. *Ipomoea angustifolia* sensu C. B. Clarke, in Hook. f., FBI 4: 205. 1883 non. Jacq. 1781-95. *Merremia hastata* Hallier f., Bot. Jahrb. Syst. 16: 552. 1893; FPM 2: 929. 1923.

Annual or biennial herb with root stock; stem twining, angular or terete, green to violet, glabrous, latex milky. Leaves simple, linear-oblong or lanceolate to oblong, 1–2 × 0.2–0.4 cm, glabrous, basally hastate and toothed at base on the auricles, margin entire, apex obtuse-mucronate; lateral veins slightly raised beneath; petiole nearly absent. Inflorescence axillary, solitary or 2–3-flowered cymes; peduncle up to 5 cm long; bracts minute, subulate; pedicels up to 1.2 cm long. Flowers yellowish-white, to 9 mm across. Calyx-lobes 5, sub-equal; outer 2 small; inner 3 large, 3 mm long, lanceolate, pale green. Extra floral nectaries on the base of the calyx, 4–5 small pore like structures. Corolla pale yellow or white with purple to brown throat, funnel-shaped, 1.5 cm long; mouth slightly 5-lobed, 1 cm across, glabrous; disk small, less than 1 mm; style inserted, glabrous, stigma biglobose. Fruit capsular, globose to ovoid 6–7 × 5–6 mm, 4-valved, glabrous, fruiting calyx exceeding the fruit, pedicels thickened above. Seeds 4, ovate, 1–2 × 2–3 mm, basally rounded, dorsal surface with a longitudinal ridge, glabrous, black.

Common, in dry deciduous-forests near moist places and rock crevices at high altitude. Fl. & Fr.: Throughout the year. Idupulapaya (KDP), *KRKS* & *BSS* 39088; Pulivendula Ghat (KDP, *KRKS* 283 (HJCB-E).

CUSCUTACEAE

CUSCUTAL.

Cuscuta reflexa Roxb., Pl. Coromandel 2: 3. t. 104. 1798; FBI 4: 225. 1883; FPM 2: 931. 1923; FAP 2: 647. 1997.

Local name: *Sitamma pounalu, Passitiga, Sitamma savaram.*

Leafless, twining, parasitic herb, stem coarse, 2–2.5 mm in diameter, yellow, profusely branched, glabrous. Inflorescence lax raceme; bracts small, broadly ovate to deltoid, fleshy, glabrous. Flowers white or cream, up to 4 mm long, sub-sessile. Calyx 5, ovate 3 mm long, obtuse at apex, margin thin. Corolla white or cream, tubular; tube up to 1.3 cm long, up to 5 mm across, deeply 5-lobed; lobes oblong, sub-acute, reflexed. Stamens 5, exserted; anthers up to 1.5 mm long, subsessile, up to 1 mm long; corolla scales 5, attached at the base of corolla, fimbriate. Ovary 1.5

mm long, conical, style 1, short, stout, stigma bilobed; lobes linear, papillate. Capsule globose, circumscissile near base, fleshy. Seeds 4, glabrous, black, 3–3.5 × 2–2.5 mm.

Occasional, on shrubs and spiny bushes on road sides and scrub forest. Fl. & Fr.: Dec. – Apr. Thimmamma marrimanu (ATP), *KRKS* & *KRS* 37936.

SOLANACEAE

Key to Genera

1a. Corolla funnel-shaped, to 12 cm long; fruit spinescent............................ **Datura**

1b. Corolla salver-shaped, to 2 cm long; fruit not spinescent................................(2)

2a. Inflorescence extra axillary cyme; corolla purple-blue; fruiting calyx accrescent at base of fruit... **Solanum**

2b. Inflorescence apical or axillary, solitary; corolla yellowish; fruiting calyx enclose the fruit .... **Physalis**

DATURA L.

Key to Species

1a. Leaf margin entire or undulate; plant sticky villous hairy; fruit spines weak, more than 7 mm long; base of spine not swollen; fruit loculicidally dehiscent; seeds dispersed...................................... **D. innoxia**

1b. Leaf margin lobed to distinctly dentate; plant glabrescent; fruit spines strong, less than 5 mm long; base of spine swollen, bulbous; fruit irregularly dehiscent; seeds not dispersed................................(2)

2a. Stem, leaves and corolla dark purple-pink tinge or coloured; fruit spines 3 mm long, sharp .... **D. metel**

2b. Stem and leaves green and corolla white; fruit spines 1 mm long, blunt.. **D. stramonium**

Datura innoxia Mill., Gard. Dict. ed. 8. n. 5. 1768; Santapau, J. Bombay Nat. Hist. Soc. 47: 659. 1948; FAP 2: 248. 1997. *D. metel* auct. non L.: FBI 4: 243. 1883; FPM 2: 941. 1923.

Local name: *Ummetta.*

Biannual, erect subshrub to 1.5 m high; branches hispid, sticky or whitish pubescent. Leaves simple, ovate, 3.5–13 × 2–9 cm, softly pubescent, base oblique, margin undulate to entire, apex acute; petiole up to 6 cm long. Flowers axillary, solitary to 13 cm long; pedicels to 1.5 cm long, pubescent. Flower funnel-shped, white, to 13 cm long. Calyx tubular, up to 7 cm long; lobes 5, ovate-oblong, acute. Corolla white, funnel-shaped; tube up to 12 cm long, slightly 5-lobed, glabrous, 6 cm across; lobes apically elongated to 5 mm long. Stamens 5, sub-exserted; filaments attached up to 7 cm to the corolla tube from the base; anthers 5 mm long, white. Ovary 0.5 cm across; style 10–12 cm long, filiform and increase in size towards the apex; stigma slightly lobed, obtuse, sticky, 2 mm diameter. Capsule pendulous, globose, 4-valved, with 2 cm long, weak spines; stalk 2–2.5 cm long; testa hard;

persistent calyx disk like at base of fruit, 3 cm across; opening of fruit loculicidal. Seeds light brown-yellow, curved, longitudinally rugose, 3 × 4 mm, easily dispersed from apex.

Common, in wastelands, road sides and villages. Fl. & Fr.: Throughout the year. Thummalapalle (KDP), *KRKS* & *KRS* 36976.

Note: Leaves are used for wound healing.

Datura metel L., Sp. Pl. 179. 1753; FAP 2; 648. 1997. *D. fastuosa* auct. non L.; FBI 4: 242. 1883.

Local name: *Errakolaku, Nalla thummaddi.*

Perennial shrub, to 2 m high; stem dark pinkish-purple; branchlets glabrescent. Leaves simple, oblong-ovate, 10–15 × 7–10 cm, sparsely tomentose on both surface, base obliquely truncate, margin deeply dentate or frequently lobed, apex acute; lateral veins 5–7 pairs; petiole 6–8 cm long. Flowers axillary, solitary, to 12 cm long; pedicel up to 2 cm long, glabrescent. Calyx tubular, up to 7 cm long, lobes 5, lanceolate, acuminate, greenish purple. Corolla white within, purple-pink without, funnel-shaped, up to 12 cm long; lobes 5, broad ovate, teeth 5, lanceolate; mouth 7 cm across. Stamens sub-exserted; filaments 10 cm long, glabrous; anthers 5 mm long, oblong-ovate. Ovary 0.7 cm across; style up to 8 cm long; stigma ovate, apex obtuse, slightly canaliculate. Capsule spinescent, 3 × 2.5 cm, subglobose, horizontally (stalk 90° curved), 4-valved, testa brown with strong tuberculate spines, breaking irregularly near the apex; spines 3 mm long; fruiting calyx persistent at base of fruit like disk, with wavy margin, 3 cm across. Seeds brown, 3 × 2 mm, rugose, curved, strongly attached each other with pulp of placenta, it does not get separated easily.

Occasional, on road sides and wastelands. Fl. & Fr.: Throughout the year. B. Kottapalli (ATP), *KRKS* 41895; Kalasamudram (ATP), *KRKS* 42045.

Note: Leaves are used for wound healing.

Datura stramonium L., Sp. Pl. 179. 1753; FBI 4: 242. 1883; FPM 2: 941. 1923; FAP 2: 649. 1997.

Local name: *Ummetta.*

Annual erect sub-shrub, up to 1.5 m high; stem and branchlets green, sparsely pubescent. Leaves simple, ovate, 4–12 × 3–7 cm, glabrescent, base oblique, margin many-toothed or sinuate with triangular teeth; lateral veins 6–7 pairs; petiole up to 6 cm long, glabrescent. Flowers axillary, solitary; pedicels to 2 cm long, glabrescent white, to 15 cm long. Calyx tubular, to 5 cm long, pubescent; lobes lanceolate, acute at apex. Corolla white, funnel-shaped; tube up to 15 cm long, 1.5 cm across; lobes 5, cuspidate; teeth 5, lanceolate. Stamens 5, sub-exserted; filaments 10 cm long, glabrous, filiform; anthers oblong-ovate, 8 mm long. Ovary 0.5 cm across, papillate like hairs; style up to 10 cm long, terete; stigma oblong-obovate, slightly canaliculated. Capsule spinescent, suberect, 2 × 2 cm, sub-globose; testa brown, with tuberculate spines; splitting regularly at the top into 4 valves; spines 1 mm

Plate 13

A. *Datura stramonium*, B. *Glossostigma diandra*, C. *Limnophila indica*, D. *Lindernia oppositifolia*, E. *Utricularia australis*, F. *Henckelia incana*

long, base swollen, apex blunt. Seeds blackish brown, suborbiculate, compressed, rugose, 3 × 2 mm, minutely dotted.

Very rare, in wastelands and village outskirts. Fl. & Fr.: Oct. – Feb. Ambakapalli (KDP), *KRKS* 41821.

PHYSALIS L.

Key to Species

1a. Stem pubescent; leaf base cordate; flowers 1 cm across, light purple tinge without; fruiting calyx 3 cm across .... **P. peruviana**

1b. Stem glabrous; leaf base obtuse; flowers 2 cm across, purple tinge absent; fruiting calyx 2.5 cm across .... **P. angulata**

Physalis angulata L., Sp. Pl. 183. 1753. var. **angulata.** *P. minima* L., Sp. Pl. 183. 1753; FBI 4: 238. 1883; FPM 2: 939. 1923; FAP 2: 650. 1997.

Local name: *Budda busara, Budda budama.*

Annual, erect herb, up to 60 cm high; stem glabrous, smooth, shiny, herbaceous, angular. Leaves simple, ovate, 3–9 × 1.5–4 cm, glabrous on both sides, base obtuse, margin wavy or shallowly toothed, apex acute-acuminate; lateral veins 5–7 pairs; petiole to 7 cm long. Inflorescence axillary, solitary, pendulous; pedicels to 2 cm long, filiform. Flowers yellow, to 2 cm across. Calyx 0.4 cm long; tube 2 × 2 mm; lobes 5, ovate, acute, 2 mm long, pubescent without on mid vein. Corolla campanulate, light yellow; tube 0.5 cm, glabrous; lobes 5, broadly ovate, apex acute; mouth 1 cm across. Stamens 5, attached to the corolla tube; filamens 0.5 cm, filiform; anthers 5, light yellow, oblong-ovate. Ovary 0.2 cm across, glabrous; style 3 mm long; stigma slightly bilobed. Fruit berry, globose, yellow, 0.7 cm across, shiny, over-topped by inflated calyx; testa transparent. Fruiting calyx 2.5 cm across, thin, acuminate. Seeds many, muricate, sub-orbicular, 1.5 mm across.

Common, weed in cultivated fields and also in wastelands. Fl. & Fr.: Aug. – Dec. Bhumaiahgari palle (KDP), *KRKS* & *KRS* 37616.

Physalis peruviana L., Sp. Pl. ed. 2. 1670. 1763; FBI 4: 238. 1883; FPM 2: 939. 1923; FAP 2: 650. 1997.

Local name: *Buddabasara.*

Annual to perennial herb, up to 1 m high; stem pubescent, herbaceous, angular. Leaves simple, ovate, 4–6 × 2–5 cm, pubescent on both sides, base slightly cordate, margin distinctly dentate, apex acute; lateral veins 4–5 pairs; petiole 1–4 cm long, pubescent. Inflorescence axillary, solitary, pendulous; pedicels 1 cm long, pubescent. Flower 1 cm across, yellow with purple tinge. Calyx tube 3 mm across; lobes oblong-ovate, 3 mm long, acute at apex, pubescent. Corolla yellowish, campanulate; tube 0.4 cm long; mouth 0.4 cm across; lobes 5, 0.5 cm across, light purple tinge without. Stamens 5, exserted, attached near the base of corolla; anthers oblong, up to 2 mm long; filaments up to 6 mm long. Ovary 2-celled, 2 mm long, style linear, 6 mm long, stigma capitate. Berry globose, yellow, testa transparent, more than 1 cm across, over-topped by calyx. Fruiting calyx 3 cm across, acuminate. Seeds many, muricate, 2 mm across, compressed.

Common, weed of cultivated fields and also in hill tops of dry deciduous forests. Fl. & Fr.: Jul. – Oct. Namalagundu (KDP), *KRKS* & *BSS* 39659.

Note: Opposite leaves unequal in size

SOLANUM L.

Key to Species

1a. Plants unarmed .... (2)

1b. Plants armed .... (3)

2a. Plants stellate-villous; flowers purple; fruit yellow-orange, to 1 cm across .... **S. pubescens**

2b. Plants glabrous; flowers white; fruit black, to 5 mm across .... **S. nigrum**

3a. Leaves oblong, margin deeply lobed; prickles straight, to 1.5 cm long, cylindrical, yellow; prostrate herb; flowers 4–5; fruit to 1 cm across .... **S. surattense**

3b. Leaves ovate, margin undulate or entire; prickles slightly curved or straight, to 5 mm long, flattened, yellow or pink; erect or decumbent herbs or shrubs; flowers 1 or 3; fruit to 2.5 cm across .... **S. anguivi**

Solanum anguivi Lam., Tabl. Encycl. 2: 23. 1794; FAP 2: 651. 1997. *S. indicum auct. non* L., FBI 4: 234. 1883; FPM 2: 938. 1923.

Local name: *Tella mulaka, Jiddu mulaga, Kaka munchi.*

Armed, much branched shrub, up to 1.5 m high; stems and branches yellowish stellate pilose, with curved, short, sharp prickles, up to 0.5 cm; prickles yellowish, flattened with stellate hairs, up to half from the base. Leaves simple, variable in size and shape, ovate, shallowly lobed, 4–8 × 3–6 cm, stellate pubescent on both sides, base obtuse-truncate, margin wavy, apex acute or subacute; lateral veins 5–7 pairs, lower side conspicuous; spines present on veins both sides; petiole to 2 cm long, stellate pubescent. Flowers extra axillary, to 1.5 cm across, solitary or often in 3-flowered cymes; peduncle up to 3 cm long; pedicels to 4 cm long with spines, curved. Calyx prickly, cupular, stellate hairy; lobes 5, triangular, ovate-oblong, acute at apex. Corolla purple-blue, 5-lobed, spreading, to 1.5 cm across; tube 0.5 cm long; lobes ovate, mucronate, stellate without on bands. Stamens 5; filaments to 2 mm long; anthers oblong, to 0.5 cm, yellow, slightly curved, equal to style. Ovary 2 mm across, glabrous; style up to 4 mm long; stigma capitate, green. Berry globose, 2.5 cm across, glabrous, green with white striations when young, shiny, yellow when mature; fruiting calyx enlarged, accrescent, persistent, cover the fruit at base. Seeds many, compressed, blackish-brown, tuberculate, shiny, 2 mm across.

Occasional, in foot hills and hill tops of dry deciduous forests. Fl. & Fr.: Aug. – Nov. K. K. Kottala (KDP), *KRKS* & *KRS* 36426; Batrepalli (ATP), *KRKS* 41880.

Solanum nigrum L., Sp. Pl. 186. 1753; FBI 4: 229. 1883; FPM 2: 936. 1923; FAP 2: 653. 1997. *S. americanum* Mill., Gard. Dict. (ed.8) no. 5. 1768.

Local name: *Kamanchi, Kosi, Gajju chettu.*

Annual, erect, much branched unarmed herb, up to 1.5 m high; branches sparsely hairy, herbaceous, shiny. Leaves simple, ovate, ovate-oblong, 3–8 × 1.5–4 cm, sparsely hairy, base attenuate, margin distinctly toothed, apex acute; lateral veins 7–8 pairs; petiole to 2 cm long, sparsely hairy. Inflorescence extra-axillary, 5–6-flowered umbellate cyme; peduncles up to 2 cm long; pedicels 0.6 cm long, sparsely hairy. Flowers white, to 4 mm across. Calyx lobes 5; tube 1.5 mm long; lobes ovate, acute, sparsely hairy without, glabrous within. Corolla white, 3 mm across; lobes 2 mm long, glabrous. Stamens 5; anthers yellow, up to 2 mm long, filaments 0.5 mm long. Ovary 1 mm across; style 2 mm long, pubescent; stigma capitellate. Berry globose, black shiny, 0.8 cm across; testa transparent; fruiting calyx slightly enlarged. Seeds many, brown, compressed, margin thin, glabrous, 1.2 mm across.

Common, in wastelands and road sides. Fl. & Fr.: Jan. – Aug. Rekkamanu (ATP), *KRKS* & *KRS* 37930.

Note: Fruits edible when ripe, leavs are used as vegetable.

Solanum pubescens Willd., Phytographia 5. t. 3. 1794; FBI 4: 230. 1883; FPM 2: 936. 1923; FAP 2: 653. 1997.

Local name: *Ushti chettu, Kasivuste.*

Perennial shrub, up to 2 m high, unarmed; branches pubescent, green when young, grey corky when mature. Leaves ovate-cordate, 4–8 × 2–5 cm, pubescent, base obliquely truncate or cordate or acute, margin entire, apex acute; lateral veins to 7 pairs; petiole up to 5 cm long, pubescent. Inflorescence extra-axillary, distinctly arranged racemose cyme; peduncle up to 4 cm long; pedicels up to 2.5 cm long. Flowers purple, to 2 cm across, villous. Calyx-lobes 5; tube 3 mm long; lobes 6–7 mm long, lanceolate, acuminate, pubescent without. Corolla purple-blue, up to 2.5 cm across; tube 0.3 cm long; lobes 5, 1.2 cm long, oblanceolate, acuminate, pubescent without and apically stellate. Stamens 5; filaments 2 mm long; anthers linear-oblong, 7–9 mm long, porate at apex, slightly curved. Ovary 1.5 × 1 mm; style 1.2 cm long; stigma capitate. Berry globose, 1.5 cm across, glabrous, shiny, red when mature. Seeds brown, compressed, 1.5 mm across, dotted, shiny.

Common, in dry deciduous forests and scrub forests. Fl. & Fr.: Jul. – Feb. Rachakuntapalle (KDP), *SSR* 33610.

Note: Fruits are used in preparation of chutney which is also served in hotels.

Solanum surattense Burm. f., Fl. Ind. 57. 1768; FAP 2: 654. 1997.

Local name: *Mulaga, Rammalaka.*

Annual or biennual, prostrate, armed herb; prickles up to 2 cm long, yellow, sharp, straight, cylindrical, basally glandular hairy, branches stellate hairy and glandular hairy, shiny. Leaves lanceolate, 3.5–8 × 1.5–5 cm, chartaceous, distinct stellete hairy on both sides, base attenuate, margin wavy or lobed, apex acute; lateral veins up to 5 pairs; spines 2 cm long on nerves. Inflorescence extra-axillary, 4–5-flowered corymbose cyme; peduncles 3 cm long, stellate hairy, shiny; peduncles 1 cm long. Flowers purple, to 2.5 cm across, stellate hairy. Calyx-lobes 5, almost free, obovate, 5.5 mm long, prickly, acuminate, glabrous within. Corolla purple, 2.5 cm across; lobes 5, broadly ovate-triangular, stellate without, 1.2 cm long, acute. Stamens 5; filaments 1 mm long; anthers liner-oblong, 8 mm long, yellow, slightly curved. Ovary globose, 2.5 × 2 mm; style 1.3 mm; stigma acute or capitate, longer than anthers. Fruit globose berry, green, with white striations when young, glabrous, yellow when mature, 1.5 cm across; fruiting calyx covered half of the fruit, spiny, stout, accrescent. Seeds brown, circular, compressed, glabrous, 1.5 mm across, shiny.

Common, weed in plains, wastelands and agriculture fields. Fl.& Fr.: Aug. – Jan. Mabbuchintalapalle (KDP), *KRKS* 36939; Batrepalli (ATP), *KRKS* 37875.

Note: It is highly medicinal plant; dried fruits with warm gingelly oil are used for muscle and joint pains.

SCROPHULARIACEAE

Key to Genera

1a. Leaves both alternate and decussate in the same plant; root parasites .... (2)

1b. Leaves decussate or whorled; not parasites .... (3)

2a. Leaf-blade entire; corolla-tube bent from above the middle; anthers free, 1-celled .... **Striga**

2b. Leaf-blade pinnatifid; corolla-tube straight; anthers connivent in pairs, 2-celled .... **Sopubia**

3a. Prostrate herbs, rooting at nodes .... (4)

3b. Erect herbs, not rooting at nodes .... (5)

4a. Stoloniferous herb, to 2 cm long; stem filiform; leaves spatulate; flowers 2 mm across, pinkish; pedicel to 3 mm long .... **Glossostigma**

4b. Stolons absent, prostrate herb, to 20 cm long; stem not filiform; leaves oblong; flowers 5 mm across, white; pedicel to 5 mm long .... **Bacopa**

5a. Leaf simple, monomorphic, margin serrate or entire .... **Lindernia**

5b. Leaf dissected; dimorphic, margin pinnatifid .... **Limnophila**

BACOPA Aubl. *nom. cons.*

Bacopa monnieri (L.) Pennell, Proc. Acad. Nat. Sci. Philadelphia 98: 94. 1946; FAP 2: 657. 1997. *Lysimachia monnieri* L., Cent. Pl. 2: 9. 1756. *Monniera cuneifolia* Michaux, Fl. Bor. Amer. 2: 22. 1803; FAP 2: 953. 1923. *Herpestis monnieri* (L.) Kunth, Nov. Gen. Sp. 2: 366. 1818, *nom. illegit.*; FBI 4: 272. 1884.

Local name: *Brahmi, Neeti sambrani, Sambranichettu, Neerugobbi.*

Prostrate, annual herb to 30 cm long, glabrous, rooted at nodes; branches cylindrical, light pink tinged, succulent. Leaves semi-succulent, decussate, sub sessile obovate-oblong, 1–2 × 0.3–0.6 cm, base acute, margin entire, apex obtuse, glabrous, veins not conspicuous; petiole 1 mm long, sometimes absent. Flowers axillary, solitary, 1 cm across; pedicels to 1 cm long, filiform, glabrous. Calyx lobes 5, unequal, 0.6 × 0.2 cm; outer 2 ovate, obtuse, glabrous, the third one some what falcate; inner two lanceolate, acute, ciliate. Corolla blue or white with purple veins, sub-bilabiate, 7–11 mm long; lobes 5, subequal, glabrous. Stamens 4, didynamous, anthers linear-oblong, yellow or white. Ovary oblong-globose, 2 mm long; style slightly deflexed, 5.5 mm; stigma flat. Capsule oblong-globose, 5 × 2.5 mm, septicidal, or septicidal and loculicidal, 2- or 4-valved. Seeds minute, many, oblong, 0.1–0.2 mm, testa striate.

Common, near water bodies and moist places in scrub forests and wastelands. Fl. & Fr.: Aug. – Nov. Mabbuchintalapalle (KDP), *KRKS* & *KRS* 36941.

GLOSSOSTIGMA Wight & Arn. *nom. cons*

Glossostigma diandrum (L.) Kuntze, Revis. Gen. Pl. 461. 1891. *Limosella diandra* L., Mant. 2: 252. 1771. *Glossostigma spathulatum* (Hook.) Arn., Nov. Act. Acad. Caes. Leop. Carol. German. Nat. Cur. 18: 355. 1836; FBI 4: 288. 1884; FPM 2: 964. 1923; FAP 2: 659. 1997.

Annual, prostrate, herb, very small, to 2 cm long; stolons interwoven, rooting at the nodes, flaccid; stem creeping. Leaves radical, fascicled, 5–7, spatulate, 1.5 × 0.2 cm, glabrous, membranous, base attenuate, margin entire, apex obtuse; veins not conspicuous. Flowers axillary, solitary, 2 mm across, pinkish; pedicel to 5 mm long; bracts absent. Calyx campanulate, cupular; lobes 5, minute. Corolla white-pinkish; tube 1 mm long; lobes subequal; upper shorter, 0.4 mm long; lower to 0.7 mm long. Stamens 2, exserted, minute; filaments filiform, to 1 mm long; anthers 1-celled, reniform, obtuse. Capsule globose, 2 mm across, enclosed by calyx, loculicidal. Seeds ellipsoid, minute, brown, shiny, rugose.

Very rare, found in marshy areas. Fl. & Fr.: Sept. – Jan. Kalasamudram (ATP), *KRKS* 41884.

Note: Hundreds of individuals grow in one place.

LIMNOPHILA R. Br. *nom. cons*

Key to Species

1a. Herbs > 30 cm high; flowers pink-white, ebracteolate .... **L. heterophylla**

1b. Herbs < 10 cm high; flowers cream-yellow, bracteolate .... **L. indica**

Limnophila heterophylla (Roxb.) Benth., Scroph. Ind. 25. 1835; FBI 4: 323. 2011. *Columnea heterophylla* Roxb., Fl. Ind. 3: 97. 1832. *Limnophila roxburghii* G. Don var. *reflexa* (Benth.) Hook. f., FBI 4: 270. 1884.

Erect or decumbent, annual herb, to 20 cm long; stem glabrous. Submerged leaves capillary multifid; intermediate leaves 4–6 in a whorl, pinnatifid; upper leaves opposite, sessile, semiamplexicaul or sometimes also alternate, oblong, margin serrulate, apex acute. Inflorescence axillary. Flowers sessile or sub sessile, solitary, 0.3 cm long. Calyx lobes ovate-lanceolate, 2 mm long. Corolla purplish, bilabiate, 2 mm across; lobes 1 mm long. Stamens 4, minute, didynamous. Capsule broadly ellipsoid, 2 mm across. Seeds minute, brown, shiny, rugose.

Common, in stagnant water, ponds, ditches. Fl. & Fr.: Sept. – Jan. Kalasamudram (ATP), *KRKS* 37685.

Limnophila indica (L.) Druce, Bot. Exch. Club. Soc. Brit. Isles 3: 420. 1914; FAP 2: 661. 1997. *L. gratioloides* R. Br., Prodr. 442. 1810; FBI 4: 271. 1884; FPM 2: 952. 1923.

Local name: *Buradabokkena, Kuttra.*

Annual, erect herb, up to 15 cm high; stems rooting at lower nodes, glabrous. Leaves pinnately dissected, 1–2 cm long, glabrous; lower ones simple, opposite, sessile, margin serrate; upper ones 3-fid, spatulate; upper most often undivided, sessile, linear-spatulate or linear-lanceolate, 7-10 × 1.5-3 mm, base tapering, margin serrate, apex subacute. Inflorescence axillary, pedicellate; pedicel to 1 cm long, pubescent. Flowers solitary or sometimes in cyme, 2–3 mm across. Calyx lobes linear-lanceolate, cupular, 2 mm across, pubescent. Corolla white, with pale purple or pink tinged, bilabiate; upper 2 larger, to 3 mm long; lower 3 smaller, to 2 mm long, apex obtuse. Stamens didynamous; anther ovate. Ovary 1.5 mm, externally 2-lobed; style 3 mm, broadened; stigma pilose. Capsule to 3 mm across, over locking in persistent calyx, pubescent. Seeds minute, brown, rugose.

Common, at moist places, sometimes submerged in low level water and grows after water dried up. Fl. & Fr.: Oct. – Jan. Nigidi forests foot hills (ATP), *KRKS* 39990.

LINDERNIA Allioni

Key to Species

1a. Calyx divided to base; stamens 2 .... (2)

1b. Calyx not divided to base; stamens 4 .... **L. nummularifolia**

2a. Leaf margin always serrate, serratures elongated into cilia .... **L. ciliata**

2b. Leaf margin shallowly serrate, serratures not elongated .... (3)

3a. Corolla less than 3 cm across .... **L. antipoda**

3b. Corolla more than 6 cm across .... **L. oppositifolia**

Lindernia antipoda (L.) Alston in Trimen, Fl. Ceylon 6 (Suppl.): 214. 1931; Sivarajan & Mathew, J. Bombay Nat. Hist. Soc. 80: 133. 1983. *Ruellia antipoda* L., Sp. Pl. 635. 1753. *Gratiola grandiflora* Roxb., Pl. Coromandel t. 179. 1805. *G. veronicifolia* Retz., Observ. Bot. 4: 8. 1786. *Ilysanthes veronicifolia* (Retz.) Urban., Ber. Deutsch. Bot. Ges. 2: 436. 1884; FPM 2: 959. 1924. *Bonnaya veronicifolia* (Retz.) Spreng., Syst. Veg. (ed. 16) 1: 42. 1824; FBI 4: 285. 1884.

Local name: *Neella pundu.*

Prostrate or decumbent, annual herb, 10–25 cm long, rooting at lower nodes, stems obtusely quadrangular. Leaves opposite, subsessile or lower shortly petiolate, obovate, linear-oblanceolate, elliptic-linear, base narrowed, margin subentire to serrate, apex obtuse or rounded, glabrous. Flowers solitary, axillary, pedicellate. Calyx lobes free, lanceolate, serrate. Corolla bluish or purple, bilipped and 2-fid each, 0.8–1.2 cm long. Stamens 2, staminodes 2, hooked, glabrescent, hooks convergent. Capsule cylindric, 1.5 cm long. Seeds numerous.

Common, in marshy places and by the side of paddy fields. Fl. & Fr.: Nov. – Feb. Batrepalli (ATP), *BR* & *ANS* 35631.

Lindernia ciliata (Colsm.) Pennell, Brittonia 2: 182. 1936; FAP 2: 664. 1997. *Gratiola ciliata* Colsm., Proda. Descr. Gratiol. 14. 1793. *Ilysanthes serrata* (Roxb.) Urban., Ber. Deutsch. Bot. Ges. 2: 436. 1884; FPM 2: 962. 1924. *Gratiola serrata* Roxb., Fl. Ind. 1: 140. 1820. *Bonnaya brachiata* Link & Otto, Icon. Pl. Select. 25. t. 11. 1820; FBI 4: 284. 1884.

Erect, annual herb, 10–25 cm high, usually branched from base; stems quadrangular. Leaves opposite, sessile, elliptic-obovate or oblong-lanceolate, 1–3 × 0.2–0.6 cm, base cuneate or obtuse, margin serrate, serratures acute and bristle tipped or ciliate, apex obtuse, veins arched, penninerved, glabrous. Inflorescence terminal raceme; peduncle, to 10 cm long; pedicel to 0.5 cm long, glabrous; bracts linear, scarious. Flowers bilipped, white with pink tinge, 5 mm across. Calyx lanceolate, 5 mm long; lobes linear-lanceolate, margins scarious. Corolla white, throat pink, 6–8 mm long; tube 3.5 mm; lobes 5; upper lip 2-lobed, oblong, 6 mm; lower lip 3-separations, 4 mm, obtuse. Stamens 2, hooks of the staminodes divergent; filaments short, filiform; anthers divaricate, unequal. Ovary glabrous, ovoid; style slender. Capsule obovoid, not exceeding calyx lobes, 1 cm long. Seeds numerous, minute, warted, black, slightly rugose.

Common, in moist localities, plains and hill tops. Fl. & Fr.: Aug. – Nov. near Talupula (ATP), *KRKS* & *BSS* 41909.

Lindernia nummulariifolia (D. Don) Wettst., in Engl. & Prantl. Pflanzenf. 4 (3b): 79. 1891; FAP 2: 666. 1997. *Vandellia nummularifolia* D. Don, Prodr. Fl. Nepal 86. 1825; FPM 2: 958. 1924.

Erect, dichotomously branched herb, 25 cm high; branches winged, 4-angular, glabrous, reddish-purple, little branches hairy towards apex. Leaves opposite, sessile or petiolate, margin crenate, apex obtuse, glabrous; palmatly 5-nerved. Inflorescence dichotomousely branched, cymose; peduncles 2 cm long, pedicels 0.5 cm long, 4-angular, winged. Flowers solitary, pinkish, to 1 cm long. Calyx tube up to 0.3 cm long; lobes unequal, lanceolate, persistent and envelop the fruit. Corolla pale violet, brown or reddish, above 0.5 cm long, bi-lipped; lobes orbicular, 7–10 mm long. Stamens 4, perfect, inserted in corolla, curved. Ovary oblong, ovoid, glabrous, many-ovuled; stigma 2-lamellate. Capsule ellipsoid or oblong-elliptic, brown, enveloped by calyx tube, 4 × 1 mm. Seeds many, minute, irregular in shape, pitted.

Rare in moist places and in open dry deciduous forests at high altitudes. Fl. & Fr.: Oct. – Feb. Kalasamudram (ATP), *KRKS* 41891.

Lindernia oppositifolia (Retz.) Mukherjee, J. Indian Bot. Soc. 24: 134. 1945; FAP 2: 666. 1997. *Gratiola oppositifolia* Retz., Observ. Bot. 4: 8. 1786-87. *Bonnaya oppositifolia* Spreng., Syst. Veg. (ed. 16) 1: 41. 1824; FBI 4: 286. 1884. *Ilysanthes oppositifolia* Urban, Ber. Deutsch. Bot. Ges. 2: 435. 1884; FPM 2: 962. 1924.

Erect, annual herb, to 18 cm high; branches 4-angular, glabrous. Leaves opposite-decussate, sessile, oblong-lanceolate, 1.5–2 × 0.3–0.5 cm, glabrous, base cuneate, margin serrate, apex acute; pinninerved; petiole 2 mm long. Inflorescence axillary or leaf-opposed, solitary or on a raceme; bracts linear; pedicel to 0.5 cm long. Flowers bluish, to 0.8 cm across. Calyx-lobes 5, free, lanceolate, 3 mm long, acute. Corolla bluish or white, with pink throat, 2.5 mm across; tube 3 mm long; lobes 5; upper lip oblong, entire, 2.5 mm long, obtuse; lower lip 3 mm. Stamens 2; filaments 1.5 mm long; anthers unequal, to 0.5 mm; staminodes linear, 0.6 mm long, acute. Ovary 1 mm across; style 3 mm. Capsule lanceolate, 7 × 2 mm, exceeding the calyx.

Rare in moist locations of dry deciduous forests, high altitudes. Batrepalli (ATP), *KRKS* 42010.

SOPUBIA Buch.-Ham. *ex* D. Don

Sopubia delphinifolia (L.) G. Don, Gen. Syst. 4: 560. 1837; FBI 4: 302. 1884; FPM 2: 970. 1924. FAP 2: 670. 1997. *Gerardia delphinifolia* L., Cent. Pl. 2: 21. 1756.

Local name: *Edintajada.*

Annual, parasitic, erect herb, up to 50 cm high; stem sparsely pubescent, grooved. Leaves pinnatifid below, 3-fid to linear above, opposite, 1.5–2.5 cm long, margin entire, apex acute. Inflorescence axillary, solitary; pedicels to 1cm long. Flowers white-red, to 3 cm long, pubescent. Calyx tube 5-ribbed, up to 0.9 cm long, sparsely hairy; lobes subulate, unequal, deltoid, base ovate decurrent, ciliate, acuminate. Corolla purple with violet streaks, 8–16 mm long, 7 mm across, tubular; tube 1 cm long; lobes 5, broadly ovate, 5 mm long, decurrent, undulate, acute at apex, glandular-hairy without, throat pubescent. Stamens 4; didynamous, one anther cell of each stamen fertile, anther-cells connate; filament 5 mm, strigose. Ovary 2-lobed, 2.5 mm long; style 1 cm long, hispid, stigma capitate. Capsule oblong, 2–4-valved, 7 × 3 mm, obtuse, ellipsoid, brown. Seeds oblong, cuneate, longitudinally striate.

Common, in moist locations at foothills of dry deciduous forest, scrub and near agricultural fields. Fl. & Fr.: Sept. – Dec. Kalasamudram (ATP), *KRKS* 40111.

STRIGA Lour.

Key to Species

1a. Plant dark-red or pinkish; calyx 5-ribbed .... **S. gesnerioides**

1b. Plant greenish-yellow; calyx 10–15-ribbed..(2)

2a. Plant to 40 cm high; calyx ribs 15; flower to 1.5 cm long .... **S. angustifolia**

2b. Plant to 25 cm high; calyx ribs 10; flower to 0.8 cm long .... **S. asiatica**

Striga angustifolia (D. Don) Sald., Bull. Bot. Surv. India 5: 70. 1963; FAP 2: 671. 1997. *Buchnera angustifolia* D. Don, Prodr. Fl. Nepal 91. 1825. *Striga euphrasioides* Benth., Companion Bot. Mag. 1: 364. 1935; FBI 4: 299. 1884; FPM 2: 967. 1924.

Annual, much branched, erect herb, to 40 cm high; stem ribbed, strigose hairy. Leaves alternate, sessile, linear, 3–5 × 0.5 cm, pubescent, base narrow, margin entire, apex acute. Inflorescence terminal, to 15 cm long, lax spike. Flowers pedicellate; bracteoles up to 0.2 cm long, linear, acute, ciliate. Calyx 15-ribbed; tube to 1 cm long; lobes linear-lanceolate, 0.3 mm long. Corolla white; tube pubescent, 1.5 cm long; lobes 4 (5); upper one elliptic, two slightly segmented; lower three oblong-obovate, glabrous. Stamens 4, didynamous. Capsule obovoid or ovoid, to 8 mm long. Seeds minute, to 0.3 mm across, rugose with concentric furrows, brown.

Vary rare in dry deciduous forests, near moist places associated with grass. Fl. & Fr.: Oct. – Nov. Kalasamudram (ATP), *KRKS* 42005.

Striga asiatica (L.) Kuntze, Revis. Gen. Pl. 446. 1891; FAP 2: 672. 1997. *Buchnera asiatica* L., Sp. Pl. 630. 1753. *Striga lutea* Lour., Fl. Cochin. 22. 1790; FBI 4: 299. 1884; FPM 2.968. 1924.

Local name: *Rathi badanika.*

Annual root parasitic herb to 30 cm high. Leaves alternate, linear-narrow, 1–2.5 cm high, base acute, margin entire, apex acute, rough hairy, hairs stiff with bulbous base, petiole minute. Flowers in axillary spikes. Calyx tube 10-ribbed, narrow, 0.6 cm long, scabrous without; lobes 10, lanceolate, apex acuminate. Corolla white; lobes bi-lipped, glabrous; tube 0.9 cm long, glabrous; lower 3 large lobes each 3 mm across, apex rounded; upper lobes 2 mm long, obtuse. Stamens 4, didynamous, filaments paired, 1.5 mm long; anthers oblong, 0.8 mm. Ovary globose, 2.5 mm long; style 7 mm long; stigma 2-lamellate. Capsule 6 mm long, oblong or ellipsoid, enclosed by calyx tube. Seeds minute, oblong, rugose, brown.

Common, in forests and cultivated lands, associated with grasses (commonly seen with *Cymbopogon* spp.), also grows on cereals like Jowar, Maize etc. Fl. & Fr.: Aug. – Nov. Somulavaripalli (ATP), *KRKS* 37827, 37891; Pulivendula ghat (KDP), *KRKS* & *BSS* 39627.

Striga gesnerioides (Willd.) Vatke, Oest. Bot. Z. 25. 11. 1875; FAP 2: 672. 1997. *Buchnera gesnerioides* Willd., Sp. Pl. 3: 338. 1800. *Striga orobanchoides* Benth., Comp. Bot. Maq. 1: 361. 1836; FBI 4: 299. 1884; FPM 2: 967. 1924.

Local name: *Errabadanica.*

Annual, erect, root parasitic herb, up to 20 cm high, with white hairs, branches reddish-pink, decumbent. Leaf scales sub-opposite, sessile, ovate-oblong, 0.3–0.6 cm, pubescent, margin entire, apex acute. Inflorescence terminal spike, up to 6 cm long; bract linear-lanceolate, 0.4 cm long, ciliate at apex. Flowers sessile, white, pubescent. Calyx tube 10-ribbed, up to 4 mm long; lobes 5, lanceolate, 2 mm long, ciliate at apex. Corolla white, tube 8 mm long, glabrous; lobes 5, unequal, 5 mm across, obtuse at apex. Stamens 4, filaments 5 mm long; anthers oblong, 0.8 mm long. Ovary oblong, 2.5 mm long; style 4.5 mm long; stigma capitate. Capsule 0.5 × 0.2 cm, oblong-ellipsoid, black. Seeds oblong, minute, rugose, brown.

Occasional, in scrub forests and waste lands grows as root parasite on *Lepidagathis cristata* and *L.subarmata*. Fl. & Fr.: Sept. – Jan. Rachakuntapalle (KDP), *KRKS* & *KRS* 37624; Mabbuchintalapalle (KDP), *KRKS* 292 (HJCB-E).

LENTIBULARIACEAE

UTRICULARIA L.

Key to Species

1a. Plants hydrophytic; foliar organs segmented, dichotomousely dissected .... **U. australis**

1b. Plants terrestrial; foliar organs simple, linear, linear-lanceolate........ (2)

2a. Scape twisted; flowers yellow; calyx acuminate at apex........ **U. scandens**

2b. Scape erect; flowers purple; calyx rounded at apex .... **U. caerulea**

Utricularia australis R. Br., Prodr. 430. 1810; Janarthanam & Henry, Bladderworts of India 34. 1992. *Utricularia flexuosa* sensu FBI *p.p.* quoad Syn. *U. australis* R. Br., non Vahl 1804.

Common name: *Yellow bladderwort.*

Aquatic, suspended herb with submerged free-floating branches, up to 30 cm in length, 0.5 mm thick, filiform, terete. Leaves absent, foliar organs equivalent of leaves, 1.5–2.5 cm long, primary segment divided into two equal segments, each divided into numerous alternate segments up to 10, pinnately arranged, corner segments capillary and spinulose-toothed utricles; segments thread like; bladders or traps 1 × 1.5 × 1 mm, obliquely obovoid, stalked, mouth laterals oblique; appendages 2, often simple setae present on lips, the bladders are hinged by a flap, which forms a door or valve that only opens inwards. Roots absent; sometimes root like organs present at the node, to 2 mm long. Scape weakly erect, emergent above water, 10–15 cm long, flexuous, the scape usually ending in a flowerhead of 3–6 flowers; axis initially short, elongating in fruit; scales 3 on peduncle, similar to the bracts in upper half of peduncle, 1st one obovate, 1.5 × 1 mm, base truncate, apex sharply acute, 2nd one ovate, 2.5 × 2.5 mm, base cordate, apex blunt, 3rd one ovate-oblong, 3 × 2 mm, base cordate, apex acute, bracts 3–3.2 × 3–3.2 mm, basifixed, ovate to suborbicular, cordate at base, acute to obtuse at apex, slightly glandular without; bracteoles absent; pedicels 2–2.5 cm long, filiform, glabrous, pendulous in fruit. Calyx lobes 2, slightly unequal, ovate to oblong; upper lobe 2–2.5 × 1.5–1.8 mm, retuse; lower 1.5–2 × 1.5–2 mm, obtuse at apex, sparsely glandular without. Corolla yellow, with reddish brown lines and spots on the darker lower lip, *ca.* 10 mm long; upper lip inserted in lower lip; upper lip *ca.* 5 × 3.5 mm, ovate to oblong, truncate at apex; lower 6 × 12 mm, transversely elliptic, auricled and prominently gibbous at base, truncate at apex. Spur 5–6 mm long, conical, obtuse at apex, glandular within. Stamens 2, 1–1.5 mm long; filaments dilated near anthers; anther thecae confluent, sterile, 1.5–1.8 mm. Ovary 1.2 mm long, globose, glandular; style short; stigma 2-lipped, unequal,

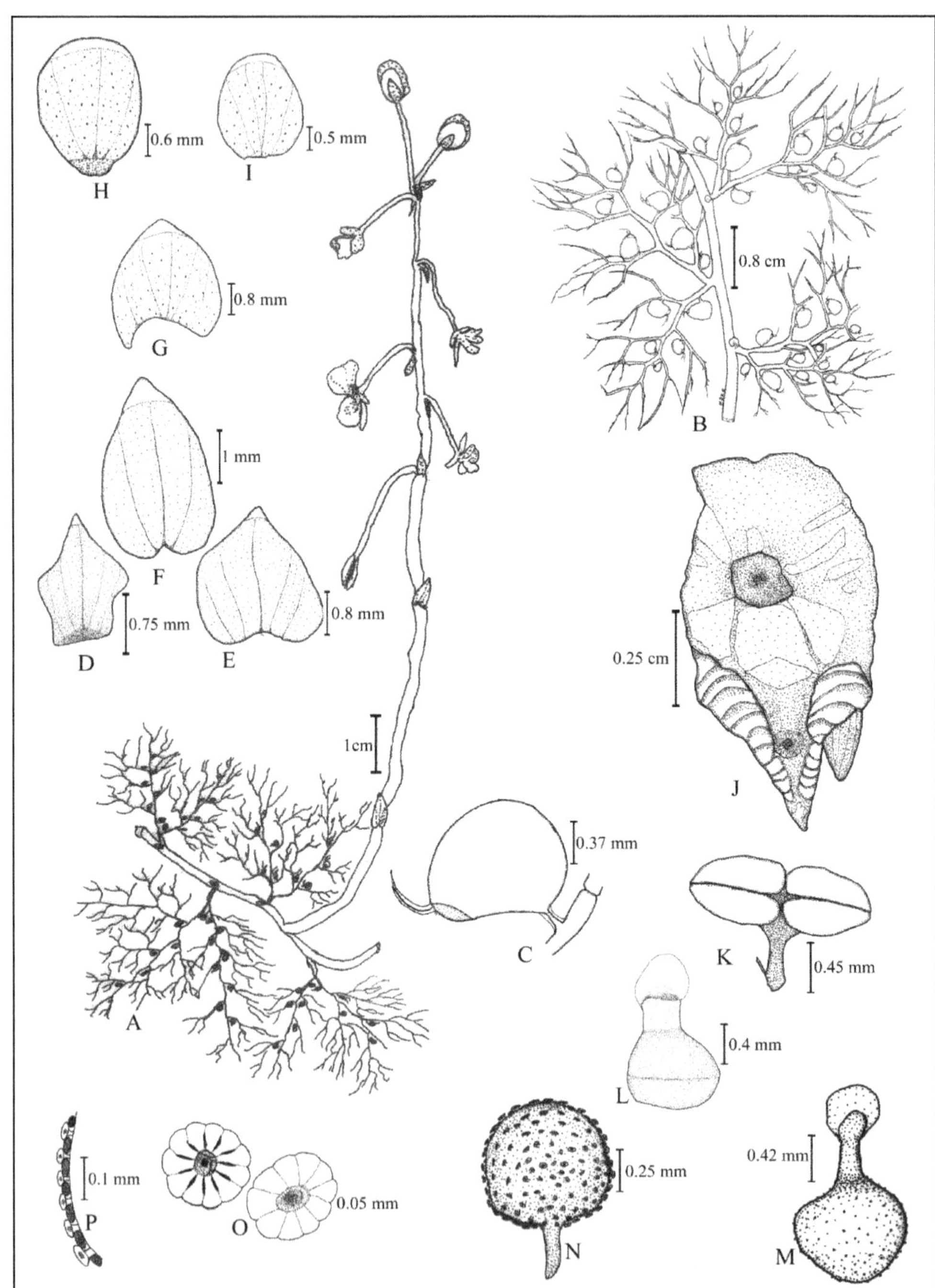

Figure 8: *Utricularia australis* R.Br.

A. Habit, B. Branching, C. Bladder, D,E & F. Scales on peduncle, G. Bract, H & I. Calyx, J. Corolla, K. Stamens, L. Ovary, M. Fruit, N. Placenta, O Seeds, P. Section of placenta.

lower lip deltoid, upper lip obovate, thin and transparent. Fruit capsule, 1.7 × 1 mm, black, obovate, capsule opens by valves, free central placentation. Seeds abortive, minute, to 0.05 mm, flattened, appendages absent, inserted between the papillae.

Very rare in stagnant water. Fl. & Fr.: Jun. – Aug. Kalasamudram (ATP), *KRKS* 41861.

Utricularia caerulea L., Sp. Pl. 118. 1753; FPM 983. 1924; FAP 2: 676. 1997. *U. racemosa* Wall. ex Walp, Meyen observ. Bot. 19: 401. 1843; FBI 4: 333. 1884.

Annual, terrestrial, erect, delicate herb, 10–15 cm high, glabrous. Rhizoids a few, filiform. Foliar organs or leaves few, simple, linear-lanceolate, 5 × 2 mm, spatulate, rounded at apex. Traps few, minute, ovoid-obovoid. Scape 5–7-flowered raceme; scales and bracts medifixed; bracteoles similar to bracts, 1 mm long, pedicels to 2 mm long. Flowers 8 cm across, purple. Calyx-lobes 2, subequal, ovate; upper lobe 2.5 mm long, broadly ovate, acute; lower lobe slightly smaller than upper one, margin incurved. Corolla deep purple, 5–7 mm long; upper lip 2-lobed, 1.3 mm long; lower lip flat, broadly suborbicular-ovate, 2 mm long; spur horizontal, slightly curved, narrowly conical, 3 mm across. Stamens 2; filaments 1.3 mm long; anthers 2-celled, 0.3 mm across. Ovary globose, 1 × 0.7 mm. Capsule 2.5 × 2 mm, globose, fruiting calyx thick, shorter than capsule. Seeds numerous, obliquely oblong; testa obscurely reticulate.

Occasional, in moist areas near foot hills of dry deciduous forests. Fl. & Fr.: Oct. – Dec. Kalsamudram (ATP), *KRKS* 42031; Nigidi forest (ATP) *KRKS* 40130.

Utricularia scandens Benj., Linnaea 20: 309. 1847; FBI 4: 332. 1884; FPM 2: 982. 1924; FAP 2: 678. 1997.

Terrestrial annual herb, twining, up to 15 cm long. Rhizoids filiform. Foliar organs linear, to 6 × 2 mm, cuneate, obtuse, decurrent into pseudopetiole; traps few, along the rhizoids. Scape slender, green, twining, 5–8-flowered raceme, terminal, pedicellate; scales lanceolate; bracts basifixed; bracteoles similar to bracts, but smaller; pedicel to 5 mm long, reflexed in fruit. Calyx-lobes 2, equal, obovate, 3 mm long, shortly acuminate; lower minutely 2-fid. Corolla yellow, 3 × 1.5 mm; upper lip erect, ovate, 1.5 mm, obtuse; lower lip orbicular, 3 mm, obtuse; palate raised, 2–3 gibbous. Spur slightly curved, facing downwards, 4 mm long. Stamens 2; filaments 1 mm; anthers confluent, 0.6 mm. Ovary globose, 1 mm long. Capsule globose, 2.5 mm across; persistent-calyx chartaceous, larger than capsule. Seed ovoid, testa reticulate.

Common, in marshy areas at foot hills and plains of dry deciduous forests associated with *Cyperus* spp. and *Eriocaulon* spp. Fl. & Fr.: Oct. – Dec. Kalasamudram (ATP), *KRKS* 42032.

GESNERIACEAE

HENCKELIA Spreng.

Henckelia incana (Vahl) Spreng., Syst. Veg. 1: 38. 1824. *Didymocarpus tomentosus* Wight, Icon Pl. Ind. Orient. t. 1349. 1848; FBI 4: 354. 1884; FPM 2: 989. 1924;

FAP 2: 680. 1997. *D. rottleriana* var. *tomentosa* (Wight) C.B. Clarke, in A & C. DC. Monogr. Phan. 5: 101. 1883.

Scapigerous, annual herb, acaulescent. Leaves arising from a rosette, radical, ovate to broadly elliptic, 3–10 × 2–6 cm, silvery pubescent-tomentose, base usually attenuate to sub-truncate into the winged petiole, margin irregular crenate, apex acute-rounded. Inflorescence axillary, dichotomously branched cyme. Peduncles 5–11 cm long, pilose; bracts linear, 3–5 mm long, pedicels to 2 cm long. Flowers 1 cm across, blue. Calyx lobes lanceolate, 2–3 mm long, tomentose; tube 2 mm long. Corolla pale blue, tubular, funnel-shaped; tube 1 cm long, gibbous, on lower side above base, narrower at the throat; limb flaring, outside sparsely pilose; lobes 5, entire. Stamens 2; staminodes 2; filaments adnate, for 1 mm long, to base of corolla-tube, 2 mm long. Ovary linear, 3 mm long, glabrous. Capsule slender, cylindric, slightly curved, 10–14 mm long, tapering at the ends. Seeds many, sub-trigonous, minutely pitted.

Occasional, in dry deciduous forests near rocky areas under rocky shadows associated with Pteridophytes and Bryophytes in Nigidi forest. Fl. & Fr.: Jun. – Dec. Batrepalli (ATP), *KRKS* & *BSS* 39578.

BIGNONIACEAE

Key to Genera

1a. Calyx spathaceous; lobes obscure; corolla white, salver-shaped .... **Dolichandrone**

1b. Calyx cupular or campanulate; lobes 5; corolla yellow, bell-shaped .... (2)

2a. Leaf margin serrate or crenate; plants shrubs, ornamental .... **Tecoma**

2b. Leaf margin entire; plants trees, wild .... **Stereospermum**

DOLICHANDRONE (Fenz.) Seem. *nom. cons.*

Key to Species

1a. Leaflets elliptic, 10 cm long, shiny; calyx glabrous; corolla creamish .... **D. atrovirens**

1b. Leaflets obovate, 4 cm long, not shiny; calyx stellate hairy; corolla white .... **D. falcata**

Dolichandrone atrovirens (Roth) Sprague, Bull. Misc. Infom. 304. 1919; FPM 2: 996. 1924; FAP 2: 682. 1997. *Bignonia atrovirens* Heyne ex Roth, Nov. Pl. Sp. 284. 1821. *Dolichandrone crispa* Seem., Journ. Bot. 8: 781. 1870; FBI 4: 379. 1884.

Local name: *Vadi, Oddi, Niru oddi, Vanka nirodi, Niruddi.*

Moderate, deciduous tree, to 12 m high; bark brown, rough; wood yellowish-brown; branches glabrous, greenish yellow. Leaves up to 30 cm long, imparipinnate; leaflets opposits, 3–5, broadly ovate, 6–10 × 4–6 cm, thin-coriaceous, base obliquely truncate, margin entire, apex acute-acuminate; lateral veins about 5–7 pairs; petiolules 0.7–1.3 cm long, apical leaflet longer than remaining; rachis to 10 cm long. Inflorescence axillary or terminal corymb or panicle; pedicel to 1.5 cm long. Flowers white, to 6 cm across. Calyx spathaceous; tubular, 1.5 cm long, acute, glabrous. Corolla white, 5 cm across, salver-shaped; tube 5 cm long; lobes 5, obtuse, 2 cm across, wavy margin. Stamens 4 (2+2), filament pairs 6 and 8 mm long; anthers 3 mm long. Ovary sessile, 4 mm across; style 2 cm long; stigma capitate, white. Fruit compressed capsule, curved, up to 30 cm long, 1.5 cm width, brown dotted, acuminate with speckles. Seeds 0.6 cm across; wings to 0.8 cm, white, transparent.

Very common, in dry deciduous forests and rare in scrub forests associated with *Croton scabiosus*. Fl.: Jul. – Set. & Fr.: Jan. – Apr. Batrepali (ATP), *KRKS* & *BSS* 37871.

Dolichandrone falcata (Wall. ex DC.) Seem., Fl. Vill. Nov. 151. 1880; FBI 4: 380. 1884; FPM 2: 996. 1924; FAP 2: 682. 1997. var. **falcata**. *Spathodea falcata* Wall. ex DC., Prodr. 9: 206. 1845.

Local name: *Chitti-niruddi, Chinna oddi, Oddi.*

Deciduous tree, 5–8 m high; young branches covered with grey stellate hairs; bark blackish, exfoliating; wood hard, white. Leaves opposite, pinnate, 5 to 6 cm long; leaflets 7, obovate or obcordate, 2–3.5 × 1.8–2.5 cm, base rounded or cuneate, margin entire, apex acute, grey stellate when young, glabrous when mature, apical leaflets larger than remaining; petiole to 3 cm long, pubescent. Inflorescence axillary corymb. Flowers white, to 3 cm across. Calyx spathaceous, obtuse, emarginated, to 1.5 cm long, glabrous. Corolla white, 3 cm long, salver-shaped; tube 2.5 cm long; lobes 5, 3 cm across, irregular, obtuse. Stamens 4, didynamous, filaments 2.5 mm long; anthers 2.5 mm long. Ovary 4 mm across; style 1 cm long; stigma capitate. Capsule flat, falcately curved, 20–40 cm long, 1.5–2 cm wide; epicarp striate, dark brown, apex beaked. Seeds 1 cm across, compressed; wing membranous at both sides.

Occasional, in dry deciduous forests and foot hills of dry deciduous forests. Kalasamudram (ATP), *KRKS* & *BSS* 40301.

STEREOSPERMUM Chamisso

Stereospermum tetragonum DC., Biblioth. Universelle Geneve n.s., 17: 124. 1838; FPM 2: 998. 1924. *S. colais* (Dillw.) Mabb., Taxon 27: 553. 1979. *Bignonia colais* Buch.-Ham. ex Dillw., Rev. Ref. Hort. Mal. 28: 1839. *S. chelonoides auct. non* (L. f.) DC.; FBI 4: 382. 1884. *S. personatum* (Hassk.) Chatterjee, Bull. Bot. Soc. Bengal 2: 70. 1948; FAP 2: 685. 1997.

Local name: *Girikaya, Magavepa, Tapada, Isikarasi, Pulakevalamu, Pisul, Kaligottu, Godalippulusu, Kala-Kod.*

Large deciduous tree, up to 20 m high, bark brown to grey, fairly smooth; branches glabrous. Leaves 30–40 cm long, pinnately compound, clustered at the ends of branches; rachis canaliculate, leaflets 5–9, opposite glabrous, elliplic oblong, 5–12 × 3–5 cm, sparsely pubescent above on midvein, base cuneate or suboblique, margin entire, apex caudate; lateral veins 5–7 pairs; petiolule 2–3 cm long. Inflorescence axillary, 15 cm long, corymbose panicle. Flowers 4 cm across, yellow, glabrous. Calyx campanulate, bilabiate, 0.7 cm across. Corolla yellow with red veins, bell-shaped, 2.5 cm long; tube 1.5 cm, gradually widened, villous in the middle from below upwards; lobes 5, subequal, 1 cm long, margin crisped. Stamens 4; filaments 8 mm long, with a tuft of woolly hairs below; anthers divaricate, 2 mm long. Ovary oblong-subacute. Capsule terete, 4-gonous, 30–40 × 1 cm, elongate, twisted; epicarp thick, with speckles. Seeds compressed, trigonous, grooved outside, ridged within; septum thick, corky, ridged, in the cavities of which lie the wings of seeds.

Common, in dry deciduous forest, on slopes. Fl.: Apr. – Sept. & Fr.: Oct. – Jan. Nigidi forest (ATP), *KRKS* 39590.

TECOMA A. L. Juss.

Tecoma stans (L.) Kunth, Nov. Gen. Sp. Pl. 3: 144. 1819; FAP 2: 686. 1997. *Bignonia stans* L., Sp. Pl. (ed. 2) 871. 1763. *Stenolobium stans* (L.) Seem., Ann. Mag. Nat. Hist. (ser. 3) 10: 30. 1862; FPM 2: 1000. 1924.

Local name: *Swarna ganneru, Suvarna ganneru, Ganta poolu, Pasupu ganneru.*

Large, shrub to small tree to 8 cm high, glabrous. Leaves imparipinnate, 5-foliolate, leaflets lanceolate, 11 × 2.5 cm, base acute, margin serrate, apex acuminate. Inflorescence terminal raceme. Flowers yellow, 5 cm across. Calyx tubular, 0.6 × 0.3 cm, campanulate; lobes 5, triangular. Corolla golden yellow, bell-shaped, 3 × 2 cm, veined; lobes 5, subequal, rounded, ciliate. Stamens 4, didynamous, filaments slender, hairy at base; staminode 1, club-shaped, 0.5 cm long. Disk cushion-shaped. Ovary glabrous, ovate-lanceolate; style filiform, slender; stigma bifid, lobes dilated, flattened. Capsule up to 18 × 0.6 cm, brown, beaked.

Common, in villages and near temples. Occasional along the road sides. Fl. & Fr.: Through the year. Batrepalli (ATP), *KRKS* 40390.

Note: Ornamental plant.

PEDALIACEAE

Key to Genera

1a. Leaves palminerved; flowers in racemes; stamens 2; ovary 1-locular; corolla purple with yellow throat; lobes pink spotted .... **Martynia**

1b. Leaves penninerved; flowers solitary; stamens 4; ovary 2- or 4-locular......... (2)

2a. Fruit drupe, indehiscent, spiny; corolla yellow; ovules 2 per locule, pendulous .... **Pedalium**

2b. Fruit capsule, dehiscent, bivalved, not spiny; corolla pink, white or purple; ovules many per locule, axile .... **Sesamum**

MARTYNIA L.

Martynia annua L., Sp. Pl. 618. 1753; FPM 2: 1003. 1924; FAP 2: 686. 1997. *M. diandra* Gloxin, Obs. Bot. 14. t. 1. 1785; FBI 4: 386. 1884.

Local name: *Thelukondikaya, Garuda mukku.*

Annual, erect, under shrub, up to 1 m high, glandular hairy. Leaves opposite, broadly cordate, 5–18 × 6–23 cm, viscid hairy, chartaceous, base cordate, margin dentate or sinuate, apex acute; palmately 5-nerved from base; petiole to 14 cm long. Inflorescence axillary raceme, to 12-flowered, at the bifurcations of the branches; peduncle to 10 cm long; pedicel up to 1 cm long. Flowers 3 cm long, glandular hairy. Calyx glandular hairy; lobes 5, free, imbricate, 1.5 cm long; anterior lobes broader, ovate; posterior lobes lanceolate, strigose, ciliate, acute. Corolla purple with yellow throat, 3 × 2.5 cm; narrow tube 1 cm long; ventricose tube 3 cm long; lobe suborbicular, 2 cm long, strigose in anterior lobe. Stamens 2; filaments 1.7 cm long; anther cells 4 mm long, longitudinally coherent with the opposite pair; staminodes 3, posterior one much shorter. Ovary globose, 2 × 1.5 mm, 1-locular, fleshy, 4-celled; ovule 1 per cell; style 2 cm. Disk larger than ovary. Drupe ovoid, with 2 apical, curved hooks, splitting into 2 valves, 3 × 1.7 cm; pericarp glandular hairy; endocarp woody, black. Seeds oblong, 8 × 6 mm.

Common, along the road sides and wastelands. Fl. & Fr.: Jun. – Sept. K. K. Kottala (KDP), *KRKS* 36441.

PEDALIUM Royen ex L.

Pedalium murex L., Syst. Nat. (ed. 10) 1123. 1759; FBI 4: 386. 1884; FPM 2: 1001. 1924; FAP 2: 687. 1997.

Local name: *Yenugu palleru, Pedda palleru, Mullagada-kaya chettu, Yenugu mullu.*

Erect annual herb, 15–30 cm high, stem glabrous, fleshy, 4-gonous. Leaves opposite, repand-angulate, 2.5–5 × 2.5 cm, succulent, base truncate-acute, margin distinctly dentate, apex truncate-emarginate; petiole to 4 cm long; veins penninerved. Flowers light yellow, to 2 cm long, axillary, solitary; pedicel 4 mm long with yellow basal glands. Calyx-lobes 5, free, valvate, linear-lanceolate, 3 mm long, pubescent without, acuminate. Corolla yellow, 2 cm across, hispid within; lobes unequal, 2 large, 3 small. Stamens 4; filaments 0.8–1 cm long; anthers 0.6 mm long, disjunct; staminodes 2.5 mm long; apex reddish. Ovary conical, 1.5 mm long, puberulent, 2-celled; style 1 cm long; ovules 2-per cell. Drupe indehiscent, 1.5 × 0.8 cm, apex conical, lower half 4-gonous, spiny along the angles; epicarp fleshy. Seeds oblong, superposed.

Common, in wastelands, road sides and foot hills of dry deciduous forests. Fl. & Fr.: Jun. – Oct. Rachakuntapalle (KDP), *KRKS* & *KRS* 37635.

SESAMUM L.

Key to Species

1a. Prostrate herb; leaves dissected; flower bud truncate at apex; corolla lobes unequal, lower one very longer than remaining 4; fruit 1 – 1.5 cm long .... **S. laciniatum**

1b. Erect herb; leaves not dissected; flower bud acute at apex; corolla lobes equal; fruit > 3 cm long .... (2)

2a. Leaves simple, ovate-oblong; flower white or pink or pinkish.-cultivated .... **S. indicum**

2b. Leaves palmately lobed, lobes linear oblong; flower pinkish .... **S. alatum**

Sesamum alatum Thonn., in Schumm. & Thonn. Beskri Guin. Pl. 284. 1827; FAP 2: 687. 1997.

Local name: *Rekkanuvvulu, Pichinuvvulu, Adavi nuvvulu.*

Annual, erect herb, up to 2 mm high, stem 4-angular, straw-coloured. Leaves dimorphic, lower leaves opposite, palmately-lobed; lobes 1–5, linear-lanceolate, 3–8 × 0.3–0.5 cm, margin entire, apex obtuse; upper leaves simple, glandular below; petiole up to 6 cm long. Flowers pink, to 2.5 cm long, axillary, solitary; pedicels short, with 2 cup-shaped glands on either side. Calyx small, 5-lobed, 2 mm long. Corolla pink with green veins, tubular, ventricose, bilipped, 5-lobed. Stamens 4, didynamous, staminode 1. Ovary bicarpellary, tetralocular, disk annular. Capsule oblong, tetragonous, beaked, oblanceolate, 4 × 0.7 cm, beak 0.7 cm. Seeds black, ovate, winged on both sides, 2 × 1.5 mm.

Occasional, in wastelands in sandy soils. Fl. & Fr.: Jun. – Nov. Kalasamudram (ATP), *KRKS* 39544.

Sesamum indicum L. Sp. Pl. 634. 1753; FBI 4: 387. 1884; FPM 2: 1002. 1924. *S. orientale* L., Sp. Pl. 634. 1753; FAP 2: 688. 1997.

Local Name: *Nuvvulu.*

Annual erect herb, to 1.5 m high, stem 4-angular, viscid. Leaves ovate or linear-oblong, 5–10 × 3–7 cm, lower leaves divided, upper leaves simple and reduced towards apex. Flowers axillary, solitary, pinkish or rose. Calyx lobes linear to 0.5 cm long, pubescent. Corolla tube 3 – 4 cm long, pubescent without; lobes unequal, lower one bigger than the remainings. Capsule densely pubescent, oblong 4-lobed, to 3 cm long, beaked. Seeds dark brown, ovate, flattened, 1 mm long.

Cultivated as main crop or boarder crop. Fl. & Fr.: Throughout the year. Gollalaguduru (KDP), *KRKS* 41976.

Sesamum laciniatum Klein ex Willd., Sp. Pl. 3: 359. 1800; FBI 4: 387. 1884; FPM 2: 1002. 1924; FAP 2: 688. 1997.

Scabrid, prostrate, annual or perennial herb, up to 1 m long. Leaves elliptic, blade sinuate, older leaves deeply pinnatifid or 3-lobed, 1 × 0.7 cm, coriaceous, base truncate-acute, margin dissected, apex sub-acute, hispid, dense below; whitish below and brownish above when dry; petiole 0.5 cm long. Flowers light pinkish, villous, to 3 cm long, solitary, axillary; pedicel to 3 mm long. Calyx valvate, unequal, lanceolate. Corolla pinkish, 2 × 1 cm, pubescent without, villose at the throat within; tube 2.5 cm long; lobes to 1 cm long. Stamens 4, didynamous; filaments pairs 1.1 and 1.4 cm long, spotted; anthers 2.8 mm long; staminode 0.8 mm long. Ovary oblong-globose, 2 × 1.5 mm long; style 1.7 cm long; stigma capitate. Capsule ovoid, with 2 apical short beaks, 1.5 × 0.7 cm, pubescent. Seeds brownish; testa reticulate, 2 × 1 mm.

Occasional, in plains of dry deciduous forests near rocky and sandy places. Fl. & Fr.: Aug. – Jan. Kalasamudram (ATP), *KRKS* & *TP* 37695.

ACANTHACEAE

Key to Genera

1a. Upper lip of corolla obsolete; corolla-tube cartilaginous .... **Blepharis**

1b. Upper lip or corolla well-developed; corolla-tube thin-fleshy or membranous .... (2)

2a. Corolla-tube as long as, or slightly longer than the lips; lobes exclusively 2+3 .... (3)

2b. Corolla-tube more than twice as long as lobes; lobes sub-equal, spreading or 4+1(stamens 4; 2 staminodes in *Barleria*) .... (12)

3a. Leaves sub-radical; corolla-lobes apically forked; ovules in two vertical rows .... **Elytraria**

3b. Leaves cauline; corolla-lobes entire; ovules in one vertical row .... (4)

4a. Stamens 4, didynamous; flowers either in condensed spikes or in axillary whorls .... (5)

4b. Stemens 2, equal; flowers in cymes, elongate spikes or panicles .... (6)

5a. Flowers in condensed spikes; calyx-lobes imbricate, free, always unequal; capsule 2–4-seeded .... **Lepidagathis**

5b. Flowers in axillary whorls; calyx-lobes valvate, basally connate, sometimes unequal; capsule 8-seeded .... **Hygrophila**

6a. Inflorescence a dense spike (except *J. glabra* cyme); corolla lip crested .... (7)

6b. Inflorescence a cyme or lax panicle .... (9)

7a. Placenta separation elastically from the base; bracts obovate .... **Rungia**

7b. Placenta not separated from the valves; bracts elliptic or lanceolate .... (8)

8a. Corolla tube 6 times longer than the calyx, greenish, slender; corolla upper lip very small than lower ones and erect, acute .... **Rhinacanthus**

8b. Corolla tube 2 times longer than the calyx, white or pink, thick; corolla upper lip almost equal to lower ones and curved, obtuse .... **Justicia**

9a. Seeds glabrous; corolla-lip blotched within; stamens connivent (not in *Andrographis*); anther cells parallel, bearded .... (10)

9b. Seeds papillose or glochidiate; corolla-lip not blotched within; stamens free; anther cells superposed, glabrous, always muticous .... **Dicliptera**

10a. Capsule > 8-seeded; flowers on all sides of the peduncle; panicles both axillary and terminal .... **Andrographis**

10b. Capsule 4-seeded; flowers 1-sided on the peduncle; panicles exclusively axillary .... (11)

11a. Racemes to 11 cm long, usually branched, exceeding the leaf; capsule obovoid, pointed at apex; leaves to 6 cm long, densely hairy .... **Erianthera**

11b. Racemes to 3 cm long, not branched, not exceeding the leaves; capsule ellipsoid or ellipsoid-oblong; leaves to 4 cm long, densely glandular hairy .... **Indoneesiella**

12a. Corolla-limb one-sided; anthers with 1 cell fertile, one sterile; bracts 4-farious .... **Crossandra**

12b. Corolla-limb equally spreading or two-sided; anthers with both cells fertile; bracts not 4-farious .... (13)

13a. Fertile stamens 4; corolla-tube ventricose .... (14)

13b. Fertile stamens 2; corolla-tube cylindric .... **Barleria**

14a. Flowers in one-sided terminal racemes; corolla-lobes imbricate in bud .... **Asystasia**

14b. Flowers in usual axillary cymes; corolla-lobes twisted to left in bud .... (15)

15a. Bracts absent; capsules basally beaked; leaf-base truncate or rounded to acute .... **Dipteracanthus**

15b. Bracts present; capsules not basally beaked; leaf-base cuneate to attenuate .... **Ruellia**

ANDROGRAPHIS Wall.

Key to Species

1a. Leaves linear-lanceolate, to 5.5 cm long; inflorescence axillary and terminal, branched panicle, glandular hairs absent .... **A. paniculata**

1b. Leaves elliptic-ovate, to 2.5 cm long; inflorescence axillary raceme, glandular hairy .... **A. glandulosa**

Andrographis glandulosa Nees, in Wall. Pl. Asiat. Rar. 3: 115. 1832; FBI 4: 505. 1884; FPM 2: 1051. 1924; FAP 2: 692. 1997.

Perennial herbs, or often under shrubs, up to 50 cm high, woody root stock; branches grey-pubescent and ciliate with long hairs; stem 4-angular. Leaves elliptic-ovate, 2–2.5 × 1–1.5 cm, pubescent on both sides, base acute, margin entire, hairy, apex acute; lateral veins 4–5 pairs, conspicuous below; sub-sessile, petiole up to 1 mm long. Inflorescence axillary, helicoid cyme; peduncle 2.5–4 cm long, brown pubescent; pedicels up to 3 mm long, glandular hairy; bracts ovate, 1.5 mm long, glandular hairy without, pubescent within; acute at apex. Flowers white, to 5 mm across. Calyx 5-lobed, tube up to 1 mm; lobes linear-lanceolate, 0.5 cm long, glandular hairy without. Corolla white with purple tinge on lower lip; tube 0.5 cm long, pubescent without; lobes bilipped, the upper slightly 2-fid, the lower usually deflexed, 3-lobed. Stamens 2, anther 2-celled, attached at base of corolla tube, bearded with a tuft of white hairs. Ovary 2-celled; ovules 2–6 in each cell;

style slender; stigma minutely bifid. Capsule 1 cm long, 4–6-seeded, linear-oblong; retinaculum acute or rounded. Seeds 4-8, hard, deeply rugosely pitted, glabrous.

Common, in dry deciduous forests in foot hills. Fl. & Fr.: Oct. – Jan. Rachakuntapalle (KDP), *TP* & *SSR* 33616; K. K. Kottala (KDP), *KRKS* & *BSS* 39038; Pulivendula ghat (ATP), *KRKS* & *BSS* 39625, KRKS 265 (HJCB-E); Chinnapalli (ATP), *KRKS* & *BSS* 40387.

Andrographis paniculata (Burm. f.) Wall. ex Nees, in Wall., Pl. Asiat. Rar. 3: 116. 1832; FBI 4: 501. 1884; FPM 2: 1048. 1924; FAP 2: 694. 1997. *Justicia paniculata* Burm. f., Fl. Indica 9. 1768. *Andrographis subspathulata* C. B. Clarke, in Hook. f., Fl. Brit. India 4: 502. 1884.

Local name: *Nela vemu.*

Perennial, erect, branched herb, up to 40 cm high; branches quadrangular, glabrous. Leaves linear-lanceolate, 5–8 × 1–2 cm, glabrous, base acute, margin entire or recurved, apex acuminate, acute; petiole to 0.3 cm long; lateral veins 5–6 pairs. Inflorescence axillary and terminal, branched, zigzag, helicoid panicle, up to 15 cm long; bracts 2, linear-lanceolate, 1.5 mm long; bracteoles 4, either side, 1 mm long, glabrous; pedicels 0.5 cm long, terete, glandular hairy. Flowers white, to 5 mm across. Calyx 3–4 mm long, lanceolate, acute; glandular hairy. Corolla white with pink tinge, 5 × 2.5 mm; bilipped, upper one slightly 2-fid, lower one deflexed, 3-lobed. Stamens 2; filaments hairy at base, bearded, anthers ovoid, 2 mm long. Ovary oblong, 1.5 mm long, puberulous; style 0.7 cm long, sparsely hairy. Capsule oblong, compressed, minutely hairy, acute, epicarp translucent. Seeds 8–12, sub-quadrate rugose, pitted, glabrous.

Common, in dry deciduous forests, especially foot hills of Nigidi forest. Fl. & Fr.: Aug. – Jan. Batrepalli (ATP), *KRKS* & *BSS* 37945.

ASYSTASIA Blume

Asystasia gangetica (L.) T. Anderson, in Thw., Enum. 234. 1860; FPM 2: 1063. 1924; FAP 2: 695. 1997. *Justicia gangetica* L., Cent. Pl. 2: 3. 1756. *Asystasia coromandeliana* Wight ex Nees, in Wall., Pl. Asiat. Rar. 3: 89. 1832; FBI 4: 493. 1884.

Erect or procumbent herb, often some what climbing, to 75 cm high; stem obtusely quadrangular, white pubescent. Leaves ovate or elliptic-ovate, 2–4 × 1.5–4 cm, glabrous above, sparsely pubescent below, more on nerves, base rounded to truncate, rarely acute in basal leaves, margin entire, apex acute to shortly acuminate; lateral veins 4–6 pairs, clearly impressed beneath; petiole to 2.5 cm long. Racemes sub-terminal; primary peduncle to 10 cm long, pubescent; secondary peduncle 5 cm long; rachis 11 cm long, white pubescent. Flowers on rachis 1.5–2 cm apart; pedicel 3 mm long, bracts lanceolate, 3 mm long. Calyx divided near the base; lobes 8–9 mm long, linear-lanceolate, pubescent. Corolla white, to 4.2 cm; tube inflated, limb ventricose; lobes rounded. Stamens 4, didynamous, connate at the base in pairs, filament pairs 8 and 12 mm long; anthers oblong, cells parallel, minutely spurred at the base. Ovary 3.5 mm long; style 2.2 cm long, linear, lower part hairy. Capsule

Plate 14

A. *Andrograpis paniculata*, B. *Barleria acuminata*, C. *B. stocksii*, D. *Justicia vahlii*, E. *Lepidagathis subarmata*, F. *Vitex pinnata*

up to 3 cm long with a solid base, to 1.6 cm long, upper part clavate, constricted above the middle. Seeds 4, angular, 4 mm across, corrugated at the edges and tubercled on the face.

Common, in dry deciduous forests along foot hills, near forest in rocky crevices. Fl. & Fr.: Sept. – Feb. Kalasamudram (ATP) *KRKS* 40428.

BARLERIA L.

Key to Species

1a. Capsules with a hollow apical beak .... (2)

1b. Capsule without an apical beak .... (4)

2a. Plants unarmed; flowers pink, purple or lilac .... **B. stocksii**

2b. Plants armed; flowers orange or golden yellow .... (3)

3a. Leaf margin entire; outer sepals with glandular hairs inside; style glabrous .... **B. cuspidata**

3b. Leaf margin ciliate; outer sepals without glandular hairs; style hirsute at base .... **B. prionitis**

4a. Leaves with a sharp apicula; inflorescence axillary .... (5)

4b. Leaves without a sharp apicula; inflorescence terminal .... **B. acuminata**

5a. Leaves spinescent at apex, obovate-spatulate; flowers solitary, axillary; corolla dull white .... **B. noctiflora**

5b. Leaves not spinescent at apex, elliptic-lanceolate; flowers in dense, ovoid spikes, axillary and terminal; corolla pink, blue or white .... **B. cristata**

Barleria acuminata Wight ex Nees, in Wall., Pl. Asiat. Rar. 3: 93. 1832; FPM 2: 1059. 1924; FAP 2: 696. 1997. *B. tomentosa* Roth, var. *acuminata* (Wight ex Nees) C. B. Clarke, in Hook. f. Fl. Brit. India 4: 485. 1884.

Perennial, erect shrub, up to 2.5 m high, tomentose, mixed with glandular hairs. Leaves ovate, 5–12 × 4–8.5 cm, base truncate or obtuse, margin entire, apex acute; lateral veins 6–8 pairs; petiole 2.5–4 cm long. Inflorescence axillary or terminal scorpioid cymes, 4.5–5 cm long; peduncle to 6 mm long; bracts linear-lanceolate, 1—2 × 0.1–0.3 cm, acuminate. Flowers sub-sessile, blue-white, to 5 cm across. Calyx scarious, hairy without; outer sepals linear, 1.2–1.8 × 0.2–0.6 cm, glandular hairy, acuminate; inner sepals triangular, 5–10 × 1–2 mm, glandular hairy, acuminate. Corolla faint blue, 3.5–5 cm long, pubescent without; tube 3–4 cm long; lobes oblong-obovate, sub-equal, 10–15 × 8–14 mm, emarginated. Stamens 2, exserted; filaments 5–10 mm long, hairs bluish; anthers 3–5 mm long; staminodes 2. Disk covering about half part of ovary at base. Ovary ovoid, 3–4 mm long, hirsute above; style terete, 3.5–4.5 cm long; stigma 1 mm long, bi-lobed. Capsule ellipsoid, 1–2 × 0.5–0.7 cm, compressed, hirsute at apex, 4-seeded. Seeds broadly suborbicular-ovate, 4–4.8 × 4.2–5 mm, sub-cordate or truncate, clothed with brownish appressed hairs.

Common, in dry deciduous forests, especially near stream sides. Fl. & Fr.: Sept. – Feb. Rachakuntapalle (KDP), *KRKS* & *SSR* 39014; Ambakkampalli (ATP), *KRKS* & *BSS* 40467; Namalagundu (ATP), *KRKS* & *BSS* 40185.

Barleria cristata L., Sp. Pl. 636. 1753; FBI 4: 488. 1884; FPM 2: 1060. 1924; FAP 2: 697. 1997.

Local name: *December puvvulu, Neelambarum, Gobbi, Kodikannu, Nirugorinta.*

Perennial, erect sub-shrub, up to 2 m high; stem appressed hairy, densely so at nodes. Leaves elliptic-lanceolate, 6–8 × 2–4 cm, sparsely hairy above, densely hairy below, base attenuate, margin entire or ciliate, apex acuminate or acute; lateral veins 5–7 pairs; petiole 0.5 cm long. Inflorescence axillary spike, 4.5–5 cm long; bracts and bracteoles linear, 1–1.5 × 0.1–0.2 cm, distantly toothed at margin, apex acute. Flowers blue, to 6 cm long. Outer calyx sub-equal, 1.7–2 × 0.6–0.8 cm, veined, broadly lanceolate, margin spinescent, apex mucronate; inner calyx linear-lanceolate, 7–8 × 2 mm, apex acute, margin ciliate, glandular-pubescent without, whitish, veined; posticous similar to outer calyx but slightly narrow, 1.6–1.8 × 0.6–0.7 cm, margin spinescent, pubescent. Corolla pink or blue, pubescent without, intermixed with glandular hairs; tube lower narrow, upper broad; lobes obovate-oblong, 1–1.8 × 0.8–1.8 cm, obtuse. Stamens 2, exserted; filaments 2–3 cm long, pubescent; anthers 2 mm long; staminodes 2, 4–6 mm long. Disk cupular, unevenly lobed at apex. Ovary 3 mm long, ovules 2 in each locule; style terete, 2.5–4.5 cm long, hairy at base; stigma 5 mm long, 2-lobed. Capsule ellipsoid, 1.2–1.8 × 0.4–0.6 cm, 4-seeded. Seeds ovate-suborbicular, 4–5 × 4–4.5 mm compressed, silky.

Occasional, in forests, common on road sides and village outskirts. Fl. & Fr.: Sept. – Mar. Talupula (ATP), *KRKS* & *BSS* 42000.

Barleria cuspidata B. Heyne ex Nees, in Wall., Pl. Asiat. Rar. 3: 93. 1832; FBI 4: 483. 1884; FPM 2: 1059. 1924; FAP 2: 698. 1997.

Perennial shrub, up to 1 m high, armed; stem terete or obscurely 4-angled; nodes bristled; thorns 2–4, axillary, up to 2 cm long. Leaves ovate or elliptic-oblong, 6–7 × 2–4 cm, glabrescent above, grey pubescent below, base attenuate, margin entire, apex acute-apiculate; lateral veins 3–5 pairs; petiole 5–10-mm long; floral leaves narrow and smaller, cuspidate. Inflorescence axillary, crowded; bracts linear-lanceolate; 1–2.2 × 0.2–0.5 cm, cuspidate, pubescent. Flowers yellow, 2.5 cm long. Calyx 5-lobed; outer sepals ovate-lanceolate, sub-equal 1.5–2.5 × 0.5–0.8 cm, cuspidate at apex, glandular-pubescent within; inner sepals linear-lanceolate, 3–4 mm, acute at apex. Corolla, yellow; tube 2 cm long, bent at base, hairy without; lobes oblong. Stamens 2, exserted; filaments 4.8–5.5 cm long, hairy at base; anthers 4 mm long; staminodes 2. Disk covering ¼th of ovary. Ovary ovoid, 3–4 mm long, glabrous; style slender; stigma ovate, 1 mm long, papillate. Capsule ovoid, 1.5–2 × 0.6–0.8 cm, 2-seeded. Seeds ovoid, 8 × 6 mm, subcordate, silky hairy, brown.

Very common, in foot hills of dry deciduous forests, scrub forests and waste lands. Fl. & Fr.: Aug. – Mar. Rachakuntapalle (KDP), *SSR* & *TP* 33634.

Barleria noctiflora L. f., Suppl. Pl. 290. 1782; FBI 4: 484. 1884; FPM 2: 1059. 1924; FAP 2: 699. 1997.

Local name: *Naga gorinta.*

Perennial under shrub, up to 90 cm high; stem terete, white-pilose, thorns leaf-opposed, often branched; spines up to 2 cm long, white. Leaves clustered, obovate or spatulate, 1–4 × 0.4–1 cm, sparsely pubescent, base cuneate, margin ciliate-entire, apex mucronate; lateral veins 4 or 5 pairs. Flowers axillary, solitary, sessile to 5 cm long; bracts spinescent, branched. Outer sepals elliptic-ovate, sub-equal, 1–2 × 0.5–1 cm, dentate and spinescent at margins, long-spinescent at apex, pubescent without, prominently veined; inner sepals elliptic-lanceolate, 5–10 × 1–2 mm, ciliate at margins, acuminate, glandular-pubescent without, reticulately veined. Corolla dull white, 3.5–5 cm long, pubescent without; tube 3–3.5 cm long; lobes obovate, sub-equal, 8–12 × 6–7 mm. Stamens 2, exserted; filaments 1.5–2 cm long, hairy at base, bluish above. Anthers 2–3 mm long, bluish; staminodes 2, scaly. Disk covering half of ovary at base. Ovary ovoid, 3 mm long, glabrous; ovules 2 in each locule; style 2.5–3.5 cm long; stigma 2-lobed. Capsules ellipsoid, 12–16 × 4–6 mm, compressed. Seeds broadly ovoid, 4.5–5 × 3.5–4.2 mm, white or brown, silky appressed-hairy.

Rare, in scrub forests and foot hills of dry deciduous forests. Fl. & Fr.: Jul. – Feb. Pedda Rangapuram (KDP), *KRKS* & *BSS* 40488.

Barleria prionitis L., Sp. Pl. 636. 1753. FBI 4: 482. 1884; FPM 2: 1058. 1924; FAP 2: 699. 1997. subsp. **prionitis**. Wood *et al.*, Kew Bull. 38: 436. 1983.

Local name: *Mullugorinta, Gobbi, Mundla gobbi, Mulla kanakambaralu.*

Perennial, erect, armed shrub, up to 1.5 m high, much branched; stem terete, glabrous; thorns simple, 2–4, axillary, 5–15 mm long, whitish. Leaves ovate-elliptic oblong, 6–15 × 4–6 cm, pubescent above when young, base attenuate, margin entire, apex cuspidate; petiole 0.5–1.5 cm long, grooved above; lateral veins 4–6 pairs. Flowers axillary, sessile, 3–4.5 cm long, often solitary or becoming spicate above; bracts linear-lanceolate, 1–1.5 cm long, apex cuspidate. Calyx 5-lobed, spine tipped; inner sepals linear-lanceolate, 1–1.5 × 0.2 cm, acuminate at apex; outer sepals ovate-lanceolate, sub-equal 1–1.5 × 0.4–0.5 cm, acuminate, hairy within. Corolla bi-lipped, pubescent outside, orange-yellow; tube 1.5 cm long, swollen and curved at base, pubescent above; lobes subequal, oblong, 1.2–2 × 0.6–1.2 cm, acute. Stamens 2, exserted; filaments 2–2.5 cm long, glandular-pubescent, yellowish; anthers 3 mm long; staminodes 2, 1–2 mm long. Disk covering basal portion of ovary. Ovary ovoid, 3 × 2 mm; style terete, 2.5–3 cm long, hirsute at base; stigma linear, 1 mm long, pinkish. Capsules ovoid with beak, 1.5–2 × 0.6–0.8 cm, compressed, 2-seeded. Seeds ovoid, 7.4–8.5 × 6–6.8 mm, sub-cordate, flattened, brown hairy.

Common, in foot hills of dry deciduous forests, scrub forests and road sides. Fl. & Fr.: Sept. – Dec. Kalasamudram (ATP), *KRKS* 42014.

Barleria stocksii T. Anderson, J. Linn. Soc. Bot. 9: 493. 1867; FBI 4: 489. 1884; FPM 2: 1060. 1924; FAP 2: 700. 1997.

Perennial, under shrub; up to 45 cm high; root stalk tuberous; stem terete, slightly 4- gonous, glabrous, blackish brown when dry, sparsely hairy; hairs 2 mm long, white. Leaves sub-sessile, elliptic-ovate, 2–4 × 0.5–2 cm, glabrous, whitish, base obtuse, margin entire, apex acute; lateral veins 5–7 pairs; petiole to 0.5 mm long. Inflorescence axillary, solitary or raceme. Flowers solitary, to 5 cm long in upper leaf axils and gradually passing into terminal raceme; bracts oblong, linear-lanceolate, 2.2–2.8 × 0.5–1.5 cm, ciliate at margins, acute at apex, hairy on veins without, longer than the calyx lobes. Outer sepals 2, subequal, 1.8–2.6 × 0.2–0.4 cm, ciliate at margin, acuminate at apex, each lobe lanceolate, 3-nerved, 1.5 cm, deeply bilobed, each side 2-nerved; inner sepals 2, 1.2–1.6 × 0.1–0.2 cm, ciliate at margin, acuminate at apex. Corolla 5 cm long, pink; tube cylindric below, funnel-shaped above; lobes broadly obovate, emarginated at apex. Stamens 2, exserted. Ovary covered by disk at base. Capsule obovoid with beak, 1.5–2 cm long, 2-seeded, beaked, silky. Seeds diskoid, 8 × 6 mm, clothed with dense appressed hairs.

Rare, in dry deciduous forests. Fl. & Fr.: Jul. – Sept. Batrepalli (ATP), *KRKS* 39639.

Note: Fruits are rarely seen because flowers are eaten by larvae before fruit formation.

BLEPHARIS Juss.

Key to Species

1a. Herb, prostrate with bristle hairs; leaves elliptic-oblanceolate, 1 cm long; flowers white, to 8 mm across .... **B. integrifolia**

1b. Subshrubs, climbing or straggling; leaves elliptic-oblong, 5 cm long; flowers white with pink tinge, to 1.5 cm across .... **B. maderaspatensis**

Blepharis integrifolia (L.f.) E.Mey. & Drege ex Schinz, Vierteljahrsschr. Naturf. Ges. Zurich 60: 416. 1915. *Acanthus integrifolius* L.f., Suppl. Pl. 294. 1782. *Blepharis repens* (Vahl) Roth, Nov. Pl. Sp. 321. 1821; FAP 2: 701. 1997. *Acanthus repens* Vahl, Symb. Bot. 2: 76. 1791. *Blepharis molluginifolia* Pers., Syn. Pl. 2: 130. 1806; FBI 4: 479. 1884; FPM 2: 1013. 1924.

Local name: *Chinna antrintapulu.*

Annual, prostrate herb, up to 30 cm long; branches diffusely spreading, rooting at nodes, cylindrical, terete, rough, strigose, stinging hairy. Leaves whorled, unequal, narrowly oblong or elliptic-oblanceolate oblanceolate, 1.5–4 × 0.5–1 cm, strigose above, whitish dotted below, base acute, margin entire, apex mucronate; lateral veins prominent above, 3–4 pairs; petiole to 1 mm. Floral leaves 4, unequal, obovate-circular, 3 × 2 mm, sessile, base obtuse, margin ciliate, apex apiculate. Inflorescence 1–few-flowered cyme, axillary, if one-flowered sessile; if cyme with many flowers then pedunculate, peduncle to 1.5 cm long, strigose. Cymose flowers smaller than solitary flowers, to 1 cm long. Bracts of axillary sessile flower 4, opposite, paired, 2.5 × 1.5, 4 × 2, 4 × 1.75, 7 × 3 mm respectvely, obovate, oblong or oblanceolate, spinescent, stellate hairy at apex; bracteoles oblong-lanceolate, 7 × 1.5 mm. Calyx lobes 4, villous at margins; outer calyx lobe forked or 2-toothed at apex; inner pair linear-lanceolate, 7 × 0.5 mm. Corolla bluish-white to 1 cm, lower lip 3-lobed at apex, tube 3 mm long. Stamens 4, to 1.5 mm; anthers 1.3 mm long. Ovary 1 × 1.2 mm; style 2 mm; stigma minutely 2-fid. Capsule 5 × 4 mm. Seeds 4, 4 × 3.5 mm.

Common, in plains, scrub forests and road sides. Fl. & Fr.: Jul. – Nov. Bhumaiahgari palle (KDP), *KRKS* & *KRS* 37605.

Blepharis maderaspatensis (L.) B. Heyne ex Roth, Nov. Pl. Sp. 320. 1821; FAP 2: 701. 1997. *Acanthus maderaspatensis* L., Sp. Pl. 639. 1753. *Blepharis boerhaviifolia* Pers., Syn. Pl. 2: 180. 1806 (*'boerhaviaefolia'*); FBI 4: 478. 1884; FPM 2: 1013. 1924.

Local name: *Anthrinta pulu.*

Annual straggling herb to under shrub, up to 1.5 mm long, branches wiry, dichotomously branched, cylindrical, scattered hairy, rough, nodes up to 10 cm. Leaves 4 in a whorl, unequal, elliptic-oblong or ovate, 2–5 × 1.1–3 cm, glaucous beneath, scabrid on both sides, base obtuse, margin distantly serrate, apex apiculate. Inflorescence axillary, solitary or few-flowered clusters cymes; bracts 3–4 pairs, vary in size 4.3, 6.5, 10 mm each, with glochidiate spinous teeth; bracteoles spatulate, 12 mm long, spinous. Flowers 2 cm across, whitish, pink tinge or nerved on lower lip. Calyx lobes 4, oblong; outer one unequel, 12 and 17 mm each; inner ones 7 mm long, reticulate. Corolla white with pink nerves; tube 4.5 mm long, cartilaginous; upper lip absent; lower lip 3-lobed at apex, 2 cm long, sericeous. Stamens 4; filamens 4 mm long. Ovary 2 × 1.5 mm; style 7.5 mm long; stigma bifid, 0.5 mm. Capsule oblong or obovoid, 8 × 4 mm, shining, glossy. Seeds 2, orbicular, 4 × 3.5 mm, echinate, flattened.

Common, in plains and near water in scrub forests. Fl. & Fr.: Oct. – Mar. Bhumaiahgari palle (KDP), *KRKS* & *KRS* 37646.

CROSSANDRA Salisb.

Crossandra infundibuliformis (L.) Nees, Pl. Asiat. Rar. 3: 98. 1833; FAP 2: 702. 1997. *Justicia infundibuliiformis* L., Syst. Nat. (ed. 10) 850. 1758. *Crossandra undulaefolia* Salisb. in W. Hooker, Parad. Lond. t. 12. 1805; FBI 4: 492. 1884; FPM 2: 1055. 1924.

Local name: *Kanakambaralu.*

Unarmed under shrub, up to 1.5 m high; branches pubescent, terete, cylindrical. Leaves crowded near the inflorescence, elliptic to oblong, 5–10 × 2–4 cm, acute at both sides; lateral veins 10–12 pairs; petiole to 6 cm long. Spikes dense, axillary; peduncles 5–10 cm; bracts imbricate, 4-farious, lanceolate, 1.5 × 0.5 cm; bracteoles linear-subulate, 1.5 × 0.3 cm, pubescent, ciliate, acuminate. Flowers sessile, orange, to 3 cm long. Calyx-lobes 5, unequal, imbricate, lanceolate, 7–9 mm, glandular-hairy; outer lobes obtuse or 2-fid, the rest acuminate, ciliate; inner ones shorter. Corolla orange-scarlet with yellow throat, 3 × 2 cm, salver form; tube narrow, 2.5 cm; lobes 5, subequal, imbricate, connate near middle, 1-sided, split at one end, 1.5 cm long, puberulous, the mid 1 or 3 lobes retuse, laterals obtuse. Stamens 4, attached at the middle of tube, included; filaments curved, 0.7 mm long; anthers 1-celled, oblong, 1.5 mm long. Ovary oblong-cylindric, 3 × 1.5 mm, 4-ovuled; style filiform, 1-2 cm; stigma fimbriate or 2-fid. Capsule oblong, 1 × 0.5 cm. Seeds 4, orbicular, 3 mm across; testa appressed with fimbriate scales.

Cultivated for its ornamental flowers. Fl. & Fr.: Oct. – Mar. Chinnapalli (ATP), *KRKS* & *KRS* 37699.

DICLIPTERA Juss.

Dicliptera paniculata (Forssk.) I. Darbysh, Kew Bull. 62: 122. 2007. *Peristrophe paniculata* (Forssk.) Brummitt, Kew Bull. 38: 451. 1983; FAP 2: 722. 1997. *Dianthera paniculata* Forssk., Fl. Aegypt.-Arab. 7. 1775. *Peristrophe bicalyculata* (Retz.) Nees, in Wall., Pl. Asiat. Rar. 3: 113. 1832; FBI 4: 534. 1885; FPM 2: 1084. 1924. *Dianthera bicalyculata* Retz., Acta Holm. 1775. 197. t. 9. 1776.

Local name: *Chebeera, Oosukada, Sannapulla, Atrial.*

Annual erect herb, up to 1.5 m high, stem hexagonal, distintly tuberculate hairy, much branched, branches rough hairy, hairs only on margins of ridges. Leaves ovate, 2–5 × 1–1.8 cm, pubescent, dense beneath, lanceolate, base rounded, margin entire, apex acuminate; lateral veins 4–6 pairs; petiole 2 cm long; upper leaves remote, subtending the cymes. Panicles of lax cymes, divaricate, terminal or axillary; peduncle to 7 mm long; pedicel 0.6–1.6 cm long; bracts 2, unequal, linear-lanceolate, larger ones 4–5 mm long, smaller ones 2–3 mm long; bracteoles linear-acuminate, longer than calyx lobes, to 1 cm long, 1.5 mm wide, mucronate at apex. Flowers, pink, to 1.5 cm long. Calyx lobes 5, 5 × 1 mm long, 1 mm wide, ends tapering, margin thin, acuminate at apex, pubescent without. Corolla pink, 1.5 cm long, tube cylindrical, 4–6 mm, lobes twisted in bud, 2-lipped, upper ones obtuse, lower 3-lobed, pubescent without. Stamens 2, free; filaments dilated, anthers extruded, slightly higher, muticous, glabrous. Ovary pubescent towards apex, style filiform, 1 cm long, glabrous; stigma bifid. Capsule 1 cm long, clavate with solid basal beak, tip acuminate. Seeds 4, flattened, orbicular, 1.5–2 mm across, papillose.

Common, in wastelands, road sides and foot hills of dry deciduous forest. Fl. & Fr.: Jul.–Dec. Chinnapalli (ATP), *KRKS* 37803.

DIPTERACANTHUS Nees

Dipteracanthus prostratus (Poir.) Nees, in Wall. Pl. Asiat. Rar. 3: 81. 1832; FAP 2: 705. 1997. *Ruellia prostrata* Poir. in Lam., Encycl. 6: 349. 1804; FBI 4: 411. 1884; FPM 2: 1017. 1924.

Local name: *Nela neelambaram.*

Prostrate or diffuse, straggling, much branched under shrub, up to 50 cm; branches terete to sub-quadrangular, glabrous or sparsely hairy; internodes to 10 cm long, nodes purple; rooting at nodes. Leaves ovate-elliptic to deltoid, 2–6 × 1–2 cm, glabrescent above, pubescent below, base acute reducing to petiole, margin entire, apex acute or subacute; lateral veins 5–6 pairs below; petiole to 1.5 cm long. Flower axillary, solitary, sub sessile; bract 0; bracteoles elliptic, 1 × 0.5 cm, pubescent. Calyx-lobes 5, linear-subulate, 7 × 1 mm, divided from base, hairy. Corolla blue-violet, 3 cm long; tube 2 cm long, narrow and cylindrical below, upper half infundibuliform; lobes 5, 6–9 × 8 mm, obovate. Stamens 4; unequal filament pairs 0.5 and 1 cm long; anthers 2.5 mm long. Ovary pubescent, 3 mm long; style 2 cm long. Capsule 1.5 × 0.4 cm, oblanceolate, pubescent, base beaked, to 7 mm long, clavate. Seeds 16, orbicular, 3 mm diameter, flattened, glabrous, hairs on the margins.

Common, in scrub forests and foot hills of dry deciduous forests. Fl. & Fr.: Oct. – Dec. Kalasamudram (ATP) *KRKS* & *BSS* 39117; *KRKS* 42036; Thummalapalle (KDP), *KRKS* & *KRS* 33691, *KRKS* 39025.

ELYTRARIA Michx. *nom. cons.*

Elytraria acaulis (L. f.) Lindau, in Engler & Prantl., Pflanzenfam. 1: 304. 1897; FPM 2: 1009. 1924; FAP 2: 707. 1997. *Justicia acaulis* L. f., Suppl. Pl. 84. 1781. *Elytraria crenata* Vahl, Enum Pl. 1. 106. 1804; FBI 4: 394. 1884.

Local name: *Yeddu adugu, Nelavemu, Kukkupan, Pogetiaku.*

Acaulescent herb, to 25 cm high (including spike); stem minute. Leaves radical or sub-radical, obovate or oblanceolate, 5.5–17 × 2.5–6.5 cm, villous along the nerves below, base cuneate to attenuate, margin crenate, apex obtuse; lateral veins 12 pairs; petiole to 1.5 cm, villose. Spike scapigerous to 16 cm long, unbranched, rarely 2-branched; basal sterile spike, up to 10 cm long; bracts 6 mm, rigid, ovate, acuminate to spinescent tipped; bracteoles 2, linear, small, 4 mm long. Calyx-lobes 4–5, lanceolate, slightly connate, unequal, imbricate, outer lobes broad, 4 mm across, sericeous, acute. Corolla pale blue or white, 7 mm across; tube 4 mm long; lobes 5, bi-lipped, imbricate, acute. Stamens 2, attached at the throat of the tube, included; filaments 0.7 mm long; anther cells oblong, equal, 1 mm, apically mucronate. Staminodes 2, each 0.3 mm long. Ovary cylindric, 1.5 × 0.4 mm; ovules 10 per cell, superposed in 2 rows; style filiform, 8 mm long; stigma lanceolate. Capsule oblong, 1 × 0.3 cm; valves recurved when dry. Seeds many, flattened, retinacula inconspicuous.

Common, in scrub forests, plains and foot hills of dry deciduous forests. Fl. & Fr.: May – Sept. Kalasamudram (ATP), *KRKS* & *KRS* 36452; Thummalapalle (KDP), *KRKS* & *KRS* 33698.

ERIANTHERA Benth.

Erianthera longipedunculata (Sreem.) M.R.Almeida, Fl. Maharashtra 4A: 41. 2003. *Neesiella longipedunculata* Sreemadh., Phytologia 15: 271. 1967. *Indoneesiella longipedunculata* (Sreemadh.) Sreemadh., Phytologia 16: 466. 1968; FAP 2: 715. 1997.

Annual, erect, villous herb, up to 70 cm high, stem quadrangular, grooved, sometimes branched. Leaves elliptic to oblong or oblanceolate, 4–5 × 1.5–2 cm, sparsely villous on both sides, base obtuse, margin entire, apex obtuse-subacute; lateral veins 6–8 pairs; petiole minute, to 1 mm long. Raceme axillary, branched, to 11 cm long, exceeding the leaves. Flowers 1 cm long in helicoid cymes; bracts 2, minute, to 1.5 mm long, with glandular hairs; bracteoles linear-lanceolate, to 2 mm long, with glandular hairs. Calyx tube 1.5 mm long; lobes 5, linear, to 8 mm long with glandular hairs mixed with non glandular hairs; pedicel 1 mm long. Corolla white with pink-purple tinge, 5.5 mm long; tube to 5 mm long; lobes unequal, 2 + 3, imbricate, linear, obtuse, hairy within. Stamens 2; filamens 7 mm long; anthers 2 mm long. Ovary obovoid, 2 mm across, pubescent; style 1.5 cm long, hairy. Capusle obovoid, 1.3 × 0.4 cm, ciliate, tapering at base, acute at apex. Seeds 4, reticulate, pitted, glabrous, 2 mm diameter.

Occasional, in riparian forests and plains. Fl. & Fr.: Oct. – Dec. Somulavaripalli (ATP), *KRKS& BSS* 40182; Gandi river (KDP), *KRKS & BSS* 40324.

HYGROPHILA R. Br.

Hygrophila auriculata (Schumach.) Heine, Kew Bull. 16: 172. 1962; FAP 2: 713. 1997. *Barleria auriculata* Schumach. in Schumm. & Thonn., Beskr. Guin. Pl. 285. 1827. *Asteracantha longifolia* (L.) Nees in Wall., Asiat. Rar. 3: 90. 1832; FPM 2: 1015. 1924. *Barleria longifolia* L., Cent. Pl. 2: t. 22. 1756. *Hygrophila spinosa* T. Anderson in Thwaites, Enum. Pl. Zeyl. 225.1860; FBI 4: 408. 1884. *H. schulli* (Buch.-Ham.) M.R. Almeida & S.M. Almeida, J. Bombay Nat. Hist. Soc. (Suppl.) 221. 1986. *Behel schulli* Buch-Ham., Trans. Linn. Soc. Lond. 14: 289. 1825.

Local name: *Mudigubbaku, Mulla gobbi, Neerugubbi, Kokilachi.*

Erect, armed, unbranched or rarely branched herb, up to 1.5 m high; stem sub-quadrangular, reddish, strigose hairy with tuberculate base; nodes thickened; spines 6 at node, up to 3 cm, dark brown. Leaves oblong-lanceolate or oblanceolate, 5.5–14 × 1.5–2.6 cm, sparsely hispid on both sides, more on midvein, base attenuate, tapering or cuneate, margin entire-ciliate, apex sub-acute; lateral veins 10 pairs, prominent beneath; petiole to 0.3 cm long. Inflorescence axillary cymes, clustered or whorled; bracts leafy, lanceolate, hispid, 8 at each node, 2 × 0.3 cm, ciliate at margin, acute at apex; bracteoles linear-lanceolate 1.5 × 0.3 cm. Flowers whorled at node, to 3.2 cm long, bluish-purple. Calyx 4-lobed; lobes unequal; upper one broader than remaining three, up to 1 cm long, lanceolate. Corolla bluish-purple, to 3.2 cm long, bilipped, lower one 3-lobed, upper one obtuse at apex. Stamens 4; filament pairs unequal, 5 and 10 mm long, each pair connate at the base, hairy; anthers unequal. Ovary oblong, 2.5 mm long, apically ciliate; style 2.5 cm long, slightly puberulous; stigma acute. Capsule 1 cm long, linear-oblong, sub-quadrangular in length, 4–8-seeded. Seeds orbicular, 3 mm across.

Common, in moist places. Fl. & Fr.: Throughout the year, particularly Sept. – Mar. Batrepalli (ATP), *KRKS* 37956; *KRKS* 289 (HJCB-E).

Note: Leaves are used as vegetable, also for joint pains.

INDONEESIELLA Sreemadh.

Indoneesiella echioides (L.) Sreemadh., Phytologia 16: 466. 1968; FAP 2: 714. 1997. *Justicia echioides* L., Sp. Pl. 16. 1753. *Andrographis echioides* (L.) Nees, in Wall. Pl. Asiat. Rar. 3: 117. 1832; FBI 4: 505. 1884; FPM 2: 1051. 1924.

Local name: *Akupootha, Pottinelavemu, Sanyasi.*

Annual erect herb, up to 50 cm high; stem 4-angular, grooved, pubescent, glandular hairy. Leaves oblong to oblanceolate, 3–9.5 × 0.7–2.1 cm, sparsely white-villous, base acute or cuneate, margin entire, apex obtuse or rounded; lateral veins 4–5-pairs; petiole very short, to 2 mm long. Raceme axillary unbranched to 3 cm long, usually not exceeding the leaves; rachis pubescent, glandular hairy; bracts lanceolate, minute, to 1 mm long. Flowers white, 1 cm long, with brown tinge, on lower lip. Calyx glandular hairy, connate at base; tube to 2 mm long; lobes 5, linear, 6 mm long. Corolla white with brown tinge; tube 8 mm long; lower lip 3-lobed; lobes obtuse. Stamens 2, adnate to the throat of corolla; filaments 8 mm long, sparsely hairy; anthers 2.5 mm long. Ovary elliptic, 1.5 mm, puberulous; style 15 mm long, hairy. Capsule ellipsoid or ellipsoid-oblong, tapering at both ends, 1 cm long. Seeds 4, ovoid-oblong, glabrous, reticulately pitted.

Common, in all wastelands and hill tops. Fl. & Fr.: Aug. – Dec. Batrepalli (ATP), *KRKS* 42003; Somulavaripalli (ATP), *KRKS* 40182.

JUSTICIA L.

Key to Species

1a. Slender herbs or under shrubs .... (2)

1b. Shrubs .... (8)

2a. Calyx 5-partite .... (3)

2b. Calyx 4-partite .... (4)

3a. Bracts longer than calyx, orbicular or broadly ovate, to 1 cm long ...... **J. glauca**

3b. Bracts as long as or smaller than calyx, linear, to 1.5 mm long .... **J. glabra**

4a. Bracts and bracteoles usually shorter than calyx-lobes .... (5)

4b. Bracteols as long as or slightly longer than calyx lobes .... (6)

5a. Leaves elliptic-oblong or ovate-oblong; petiole to 1.5 cm long; flowers laxly arranged in spikes .... **J. diffusa**

5b. Leaves linear-lanceolate; petiole to 0.5 cm long; flowers compactly arranged in spike .... **J. vahlii**

6a. Erect herbs, branches and spikes sparsely hairy.-hairs bulbous-based .... **J. micrantha**

6b. Procumbent or diffused herbs; branches and spikes conspicuously hairy ..... (7)

7a. Branches straight, rooting at nodes .... **J. procumbens**

7b. Branches zig-zag, not rooting at nodes .... **J. japonica**

8a. Plants more than 2 m high; bracts leafy .... **J. adhathoda**

8b. Plants less than 2 m high; bracts scarious with green nerves .... **J. betonica**

Justicia adhatoda L., Sp. Pl. 15. 1753; FAP 2: 715. 1997. *Adhatoda vasica* Nees in Wall., Pl. Asiat. Rar. 3: 103. 1832; FBI 4: 540. 1885; FPM 2: 1082. 1924.

Local name: *Addasaram, Vasakam.*

Dense shrub, up to 2.5 m high, branches many from base, ascending, grey-pubescent, 4 angular at young. Leaves thin-coriaceous, elliptic-lanceolate or oblanceolate, 10–18 × 3–6 cm, base acute to attenuate, margin entire, apex acuminate; lateral veins 12–14 pairs, prominent; petiole stout, to 3.8 cm, canaliculated. Spikes axillary or terminal, to 10 cm long with leafy bracts; bracts 2.5 × 5 cm, elliptic or elliptic-ovate, lanceolate, 5–9-veined, closely reticulate; bracteoles oblong-lanceolate, 1.8 × 0.7 cm, ciliate along margins. Flowers white, to 2.5 cm long. Calyx lobes 5, slightly connate, imbricate, lanceolate, 1.2 × 0.3 cm, thick, puberulous, acute, 3-veined. Corolla white, yellowish at throat, puberulous without, 2-lipped, 2 + 3; upper lip ovate-oblong, 2-fid, curved, obtuse at apex, notched; lower lip oblong, rounded, 3-fid, midlobe broad, crested with purple lines, laterals recurved, obtuse. Stamens 2, attached at the throat of corolla tube, exserted, curved; filaments to 2 cm, hairy below, declinate; anthers oblong, 4 mm long, parallel, basally apiculate. Ovary oblong, 3 × 2 mm, pubescent, 4-ovuled; style 2.5 cm long; stigma minutely 2-fid. Capsule clavate, basally beaked, 1.8 cm long, pubescent, 4-seeded. Seeds orbicular-oblong, tubercled or rugose.

Common, on road sides and foot hills of dry deciduous forests. Fl.& Fr.: Oct. – Mar. Vempalli (KDP), *KRKS* & *SSR* 39047.

Justicia betonica L., Sp. Pl. 15. 1753. var. **betonica**; FBI 4: 525. 1885; FPM 2: 1078. 1924; FAP 2: 716. 1997.

Local name: *Telarantu, Tella-addasaramu, Velimungilu.*

Perennial shrubs, to 1.5 m high; branches terete, striate, puberulous, nodes swollen, internode 8-angled. Leaves ovate or lanceolate, linear-elliptic, 2–8 × 1–2 cm, glabrous, base attenuate or cuneate, margin entire to subcrenate, apex acute or shortly acuminate; lateral veins 5 pairs, conspicuous beneath; petiole 1–2 cm long. Spikes 1–3, terminal or sub-terminal, 8–15 cm long; bracts closely imbricate, ovate, elliptic-oblong, lanceolate, leafy, 1.5 × 0.7 cm, white scarious with green veins, ciliate, shortly acuminate; bracteoles similar to bracts. Calyx lobes 5, 4 mm long, linear-lanceolate, scarious, pubescent, 3-ribbed. Corolla white, 2 cm long, 2-lipped; upper 2-lobed, lower 3-lobed; tube cylindric, 5 mm long, pubescent without; lobes 9 mm long. Stamens 2, filaments slightly exserted, to 6 mm long, curved; anthers spurred, 1.5 mm long. Ovary oblong, 2 mm long, pubescent; style 1 cm long, hairy. Capsule clavate, 0.7–1.2 × 0.3–0.5 cm, pubescent, base beaked, apex mucronate. Seeds 2–4, spinulose (or) corrugated or rugose.

Occasional, on road sides and field bunds. Fl. & Fr.: Throughout the year. Batrepalli (ATP), *KRKS* 37976.

Justicia diffusa Willd. Sp. Pl. 1: 87.1797; FBI 4: 538. 1885; FPM 2: 1081. 1924. *Rostellularia diffusa* (Willd.) Nees in Wall., Pl. Asiat. Rar. 3: 100. 1832 var. *diffusa*; FAP 2: 727. 1997.

Slender diffused herb, branchlets erect to 30 cm high. Leaves oblong to broadly elliptic-ovate, pubescent, long hairy, 2–4 × 2–3 cm, base truncate to rounded, margin entire, apex subacute; raphides irregularly scattered; petiole to 1.5 cm long. Flowers in terminal spikes. Spikes linear, forks of dichotomy, to 6 cm long; bracts and bracteoles ovate, 2 mm long, scarious, midrib green, acute. Calyx-lobes 4, subequal, lanceolate, 3.5 mm, flat, margin scarious, minutely ciliate, acute. Corolla 5 × 2 mm across; tube 3 mm long; upper lip 2 mm long; lower lip broadly ovate, 4 mm long. Stamens 2, to 2.5 mm long. Ovary 1 mm long; style 3 mm long, hairy. Capsule 4 mm long, minutely hairy above, ends tapering. Seeds ovate, minutely papillose when wet.

Occasional, in plains and scrub jungles. Fl. & Fr.: Oct. – Jan. Gollalaguduru (KDP), *KRKS* 41971.

Justicia glabra K.D. Koen. ex Roxb., Fl. Ind. 1: 132. 1820; FBI 4: 535. 1885; FPM 2: 1079. 1924; FAP 2: 717. 1997.

Slender, erect or decumbent herb, to 40 cm high; branches sparsely villous, 8-angular; internodes up to 10 cm long. Leaves ovate-oblong, 2–6 × 1–3 cm, glabrous, base obtuse, sometimes acute, margin entire, apex acute; lateral veins 5–7 pairs, mid vein raised beneath; petiole 1–2.5 cm long, sparsely hairy. Inflorescence axillary or terminal cyme; bracts and bractoles very small, to 1.5 mm long, linear. Flowers white, to 7 mm long. Calyx lobes 5, setaceous, to 5 mm long, villous, lanceolate. Corolla white, pinkish or pale yellow with purple spots, 7 × 3 mm across, puberulous without; tube cylindrical, 4 mm long; upper lip 5 mm long; lower lip 7 mm long. Stamens 2 to 4 mm, twisted; anthers cells 1 mm long. Ovary 1.5 mm, puberulous above; style 6 mm, hairy. Capsule 4-angled, 1 × 0.4 cm, puberulous, apically pointed, tapering. Seeds 1.5 mm across, echinate; retinacula jointed.

Common, in agriculture lands and foot hills of dry deciduous forests. Fl. & Fr.: Aug. – Nov. Kalasamudram (ATP), *KRKS* 36481, *KRKS* & *BSS* 39515.

Justicia glauca Rottler, Ges. Naturf. Freunde, Berlin Neue Schriflen 4: 219. 1803; FBI 4: 529. 1885; FPM 2: 1079. 1924; FAP 2: 717. 1997.

Local name: *Addasaram*.

Diffuse or decumbent herb, to 50 cm high; branches terete, grooved, puberulous, internodes up to 4 cm long. Leaves broadly ovate, elliptic-ovate, 4–7 × 2–4 cm, glabrous, base acute, margin entire, apex acute; lateral veins 4–6 pairs, prominent beneath; petiole to 1 cm long. Inflorescence terminal and axillary spike, sometimes solitary; spikes up to 6 cm long; bracts orbicular or broadly ovate, 1 × 0.8 cm, pubescent, base acute, margin entire, apex acute; bracteoles linear, lanceolate, 4 × 1 mm. Flowers 0.7 cm across. Calyx lobes 5, lanceolate, 6 mm long, acuminate, ciliate at margin, base connate. Corolla white with purple spots, 8 × 3 mm, puberulous without; tube 4 mm long; upper lip 5 mm long, lower lip 7 mm long. Stamens 2, to 4 mm long; upper anther cell 1 mm long, lower one 1.5 mm long. Ovary 1.5 mm long, style 7 mm long, pubescent below. Capsule oblong, 1 × 0.5 mm, apically pointed, tapering into a solid beak below. Seeds 4, spinulose, tubercled.

Common, in wastelands, abundant in foot hills of dry deciduous forests. Fl. & Fr.: Aug. – Nov. Thummalapalle (KDP), *KRKS* 36914.

Justicia japonica Thunb., Fl. Jap. 20. 1784. *J. simplex* D. Don., Prodr. Fl. Nep. 118. 1825, non *Rostellularia simplex* Wight, 1850; FBI 4: 539. 1885; FPM 2: 1080. 1924. *Rostellularia japonica* (Thunb.) J. L. Ellis, Bull. Bot. Surv. India 22: 196. 1980; FAP 2: 728. 1997.

Softly hairy, diffuse, annual herb, up to 35 cm high; branchlets striate, angled, zig-zag, strigose. Leaves elliptic-oblong, 2–3.5 × 0.7–1.8 cm, pubescent on both sides, base acute-obtuse, margin entire, apex acute; lateral veins 3–4 pairs; petiole to 0.5 cm long. Inflorescence terminal or axillary, dense villous spike, up to 3 cm long; bracts linear-lanceolate, 3 mm long, margin ciliate with white hairs, apex acuminate; bracteole oblong, ciliate at margin and apex with white 1 mm long hairs. Flowers bilipped, light pinkish, to 5 mm across. Calyx lobes 4; outer broadly oblong 2 mm long with ciliate margin from the middle apex acuminate, ciliate; lower narrowly oblong, 2 mm long, base acute, margin short ciliate, apex long ciliate, to 0.5 mm long. Corolla bilipped, lower 3-lobed, lobes obtuse, villous within on mid lobe; upper slightly 2-lobed, pubescent within; tube up to 1 mm, ciliate within. Stamens 2; filaments to 1 mm long; anthers 0.5 mm long. Ovary 2 mm long, glabrous; style 3 mm long; stigma entire. Capsule oblong, 3 mm long, glabrous, 4-seeded. Seeds brown, cordate at base, 1 mm across, folding from middle.

Common, in wastelands and forests, foot hills and agriculture lands. Thummalapalle (KDP), *KRKS* & *KRS* 36981.

Justicia micrantha Wall. ex. C.B. Clarke, in Hook. f., Fl. Brit. India 4: 536. 1885; FPM 2: 1080. 1924. *Rostellularia crinita* (Nees) Nees, in DC. Prodr. 11: 373. 1847; FAP 2: 726. 1997. *Rostellaria crinita* Nees, in Wall., Pl. Asiat. Rar. 3: 101. 1832.

Erect herb, up to 60 cm high, branches trichotomous, pale, zig-zag, striate and grooved, thinly pubescent. Leaves ovate or elliptic, 2.5–5 × 1.5–2.5 cm, sparsely hairy above, less hairy below, base attenuate, margin distantly crenate, apex acute or subobtuse; lateral veins 4–6 pairs; petiole to 2 cm long. Spike terminal, to 5 cm long; bracts and bracteoles linear-lanceolate, blunt with bulbous based hairs, margins scarious, 4 mm long. Flowers purple, to 6.5 mm across. Calyx lobes 4, unequal, 4–5 mm long, scarious, midrib green, apex blunt. Corolla 2-lipped, pale purple, 6.2 mm long, deltoid-oblong, tube 2 mm long. Stamens 2; filaments glabrous, 2.5 mm long; anthers 0.6 mm. Ovary 1.5 mm, pubescent at the tip; style 5 mm long, lower part hairy. Capsule 4 × 2 mm, beaked at apex, 4-seeded. Seeds 1 mm, very scarcely papillose.

Occasional, in foot hills of dry deciduous forests. Fl. & Fr.: Jul. – Oct. Batrepalli (ATP), *KRKS* & *BSS* 39638.

Justicia procumbens L., Sp. Pl. 15: 1753; FBI 4: 539. 1885; FPM 2: 1080. 1924. *Rostellularia procumbens* (L.) Nees, in DC. Prodr. 11: 371. 1847; FAP 2: 728. 1997.

Slender, diffuse herb, to 30 cm high; branches glabrous, rooting at nodes, angular. Leaves sub-sessile, lanceolate or oblong, 2–3.5 × 0.5–0.8 cm, sparsely pubescent, base acuminate, margin entire, apex long acuminate; lateral veins 5–7 pairs; petiole to 1.5 mm long. Inflorescence terminal spike, to 7 cm long; bracts obovate, 4 × 2 mm, scarious, midrib green, broad, apex acute; bracts and bracteoles linear-lanceolate, 4.5 × 0.7 mm, midrib green, remaining white, apex acute. Flowers pinkish, to 4 mm long. Calyx lobes 4, sub-equal, lanceolate, 4 mm long, midrib ridged, scarious, hairy, acute at apex. Corolla pinkish or lilac, 6 × 3 mm across; tube 3.5 mm, bilipped; upper lip 3.5 mm, truncate, shortly emarginated; lower lip suborbicular, 6 × 5.5 mm, puberulous, without. Stamens 2, to 3 mm; anthers 0.7 mm. Ovary 1.5 mm, pubescent, oblong; style 2.5 mm long, sparsely hirsute. Capsule 3 × 1.2 mm, sparsely hairy, pointed. Seeds 1 mm across.

Rare in foot hills of dry deciduous forests and road sides of clay soils. Fl. & Fr: Sept. – Dec. Idupulapaya (KDP), *KRKS* & *BSS* 40323.

Justicia vahlii Roth, Nov. Pl. Sp. 14. 1821; FPM 2: 1081. 1924. *J. diffusa* Roth var. *vahlii* (Roth) C. B. Clarke, in Hook. f. Fl. Brit. India 3: 538. 1885. *Rostellularia vahlii* (Roth) Nees, in Wall., Pl. Asiat. Rar. 3: 102. 1832; FAP 2: 730. 1997.

Straggling or spreading glabrous herb, to 1.5 m high, from a stout root stock; branches slightly angular, internode to 20 cm long. Leaves chartaceous, oblong-lanceolate, 2–3.5 × 0.3–0.5 cm, glabrous, base acuminate, margin entire, apex acute or acuminate; lateral veins 4–6 pairs; petiole 2 mm long. Spikes terminal, distantly crowded on axis, to 8 cm long; bracts ovate or oblong-lanceolate, 2.5 mm long, midrib green, scarious at margin, base ciliate at apex; bracteole similar to bract. Flowers whitish pink, to 5 mm across. Calyx lobes 4, subequal, 4.4 mm long, linear, midrib scarious, acuminate at apex. Corolla 6.6 mm long, pale with reddish streaks, sparsely hairy without; tube 2 mm long. Stamens 2, exserted, adnate at the throat of corolla; anthers 0.8 mm long; lower anthers cell spurred; filaments 3 mm long, flattened. Ovary puberulous, style slender, 5 mm long; stigma slightly bifid. Capsule 4 mm, clavate, pubescent. Seeds brown, 1.2 mm across, papillate, flattened.

Common, in high altitudes of dry deciduous forests along rocky crevices in moist areas. Fl. & Fr.: Nov. – Dec. Pulivendula ghat, near Anjaneyaswamy temple (KDP), *KRKS* & *SSR* 39029.

LEPIDAGATHIS Willd.

Key to Species

1a. Bracts and calyx lobes with stiff dense hairs, with acute tip .... **L. cristata**

1b. Bracts and calyx lobes glabrescent but ciliate with long, stiff spinous hairs, with spinescent tip .... **L. subarmata**

Lepidagathis cristata Willd., Sp. Pl. 3: 400. 1800; FBI 4: 516. 1885; FPM 2: 1066. 1924; FAP 2: 718. 1997.

Local name: *Mullabanti, Nakkapintuka, Karapani, Karapani poudu.*

Perennial, procumbent herb, up to 30 cm long, main stem erect; branches spreading from inflorescence, quadrangular, glabrous. Leaves linear-oblong or oblanceolate, 3–6 × 0.5–1 cm, nearly glabrous or scabrid, base obtuse, margin entire to serrulate, sub-sessile. Spikes condensed, ovoid, pilose, densely crowded at the base axillaries of the stem, 4 cm across; bracts ovate or elliptic, spinescent; bracteoles lanceolate, membranous hairy, spinescent, 7 × 2 mm. Flowers white, to 5 mm across. Calyx-lobes 5, imbricate densely hairy; outer lobe 8 × 4 mm, obtuse at base; mid ones 2, united above the middle; inner lobes lanceolate, 8 × 2 mm, densely sericeous throughout, acute at apex. Corolla white, pinkish or brown dotted, 8 × 2 mm across, hairy without, tube 4 mm long; upper lip 5 mm long, oblong, obtuse, notched at apex with pink tinge; lower lip 3-lobed, 7 mm long, divided nearly to the base. Stamens 4, exserted; filament pairs 1.5 and 2.5 mm long; anther cells unequal, 1.2 mm long. Ovary ellipsoid-globose, 1.5 × 1 mm; style 7 mm long, sparsely hairy. Capsule oblong, compressed, 6 × 2 mm, beaked. Seeds 2, ovate, densely hairy, 3 × 1.5 mm.

Occasional, in scrub forest, wastelands and foot hills of dry deciduous forests. Fl. & Fr.: Aug. – Jan. Kalasamudram (ATP), *KRKS* 39905; Mabbuchintalapalle (KDP), *KRKS* 288 (HJCB-E).

Lepidagathis subarmata (C. B. Clarke) Gamble, Fl. Madras 2: 1067. 1924; FAP 2: 720. 1997. *L. mitis* Dalz. var. *subarmata* C.B. Clarke, in Hook. f. Fl. Brit. India 4: 516. 1885.

Local name: *Kodabanti.*

Perennial stiff under shrub up to 1 mm long with radiating branches raised from basal nodes; branches decumbent, quadrangular, pinkish, sparsely hairy. Leaves obovate-oblong, obovate, to 4–6 × 0.4–1.5 cm, glabrous, scabrous on nerves beneath, base acute, margin entire, apex obtuse; lateral veins 5–6 pairs; petiole to 1 mm long. Spikes clustered, cushion like at the base, globose head up to 5 cm across; bracts 5 mm, ovate, long spinous acuminate, densely sericeous, ciliate along margins. Flowers white, to 10 mm across, dark brown dotted within corolla throat. Calyx 5 mm long, imbricate; lower lobes connate almost at tip, sericeous, ciliate at margins, spinescent at apex. Corolla light pink with maroon streaks near throat, 1 cm long, 2-lipped; upper 2-fid, lower 3-lobed, hairy without. Stamens 4, filaments 3 mm long; anthers 1 mm long. Ovary glabrous, 1 mm long. Capsule 5 mm long, flattened. Seeds 2, faintly rugose.

Occasional, in plains of black cotton soils and wastelands. Fl. & Fr.: Sept. – Jan. Kalasamudram (ATP), *KRKS* 39927; Mabbuchintalapalle (KDP), *KRKS* 273 (HJCB-E).

RHINACANTHUS Nees

Rhinacanthus nasutus (L.) Kurz., J. Asiat. Soc. Bengal, Pt. 2, Nat. Hist. 39: 79. 1870. *Justicia nasuta* L., Sp. Pl. 16. 1753. *Rhinacanthus communis* Nees in Wall., Pl. Asiat. Rar. 3: 109. 1832; FBI 4: 541. 1885; FPM 2: 1083. 1924.

Local name: *Naga malli, Vontikommu, Rhinachettu, Pamu malli.*

Erect herb or subshrub, divaricately branched, to 1.5 m high; branches striate, subterete or obscurely angled, young shoot pubescent. Leaves elliptic or elliptic-lanceolate, 4–20 × 2–10 cm, puberulous, base acute to narrowed, margin entire or rarely very scarcely shallowly serrate, apex acute or shortly acuminate; lateral nerves 6-10 pairs, clearly impressed beneath; petiole 1–2 cm long, pubescent. Flowers sessile or shortly pedicellate, solitary or 2-3 together, distant on the divaricate branches of large, lax, terminal panicles; bracts and bracteoles ovate-lanceolate, 2 mm, glandular hairy, pubescent; calyx deeply partite, 3 mm, pubescent with glandular hairs. Sepal lobes subequal, linear, subacute. Corolla white, 3.2 cm, tube narrowly cylindric, 2.5 cm long, imbricate, 2-lipped, 2+3, upper lip linear-oblong, recurved, 9 × 1.5 mm, lower lip 1.3 cm long, lobes broadly oblong, obtuse, hispid without. Stamens 2, attached at the throat, exserted, filaments 2 mm long, anther 2-celled, cells superposed, 2 mm. Ovary 2 mm long, style 2.3 cm long, hairy, stigma minutely 2-fid. Capsule linear-ellipsoid, 1.8–2 cm, narrowed, pointed with a solid basal beak. Seeds 4, orbicular, 4.3 mm diam, tuberculate, black.

Common, in hill slopes above 500 m, among crevices. Fl. & Fr. Most part of the year. Kalasamudram RF (ATP), *NY* 1192.

RUELLIA L.

1a. Roots tuberous; leaves 4–12 × 3–4.5 cm .... **R. tuberosa**

1b. Roots not tuberous; leaves 1.5–4 × 0.6–2. 5 cm .... **R. patula**

Ruellia patula Jacq., Misc. 2: 358. 1781; FBI 4: 412. 1884; FPM 2: 1017. 1924. *Dipteracanthus patulus* (Jacq.) Nees, in Wall., Pl. Asiat. Rar. 3: 82. 1832; FAP 2: 704. 1997.

Erect or suberect herb, to 50 cm high, branches sub-quadrangular, hoary pubescent or villous, terete, glandular hairy. Leaves elliptic or ovate, 1.5–4 × 0.6–2 cm, appressed pubescent, glandular hairy, base obtuse or acute, margin entire to coarsely crenate, apex acute-obtuse; lateral veins 4 pairs; petiole 1.8 cm long, pubescent. Inflorescence axillary, solitary or cyme; bracteoles foliaceous, ovate or elliptic, 1 cm long, densely hairy. Flowers solitary, whitish-blue or violet, villous. Calyx tube 1 mm long; lobes 5, lanceolate, 5 × 2 mm, strigose, ciliate at margin, acute at apex. Corolla blue-violet, 2.5–3 cm long, pubescent without; tube 1.5 cm, companulate, infundibuliform, lower half cylindrical and upper half swollen; lobes 5, 6 mm long, broadly elliptic. Stamens 4, filaments included, glabrous, 4 and 5.8 mm long; anthers 2 mm, ovary glabrous, 1.5 × 0.5 cm. Seeds 12, suborbicular, flattened, 3 mm across, margin thick brown.

Common, in wastelands, scrub forests and foot hills of dry deciduous forest near rocky crevices. Fl. & Fr.: Oct. – Dec. Kalasamudram (ATP), *KRKS* & *KRS* 36485.

Ruellia tuberosa L., Sp. Pl. 635. 1753; FPM 2: 1017. 1924; FAP 2: 730. 1997.

Local name: *Jurbula gadda.*

Perrennial, erect, tuberous herb to 50 cm high; much branched, branches quadrangular, tender parts hispid. Leaves elliptic-obovate or elliptic-oblong, 4–12 × 3–4.5 cm, base cuneate, tapering, margin entire to undulate or crenate, apex obtuse; lateral veins 8–10 pairs; petiole 1–2.5 cm long. Inflorescence axillary dichasial cyme; bracts and bracteoles linear-lanceolate. Flowers bluish-violet, to 5 cm long, funnel-shaped. Calyx lobes 5, subequal, 1.5 cm long, setaceous, puberulous. Corolla blue to violet; tube 5 cm long, lobes subequal. Stamens 4, included, filament pairs 4 and 8 mm long, hairy at base, anthers 4 mm long, sagittate, staminode narrowed. Ovary 4 mm long, style 2 cm long, stigma bilamellate, ciliate. Capsule oblong, 2.8 cm long, linear-oblong. Seeds flattened, brown, hairy, more dense on margins.

Common, in village outskirts near temples, field bunds and foot hills. Fl. & Fr.: Through the year. Kalasamudram (ATP), *KRKS* 42046.

RUNGIA Nees

Rungia repens (L.) Nees, in Wall., Pl. Asiat. Rar. 3: 110. 1832; FBI 4: 549. 1885; FPM 2: 1070. 1924; FAP 2: 731. 1997. *Justicia repens* L., Sp. Pl. 15. 1753.

Local name: *Palakavelli, Kharmor.*

Annual, erect or decumbent herb, up to 40 cm high; branches slender, sub-erect, strigose, often rooting at nodes. Leaves sub-sessile, oblong-lanceolate, 2–3.5 × 0.4–0.6 cm, scabrous, hairy on both sides, base acute, margin entire, apex acute; lateral veins 3–4 pairs; petiole to 2 mm long. Spikes terminal, to 7.3 cm long, quadrangular; sterile bracts 5 × 3 mm, elliptic-oblong, base acute, margin scarious, apex acute, midrib green, pubescent without; fertile bracts similar to sterile bracts but larger in length and width; bracteoles 2, 3.8 × 1 mm, lanceolate, margins scarious, glandular ciliate. Flowers pinkish-white, densely arranged on spike, to 4 mm long. Calyx 5-lobed, subequal, shortly connate, lanceolate, 3.2 mm long, ciliate, glandular hairs. Corolla pinkish, 5 × 2 mm; tube 3 mm long; lobes 5, unequal, 3 + 2, imbricate; upper lip 4 mm, shortly 2-fid; lower lip 5 mm, crested, 3-lobed, obtuse. Stamens 2, at the junction of 2 lips, exserted; filaments 2.5 mm, puberulous; anther cells oblong, divergent, unequal, to 1.2 mm long; lower cell shortly appendaged below; connectives broad. Ovary globose, 1 mm across, puberulous; style 4.5 mm, hairy; stigma minutely 2-fid. Capsule elliptic-oblong, flat, 5 mm, pubescent. Seeds orbicular, compressed, concentrically tuberculate, elastically separating from the valves.

Common, near water and moist places of scrub forests and wastelands. Fl. & Fr.: Nov. – Jan. Mabbuchintalapalli (KDP), *KRKS* & *BSS* 41806, *KRKS* 287 (HJCB-E).

VERBENACEAE

Key to Genera

1a. Leavs digitately 3–5(7)-foliolate .... **Vitex**

1b. Leaves simple .... (2)

2a. Inflorescence condensed or elongated spike; fruiting bracts conspicuous; herbs or subshrubs; stamens included .... (3)

2b. Inflorescence a raceme or cyme; fruiting bracts inconspicuous or 0; shrubs or trees .... (6)

3a. Calyx cupular or truncate, not ribbed; stigma lateral; spikes condensed or slightly elongate, terminal and axillary; ovary 2-celled, 2-ovuled .... (4)

3b. Calyx tubular, ribbed; stigma terminal; spikes elongate, cylindric, exclusively terminal .... (5)

4a. Drupe fleshy; calyx truncate; lobes obscure .... **Lantana**

4b. Drupe dry; calyx cupular; lobes 2 .... **Phyla**

5a. Stamens 2; anthers divaricate; flowers embedded in excavations of spike .... **Stachytarpheta**

5b. Stamens 4; anthers parallel; flowers not embedded, but attached only by their base .... **Priva**

6a. Corolla regular; lobes either 4 or 6–8; stamens as many as corolla-lobes, equal .... **Tectona**

6b. Corolla bilabiate; lobes exclusively 5; stamens 4; didynamous .... (7)

7a. Stamens inserted well below tube, involute in bud, clearly exserted later; inflorescence cymose .... **Clerodendrum**

7b. Stamens inserted near the middle of tube, not involute in bud, included or subexserted; inflorescence racemose or corymbose .... (8)

8a. Racemes paniculate; corolla yellow, > 2 cm across .... **Gmelina**

8b. Cymes corymbose; corolla cream or white, < 0.7 cm across .... **Premna**

CLERODENDRUM L.

Clerodendrum phlomides L. f., Suppl. Pl. 292. 1781; FBI 4: 590. 1885; FPM 2: 1099. 1924; FAP 2: 741. 1997.

Local name: *Thakkalaku, Konda takai, Thakkali.*

Large bushy shrub or small tree, to 3 m high, mild aromatic; branches pale yellowish pubescent, 4-angular. Leaves ovate-rhomboid or deltoid-ovoid, chartaceous, 2–6 × 1–3.5 cm, densely hairy, base truncate or subcordate, margin coarsely crenate-dentate, apex acute or obtusely acute; lateral veins 4 pairs, rised beneath; petiole to 2.3 cm long. Inflorescence axillary or terminal dichotomously branched cymose panicle; bracts lanceolate, leafy, 5 mm long. Flowers moderate scented, white, to 5 cm long. Calyx campanulate; tube to 4 mm long; lobes ovate-lanceolate, to 1 cm long, acuminate. Corolla white or purplish, salver-shaped; tube to 2.3 cm long, hairy without, cylindrical; lobes 5, elliptic, obtuse, prominently nerved, subequal, to 8 mm long; stamens 4, inserted on the corolla tube, filament pairs 2 and 2.3 cm long; anthers oblong, 2.5 mm, greyish. Ovary globose, 2 mm across, style to 3.7 cm long; stigma 2-fid. Drupe obovoid, 4-lobed, black, to 1 cm long, with accrescent calyx lobes.

Common, in hedges of villages and agriculture lands. Fl. & Fr.: Feb. – May. Chinnakudala (KDP), *KRKS* & *BSS* 40461.

Note: Leaf extract is used as medicine for fever and muscle pains.

GMELINA L.

Gmelina asiatica L., Sp. Pl. 626. 1753; FBI 4: 582. 1885; FPM 2: 1098. 1924; FAP 2: 744. 1997.

Local name: *Adavigummadi, Kavva gummidi, Chiru gummudu, Salla gummudu, Chiru nelli, Nela gummudu.*

Large armed shrub to small tree, often struggling on the other shrubs, up to 3 m high; arms up to 2 cm long, having leaves; bark brownish-white, thin; wood grey; branches pubescent with light yellow hairs. Leaves chartaceous, elliptic, ovate or rhomboid, 1.6–3.5 × 1.4–2.5 cm, glabrous, glaucous beneath, base acute, margin entire or distantly sinuate, apex obtuse to subacute; secondary veins 3–4 pairs, impressed beneath; petiole to 1.4 cm long. Inflorescence axillary and terminal raceme, to 8 cm long. Flowers yellow, bell-shaped, to 4.3 cm long. Calyx 4-lobed, 6 mm long, puberulous without, lower part of the tube narrow; upper portion cup-shaped, hairy, glandular without. Corolla yellow, 4 cm long, 2.5 cm across; tube 1.5 cm long, cylindrical, narrow at base, widen towards apex, to 1 cm across; lobes 5, acute, glandular pubescent without. Stamens 4, with a 5th rudimentary one; filament pairs 1 and 1.7 cm long; anthers 2.5 mm long. Ovary 4 mm across; style slender, 3 cm long; stigma shortly bifid. Drupe yellow, obovate, 1.8 × 2.5 cm, dotted, shiny, pulpy. Seeds 2, inserted in pulp, hard.

Common, in scrub forests and foot hills of dry deciduous forests. Fl. & Fr: Jun. – Sept. Kalasamudram (ATP), *KRKS* & *KRS* 40072.

LANTANA L.

Key to Species

1a. Unarmed shrubs; spikes elongate; flowers white .... **L. veronicifolia**

1b. Armed shrubs; spikes condensed; flowers variable in colour (pinkish red-yellow) .... **L. camara**

Lantana camara L., Sp. Pl. 627. 1753. var. **aculeata** (L.) Mold., Torreya 34: 9. 1934; FAP 2: 745. 1997. *L. aculeata* L., Sp. Pl. 627. 1753; FBI 4: 562. 1885; FPM 2: 1087. 1924.

Local name: *Gabbuseeki, Beliaku, Belichettu.*

Armed straggling or sub-scandant, aromatic shrub, up to 2.5 m high; branches 4-angular, scabrous, rough hairy, brownish, armed; prickles recurved, 1 to 2 mm long,. Leaves ovate or ovate-oblong, 5–8 × 2–5 cm, scarbrid above, pubescent beneath, base truncate, rounded, margin crenate-serrate, apex acute; lateral veins 5–6 pairs; petiole to 2.2 cm long. Inflorescence short capitate umbellate spike, axillary and terminal; bracts lanceolate, hispidulous, to 0.5 cm long, acute at apex. Flowers scented, yellow-white, tinged with red, salver-shaped, to 1 cm long. Calyx membranous, truncate, shortly or obscurely toothed, connate, 0.2 cm across. Corolla variable in colour, hypocrateriform; tube slender, 7 mm long; limb 7.3 mm across; obscurely 2-lipped; lobes 5, spreading. Stamens 4, didynamous, included, anthers oblong, cells parallel. Ovary 2 mm across; style short, stigma oblique. Drupe fleshy, globose, shiny, black, 1-seeded pyrenes. Seed brown, seed coat hard.

Common, in wastelands, road sides, scrub forests and foot hills of dry deciduous forests. Fl. & Fr.: Through the year. Kalasamudram (ATP), *KRKS* 41809; Batrepalli (ATP), *KRKS* 42048.

Lantana veronicifolia Hayek, Repert. Spec. Nov. Regni Veg. 2: 163. 1906. *L. wightiana* Wall., ex Gamble, Fl. Madras 1087. 1929; FAP 2: 745. 1997. *L. indica* Roxb. var. *albiflora* Wight ex Hook. f., Fl. Brit. India 4: 562. 1885.

Local name: *Thellabeli, Podabeli.*

Unarmed shrub, up to 2 m high; branches brown, some what quadrangular, scandent. Leaves elliptic-ovate, 2–4.6 × 1–2.5 cm, scabrous, more rugose above, pubescent below, base rounded, acute, margin crenate, apex acute or subacute; lateral veins 6–7 pairs; petiole to 9 mm long. Inflorescence elongated, spike to 3 cm long; bracts ovate, 3 mm long. Flowers white, to 8 mm long. Calyx cupular, 2 mm across, obscurely 5-lobed. Corolla white, 8 mm long; tube slender; lobes 5, 4 mm across. Stamens 4, didynamous, included; filament pairs 0.7 and 1 mm long; anthers 0.5 mm long. Ovary 0.5 mm across; style 9 mm long. Berries brown or black on stalk, shiny, 3 mm across.

Occasional, in scrub forests and wastelands. Fl. & Fr.: May – Sept. Gajjalavaripalli near Vemana temple (ATP), *KRKS* 42047.

PHYLA Lour.

Phyla nodiflora (L.) Greene, Pittonia 4: 46. 1899; FAP 2: 746. 1997. *Verbena nodiflora* L., Sp. Pl. 20. 1753. *Lippia nodiflora* (L.) A. Rich., in Michaux, Fl. Bor. Amer. 2: 15. 1803; FBI 4: 563. 1885; FPM 2: 1088. 1924.

Local name: *Bokkena, Bokkenaku, Mosalipappu.*

Prostrate perennial herb, up to 50 cm long, much branched, branches glandular, angular, grooved, glabrous; rooting at nodes. Leaves thick-chartaceous, obovate-spatulate, sometimes orbicular, 5–18 × 5–8 mm, sparsely hairy, base decurrent, margins serrate at upper half, apex obtuse, 2–4-nerved. Spikes condensed, cylindrical, to 1 cm long, axillary; peduncle 3–6 cm; bracts broadly elliptic or obovate, 2.4 mm long. Flowers numerous, sessile, aggregated to 2 mm across. Calyx membranous, 2-lobed, lanceolate, scarious, pink. Corolla white or pale pink, salver form; tube 1.5 mm long; lobes 5, spreading, 2-lipped, 2 + 2 +1, lower one 2-fid at apex. Stamens 4, included, attached in pairs at throat of tube; filaments curved, 0.4 mm long, anthers ovate, cells parallel. Ovary glabrous, 2-celled, ovule 1 per cell; style short, 0.5 mm long; stigma oblique, subcapitate. Drupes subglobose, 1 mm long, splitting into two when dry. Seed 1, pyrenes, minute.

Common, near moist locations in all low altituds. Fl. & Fr.: Through the year. Somulavari palli (ATP), *KRKS* & *KRS* 37837.

PREMNA L. *nom. cons.*

Premna tomentosa Willd., Sp. Pl. 3: 314. 1800; FBI 4: 576. 1885; FPM 2: 1095. 1924; FAP 2: 748. 1997.

Local name: *Peddanarava, Kampu gummadi, Kokkiti chettu, Nagoora, Pomanti, Nela neredu, Adavinela, Nagal, Narvu.*

Moderate sized deciduous tree, up to 10 m high; bark grey or light brown; wood light brown, smooth; branches with densely soft, tawny tomentose, stellate hairy, yellowish, quadrangular when young. Leaves thick, chartaceous, ovate, 2.5–16 × 2–10 cm, tawny yellow, stellate tomentose, dense below than above, base truncate-obtuse, margin entire, apex acuminate; lateral veins 5–8 pairs, prominent below; petiole to 3.8 cm, yellow tawny-tomentose, canaliculate. Inflorescence corymobose cyme, terminal and axillary, up to 12 cm long; bracts to 1 mm long; pedicel to 2 mm long; corymbose axis tomentose with yellowish hairs. Flowers yellowish, to 4 mm long. Calyx cupular, 4–5-toothed, 0.4 cm, across, villose. Corolla greyish-yellow, 4 mm across; tube 3 mm long, villous within along throat; lobes 5, orbicular, to 2 mm across, obtuse. Stamens 4, didynamous; filament pairs 1.5 and 2 mm long; anthers rounded, 0.5 mm long. Ovary truncate, puberulous; style 2.5 mm long; stigma unequally 2-fid. Drupe ovoid or sub-globose, 5 mm across, black, smooth, stellate at apex; mesocarp hard, brown. Seeds 4, ovate 3 × 1.5 mm, brown.

Common, in foot hills of dry deciduous forests and scrub forests. Fl.: May – Jul. & Fr.: Jul. – Sept. Kalasamudram (ATP), *KRKS* & *KRS* 36499.

PRIVA Adans.

Priva cordifolia (L. f.) Druce, Bot. Soc. Exch. Club. Brit. Isles 1916; 641. 1917; FAP 2: 749. 1997. *Buchnera cordifolia* L. f., Suppl. Pl. 287. 1781. *Priva leptostachya* Juss., Ann. Mus. Hist. Nat. 7: 70. 1806; FBI 4: 565. 1885; FPM 2: 1091. 1924.

Local name: *Pachareni.*

Erect herb, up to 1 m high; stem 4-angular, sticky-pubescent, hairs hooked, with bulbous base, internodes to 10 cm long. Leaves broadly ovate-elliptic, 2–4 × 1–2.8 cm, chartaceous, sticky hairs hooked at tip, base truncate-acute or sub-cordate, margin dentate, apex acute; lateral veins 5–7 pairs, prominent below; petiole 0.5–1.8 cm long, pubescent with sticky hairs. Spikes terminal, elongate, 15–30 cm long; bracts linear-lanceolate, persistent, hairy. Flowers distantly arranged, subsessile, white, to 5 mm across, pubescent. Calyx tubular, 5-ribbed, 5-toothed, 6 mm long, sticky hairs. Corolla white, tube 8 mm long, twisted, scarcely enlarged above, limb 2-lipped; upper lip 2-lobed; lower lip larger than the upper lip, deeply 3-lobed. Stamens 4; filament pairs 1.5 and 2 mm long, inserted in the middle of the corolla tube, included; anther cells parallel or slightly divergent. Ovary oblong-cylindrical, 1.5 mm long, 2-locular, each 2-ovuled; style 4 mm long, filiform, persistent; stigma oblique. Drupe of 2, 2-seeded, pyrenes enclosed in dilated calyx-tube, separating when ripe, into two as 2-seeded pyrenes, sticky pubescent, laterally compressed, apically beaked. Seeds laterally ribbed, apex hooked, spiny without, 3-angular, 4 × 2 mm, light brown.

Common, in wastelands, road sides and foot hills of dry deciduous forests. Fl. & Fr.: Jun. – Sept. Kalasamudram (ATP), *KRKS* & *KRS* 36500.

STACHYTARPHETA Vahl

Stachytarpheta jamaicensis (L.) Vahl, Enum. Pl. 1: 206. 1804; FAP 2: 750. 1997. *Verbena jamaicensis* L., Sp. Pl. 19. 1753. *Stachytarpheta indica* sensu, Hook. f., FBI 4: 564; *p.p.*; auct. non (L.) Vahl 1804. *S.indica* var. *jamaicensis* (L.) Trimen, Cat. Pl. Ceylon 68. 1885 & Handb. Fl. Ceylon 3: 348. 1895; FPM 2: 1090. 1924.

Local name: *Medabalaku.*

Erect, woody, biannual herb, up to 80 cm high; branches slightly quadrangular, almost glabrous when mature, distantly with bulbous base hairs when young. Leaves variable, elliptic, obovate-elliptic, 4–8 × 2.5–5.3 cm, sparsely hairy below on midvein, glabrescent above, base tapering and decurrent, margin dentate, scabrous with bulbous hairs, apex obtuse or acute; lateral veins 5–6 pairs, prominently veined; petiole to 1 cm long. Spikes terminal, 30 cm long; rachis glabrous, hollowed; bract 4.8 mm long, lanceolate subulate. Flowers sessile, bluish, to 7 mm across. Calyx 5–6 mm long, membranous, tubular, teeth 4, small. Corolla blue, 7 mm long; tube slender, cylindric, 5 mm long, limb oblique; lobes 5, subequal, flat, spreading, 5 mm diameter (3 + 2). Stamens 2; filaments hairy, slender, 1.5 mm long; staminodes 2; anthers 0.8 mm long. Ovary glabrous, 2 mm long; style 7 mm long, filiform. Fruit oblong, ribbed, 3–4 mm long, splitting into 2 pyrenes, dark chocolate-brown. Pyrenes oblong, 6 mm across.

Occasional, in plains of dry deciduous forests, road sides near moist places and water bodies. Fl. & Fr.: Sept. – Jan. Gandi river (KDP), *KRKS* 40325.

Note: There are variants in populations and no specieman exactly matching with protologue from our collection.

TECTONA L. f. *nom. cons.*

Tectona grandis L. f., Suppl. Pl. 151. 1781; FBI 4: 570. 1885; FPM 2: 1092. 1924; FAP 2: 752. 1997.

Local name: *Tekuchettu, Pedda teku.*

Large deciduous tree, up to 20 m high; bark grey or light brown, fibrous with shallow longitudinal fissures, exfoliating in long, thin, narrow, some what corky flakes, wood golden yellow, turning brown, branchlets quadrangular, stellate hairy. Leaves broadly elliptic, 15–30 × 9–29 cm, rough hairy above, with tuberculate hairs, grey tomentose below; base rounded or acute, margin entire or wavy, apex acute or obtuse; lateral veins 10–12 pairs; petiole to 3 cm long. Panicle terminal, dichotomousely branched, 10–30 cm long; pedicels to 2 mm long; bracts minute. Flowers 6–8-merous, yellowish-white, to 9 mm long. Calyx stellate, tomentose, campanulate, 5–7-lobed, spreading, subequal, 2–3.5 mm long. Corolla yellowish white, glabrous, 1 cm across, sub-rotate; lobes 5 or 7, speading, broadly oblong, 5–6 mm long. Stamens 5–7, inserted near the base of the corolla, equal, exserted; filaments 3 mm long; anthers ovate or oblong, cells parallel. Ovary globose, 2 mm across, hairy without; style linear, 4 mm long; stigma shortly bifid. Drupe globose,

1.2 cm in diameter, slightly 4-lobed, epicarp stellate hairy, velvety, endocarp bony, accrescent calyx to 2.5 cm across, bladder like, enclosing fruit. Seeds oblong, dark brown, to 5 mm across.

Common, in dry deciduous forests, on slopes and to hill tops. Fl.: Jul. – Sept. & Fr.: Oct. – Jan. Batrepalli (ATP), *KRKS* 39614.

VITEX L.

Key to Species

1a. Inflorescence axillary; leaves glabrous; petiolule *ca.* 1.8 cm long in all leaflets .... **V. leucoxylon**

1b. Inflorescence terminal; leaves hairy or glabrous; petiolule *ca.* 4 mm long in all leaflets (except *V. negundo*) .... (2)

2a. Lower surface of the leaflets grey tomentose; shrubs; petiolule of middle leaflet longer than the remaining ones .... **V. negundo**

2b. Lower surface of the leaflets minutely pubescent; trees; petiolule of all leaflets equal .... (3)

3a. Petiole completely winged; wing to 3 mm width .... **V. altissima**

3b. Petiole basally winged or absent; wing to 1 mm width .... **V. pinnata**

Vitex altissima L. f., Suppl. Pl. 294. 1781; FBI 4: 584. 1885; FPM 2: 1102. 1924; FAP 2: 754. 1997. *V. alata* Heyne ex Roth., Nov. Pl. Sp. 316. 1821; FBI 4: 584. 1885.

Local name: *Nemali adugu, Busi chettu, Mayurapaku, Gandarapur chettu.*

Deciduous tree, up to 20 m high; crown dense; bark yellowish-grey, scaly fibrous; branches villous at young with soft hairs. Leaves 3–5-foliolate; petiole to 5 cm long, winged throughout, with an auricle base; leaflets thick, chartaceous, middle leaflet obovate, 10–12 × 2–3 cm; lateral ones oblanceolate, 4–9.5 × 1.5–3 cm, all glabrescent above, thinly pubescent beneath, more on veins, base acute, margin entire, apex acuminate; lateral veins 9–15 pairs; petiolule to 3 mm long. Inflorescence terminal, 10–25 cm long panicle, branches fulvous-pubescent; bracts 3–4 mm, oblong, pubescent. Flowers 4 mm across, shortly pedunculate cymes, blue, bilipped. Calyx 3 mm across, cupular; tube to 4 mm across; teeth or lobes triangular, 1.2 mm across, pubescent without. Corolla white, tinged with blue, to 6 mm across, 2-lipped; tube hairy, cylindrical; limb pubescent; upper lip 2.4 mm long, 2-lobed, lobes triangular; lower lip 4 mm long, 3-lobed, mid-lobe larger than lateral ones. Stamens 4, exserted; filament pairs 3 and 3.5 mm long; anthers 0.6 mm long. Ovary globose, to 1 mm across; style 4.6 mm long; stigma bilobed. Drupe globose, 5 mm across, dark purple, smooth, hairy at apex, one-seeded. Seed brown, testa hard, 3 mm across, longitudinally striated.

Rare in dry deciduous forests. Fl.: Apr. – Oct. & Fr.: Aug. – Dec. Rachakuntapalli (ATP), *KRKS* & *BSS* 41856.

Note: Petiole wing larger when it is young and reduces its size in old tree.

Vitex leucoxylon L. f., Suppl. Pl. 293. 1781; FBI 4: 587. 1885; FPM 2: 1103. 1924; FAP 2: 754. 1997.

Local name: *Lukki, Kondavavili, Gajavavili, Nakkaneredu, Nevaledi.*

Large, deciduous tree, to 20 m high; crown spreading; bark grey, smooth; wood dark purple, straight grained; branchelets puberulous or glabrous. Leaves 3–5-foliolate; petiole to 7 cm long; leaflets thin, coriaceous, narrowly elliptic-oblong; middle leaflet larger than laterals, 7–10 × 2–3 cm, glabrous, base acute, margin entire, apex narrowly obtuse or acute; lateral veins 8–11 pairs, petiolule to 1.8 cm long; lateral leaflets 2–4, gradually decreasing in size from middle leaflet, base oblique. Inflorescence axillary, dichotomously branched cyme, corymbiform, panicles 13 cm long; peduncle 4–7 cm long, slender, glabrous; bracts 2 mm long. Flowers white-yellow, to 1.5 cm across in cymes. Calyx 3–5 mm across, cupular, puberulous, appressed without, 5-toothed, teeth 1 mm long. Corolla whitish, purplish within, 1.5 cm across; tube 5 mm long, puberulous without; lobes 5; upper lip 3.5 mm long; lower lip 8 mm long, obtuse. Stamens 4; filament pairs 4 and 6 mm long; anthers 0.5 mm long. Ovary 2 mm across, pubescent; style 8 mm long; stigma 2-lobed. Drupe ellipsoid-obovoid, dark purple, 1.5 × 0.8 cm, spiny at apex. Seeds 4, oblong to 1 cm long, connate; testa hard.

Occasional, in plains and foot hills of dry deciduous forests. Fl.: Dec. – Mar. & Fr.: Apr. – Jul. Thimmamma marrimanu (ATP), *KRKS* 37969.

Vitex negundo L., Sp. Pl. 638. 1753; FBI 4: 583. 1885, incl. var. *incisa*; FPM 2: 1102. 1924; FAP 2: 754. 1997.

Local name: *Tella vavili, Vavili.*

Large shrub, white or greyish pubescent, to 3 m high; branchlets grey pubescent, 4-angular, blackish when dry. Leaves 3(-5)-foliolate; leaflets ovate-oblong, lanceolate, middle one larger than lateral ones, to 10 × 2.5 cm, blackish pubescent above, grey tomentose below, base short acute-rounded, margin entire, apex acuminate; lateral veins 16-19 pairs; petiolule 2.2 cm long; lateral leflets 2, smaller than middle, base oblique; petiolule very small, 0.3–0.5 cm long; petiole to 7 cm long, grey pubescent, canaliculated. Inflorescence pyramid-shaped, terminal, 10–25 cm long, cymose panicle, cymes dichotomously branched on main axis, basally helicoid cyme dichotomously brached; bract ovate, to 1 mm long, villose. Flowers sessile, purple, to 8 mm across, pubescent. Calyx bell-shaped, tomentose without, to 3 mm across, teeth 5, 0.5 mm ovate. Corolla purple to violet, 7 mm across, hairy inside at the inception of the stamens; upper lip 2 mm long, 2-lobed, obtuse; lower lip 5 mm long, 3-lobed; lateral lobes 2, oblong-obtuse, terminal lobe 3 mm, crenulate, obovate. Stamens 4; filaments hairy at the base, 3.5 and 5 mm long; anthers 1 mm long. Ovary glabrous; style 7 mm long; stigma 2-lobed, forked. Drupe globose, black, 3 mm diameter. Seed 1, subglobose, 2–2.5 mm across.

Common, on road sides and wastelands. Fl. & Fr.: Mar. – Jul. Mobbuchintalapalle (KDP), *KRKS, SSR, KRS* & *TP* 33656.

Vitex pinnata L., Sp. Pl. 638. 1753; FAP 2: 755. 1997. *V. pubescens* Vahl, Symb. Bot. 3: 85. 1794; FBI 4: 585. 1885; FPM 2: 1103. 1924.

Local name: *Nemaliadugu, Rachangi.*

Small tree, to 5 m high; trunk fluted, dark greyish, somewhat fissured, flaking in long thin pieces, bluish lower side when flaking off; wood reddish brown, close-grained, smooth; branchlets 4-angled with tawny pubescence. Leaves 3-foliolate, coriaceous, ovate or elliptic to 16 × 7 cm, pubescent, base rounded, margin entire, apex acuminate; lateral veins 6–7 pairs, petiole to 5 cm long, slightly winged at base; petiolule minute, to 3 mm long. Panicles terminal, pyramidal, subcorymbose, minutely tomentose; branches helicoid cymes; bracts foliaceous, ovate or obovate, 6–12 mm long, persistent. Flowers 5 mm across, pale blue, pubescent. Calyx 7.5–10 mm diameter, 5-toothed, tomentose, often unequally split in fruit. Corolla 1–1.2 cm, pale blue, mildly fragrant, upper lip 2-lobed, lower lip 3-lobed, obtuse at apex. Stamens 4, didynamous, dark blue; filaments to 3 mm long; anther cell horizantal. Drupe black, depressed globose, 8–10 mm diameter, shiny, enclosed with calyx up to ¾ th, apex pubescent, endocarp thick, pink pulpy. Seed solitary, brown, with longitudinal striations, globose, hard, to 0.5 cm across.

Occasional, in dry deciduous forests. Fl.: Jun. – Jul. & Fr.: Aug. – Sept. Rachakuntapalle (KDP), *KRKS* 40191.

Note: Scorpoid cymose flowers also seen in a population. Bark light blue to dark blue when cut open, also seen the same population in different individuals.

LAMIACEAE

Key to Genera

1a. Stamens ascending to the upper lip, or spreading and projecting straight forward .... (2)

1b. Stamens descending, lying upon or enclosed in the lower lip........................ (5)

2a. Calyx 8 or 10-toothed; upper lip of corolla 1-lobed, hooded; verticils usually single, axillary or terminal; one arm of stigma obsolete .... (3)

2b. Calyx 5-lobed; upper lip of corolla 2 or 4-lobed, spreading; verticils or second spikes aggregated into a raceme; stigma unequally 2-fid.............. (4)

3a. Corolla orange; upper lip twice as long as lower lip; calyx teeth spinescent .... **Leonotis**

3b. Corolla white; upper lip less or as long as lower lip; calyx-teeth acute or acuminate.-stamens 4, didymamous, included in upper lip of corolla .... **Leucas**

4a. Stamens spreading and projecting out; corolla-lobes 4, < 5 mm long, obscurely 2-lipped .... **Pogostemon**

4b. Stamens ascending and not projecting; corolla-lobes 5, > 1 cm, distinctly 2-lipped .... **Anisomeles**

5a. Calyx campanulate, equally 5-toothed; corolla 2 + 3; lower lip shorter than upper lip; flowers in short cymes or verticils....................... **Hyptis**

5b. Calyx distinctly 2-lipped; corolla 4 + 1; lower lip as long as, or longer than upper lip; flowers in verticils aggregated into trichotomous or paniculate racemes ....(6)

6a. Thyrses > 1 cm across; fruiting calyx, 2.5 mm .... **Basilicum**

6b. Thyrses > 1.5 cm across; fruiting calyx > 4 mm long...(7)

7a. Lower lip of calyx truncate; thyrses of a compact, cylindric spike ....**Anisochilus**

7b. Lower lip of calyx 2 or 4-toothed; thyrse of a lax, spicate raceme...................(8)

8a. Lower lip of corolla not longer than upper lip; flowers 6 – 10, in distinct whorls, aggregated to 3-chotomous racemes...................................(9)

8b. Lower lip of corolla twice as long as upper lip; flowers in decussate cymes or 6–9-flowered whorls, aggregated to 3-chotomous thyrses or racemes ....**Plectranthus**

9a. Corolla-tube clearly exserted; stamens included; filaments not toothed; stigma capitate or minutely clavate; fruiting pedicels spreading .... **Orthosiphon**

9b. Corolla-tube hardly exserted; stamens exserted; upper pair of filaments toothed near base; stigma 2-fid; fruiting pedicels usually recurved ....**Ocimum**

ANISOCHILUS Wall.

Anisochilus carnosus (L. f.) Wall. ex Benth., in Wall., Pl. Asiat Rar. 2: 18. 1830-1831; FBI 4: 627. 1885; FPM 2: 1126. 1924; FAP 2: 759. 1997.

Local name: *Ritchu-rodda, Karpuravalli.*

Aromatic, erect herb, to 70 cm high; stem stout, 4-gonous, bluntly quadrangular, finely pubescent, often tinged with red. Leaves ovate, fleshy, 3–6 × 4–7 cm, verrucose above, pubescent below, glandular when mature, base cordate or rounded, margin serrate-crenate, apex acute; lateral veins 3–5 pairs; petiole to 5 cm long. Spikes 2–4 cm long, dense, compact, quadrangular, elongating and cylindric in fruit, peduncled, terminal; bracts broadly ovate, 15 mm long. Flowers dark purple and with white tube. Calyx-tube 1.5 mm long; upper lip 1.5 – 2 mm long, pubescent without, acute; lower lip reduced, truncate. Corolla-lobes 5, purple, 7 × 1.5 mm, bilipped, 4 + 1; upper lip short, erect, entire, rounded at apex; lower lip concave, 3–4.5 mm long, hairy without; tube white, narrow, truncate below, 5 mm long, inflated above. Stamens 4, didynamous, declinate in lower lip, filament pairs of 2 and 3 mm long; anthers 0.5 mm long. Ovary 0.6 mm across; style 3.2 mm long, filiform; stigma bifid. nutlets suborbicular, compressed, smooth, inner face angular, shining brown.

Common, in dry deciduous forests on exposed slopes near moist locations. Fl. & Fr.: Oct. – Jan. Rachakuntapalle (KDP), *SSR &TP* 33623; Talupula (ATP), *KRKS & BSS* 41922.

ANISOMELES R. Br.

Key to Species

1a. Branchlets tawny-tomentose, obtusely 4-gonous; leaves lanceolate, coriaceous, margin serrate; corolla purple .... **A. malabarica**

1b. Branchlets pubescent, acutely 4-gonous; leaves ovate, chartaceous, margin crenate-srreate; corolla pink-violet .... **A. indica**

Anisomeles indica (L.) Kuntze, Revis. Gen. Pl. 2: 512. 1891; FPM 2: 1140. 1924; FAP 2: 761. 1997. *Nepeta indica* L., Sp. Pl. 571. 1753. *Anisomeles ovata* R. Br. in Ait., Hort. Kew. ed. 2. 2: 364. 1811; FBI 4: 672. 1885.

Local name: *Adabeera, China ranaberi.*

Suffruticose under shrub, scarcely aromatic, softly pubescent, 1–2 m high; stem and branches acutely tetragonous. Leaves chartaceous, ovate, 3–9.5 × 2–5 cm, softly pubescent above, soft velvety below, dark brown when dry above, grey below, base subtruncate or cuneate, margin crenate-serrate, apex acute to acuminate; lateral veins 4, conspicuous towards base and 4 inconspicuous towards apex; petiole to 4.3 cm long. Spikes 3.5–20 cm long, axillary and terminal, greenish. Flowers in dense whorls, pinkish towards apex, whitish below, to 1.7 cm long. Calyx 0.5 cm long; tube to 7 mm long; lobes 5, equal, triangular, acute at apex, villous without. Corolla bilipped, pink-violet, 1.6 cm long; tube whitish, annulate within, to 9 mm long; upper lip oblong, 9 mm long, rounded at the apex; lower lip much longer, to 1.3 cm long, 3-lobed, the lateral lobes small, rounded, mid one deeply divided into 2 oblong lobes, densely hairy within. Stamens 4, filaments to 1.2 cm long, hirsute, covered by upper lip of corolla; anthers 0.4 mm long, dark brown. Ovary 0.5 mm long; style 1.6 cm long. Nutlets ovoid, 1.8–2 mm across, black, polished, smooth, inner face obtusely angled, dorsal face rounded.

Common, in foot hills of dry deciduous forests and field bunds. Fl. & Fr.: Sept. – Mar. *KRKS, KRS, SSR &TP* 37826.

Anisomeles malabarica (L.) R. Br. ex Sims, Curtis, Bot. Mag. t. 2071. 1819; FBI 4: 673. 1085; FPM 2: 1140. 1924; FAP 2: 761. 1997. *Nepeta malabarica* L., Mant. Pl. 566. 1771.

Local name: *Magabeera, Mogabheri.*

Under shrub, up to 2 m high, clothed with dense tomentose or whitish woolly, light grey when dry; stem obtusely quadrangular, slightly branched. Leaves ovate-lanceolate, thick, subcoriaceous, 4–9 × 1.6–2 cm, softly tomentose, base rounded or shortly cuneate, margin crenate-serrate, apex acute; lateral veins 6–8 pairs, prominent beneath; petiole to 1.7 cm long; apical leaves reduced along spike. Spikes to 18 cm long, interrupted; bracts to 1.8 cm long, linear, soft woolly; bracteoles to 9 mm long. Calyx 8.5–10 mm long, densely villous without; tube 6 mm long; lobes 5, narrowly lanceolate, to 4 mm long. Corolla purple, hairy without, 1.4 cm long, bilipped, 2 + 3; tube 7 mm long, white; upper lip flat, obtuse at apex, to 5 mm long;

lower lip 3-lobed, pink, mid lobe emarginated, lateral lobes short, white hairy within. Stamens 4, didynamous, ascending to the upper lip of corolla, exserted, filaments hairy, to 1.2 cm; anther 0.7 mm long. Ovary 3 mm long, strigose; style bifid, to 0.4 mm long. Nutlets pale, to 4 mm long, ellipsoid, compressed, smooth, polished, brown.

Common, in wastelands, road sides and scrub forests. Fl. & Fr.: Through the year. Thummalpalle (KDP), *KRKS* & *KRS* 33694.

BASILICUM Moench

Basilicum polystachyon (L.) Moench, Suppl. Menth. Pl. 143. 1802; FAP 2: 762. 1997. *Ocimum polystachyon* L., Mant. Pl. 567. 1771. *Moschosma polystachyum* (L.) Benth., in Wall., Pl. Asiat. Rar. 2: 13. 1830-1831; FBI 4: 612. 1885; FPM 2: 1117. 1924.

Slender, annual, erect herb or subshrub, to 1 m high; branches slender, sharply 4-angled, glabrous, brownish, pubescent. Leaves ovate-deltoid, 1–4.5 × 1–3.2 cm, glabrous below, pubescent above on veins, base cuneate or truncate, margin deeply serrate or crenate, apex acute; lateral veins 6 pairs; petiole to 2.5 cm long. Thyrses terminal, rachis slender, to 13 cm long, pubescent. Flowers in whorls of 6 at a point, to 8 mm long, bracts ovate, to 2 × 1, apex apiculate, pubescent; pedicel 5 mm long. Calyx tubular; tube 5 mm long, ovoid, curved upwards, 2-lipped, 4 + 1; upper lip boat-shaped, 2 mm across; lower 4-lobed, 3-angular, acuminate, spinescent, to 2 mm long. Corolla 1 mm long, white, bi-lipped, 3 + 2. Stamens 4, didynamous, descending into the lower lip; filaments slightly exserted, to 1 mm long; anthers 2-celled, 0.2 mm long. Ovary 1 mm across; style gynobasic, 2.3 mm long; stigma forked, disk anteriorly developed. Nutlets ellipsoid, 0.5 mm across, erect, basilar, smooth, acute, dark brown, rugose.

Occasional, in dry deciduous forests near rocks. Fl. & Fr.: Oct. – Feb. Rachakuntapalle (KDP), *KRKS* 37914.

HYPTIS Jacq.

Hyptis suaveolens (L.) Poit., Ann. Mus. Natl. Hist. Nat. 7: 472. t. 29. f. 2. 1806; FBI 4: 630. 1885; FPM 2: 1129. 1924; FAP 2: 766. 1997. *Ballota suaveolens* L., Syst. Nat. ed. 10. 1100. 1759.

Local name: *Seemathulasi.*

Subshrub to 3 m high; branches terete, 4-gonous, ovate, 2–7 × 1–6 cm, chartaceous, strigose, base obliquely truncate to acute, margin irregularly serrate, apex acute; petiole to 2 cm long. Flowers in verticils or in short, stalked cymes, or both at the same node, 6-flowered, axillary; pedicel to 1 cm long. Calyx campanulate; tube 4 mm long, 10-ribbed; lobes 5, subequal, linear, 4 mm, acuminate. Corolla blue, 5 × 2 mm across; tube 4 mm long; lobes 5, bilipped, 2 + 3; upper lip erect, 3 mm across; lower lip deflexed, saccate; mid lobe folded, short and bifid, 2.5 mm long, membranous, finely hispid without, pubescent below the stamens within. Stamens 4, didynamous, descending into the lower lip, partly exserted; filaments 3 mm long, bearded; anthers 2-celled. Ovary 1 mm long, one lobe slightly smaller; style gynobasic, 4.5 mm long; stigma minutely bifid. Disk anteriorly developed.

Plate 15

A. *Basilicum polystachyon*, B. *Leucas lavandulifolia*, C. *L. mathewiana*, D. *Orthosiphon rubicundus*, E. *Achyranthes aspera* var. *sicula*, F. *Allmania longipedunculata*

Nutlets 2, ovoid or oblong, 2 aborted and seen as rudimentary white structures at base, erect, basilar.

Very Common, on road sides and wastelands. Fl. & Fr.: Aug. – Nov. Kalasamudram (ATP), *KRKS* 36448.

LEONOTIS (Pers.) R. Br.

Leonotis nepetiifolia (L.) R. Br., Prodr. 504. 1810; FBI 4: 691. 1885; FPM 2: 1155. 1924; FAP 2: 767. 1997.

Local name: *Ranabheri, Mulla golimidi, Hanumantha beera, Rama chettu, Errathummi.*

Erect herb, to 1.5 m high; branchlets pubescent. Leaves ovate, 3–7 × 1.5–5 cm, chartaceous, base truncate to cuneate, margin deeply crenate to serrate, apex acute; petiole to 7 cm long. Verticils axillary; flowers many; bracts acuminate. Calyx tubular; mouth oblique, 8–10-toothed, posterior tooth wider and longer, valvate, 1.5–2 cm, coriaceous, spinescent. Corolla orange, 1.5 × 0.4 cm across, tubular, bilipped; tube 1 cm long, with 5 annular rings within; upper lip hooded, 1.3 cm long, densely velutinous without; lower lip 3-lobed; midlobe bifid, 0.5 cm long, velutinous without, acute. Stamens 4, didynamous, hooded by the upper lip, exserted; filaments thick, minutely bearded, to 1 cm long; anthers divaricate, at length confluent, 2.3 mm across. Ovary 1 mm long, cells trigonous; style gynobasic, 2 cm long; stigma unequally 2-fid, capitellate. Disk cupular, anteriorly developed. Nutlets erect, basilar, 4 mm, trigonous. Seeds oblong.

Common, in wastelands and road sides. Fl. & Fr.: Oct. – Feb. Chinnapalli (ATP), *KRKS* & *KRS* 37698.

LEUCAS R. Br.

Key to Species

1a. Calyx mouth oblique, throat villi with white hairs which are longer than calyx lobes.......... **L. hirta**

1b. Calyx mouth straight or truncate, throat hairy with colourless hairs, not as above .... (2)

2a. Calyx-tube bent in the middle .... (3)

2b. Calyx-tube curved, not bent in the middle .... (4)

3a. Leaf-apex acuminate; calyx-tube sigmoidly curved, pubescent, posterior tooth twice as long as the rest; corolla-lips subequal; upper lip of corolla hairs red.......... **L. martinicensis**

3b. Leaf-apex acute; calyx-tube abruptly deflexed, strigose, teeth subequal, lower lip of corolla longer than the upper; upper lip of corolla hairs brown.......... **L. nutans**

4a. Mouth of calyx truncate, upper teeth to 3 mm long, with bristle-like hairs on the teeth and tube outside .... **L. stricta**

4b. Mouth of calyx oblique, upper teeth to 1 mm or < 1 mm long, pubescent all over calyx (not within at base)..(5)

5a. Verticil 1 at apex; flowers loosely spreading; bracts half of the calyx length; pedicel 1 – 2 mm long; calyx mouth suboblique; corolla 16 – 17 mm long ....**L. mathewiana**

5b. Verticils 2 – 3 at apex; flowers compact; bracts equal to calyx lobes in length; pedicel *ca.* 0.5 mm long; calyx mouth oblique; corolla 13 – 15 mm long ....(6)

6a. Base of calyx tube thin, membranous; bracts linear, to 8 mm long, with bristle-like hairs along margin..**L. aspera**

6b. Base of calyx tube not thin; bracts lanceolate, spinescent, pubescent ....**L. lavandulifolia**

Leucas aspera (Willd.) Link, Enum. Pl. Hort. Berol. 2: 113. 1822; FBI 4: 690. 1885; FPM 2: 1150. 1924; FAP 2: 768. 1997. *Phlomis aspera* Willd., Enum. Pl. Hort. Berol. Alt. 621. 1809.

Local name: *Thummi, Tella thummi.*

Annual herb, up to 60 cm high; stem and branches 4-gonous, hispid with spreading hairs. Leaves subsessile, linear, linear-lanceolate, 2–10 × 0.3–1.5 cm, sparsely hairy to thinly tomentose on both sides, base obtuse, attenuate, margin sub-entire, remotely crenate, apex acute or obtuse; petiole up to 1 cm long. Inflorescence terminal and axillary verticil, 1–2.8 cm in diameter, forming globose heads, sometimes an additional head develops in the axil of sub terminal leaves; bracts 7–10 mm long and 1.5 mm width, linear, acuminate or spinulose at apex, mid nerve prominent, ciliate with bristle-like hairs. Flowers white, whorled, densely, many in the head of verticil, to 1.3 cm long, pubescent. Calyx-tube 8–12 mm long, tubular, shortly curved or apparently straight, constricted above the nutlets, prominently 10-nerved or ribbed and pilose with bristle-like, bulbous-based hairs above the constriction, glabrous and membranous in lower part; mouth strongly oblique, produced above the mouth; true villi absent; teeth 10, 1–1.5 mm long, ciliate with bristle-like hairs, triangular, spine-tipped. Corolla white; tube 1.5–2.8 cm long, strongly curved, exserted; upper lip bearded with white hairs, smaller than lower one; lower lip about twice as long as upper, 3-lobed, middle lobe obovate, truncate. Stamens 4, didynamous; filaments to 1 and 1.1 cm long, ciliate; anthers to 1 mm across, curved, light brown, in upper lip of corolla. Ovary 2 mm across, style to 1.5 cm long, stigma bilobed, lobes unequal. Nutlets 2 mm long, oblong, truncate at apex, smooth or granulate, brown.

Very common, in wastelands, agriculture lands and road sides. Fl. & Fr.: Aug. – Nov. Rachakuntapalle (KDP), *KRKS* 37918; Sanevaripalli (ATP), *KRKS* 40141, 40143; Chinnapalli (ATP), *KRKS* 37812; Thummalapalle (ATP), *KRKS* & *KRS* 36983.

Leucas hirta (Heyne ex Roth) Spreng., Syst. Veg. 2: 743. 1825; FBI 4: 687. 1885; FPM 1153. 1924; FAP 2: 771. 1997. *Phlomis hirta* Heyne ex Roth, Nov. Pl. Sp. 264. 1821.

Erect herbs to 80 cm high with stout root stock; branches pubescent with reflexed hairs. Leaves 5–6 × 1.5 cm, elliptic-oblong, base acute, margin distantly crenate, apex acute or obtuse, hirtus above, densely tomentose beneath with scabrid hairs; petiole 8 mm long. Heads 1–2, terminal, many- flowered, 2–2.5 cm diam. Bracts 10 × 1 mm, linear, bristly. Flowers 15–30. Calyx-tube *ca.* 7 mm long, not oblique, whitish, villous at the throat conspicuous, hairs longer than the tooth, densely hairy outside. Corolla tube 5 mm long, upper lip emarginate, white, hairy, slightly shorter than the lower, mid-lobe of lower lip emarginate.

Rare in dry forests. Fl. & Fr.: Dec. – Apr. Kalasamudram (ATP), *EC* 6528.

Leucas lavandulifolia Sm., in Rees. Cycl. 20. 1812; Mukerjee, Rec. Bot. Surv. India 14 (1): 167. 1940. *L. linifolia* (Roth) Spreng. Syst. Veg. 2: 743. 1825; FBI 4: 690. 1885; FPM 2: 1149. 1924. *Phlomis linifolia* Roth, Nov. Pl. Sp. 260. 1821.

Local name: *Pula pummi, Pula thummi.*

Erect, foetid, annual herb, up to 70 cm high; branched from the base, sometimes woody at base; stem and branches acutely 4-gonous, puberulous with fine erect hairs. Leaves sessile or sub-sessile, linear, linear-lanceolate or lanceolate, 3–10 × 0.2–1.5 cm, glabrate above, puberulous beneath or on both sides, base acute, margin subentire or distantly serrulate, apex acute; petiole very short, narrowed. Inflorescence whorls 1.5–2.5 cm in diameter, many-flowered, terminal and axillary, often internodes are too short so that whorls become spicate towards the end of branches; bracts 3–8 mm long, shorter than calyx, linear, puberulous, ciliolate with fine short hairs, spinulose at apex. Flowers white, in verticil, to 1.8 cm long, hairy. Calyx-tube 6–9 mm long, turbinate, straight or inconspicuously bent, finely puberulous, veins inconspicuous; mouth very oblique, distance of projected upper and lower depressed rim of mouth varies from 3–5 mm, pubescent within; villi absent; teeth irregular, spinulose; upper most largest, 2 mm long; smaller ones 0.5–1 mm long, puberulous. Corolla white; tube 10–12 mm long, exserted, pubescent outside, annular within near the middle; upper lip densely bearded with white hairs, half or less the length of lower lip; lower lip horizontal, very long and broad. Stamens 4, didynamous; filaments glabrous, to 1 cm long; anthers light brown, two-lobed. Ovary 2 mm across; style to 1.3 cm long; stigma bilobed; lobes unequal. Nutlets 2–2.5 × 1 mm, oblong or ovoid, rounded at apex, inner surface angular, rounded at apex, dark brown.

Common, in dry deciduous forests. Fl. & Fr.: Aug.–Dec. Batrepalli (ATP), *KRKS* 42050; Batrepalli (ATP), *KRKS* 37880; Kalasamudram (ATP), *KRKS* 40084.

Leucas martinicensis (Jacq.) R. Br., Prodr. 504. 1810; FBI 4: 688. 1885; FPM 2: 1149. 1924; FAP 2: 773. 1997. *Clinopodium martinicense* Jacq. Enum. Syst. Pl. 25. 1769.

Local name: *Pedda thummi.*

Annual herb, up to 1 m high; stem and branches 4-gonous, grooved, hirsute with short deflexed hairs, blackish when dry. Leaves ovate-lanceolate or lanceolate, 4–9 × 1.5–4 cm, appressed hirsute or pubescent on both sides, base cuneate, margin entire towards base, crenate-serrate upwards, apex obtuse or subacute to acuminate;

petiole up to 1 cm long. Inflorescence axillary verticil, to 2.6 cm across; bracts up to 1 cm long, about as long as calyx or longer than calyx, linear or linear-lanceolate, ciliate, pubescent, spinescent. Flowers whorled, white, to 1.5 cm long, with bristle hairs. Calyx-tube 6–8 mm long, tubular, sigmoidly curved, hispid outside above the nutlets, prominently ribbed and reticulately veined between the ribs, glabrous within, membranous in fruit; mouth inconspicuously oblique; villi absent; teeth 10, 1.5–4 mm long, upper most tooth the longest, nearly twice as long as rest, spinescent. Corolla white; tube included within the calyx or very slightly exserted, not annulate within; upper lip bearded with white hairs, apically with red hairs, almost as long as lower; lower lip subpatent, with oblong lobes. Stamens 4, didynamous; filaments to 5 and 7 mm long, glabrous; anthers 0.5 mm across, red. Ovary 1.5 mm across; style to 4 mm long; stigma bilobed. Nutlets 2 mm long, oblong-obovoid, dark brown, shining.

Common, in dry deciduous forests, laterite soils, near moist places. Fl. & Fr.: Oct. – Dec. Batrepalli (ATP), *KRKS* 42006.

Leucas mathewiana Sunojk., Candollea 64: 46. 2009.

Erect, annual herb, 20–40 cm high, stem obtusely quadrangular and grooved, pubescent. Leaves membraneous, narrowly linear-lanceolate, 3–5.5 × 0.4–0.6 cm, base long cuneate to angustate, margin almost entire or remotely 2-3-serrate, apex obtuse, lateral veins not prominent, sparsely pubescent on both sides with extrorse hairs; petiole 3-5 mm long. Inflorescence a terminal solitary verticillaster, subtended by a pair of leaves, fewer numbers of flowers, spreading horizontally; bracteoles 4–7 mm long, 0.05 mm broad at base, narrowly linear, spinulous tip, slightly bend toward flowers, hispid outside and margins. Flowers10–25, pedicel 1–2 mm long. Calyx tubular, side *ca.* 10 mm long (up to 14 mm long at seeding stage), slightly bent at middle, 10-nerved, glabrous within, fully hispid without, hairs 1–2 mm long, scattered and prominent towards upper half, each nerve excurrent in a teeth, mouth semi-oblique, rim annulate with thick veins; teeth 10, equal sized, 1–1.5 mm long, tip subulate, spinulose. Corolla white, 16–17 mm long; tube 6–8 mm long, included in calyx tube, outside hairy above, inside middle annulate with a ring of fleshy hairs; lower lip *ca.* 10 mm long, middle lobe *ca.* 7 mm wide, fan shaped, side lobes 2.5 mm broad at free end, base narrowed, margin slightly revolute; upper lip 6 mm long, concave, bent, white spreading hairs outside, margins ciliate, tip of upper cilia slightly dark brown. Stamens included in the upper lip, light reddish brown, upper pair shorter than the lower; filaments hairy, anther lobe 0.9 mm long, 0.5 mm broad, confluent theca; disk slightly-cup shaped, four lobes equal. Carpel locules 1 mm high, with rounded top; style 11 mm long, bent towards upper side; stigma bilobed, lobes unequal. Nutlets 3.5–4 × 1.5 mm, almost cylindrical, dark brownish, glabrous, shining, basal attachment flat, apex rounded.

Rare, found in the open, rocky wastelands, near deciduous forest at an altitude between 450-500 m in Nigidi forest (Type). The plant is found growing among grasses. Fl. & Fr.: Sep. – Dec. Palakonda hill ranges, Rachakuntapalle (KDP), *KRKS* 37631; Somulavaripalli (ATP), *KRKS* 37833.

Leucas nutans (Roth) Spreng., Syst. Veg. 2: 743. 1925; FBI 4: 688. 1885; FPM 2: 1149. 1924; FAP 2: 774. 1997.

Diffuse, annual herb, up to 30 cm high; branches 4-gonous, tomentose with spreading hairs. Leaves ovate-oblong or oblong-lanceolate, obtuse, 2–5 × 1–2 cm, pubescent on both sides especially on nerves, gland-dotted beneath, base obtuse, margin serrulate, apex obtuse; petiole to 4 mm long. Inflorescence axillary globose verticils, to 2.5 cm diameter; bracts sometimes as long as calyx or short, usually 5–8 × 1.5–2 mm, lanceolate, foliaceous, narrowed at base into a short stalk, long acuminate at apex, ciliate. Flowers whorled, many in a head, white, to 1.8 cm long, pubescent. Calyx-tube 8–10 mm long, tubular, abruptly falcately decurved in the middle, strongly ribbed and hispid outside, hairy or pubescent within; mouth oblique, produced on the upper side; villi shorter than teeth; teeth 10, 1.5–3 mm long, subequal, triangular at base, long acuminate at apex, ciliate, upper one tooth little longer than others. Corolla-tube little exserted from the calyx, annulate near the base; upper lip bearded with white hairs, shorter than lower; lower lip 3-lobed, middle one longer than rest, truncate. Stamens 4, filaments didynamous, included in upper lip of corolla, 7 and 8 mm long; anthers brown. Ovary 2 mm across; style to 8 mm long; stigma unequally lobed. Nutlets 2 – 3 mm long, obovoid-oblong, 3-gonous, obliquely rounded or truncate at apex, brown, smooth.

Occasional, in dry deciduous forests. Fl. & Fr.: Oct. – Dec. Somulavaripalli (ATP), *KRKS* 40142; Rachakuntapalle (KDP), *KRKS* 39037; Mabbuchintalapalle (KDP), *KRKS* 271 (HJCB-E).

Leucas stricta Wall. ex Benth., in Wall., Pl. Asiat. Rar. 1: 61. 1830; FBI 4: 688. 1885; FPM 2: 1150. 1924; FAP 2: 774. 1997.

Erect, annual herb, to 40 cm high; stem and branches 4-gonous, pubescent and sparsely hispidly hairy with spreading hairs. Leaves oblong-lanceolate or linear, 2.5–5 × 0.3–1 cm, pubescent on both sides with scattered hairs on the veins, base acute, margin entire to distinctly serrate, subsessile, petiole to 2 mm long. Inflorescence globose verticil in terminal axils and sometimes lower axils, to 2 cm diameter; bracts as long as calyx, to 1.2 cm long, linear-lanceolate, bristle hairy at magin or ciliate, hispid all over the surface, base reduced. Flowers white in many-flowered heads, pubescent, to 2.5 cm long. Calyx-tube 7–10 mm long, tubular to funnel-shaped, straight or slightly curved, ribbed, pubescent and with scattered long white hairs ouside especially near the mouth, pubescent within; mouth suboblique; villi shorter than teeth; teeth 10, 2–4 mm long, subequal, subulate-acicular, ciliate; ribs 10, prominent within rather than outside, in continuation with 10 teeth. Corolla-tube included within the calyx tube, to 8 mm long; upper lip bearded with white hairs, shorter than lower; lower lip twice as long as upper, 3-lobed, middle lobe 5 mm wide and longer than laterals. Stamens 4; filaments didynamous 9 and 10 mm long; anthers brown, to 1 mm long. Ovary 2 mm across; style to 1 cm long; stigma bilobed. Nutlets 3 mm long, oblong, rounded at top, smooth, brown.

Occasional, in dry deciduous forests. Fl. & Fr.: Oct. – Dec. Batrepalli (ATP), KRKS & *BSS* 40346; Kalasamudram (ATP), *KRKS* 40139.

OCIMUM L.

Key to Species

1a. Lower lip of calyx middle 2 long and side 2 short-toothed; corolla pinkish white; filaments > 1 cm; scarcely aromatic .... **O. filamentosum**

1b. Lower lip of calyx eqully 4-toothed; corolla cream or purplish white; filaments < 0.5 cm long; strongly aromatic .... (2)

2a. Pedicels as long as calyx .... **O. tenuiflorum**

2b. Pedicels shorter than the calyx.......... (3)

3a. Calyx with long hairs, upper lip of calyx lobe acute; corolla 4 – 6 mm long ....**O. americanum**

3b. Calyx almost glabrous, upper lip of calyx lobe acuminate or mucronulate; corolla 7–12 mm long.......... **O. basilicum**

Ocimum americanum L., in Juslenius, Cent. Pl. 1: 15. 1755; FAP 2: 776. 1997. *O. canum* Sims, Curtis Bot. Mag. t. 2452. 1823; FBI 4: 607. 1885; FPM 2: 1111. 1924.

Local name: *Kukka thulasi, Bhuthulasi.*

Erect, much branched strongly aromatic herb, up to 50 cm high; stem quadrangular, grooved, striate, puberulous. Leaves elliptic-ovate or elliptic-lanceolate, 1–2.5 × 0.5–1 cm, glabrous or nearly so, gland dotted, acute at both ends, margin entire or shallowly serrate; lateral veins 4–5 pairs, impressed beneath; petiole 1.2–2 cm long. Thyrses 10–15 cm long, flowers in rather close whorls, *ca.* 6 in a whorl; bracts elliptic-lanceolate, stalked; pedicel 2 mm long, recurved. Calyx not much enlarged, pubescent with long hairs, 1 + 4, upper lip suborbicular, rather broader than long, flat, shortly mucronate, reticulately veined, ciliate; lower lip with lanceolate-subulate teeth, 2–2.5 mm, central tooth of lower lip longer than the lateral and exceeding the upper lip of the calyx. Corolla white, 4–6 mm long, hairy without; upper lip broadly oblong, 4-toothed at the subtruncate apex; lower lip longer than the upper, oblong, obtuse. Stamens much exserted; filaments slender, 5 mm long; anthers 0.8 mm long. Ovary 0.5 mm long; style 5.5 mm long, exserted beyond the filaments. Fruit 3–3.5 mm across; nutlets ellipsoid, black, 1.4 mm across, mucilaginous when wet.

Common, in wastelands and foot hills of all types of forests. Fl. & Fr.: Nov. – Jan. Kalasamudram (ATP), *KRKS* & *KRS* 36483; Batrepalli (ATP), *KRKS* 306 (HJCB-E).

Ocimum basilicum L., Sp. Pl. 597. 1753; FBI 4: 608. 1885; FPM 2: 1111. 1924; FAP 2: 776. 1997.

Local name: *Sabja, Kammagaggiraku, Bhoothulasi, Rudrajada, Vibhudi patri.*

Erect herb or undershrubs, scented, 60–100 cm high; stem quadrangular, striate, yellowish brown, pubescent in young branchlets. Leaves elliptic-lanceolate, 2–7 × 1–3 cm, glabrescent, base cuneate, margin entire to shallowly serrate, apex acute; lateral veins 5 pairs, raised beneath. Thyrses 10–15 cm long, peduncles stout, purplish green; bracts oblanceolate, petiole 2–3 mm long; pedicel 3 mm long, shorter than calyx.

Calyx 2–3 mm long, in fruit 5–8 mm long, almost glabrous; upper lip 3 mm long, sub-orbicular, ciliate at margins, becoming horizontal in fruit from the deflection of the pedicel; lower lip 4 mm long, hairy within, central pair of teeth longer than the upper lip. Corolla white or slightly pale purple, 7–12 mm long; tube inflated below, 4 mm long; lower lip fimbriate. Stamens 4; filament pairs 6 and 7 mm long, toothed at the base; anthers 0.6 mm long. Ovary 0.7 mm across; style glabrous, 8.5 mm long; fruiting pedicel recurved. Nutlets oblong or ellipsoid, 1.8 mm across, tuberculate; seeds mucilaginous when wet.

Occasional, in forests, cultivated near temples and agriculture fields. Fl. & Fr.: Through the year. Velpula (KDP), *KRKS* & *KRS* 37674.

Ocimum filamentosum Forssk., Fl. Aegypt.-Arab. 108. 1778. *O. adscendens* Willd., Sp. Pl. 3: 166. 1800; FBI 4: 609. 1885; FPM 2: 1112. 1924. *Becium filamentosum* (Forssk.) Chiov., Nuovo Giorn. Bot. Ital. (n.s.) 26: 162. 1919; FAP 2: 762. 1997.

Annual, prostrate herb, aromatic. Branches ascending, to 50 cm high; branchlets puberulous, subtetragonous. Leaves ovate or elliptic-oblong to lanceolate, 1–2.5 × 0.7–1.8 cm, glabrous, chartaceous, base attenuate, margin entire or serrulate, repand, apex obtuse; lateral veins 4 pairs, impressed below; petiole 1–2 cm long. Inflorescence of a few-flowered verticil, aggregated in terminal, 3-chotomous racemes; peduncle to 15 cm long; bracts lanceolate, acute; pedicel to 2 mm long. Calyx, 2-lipped enlarged in fruit; lobes 5, imbricate, 1 + 4; upper lip obovate, 4 mm, mucronate; lower lip 3 mm; middle 2 lobes longer, mucronate, lateral lobes reduced, indistinct, extended in the form of a sheath, puberulous without, glabrous within. Corolla white, 3.5 × 1 mm across, tubular, bilipped; tube 4.5 mm long; lobes 5, unequal, 4 + 1; upper lip spreading, 5 mm; mid lobes 2; lateral lobes obsolete; lower lip deflexed, 3.5 mm long, obtuse. Stamens 4, didynamous; filaments declinate, 1.7 cm long; upper pair toothed above the base; anthers 1-celled, oblong, 0.7 mm across. Ovary 0.5 mm; style gynobasic, 2 cm long; stigma equally 2-fid. Disk 4-lobed. Nutlets orbicular, compressed, erect, basilar; calyx enlarged in fruit. Seeds reddish-brown, mucilaginous when wet.

Occasinal in wastelands of foot hills. Fl. & Fr.: Oct. – Jan. Bhumaiahgari palle (KDP), *KRKS* & *KRS* 37607.

Ocimum tenuiflorum L., Sp. Pl. 597. 1753; FAP 2: 777. 1997. *O. sanctum* L., Mant. Pl. 85. 1767; FBI 4: 609. 1885; FPM 2: 1111. 1924.

Local name: *Tulasi.*

Erect, much branched aromatic subshrub, up to 1 m high, branches hispid, obtusely quadrangular. Leaves elliptic-oblong, 2–4 cm, softly, minutely pubescent and glandular beneath, base acute, margin serrate, apex obtuse with mucro; lateral veins 5 pairs, impressed above; petiole to 1.5 cm long. Thyrses to 13 cm long, bracts ovate, pedicels slender, 3 mm long. Calyx purplish, hispid without and glandular, upper lip ovate or suborbicular, reflexed, apiculate; lower lip larger than the upper; teeth 4, lanceolate, aristate. Corolla white, purplish within, 3.5 mm long; tube 1 mm long; lobes 5, sericeous. Stamens 4, exserted, filaments slender, villous at the base of anther, 0.5 mm long. Ovary 1 mm long; style 4–5 mm long. Fruit with calyx 3–3.5 × 1.5 mm. Nutlets brown with minute tubercles, nearly smooth, ellipsoid, 1 mm diameter; seeds brown, mucilaginous when wet.

Occasional, in wastelands, agriculture lands and cultivated. Fl. & Fr.: Through the year. Thummalapalle (KDP), *KRKS* & *KRS* 33697.

ORTHOSIPHON Benth.

Orthosiphon rubicundus (D. Don) Benth., in Wall., Pl. Asiat. Rar. 2: 41. 1830-1831. var. **rubicundus**; FBI 4: 614. 1885; FPM 2: 1114. 1924; FAP 2: 778. 1997. *Plectranthus rubicundus* D. Don, Prodr. Fl. Nep. 116. 1825.

Local name: *Nela thappidi, Rubicunda.*

Erect, much branched herb, up to 70 cm high; root stock woody, stout, the roots often furnished with fusiform root tubers. Leaves ovate-oblong, elliptic-oblong, 4–6 × 2–3 cm, pubescent above, more below, base cuneate, margin serrate-dentate, apex obtuse; petiole 0.6–2 cm long. Thyrses terminal, 10–20 cm long, verticils distant and 6-flowered, bracts ovate-oblong, reflexed, 1.2–2.5 mm long. Calyx 5 mm long, minutely pubescent; lobes 5, upper lobe subpetaloid, purple, slightly decurrent on the tube, 2.5–3 mm broad, lower lip 2-awned, lateral lobes with short awns. Corolla twice as long as calyx, up to 1.2 cm long; tube 8 mm long, longer than the calyx. Stamens 4, declinate; filaments glabrous; anther cells confluent; disk gibbous. Ovary 6 mm long; style slender, to 1 cm long; stigma capitate. Nutlets suborbicular, compressed, calyx enlarged in fruit, to 8 mm long, 12-nerved.

Common, in dry deciduous forests, slopes and rocky areas. Fl. & Fr.: Sept. – Jan. Kalasamudram (ATP), *KRKS* 39557, 40386; Batrepalli (ATP), *KRKS* 39647.

PLECTRANTHUS L' Herit. *nom. cons.*

Plectranthus caninus Roth, Nov. Pl. Sp. 279. 1821; FAP 2: 782. 1997. *Coleus spicatus* Benth. in Wall., Pl. Asiat. Rar. 2: 15. 1830-1831; FBI 4: 624. 1885; FPM 2: 1123. 1924.

Fleshy herb, up to 60 cm high, strongly aromatic; stem subquadrangular, pale yellowish, sulcate, clothed with long hirsute hairs. Leaves ovate or obovate, fleshy, 2–3 × 1.5–2 cm, pubescent, base narrowed, margin crenate, apex obtuse; petiole to 7 mm long. Spikes terminal, to 15 cm long, flowers densely arranged. Flowers whorls close in flowering, some what separating in fruiting; bracts closely imbricate, ovate, 6 mm long. Calyx 4 mm long, woolly without, rings of hairs within the throat; upper lip shorter than the lower, orbicular, ciliate, lower lip 6 mm long, four triangular toothed; teeth ciliate, the 2 middle lobes broader than the 2-lateral ones. Corolla bluish or violet, 1.3 cm long; tube bent downwards almost at right angle; upper lip broadly obovate, rounded, 4-lobed, the middle lobes broader than the 2-lateral ones; lower lip 6 mm long, stipitate, narrowed into a neck at the base, boat-shaped part is nearly horizontal. Stamens 4, included in lower lip of corolla; filaments connate below, 5.5 mm long; staminal sheath 8 mm long; anther 0.8 mm long. Ovary 0.5 mm long; style 1.5 cm long, with 2 acute rather deep lobes. Nutlets subglobose, 1.4 mm diameter, smooth, black, shiny.

Occasional, in scrub forest and foot hills of dry deciduous forests. Fl. & Fr.: Nov. – Feb. Kalasamudram (ATP), *KRKS* & *KRS* 37959; Thummalapalle (KDP), *KRKS* & *SSR* 39022.

POGOSTEMON Desfon.

Pogostemon myosuroides (Roth) Kuntze, Revis. Gen. Pl. 2: 530. 1891; FAP 2: 783. 1997. *Mentha myosuroides* Roth, Nov. Pl. Sp. 257. 1821. *Dysophylla myosuroides* (Roth) Benth., in Wall., Pl. Asiat. Rar. 1: 30. 1830; FBI 4: 638. 1885; FPM 2: 1136. 1924.

Erect, much branched herb, root stock woody, stem terete, pale tawny or white silky tomentose. Leaves thick, wrinkled, linear-oblong or elliptic-oblong, 2.5–7 × 1–2 cm, silky tomentose, base narrowed, margin denticulate, entire at the base, apex obtuse or subacute; lateral veins 5 pairs, very conspicuous beneath, oblique; petiole to 5 mm long. Spikes slender, to 9 cm long, 4 mm diam., tomentose. Flowers very small, whorls close and compact; bracts minute, linear-oblong, ciliate. Calyx minute, 1.2–1.5 mm long, densely silky, teeth equal, triangular, shorter than the tube. Corolla reddish, 1.5 mm long, subequally 4-fid; lobes subequal, oblong, rounded, tube included. Stamens 4, exserted, subequal or slightly unequal, straight or declinate, filaments bearded, anther cells confluent. Ovary 4-partite, style arms subulate. nutlets ellipsoid, compressed, granulate.

Occasional, in dry deciduous forests near rocky areas. Fl. & Fr.: Dec. – Mar. Rachakuntapalle (KDP), *KRKS* & *KRS* 37904.

NYCTAGINACEAE

Key to Genera

1a. Straggling shrub, spinescent; stipules leafy, coloured, reticulately veined, thin .... **Bougainvillea**

1b. Erect or herbs, not spinescent; stipules linear, ovate, not coloured, not as above .... (2)

2a. Stem semisucculent, nodes swollen, glabrous; flowers salver-shaped.(-root tuberous) .... **Mirabilis**

2b. Stem non succulent, slender, pubescent; flowers not salver-shaped .... **Boerhavia**

BOERHAVIA L.

Key to Species

1a. Anthocarp 10-ribbed, > 7 mm long, with wart-like appendages; flowering pedicel > 1.5 cm long .... **B. chinensis**

1b. Anthocarp 5-ribbed, < 3 mm long, without wart-like appendages; flowering pedicel < 0.5 cm long .... (2)

2a. Perennial diffuse herb; flowers pink; anthocarp ellipsoid, sticky glandular hairy .... **B. diffusa**

2b. Annual erect herb; flowers white; anthocarp obconical, glabrous .... **B. erecta**

Boerhavia chinensis (L.) Rottb., Pl. Hort. Univ. Rar. Prog. 4. 1773. *Commicarpus chinensis* (L.) Heimerl, in Engl. & Prantl, Nat. Pflanzenfam. ed. 2. 16 C: 117. 1934; FAP 2: 788. 1997. *Valeriana chinensis* L., Sp. Pl. 33. 1753. *Boerhavia repanda* Willd., Sp. Pl. 1: 22. 1797; FBI 4: 709. 1885; FPM 2: 1163. 1925.

Local name: *Teega mamidi.*

Assending or sub-erect herb, to 1.5 m high. Leaves green, triangular-ovate, 3–7 × 4–6 cm, base truncate or rounded-cordate, margin undulate-wavy, apex acute, glabrous above, pubescent below; petiole 1.5–2 cm long. Umbels terminal on capillary pedicels; bracts linear or lanceolate, to 1.5 mm long. Flowers pink, to 4 mm across. Perianth 4 mm long, with slender tube, funnel-shaped limb; lower part to 2 mm long, glandular, 10-ribbed, appendages on each rib; upper part of perianth pink, 3 mm across. Stamens 3, white. Anthocarp oblanceolate, apex obtuse with long glandular hairs, to 4 mm long, 5-ribed, glandular hairy below.

Occasional, on road sides, borders of agriculture fields. Fl. & Fr.: Jul. – Nov. Kalasamudram (ATP), *KRKS* & *TP* 37686.

Boerhavia diffusa L., Sp. Pl. 3. 1753; FPM 2: 1162. 1925; FAP 2: 787. 1997. *B. repens* L., Sp. Pl. 3. 1753; FBI 4: 709. 1885.

Local name: *Atika mamidaku.*

Perennial diffuse herb, hirsute, up to 1 m long; branches slender, pubescent. Leaves light green, pinkish below, opposite, sub-equal, ovate, lanceolate, sub-orbicular, thick, 1–4 × 0.5–3 cm, base rounded or sub-cordate, margin some what undulate, apex obtuse, apiculate; greenish pubescent above, white beneath; lateral veins 4–5 pairs; petiole slender, up to 1.8 cm long, pubescent. Umbels axillary and terminal, glandular hairy; peduncle to 5 cm long, secondary peduncles 5, to 1.5 cm long; pedicel < 1 mm long; bracts 1 mm long; bracteole similar to bract. Perianth tube 4 mm long; lower part prominently 5-ribbed; upper part deep pink, 4 mm across; 5-lobed. Stamens 2 or 3; filaments 3 mm long; anthers 0.5 mm. Ovary 0.6 mm long; style 3 mm long; stigma diskoid. Anthocarp club-shaped, glandular-hairy, ellipsoid, 3 × 1 mm, rounded at apex.

Common, in all places, weed in agriculture fields. Fl. & Fr.: Jun. – Jan. Thummalapalle (KDP), *SSR* & *TP* 33636.

Note: Leaves are used for curing Jaundice and as leafy vegetable.

Boerhavia erecta L., Sp. Pl. 3.1753; FAP 2: 787. 1997. *B. punarnava* Saha & Krishn., J. Sci. Industr. Res. 21C: 249. 1962.

Local name: *Punarnava, Tella atikamamidi, Thikka jilakara.*

Annual erect herb, up to 70 cm high; stem pubescent, terete. Leaves light green-pinkish below, opposite, sub-equal, ovate, 2–5 × 1.5–2 cm, glabrous above, pubescent below, base truncate-obtuse, margin undulate, apex apiculate-acute; lateral veins 5–6 pairs; petiole 0.5–1.5 cm long. Paniculate umbels axillary and terminal, 10–20 cm long, zigzag, slender, forked; pedicel to 0.6 cm long; bract and bracteole 1 mm long, lanceolate. Flowers white, to 3 mm long. Perianth tube 2.5 mm long; lower parts with 5 faint ribs; upper part white, with pink tinge, 3 mm across. Stamens 1 or 3; filaments 2.5 mm long; anthers 0.4 mm across. Ovary 0.5 mm long; style 2.5 mm long. Anthocarp obconical, 2.5 mm long, glabrous, top truncate, the grooves between the ribs some what undulate.

Very common, in waste lands and as weed in agriculture fields. Fl. & Fr.: Throughout the year. Thummalapalle (KDP), *KRKS* & *KRS* 33686.

BOUGAINVILLEA Juss.

Bougainvillea glabra Choisy, in DC. Prodr. 12(2): 437. 1849; FPM 2: 1164. 1925.

Local name: *Kagithapoolu.*

Large, thorny, climbing shrub, spines axillary, up to 1.5 cm long. Leaves ovate, 3–8 × 3–10 cm, base truncate-acute, margin entire, apex acute; lateral veins 5–7 pairs. Flowers axillary in dichotomous cyme; bracts 3, white or red, thin, reticulately veined, 4 × 3 cm; pedicels 3–4 cm long. Calyx-tube 2 cm long, pink; lobes 5, pink below, white above, 0.5 cm long. Corolla white, 5-lobed, each 2-fid, 0.3 cm across. Stamens 8, in different lengths, 1.5–2 cm long; anthers yellow, 0.3 mm across. Ovary oblong, to 0.5 cm long; style hairy, curved; stigma linear, forked.

Grown as ornamental. Fl. & Fr.; Throughout the year. Chinnapalli (ATP), *KRKS* 40389.

MIRABILIS L.

Mirabilis jalapa L., Sp. Pl. 177. 1753; FPM 2: 1164. 1925.

Local name: *Nityamalle.*

Perennial herb, to 75 cm high; root tuber black, cone-shaped, 20 × 6 cm; branches glabrous, succulent, swollen at nodes. Leaves opposite, triangular-ovate to elliptic, 4–10 × 3–6 cm, thick-chartaceous, glabrous, base truncate to shortly decurrent, margin entire, apex acute to acuminate; lateral veins 5 – 7 pairs; petiole to 3 cm long. Flowers sessile, axillary or terminal corymbs, subtended by an involucre of 5, ovate-elliptic, acute, shortly connate bracts, accrescent in fruit. Perianth salver-shaped, 5-lobed, 5 cm long; lower part green, glabrous, 0.5 cm across; upper part pink or yellow or white, to 3 cm across. Stamens 3–6, exserted; filaments to 6.5 cm long; anthers 1.5 mm across. Ovary globose, 1.5 mm, sub-sessile; style 6 cm long; stigma shortly lobed, fimbriate. Anthocarp globose, to 6 mm across, ribbed and with knobs, black when mature, hard, shiny.

Grown as ornamental near temples, also as an escape. Fl. & Fr.: Jul. – May. Bhumaiahgaripalli (KDP), *KRKS* & *KRS* 42053.

AMARANTHACEAE

Key to Genera

1a. Anthers one-celled .... (2)

1b. Anther two-celled .... (3)

2a. Bracteoles with a vertical dorsal crest; anther-filaments fused almost to the top; stigmas 2, erect or spreading, sometimes minute; fruit dehiscent .... **Gomphrena**

2b. Bracteoles not dorsally crested; anther-filaments fused at the base only; stigma 1, capitate, entire; fruit indehiscent .... **Alternanthera**

3a. Ovary one-ovuled .... (4)

3b. Ovary two-ovuled .... **Celosia**

4a. Leaves alternate, often basal leaves opposite .... (5)

4b. Leavs opposite, sub opposite or whorled .... (8)

5a. Pseudostaminodes present, subulate; fruiting spike densely white-woolly .... **Aerva**

5b. Pseudostaminodes absent; fruiting spikes not as above .... (6)

6a. Flowers unisexual; style very short (< 1 mm long) or absent .... **Amaranthus**

6b. Flowers bisexual; style 2 – 3 mm long .... (7)

7a. Flowers spicate; fertile flowers each subtended by 2 lateral sterile scales (modified flowers); tepals falling off with fruits; flowers in spicate, often long racemes; filaments free; stigmas 2, recurved; fruit crustaceous, indehiscent .... **Digera**

7b. Flowers capitate; fertile flowers not subtended by scales; tepals persistent on fruiting spikes; flowers in short globose heads, filaments connate at their base; stigma 1, entire or faintly 2-lobed; fruit thin walled, dehiscent .... **Allmania**

8a. Fertile flowers subtended by modified sterile flowers bearing strongly accrescent .... **Pupalia**

8b. Fertile flowers not subtended by modified sterile flowers .... (9)

9a. Pseudostaminodes absent .... **Nothosaerva**

9b. Pseudostaminodes present .... (10)

10a. Prostrate herbs; leaves linear, filiform; bracteole wings divergent; capsule circumscissile, longitudinally split at maturity; tepals persistent with fruiting spikes; pseudostaminodes entire .... **Trichurus**

10b. Erect herbs; leaves other than filiform; bracteole wings entirely adnate to midrib; capsule indehiscent or irregularly rupturing; tepals falling off with fruits; pseudostaminodes fimbriate .... **Achyranthes**

ACHYRANTHES L.

Achyranthes aspera L., Sp. Pl. 204. 1753; FBI 4: 730. 1885; FPM 2: 1176. 1925.

Key to Varieties

1. Erect herb, stem 4-angular, glabrescent; leaves ovate, acuminate, silky villose below; flowers distinctly arranged on glabrous spike axis, 10–12 flowers per 5 cm length of spike axis at base, 21–24 flowers per 3 cm length of spike axis below the opening flowers; perianth 0.6 cm, staminal sheath white .... var. **sicula**
2. Erect-decumbent herb, stem 4-angular-rounded, pubescent; leaves obovate or elliptic-ovate, , densely breadth; flowers compactly arranged 19–22 flowers per 5 cm length of spike axis at base, 30-32 flowers per 3 cm length of spike; perianth 0.3 cm; staminal sheath pink .... var. **aspera**

Achyranthes aspera L. var. **aspera**; FAP 2: 791. 1997.

Local name: *Uttareni.*

Annual erect herb, up to 1.5 m high; branches erect to sub-scandent or straggling, much branched, 4-angular, sulcate, densely tomentose. Leaves opposite, elliptic-ovate or obovate, 2–6 × 2–4 cm, sparingly pubescent above, densely so below with white hairs, base rounded or acute, margin entire, apex obtusely acute; lateral veins 5–7 pairs; petiole 0.5–1.5 cm long. Spike terminal, 10–30 cm long, axis cylindrical, pubescent. Flower 0.3 cm long, 19–22 flowers per 5 cm length of spike axis, sessile, light pinkish-green; bracts lanceolate or narrowly deltoid-lanceolate, 3 × 1–1.5 mm, glabrous; bracteoles detoid-lanceolate, 2–3 × 1–1.5 mm, glabrous, basal wings 0.2–0.3 mm long; spines 1 mm long. Perianth segments 5; outer one longer than remaining, 3 × 1–1.5 mm, spiny acuminate, midrib distinct, lateral veins 2, distinct, margin pale; inner ones lanceolate, 2–3 × 1 mm, long acuminate, midrib distinct, margin pale. Stamens 5, 1 mm long, alternating with subquadrate pseudo-staminodes, staminal sheath fimbriate at apex, prominently appendaged within; anthers 0.5 mm long. Ovary obconical, 0.7 mm across; style 1 mm long; stigma capitellate. Utricle oblong to ovoid, 1.5 mm long; operculum cup-shaped. Seed oblong, testa coriaceous, white, 1–2 mm long.

Common, in plains, agriculture lands and road sides. Fl. & Fr.: Aug. – Dec. Rachakuntapalle (KDP), *SSR* & *TP* 33603.

Note: Flower opens between 7 and 10 AM. Leaves are used for skin diseases and root is used for dental problems. Seeds are cooked and eaten.

Achysanthes aspera L., Sp. Pl. 204. 1753. var. **sicula** L. Sp. Pl. 204. 1753. *A. aspera* var. *argentea* (Thwaites) C.B.Clarke in Hook. f., Fl. Brit. India 4: 730. 1885; FAP 2: 792. 1997. *A. argentea* Thwaites, Enum. Pl. Zeyl. 249. 1861.

Annual, erect herb, up to 1 m high; branches erect, quadrangular, glabrescent. Leaves opposite, ovate-elliptic, 3–4.5 × 2–3 cm, pubescent above, silky white villose below, base acute, margin entire, apex acuminate or acute; lateral veins 6–7 pairs; petiole to 1 cm long, pubescent. Spike terminal, to 40 cm long, glabrescent. Flowers 0.5 cm long, distantly arranged on spike axis, 10–12 flowers per 5 cm length of spike axis, sessile, white-green; bracts lanceolate, 3 × 1 mm, basally winged, wing 0.2–0.3 mm long, apically spiny; bracteoles 2, 4 × 1 mm, winged, wing 0.3–0.4 mm long. Perianth segments 5, 5–6 mm across; outer one larger than innerones, 4–6 × 1–1.5 mm, lanceolate, acuminate, midrib distinct, spiny, margin thin; inner ones 3-4 × 1-1.2 mm, lanceolate, acuminate, midrib distinct, spiny, margin thin. Stamens 5; filaments 1 mm long, alternating with thin pseudo-staminodes; staminal sheeth filiform at the tip to 5 mm across, 1 mm height, prominently appendaged within; anthers white, 0.5 mm long. Ovary obconical, 0.7 mm across; style 1 mm long; stigma capitate. Utricle oblong to elliptic, 2 mm across, operculum cap like. Seed oblong, elliptic, 2.2 mm long, testa coriaceous, brown.

Very rare, near agriculture lands and field bunds, road sides and scrub forests. Fl. & Fr.: Sept. – Jan. Thalupula (ATP), *KRKS* & *BSS* 41940.

AERVA Forssk. *nom. cons.*

Key to Species

1a. Sub-shrub; stem strong, to 4 mm diam.; leaves oblanceolate to spathulate, to 9 cm long; spike terminal, to 25 cm long .... **A. javanica**

1b. Herb; stem weak, to 2 mm diam.; leaves elliptic-ovate, to 2 cm long; spikes axillary, to 1 cm long .... **A. lanata**

Aerva javanica (Burm. f.) Juss. ex Schult., Syst. Veg. ed. 15.5: 565. 1819; FBI 4: 727. 1885; FAP 2: 793. 1997. *Iresine javanica* Burm. f., Fl. Indica 212. t. 65. f. 2. 1768. *Aerva tomentosa* Forssk., Fl. Aegypt.-Arab. 122: 170. 1775; FPM 2: 1178. 1925.

Local name: *Konda pindi, Pindikura, Pedda pindi konda.*

Annual or biannual, erect sub shrub, to 1.5 m high; frequently woody; branches terete, striate, densely whitish tomentose. Leaves opposite at base, alternate towards apex, oblanceolate to spatulate, 5–9 × 1–2.5 cm, densely whitish tomentose with stellate hairs, base cuneate to attenuate, margin entire, apex obtusely apiculate; lateral veins 5–6 pairs; petiole 2–3 mm long, tomentose. Spikes paniculate, terminal, 18–25 cm long. Flowers unisexual. Male flower: bracts and bracteoles deltoid, ovate, 0.75–2.25 × 0.5 mm, hyaline, acute or obtuse, midrib obscure, 3 mm long; stamens 5, 1.5 mm long, pseudostaminodes alternating the fertile stamens, gynoecium rudimentary. Female flower: bracts and bracteoles lanceolate, 1–2.5 × 0.5 mm; tepals 5, obovate-oblanceolate to spatulate, 2.5 mm long, tomentose without, obtuse; outer 2 tepals with faint midrib, 2–3 × 0.7–1 mm; inner 3 tepals with thick green midrib, 1–1.5 mm, pseudostaminodes in the place of stamens, subulate. Ovary 0.5 mm across; style 1.8 mm long; stigma deeply forked. Seeds rounded, slightly compressed, black, shining, very faintly reticulate; testa coriaceous, 0.5–1 mm across.

Common, in wastelands, road sides and scrub forests. Fl. & Fr.: Jul. – Oct. Velpula (KDP), *KRKS* & *KRS* 36996.

Aerva lanata (L.) Juss., Ann. Mus. Nat. Hist. Paris 11: 131. 1808; FBI 4: 728. 1885; FPM 2: 1178. 1925; FAP 2: 793. 1997. *Achyranthes lanata* L., Sp. Pl. 204. 1753.

Local name: *Teega pindikura, Chinna pindi konda.*

Erect or prostrate, sometimes straggling herb, to 2 m high; branches slender, 1–3 mm across, weak, white lanate or villous. Leaver alternate, variable, reniform to elliptic-ovate, basally large, to 2.5 × 2 cm and small towards apex, to 0.4 × 0.3 cm, pubescent above, tomentose below, base acute-attenuate, margin entire, apex orbicular or acute-apiculate; lateral veins 5–7 pairs; petiole, 1–2 cm long sessile in apical leaves, . Spikes axillary, to 1 cm long. Flowers bisexual; bracts deltoid-ovate to oblong-ovate, 0.7–1 × 0.5 mm; membranous with a short arista formed by the excurrent midrib, pilose, persistent; bracteoles similar to bracts. Perianth lobes 5; outer 2 tepals ovate-oblong, 1.5–2 × 0.5–0.7 mm, hyaline, abruptly connate at apex to a distinct mucro formed by the excurrent mid vein; inner 3 tepals slightly shorter and narrow with a broad central green vitta along the midrib. Stamens 5, 1 mm long, pseudostaminodes present at anthesis, attaining about half the length of the style; anthers 0.2 mm across. Ovary 0.4 mm long; style 0.2 mm long; stigma 2-fid. Utricle rotund, 1 mm across, compressed. Seed reniform, 0.6–0.8 mm across, black, shining, testa smooth in the centre, faintly reticulate around the margin, coriaceous.

Common, along bushes on road sides and scrub forests. Fl. & Fr.: Jul. – Oct. Rachakuntapalle (KDP), *TP* & *SSR* 33612, Kalasamudram (ATP), *KRKS* 39103.

ALLMANIA R. Br. ex Wight

Key to Species

1a. Leaves extremely variable; spikes globose, axillary, sessile or subsessile .... **A. nodiflora**

1b. Leaves narrow-spatulate or lanceolate; spikes oblong or ovoid; terminal; pedunculate to 2–4 cm long .... **A. longipedunculata**

Allmania longipedunculata (Trimen) Gamble, Fl. Madras 2: 1168. 1925; FAP 2: 794. 1997. *A. nodiflora* (L.) R. Br. ex Wight, var. *longepedunculata* Trimen, Handb. Fl. Ceylon 3: 394. 1895. *Celosia nodiflora* L., Sp. Pl. 205. 1753. *Allmania albida* (Wight) R. Br., Numer. List (Wallich) n. 6891. 1832; FBI 4: 717. 1885. *Chamissoa albida* Wight., Ic. t. 1853.

Local name: *Pendlibadde aku.*

Annual ascending or prostrate herb, to 30 cm tall; branches angular, pilose with whitish hairs, sulcate. Leaves alternate, spathulate or oblanceolate, 2–5 × 0.5–2 cm, thinly pilose below, base attenuate, margin entire, apex widely retuse to apiculate or obtuse; lateral veins 4–5 pairs; petiole to 1.5 cm long. Spike terminal, white or yellowish, to 2 cm long; peduncle to 4 cm long; bracts deltoid-ovate, 2–3 × 1–2 mm, acuminate, aristate with excurrent dark midrib; bracteoles lanceolate, 2–3 × 1.5–2 mm, silvery membranous with a reddish or greenish midrib, excurrent in a very short mucronate tip, glabrous, or less pilose. Perianth calycine, 5-lobed; lobes oblong-lanceolate, 0.5–0.7 × 0.1–0.2 cm, glabrous or pilose without. Stamens 5, 3-3.5

mm long, subulate, filaments alternating the sheath. Ovary sub-globose, 1 mm long; style 2 mm long; stigma capitate. Fruit subglobose or ovoid, membranous, utricle circumscissile dehiscent. Seed lenticular, black, reticulate, more or less shining, areolate, dull, testa crustaceous.

Common, along with grasses in open lands and foot hills of dry deciduous forests, near water bodies and wetlands. Fl. & Fr.: Jul. – Nov. Kalasamudram (ATP), *KRKS* & *KRS* 36475.

Allmania nodiflora (L.) R. Br. ex Wight, in Hook. J. Bot. 1: 226. t. 128. 1834.

Key to Varieties

1a. Leaves spatulate, glabrous .... var. **roxburghii**

1b. Leaves lanceolate or linear, pubescent or scabrid .... (2)

2a. Leaves linear, margin entire, scabrid; heads 1.5 cm across, brick red .... var. **nodiflora**

2b. Leaves lanceolate, margin undulate, pubescent; head 2 cm across, white .... var. **procumbens**

Allmania nodiflora (L.) R. Br. ex Wight var. **nodiflora**; FBI 4: 716. 1885; FPM 2: 1167. 1925; FAP 2: 796. 1997. *Celosia nodiflora* L,. Sp. Pl. 205. 1753.

Local name: *Errabaddi aku, Pendly meddi aku, Erra banthi aku.*

Annual spreading, prostrate or ascending herb, up to 30 cm long; branches reddish, flexuose below, angular, sulcate, moderately pilose. Leaves alternate, variable in shape and size, linear, oblong, 1.2–5 × 0.5–2 cm, glabrous above, pilose below, base alternate, margin entire, apex rounded or widely retuse to apiculate or shortly acuminate; lateral veins 4–5 pairs; petiole 1–2 mm long, pilose. Head terminal or axillary, globose, 1–1.5 cm across; peduncle 0.5–1 cm long. Flowers sessile, pale reddish to stramineous; bracts deltoid ovate, 2–3 × 1–1.5 mm, membranous with excurrent dark midrib; bracteoles linear, 3–4 × 1–1.5 mm, pubescent along the keel. Perianth lobes 5, elliptic-lanceolate, 4–5 × 1.5–2 mm, silvery membranous with reddish or greenish midrib, excurrent in very short mucro, glabrous or less pilose without. Stamens 5, 3.5–4 mm long, subulate; anther 0.5 mm across. Ovary depressed-globose, 1 mm across; style slender, 2.5–3 mm long; stigma capitate. Urticle ovoid, 3–4 mm. Seed lenticular, 0.5–1 mm, black, reticulate, shining dull and areolate, testa crustaceous.

Common, in scrub forest, agriculture field borders and wastelands. Fl. & Fr.: Aug. – Nov. Rachakuntapalle (KDP), *TP* & *SSR* 33635.

Allmania nodiflora (L.) R. Br. ex Wight var. **procumbens** Hook. f., Fl. Brit. India 4: 717. 1885; FPM 2: 1167. 1925; FAP 2: 796. 1997. *Celosia nodiflora* L., Sp. Pl. 205. 1753.

Diffuse herb, to 70 cm long; branches prostrate, from root stock, root stout, woody. Leaves alternate, lanceolate, 3–6 × 0.5–1.5 cm, pilose on both sides, base acute, margin entire, apex acute; petiole to 2 mm long, pilose. Inflorescence axillary head, white, 2 cm across, peduncle 3 mm long. Flowers sessile, pale green or white, 0.5 cm long; bracts ovate, 2–3 × 1–1.5 mm, membranous, pilose, acuminate; bracteole 1–2 × 1 mm, membranous, pilose at margin, acuminate. Perianth lobes 5, pilose without. Anthers 5, 3 mm long. Ovary sub globose. Seed black, 2 mm across, slightly compressed.

Common, in foot hills of dry deciduous forests among grasses and near rocks. Fl. & Fr.: Aug. – Dec. Kalasamudram (ATP), *KRKS* 39684.

Allmania nodiflora (L.) R. Br. ex Wight, var. **roxburghii** Wight, in Hook. J. Bot. 1. 226, t. 128. 1834; FAP 2: 796. 1997. *Achyranthes nodiflora* Roxb., Fl. Ind. 678. 1820. *Celosia nodiflora* L., Sp. Pl. 205. 1753.

Straggling, many-branched herb, with stout root-stock, up to 60 cm long; branches glabrous or pubescent. Leaves spatulate, 3–6 × 1–1.5 cm, glabrous above, pilose or pubescent below, base rounded, margin entire, apex acute, apiculate; lateral veins 5–6 pairs; petiole 0.3–0.5 cm long, pilose. Inflorescence axillary head, 1.5 cm across. Flowers sessile, 0.4 cm long; bracts thin, lanceolate, pilose without, 0.3 cm long; bracteole 0.2 cm long with excurrent green midvein. Perianth lobes 5, 0.4 cm long, pilose above, thin at margin, midvein green. Stamens 5, 3 mm long. Ovary sub globose; style 2 mm, stigma capitate. Capsule ovoid, covered with cup-shaped operculum. Seed black, 0.2 mm across.

Rare in dry deciduous forests, near streams and among rocks. Fl. & Fr.: Aug. – Dec. Rachakuntapalle (KDP), *KRKS* & *KRS* 37633.

ALTERNANTHERA Forssk.

Key to Species

1a. Creeping herb, rooting at nodes; leaves elliptic-orbicular, shiny; bracts spiny .... **A. pungens**

1b. Erect herbs, not rooting at nodes; leaves not as above; bracts apiculate, but not spiny .... (2)

2a. Leaves linear, oblanceolate-spatulate, sessile or sub-sessile; heads 7 mm across; bracts acute; hydrophyte .... **A. sessilis**

2b. Leaves elliptic-oblong or obovate-oblong, petiole to 1 cm long; heads to 4 mm across; bracts acuminate; mesophyte .... **A. ficoidea**

Alternanthera ficoidea (L.) Sm., Cycl. 39 (2Suppl.): n. 8. 1819. *Gomphrena ficoidea* L., Sp. Pl. 1: 225. 1753. *Alternanthera tenella* Colla, Mem. Acad. Sci. Terime 33: 131. t. 9. 1829; FAP 5: 57. 2008.

Annual, decumbent or erect herb, to 1.5 m high; stem slightly angular, with appressed hairs. Leaves opposite, elliptic-oblong or obovate-oblong 1.5–6.5 × 0.3–3.25 cm, densely clothed with appressed long hairs below, less above, base attenuate, margin entire, apex acute-apiculate, dotted on both sides; lateral veins 5–7 pairs; petiole to 1 cm long. Flowers silvery or light yellowish in terminal axils of branches, 2–3 spikes together on common peduncle; bracteole lanceolate, 2–2.5 × 0.5–0.7 mm, very acute, densely hairy on dorsal side. Perianth 5-lobed, oblong or ovate-oblong, 3–3.5 × 0.8–1 mm, 1-nerved, in lower half rather densely hairy on dorsal side. Stamens 5, 1.5–2 mm long, alternating with pseudo-staminodes, strap-shaped, about as long as stamens, divided into a few narrow strips at apex. Ovary oblong, 2 mm across; style broader than long, 0.25–0.5 mm long; stigma capitellate. Utricle oblong, 1–1.5 mm, compressed, lenticular, testa crustaceous.

Common, on road sides. Fl. & Fr.: Jul. – Dec. Thimmammamarimanu (ATP), *KRKS* & *BSS* 41939.

Alternanthera pungens Kunth, Nov. Gen. Sp. 2: 206. 1817; FAP 2: 798. 1997. *A. repens* (L.) Link, Enum. Pl. Hort. Berol. 1: 154. 1821. *Achyranthes repens* L., Sp. Pl. 205. 1753.

Local name: *Mullu ponnaganti, Mulla tappadi.*

Prostrate-creeper, herb to 50 cm long; rooting at nodes; branches villous with white hairs. Leaves opposite, unequel, obliquely elliptic to orbicular, 1.5–3 × 1–2 cm, distinctly villous above, more below, base attenuate, margin entire, apex mucronate; lateral veins 6–7 pairs; petiole to 0.5 cm long. Inflorescence axillary, clustered head, to 0.5 cm across; bracts oblanceolate, to 5 mm long, midnerve shortly awned; margin scarious; bracteoles 3 mm long, similar to bract. Flowers white. Tepals 5; outer 2 lobes oblanceolate, 4–5 × 1.5–2 mm, very rigid, 5-nerved, mid nerve and outer pair stronger, jointed above into a pungent awn; inner one faintly visible, lobes oblong, 3–3.5 × 1–1.5 mm, 3-nerved, margin serrate, obtusely apiculate, with a tuft of glochidiate, barbellate bristles at base; inner 2 lobes linear, curved with a tuft of glochidiate, barbellate bristles at middle, apex acute. Stamens 5; filaments 0.3 mm long; anthers oblong, 0.3 mm long; staminodes membranous, dentate, alternating to the stamens. Ovary depressed-globose, 2 mm long; style broader than long, 0.1 mm. Seed diskoid, brown shining, faintly reticulate; testa crustaceous.

Common, weed in all open areas, agriculture fields and wastelands. Fl. & Fr.: Through the year. Batrepalli (ATP), *KRKS* & *BSS* 37974; Kalasamudram (ATP), *KRKS* 40038.

Note: Leaves are used as vegetable.

Alternathera sessilis (L.) R. Br. ex. DC., Cat. Hort. Monsp. 4: 77. 1813; FBI 4: 731. 1885; FAP 2: 798. 1997. *Gomphrena sessilis* L., Sp. Pl. 225. 1753. *Alternanthera triandra* Lam., Encycl. 1: 95. 1783; FPM 2: 1179. 1925.

Local name: *Ponaganti aku.*

Annual, prostrate to erect herb, to 50 cm high, sometimes rooting at basal nodes; branches 5–6-angular, glabrescent, villous at nodes or sparsely hairy. Leaves opposite, variable in shape and size, linear-lanceolate or ovate-oblong or oblanceolate to spatulate, 1–9 × 0.2–2 cm, glabrous or thinly pilose below, base cuneate to attenuate, margin distinctly serrate, apex obtuse or acute; lateral veins 5–7 pairs; petiole to 2 mm long or sessile. Spike axillary, white, to 0.8 cm long; bracts deltoid-ovate, 1 × 0.4 mm, scarious, white, mucronate with excurrent midrib, glabrous; bracteoles similar to bracts, 1–1.5 × 0.5–0.8 mm, persistent. Perianth segments 5, ovate-elliptic, 2.5–3 × 0.5–1 mm, 1-nerved, acuminate to rather blunt, white, glabrous, apex shortly mucronate with midvein, margins obscurely denticulate. Stamens 3; anthers fertile, staminodes 2, 1–1.5 mm long, alternating with pseudostaminodes resembling the filaments, shorter than fertile ones. Ovary depressed-globose, 1 mm across; style 0.2 mm long; stigma simple, fimbriate. Utricle cordiform, strongly compressed, 1.2 × 2 mm. Seed diskoid-orbicular, 0.5–0.75 mm, brown, shining, faintly reticulate, testa crustaceous.

Common, near water bodies in dry deciduous forests. Fl. & Fr.: Aug. – Feb. Velpula (KDP), *KRKS*& *KRS* 37678; Kalasamudram (ATP), *KRKS* 40435.

Note: Leaves are used as vegetable.

AMARANTHUS L.

Key to Species

1a. Tepals and stamens 5 each in the flower .... (2)
1b. Tepals and stamens 3 each in the flower .... (3)
2a. Leaf axils with paired spines, midrib of the tepals green or purple .... **A. spinosus**
2b. Leaf axils without spines, midrib of the tepals green .... **A. hybridus**
3a. Spike elongated, axillary and terminal; leaves to 7 cm long .... **A. viridis**
3b. Spike axillary globose; leaves to 4 cm long .... **A. tricolor**

Amaranthus hybridus L., Sp. Pl. 990. 1753. subsp. **hybridus**; FAP 2: 800. 1997. var. **erythrostachys** Moq., in DC., Prodr. 13(2): 259. 1849; Towns. in Dassan. & Fosb., Rev. Handb. Fl. Ceylon 1: 13. 1980. *A. hypochondriacus* L., Sp. Pl. 991. 1753. *A. paniculatus* L., Sp. Pl. 2: 1406. 1763; FBI 4: 718. 1885; FPM 2: 1170. 1925.

Local name: *Molaka kura, Thota kura.*

Annual, erect herb, up to 2 m high; branches stout, angular, glabrous or thinly to moderately furnished with short or long muticellular hairs. Leaves alternate, broadly lanceolate to ovate, 3–20 × 1.5–10 cm, gradually narrowed, thinly pilose below on veins, base cordate, margin entire, apex blunt to mucronulate; lateral veins 5–15 pairs; petiole up to 15 cm long. Spikes red or purple, axillary and terminal, varying from a single spike to a broad, much-branched panicle, 10–30 × 3–5 cm; bracts and bracteoles unequal in size, deltoid-ovate 2.5–3 × 0.5–1 mm, pale membranous, scarious, acuminate with a long tip, arista erect and formed by the stout excurrent midrib, exceeding the perianth. Flowers androgynous formed of cymose clusters, increasing upwards. Perianth segments 5, lanceolate or oblong, 3–3.5 × 1–1.5 mm, acute-aristate or the inner sometimes blunt in the female flower, 3-nerved. Stamens 5, 1.5 mm long, dorsifixed, X-shaped. Ovary conic or pyriform; styles 3, fimbriate, 1 mm long, curved. Utricle subglobose to ovoid, 2–3 mm, capsule lid swollen with a distinct neck. Seed ovoid, 0.75–1.25 mm, compressed, almost smooth, faintly reticulate around the margin, testa crustaceous.

Cultivated and escape. Fl. & Fr.: Throughout the year. Batrepalli (ATP), *KRKS* 42051.

Amaranthus spinosus L., Sp. Pl. 991. 1753; FBI 4: 718; FPM 2: 1170. 1925; FAP 2: 800. 1997.

Local name: *Mulluthota kura, Mullu koytaku, Teeta ganjaraku.*

Armed, annual, erect herb, up to 1 m high; stem stout, angular, reddish, glabrous basally, multicellular long hairy apically. Spines axillary, 1.5 cm long, paired or clustered, subtending inflorescence. Leaves alternate, ovate to rhomboid-ovate, rarely oblong-lanceolate, 1.5–7 × 0.8–4 cm, glabrous or thinly pilose on the veins on lower side, base cuneate or attenuate, margin entire, apex obtuse-apiculate with fine mucro; lateral veins 5–7 pairs; petiole up to 9 cm long. Spikes clustered, axillary or terminal, up to 7 cm long, 0.8 cm diameter; lower clusters entirely female; upper flowers male and bisexual; bracts deltoid-ovate, 1–1.5 × 0.5–0.75 mm, pale scarious margin, pale or reddish awn formed by the excurrent green midrib, to 0.5 cm; bracteoles deltoid-ovate, 2–2.5 × 1–1.5 mm, sub-equaling or shorter than perianth lobes. Flowers pale greenish-white, up to 0.5 cm across. Perianth segments 5; those of female flowers narrowly oblong or spatulate-oblong, 1.5–2.5 × 0.5–1 mm, obtuse or acute, mucronulate, frequently with a greenish dorsal midvein; those of male flowers broader than the female, oblong-lanceolate, 1.5–3 × 0.75–1 mm, acute or acuminate, midrib excurrent. Stamens 5; filaments 1–1.5 mm long; anthers oblong, 0.2 mm long, sagittate. Ovary 0.5 mm across; style 0.5 mm long; stigma lobes 3, flexuose or reflexed, 1–1.5 mm. Utricle ovoid, urceolate, 1.5–2 mm, with a short inflated neck below the style base, circumscissile, rarely indehiscent. Seed globular, 0.75-1 mm, compressed, black, shining, very faintly reticulate, testa crustaceous.

Common, in wastelands, borders of agriculture lands and foot hills of forests. Fl. & Fr.: Aug. – Dec. Bhumaiahgari palle (KDP), *KRKS* & *KRS* 37618.

Amaranthus tricolor L., Sp. Pl. 989. 1753; FAP 2: 801. 1997. *A. mangostanus* L., Cent. Pl. 1. 32. 1755; FBI 2: 1171. 1925. *A. gangeticus* L., Syst. Nat. ed. 10. 1268. 1759. var. *tristis*; FBI 4: 719. 1885; FPM 2: 1171. 1925.

Local name: *Sirraku, Peruguthotakura.*

Annual erect or decumbent herb, to 50 cm high; stem angular, glabrescent, upper parts pilose with less crisped hairs. Leaves alternate, rhomboid-ovate, broadly elliptic, 1–4 × 0.5–3 cm, glabrous, white dotted below, base attenuate, margin entire, apex emarginated to apiculate; lateral veins 4–7 pairs; petiole to 7 cm long, longer than lamina. Spikes axillary and terminal, variable in length. Flowers green to crimson in globose clusters on spike, male and female flowers intermixed; bracts deltoid ovate, 1–1.5 × 0.5–0.7 mm, bracteoles broadly ovate, 2.5–3 × 0.5–1 mm, sub-equal to the perianth, membranous, lanceolate, apex apiculate, midrib green. Perianth segments 3, elliptic or oblong-elliptic, 3–3.5 × 1–1.5 mm, narrowed above, pale membranous, midrib excurrent into a long, pale-tipped awn, perianth segments slightly accrescent in fruit. Stamens 3; filaments 2–2.5 mm long; anthers

0.3 mm long. Ovary compressed, 1 mm across; style 3-fid, minute. Utricle ovoid-urceolate, 2.25–2.75 mm, with a short neck below, circumscissile, membranous. Seed ovoid 1–1.5 mm, black or brown, shining, very faintly reticulate, lenticular, testa crustaceous.

Common, weed in agriculture fields, road sides and foot hills of forests. Fl. & Fr.: Throughout the year. Thummalapalle (KDP), *KRKS* & *KRS* 36415; K.K. Kottala (KDP), *KRKS* & *KRS* 36417.

Amaranthus viridis L., Sp. Pl. 2: 1405. 1763; FBI 4: 720. 1885; FPM 2: 1171. 1925; FAP 2: 801. 1997.

Local name: *Chilakamukku aku, Chinna thotakura.*

Annual, erect herb, up to 75 cm high; stem glabrous when mature, pilose when young, with floccose multicellular hairs, angular. Leaves alternate, deltoid-ovate to rhomboid-oblong, 2–7 × 1.5–5.5 cm, glabrous or shortly to fairly long-pilose when young and on the veins of lower surface, base shortly cuneate or sub-truncate, margin entire, apex obtuse or emarginate-apiculate; lateral veins 7–9 pairs; petiole to 6 cm long. Spike axillary and terminal, often paniculate, 2.5–12 cm long. Flowers intermixed but later more numerous; bracts and bracteoles deltoid-ovate to ovate-lanceolate, 1–2 × 0.5–0.7 mm, whitish membranous with a very short, pale awn formed by excurrent midrib. Perianth lobes 3, very rarely 4; male flower segments ovate-oblong, 2–2.5 × 0.5–1 mm, margin scarious, apex mucronate, midrib green. Stamens 3, filaments 1.5 mm long; anthers 0.3 mm long. Ovary 0.7 mm across; style 0.2 mm, stigmas 2–3, erect. Utricle subglobose, 1.25–1.5 mm, not exceeding the perianth, indehiscent. Seed ovoid, 1–1.25 mm, slightly compressed, dark brown to black, reticulate and verrucose, testa crustaceous.

Common, weed in agriculture lands, road sides and wastelands. Fl. & Fr.: Throughout the year. Thummalapalle (KDP), *KRKS* 36414; Kalasamudram, (ATP), *KRKS* & *BSS* 39619.

Note: Leaves are used as vegetable.

CELOSIA L.

Key to Species

1a. Erect herb, to 2 m high; leaves linear-lanceolate, glabrous, shiny; inflorescence terminal, conical spike, to 10 cm long; flowers dense on axis, white .... **C. argentea**

1b. Erect herb, to 40 cm high; leaves deltoid-ovate, elliptic, pubescent, not shiny; inflorescence terminal, 2-few-flowered clusters on receme .... **C. polygonoides**

Celosia argentea L., Sp. Pl. 205. 1753; FBI 4: 714. 1885; FPM 2: 1166. 1925; FAP 2: 802. 1997. var. **argentea**.

Local name: *Gurugu aku, Gurugu.*

Annual, erect much branched variable, tall herb, up to 2 m high; branches ascending, strongly ridged and sulcate, glabrous, reddish or greenish. Leaves alternate, oblong-lanceolate to narrowly linear, 2–15 × 1–3.5 cm, glabrous, sometimes reddish, base attenuate, margin entire, apex acuminate or obtuse, apiculate; lateral veins 7-9 pairs; petiole 2–10 mm. Inflorescence terminal, dense spike, 2.5–10 × 1.5–3 cm, conical at first but becoming cylindrical; peduncle sulcate, up to 20 cm long; bracts and bracteoles lanceolate, the lower deltoid, 3–4 × 2–3 mm, much shorter than tepals, hyaline, aristate with excurrent midrib, margins scarious, persistent. Flowers white to pinkish, to 5 mm across. Perianth segments 5, elliptic-oblong, 6–8 × 2–3 mm, acute to rather blunt, mucronate, lateral veins 2-4, margin widely hyaline. Stamens 5, 6–7 mm long; filaments delicate; anthers creamy. Ovary ovoid, 2–3 mm long; style filiform, 5–6 mm; stigmas 2–3. Urticle two-seeded, ovoid, 3–4 mm across. Seeds lenticular, 1–1.5 mm diameter, black, shining; testa very finely reticulate, crustaceous.

Common, weed in agriculture fields, wastelands, roadsides and foot hills of forests along with grasses. Fl. & Fr.: Nov. – Feb. Chinnapalli (ATP), *KRKS* 37804.

Note: Leaves are used as vegetable when young. Seeds used for eye diseases.

Celosia polygonoides Retz., Observ. Bot. 2: 12. 1781; FBI 4: 715. 1883; FPM 2: 1166. 1925; FAP 2: 802. 1997.

Local name: *Eluka-uttareni.*

Biannual, erect herb, to 40 cm high; root stock woody; branches angular, villose with yellowish hairs. Leaves alternate, broadly ovate to deltoid-ovate, 1–5 × 1.5–3 cm, pilose on lower side, glabrescent above, base cordate-truncate or cuneate, margin entire, apex obtuse to acute; lateral veins 4–7 pairs; petiole to 20 cm long. Inflorescence terminal raceme; peduncle densely patent-hairy; bracts and bracteoles glabrous, ovate to deltoid-ovate, 15–2 × 1–1.5 mm, whitish, acute, mucronate with the excurrent midrib. Flowers distant to approximate, 2-few in sessile cluster, to 4 mm across. Perianth segments 5, elliptic-oblong, 2.5–3 × 1.5–2 mm, mucronate with the shortly excurrent midrib, midrib green. Stamens 5, 2.5 mm long; anthers 0.2 mm long. Ovary ovoid, 1 mm long; style 0.5 mm, embedded; stigma lobess 2–3, circinately recurved, sub-equalling or longer than the style, 0.5–0.75 mm. Urticle urceolate, 2.5–3 mm long. Seeds reniform, compressed, 0.6 mm across, black, uniformly verruculose, testa crustaceous.

Common, in wastelands, foot hills of scrub forest and road sides. Fl. & Fr.: Oct. – Jan. Kalasamudram (ATP), *KRKS* & *KRS* 37696.

DIGERA Forssk.

Digera muricata (L.) Mart., Beitr. Amarantac. 77. 1825; FAP 2: 804. 1997. *Achyranthes muricata* L., Sp. Pl. ed. 2. 1: 295. 1762. *Digera arvensis* Forssk., Fl. Aegypt. Arab. 65. 1775; FBI 4: 717. 1885; FPM 2: 1169. 1925.

Local name: *Chenchali aku.*

Annual, prostrate or decumbent, sometimes erect herb, to 40 cm high; stem glabrous or pilose, angular. Leaves alternate, broadly ovate or rarely sub-elliptic, 2–5 × 0.5–3 cm, glabrous or hairy below, base attenuate or narrowed, margin entire, apex acute or acuminate-apiculate; lateral veins 7–10 pairs; petiole slender, 4–5 cm long. Spike axillary, slender, 10 cm long. Flowers whitish-pink, distantly arranged. Fertile flowers each subtended by 2 lateral, sterile, modified flowers consisting of scales; bract obovate, 2.5–3 × 0.75–1 mm, midrib prominent; bracteole lanceolate, 2–2.5 × 0.5–0.75 mm, glabrous, membranous with a green or brownish percurrent midrib. Perianth segments 5, ovate-oblong, 3–4.5 × 1–1.5 mm, acute, the inner perianth lobes slightly shorter, more delicate, blunt or erose, 1–3-nerved, margin hyaline with a darker central midrib. Stamens 5, 3 mm long. Ovary globose, 1 mm across; style slender, to 2 mm long; stigma bifid, finally recurved. Fruit subglobose, 2–2.5 mm, indehiscent, slightly compressed, bluntly keeld, furnished throughout with verrucose, testa crustaceous.

Weed in agriculture fields, wastelands and road sides. Fl. & Fr.: Through the year. Thummalapalle (KDP), *KRKS* & *KRS* 33685.

Seeds are given for urinary disorders. The plant is laxative in large doses.

GOMPHRENA L.

Gomphrena serrata L., Sp. Pl. 224. 1753; FAP 2: 805. 1997. *G. decumbens* Jacq., Pl. Rar. Hort. Schoenbr. t. 482. 1805; FPM 2: 1179. 1925.

Local name: *Tella pagada banthi.*

Annual, ascending or erect herb, to 20 cm high; much branched, spreading from base, densely furnished with long, white lanate hairs, villose. Leaves opposite, elliptic-oblong to oblanceolate 1.5–5 × 0.5–1.5 cm, pilose with lanate hairs, base attenuate, margin entire, apex obtuse-mucronate; lateral veins 5–7 pairs; petiole to 0.5 cm long, pilose. Spike white, terminal, to 2.5 cm long; peduncle to 6 cm long with pilose hairs and having terminal pair of leaves. Flowers silvery white in spikes; basal region (to 1 cm) sterile; bracts deltoid-ovate, 2.5–3 × 0.5–1 mm, shortly mucronate with excurrent midrib; bracteoles lanceolate with a vertical dorsal crest, 3–3.5 × 0.7–1 mm, laterally compressed, hyaline, villose. Perianth segments 5, narrowly lanceolate, 4.5–5 × 1–1.5 mm, 1-nerved; the outer 3 more or less flat, lanate at the base, apex mucronate; the inner ones densely lanate almost to the tip, slightly longer than the outer ones. Stamens 5, staminal tube subequalling the perianth; anthers yellow. Ovary compressed-globose, 1.3 mm; style and stigma together 1 mm, stigma divergent. Capsule shortly compressed pyriform, to 1.75 mm long. Seed ovoid, compressed, brown, faintly reticulate, shining, testa coriaceous.

Common, in wastelands, road sides and scrub forests. Fl. & Fr.: Through the year. Kalasamudram (ATP), *KRKS* & *KRS* 36478.

NOTHOSAERVA Wight

Nothosaerva brachiata (L.) Wight, Icon. Pl. Ind. Orient. 6: 1. 1853; FBI 4: 726. 1885; FPM 2: 1175. 1925; FAP 2: 805. 1997. *Achyranthes brachiata* L., Mant. Pl. 50. 1767.

A. *Nothosaerva brachiata*, B. *Dendrophthoe falcata* var. *pubescens*, C. *Scurrula parasitica*, D. *Viscum articulatum*, E. *Acalypha fruticosa*, F. *Croton scabiosus*

Annual, erect herb, up to 50 cm high, branches spreading from base upwards; stem subterete, striate, sparsely hairy. Leaves decussate, lower leaves on main stem ovoid-elliptic, 1–3 × 0.5–2 cm, upper leaves on main stem and all leaves of branches becoming shorter and narrower towards apex, reduced in flowering branches, glabrous to thinly hairy above, pubescent below, more on nerves, base acute to attenuate, margin entire, apex acutely apiculate; petiole about half the length of the lamina, to 1.5 cm long on main stem, shorter on branches, reduced towards apex and finally sessile. Paniculate spikes dense, axillary, clustered, 2 × 0.3 cm, subtended by reduced leaves; peduncles short, 3 mm long, axis thinly to rather densely pilose. Flowers bisexual, 0.8 mm across, pale whitish woolly; bracts ovate, 0.5 mm, hyaline, acute or shortly acuminate, glabrous or very thinly hairy, nerveless; bracteoles 2, broadly ovate, concave, 1–1.25 mm, hyaline, acuminate, persistent. Perianth segments 5, sub equal, oblong, deltoid-ovate, 1.25–1.5 × 0.5–0.75 mm, hyaline, membranous, 1-nerved, sub acute to shortly acuminate, villous on the outer suface, with a dark greenish midrib, extends up to $^{2}/_{3}$ of the segment, truncate, acute. Stamens 2; filaments 1–2.5 mm long; anthers 2-celled, hemispheric, 0.1 mm across. Ovary obconic to depressed-globose, 0.3 mm across, 1-locular; ovule 1 on short, erect funicle; style short, 0.5 mm long; stigma capitellate. Urticle pyriform, oblong, 0.5–0.75 mm, compressed, epicarp membranous, irregularly rupturing, falling off with the perianth-lobes. Seed lenticular, 0.5 mm, brown, smooth and shining, testa crustaceous.

Occasional, in dry deciduous forests, near ponds at Muniswara temple. Fl. & Fr.: Aug. – Dec. Kalasamudram (ATP), *KRKS* 42027.

PUPALIA A. L. Juss. *nom. cons.*

Pupalia lappacea (L.) Juss., Ann. Mus. Nat 1. Hist. Nat. 2: 132. 1803; var. **lappacea**; FBI 4: 724. 1885; FPM 2: 1173. 1925; FAP 2: 807. 1997. *Achyranthes lappacea* L., Sp. Pl. 204. 1753. *Pupalia atropurpurea* (Lam.) Moq., in DC., Prodr. 13(2): 331. 1849; FBI 4: 723. 1885; FPM 2: 1173. 1925. *Achyranthes atropurpurea* Lam., Encycl. 1: 546. 1785.

Local name: *Erra uttareni, Tella uttareni.*

Annual or perennial, erect herb to straggling undershrub, up to 2 m high; stem much branched and swollen at nodes; branches opposite, slender, obtusely 4-angled to almost terete, thinly pilose to tomentose. Leaves opposite, variable in shape and size, ovate-elliptic to oblong or orbicular, 2–7 × 1–4 cm, sparsely tomentose or pilose on the veins in lower side, base attenuate gradually, margin entire, apex acuminate to apiculate; lateral veins 7–9 pairs; petiole to 1 cm long, pilose. Spike terminal, unbranched, to 50 cm long; peduncle tomentose. Flowers 3 in a cluster, central one 1-bracteate, lateral flowers 2, bracteolate; sterile flowers modified into hooked bristles in fruit; bracts lanceolate, 1.5–2.5 × 0.5–0.7 mm, persistent, more or less deflexed after the fall of the fruit, sub-glabrous or pilose, sharply awned with percurrent midrib; bracteoles broad, subcordate-ovate, 3–5 × 0.5–1 mm, abruptly narrowed, dark arista formed by the excurrent midrib, membranous with pale margin, densely hairy; bracteoles of sterile flowers ovate-lanceolate, pilose. Perianth segments oblong-ovate to lanceolate, 3–6 × 1–1.5 mm, narrowed to obtuse-mucronate at apex, aristate; outer tomentose, dorsally 3-nerved; inner 2-nerved, mucronate at apex with arista; inner most 3 slightly shorter and more densely pilose. Stamens 5, 2.5–3 mm long, anthers yellowish. Ovary depressed-globose, basally narrowed, 1 mm across; style erect, slender, 0.8–2 mm long; stigma capitate. Capsule depressed globose, circumscissile, 2.5 mm long, top areolate. Seed ovoid-oblong, 1.5–2 mm, lenticular, dark brown, shining; testa at first faintly reticulate but finally smooth or punctulate, thinly crustaceous.

Common, along road sides, wastelands, borders of agriculture fields and foot hills of forests. Fl. & Fr.: Oct. – Apr. Thummalapalle (KDP), *SSR* & *TP* 33608.

TRICHURIELLA Bennet

Trichuriella monsoniae (L. f.) Bennet, Indian J. Forest. 8: 86. 1985. *Trichurus monsoniae* (L.f.) C. Towns., Kew Bull. 29: 466. 1974 & in Dassan & Fosb., Rev. Handb. Fl. Ceylon 1: 36. 1980; FAP 2: 808. 1997. *Illecebrum monsoniae* L.f., Suppl. Pl. 161. 1781. *Celosia monsoniae* Retz., Observ. Bot. 2: 13. 1781. *Aerva monsoniae* (Retz.) Mart., Nova. Act. Phys.-Med. Acad. Caes. Leop.-Carol. Nat. Cur. 13: 291. 1826; FBI 4: 728. 1885; Gamble 2: 1178. 1925.

Annual, tufted, prostrate herb, to 90 cm long; stem wiry, slender with ridges, branches horizontally spreading, opposite or whorled, pilose with white hairs, villous. Leaves whorled, linear-filiform, up to 2.5 cm long, crowded at nodes, more commonly thinly pilose, with multicellular hairs, sessile, acuminate. Spikes pink, 0.5–3 × 0.5 cm, terminal and axillary. Flowers in spike; bracts and brateoles lanceolate, 1–1.5 × 0.5 mm, hyaline, with a green midrib. Perianth lobes 4, lanceolate, sub-clavate, 2–2.5 × 0.7 mm, hyaline with a green midrib, dorsally lanate, curving towards the apex of the spike, persistent. Stamens 4; filaments 1–1.5 mm long, alternating with triangular to sub-quadrate, pseudo-staminodes. Ovary depressed globose, 0.2 mm across; style simple; stigma capitate. Utricle ovoid, to 1 mm, circumscissile, longitudinally split at maturity, lid much shorter than the base. Seed ovoid, compressed, 0.6–0.7 mm, shining, reddish-brown, testa crustaceous.

Common, in wastelands in sandy soils and scrub forests. Fl. & Fr.: Oct. – Feb. Kalasamudram (ATP), *KRKS* & *TP* 37694.

CHENOPODIACEAE

CHENOPODIUM L.

Chenopodium murale L., Sp. Pl. 219. 1753; FBI 5: 4. 1886; FPM 2: 1181. 1925; FAP 2: 811. 1997.

Annual erect herb, to 1 m high, upright or spreading, normally much branched, green, rarely red-tinged, with mealy hairs especially on young parts but rarely densely clothed. Leaves variable alternate, rhomboid to deltoid-ovate, rarely narrower, 2–7 × 1–4.5 cm; petiole 3 cm, base attenuate, margin sharply toothed, apex acute. Inflorescence leafy axillary spike or composed of divaricately branching cyme up to 5 cm. long, terminal and from upper and sometimes median axils. Flowers greenish, minute, about 1–1.5 mm. in diam. Perianth segments 5, papillose on margins and outside, each with a blunt raised green keel towards apex only. Stamens

5. Pericarp very difficult to detach from seed. Seeds black, somewhat shining, 1.2–1.5 mm. in diam., acutely keeled; testa marked with very close minute rounded pits.

Occasional, in agricultural fields. Fl. & Fr.: Sep. – Nov. Kalasamudram (ATP), TP & *EC* 2924.

ARISTOLOCHIACEAE

ARISTOLOCHIA L.

Key to Species

1a. Perennial herb, to 50 cm high; leaves reniform-cordiform, whitish, basally 5-nerved; flowers solitary, bracts large at the base of the pedicel; fruit ellipsoid-oblong .... **A. bracteolata**

1b. Perennial climber, to 4 m long; leaves oblong-elliptic, dark green, basally 3-nerved; flowers axillary racemes, 5–8, bracts small, opposite the pedicel; fruit ellipsoid-oblong .... **A. indica**

Aristolochia bracteolata Lam., Encycl. 1: 258. 1783; FAP 2: 818. 1997. *A. bracteata* Retz. Observ. Bot. 5: 29. 1788; FBI 5: 75. 1886; FPM 2: 1201. 1925.

Local name: *Gadida gadapa.*

Perennial, slender, prostrate herb, to 50 cm high; branches angular with ridges, white glaucous. Leaves reniform-cordiform, 3–5 × 4–6 cm, glaucous below, glabrous above, base cordate to broadly sagittate, margin entire to undulate, apex obtuse to sub acute; lateral veins 5 at base, palmately divergent; petiole to 3 cm long. Flowers axillary, solitary, 5 cm long, purple; bract cordate-orbicular, 1 × 1 cm; pedicel to 0.5 cm long, tube 2 cm long, curved, glabrous without, hirsute within, base inflated, 0.5 cm across; lip dark purple, linear, 2 cm long, with revolute margin, apex obtuse; mouth 0.4 cm across. Stamens 6; anthers oblong, 0.6 mm long. Ovary 8 × 0.3 mm; style 2 mm long; stigmatic lobes 6, connivent, acute. Capsule about 2.5 cm long 1.5 cm across, glabrous, ellipsoid-oblong, 12-ribbed, septicidally opened. Seeds cordate, 4 mm across, partly covered by the remains of the placenta, glands dotted; testa thin, albumen fleshy.

Common, in wastelands and cultivated fields in black soils. Fl. & Fr.: Aug. – Jan. Velpula (KDP), *KRKS* & *KRS* 37673.

Aristolochia indica L., Sp. Pl. 960. 1783; FBI 5: 75. 1886; FPM 2: 1202. 1925; FAP 2: 819. 1997.

Local name: *Isora, Eswari, Nalla eswari.*

Perennial, slender twiner, to 4 m long, woody at base, grooved, glabrous. Leaves alternate, oblong-elliptic, 3–14 × 2–4 cm, basally 3-nerved, glabrous, base truncate or sub-cordate, margin entire, apex acute or obtuse, sometimes acuminate; lateral veins 6–7 pairs; petiole 0.6–0.8 cm long. Racemes axillary or terminal, to 3 cm long, 8–15-flowered. Pedicel to 1.5 cm long. Flowers 6 cm long; bract 2 mm long; tube 2.5 cm long, glabrous without, puberulous within, mouth 0.7 cm across; base inflated,

0.8 cm across; limb 1-lipped, lip 3.5 cm long, 0.5 cm wide, dark purple, greenish, gradually narrowed, margin revolute. Stamens 6; anthers oblong, 1 mm. Ovary 1 × 2.5 cm, glabrous; style 3 mm; stigmatic lobes 6, erect, 1.5 mm, acute. Capsule 4 × 2 cm, oblong or globose-oblong, opening from below upwards, 6-valved, hanging when open. Seeds oblong, 5 mm across, obtuse at apex, laterally winged; testa thin.

Common, in dry deciduous forest and scrub forests. Fl. & Fr.: Nov. – Feb. Rachakuntapalle (KDP), *KRKS* & *KRS* 37908.

Note: Roots are said to be used for snake bites.

LAURACEAE

CASSYTHA L.

Cassytha filiformis L., Sp. Pl. 35. 1753; FBI 5: 188. 1886; FPM 2: 1241. 1925; FAP 2: 824. 1997.

Local name: *Sithamma savaralu, Sithammavari jadalu, Antara valli, Pachi teega, Nulu teega.*

Perennial, parasitic, green, glabrous, leafless, branches wiry, twining herb, branches slender, green, attached to hosts by suckers or haustoria, younger ones yellow. Flowers in spikes, white, 3-merous, bisexual, 2 mm across. Tepals 6, free, unequal, 3 + 3, truncate, obtuse at apex; outer lobes ovate-orbicular, 1 mm long, ciliate; inner lobes obovate, 1.5 mm long. Fertile stamens 9; filaments 1.5 mm; glands sessile; anthers 2-celled, 1 mm long; staminodes 3, to 1 mm long. Ovary 1.5 mm; style 0.5 mm long; stigma capitate. Drupe globose, enclosed within inflated perianth, crowned by lobes, white when mature, 0.5 across. Seed 1, glabrous, globose; testa thin, hard, black, 3 mm across.

Common, in scrub forests and agriculture fields, grows on herbs, shrubs and trees. Fl. & Fr.: Jul. – Jan. Mabbuchintalapalle (KDP), *KRKS* & *KRS* 36906; N. P. S. Kottala (KDP), 40486.

Note: The juice of the plant is used for treating inflamed eyes.

HERNANDIACEAE

GYROCARPUS Jacq.

Gyrocarpus americanus Jacq., Select. Strip. Amer. Hist. 282. t. 178. f. 80. 1763; FPM 2: 1242. 1925; FAP 2: 828. 1997. *G. jacquinii* Roxb., Pl. Coromandel. t. 1. 1795; FBI 2: 461. 1878.

Local name: *Kumari poliki, Tanaku.*

Deciduous tree, to 20 m high; trunk cylindric, bark whitish; wood light, soft, white; young shoots pubescent. Leaves broadly ovate, 3-lobed, 5–10 × 8–15 cm, chartaceous, glabrous above, thinly pubescent below with appressed hairs, base truncate to sub-cordate, margin entire, apex acute to acuminate; lateral veins forking, extending to $^{2}/_{3}$ of the leaf; veins flattened to impressed above, slightly raised below; petiole 8–12 cm long. Cymes compound in terminal clusteres, to 20 cm wide, pubescent, reduced panicles. Flowers white, small, 2 mm across, without bracts, unisexual. Male flowers 2–3 mm across; tepals 4–7, unequal, ovate, concave, 2 mm, thinly pubescent without, acute; stamens 4–7; filaments 3 mm, sparsely hairy; anthers 0.5 mm long, valves opening from base upwards; staminodes flat, obovate-spatulate, 1 mm across, obtuse. Female flowers 3 mm across; tepals 2, basally adnate to ovary, 2-winged above, appressed-pubescent, obtuse. A few of bisexual flowers scattered among a mass of male flowers, yellow or creamish. Gland white, disk-shaped, 1 mm across. Drupe obovoid-globose, longitudinally ribbed, crowned by 2, spatulate wings; wings 7 × 0.7 cm, apex obtuse, divaricate. Seed 1; cotyledons large, convolute.

Occasional, in dry decdiduous forests near rocky boulders. Fl.: Jan. – Mar. Fr.: Jul. – Aug. Thimmamma marrimanu (ATP), *KRKS* & *KRS* 37932.

LORANTHACEAE

Key to Genera

1a. Leaves linear or oblong to linear, glabrous; flowers in racemes up to 8 cm long .... **Dendrophthoe**

1b. Leaves elliptic or elliptic-ovate, pubescent or tomentose; flowers not in racemes .... (2)

2a. Leaf base acute, apex acute or emarginate; petiole to 1.2 cm long; inflorescence recemose fascicle; corolla orange-yellow, tube 1.8 cm long, inflated below .... **Scurrula**

2b. Leaf base truncate, apex obtuse; petiole to 0.5 cm long; inflorescence umbellate receme or cyme; corolla greenish yellow, tube 2.4 cm long, inflated above the middle .... **Taxillus**

DENDROPHTHOE Mart.

Dendrophthoe falcata (L. f.) Ettingsh., Akad. Wiss. Wien. Math.-Naturwiss. Kl. Denkschr. 32: 68. t. 13. ff. 14 & 15. 1871; FAP 2: 830. 1997. *Loranthus falcatus* L. f., Suppl. Pl. 211. 1781. *L.longiflorus* Danser, in Lam., Encycl. 3: 598. 1792; FBI 5: 214. 1886; *L. longiflorus* Danser var. *falcata* (L. f.) Kurz, Forest Fl. Burma 2: 321. 1877; FBI 5: 215. 1886; FPM 2: 1253. 1925.

Key to Varieties

1a. Leaves linear-lanceolate, curved; corolla creamish yellow, green band at apex .... var. **falcata**

1b. Leaves elliptic-linear, straight; corolla green-yellow, light yellow at middle, dark green at apex .... var. **pubescens**

Dendrophthoe falcata (L. f.) Ettingsh., var. **falcata**.

Local name: *Badanika, Jiddukai, Bajinika.*

Large, semiparasitic shrub with haustoriferous runners; stem to 2 m high; branches cylindric, glabrous, slender, coriaceous, corky. Leaves alternate, sub-opposite, simple, linear-oblong, straight or curved, 3–10 × 0.4–1 cm, glabrous, coriaceous, base attenuate, margin entire, apex obtuse; lateral veins indistinct, midvein prominent; petiole 0.5–1 cm long. Inflorescence axillary raceme, up to 5 cm long; peduncle up to 1 cm long, glabrous. Flowers in fascicles of about 25; pedicel to 0.5 cm long; bracteole 2 mm across, light coloured, orbicular. Calyx cylindrical, up to 0.4 cm long, connate. Corolla creamish yellow; tube up to 4 cm long, slender, narrow at base, gradually wider, green band at apex, followed by constriction and swollen portion; perianth lobes apically connate and forms club shaped stiff structure; corolla lobes 5, green, recurved at mature. Stamens 5, filaments adnate and opposite to the perianth lobes, up to 3.8 cm long; anthers red, basifixed, up to 3 mm long. Ovary inferior; style to 4 cm long, green; stigma capitate, dark green. Fruit oblong, ellipsoid, red or light pink, glabrous, 1 × 0.5 cm, annulate above, fleshy, gelly pulp. Seed oblong, white, 0.8 × 0.4 cm, sticky nature (gummy), sweet; caruncle green, 0.2 cm across.

Common, parasite on trees, especially on *Maerua oblongifolia, Albizia amara* and *Croton scabiosus*. Fl. & Fr.: Jan. – Jul. Thummalapalle (KDP), *KRKS, KRS, TP* & *SSR* 33647.

Note: Because of sticky nature of seed, it sticks on host stem and germinates. In the germination process first haustoria comes out and penetrate into host tissue then it develops into adult plant.

Dendrophthoe falcata (L. f.) Ettingsh., var. **pubescens** Hook. f., Fl. Brit. India. 5: 215. 1886.

Local name: *Badanika, Bajanika.*

Large, semi parasitic shrub with haustoriferous stem, to 2 m high; branches cylindric, glabrous, slender, coriaceous, corky. Leaves opposite or alternate or subopposite, simple, variable in shape, ovate-oblong, obovate or elliptic-oblong, 4–12 × 1.5–5 cm, glabrous, coriaceous, base acute-attenuate, margin entire or undulate, tip obtuse; lateral veins 5–7 pairs, midvein prominent; petiole 0.5 cm long. Raceme axillary, to 8 cm long, pubescent; peduncle 0.6 cm long; pedicel 0.6 cm; bract ovate-elliptic, orbicular, 0.2 mm across. Flowers to 3 cm long, pubescent. Calyx cupular, to 3 × 2 mm, pubescent with white minute hairs, hairs deciduous; lobes 5, dentate, minute. Corolla tubular, to 2.5 cm long, sparsely hairy, green-yellow at base, light yellow at middle, dark green at apex; lobes 5, green, base of lobes swollen, apex obtuse. Stamens 5, alternating the corolla lobes, red; filaments 0.5 cm long; anther cells 0.3 cm long. Ovary oblong-subglobose, glabrous; style 2.5–3 cm long; stigma capitate, green, ovary covered by calyx tube. Fruit green, glabrous, sub-cylindric, ellipsoid, 1.8 × 0.8 cm, annulate above with prominent calyx-tube, light red or pink at maturity, gummy pulp. Seed 1, light brown-yellow, 1 × 0.5 cm, sticky nature, gel sweet; caruncle green, 0.2 mm across,.

Common, on tree species in dry deciduous forests especially on *Ziziphus* spp., *Croton scabiosus* and *Dalbergia lalifolia*. Fl. & Fr.: Jun. – Dec. Rachakuntapalle (KDP), *KRKS* & *SSR* 39034; near Rathnachalam Anjaneyaswamy temple, Pulivendula Ghat (ATP), *KRKS* & *BSS* 39672.

Note: Because of sticky nature of seed, it sticks on host stem and germinates. In the germination process, haustoria first comes out and penetrate into host tissue then develop into adult plant as that of dicots. This species almost as var. *falcata* but vary in leaf shape and colour of the flower.

SCURRULA L.

Scurrula parasitica L., Sp. Pl. 110. 1753; FAP 2: 831. 1997. *Loranthus buddleioides* Danser in Lam., Encycl. 3: 600. 1792; FPM 2: 1251. 1925.

Large semi parasitic shrub, up to 1.5 m high; young branches rusty tomentose, dark brown; stem corky, glabrous, grey, rough. Leaves opposite, simple, elliptic-ovate or obovate, 3.5–5.5 × 2–35 cm, glabrescent above, tawny-tomentose below, base acute, margin entire, apex obtuse-acute or emarginated, lateral veins 5–6 pairs, prominent; petiole 0.5–1.2 cm, glabrous. Inflorescence axillary fascicle, 6–7-flowered; pedicel 0.3 cm long, tomentose; bract ovate, 1.2 × 1 mm, tomentose. Flower bisexual, reddish brown, tomentose, zygomorphic, to 2.5 cm long. Calyx 3 mm long, apex truncate, base narrow, pulverulent. Corolla orange-yellow, 4 mm across, split length wise towards one side, pulverulent without; tube 1.8 cm long, inflated below, gradually narrowed above; lobes 4, lanceolate 3 × 0.3 mm, acute. Stamens 4; filaments 1 mm long; anthers linearly-oblong, 0.8 mm long. Ovary 2 mm long; style 1.7 cm long; linear; stigma obovoid-capitate, reddish. Fruit conical-pyriform, up to 0.8 cm long, base acute, apex truncate, mucronate at tip, pulp gummy, pubescent, red. Seed 1, dark green, 4 mm across.

Very rare, in forests on *Dalbergia paniculata*. Fl. & Fr.: Jan. – Apr. Kalasamudram (ATP), *KRKS* 41886.

TAXILLUS Tieghem

Taxillus bracteatus (Heyne ex Wall.) Tieghem, Bull. Soc. Bot. France 42: 256. 1895; FAP 2: 831. 1997. *Loranthus bracteatus* Heyne ex Wall. in Roxb., Fl. Ind. ed. Carey & Wall. 2: 220. 1824; FBI 5: 213. 1886; FPM 2: 1252. 1925.

Large, semiparasitic shrubs with haustoriferous runners, to 1 m high on trees; stem brown tomentose, corky; branches whitish, tawny-tomentose. Leaves alternate, elliptic, sub-orbicular, 1.5–2.5 × 1–2 cm, pubescent, base truncate-obtuse, acute, margin entire, apex obtuse; lateral veins 3–5 pairs, mostly raised from base; petiole to 0.5 cm long, tometose. Inflorescence axillary umbellate cyme. Floweres 2.5 cm long, subtended by a lateral bract, bract oblanceolate, 7 × 3 mm. Calyx 4 mm long, margin obscurely 2–4-toothed, white-pilose without. Corolla greenish yellow, red inside, 0.4 cm across; tube 2.4 cm long, inflated above middle, pilose without; lobes 5, linear lanceolate, reflexed, 7 × 1.5 mm, subacute, red within. Stamens 5; filaments 3 mm long; anthers 2 mm long, basifixed. Ovary oblong-cylindric, 2 × 1.5 mm; style 2.2 cm long; stigma capitate, light yellow. Fruit oblong or ellipsoid, up to 2 cm long, red, with persistent bract, annulate above by persistent calyx. Seed 1; yellowish, oblong, to 1.7 cm long, sticky nature.

Occasional, in dry deciduous forests on *Grewia* spp. Fl. & Fr.: Jul. – Aug. Batrepalli (ATP), *KRKS* & *BSS* 37977.

VISCACEAE

VISCUM L.

Key to Species

1a. Stem rounded, many branches at node; nodes swollen.........**V. ramosissimum**

1b. Stem flattened or 3 angular, dichotomously branched, nodes not swollen .... **V. articulatum**

Viscum articulatum Burm. f., Fl. Ind. 311. 1768; FBI 5: 226. 1886; FPM 2: 1257. 1925; FAP 2: 832. 1997.

Local name: *Dabba badanika.*

Subshrub, moderate sized, semiparasite, evergreen, up to 1 m long, drooping; branches usually dichotomously branched, sometimes more than 2 branches at each node, yellowish green, internodes flattened, at right angle to the preceding one, gradually broadening from the base to tip, twisted near the base, longitudinally prominently grooved. Leaves absent. Flowers sessile triads in axillary fascicles, unisexual, actinomorphic, central flower of the triad female and lateral male. Male flowers sessile, usually 2 in a triad on either side, up to 3 mm across, actinomophic, epigynous. Perianth lobes 4, green, triangular, 1.5 mm long, valvate. Stamens 4, 1 × 0.7 mm, opposite and adnate to the perianth lobes; filaments absent; pollen sacs many, dehisce introrsely. Female flower sessile, centre in the triad, up to 4 mm across, actinomorphic. Perianth lobes 3 or 4, green, triangular, valvate, up to 2 mm across. Ovary inferior; style absent; stigma nipple-shaped, truncate. Fruit berry, globose, green when young, yellowish when mature, up to 4 mm across, apex truncate. Seed 1, globose, 2.5 mm across.

Common, in dry deciduous forests and scrub forests, on trees like *Albizia* sp., *Dalbergia* sp., *Croton scabiosus, Ziziphus* sp., etc. Fl. & Fr.: Sept. – Apr. Kalasamudram (ATP), *KRKS* 41885; Thimmamma marimanu (ATP), *KRKS* 37967; Rachakuntapalli (KDP), *KRKS* & *SSR* 39035.

Viscum ramosissimum Roxb. ex DC., Prodr. 4: 278. 1830; FBI 5: 225. 1886; FPM 2: 1258. 1925; FAP 2: 834. 1997.

Local name: *Gundu badanika.*

Large, semi-parasitic, branches many, slender, sub-shrub, up to 1.5 m high, hanging broomlike. Branches slender, cylindrical, numerous; nodes swollen, internodes terete, basally larger, up to 5 cm long, apically smaller, up to 2 cm long. Leaves usually absent, when present only on lower branches at young stage, obovate, spatulate, flattened, 1.5–2.5 × 0.4–0.8 cm, faintly 3-nerved from base. Flowers triads in axillary fascicles, small, sessile, unisexual, central female and laterals male. Male flowers compressed, up to 3 mm across, actinomorphic. Perianth lobes 4, free, ovate, valvate, up to 2 mm long. Stamens 4, opposite and adnate to the perianth lobes; staminal filaments absent; pollen sacs many, dehisce introrsely, pistillode absent. Female flowers oblong, larger than male, present in the center of the triad, 0.4 cm long, actinomorphic, epigynous. Perianth lobes 4, free, conical, staminodes absent. Ovary inferior; style reduced; stigma capitate, minute, nipple-shaped. Fruit berry, small, globose, up to 5 mm across, green, apex truncate. Seed 1, 0.4 cm across, globose, whitish.

Occasional, in scrub and dry deciduous forests, on *Ziziphus* spp. Fl. & Fr.: Apr. – Aug. Thimmamma marrimanu (ATP), *KRKS* 37966.

SANTALACEAE

SANTALUM L.

Santalum album L., Sp. Pl. 349. 1753; FBI 5: 231. 1886; FPM 2: 1261. 1925; FAP 2: 834. 1997.

Local name: *Chandanam, Srigandham, Harichandanam.*

Small, semiparasitic, evergreen tree, up to 8 m high; bark dark brown; wood outerside white and centre yellowish brown, strongly scented when mature; branches glabrous, pinkish green. Leaves opposite-alternate, elliptic-ovate, lanceolate, 4–6.5 × 2–3 cm, glaucous beneath, green above, base rounded to acute, margin entire, apex gradually acute; lateral veins 10–12 pairs, faint, irregularly arching near margin; petiole to 1.5 cm long. Inflorescence of paniculate cyme, 3-chotomous, to 2.5 cm long, terminal and axillary. Flowers bisexual, 5-merous, brownish purple or maroon red, 6 mm across. Tepals 5, equal, basally connate into a campanulate tube, 2 mm long, shortly connate to the basal part of ovary; lobes ovate, thin-fleshy, 2.5 × 1.5 mm, glaucescent without, minutely ciliate, acute. Stamens 5; filaments 1 mm long; anthers 0.7 mm long. Ovary superior, later becoming half-inferior, globose, 1 mm across; ovules 2–3, pendulous from below the long, acuminate, central column; style 1.5 mm long; stigma 3-lobed. Haustoria 3, connate at base to the ovule, to 1.5 mm long. Disk of 5, fleshy, ovoid, obtuse scales, alternating with stamens. Drupe globose, 1 × 0.8 cm, annulate above, beaked with the basal part of style, reddish when young, black when ripe. Seed sub-globose, whitish, endocarp hard, 4 × 3 mm.

Common, in foot hills and open dry deciduous forests. Fl.: Dec. – Apr. Fr.: Aug. – Mar. Batrepalli (ATP), *KRKS* 36954.

Note: Heart wood strongly, characteristically scented, lasts for many years and in great demand for carvings.

EUPHORBIACEAE

Key to Genera

1a. Inflorescence a cyathium; tepals 0; stamens 1 per male floret; filament jointed to a pedicel; pistil stalked .... **Euphorbia**

1b. Inflorescence other than a cyathium; tepals usually conspicuous; stamens 2 – 10 or many; filaments not jointed; pistil subsessile .... (2)

2a. Cells of ovary 2-ovuled; stamens opposite to tepals .... (3)

2b. Cells of ovary 1-ovuled; stamens alternating with tepals .... (9)

3a. Stamens connate .... (4)

3b. Stamens free...(7)

4a. Tepals gland-dotted; stamens born on a gynophore .... **Sauropus**

4b. Tepals not gland-dotted; stamens sessile or born on a gynophore ....(5)

5a. Tepals of male flower turbinate; ovary truncate .... **Breynia**

5b. Tepals of male flowers spreading; ovary globose.
– stamens opposite tepals ....(6)

6a. Leaves coriaceous; flowers silky; drupe globose, green–light
brown-dark brown .... **Cleistanthus**

6b. Leaves not coriaceous; flowers glabrous; capsule or drupe
globose, subglobose, if drupe green only... **Phyllanthus**

7a. Stamens monadelphous, borne on a gynophore; tepals 10, biseriate;
disk of female flower with an inner, fimbriate corona .... **Bridelia**

7b. Stamens free, not borne on a gynophore; tepals 4 – 9, one or
biseriate; disk of female flower without an inner corona ....(8)

8a. Ovary 1-locular; inflorescence a spike; stamens inserted within
or in the hollows of disk; fruits hardly enlarged................................ **Antidesma**

8b. Ovary 2 or 3-locular; inflorescence a fascicle; stamens not
inserted in disk; fruits notably enlarged .... **Flueggea**

9a. Seeds carunculate...(10)

9b. Seeds not carunculate ....(13)

10a. Leaves palmately lobed...(11)

10b. Leaves simple ....(12)

11a. Large shrubs atleast 3 m high; stem and petiole hollow; capsule
papillate with smooth spines.-stamens many in male flowers ....**Ricinus**

11b. Small shrubs up to 2 m high, stem and petiole solid; capsule
without papilla ....**Jatropha**

12a. Herbaceous; leaves linear-lanceolate, capsule barbellate or softly
prickled on the outer side of cocci.. **Microstachys**

12b. Undershrubs or small trees, leaves ovate or ovate-oblong;
capsule without prickles ....**Croton**

13a. Perianth 2-seriate, inner whorl longer, petaloid...........................**Chrozophora**

13b. Perianth 1-seriate, sepaloid ....(14)

14a. Climbers or erect herbs; leaf and fruit with stinging hairs;
female flower bracts not cupular ....**Tragia**

14b. Erect herbs; leaves and fruit glabrous or pubescent not as above;
female flower bracts cupular.-stamens many .... **Acalypha**

ACALYPHA L.

Key to Species

1a. Spike unisexual; female spikes terminal; male spike axillary .... **A. alnifolia**

1b. Spike bisexual; female and male flowers in the same axillary or terminal spike .... (2)

2a. Female bracts very few, 1–5 in each spike .... **A. fruticosa**

2b. Female bracts many, 8–20 in each spike .... (3)

3a. Lower leaves and upper leaves on stem with almost equal length of petiole, lamina ovate-lanceolate; bracts dentate with glandular tip, smaller than capsule .... **A. lanceolata**

3b. Lower leaves with long petiole, reduced towards apex of the stem, lamina ovate or rhomboid; bracts larger than capsule, dentate without glandulat tip .... **A. indica**

Acalypha alnifolia Klein ex Willd., Sp. Pl. 4: 525. 1805; FBI 5: 415. 1887; FPM 2: 1330. 1925.

Local name: *Chiru kuppinta.*

Perennial hispid shrub, up to 1.5 m high; branches slender, hispid, brown, hard. Leaves alternate, ovate-elliptic, 1.5–2 × 1–1.5 cm, hispid, dense below, base cordate, rounded, margin crenate to serrate, apex acute to obtuse; lateral veins 4–5 pairs; petiole 0.3–0.7 cm, hispid. Inflorescence axillary spike, unisexual. Male spike axillary, slender, elongate, up to 4 cm long, terete, usually in lower axils. Flowers sessile, 0.2 mm across, arranged in clusters on hispid axis; bracts lanceolate, 1 mm long, hispid. Perianth lobes 4, sub-elliptic, 0.5 mm across, hispid. Stamens many, anthers 0.2 mm across, vermiculiform, sessile. Female spike short, to 1 cm long, terminal or sub terminal; bracts cupular, dentate at margin, enclosing ovary, hispid. Ovary 2 mm across, tomentose; styles 3, to 0.5 cm long, ciliate, filiform. Capsule 0.5 cm, globose, hirsute. Seeds globose, smooth, shiny, brown, 2 mm across.

Common, in foot hills of dry deciduous forests, scrub forests and field bunds. Fl. & Fr.: Jul. – Feb. Thimmamma marrimanu (ATP), *KRKS* & *KRS* 37935.

Acalypha fruticosa Forssk., Fl. Aegypt.-Arab. 161. 1775; FBI 5: 414. 1887; FPM 2: 1331. 1925; FAP 2: 839. 1997.

Local name: *Chinni.*

Perennial bushy hispid under shrub with woody base, to 2.5 m high, monoecious, with waxy glands; branchlets finely pubescent, terete, striate, young parts greyish, puberulous. Leaves ovate-oblong, 1.5–3 × 1.5–2.5 cm, membranous to chartaceous, pilose or pubescent on both sides, yellowish glandular below, base rounded-truncate or obtuse, margin serrate to crenate, apex acute-acuminate; basal veins 2–4 pairs; lateral veins 2–4 pairs; petiole 1–2 cm long; stipules subulate or linear, up to 2 mm long. Spikes axillary or terminal, 1–3 cm long, solitary or paired, bisexual or unisexual. Male flowers many, clustered along the spikes, 0.4 mm across followed by sterile flowers; pedicels up to 0.3 mm long, pubescent. Sepals 4, ovate-lanceolate, 0.5 mm long, pubescent; stamens 10–15; filaments 0.2 mm long; anthers vermiculiform, 0.3–0.5 mm long. Female flowers 2–3 mm across, few at the very base of the spike or in separate, small clusters resembling few-flowered spikes in the axils of young leaves; bracts leafy, folded, dentate along margins, 5–6 mm across, glandular, 1 or 2-flowered. Pedicel 1 mm long; sepals 3, ovate to ovate-lanceolate, 0.7–1 mm long, pubescent, pistils 2, one sterile. Ovary 2–3 mm across, glandular, pubescent; styles 3, filiform, laciniate, up to 5 mm long. Capsules ovoid, trilobed, 3–4 mm across, tomentose. Seeds sub-globose or ovoid, 2 mm across, smooth.

Common, in scrub forests and wastelands in the Core Zone of Mining Area. Fl. & Fr.: Mar. – Jul. Thummalapalle (KDP), *KRKS, KRS, SSR* & *TP* 33654.

Acalypha indica L., Sp. Pl. 1003. 1753; FBI 5: 416. 1887; FPM 2: 1330. 1925; FAP 2: 839.

Local name: *Kuppichettu, Pippinta, Muripinda.*

Annual erect herb, up to 90 cm high, monoecious; stem simple or branched, finely pubescent. Leaves elliptic-ovate, rhombate or broadly ovate, 2–7 × 1.5–5 cm, membranous to sub-coriaceous, sparsely pubescent above, finely pubescent below, more on veins, base cuneate-rounded or attenuate, margin crenate or serrulate, apex acute or abruptly acuminate; basal veins 3–5 at base; lateral veins 3–5 pairs; petiole 3–8 cm long, longer than the leaf-blade, decreasingly at apex of stem, stipule subulate, up to 2 mm long. Spike axillary 1.5–9 cm long, finely pubescent, nearly as long as petiole. Male flowers clustered at the upper portion of the spike to 0.5 mm across, spike to 0.6 cm long, terminating with long-stalked allomorphic flowers, narrow toothed, tubular, inter mixed with sterile flowers, sub-sessile; pedicels 0.3 mm long. Sepals 4, obovate, 0.5 mm, across. Stamens 10–15; filaments 0.2 mm long; anthers vermiculiform, 0.2-0.3 mm long. Female flowers many from base to above the middle of the spike, each in axils of bracts; bracts alternate, cup-shaped, larger than capsule, denticulate along margins, 0.5–1.3 cm across; flowers sessile, 1–2 mm across. Sepals 3, ovate-lanceolate, 0.5–0.7 mm long, pistils 2, one sterile. Ovary 1–1.5 mm across, densely whitish tomentose; styles 3, filiform, laciniate, 2–4 mm long. Capsules completely enclosed by bracts, ovoid, trilobed, 3–4 mm across, minutely pubescent, mixed with a few short stipitate pale brown glands.

Very common, in wastelands, road sides and weed in agriculture fields. Fl. & Fr.: Throughout the year. Thummalapalle (KDP), *KRKS* & *KRS* 33680; Chinnapalli (ATP), *KRKS* & *KRS* 37815.

Acalypha lanceolata Willd., Sp. Pl. ed. 4, 4: 524. 1805; FPM 2: 1331. 1925; FAP 2: 840. 1997.

Annual erect herb, to 50 cm high; sometimes ascending; branches hirsute or finely pubescent, angular. Leaves crowded, broadly ovate or ovate-lanceolate, 1–5 × 0.5–4 cm, membranous to thinly coriaceous, finely pubescent, gland-dotted, base cuneate, rounded or truncate, margin serrate or dentate, apex acute to subacuminate;

Plate 17

A. *Acalypha lanceolata*, B. *Euphorbia clarkeana*, C. *E. fusiformis*, D. *E. kadapensis*, E. *E. longistyla*, F. *E. prostrata*

basally 3–4-nerved; lateral veins 4–6 pairs; petiole 0.5–8 cm long, reducing towards apex of stem; stipule filiform, 2 mm long. Spikes axillary, sessile or peduncled, 1.5–8 cm long. Male flowers 0.5–0.8 mm across, few, clustered in a small head, concealed in upper most bract; bracts ovate, 0.5–0.7 mm across. Sepals 4, ovate to ovate-lanceolate, 0.5 mm across. Stamens 8–15; filaments 0.1 mm long; anthers vermiculiform, 0.2–0.3 mm long. Female flowers 1–2 mm across, solitary in each bract; bracts reniform or campanulate, smaller than the capsule, folded, 3–4 mm across, shallowly dentate, glandular hairs along margin. Sepals 3, ovate to ovate-lanceolate, 0.5–0.8 mm across, pubescent. Ovary 0.8 mm across, densely pubescent; style 3, filiform, 2–4 mm long, fimbriate. Capsules 3-lobed, 2–2.5 × 4–5 mm, hirsute hairs with glandular bases. Seeds ovoid to globose-ovoid, 1.5–2 mm across.

Occasional, near water, streams and dry deciduous forests in high altitudes. Fl. & Fr.: Aug. – Dec. Kalasamudram (ATP), *KRKS* & *BSS* 40074, Rachakuntapalle (KDP), *KRKS* & *BSS* 40177; Batrepalli (ATP), *KRKS* 42016.

Acalypha malabarica Müll.Arg., Linnaea 34: 42. 1865; FBI 5: 416. 1887; FPM 2: 1330. 1925; FAP 2: 840. 1997.

Annual herb, up to 50 cm high, monoecious; stem and branches striate, thinly pubescent. Leaves alternate, ovate to ovate-lanceolate or elliptic-lanceolate, 1.5–6 × 1.2–3 cm, membranous to thinly cariaceous, pilose or hispid on both sides, gland-dotted beneath; base cuneate to subtruncate, margin crenate-serrate, apex acute-mucronate, basal veins 3–5; lateral veins 4–6 pairs; petiole 1.5–5.5 cm long; stipules subulate to linear, up to 3 mm long. Spike axillary, up to 3 cm long, 2–3 clustered, androgynous, finely pubescent, glandular based hairs. Male flowers 0.5 mm across, very few, clustered near the apex of spike at axils of 5 mm long bracts. Sepals 4, ovate-lanceolate. Stamens 10–15; filaments 0.1 mm long; anthers vermiculiform, 0.2 mm across. Female flowers 2–2.5 mm across, numerous. Bract 1 or 2-flowered, leafy, subcampanulate, folded, 5 mm across, glandular along margin. Sepals 3 or 4, ovate to ovate-lanceolate, 0.5–0.8 mm across, hispid along margins. Ovary 0.5 mm across, glandular-pubescent; styles 3, filiform, laciniate, 1.5–2 mm long. Capsules 3-lobed, 3–4 mm across, pubescent with glandular based hairs or nearly glabrous. Seeds 3 in capsule, 1 mm across, elliptic-globose, brown, smooth.

Occasional, in moist and shady places in wastelands, foot hills of forests and road sides. Fl. & Fr.: Aug. – Jan. Bhumaiahgari palle (KDP), *KRKS* & *KRS* 37615; Rachakuntapalle (KDP), *KRKS* & *BSS* 40190.

ANTIDESMA Burm. ex L.

Antidesma acidum Retz., Observ. Bot. 5: 30. 1788 (*'acida'*); FAP 2: 842. 1997. *A. diandrum* Roth, Nov. Pl. Sp. 369. 1821; FBI 5: 361. 1887; FPM 2: 1298. 1925. *Stilago diandra* Roxb., Pl. Coromandel t. 166. 1802.

Local name: *Pella-gumudu, Pulleru, Gummadi pella.*

Shrub or small tree, up to 3 m high, deciduous; branchlets glabrescent with age. Leaves alternate, obovate to oblanceolate or oblong to elliptic, 5–12 × 1.5–7 cm, membranous to chartaceous, glabrous above, sparsely pilose below on midrib; base cuneate or acute, margin entire, apex acute to apiculate; lateral veins 3–8 pairs; petiole

1–7 mm. Male inflorescence spike, axillary and terminal, simple, 1 or 2-branched, 2–12 cm long; bracts 0.5–1 mm long, ciliate. Male flowers pedicels 0.5–1.5 mm long. Calyx cup-shaped, 0.8–1 × 1–1.3 mm; lobes 4 or 5, deltoid or suborbicular, 0.3 × 0.5 mm; disk cushion-shaped, enclosing the base of filaments, pubescent. Stamens 2, 1.2–2.5 mm long; anthers 0.6 mm broad. Female inflorescence simple or rarely once-branched, 2–5 cm long; bracts as in male. Female flowers pedicels 0.5–1.5 mm long. Tepals cupular, urceolate, 1–1.5 × 1.2 mm; lobes 4, triangular or deltoid, 0.5 mm long; disk annular. Ovary obovoid, 1 × 1 mm, glabrous; styles 2, terminal, 0.5-1 mm long, 2-fid. Fruit ellipsoid to suborbicular or often broadly oblong, prominently ribbed. Some what laterally compressed, with short terminal, persistent style at apex, 4–6 × 3–5 mm, glabrous. Fruiting pedicels 2–3 mm long.

Occasional, in exposed slopes of dry deciduous forests, usually near the streams. Fl.: Aug. – Oct. & Fr.: Oct. – Feb. Batrepalli (ATP), *KRKS* & *BSS* 39593.

BREYNIA J. R. Forst. & G. Forst. *nom. cons.*

Breynia retusa (Dennst.) Alston, Ann. Roy. Bot. Gard. (Peradeniya) 11: 204. 1929; FAP 2: 488. 1997. *Phyllanthus retusus* Dennst., Schluss. Hort. Malab. 1. Register: 15; 2. Register: 24; 3. Register: 31. 1818. *Breynia patens* (Roxb.) Rolfe, J. Bot. 20: 359. 1882; FBI 5: 329. 1887; FPM 2: 1304. 1925. *Phyllanthus patens* Roxb., Fl. Ind. 3: 667. 1832.

Local name: *Devadia, Dawadaree.*

Shrub, monoecious, 1–3 m high; very slender, branches brown, swollen at nodes; branchlets very slender, clustered at nodes. Leaves ovate, oblong, elliptic to suborbicular, 1–2 × 0.5–1.2 cm, membranous to chartaceous, often glaucescent beneath; leaf beariang branchlets 5–16 cm long, base acute, obtuse to rounded, margin entire, somewhat scarious-reflexed, apex obtuse, rounded, emarginate or retuse; lateral veins 3–6 pairs; petioles 1–2 mm long; stipules triangular-acuminate, up to 2 mm long. Male flowers terminal and axillary; pedicels 2–7 mm long. Calyx 1.5–5 × 1.5–3 mm. Stamens 3; anthers 0.6–2 mm long. Female flowers subterminal, pedicels 1.5–5 mm long or sessile. Tepals 3, initially obconic, cupular or often campanulate, 1–4 × 3.5–6 mm, finally flattened and up to 8 mm in diam.; lobes flabellate, almost free, 1–3 × 1.5–4 mm, thinly pubescent without, accrescent. Ovary obconic, globose or turbinate, 1–2 mm in diam., excavate; styles 3, 1–2 mm long, connate at base into a column, bifid above. Fruits depressed, subglobose, often shallowly lobed, 4–5 × 8–10 mm; pedicels 4–6 mm long; persistent calyx patelliform, 9–12 mm across. Seeds with membranous testa, 0.3 cm across, thick brownish-black.

Occasional, near streams in dry deciduous forests. Fl. & Fr.: Apr. – Aug. Batrepalli (ATP), *KRKS* 39595.

BRIDELIA Willd. *nom. & orth. cons.*

Key to Species

1a. Fruit with 3–4 mm long pedicel; inflorescence mostly borne on terminal and leafless branches, up to 35 cm long; leaves with 12-20 pairs of nerves .... **B. retusa**

1b. Fruit sessile to sub-sessile; inflorescence born on main leafy branches, clustered; leaves with 4-10 pairs of nerves........................ **B. montana**

Bridelia montana (Roxb.) Willd., Sp. Pl. ed. 4, 4: 978. 1806; FBI 5: 269. 1887; FPM 2: 1281. 1925; FAP 2: 846. 1997.

Local name: *Pantenga, Gundubigalu.*

Erect or scandant shrub or small tree, up to 4 m high, much branched, branches spreading, glabrous; branchlets shining, green. Leaves obovate-oblong to rhombate, 2.5–4.5 × 1–2.8 cm, chartaceous to thinly coriaceous, glabrous, base cuneate, acute-rounded, margin entire, apex obtuse or apiculate, acute, lateral veins 4–10 pairs, mostly bifurcating midway; petiole 3–5 mm long; stipules linear-subulate to filiform, 2.5–7 × 0.3–1 mm. Inflorescence on main leafy branches. Flowers 5–8 mm in diameter, in axillary clusters. Flowers 5–8 mm across, mostly bisexual. Bracts broadly triangular to deltoid, 2.5 mm long. Male flowers sessile or pedicellate; pedicels up to 0.5 mm long. Calyx 2.5 × 3–4 mm; sepals triangular, 1.5–2 × 1–1.5 mm; petals flabellate, 1 × 1.5 mm; disk flat, pulpy, entire, 1.5 mm long. Stamens 5, on gynophores; anthers 0.6 × 0.5 mm. Female flowers sessile to sub sessile. Calyx 3–3.5 mm across; sepals rarely 6, triangular-oblong, 2 × 1 mm; petals 0.5–2 mm long; outer disk patelliform, 1.5 mm across; inner disk urceolate, lobulate, covering the ovary. Ovary subglobose or ovoid, 1 × 1.2 mm, 2-locular; styles 2, 1 mm long, almost free, deeply bifid. Fruit sessile to subsessile, subglobose, 4–9 × 5–10 mm, shallowly to prominently bilobed, 2-seeded or often 1-seeded by abortion; fruiting calyx 4 mm across; pyrene furrows shallow, ridges rounded, curved from a central one. Seed with fleshy albumen, brown, 0.5 cm across.

Common, in open dry deciduous forests near water streams, foot hills of forests and occasional in scrub forests. Fl.: Sept. – Nov. & Fr.: Dec. – Mar. Rachakuntapalle (KDP), *KRKS* & *KRS* 37630; *KRKS* & *SSR* 39019.

Bridelia retusa (L.) A. Juss., Euphorb. Gen. 109. t.7, f. 22. 1824; FAP 2: 847.1997. *Clutyia retusa* L., Sp. Pl. 1042. 1753. *Bridelia cinerascens* Gehrm., Bot. Jahr. Syst.41(95): 30. 1908; FPM 2. 1280. 1925; FAP 2: 845. 1997.

Local name: *Bontha-yepi, Kora maddi, Verri karaka.*

Small tree, 8–5 m high; branchlets pubescent when young, glabrescent when old. Leaves alternate, broadly oblong, obovate, 5–13 × 2–8 cm, thinly coriaceous, glossy above, glaucous beneath, puberulous on midrib above, base rounded to acute, margin entire, apex rounded, acute or acuminate; lateral veins 12–20 pairs, terminating at the margins; petiole 5–15 mm long, pubescent. Stipules lanceolate to linear-lanceolate. Leafless panicles or in axillary fascicles, unisexual. Male flowers pedicellate; pedicels 1.5–2 mm long. Calyx 3.5–6 mm across; sepals triangular 1.5–2.5 × 1–1.6 mm; petals suborbicular, toothed or lobed, 1 × 1 mm; disk flat, 2 mm in diameter. Staminal column 1.5 mm long; filaments 1 mm long; anthers ovoid, 0.8 mm long; pistillode conic, 1 mm long. Female flowers pedicellate; pedicels 0.5–2 mm long, scattered, pilose to glabrous. Calyx 4.5–5 mm across; sepals narrowly triangular, acuminate, 1–2 mm across; outer disk diskiform, up to 2.5 mm across;

inner disk conical-cupular. Ovary ovoid, depressed, 1.5–2 mm across, 2-locular; styles 2, 0.8–1.5 mm long, connate below bifid above. Fruits subglobose, slightly depressed, 5–8 × 6–9 mm, short beaked at apex, shallowly 2-lobed, glabrous, 2-locular. Seeds 2; pedicels 2–4 mm long; fruiting calyx up to 7 mm across.

Rare in dry deciduous forests near and along the streams and hill slopes. Fl. & Fr.: Jan. – Nov. Kalasamudram (ATP), *KRKS* & *BSS* 39195.

CHROZOPHORA Neck. *ex* A. Juss.

Chrozophora rottleri (Geiseler) A. Juss. ex Spreng., Syst. Veg. 3:850. 1826; FPM 2: 1317. 1925; FAP 2: 848. 1997. *Croton rottleri* Geiseler, Croton. Monogr. 54. 1807. *C. tinctoria* auct. non (L.) A. Juss., 1824; FBI 5: 408. 1887.

Local name: *Erri miriyam.*

Erect herbs or subshrubs, up to 60 cm high, monoecious, densely hispid or stellate-tomentose; branches terete, fulvous-tomentose. Leaves alternate, broadly ovate-orbicular to oblong coriaceous, 2.5–7.5 × 2–6 cm, densely tomentose, with stellate hairs on both sides, base rounded or truncate-cuneate, margin undulate-toothed or crenate, apex obtuse or sub acute, 2-glandular beneath at the base of lamina; basal veins 3(5); lateral veins 4–5 pairs; petiole 1.5–6 cm long, stipules deciduous. Inflorescence axillary or terminal raceme, 2.5–9 cm long, bisexual, densely tomentose; bracts linear, 3 mm long, caducous. Male flowers sub sessile, crowded at upper part of raceme, 5 mm across. Calyx-lobes 5, 3–4 × 1.2 mm; connate at base, ovate-lanceolate, concave, acute-obtuse; petals 5, elliptic-oblong, lanceolate, obtuse 3.6–3.8 × 1.5–1.6 mm, free; disk obscure, glands 5, free. Stamens 5–15, biseriate; filaments connate below into 3.2–3.7 mm long column, with 1–3 series of free filaments, the outer ones short, inner ones long; anthers oblong, 1 mm across. Female flowers green, 3 mm across, 4–8 arranged at lower part of racemes, pedicellate, 1.5–3 mm long, elongating in fruit, up to 11 mm long. Calyx-lobes 5, 1.5–2 × 0.5–0.8 mm, valvate, connate at base, lanceolate-triangular, acute, stellate without. Petals 5, 1.4–2 × 0.5 mm, connate at base, valvate, similar to calyx; disk of 5 short broad glands alternating with petals. Ovary ovoid, 2.5–3 × 2–3 mm; styles 3, each bifid, 2–2.3 mm long, more than half way down; stigma red. Fruit capsule, subglobose, 7–9 × 5–8 mm, depressed, 3-lobed, densely stellate, reddish-purple when mature, tinctorial when wet; stalk 5–10 mm long. Seeds estrophiolate, globose, 3 mm in diameter; testa smooth, shiny, brown, hard.

Common, weed in agriculture fields and wastelands especially black soils. Fl. & Fr.: Through the year. Kalasamudram (ATP), *KRKS* 39527.

CLEISTANTHUS Hook. f. *ex* Planch.

Cleistanthus collinus (Roxb.) Benth. ex Hook. f., Fl. Brit. India 5: 1887; FPM 2: 1282. *Clutia collina* Roxb., Pl. Coromandel 2:37, t. 169. 1802 (*'Cluytia'*).

Local name: *Lodisa, Vodisa, Nalla-kordtha.*

Deciduous, monoecious tree, up to 8 m high; bark dark brown, rough; wood dark reddish brown; branchlets terete, glabrous. Leaves coriaceous, elliptic, orbicular or oblong-obovate, 4-12 × 3-7.5, base obtuse, margin entire, apex obtuse or retuse, glabrous, bright green; secondary nerves 5-6 pairs; petiole to 0.7 cm. Flowers yellowish-green in small, axillary, silky clusters. Male flowers: pedicels 1 – 2 mm long. Sepals occasionally 6, triangular-oblong or lanceolate-oblong, pilose outside. Petals linear; disk annular. Staminal column *ca.* 11.5 mm long; pistillode ovoid. Female flowers: pedicels 0.5-1.5 mm long. Calyx-tube *ca.* 1 mm long; sepals triangular-lanceolate. Petals subulate; disk shortly cupular-annular. Ovary subglobose, styles 3-4 mm long, almost free or basal column *ca.* 1.5 mm long, shortly to deeply bifid above. Fruit drupaceous, globose, 2 cm across, 3-valved, green-dark brown, shining; seeds 3, globose, testa smooth, shiny.

Common, in dry forests. Fl.: Feb., Fr.: May onwards dehiscing by next February. Kalasamudram (ATP), *EC* 2885.

Note: Leaves, fruit and bark are poisonous and used as fish poison.

CROTON L.

Key to Species

1a. Herbs, to 60 cm high; leaves narrowly ovate, oblong, to 4 cm long.......... **C. bonplandianus**

1b. Shrubs or small trees, to 5 m high; leaves ovate, elliptic oblong, to 9 cm long .... **C. scabiosus**

Croton bonplandianus Baill., Adansonia 4: 339. 1864; FAP 2: 850. 1997. *C. sparsiflorus* Morong, Ann. New York Acad. Sci. 7: 221. 1893; FPM 2: 1316. 1925.

Local name: *Kukka mirapa, Galivana chettu.*

Erect herb or subshrub, up to 60 cm high, monoecious, perennial by root stock; much branched; branchlets terete, stellate with white hairs. Leaves narrowly ovate, oblong, 2–4 × 1–2.5 cm, membranous to thinly chartaceous, stellate, base acute, obtuse, margin crenulate or serrate-dentate, apex gradually acute; lateral veins 3–8 pairs; basal glands sessile; petiole 0.1–2 cm long, sparsely pubescent to glabrous. Inflorescene 3–10 cm long, in terminal or subterminal racemes; bract 0.3 mm long; pedicel 1 mm long; glands 2 on either side at base of lamina. Male flowers 4 mm across, white. Sepals elliptic oblong to ovate, 1–2.5 × 0.5–1.5 mm. Petals narrowly oblong to spatulate, 1.5–2.5 × 0.5–1 mm, villous. Stamens 14–16 (20), 1.5–2.5 mm long; filaments 2 mm long; anthers 0.5 mm long. Female flowers 2.5 mm across, pedicels 0.5–1.5 mm long, with 1 or 2 sessile glands at base. Sepals 5, triangular to oblong, 1–2 × 0.4–1 mm, valvate, connate at base. Petals obsolete. Ovary oblong 1–2 × 1–1.5 mm, 3-loculed, white tomentose, styles 3, 2–3 mm long, bifid. Capsule oblong to ellipsoid, 4–7 × 3–5 mm, shallowly lobed, scattered puberulous to glabrous, bivalved cocci, epicarp warty. Seeds oblong, 0.5 × 0.3 cm, dotted brown.

Very common, in wastelands and road sides. Fl. & Fr.: Throughout the year. Mabbuchintalapalle (KDP), *KRKS* & *KRS* 33661.

Croton scabiosus Bedd., Fl. Sylv. S. India t. 283 B. 1872; FBI 5: 386. 1887; FPM 2: 1314. 1995; FAP 2: 851. 1997.

Local name: *Verrichilla, Errachillaku.*

Shrub or small tree, up to 5 m high, monoecious; bark blackish, scaly, corky; wood yellowish white, hard; branchlets striate, silvery shiny, stellate. Leaves ovate, elliptic-oblong, 4–9 × 3–7 cm, thinly coriaceous, stellate on both sides, base cordate, obtuse-truncate, margin crenate, apex acute; basally 3-veined or absent; lateral veins 5–7 pairs; basal glands 2, sessile; petiole 2–5 cm long, striate; stipules lanceolate, to 3 mm long. Inflorescence 3–10 cm long, terminal or subterminal, stellate; bracts ovate, subulate or linear-lanceolate, 0.5–3 mm long. Male flowers 4–4.5 mm across. Pedicels 2.5–3.5 mm long. Sepals 4, ovate to oblong-elliptic, 1.5–3 × 1–2 mm, slightly connate at base, ovate, acute, granulate with stellate hairs. Petals 4, obovate or oblong, 2–3 × 1–2 mm, margin ciliate. Stamens 10–15; filaments 1.6–2 mm long, hairy at base; anthers 0.9 × 0.7 mm. Female flowers 5 mm across, pedicel 2.5–6 mm long. Sepals 4, oblong, elliptic to obovate, 3–5.5 × 1.5–3 mm, connate at base, tomentose without, margin ciliate. Petals 4, ovate, acute, 2 × 1.5 mm, free, granulate with hairs without, margin ciliate. Ovary 2–3 × 2.5–3 mm across, strongly 3-lobed, tomentose; styles 3, 4 mm long, bifid near the base. Capsule deeply 3-lobed, intruded at apex, 5.5–7.5 × 7–8 mm, subglobose, granulate with stellate hairs. Seeds carunculate, dark brown, cylindrical, curved, 3–5 × 3–4 mm; testa hard, shiny, formation of seeds very low, most of them abortive, germination capacity very low.

Common, in Core Zone of Thummalapalle Uranium Mining Area, Kadapa district. It extends up to Idupulapaya forest in Palakonda hill ranges, also found in Guvvlacheruvu forest of Kadapa district. Fl.: May – Sept. & Fr.: Aug. – Dec. K. K. Kottala (KDP), *KRKS* & *KRS* 36431; Rachakuntapalle (KDP), *KRKS* & *BSS* 40192.

Note: Seed setting is very less because of fungal infection and insect attack. Most of the seeds are abortive, < 3 per cent of seeds are having embryos, but germination ability is very low.

Endemic to Andhra Pradesh.

EUPHORBIA L.

Key to Species

1a. Plant geophytic, root tuberous, normally not developed above the ground; leaves in rosettes at the ground level .... **E. fusiformis**

1b. Plants not geophytic, root not tuberous, well developed above the ground level; leaves distributed along the stems and branches .... (2)

2a. Plants succulent; leaves early caducous .... (3)

2b. Plants not succulent; leaves persistent (not early caducous)........................... (6)

3a. Plant spiny .... (4)

3b. Plant not spiny .... **E. tirucalli**

4a. Branches terete; spine shields rounded .... **E. caducifolia**

4b. Branches winged; spine shields conical....................(5)

5a. Angles of branches twisted; capsules always acutely keeled.............. **E. tortilis**

5b. Angles of branches straight; capsules obtusely keeled .... **E. antiquorum**

6a. Stipules usually present, interpetiolar; involucral glands 4–5 with petaloid limbs, though often very small or minute ....(8)

6b. Stipules absent; involucral gland single without petaloid limbs....................(7)

7a. Floral gland single, compressed mouth, leaves orange-blotched, margin lobed (denticulate).- often grown as ornamental .... **E. cyathophora**

7b. Floral gland single, circular mouth; leaves only green, margin serrulate.-weed .... **E. heterophylla**

8a. Root primordia present at all nodes; leaves retuse at apex; stipules connate.................... **E. serpens**

8b. Root primordia absent; leaves rounded, obtuse, subacute or acute at apex; stipules free....................(9)

9a. Leaves of main stems distinctly larger than those elsewhere (microphyllous branches) .... **E. clarkeana**

9b. Leaves of stems and branches almost equal in size....................(10)

10a. Limb of glands indistinct or same size as the glands ....(11)

10b. Limb of gland distinct and a little to much larger than the glands ....(14)

11a. Plants prostrate, only branches or tips of branches decumbent; cyathia solitary, paired or 3 together ....(12)

11b. Plants erect, decumbent or ascending; cyathia more than 3 in glomerules ....(13)

12a. Stems, branches and leaves pinkish grey; mature capsules remaining within the involucres, hairs on entire capsule............. **E. thymifolia**

12b. Stems, branches and leaves green; mature capsules protruding out from the involucres, hairs on keels of the capsule .... **E. prostrata**

13a. Stem hirsute with yellowish hairs; leaf ovate-lanceolate, serrate margin; cyathia more than 10; limb of gland inconspicuous, glands cupular .... **E. hirta**

13b. Stem pubescent, leaf elliptic, serrulate margin; cyathia less than 10; limb of gland conspicuous, glands oblong .... **E. indica**

14a. Cyathia many.................... **E. corrigioloides**

14b. Cyathia solitary to 2 or 3....................(15)

15a. Plant hispid; floral leaves imbricate and concealing the cyathia .... **E. cristata**

15b. Plant glabrous; floral leaves not imbricate but concealing the cyathia ....(16)

16a. Limb of gland 2- or 3-fid from the middle; capsule along the sutural ridges between cocci. -low elevations .... **E. kadapensis**

16b. Limb of gland multi-fid from the base; capsule without sutures. -high elevations .... **E. longistyla**

Euphorbia antiquorum L., Sp. Pl. 450. 1753; FBI 5: 255. 1887; FPM 2: 1277. 1925; FAP 2: 857. 1997.

Local name: *Bontha kalli, Jamudu.*

Succulent, arborescent tree, up to 5 m high, glabrous; latex milky, stem woody; branches jointed, 3-winged, spreading; wings broad, wavy margined; podaria in rows, 1.5–2 cm apart, slightly lobulate; spines 2, divaricate, stout, 5 mm long, blackish. Leaves sessile or subsessile, broadly ovate or suborbicular, 4–12 × 3–6 mm, glabrous, fleshy, base subcordate, rounded, margin entire, apex apiculate, early caducous; veins obscure; petioles 1 mm long. Cyathia in simple, axillary, dichasial cymes, central one sessile, lateral two pedunculate; primary peduncle 1–1.5 cm long, stout; secondary peduncles 0.5–1 cm long; bracts broadly triangular, obtuse, 3 × 2.5 mm. Involucre turbinate, 6–10 × 4–8 mm, pinkish outside; lobes 5, 2 mm across, laciniate. Glands 5, transversely oblong, fleshy, spongy, pitted, 2 × 5 mm, pale yellow. Male florets 5 groups or fascicles, 5 to 6 flowers in each; pedicels 1 mm long; anthers subglobose transversely dehiscent, yellowish; bracteoles broadly setaceous, deeply laciniate, 2.5 mm long, overtopping anthers. Female floret 5 mm across; gynophores 2 mm long; ovary 4 mm across; styles 3, connate up to middle, deeply lobed; stigma papillose, bifid. Fruit trilobed, 7–9 mm across, obtusely keeled, separating into 3, bivalved, 1-seeded cocci, yellowish-orange. Seeds globose, 3 mm across, greyish brown; testa smooth.

Common, in foot hills of dry deciduous forests and scrub forests. Fl.: Nov. – Dec. & Fr.: Dec. – Feb. Somulavaripalli (ATP), *KRKS* & *BSS* 40436.

Euphorbia caducifolia Haines, Indian Forester 40: 154. 1914; FPM 2: 1276. 1925; FAP 2: 858. 1997.

Local name: *Kolajamudu, Bontha jamudu.*

Thorny shrub, up to 3 m high, latex milky, glabrous, fleshy; branches many, bushy, dendroid, suberect, cylindrical, smooth, pinkish-green with silver-white coating peels when mature, 3–5 cm in diameter; podaria spirally arranged, distant; spines 2, divaricate, rigid, 5–8 mm long, slightly curved at tip, reddish when young, blackish when old. Leaves subsessile, located singly just above the spines, variable, ovate, ovate-elliptic to orbicular, 2–8 × 1–5 cm, fleshy, glabrous, with prominent midrib and indistinct lateral veins, base cuneate-truncate, margin undulate-entire, apex acute, early deciduous. Cyathia axillary in small fascicles, 1–1.5 cm long, fleshy, brown; bracts opposite, broadly triangular. Involucre turbinate, 3–4 × 2–3 mm; lobes 5, broadly cuneate, fimbriate at apex; glands 5, transversely oblong, fleshy, pale yellow. Male florets in 5 fascicles of 6 or 7 flowers each; pedicels 1 mm long; anthers subglobose, dithecous, 1 mm across, yellow, transversely dehiscent;

bracteoles 1.5–4 mm long, broad, laciniate at apex, covering florets. Female flower 1.5 mm long; gynophores 1.3 mm long; perianth lobes 3, minute, triangular. Ovary trigonous, minutely papillose; styles 3, connate up to middle, simple; stigma simple, papillose. Capsule trilobed, with laterally compressed and sharply keeled cocci, 7–9 mm in diameter, smooth, glabrous, reddish. Seeds ovoid to spherical or globose; 2 mm across; testa brownish, smooth, shiny.

Common, in foot hills of dry deciduous, scrub forest and dense in rocky hillocks. Fl.: Jan. – Feb. & Fr.: Mar. – Apr. Thummalapalle (KDP), *KRKS, KRS, SSR* & *TP* 33648; Mabbuchintalapalle (KDP), *KRKS* & *BSS* 39012.

Euphorbia clarkeana Hook.f., Fl. Brit. India 5: 253. 1887; Raja Kullayisw. *et al.*, J. Econ. Taxon. Bot. 37 (2): 305. 2013.

Annual prostrate herb, up to 25 cm long; stems many from a woody crown; branches pink to purplish, pubescent, nodes thickened, internodes 0.5–0.7 cm (floral branches 0.5–2 mm) long. Leaves two types, subsessile, elliptic-oblong, 0.4–0.5 × 0.2–0.3 cm, (microphyll 1.2–2.9 × 0.6–1.4 mm) glabrous, coriaceous, base oblique, margin entire, apex apiculate, midnerve prominent below; petioles 0.36 mm (microphyll 0.32 mm), glabrous; stipules interpetiolar, linear-lanceolate, 0.96 mm, (microphyll 0.6 mm) with glandular hairs. Oval shaped glands present at nodes, 0.45– 0.3 mm (microphyll 0.15 mm), pinkish, puberulous. Cyathia solitary, axillary, 3–5, congested by reduction of subtending leaves, often arranged in microphyllous branches or lateral branchlets, 1.26–1.89 mm, glabrous; peduncle 0.45–0.36 mm; involucre campanulate, 1.26 × 0.81 mm, glabrous; lobes 5, oblong, laciniate, glabrous; glands 4, shortly stalked, 0.11 × 0.23 mm, oblong, cupular, pink, limb of gland inconspicuous, 0.32–0.51 mm diameter, white, margin wavy. Staminate florets few, up to 4, 0.52–1 mm long, pedicels 0.32–0.54 mm long, glabrous; anther subglobose, 0.23 × 0.3 mm, transversely dehiscing; bracteoles lanceolate, 0.3–0.54 mm long. Pistillate florets glabrous; gynophore 1.26–1.51 mm long; ovary 0.53–0.63 mm long, glabrous; style 3, 0.36–0.58 mm long, pink, bifid from the middle. Capsule sub globose, 3-angled, 1.36 × 1.06 mm. Seeds quadrangular, subovoid-oblong, ovoid, 1.15– 1.21 × 0.69–0.72 mm, transversely 4-furrowed, reddish brown, white-coated, all facets with 5–7 transverse ridges.

Rare in cultivated fields, roadsides and open wastelands. Fl. & Fr.: Apr. – Dec. Boggudupalli (KDP), *KRKS* 41960; Thalupula (ATP), *KRKS* 41855.

Euphorbia corrigioloides Boiss., Cent. Euphorb. 11. 1860; FBI 5: 251. 1887; FPM 2: 1275. 1925; FAP 2: 859. 1997.

Local name: *Erra pala-alumu.*

Prostrate herb, up to 50 cm long, pubescent, puberulous; branches arise from a stout root, terete; nodes annular, thickened; internodes 0.5–1 cm long, sparsely hairy. Leaves elliptic-oblong, often with reddish-black in the middle, 0.5–1.5 × 0.4–0.5 cm, sparsely hairy or glabrous on both sides, base oblique, margin serrulalte, apex acute or obtuse; veins prominent; petiole 1 mm long; stipules sheathing, laterally

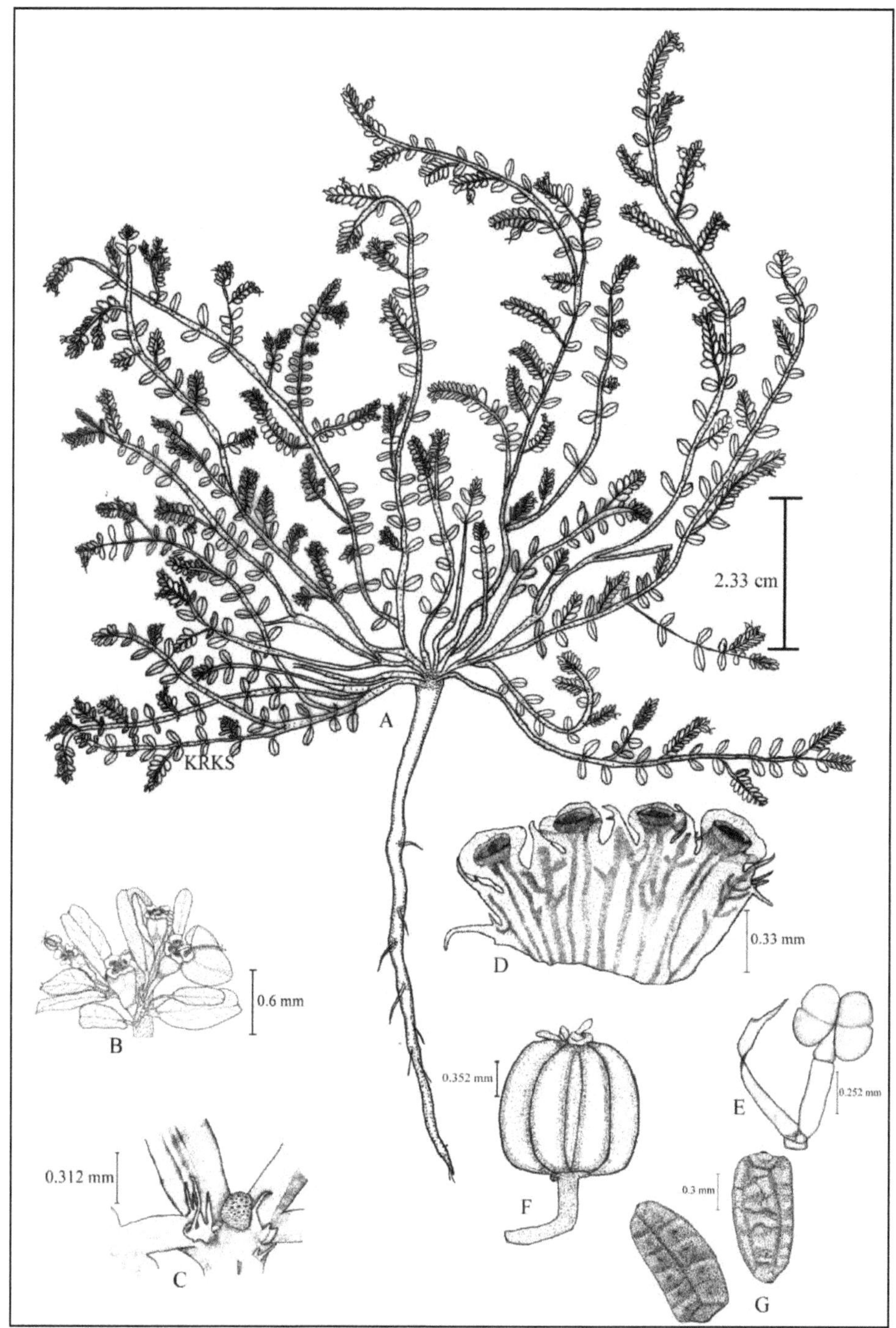

Figure 9: ***Euphorbia clarkeana*** **Hook.f. -**

A. Habit, B.Cyathial branch at apex, C. Gland and stipule at node, D. Involucral bracts. E. Male floret, F. Fruit, G. Seeds.

fimbriate, 2 mm long, laciniate along margins. Cyathia in axillary, condensed cymes or clusters, 0.6 cm long, pubescent; pedicels 0.2 mm long. Involucre turbinate, 2.3 × 2 mm; lobes triangular, trifid, exceeding the glands; glands 4, suborbicular to transversely oblong, 1 × 1.5 mm, wavy along apical margin. Staminate florets 10–18 in a fascicle; pedicels 1–2 mm long; bracts simple, filiform, 1.5 mm long; anthers subglobose, dithecous, yellow, transversely dehiscent. Female floret 4 mm across; gynophores 1.5–2 mm long, sparsely hairy in keels; ovary ovoid, 1.5 × 1 mm; styles free, 1 mm long, each bifid; stigma capitate, pink. Fruit globose, 4 mm in diameter, keeled, glabrous, puberulous along keels. Seeds ovoid, smooth, brown, 2 mm across.

Occasional, in black soils, wastelands and agriculture fields. Fl. & Fr.: Aug. – Jan. Nuthnakaluva, Vemula (KDP), *KRKS* & *BSS* 41954.

Euphorbia cristata Heyne ex Roth., Nov. Pl. Sp. 226. 1821; FBI 5: 247. 1887; FPM 2: 1274. 1925; FAP 2: 859. 1997.

Ascending-erect herb, up to 50 cm high, hirsute; branches pseudo-dichotomously branched, pinkish, clothed with flexuous hairs; internodes 2–3 cm long. Leaves broadly ovate or ovate-oblong, 1.5–2.5 × 0.7–1.5 cm, hispid on both surfaces, base obliquely cordate, margin serrulate towards apex, apex acute or obtuse; lateral veins 3–6 pairs, prominent; upper leaves smaller, closer, imbricate and concealing the cyathia; petiole 1.5 mm long. Cyathia subterminal or axillary, solitary; peduncle 5 mm long, hirsute, involucres campanulate, 1.5–4 × 1–2 mm; lobes triangular; glands 4, transversely oblong, green-red, 1.5 × 1 mm, hairy; limbs 4, laciniate or palmately 7–11-lobed for half way, 1.5 × 2.5 mm, sparsely pubescent on lobes, with 2-celled hairs, crimson-red. Staminate florets 6–8; pedicels, 1.5 mm long; bracteoles simple, setaceous, 1–1.5 mm long; anthers subglobose, transversely dehiscent, yellow; peripheral bracteoles many, laciniate at apex, adnate to involucral cup, 1 mm long. Female florets pendulous; gynophores 3 mm long, hairy; ovary subglobose, hirsute, 1.5–2.5 × 1–1.7 mm; styles 3, free, bifid up to middle; stigma capitate. Fruit subglobose, 2.5 × 2.5 mm, hirsute, pale pink. Seeds 2 mm long, obtusely quadrangular, transversely 3 or 4-furrowed; testa reddish brown or white.

Common, in slopes of dry deciduous forests and scrub forests. Fl. & Fr.: Aug. – Dec. K. K. Kottala (KDP), *KRKS* & *KRS* 37658; Batrepalli (ATP), *KRKS* 42017, 302 (HJCB-E).

Euphorbia cyathophora Murray, Comment. Soc. Regiae Sci. Gott. 7: 81 1786.

Annual herb or subshrub erect to 1 m high, with hollow glabrous stems and branches. Leaves to 10 × 5 cm, markedly panduriform, base cuneate, apex obtuse, margin shallowly and irregularly toothed, glabrous on upper surface, sparsely pilose with septate hairs on lower surface, glabrescent; petiole to 1.5 cm long, pilose; stipules modified as brownish glands. Cymes terminal, each forking *ca.* 4 times, with rays progressively shorter from 5 cm to 2 mm long and cyathia densely clustered. Basal bracts similar to the leaves but with a bright orange-red blotch at the base, the upper bracts progressively smaller, lanceolate, subsessile and entirely red. Cyathia *ca.* 3.5 × 3 mm with barrel-shaped involucres, glabrous; gland peltate, 1.5 mm long, funnel-shaped, the opening transversely elliptic, 2 mm wide; lobes 1.5

× 1.5 mm, rounded, margin deeply and bluntly toothed. Male flowers: bracteoles ligulate, feathery; stamens 4 mm long. Female flower: ovary pedicellate, glabrous; styles 2 mm long, bifid almost to the base. Capsule exserted on a reflexed pedicel to 5.5 mm long, 4 × 5 mm, deeply 3-lobed. Seeds 2.8 × 2.2 mm, ovoid-conical with acute apex and warts sharply pointed, blackish-brown.

Occasional, on roadsides and often cultivated in gardens. Fl. & Fr.: Almost throughout the year. Roadside, near Kadiri, Pulivendula cross (ATP), *KRKS* 439 (HJCB).

Euphorbia fusiformis Buch.-Ham. ex D. Don, Prodr. Fl. Nepal. 62. 1825; FBI 5: 257. 1887. *E. acaulis* Roxb., Fl. Ind., ed. Carey 2: 472. 1832; FPM 2: 1276. 1925; FAP 2: 856. 1997.

Local name: *Barresepu gadda.*

Geophytic herb, glabrous; root-stout, 10–70 cm long, tuberous, cylindric, latex milky white. Leaves radical, succulent, oblanceolate or elliptic-lanceolate, 7–20 × 1.5–7 cm, glabrous, often pinkish along margins, base attenuate, margin entire or curled, undulate at young stage, apex acute to subacute; midvein distinctly raised and pinkish beneath, lateral veins obscure; petiole 2–5 cm long, canaliculated, pinkish. Cyathia in diads or triads; primary peduncles 1–3 from the apex of the root stock, 5 cm or more long; secondary or tertiary peduncles 2–5 cm long, glabrous. Involucre turbinate, 3 mm across; lobes 5, subobovate, laciniate at apical margin; glands 4, transversely oblong, 1.5 × 1 mm; bracts ovate-lanceolate or triangular, obtuse or acute-mucronate. Staminate florets 3 or 4 fascicled, each with 4–6 flowers; pedicels 1.5 mm long; bracteole laciniate; anthers subglobose, longitudinally dehiscing. Female florets glabrous; gynophores 2–4 mm long; ovary subglobose; styles connate at base; stigma papillose. Fruits 4–6 mm long, 6–8 mm across, with compressed cocci; pedicels 1 cm long. Seeds broadly ovate to globose, 2–3 × 1.5–2.5 mm, whitish, variegated.

Very rare, in hill tops and slopes of dry deciduous forests along rocks and in gravelly soil. Fl. & Fr.: Feb. – Jul. Batrepalli (ATP), *KRKS* 39676.

Uses: Tubers are used for Lactation in animals.

Euphorbia heterophylla L., Sp. Pl. 453. 1753; FAP 2: 861. 1997. *E. geniculata* Ortega, Nov. Pl. Rar. Descr. Dec. 2: 18. 1797; FBI 5: 239. 1887; FPM 3: 1884. 1936.

Local name: *Pedda palaku.*

Erect herb, up to 70 cm high, latex milky, stem terete, hollow, puberulous, when young, dichotomously branched at apex. Leaves alternate below, opposite or rarely whorled above, elliptic to oblong, 3–8 × 2–5 cm, glabrescent above, pubescent, white glaucous below, base attenuate, acute, margin denticulate, crenate, apex acute-acuminate; floral leaves 1–3 cm long; lateral veins 10–12 pairs, distinct beneath; petiole 2–4 cm long; stipules scaly, sessile, 1 mm long, glandular, deciduous. Cyathia in terminal clustered cymes; peduncles 4–5 mm long, glabrous; bracts of cyathia 2, lanceolate, 2–2.5 cm long, crenate at margin. Involucre campanulate, 3–5 × 2–3 mm, glabrous; lobes triangular, laciniate; gland 1, extra-involucral, stipitate, turbinate with circular mouth, 0.4–0.5 × 0.3–0.4 mm, appendages 0. Staminate florets 12–15 in 2 or 3 fascicles; pedicels 2 mm long; bracteoles filiform, 2 mm long; peripheral bracteoles laciniate, adnate to involucral wall; anthers subglobose, 0.5 mm long, longitudinally dehiscent, yellow. Female floret laterally pendulous, 6 mm long; ovary glabrous, 2 × 3 mm; style 3, 1 mm long, connate at base, bifid at apex; stigma capitate. Fruit subglobose, 4–5 × 4.5–5.5 mm, smooth, glabrous, white coated. Seeds angular, 2 mm across, 3-angled, oblong-ovoid, truncate at base, transversely wrinkled, bluntly tuberculate; testa pale brownish black.

Common, weed in agriculture fields, road sides and foot hills of all forests. Fl. & Fr.: Through the year. Thummalapalle (KDP), *KRKS, KRS, SSR* & *TP* 33668.

Euphorbia hirta L., Sp. Pl. 454. 1753; FPM 2: 1275. 1925; FAP 2: 862. 1997. *E. pilulifera* L., Sp. Pl. 454. 1753; FBI 5: 250. 1887.

Local name: *Palaku, Pala alumu, Reddivari nanabalu, Pachabotlu.*

Prostrate or erect herb, up to 25 cm high, latex milky, all parts hirsute with yellow patent spreading multicellular hairs; stem stout; branches arise from root-stock; nodes slightly thickened; internodes 2–4 cm long, slightly reddish. Leaves opposite, asymmetric, oblong-ovate, 1–2.5 × 1–2 cm, hirsute on both sides; base oblique, margin serrate, apex acute, membranous; lateral veins 3–5 from base; petiole 1–6 mm long; stipules linear-lanceolate, 2 mm long, laciniate, glandular. Cyathia glomerulate in axillary globose heads; pedicels 1–2 cm long, dichotomously branched, 2–7 mm long, heads 0.5–1 cm in diameter. Involucre turbinate, 1 × 0.5 mm; lobes 5, deltoid, acute, 1 mm long; glands 4, cupular, reddish, 0.5–0.8 mm across; limbs minute or as large as the gland, rounded; hyaline, white with purple tinge. Staminate florets 4–6; pedicels 0.8 mm long; anthers subglobose; bracteoles setaceous, 0.5 mm long. Female floret pendulous; gynophores 1–2 mm long; ovary globose, 3-lobed, 0.8 mm in diameter; style 0.4 mm long, free, each bifid, half way; stigma capitate. Fruit subglobose to trigonous, 1.2–2 mm in diameter, pubescent. Seeds 3, oblong, 1 mm long, with 3 or 4 obscure, transverse ridges, purplish or reddish grey.

Very Common, in wastelands and agricultural fields. Mabbuchintalapalle (KDP), *KRKS, KRS, SSR* & *TP* 33640.

Note: This species shows ecological variations like reddish or purplish shoots during dry conditions, green and luxurious shoots during normal monsoon season.

Euphorbia indica Lam., Encycl. 2: 423. 1788; FBI 5: 250. 1887; FAP 2: 862. 1997. *E. hypercifolia* auct. non L., 1753; FBI 5: 249. 1887; FPM 2: 1275. 1925.

Erect or decumbent herb, up to 50 cm high, sparsely puberulous or hirsute, main stem arising from root-stock; internodes 1–2 cm long, rounded, terete. Leaves elliptic-oblong, 0.5–2 × 0.3–0.7 cm, pubescent on both sides or glabrous, base obliquely cordate, margin serrate above from the middle, apex obtuse; basally 3–4-veined, lateral veins 5–7 pairs, obscure above, prominent beneath; petiole 2–3 mm long; stipules triangular to setaceous or lanceolate, fimbriate, 1–2 mm long. Cyathia clustered on 1–2 cm long, axillary, microphyllous shoots, usually 1–2 pairs of small leaves; cymes dichotomously branched. Involucre turbinate, 1.2–2 × 1–1.2 mm, hirsute without; lobes 5, triangular, 0.5 mm long, lacerate; glands 4, pink, rounded, 0.1 mm across; limbs light pink or white, obovate or orbicular, as long as or slightly longer than the gland, entire or slightly sinuate. Staminate florets 2 or 3 fascicles of 3 or 4 flowers each; pedicels 0.5–1 mm long; anthers subglobose, yellow, transversely dehiscent; bracteoles filiform, 1 mm long; peripheral bracteoles laciniate. Female floret 1.5 mm long; gynophores 1 mm long; ovary 1 mm long, hairy; style 0.2 mm long, free, each bifid half way; stigma capitate. Fruits subglobose, 1.5 mm across, dorsally keeled, pubescent. Seeds oblong-ovoid, with 5 or 6 transverse broken ridges and shallow furrows, greyish brown.

Common, in cultivated fields, wastelands, road sides and scrub forests. Fl. & Fr.: Throughout the year. Thummalapalle (KDP), *KRKS* 36916.

Euphorbia kadapensis Sarojin. & R.R.V. Raju, Phytotaxa 181 (3): 179. 2014

Erect, decumbent to ascending herb (hemicryptophyte) up to 30 cm tall. Latex milky. Stem sympodial, dichotomously branched. Branches terete, glabrous, pink, filiform towards apices, sparsely hairy, hairs hyaline; internodes 1–2.5 cm long, thickened at nodes; stipules scaly, connate, 0.1 cm long, shortly laciniate. Leaves simple, opposite, equal, inequilateral, subsessile; petiole 0.1 cm long, sparsely hairy; lamina ovate to elliptic, 1.0–2.5 × 0.4–0.6 cm, glabrous, oblique at base, margin serrulate and more so distally, apiculate, midrib prominent, nerves slightly distinct on adaxial surface, in 6–8 pairs. Cyathia terminal, at times axillary, 1–2, 0.6–0.9 cm long; peduncles 0.2 cm long; involucre turbinate, 4 × 2 mm, glabrous, alternately pink and green; involucral lobes 5, laciniate; appendages of glands 4, 2–3-fid up to half of its length and occasionally to the base, 2 × 1 mm, pink; glands 4, oblong, yellowish green to brown, stipitate, stalks hairy; staminate florets 15–18, exserted, 1.5 mm long, pink, bracteate; bract simple, 2 mm long; anther lobes subglobose, dehiscing transversely, yellow; pistillate flowers laterally pendulous, 0.5 × 0.3 cm, glabrous, pedicel 0.5 cm long, knobbed distally; ovary tricarpellate; styles 3, free from base, 0.2 cm long; stigmas 3, simple, pink. Fruiting pedicels recurved, capsules schizocarpic, trigonous, distinctly keeled, prominently ridged along sutures, glabrous; cocci obscurely venulose. Fruits 3.5 mm long. Seeds 3 per capsule, ecarunculate, 1.5 × 1.0 mm oblong-ovoid, tetragonal, transversely and obscurely furrowed, truncate at base, not tuberculated, brown.

Occasional, in the calcified rocky areas. Fl. & Fr.: Oct. – Jan. Mabbuchintalapalle (KDP) *KRKS* 36936, *KRKS* & *BSS* 41844.

Euphorbia longistyla Boiss., Cent. Euphorb. 9. 1860; FBI 5: 247. 1887; FPM 2: 1275. 1925. *E. senguptae* sensu FAP 2: 867.1997. non N.P.; Balakr & Subr.

Erect herb, up to 30 cm high, glabrous; stem pseudo-dichotomously branched, terete, yellowish brown; nodes thickened, internodes 2–3 cm long; branches filiform towards the apex, minutely velutinous at base. Leaves elliptic to oblong-elliptic, 1–2 × 0.5–1 cm, glabrous, base oblique, margin entire towards base, serrate towards apex, apex apiculate-obtuse, midvein prominent; petiole 1–2 mm long; stipules linear-lanceolate, acuminate, shortly laciniate at upper margins. Cyathia terminal, solitary; peduncle 2 mm long, glabrous. Involucre turbinate, 1–2 × 1–1.5 mm, glabrous; lobes 5, laciniate, pubescent towards margins; glands 4, shortly stalked, rounded, highly velutinous at base; limbs of glands 1 × 1.5 mm, multifid-pectinate, minutely or bristly-tuberculate. Staminate florets few; pedicels 1 mm long, glabrous; anthers subglobose, transversely dehiscing; bracteoles filiform, laciniate, 1.5 mm long. Female flowers 4 mm long; gynophores 2 mm long, glabrous; ovary subovoid, 2 mm across, glabrous; styles filiform 2.5 mm long, free from base; stigma capitate. Fruit subglobose, keeled, 4 mm across, glabrous with clearly distinct veins on the cocci. Seeds quadrangular, subovoid to oblong-ovoid, 1.5 × 1 mm, transversely 3 or 4-furrowed, reddish brown.

Occasional, on slopes of dry deciduous forests. Fl. & Fr.: Jul. – Dec. Rathnachalam Anjaneyaswamy temple, Pulivendula Ghat (ATP), *KRKS* & *BSS* 39665; Batrepalli (ATP), *KRKS* & *SSR* 37886, 37890; Somulavaripalli (ATP), *KRKS* 41902.

Euphorbia prostrata Aiton, Hort. Kew. 2: 139. 1789; FBI 5: 266. 1887; FPM 2: 1276. 1925.

Annual prostrate herb, up to 20 cm long, branches many from base of root-stock, filiform, terete on lower surface, pubescent, nodes thickened, internodes 0.5–1 cm long. Leaves obliquely obovate to elliptic or suborbicular, 3–8 × 1–5 mm, glabrous or sparsely hairy beneath; base oblique, margin entire towards base, serrate towards apex, apex rounded or acute, veins obscure; petiole 1 mm long; stipules laciniate, 0.5 mm broad, lanceolate, glandular trichomes at nodal part merged with stipules. Cyathia axillary, 2–3 together on microphyllous branches; peduncles 3 mm long. Involucre campanulate, 1 × 0.7 mm; lobes 5, triangular, ciliate; glands 4, stalked, rounded, 0.2 × 0.2 mm, purple; limbs as large or a little larger than the glands. Staminate florets 2 to 3, pedicels 1 mm long; anthers subglobose, pinkish. Female florets 1.5 mm long; gynophores 1 mm long; ovary subglobose, hairy at keels; styles short, 1 mm long, free from base, each bifid at apex; stigma capitate. Fruit long exserted from the cyathial cup, 1 mm long, ovoid to subglobose, acutely keeled, 1 × 1.5 mm, with 2–3 rows of hairs restricted to the keels only. Seeds ovoid, tetragonous, transversely rugulose with 5–7 sharp ridges, 0.8 × 0.5 mm, pale brown to grey.

Common, in plains and foot hills of forests. Fl. & Fr.: Through the year. Kalasamudram (ATP), *KRKS* & *BSS* 39111, 39142; Bhumaiahgari palle (KDP), *KRKS* 37603; Somulavaripalli (ATP), *KRKS* 37834; Gollala guduru (KDP), *KRKS* & *BSS* 41960.

Euphorbia serpens Kunth, Nov. Gen. Sp. 2: 52. 1817; FAP 2: 867. 1997. *E. microphylla* auct. non B. Heyne ex Roth, 1821; FBI 5: 252. 1887.

Local name: *Chitti palaku.*

Annual prostrate herb, up to 40 cm long, glabrous; branches many, dichotomousely branched, nodes thickened with roots; internodes 0.5–1.5 cm, pale pink or pale green. Leaves opposite, suborbicular, subquadrate or ovate-elliptic, 3–6 × 2–5 mm, glabrous, base cordate, margin entire, apex obtuse, whitish green below, green above; veins indistinct, 3-nerved at base, 3–4 pairs of lateral veins;

petiole 0.5–1 mm long; stipules united above, scaly, broadly triangular, 0.5 mm long. Cyathia axillary, solitary; peduncles 1–2 mm long, glabrous. Involucre turbinate, 1 × 0.5 mm; lobes 5, laciniate, minutely hairy; glands 4, sessile, transversely oblong, 0.1–0.2 mm across, purplish; limbs reniform, subentire or 2–3-lobed, whitish, slightly broader than the gland, 0.3 mm across. Staminate florets few, ebracteolate; pedicels 1 mm long; anthers subglobose, transversely dehiscing, pinkish. Female floret 1.8 mm long; gynophores 0.8 mm long; ovary glabrous, 1 × 1.5 mm; style 0.5 mm long, each bifid from middle; stigma capitate. Fruit 1.5 mm long, pediceled, subglobose, trigonous with a faintly carinate keels, 1.5–1.8 mm across, glabrous. Seeds ovoid, sub-tetragonal, 1 × 0.7 mm, minutely tuberculate, pinkish brown.

Common, in foot hills of forests and cultivated fields, especially in black cotton soils. Fl. & Fr.: Through the year. Chinnapalli (ATP), *KRKS* & *BSS* 37998.

Euphorbia thymifolia L., Sp. Pl. 454. 1753; FBI 5: 252. 1887; FPM 2: 1276. 1925; FAP 2: 867. 1997.

Local name: *Palaku, Sanna palaku; Reddivari-nanubala.*

Prostrate annual herb, up to 20 cm long, puberulous; stem spreading from root stock; branched terete, pubescent, pinkish, nodes thickened, internodes 0.5–2 cm long. Leaves distichous, ovate-oblong, 3–8 × 2–5 mm, glabrous above, scattered hairy below, pinkish above, base obliquely subcordate, margin crenulate-serrulate or entire up to middle, apex obtuse or subacute; 3-veined at base; petiole 0.5–1 mm long; stipules united or slightly parted, narrowly triangular, linear-lanceolate to subulate, fimbriate, 0.5–1.2 mm long, hairy. Cyathia terminal or on short axillary microphyllous shoots, solitary or 2–3 together, reddish; peduncles 1 mm long. Involucre turbinate, 1 × 0.7 mm; lobes 5, triangular, subequal, ciliate; glands 4, suborbicular, 0.2 mm across, purple-red; limbs as large as glands, white or pink. Staminate florets 3–5; pedicels 1 mm long, glabrous; bracteoles filiform, 1.5 mm long; anthers subglobose, transversely dehiscing. Female florets pendulous; gynophores 0.5 mm long, strigose above; ovary 1 mm across, densely hairy throughout, pink; styles 3, free from base, spreading, 1.5 mm long, each deeply bifid; stigma usually clavate. Fruit shortly stalked, erect, scarcely exserted from involucral cup, maturing inside, trigonous, subglobose, obtusely keeled, 1–1.5 × 0.7–1.2 mm, hispid. Seeds oblong, tetragonal, blunt, 0.7–1 × 1–0.6 mm, transversely, shallowly 2–3-furrowed on facets, reddish brown or pinkish.

Occasional, in foot hills, plains, wastelands and village outskirts. Fl. & Fr.: Through the year. Thimmamma marrimanu (ATP), *KRKS* 37970,41935. Kalasamudram (ATP), *KRKS* & *BSS* 39965.

Euphorbia tirucalli L., Sp. Pl. 452. 1753; FBI 5: 254. 1887; FPM 2: 1276. 1925; FAP 2: 867. 1997.

Local name: *Calli, Ooracalli, Callimanu, Kada jamudu.*

Small tree, up to 6 m high, green; stem woody, 10–80 cm in diameter; bark blackish, wood white, latex milky; branches spreading, unarmed, terete, glabrous, bushy, apparently leafless. Leaves alternate, sessile or subsessile, shortly deflexed, early deciduous, obovate-linear, oblong-linear to linear-lanceolate, 7–15 × 2–6 mm, fleshy, sparsely hirsute, base cuneate and sessile, margin entire, apex obtuse, veins obscure; stipules minute, circular-diskoid, glandular, dark purple. Cyathia at the forks and tips of the branches, subsessile, solitary or in small capitate glomerules, congested at the apex of branchlets; bracts rounded, sharply keeled, 2 × 15 mm, usually glabrous. Involucre turbinate, 4 × 7 mm, glabrous or tomentose with light brown hairs; lobes 5, triangular or rhombic, laciniate at margin, scarious; glands 5, peltate, circular or a little angular, entire, 1–3 mm in diameter (except limb). Staminate florets pedicellate, to 3 mm long; bracteoles filiform, laciniate, 2 mm long; anthers subglobose, transversely dehiscing. Female florets pubescent; gynophore 1 mm long, erect; ovary 2 × 1.5 mm, pubescent; styles connate up to middle, 3 mm long, each bifid at apex; stigma capitate. Fruits exserted, subglobose, 6 × 8 mm, sparsely crisped pubescent to glabrous. Seeds ovoid, 3–4 mm in diameter, speckled brown with dark brown ventral line; caruncle minute, white.

Occasional, in forest, road sides and village outskirts. Fl. & Fr.: Feb. – Jul. Kalasamudram (ATP), *KRKS* & *BSS* 39124.

Euphorbia tortilis Rottler ex Ainslie, Materia Indica 2: 424. 1826; FBI 5: 256. 1887; FPM 2: 1277. 1925; FAP 2: 868. 1997.

Local name: *Pedda jamudu, Tirgudu jamudu.*

Shrub or small tree, up to 4 m high; bark black, scaly, loosely peeling off, shining, hard; wood white; stem green, spiny, coiled, latex milky, 3-4-winged; podaria 3 mm, projecting; spines 2, divaricate, 4–7 mm long, hard, black or brown. Leaves tiny, sessile or subsessile, ovate, suborbicular, 4–6 × 4–5 mm, glabrous, fleshy, early caducous, base obtuse, margin entire, apex acute; nerves indistinct, petiole minute or absent. Cyathia in fascicles, triads or solitary; primary peduncles in fascicles, 5 mm long; secondary peduncles 3 mm long; bracts broadly triangular, obtuse, 2 × 2.5 mm. Involucre cup-shaped, 4 × 3 mm; lobes 5, laciniate-fimbriate at apex; glands 5, transversely oblong or suborbicular, fleshy, 2 × 1.5 mm; bracteoles filiform, deeply laciniate, 2.5 mm long; anthers subglobose, transversely dehising. Female floret glabrous, 5 mm long; gynophore 1 mm long, glabrous; ovary 4 mm in diameter, glabrous; styles connate; stigma repand. Fruit 5–7 mm in diameter, sharply keeled. Seeds subglobose, obtusely tetragonous, 2 × 1.5 mm, smooth, pale black.

Rare in foot hills and hill tops of dry deciduous forests, near Mining area in Core Zone. Fl. & Fr.: Sept. – May. Rachakuntapalle (KDP), *KRKS* 39074.

FLUEGGEA Willd.

Flueggea leucopyrus Willd., Sp. Pl. ed. 4, 4: 757. 1806; FBI 5: 328. 1887; FPM 2: 1296. 1925. *Securinega leucopyrus* (Willd.) Müll. Arg., in DC., Prodr. 15 (2): 1866; FAP 2: 887. 1997.

Local name: *Tella purugudu, Tella puli, Tella-pulugu.*

Plate 18

A. *Euphorbia tortilis*, B. *Phyllanthus indofischeri*, C. *P. rheedei*, D. *P. rotundifolius*, E. *Ficus tsjakela*, F. *Eulophia epidendraea*

Shrub, 1.5–4 m high, bark whitish, entirely glabrous; leaf bearing lateral branchlets spinescent at tip. Leaves alternate, obovate-elliptic or orbicular 0.8–3 × 0.3–2 cm, glabrous, chartaceous to coriaceous, glaucescent below, base rounded, cuneate, margin entire, apex obtuse or obcordate or truncate; lateral veins 4–6 pairs; petiole 1–5 mm long; stipules triangular, persistent. Inflorescence arising precociously, axillary or at leafless nodes, male flowers many and glomerulate, female flowers 1–3 in fascicles at each node. Male flowers pedicellate; pedicels 2–4 mm long; sepals 5, ovate elliptic to suborbicular, 0.6–1 × 0.5–1 mm; disk glands 5, angular, 0.3 mm in diameter. Stamens 5, 1–2 mm long; anthers suborbicular, 0.4 mm long; pistillode 1 mm long, trifid at apex. Female flowers pedicellate; pedicels 1.5–2.5 × 0.3–0.4 mm. Sepals 5, ovate-triangular, 0.8–1.5 × 0.6–1 mm, persistent in fruit, disk flat, 0.8 mm in diameter. Ovary ovoid, 1.2 × 0.8 mm, 3-locular; styles 3, 0.8 mm long. Fruits subglobose, capsular or baccate, 2–3 × 4–4.5 mm, white, unlobed, prominent reticulate venation, sweet. Seeds plano-convex, 2–2.5 mm across, smooth, brown.

Common, in scrub forests and foot hills of dry deciduous forests. Fl. & Fr.: Mar. – Dec. Kalasamudram (ATP), *KRKS* & *KRS* 36491.

GIVOTIA Griff.

Givotia moluccana (L.) Sreem., Taxon 24: 696. 1975; FAP 2: 868. 1997. *G. rottleriformis* Griff., Calcutta J. Nat. Hist. 4: 88. 1844; FBI 5: 1887; FPM 2: 1342. 1925. *Croton moluccana* L., Sp. Pl. 1005. 1753.

Local name: *Tella-puliki, Tella-punki.*

Deciduous tree, up to 20 m high, dioecious, stellate-pubescent or fulvous tomentose; branchlets marked with prominent leaf scars; bark brown, smooth, peeling off in circular scales; wood white, very light and soft. Leaves alternate, broadly ovate to orbicular-deltoid, 8–18 × 6–15 cm, minutely pubescent with stellate hairs above and tomentose with stellate hairs below, base widely cordate, margin coarsely and irregularly toothed, glandular, apex acute, acuminate; palmately 5-veined at base; petiole 2–8 cm long, sometimes black, round, glands 2 at the base of lamina and several above the middle; stipules up to 5 mm long. Inflorescence axillary or terminal paniculate cymes, unisexual; bracts linear, up to 12 mm long. Male inflorescence 12–20 cm long, branched. Flowers yellow; sepals oblong to obovate, cucullate, 4–6 mm long, densely stellate pubescent; petals campanulate, 5-lobed with cuneate and truncate lobes, longer than sepals; disk of 5 free orbicular glands. Stamens 8–20; outer ones 6, with free filaments and inner ones 10, with filaments connate at base and exserted; filaments unequal, 5 mm long; anthers oblong, 1 mm long. Female inflorescence shorter and smaller than in male; sepals and petals as in male flower. Ovary globose, stellate hairy; styles 2, forked from base; stigma capitate. Drupe subglobose ovoid or ellipsoid, 2–2.5 cm long, indehiscent, smooth, green, densely fulvous-tomentose. Seed globose or ellipsoid; testa bony.

Occasional, in dry deciduous forests in all areas. Fl.: Mar. – Aug. Fr.: May – Sept. Kalasamudram (ATP), *KRKS* 39168 (♀), 39193 (♂).

JATROPHA L.

Key to Species

1a. Stipules well-developed, divided into narrow lobes or often modified into decurrent series of stalked glands .... (2)

1b. Stipules absent or minute and subulate .... **J. curcas**

2a. Lobes of leaves entire, glandular hairy along margins; flowers reddish or purplish red .... **J. gossypifolia**

2b. Lobes of leaves serrate, each tooth ending in gland-tipped bristle; flowers green with pinkish tinge .... **J. glandulifera**

Jatropha curcas L., Sp. Pl. 1006. 1753; FBI 5: 383. 1887; FPM 2: 1340. 1925, FAP 2: 872. 1997.

Local name: *Adavi amudamu, Nepalam.*

Shrub or small tree, up to 3 m high; branches sticky, gummy and latex watery, viscous, soapy latex, glabrous or rarely more or less flocculent, bark greenish white, smooth, peeling off as thin flakes; wood white, very soft. Leaves alternate, triangular-ovate or orbicular, 8–15 × 10–15 cm, membranous, glabrous, base broadly cordate, margin entire or 3–5-angled or lobed, caudate-acute; lobes 5–15 cm long, apex acute; digitately 5–7-nerved from base, petiole 4–15 cm long; stipules small or absent. Inflorescence laxly flowered cyme; peduncle 3.5–5 cm long. Male flowers 4 mm across; sepals ovate or elliptic, 3–4 × 1–2.5 mm; petals 5–10, united up to middle, oblong or obovate, coherent up to middle, 7–8 mm long, yellowish green; disk columnar, deeply 5-lobed. Stamens 8–10; outer filaments free; inner 5 connate at base; anthers 1.5 mm long. Female flowers 5 mm across; sepals longer than in male flowers; petals absent or same as in male flowers, but free and smaller. Ovary 3-locular 2 × 1.5 mm; style 5 mm, connate at base. Fruits fleshy, drupaceous, drying to capsule, subglobose, slightly 3-lobed, 2.5–3.5 cm in diameter, glaucous, pale yellow, turning to black. Seeds ovoid or oblong; testa crustaceous, albumen fleshy, carunculate.

Occasionally planted along road sides and foot hills of forests. Fl. & Fr.: Aug. – Mar. Somulavaripalli (ATP), *KRKS* & *BSS* 41901.

Jatropha glandulifera Roxb., Fl. Ind., ed. Carey 3: 688. 1832; FBI 5: 382. 1887; FPM 2: 1340. 1925; FAP 2: 872. 1997.

Local name: *Neela-amudam, Dundagaku.*

Undershrub, 1–2 m high, with pale yellowish juice; stem stout, dichotomously branched, glabrous, shiny and marked by prominent leaf scars; branches glabrous, somewhat sticky. Leaves alternate, orbicular, deeply palmately 3–5-lobed, 3.5–8.5 × 1.5–5.5 cm, glabrous, base widely subcordate, 8–16 cm in diameter, margin serrate with tipped-glands, apex acute to acuminate; lobes obovate or oblanceolate, oblong, laterals smaller than middle, basally narrowed, main nerves 3–5 at base, palmate, grooved above, raised beneath; secondary veins many, oblique; petiole 5–15 cm long;

stipules, 1.5–2 mm diameter, deeply divided into several, long, capillary tipped-gland segments, gland at base of petiole, attachment of petiole to the stem pointed and horizontal. Inflorescence axillary, corymbose cyme, with central flower of cyme usually female; main peduncle 8–15 cm long; branches alternate, up to 5 cm long. Male flowers pedicellate; pedicels 2–3 mm long, green. Calyx lobes oblanceolate, 4 × 2 mm, glabrous; disk of 5 rounded glandular lobes. Stamens 5 × 3; inner longer; filaments connate for half the length, 2–2.5 mm long; outer smaller; anthers oblong, 1 mm long. Female flowers pedicellate; pedicels 3–4 mm long. Calyx-lobes lanceolate, 2–3 mm long; petals free up to the base, oblong-lanceolate, 5 × 2 mm; disk consisting of 3 rounded glandular lobes. Ovary oblong-ovoid, 2 × 1.5 mm, glabrous; styles 3, 1.5 mm long, connate at base; stigma club-shaped. Fruit oblong-ovoid or subglobose, faintly 3-lobed, 13 × 8 mm, glabrous, rugulose with persistent sepals. Seeds ovoid oblong, testa crustaceous, carunculate, brownish black.

Occasional, in wastelands, road sides and scrub forests. Fl. & Fr.: Jun. – Dec. Mobbuchintalapalle (KDP), *KRKS* 36907, 272 (HJCB-E).

Jatropha gossypifolia L., Sp. Pl. 1006. 1753 (*'gossipifolia'*); FBI 5: 363. 1887. var. **elegans** (Pohl) Müll. Arg., in DC., Prodr. 15 (2): 1087. 1866; FAP 2: 873. 1997. *Adenoropium elegans* Pohl, Pl. Bras. Icon. Descr. 1(1): 12. t. 15. 1826.

Local name: *Dundaga chettu, Nepalamu.*

Shrub, somewhat succulent, up to 3 m high; young stem and leaves purplish red; branchlets glandular hairy. Leaves alternate, closely spirally arranged, palmately 3–5-lobed, up to the middle, 5–13 × 10–15 cm, densely hairy on abaxial on midvein and lateral veins, base subcordate, margins with viscid glands, apex acute-acuminate, gland tipped; lobes obovate, broadest at the middle; main nerves 3–5, basally palmate; petioles 4–10 cm long, attachment of petiole to the stem wide, 45° angle to stem; stipules divided into 3-fid, each lobe laciniate, diskoid, glandular hairy; glands 2 mm across at the base of petiole; petiole 4–8 cm long, purplish, glandular hairy. Inflorescence corymbose cyme, villous; main peduncle 3–7 cm long, with glandular and normal hairs; secondary peduncle smaller in length than main peduncle, each branch with ovate-lanceolate bract, first one larger and final one smaller, 0.5-0.8 × 0.2-0.4 cm, margin glandular, ciliate, apex acuminate. Flowers 7–8 cm across, brilliant red; pedicels up to 5 mm long. Male flowers pedicellate; sepals oblong-ovate, 6 × 2 mm, glandular ciliate; petals free, reddish. Stamens 8–12, monadelphous; inner filaments longer, to 2.5 mm long; anthers 1 mm long. Female flowers pedicellate; sepals and petals as in male. Ovary globose, 2 mm in diameter; style shortly connate at base, 1.5 mm long. Fruits oblong-ovoid, 3-lobed, 3-coccous, sparingly hirsutulous, 1.5 mm long, reticulate-verruculose, reddish green, truncate at both ends. Seeds 3, oblong, brownish, carunculate, 10 × 5 mm across.

Common, in wastelands and plains. Fl. & Fr.: Jul. – Dec. Mabbuchintalapalle cross (KDP), *KRKS* & *KRS* 36942.

MICROSTACHYS A. Juss.

Microstachys chamaelea (L.) Müll. Arg., Linnaea 32: 95. 1893. *Tragia chamaelea* L., Sp. Pl. 981. 1753. *Sebastiana chamaelea* (L.) Müll. Arg., in DC., Prodr. 15 (2): 1175. 1866; FBI 5: 475. 1866; FAP 2: 886. 1997.

Perennial or annual subshrubs or herbs, erect, up to 75 cm high; latex milky; branchlets angled, puberulous, white vertical lined. Leaves linear-oblong to lanceolate, 1.5–8 × 0.4–1 cm, scabrid above, glabrous below, base cuneate, margin minutely crenulate-dentate with blackish glands, apex acute or obtuse; lateral veins 15–20 pairs. Petiole up to 3 mm long; stipules 0.5–2 × 0.2–0.5 mm, densely pilose. Inflorescence up to 3 cm long; bracts triangular to lanceolate. Male flowers 1.5 mm across, sessile; bracts minute, campanulate, covered by glands. Calyx 5-lobed, lobes ovate, fimbriate, 1.5–2 mm long. Stamens 3, exserted. Female flowers solitary, axillary in upper leaf-axils, below the male flowers in the spikes, 1 mm across. Calyx 3-lobed, ovate. Ovary 3-lobed, echinate; styles 3, shortly connate, spreading, 2.5 mm long. Fruits ellipsoid, 3-lobed, with 2 vertical rows of soft prickles, 4–5 mm across, light green with a flush of red between warts. Seeds oblong-ellipsoid, barrel-shapped with parallel sides, truncate at base, rounded at apex, 2.5 × 2 mm, brown, yellowish mottled.

Occasional, in foot hills of forestss and weed in agriculture fields. Fl. & Fr.: Jul. – Feb. Idupulapaya (KDP), *KRKS* & *SSR* 39065.

PHYLLANTHUS L.

Key to Species

1a. Trees; fruit indehiscent .... **P. indofischeri**

1b. Other than trees; fruit dry or drupaceous or baccate .... (2)

2a. Fruits more or less fleshy, drupaceous or baccate .... **P. reticulatus**

2b. Fruit dry dehiscent, capsular .... (3)

3a. Annual herbs .... (4)

3b. Perennial shrubs .... **P. pinnatus**

4a. Venation of leaves immersed; plants occurring mostly in rocky areas .... **P. rotundifolius**

4b. Venation of leaves visible at least on the under surface; plants occurring in plains and forests .... (5)

5a. Cymules mostly consisting of one male and one female flower .... **P. amarus**

5b. Cymules otherwise .... (6)

6a. Stipules peltate, cordate or sagittate at base; stem not winged .... (7)

6b. Stipules neither peltate nor cordate or sagittate at base; stem winged .... **P. rheedei**

7a. Anthers connate into a subglobose mass, vertically dehiscent; ovary and fruits smooth .... **P. maderaspatensis**

7b. Anthers free, horizontally dehiscent; ovary and fruits verrucose or muriculate .... **P. simplex**

Phyllanthus amarus Schumach. & Thonn., Beskr. Guin. Pl. 421. 1827; FAP 2: 879. 1997. *P. niruri* auct. non L., 1753; Müll. Arg., in DC., Prodr. 15 (2): 406. 1866; FBI 5: 298. 1887 *p. p.*; FPM 2: 1290. 1925. *P. nanus* Hook. f., Fl. Brit. India 5: 298. 1887.

Local name: *Nelausiri, Nelausirika, Nelavusari.*

Annual erect herb, 15–75 cm high, glabrous; stem herbaceous, simple or branched; leaf-bearing branchlets 2–10 cm long, often scabridulous. Leaves opposite, oblong to elliptic-oblong, 2–10 × 1.2–5 mm, membranous or thinly chartaceous; base obtuse to subtruncate, margin entire, apex acute or subtruncate; lateral veins 4–6 pairs; petiole 0.3–1 mm long; stipules lanceolate, triangular, 0.8–1 × 0.2–0.4 mm, entire. Inflorescence axillary cyme. Flowers unisexual, each cyme consisting of one male and one female flowers; bracteoles linear-lanceolate, 0.5–1 × 0.2–0.4 mm, entire, scarious. Male flowers pedicellate; pedicels 0.3–1 mm long, slender; sepals 5, ovate-oblong or elliptic, 0.2–0.6 × 0.1–0.3 mm, entire, scarious; disk glands 5, minute, stipitate. Stamens 3; filaments united into a slender column, 0.1–0.3 mm long; anthers connate in a mass, vertically dehiscent. Female flowers pedicellate; pedicels 0.5–2 mm long, slender; sepals 5, oblong to ovate-oblong, 0.5–1 × 0.3–0.5 mm, entire and scarious; disk annular, flat, 0.4 mm across, deeply lobed. Ovary 5-lobed, subglobose 0.4–0.6 mm, smooth; styles 3, free, 0.3 mm long, bifid, spreading. Fruits smooth, 2 mm in diameter, 6-lobed, fruiting pedicels 1–2 mm long. Seeds 6, trigonous, with 5 or 6 longitudinal ribs on the back, 1 × 8 mm, brown.

Common, on road sides, wastelands and agriculture fields. Fl. & Fr.: Throughout the year. Thummalapalle (KDP), *KRKS* & *KRS* 33687; Kalasamudram (ATP) *KRKS* & *KRS* 36484.

Phyllanthus indofischeri Bennet, Indian Forester 109: 221 1983. *Emblica fischeri* Gamble, Bull. Misc. Inform. Kew 1925: 330 1925.

Monoecious trees, up to 12 m high; bark smooth when young, fissured with square or rectangular flakes when old, grey; branches terete, glabrous. Scales puberulous, 1.5–2.2 mm long. Floriferous part of branchlets thick with prominent floral and bract scars restricted to lower portion of branchlets. Young branchlets 5–9 cm long, with floriferous portion up to 3.5 cm long. Stipules scaly,1–1.5 mm long, triangular, fimbriate to laciniate along margin, acuminate, pinkish. Leaves alternate, sub-sessile, oblong,1–2.8 × 0.5–1.3 cm, cordate to obtuse at base, entire, acute, coriaceous, glabrous, glaucous beneath; lateral nerves 5–7-pairs, looping along margin, obscure above, prominent beneath. Flowers fascicled, in leafless portion of branchlets; bracts triangular, acuminate at apex, laciniate to fimbriate along margins. Male flowers pedicellate, many, restricted to lower portion of floriferous branchlets. Pedicels filiform, 2–3 mm long. Perianth-lobes 6, imbricate, oblong to obovate,1.5–2 mm long, papery, laciniate along margins, obtuse at apex, glabrous, pale green, with a globose gland at base. Stamens 3, yellow; filaments united into a column, 1–2 mm long; anthers 1–2 mm long, dehiscing vertically. Female flowers sessile, usually 1–3, surrounded by numerous male flowers, restricted to upper portion of floriferous branchlets, protogynous. Perianth-lobes 6, oblong, 2.5–3 × 1.5–2 mm, entire along margins, obtuse at apex, green. Disk fleshy, 6-angled, enclosing three-fourth of ovary, irregularly lobed. Ovary globoid, *ca.* 1 mm across; stylar column short; stylar arms 3, each 6–8 mm long, flat, curved, bifid at apex, glandular puberulous above, glabrous beneath. Fruits drupaceous, globose, 2.5 cm across, pale green; seeds 6, grey.

Rare in dry deciduous forest. Fl.: Jun. – Aug.; Fr.: Sep. – Dec. Batrepalli (ATP), *KRKS*, 441 (HJCB).

Phyllanthus maderaspatensis L., Sp. Pl. 2: 982. 1753; FBI 5: 292. 1887; FPM 2: 1289. 1925; FAP 2: 880. 1997.

Local name: *Nala-usirike.*

Herb or undershrub with woody base, 15–50 cm high, erect or ascending, glabrous branches; branches flattened to terete or often 4-angled. Leaves simple, obovate to oblanceolate, spatulate, 0.5–2 × 0.3–1 cm, chartaceous to thinly coriaceous, often glaucous below, base cuneate-acute, margin entire, apex rounded, truncate or emarginated; lateral veins 4–8 pairs; petiole 1 mm long; stipules triangular-lanceolate, subpeltate at base, entire, 1–2 × 0.5–0.8 mm, scarious and white along margins. Inflorescence axillary cymules. Cymules consisting of 1-4 male flowers and 1 female flower; bracts ovate-lanceolate, 1–1.5 × 0.3–1 mm, peltate, entire. Male flowers pedicellate; pedicels 0.5–1 mm long, filiform; sepals 6, (3 + 3), obovate to obovate-elliptic, 0.7–1 × 0.3–0.6 mm, entire and white-margined, scarious; disk glands 6, up to 0.2 mm across. Stamens 3; column minutely trifid at apex; anthers oblong, 0.1–0.3 mm long, connate into a subglobose mass, vertically dehiscent. Female flowers pedicellate; pedicels 1–1.5 mm long; sepals 3 + 3, orbicular to obovate, 1–1.2 × 0.8–1 mm, entire and scarious at margins; disk star-shaped or annular and lobulate. Ovary subglobose, 0.5 mm in diameter; styles 3, free 0.3–0.5 mm long, shortly bifid at apex. Fruit subglobose, slightly depressed, 1.5–2.5 × 3–3.5 mm, shallowly 3–6-lobed, brown or black; fruiting pedicels 1.5–3 × 0.3–1 mm. Seeds trigonous, verruculose with concentric lines of tubercles.

Common, weed of wastelands, agriculture fields and foot hills of forests. Fl.& Fr.: Mar. – Dec. Mabbuchitalapalle (KDP), *KRKS, KRS, SSR* & *TP* 33671.

Phyllanthus pinnatus (Wight) G. L. Webster, J. Arnold Arbor. 38: 52. 1957; FAP 2: 881. 1997. *Chorisandra pinnata* Wight, Icon. Pl. Ind. Orient. 6: 13. t. 1944. 1853; FPM 2: 1285. 1925.

Local name: *Nalla puli.*

Shrub, bushy, 1–2 m high, monoecious, entirely glabrous; main branches rough, brown with stipule-scars and reduced scaly leaves; leaf-bearing branchlets fascicled, 4–7 cm long, few-leaved, often deciduous. Leaves elliptic, ovate-elliptic, 0.5–2 × 0.4–1.7 cm, membranous, glabrous, glaucous below, base obliquely cuneate, acute, obtuse, margin entire, apex acute, obtuse and apiculate; lateral veins 5–6 pairs, forked near margins; petioles 1–3 mm long, hirsute; stipules linear, subulate,

2 mm long, scarious, deciduous. Inflorescence ramiflorous, unisexual; glomerules many-flowered, often pedunculate; bracts tufted, oblong, obovate-oblong, flabellate or orbicular, 1–2 × 1–1.3 mm. Male flowers at proximal axils; pedicels 3–9 mm long, filiform; sepals 3 + 3, obovate, ovate-orbicular, 0.6–2 × 0.4–1 mm, fimbriate-dentate towards apex, thin; disk annular, 0.5 mm across, 6-lobed. Stamens 6, exserted; filaments 1.5–2.8 mm long, free, filiform; anthers oblong, 0.5–0.8 mm long, longitudinally dehiscent. Female flowers at distal axils; pedicels 0.8–3.7 cm long, filiform. Sepals 6, broadly ovate to elliptic, 2–2.2 × 1.6–1.8 mm, scarious along margins; disk subentire to 6-lobed, 2 mm across, flat. Ovary globose, 1.8 mm in across, shortly stipitate; styles 3, very shortly connate at base, bifid, 1.8–2.2 mm long. Fruit subglobose to globose, 6–8 × 10–11 mm, 3-lobed, crustaceous, wrinkled. Seeds globose, plano-convex, 3 × 2.5 mm, orange-brown.

Occasional, in scrub forests and foot hills of dry deciduous forest. Fl.& Fr.: Feb. – Oct. Mabbuchintalapalle cross (KDP), *KRKS* & *KRS* 36905.

Phyllanthus reticulatus Poir., in Lam. Encycl. 5: 298. 1804; FBI 5: 288. 1887; FAP 2: 882. 1997. *Kirganelia reticulata* (Poir.) Baill., Etud. Gen. Euphorb. 613. 1858; FPM 2: 1294. 1925.

Local name: *Nalla purugudu, Nalapuliki, Phulsar.*

Straggling shrub, bushy, scandent, 1–5 m high, glabrous or pubescent; stem brown, longitudinally striated, wood whitish, hard; leaf-bearing branchlets 3–20 cm long. Leaves alternate, oblong-elliptic, suborbicular or obovate, 1–2.5 × 0.4–2 cm, thinly coriaceous, chartaceous or sometimes membranous, sparsely pilose on major nerves, glabrous above, base rounded, acute or sometimes truncate or subcordate, margin entire, apex obtuse; lateral veins 4–7 pairs; petiole 1–3 mm long, tomentellous; stipules linear, subulate to lanceolate, truncate at base, 0.8–1.5 × 0.3–1 mm, entire, often hardened, spiniform and recurved on older branches. Inflorescences axillary or sometimes on leafless branches, racemes up to 6 cm long; male cymules 2–6-flowered; female cymules 1–6-flowered; proximal cymules often bisexual; bracts triangular or linear-lanceolate, 1–1.5 mm long. Male flowers pedicellate; pedicels 4–10 mm long, glabrous; sepals 5 or 6, entire, 1–2.5 × 8–1.5 mm, thin, the outer broadly elliptic or oblong to obovate, 1 mm broad, the inner suborbicular, 1.5 mm broad; disk glands 5 or 6, pitted, 0.2 mm across. Stamens 5 or 6, unequal; outer filaments 1 mm long, free; inner filaments 1.2 mm long, free above. Anthers oblong-ellipsoid, 0.3–0.5 mm long, erect, vertically dehiscent. Female flowers pedicels 3–8 mm long; sepals 5 or 6, in 2 or 3 series, 1.2–2.5 × 0.7–1.5 mm, entire to wavy and white along margins; inner suborbicular, 1.5 mm across; disk annular, crenate or consisting of 5 or 6 distinct glands, obscurely pitted. Ovary subglobose, 1–1.5 mm in diameter, 5–10 locular, smooth; styles 5–10, up to 0.5 mm long, triangular, free or very shortly connate at base, shortly bifid at apex. Fruits baccate, subglobose to globose, 2–5 × 3.5–5.5 mm, often slightly depressed, unlobed, fleshy, smooth, glabrous. Seeds trigonous, smooth, brown, curved, to 3 mm long.

Common, in foot hills of forests, wastelands and road sides. Fl. & Fr.: Aug. – Apr. Thummalapalle (KDP), *KRKS* & *KRS* 33675.

Phyllanthus rheedei Wight, Icon. Pl. Ind. Orient. 5: 25, t. 1895, f. 1. 1852 (as *rheedii*); FBI 5: 293. 1887; FPM 2: 1298. 1925; FAP 2: 882. 1997. *P. kozhikodianus* Sivar. & Manilal, J. Indian Bot. Soc. 56: 165. 1977.

Local name: *Nalaballi.*

Annual erect herb, 15–60 cm high, entirely glabrous; leaves bearing branchlets alternate, 5-20 cm long, stipules of branchlets 3–4-lobed; lobes lanceolate, to 2.5 mm long, glabrous, acuminate. Leaves alternate, elliptic-oblong, obovate-oblong 0.5–2.5 × 0.4–1.5 cm, glabrous, whitish below, greenish above, thin, base acute-obtuse, margin entire, apex acute; lateral veins 5–7 pairs; petiole 1–1.5 mm; stipule lanceolate or subulate, entire, 1–2 mm long. Inflorescence cyme. Flowers unisexual, male at basal axils on branchlets, female flowers in apical axils; bracts lanceolate, linear to filiform, 1–3 × 0.3–0.6 mm, thin, entire. Male flowers pedicellate; pedicels 0.5–3.5 mm long, filiform; sepals 3 + 3, ovate-oblong or obovate, 0.2–0.8 × 0.1–0.2 mm, margin scarious, midvein green; disk glands 6, minute. Stamens 3; filaments 0.2 mm long, free or connate below for about half way; anthers orbicular, 0.2 mm across, vertically or obliquely dehiscent. Female flowers pedicellate; pedicels 2.5–4 mm long, winged; sepals 3 + 3, ovate-oblong or obovate, 1–1.5 × 0.3–0.5 mm, thin; entire, scarious; disk annular, 0.3 mm across, lobed. Ovary subglobose, 0.5 mm in diameter, smooth; styles 3, free, 0.3–0.6 mm long, bifid, spreading. Fruits subglobose or depressed, 1–1.2 × 1.8–2.5 mm, unlobed or shallowly lobed, smooth; fruiting calyx 3–4 mm across; fruiting pedicels 6–9 mm long. Seeds trigonous, with 5–7 longitudinal ribs on back.

Common, in slopes and foot hills of dry deciduous forests. Fl. & Fr.: Jul. – Dec. Batrepalli (ATP), *KRKS* & *BSS* 39633; 40358; Kalasamudram (ATP), *KRKS* 41871.

Phyllanthus rotundifolius Klein ex Willd., Sp. Pl. ed. 4, 4: 584. 1805; FBI 5: 299. 1887; FPM 2: 1290. 1925; FAP 2: 882. 1997.

Local name: *Gajoppulaku.*

Subsrhub, erect, perennial, sometimes ascending, 0.5–1.5 m high, rootstock thick, branches slender, corky, plant woody, reddish; leaf bearing branchlets arising singly, terete, up to 10 cm long. Leaves orbicular-obovate, 2–14 × 2–10 mm, coriaceous, often glaucous beneath; base acute, margin entire, apex acute to apiculate or mucronate; venation immersed; petioles 0.5–1.5 mm long; stipules subulate-filiform, deltoid or ovate-triangular, 1–2 × 0.4–0.6 mm, scarious at margin. Inflorescence axillary cyme. Male cymules 1–3-flowered at proximal axils; female cymules 1-flowered at distal axils; bracts ovate to ovate-lanceolate, 1 × 0.4 mm, slightly undulate to entire along margins. Male flower pedicel 0.5–1 mm long; sepals 3 + 3, oblong-ovate, entire, 0.5–0.8 × 0.4–0.6 mm; disk glands 6, subglobose, 0.2 mm across, pitted. Stamens 3; filaments united into a column, 0.2 mm long, trifid at apex; anthers subglobose 0.2 mm across, horizontally dehiscing. Female flowers pedicels 1–1.5 mm long; sepals 3 + 3, ovate-oblong to oblong-elliptic, entire, 1.5–2 × 1–1.2 mm, scarious; disk cupular, annular, 0.5 mm across, slightly lobed. Ovary subglobose, 0.5 mm in diameter, smooth; style 3, free, 0.2 mm long, bifid at apex. Fruits subglobose, slightly depressed, 1–2 × 1.5–2.5 mm, shallowly to deeply

3-lobed, smooth; fruiting pedicels 3 mm long. Seeds trigonous, 1 mm long, with 5–7 longitudinal ribs on back, brown.

Common, in foot hills, rocky areas of dry deciduous forests. Fl. & Fr.: Oct. – Jan. Rachakuntapalli (KDP), *KRKS* 37632.

Phyllanthus simplex Retz., Observ. Bot. 5: 29. 1789; FBI 5: 295. 1887; FPM 1289. 1925. *P. virgatus* auct. non G. Forst., 1786: Chakrab. & N. P. Balakr., J. Econ. Taxon. Bot. Addit. Ser 9: 98. 1992.

Local name: *Uchichi, Usirika.*

Erect or decumbent herb, 15–60 cm high, glabrous; stem with normal distichous leaves, sparsely branched, branches winged, flattened, smooth; root somewhat stocked. Leaves alternate on branches, linear-oblong to lanceolate, 0.5–2.5 × 0.2–0.9 cm, glabrous, dark green or pinkish above, base rounded to truncate, somewhat cordate, margin entire, minutely revolute, apex rounded; lateral veins 4–10 pairs; petiole 0.4–1 mm long; stipules ovate, linear-lanceolate, triangular-acuminate or deltoid, acuminate, 1–2 × 0.5–1 mm, scarious, entire or denticulate along margin. Inflorescence axillary cyme; male cymules 2–4-flowered; female cymules 1-flowered; bracts 0.5–1 mm long, cordate. Male flowers pedicellate; pedicel 1.5–5 mm long; sepals 3 + 3, elliptic-obovate to oblong or orbicular, 1–1.5 × 0.5–1.2 mm, scarious and entire or denticulate at margin; disk glands 6, obovate-oblong or flabellate, 0.1–0.3 mm across. Stamens 3; filaments free, 0.3–0.5 mm long; anthers subglobose, 0.2 mm across, horizontally dehiscing. Female flowers pedicellate; pedicels 5–12 mm long, sepals 3 + 3, oblong-elliptic to ovate or obovate, 0.6–2.5 × 0.3–1.5 mm, scarious, entire or denticulate; disk annular, slightly cupular, 0.8–1 mm across, entire or undulate. Ovary subglobose, 0.6–1.5 mm, 3-lobed, papillate; styles 3, free, 0.5–1 mm long, bifid. Fruit subglobose, slightly depressed, 1.8–2.5 × 2.5–5 mm, 3-lobed, papillate-verruculose, fruiting pedicels up to 12 mm long, often reflexed. Seeds trigonous, 1.2–2 mm long, verruculose with concentric line of tubercles, brownish black.

Common, in scrub forests and dry deciduous forests. Fl. & Fr.: Jun. – Nov. K. K. Kottala (KDP), *KRKS* & *KRS* 36424.

RICINUS L.

Ricinus communis L., Sp. Pl. 1007. 1753; FBI 5: 457. 1887; FPM 2: 1335. 1925.

Local name: *Amudam.*

Monoecious, evergreen shrub, to 3 m high, glabrous; stem hollow, white powdery. Leaves alternate, palmatifid, 6–10-lobed, each 1-nerved with many lateral veins, 10–18 × 3–7 cm; lobes lanceolate; base peltate; margin serrate, apex acuminate; petiole to 20 cm long, hollow. Racemes terminal, paniculate, 15–30 cm long. Male flowers at base of raceme, 1.5 cm across; petals 3–5-lobed, yellowish green. Stamens numerous, crowded and connate into 8 mm long cluster of stipitate phalanges; anthers yellow, ovoid. Female ones at apical region of raceme, 6 mm across, green; pedicels 2.5–4 mm long, elongate in fruits; sepals spathaceous, split into 5–7 unequal lobes. Ovary globose, 4–5 mm across, densely echinate with soft spines, each ending in a translucent stinging hair at apex; styles 3, entire or bifid, stigma 4–6 mm long, papillate, red. Fruit schizocarpic, trilobed, ellipsoid to subglobose, 1.2–2.5 × 1.5–2 cm, reddish brown to blackish purple, marked with various shades of white, grey or brown, covered with 3–6 mm long, brown soft spines. Seeds ellipsoid, 9–14 × 5–6 mm, shiny, dark brown or grey with a large caruncle.

Cultivated for its oil seed, often as an escape. Fl. & Fr.: Oct. – Feb. Somulavaripalli (ATP), *KRKS* & *KRS* 37819.

SAUROPUS Blume

Sauropus bacciformis (L.) Airy Shaw, Kew Bull. 35: 685. 1980; FAP 2: 885. 1997. *Phyllanthus bacciformis* L., Mant. Pl. 294. 1771. *Agyneia bacciformis* (L.) A. Juss., Euphorb. Gen. 24, t. 6. 1824; FBI 5: 285. 1887. FPM 2: 1284. 1924.

Prostrate or ascending herb, up to 50 cm long, entirely glabrous, branchlets quadrangular to semiterete. Leaves alternate, variable, succulent, broadly to narrowly oblong, elliptic or elliptic-oblong, 0.5–2.6 × 0.3–1 cm, chartaceous, glabrous, base acute, rounded, margin entire, apex acute to apiculate; lateral veins 3–6 pairs; petioles 0.2–0.6 mm long; stipules linear to triangular, 1–2 mm long. Inflorescence axillary cyme. Male flowers pedicellate; pedicels 2.5–6 mm long. Calyx flat, star-shaped, 1–1.5 mm in diameter; segments 6, shortly connate, subflabellate, emarginated to bilobulate. Stamens 3, connate; filaments 0.5 mm long; anthers minute, horizontal. Female flowers pedicellate; pedicels 0.5–1 mm long. Calyx 3–3.5 mm across; segments elliptic, ovate, spatulate or obovate, acute or acuminate, apiculate, 1–2.5 × 0.7–1.3 mm. Ovary ovoid, 1 mm in diameter; style broad, erect, minute, bifid. Fruits ovoid, ellipsoid or subglobose, 4.5–6 × 3.5–7 mm, unlobed, thin-walled; pedicels 0.5–1 mm long. Seeds 3-angular oblong, acuminate, thin lower side, curved, whitish brown, glabrous.

Rare near water bodies, especially in saline soils at riparian areas. Fl. & Fr.: Jub. – Sept. Kalasamudram (ATP), *KRKS* 39531.

TRAGIA L.

Key to Species

1a. Leaves palmately 3-lobed; plant climbing or twining .... **T. plukenetii**

1b. Leaves unlobed; plant erect .... **T. involucrata**

Tragia involucrata L., Sp. Pl. 980. 1753; FBI 5: 465. 1888; FPM 2: 1332. 1925; FAP 2: 888. 1997.

Local name: *Chinna-duradagunta, Doola-gondi, Durada-gondi, Teega-durada gunta.*

Herbs, erect, up to 1.5 m high; branches densely hispid. Leaves successively smaller upwards, ovate to sub orbicular or ovate-lanceolate, elliptic-lanceolate, 5–10 × 1.5–5 cm, membranous to chartaceous, scattered hispid or hirsute above, densely below, base obtuse or acute, margin irregular serrate or dentate or denticulate, apex acute or acuminate, caudate; basally 3-veined, lateral veins 3–7 pairs, predominantly semicraspedromous or often craspedromous; petiole 2–6 mm long; stipules lanceolate, subhastate at base, 2–4 × 1–2 mm, caducous. Racemes 1.5–3 cm; peduncles 0.5–2 cm long; sepals 3, filaments dilated at base; anthers oblong to suborbicular, 0.3 mm across. Female flowers subsessile; sepals 3 × 1 mm long, 3-fid at apex. Fruits subsessile, depressed-globose, deeply 3-lobed, 3–5 × 7–10 mm, evanescently hispid; fruiting sepals 5–8 × 2–3 mm; lobes 4–6, linear, 1.5–3 mm long, hispid. Seeds globose, testa crustaceous.

Common, along road sides, foot hills of forests and wastelands. Fl. & Fr.: Aug. – Feb. Kalasamudram (ATP), *KRKS* & *BSS* 39132; Batrepalli (ATP), *KRKS* & *BSS* 39584.

Tragia plukenetii Radcl.-Sm., Kew Bull. 37: 688. 1983; FAP 2: 889. 1997. *T. cannabina* L. f., Suppl. Pl. 415. 1781, *nom. illeg.*; FPM 2: 1333. 1925. *T. involucrata* var. *cannabina* (L. f.) Müll. Arg., in DC., Prodr. 15: 944. 1866; FBI 5: 465. 1888.

Local name: *Tella-duradagunta.*

Climbing shrub, 2 m long, sparsely hispid with stinging hairs; branches scattered hispid, glabrous with age. Leaves palmately 3-lobed, 3–10 × 1.5–7 cm, chartaceous, sparsely hispid above, hirsute on nerves below; lobes oblong or ovate-lanceolate, margin serrate-denticulate, apex acute-acuminate; basal veins 3; central lobes 0.5–4 × 0.3–1.5 cm, lateral primary veins entering into lateral lobes; lateral veins 3–7; predominantly craspedromous; petiole 3–20 mm long. Racemes terminal on lateral shoots or leaf-opposed, 2–5 cm long, 1–3 female flowers at base; peduncles up to 3 cm long. Male flower pedicels 1 mm long; sepals widely ovate to suborbicular, 1 × 0.7 mm, glabrous or thinly pubescent outside. Stamens 3; filaments 0.3 mm, long; anthers oblong, 0.3 mm long. Female flower pedicels 1 mm long; sepals 1–2 mm long, fimbriate, thinly hispid outside. Ovary 2 mm in diameter, hirsute; styles 3, up to 3 mm long, trifid at apex. Fruits depressed, 3–5 mm across, deeply 3-lobed, evanescently hispid; lobes 4–10, linear, 1–2 mm long, fulvous, hispid or hirsute. Seeds globose, testa crustaceous.

Common, in dry deciduous forests near streams. Fl. & Fr.: Dec. – Mar. Mabbuchintalapalle (KDP) *KRKS* & *KRS* 36940; Bhumaiahgaripalli (KDP), *KRKS* & *KRS* 37612.

ULMACEAE

HOLOPTELEA Planch.

Holoptelea integrifolia (Roxb.) Planch., Ann. Sci. Nat. Bot. (Ser. 3) 10: 266. 1848; FBI 5: 481; 1888; FAP 2: 892. 1997. *Ulmus integrifolia* Roxb., Pl. Coromandel. t. 78. 1798.

Local name: *Nemalinara, Thapasi, Nauli, Pulari, Pedda navili.*

Polygamous, deciduous tree, up to 15 m high; bark whitish-grey, pustular, smooth, exfoliating in quadrilateral scales; branchlets terete, tender parts pubescent. Leaves chartaceous, elliptic-ovate, 8–10 × 4–8 cm, coriaceous, appressed-pubescent below, puberulous along nerves above when young, base rounded, subcordate, margin entire, apex abruptly acuminate; secondary veins 5–6 pairs; petiole to 1.3 cm long; stipules scarious. Inflorescence axillary fascicle or raceme, 2 cm long, at the scars of the old shoots; bracts scaly. Flowers yellow. Male flowers and bisexual flowers mixed in raceme. Tepals 4, free, concave, 1.5 mm, imbricate, unequal, pubescent, ciliate, obtuse. Stamens 7–9, biseriate, filament 1.7 mm; anthers slightly curved, 1.3 mm, hairy, introrse. Ovary compressed-ovate, 2 × 1.5 mm, stalk elongating in fruit; style-arms 2, stigmatose. Fruit a dry, winged, compressed samara, 1 × 0.7 cm, wings membranous, 3 cm across, yellow, reticulate. Seed flat, exalbuminous.

Common, in dry deciduous forests, open lands and foot hills of forests. Fl. & Fr. Dec. – Aug. Thimmammamarimanu (ATP), *KRKS* & *KRS* 37933.

MORACEAE

FICUS L.

Key to Species

1a. Receptacles in dense clusters or fascicles on tubercles or on leafless branches from the trunk or larger branches when mature .... (2)

1b. Receptacles in pairs or 2–6 or solitary in the axils of leaves .... (3)

2a. Receptacles in fascicles on long peduncles; figs red when ripe ...... **F. recemosa**

2b. Receptacles dense on leafless branches on short peduncles; figs yellowish .... **F. hispida**

3a. Leaf apex caudate .... (4)

3b. Leaf apex rounded-obtuse, acute .... (7)

4a. Leaf base wide, truncate-subcordate; figs <8 mm across .... (5)

4b. Leaf base not wide, acute; figs > 8 mm across.- apex cusp to 1 cm long .... **F. amplissima**

5a. Leaf apex produced into a linear lanceolate cusp; figs other than yellow and not dotted .... (6)

5b. Leaf apex not produced into a long cusp, abruptly acuminate; figs whitish-yellow, dotted when ripe; leaf ovate-oblong, base truncate .... **F. tsjakela**

6a. Leaf cusp to 8 cm long; figs sessile, pink-purple or black when ripe; bracts cupular, lobes inconspicuous .... **F. religiosa**

6b. Leaf cusp to 1.5 cm long; figs pedunculate to 2 mm long, reddish when ripe; bracts cupular, lobes ovate .... **F. arnottiana**

7a. Leaf tomentose, pubescent, basal veins 5 .... (8)

7b. Leaf glabrous, basal veins 3 .... **F. microcarpa**

8a. Figs to 2 cm across, red when ripe, puberulous; aerial roots strong .... **F. benghalensis**

8b. Figs to 0.9 cm across, brown, tomentose; aerial roots not as above ..... **F. mollis**

Ficus amplissima Sm., in Rees, Cyclop. 14: n. 68. 1810; FAP 2: 896. 1997. *F. tsiela* Roxb. ex Buch.-Ham., Trans. Linn. Soc. London 15: 149. 1827; FBI 5: 515. 1888; FPM 3: 1362. 1928.

Local name: *Juvvi, Pedda juvvi*

Spreading, deciduous tree, up to 20 m high; trunk cylindrical, bark greenish-grey, smooth, latex milky; wood whitish. Leaves alternate, ovate-oblong, 7–12 × 5–8 cm, glabrous, coriaceous, base truncate to rounded, acute, margin entire to undulate, apex caudate acuminate; lateral veins 8–12 pairs, flattened above, slightly raised below, inconspicuous intercostal veins; petiole 3.5–5 cm, gland at apex on lower side or at base of lamina; stipule lanceolate, to 2.5 cm long, covering leaf at young stage, 2–2.5 × 0.5–0.8 cm. Figs monoecious, axillary, subsessile, 1–1.8 cm diameter, in leaf axils, depressed globose or pyriform, apex truncate to broadly rounded, glabrous, smooth, purple when ripe; basal bracts 3, broadly ovate, 3 mm across, obtuse; orifice circular, closed by 3, flat, apical scales; internal bristles 0. Tepals 3, broadly lanceolate, 1 mm long, brownish, margin scarious. Male flowers distinct. Stamen 1, included; filament 0.4 mm long; anther oblong, parallel, unequal, to 0.5 mm across. Female flowers sessile. Ovary bean-shaped, white, 1 mm across; style filiform, 2 mm long. Gall flowers similar, pedicellate. Achenes smooth, 2 mm long.

Common, in foot hills of forests, village outskirts and road sides. Syconia: Jan. – May. Mabbuchintalapalle (KDP), *KRKS, KRS, SSR* & *TP* 33660.

Ficus arnottiana (Miq.) Miq., Ann. Mus. Bot. Ludge.-Bat. 3: 287. 1867; FBI 5: 513. 1888; FPM 3: 1363. 1928. *Urostigma arnottianum* Miq., Hook. Lond. J. Bot. 6: 564. 1847.

Local name: *Kalla-ravi.*

Tree, up to 7 m high, shining, glabrous; bark pale, smooth; wood white. Leaves alternate, ovate or elliptic-ovate, 7–15 × 5–12 cm, subcoriaceous, glabrous, shiny, base cordate, margin entire to slightly undulate, apex acuminate or sub caudate-cuspidate, 1–1.5 cm; veins 5 from base, laterally 6–8 pairs, intercostals transverse, zigzag in the middle, flattened above, raised below; petiole 7–12 cm, glands below the base of lamina; stipules 4–6 cm long. Figs in axillary pairs or crowded near apex, subsessile or pedunculate, to 2 mm long, globose, 5–7 mm across, reddish when ripe; orifice plane; bracts 3, ovate, concave, 1 × 2.5 mm, obtuse, caducous; internal bristles 0. Tepals gamophyllous, 3–4-lobed, brownish red, glabrous. Male flowers around the orifice, sparsely scattered in the interior of fig. Stamen 1, subsessile; anthers 2-celled, ovate-oblong, 0.3 mm across, dehiscing longitudinally. Female flowers creamy, some what reddish on stylar side. Ovary depressed-globose, 0.4 mm; style filiform, 1.5 mm long. Gall flowers stalked. Achenes smooth.

Rare in rocky areas in dry deciduous forests. Syconia: Aug. – Dec. Kalasamudram (ATP), *KRKS* & *BSS* 39198; Near Rathnachalam Anjaneyaswamy temple (ATP), *KRKS* & *BSS* 39662.

Ficus benghalensis L., Sp. Pl. 1059. 1753; FBI 5: 499. 1888; FPM 3: 1361. 1928; FAP 2: 897. 1997.

Local name: *Marri manu, Udala marri.*

Evergreen tree, up to 25 m high, 50 m canopy, branches spreading, aerial roots form as pillars from branches, bark greyish white, smooth; wood grey-white; young branches pubescent, latex milky. Leaves alternate, elliptic-ovate, 10–17 × 7–12 cm, coriaceous, glabrous above, pubescent below, base rounded, margin entire, apex obtuse, to subacute; laterally 5 or 6 veins arise from base, lateral veins 3–6 pairs, flattened above, raised below, intercostals transverse, zigzag in the middle; petiole to 4 cm long, glabrous, glands below the base of lamina; stipules deciduous, leaving annular scars, 2 × 1.5 cm. Figs monoecious, 1 or 2, axillary, depressed-globose, 1.5–2 cm across, puberulous without, sessile, dark pink or red when ripe; bracts 4–5, cupular, 6 mm across, shortly connate, obtuse, persistent; orifice subumbonate; internal bristles 0. Tepals 3–5, shortly connate, 2 mm across, brownish, glabrous. Male flowers dispersed with female. Stamen 1; filament 0.8–1 mm long; anther oblong, parallel, unequal, to 1 mm across, shortly mucronate; connective brown. Female flowers sessile. Ovary obovoid, glabrous, 1.5 × 1 mm, dark brown on stylar side; style erect or curved, 2 mm long, tapering. Gall flowers similar to female flowers, pedicellate. Achenes globose-ellipsoid, 2 × 1.5 mm, dark brown.

Common, on road sides, villages and near temples. Syconia: May – Oct. Bhumaiahgaripalle (KDP), *KRKS* & *KRS* 37601.

Ficus hispida L. f., Suppl. Pl. 442. 1781; FBI 5: 522. 1888; FPM 3: 1367. 1928; FAP 2: 899. 1997.

Local name: *Kakimarri, Brahma medi, Bomma medi.*

Small tree, up to 5 m high, bushy; branchlets hispid, bark dark grey, rough; wood grey, latex milky to reddish. Leaves alternate, obovate-elliptic, oblong, 10–18 × 5–10 cm, coriaceous, glaucescent above, scabrid on either side, hispid below, base truncate, cordate; margin entire, apex acute or obtuse; basal veins 4–5, lateral veins 6 pairs, intercostals transverse, prominent below; petiole with a sub nodal gland, to 4 cm long; stipules lanceolate, 1.5 cm long. Figs dioecious on special shoots and on branchlets, depressed-globose, base narrowed, 1.5–2.5 cm across, ripening yellow, faintly ribbed, hispid to sticky-pubescent without; peduncle stout, 0.5 cm long; bracts triangular, to 2 mm long; orifice slightly raised, closed by 5–6 apical bracts; internal bristles 0. Perianth cupular, truncate, to 2 mm across. Tepals 0, glabrous. Male flowers ostiolar, 2-seriate. Stamen 1, subsessile or stalked. Ovary depressed-globose, 0.7 mm across; style 1.5 mm long, hairy, tip clavate. Gall flowers similar, but larger, distinctly stalked. Achenes lenticular, 1.5 mm, keeled, with prominent hilum.

Common, on road sides, foot hills of forest and field bunds. Syconia: Oct.- Mar. Kalasamudram (ATP), *KRKS* & *BSS* 39522; Vempalli (KDP), *KRKS* & *SSR* 39048.

Ficus microcarpa L. f., Suppl. 442. 1781; FAP 2: 900. 1997. *F. retusa* auct. non L.; FBI 5: 511. 1888.

Local name: *Juvi, Konda-juvi, Verra-juvi.*

Evergreen tree, up to 15 m high; aerial roots few; bark brown, nearly smooth; wood light reddish-grey; branchlets glabrous. Leaves coriaceous, elliptic or obovate, 4–8 × 2.5–4.5 cm, glabrous, shining, base cuneate-acute, margin entire, apex rounded to retuse; basal veins 3, lateral veins 12–19 pairs, closely pinnate, raised on either side, intercostals 0, secondary lateral veins as prominent as primary veins; petiole to 1.5 cm long, glandular at apex below, canaliculated, glabrous; stipules ovate-lanceolate, to 1 cm long. Figs monoecious, 6–8 mm across, sessile, axillary, paired or solitary, globose, glabrous, yellowish-red when ripe; bracts 3, ovate, elliptic, boat-shaped, 2 mm across, obtuse, persistent; orifice plane or slightly raised, closed by 3, flat, apical bracts; internal bristles minute, sparse. Tepals 3–4, free, ovate-lanceolate, 1.5 mm long, brownish, glabrous, Male flowers dispersed. Stamen 1, hardly exserted; filament 0.3 mm; anther unequal, ovate-oblong, 0.7 mm, mucronate. Female flowers sessile. Ovary ovoid-globose, red-brown, 1 mm across; style 1.5 mm long, tapering. Gall flowers similar to female flowers, pedicellate. Fig wall fleshy and smooth. Achenes smooth.

Occasional, in dry deciduous forests. Synconia: Dec. – Feb. Kalasamudram (ATP), *KRKS* 52026.

Ficus mollis Vahl, Symb. Bot. 1: 82. 1790; FAP 2: 900. 1997. *F. tomentosa* Roxb. ex Willd., Sp. Pl. 4: 1136. 1806; FBI 5: 501. 1888; FPM 3: 1361. 1928.

Local name: *Jivi, Kal-juvi.*

Evergreen tree, up to 15 m high; bark greenish-white, smooth, scaly; wood white; young branches densely, yellowish tomentose; aerial roots several from spreading branches, not used for support. Leaves subopposite, often in unequal pairs, broadly oblong to elliptic-ovate to pandurate, 10–15 × 6–10 cm; coriaceous, glaucescent above, scabrid on either side, hispid below or densely rust-grey tomentose beneath, base cordate with a gland at the base of the midrib beneath, margin entire, apex acute or obtuse; basal veins 3, lateral veins 6 pairs, impressed above, prominent below; petiole with a subnodal gland, to 4 cm long; stipules ovate-lanceolate, 1 cm long, hispid. Figs monoecious on branchlets, sessile, axillary in pairs, rusty tomentose, depressed-globose, base narrowed, 1.5–2 cm across, ripening yellow, faintly ribbed; bracts 3, orbicular, concave 4 mm across, velvety-tomentose; orifice circular, slightly raised, closed by 3 apical bracts; internal bristles sparse to abundant. Perianth free, truncate, to 2 mm across. Tepals 3–4, ovate-lanceolate, 1 mm long, brown, glabrous. Male flowers dispersed subsessile. Stamen 1, filament 0.3 mm long; anther oblong, 0.7 mm across. Female flower sessile. Ovary globose, 1 mm across, brown; style 1.5 mm, tapering. Achenes lenticular, 1.5 mm smooth.

Common, in dry deciduous forests near rocks and streams. Syconia: Through the year. Rachakuntapalle (KDP), *KRKS* & *SSR* 39015.

Ficus racemosa L., Sp. Pl. 1060. 1753; FAP 2. 902. 1997. *F. glomerata* Roxb., Pl. Coromandel 2: 13. t. 123. 1799; FBI 5: 535. 1888; FPM 3: 1364. 1928.

Local name: *Medi, Moida.*

Evergreen tree, up to 20 m high; latex milky; bark greenish smooth, brownish-red, scaly, peeling off; wood soft, smooth, branchlets white-pilose. Leaves alternate, oblanceolate, 9–20 × 3–6 cm, subcoriaceous, glaucous above, glabrous below, base rounded to acute, margin entire, apex acute; lateral veins 8–10 pairs, somewhat flattened and prominent on both sides, intercostals obscure; petiole not articulate, glandular at the basal nerve axils, 3–5 cm long; stipules lanceolate, 1–1.5 cm long. Figs monoecious on separate branches, cauliflorous, subglobose, 1.5–2.5 cm across, white-pilose, walls thick, soft, ripening purplish red, ribbed, base tapering, apex truncate or broadly rounded; peduncle to 7 mm long; basal bracts ovate-triangular, to 2 mm across, obtuse, persistent; orifice subumbonate or slightly sunken, closed by 5–6 bracts; internal bristles 0. Perianth shortly cupular. Tepals 3–5, ovate-lanceolate, 1.5 mm across, scarious, lanceolate, acute. Male flowers ostiolar; peduncle to 3 mm long. Stamens 2, exserted; filaments 1 mm, connate below; anthers oblong, parallel, 0.8 mm across. Female flowers stalked. Ovary stalked, brown, 1.5 mm long; style to 3 mm long. Gall flowers similar to female, long-stalked; style short. Achenes lenticular, 1 mm, obscurely keeled.

Common, in villages, outskirts and agriculture lands. Syconia: Through the year, especially Mar. – Jul. Mabbuchintalapalli (KDP), *KRKS, KRS, SSR* & *TP* 33659.

Ficus religiosa L., Sp. Pl. 1059. 1753; FBI 5: 513. 1888; FPM 3: 1363. 1928; FAP 2: 902. 1997.

Local name: *Raavi.*

Deciduous, fast growing tree, up to 20 m high; often epiphytic when young; bark greenish, smooth; wood greyish-white, fibrous; branchlets yellowsh grey, glabrous. Leaves alternate, broadly ovate, cordate, 5–13 × 4.5–12 cm, coriaceous, glabrous, base truncate or somewhat cordate, margin entire, apex abruptly acuminate, cuspidate to 8 cm long, linear-lanceolate; basal nerves 3, lateral nerves 8–10 pairs, raised on either side, inter costals zigzag; petiole to 12 cm long, articulate, glandular at the apex below; stipules ovate-lanceolate, to 1.5 cm long. Figs monoecious, axillary, paired, sessile, obovoid or globose, 4–6 mm across, ripening pink, purple or black; fig-wall thick; basal bracts cupular, to 4 mm long; margin truncate, puberulous, persistent; orifice umbonate, closed by 3 apical bracts; internal bristles 0. Male flowers ostiolar, sessile. Tepals 2, free, ovate-lanceolate, 0.7 mm across. Stamens 1; filament 0.2 mm long; anthers oblong, parallel, 0.7 mm long. Female flower sessile. Tepals 3–4, free, linear lanceolate, 1 mm long, brownish, glabrous. Ovary ovoid-oblong, 1 mm across, red-brown; style 1.5 mm long, dilated above. Gall flowers similar to female. Achenes smooth.

Common, along road sides, near temples and foot hills of forests. Synconia: May – Aug. Batrepalli (ATP), *KRKS* 42019.

Ficus tsjakela Burm.f., Fl. Indica 227. 1768; FAP 2: 904. 1997; FBI 5: 514. 1888; FPM 3: 1362. 1928.

A large tree without aerial roots, bark rough, dark-colored. Leaf scales of the young branches large, linear-lanceolate, flaccid, caducous, 7.5 – 13 cm long. Petiole 2.5–5.5 cm long, slender, stipule small, ovate-lanceolate, 1.3 cm long. Leaves reddish at young, later dark green, glossy above, coriaceous, ovate-oblong, 7.5–19 × 4–7.5 cm, nerves 3–5 at base, 7–10 laterals, base rounded or truncate, margin entire, apex shortly and abruptly acuminate. Figs in clusters or 2–6, on very short crowded tubercles, in the axils of the leaves, depressed-globose, 6 mm dia., whitish-yellow, dotted when ripe. Basal bracts 3, broad, deeply 2-fid. Male flowers few, near the mouth of the receptacle, sessile. Sepals 3, ovate, shorter than the single stamen. Filaments as long as the anther (or longer), anther ovate. Gall and fertile flowers alike except as regards the contents of the ovary. Sepals 3–4. Fertile achenes obovate. Style elongate; stigma cylindric.

Very rare, in deciduous forests. Kalasamudram (ATP), *KRKS* 39110.

CERATOPHYLLACEAE

CERTOPHYLLUM L.

Ceratophyllum demersum L., Sp. Pl. 2: 992. 1753; FAP 2: 912. 1997; FBI 5: 639. 1888; FPM 3: 1391. 1928.

Submerged, aquatic herb to 30 cm long, stem free floating, alternate branching, branches brittle, green, glabrescent, internodes to 5 cm long. Leaves 7–12 in a whorl at node, linear lanceolate, 2–3 cm long, margin denticulate, apex acute. Flowers not seen. Nutlets ellipsoid, black when dry, 4 × 3 mm, furnished with 3 spurs. Seed globose, 3 mm across; testa stony.

Occasional in small ditches of dry deciduous forest. Fr.: Feb. – Mar. Batrepalli forest (ATP), *KRKS* & *SSR* 37900.

MONOCOTYLEDONS

HYDROCHARITACEAE

Key to Genera

1a. Plants caulescent, spathes sessile .... **Hydrilla**

1b. Plants acaulescent, spathes peduncled .... (2)

2a. Leaves linear, ribbon-like; stamens 1–3 .... **Vallisnaria**

2b. Leaves oblong-orbicular; stamens 6–9 .... **Ottelia**

HYDRILLA L. C. Rich.

Hydrilla verticillata (L. f.) Royle, Ill. Bot. Himal. t. 376. 1839; FBI 5: 659. 1888; FPM 3. 1396. 1928; FAP 3: 926. 1997. *Serpicula verticillata* L. f., Suppl. Pl. 416. 1781.

Acaulescent, submerged herb, to 30 cm long. Leaves various, whorled, 3–6 per node, basal linear-oblong, upper ovate-lanceolate, 0.6–1.5 × 0.2–0.4 cm, membranous,

Plate 19

A. *Ceratophyllum demersum*, B. *Molineria trichocarpa*, C. *Commelina diffusa*, D. *C. subulata*, E. *Najas graminea*, F. *Eriocaulon quinquangulare*

base rounded, margin entire, apex apiculate, scaberulous. Flowers unisexual; sepals 3; petals 3. Male flower: solitary, pedicillate; stamens 3. Female flower: Solitary, sessile; ovary oblong, beaked, to 3 cm long, style 3-fid. Fruit subulate. Seeds 5, oblong.

Common, in stagnant water. Fl. & Fr.: Sep. – Jan. Kalasamudram (ATP), *KRKS* 41832.

OTTELIA Pers.

Ottelia alismoides (L.) Pers., Syn. Pl. t. 400. 1805; FBI 5: 662. 1888; FPM 3:1398. 1934; FAP 3:927. 1997. *Stratiotes alismoides* L., Sp. Pl. 535. 1753.

Perennial, acaulescent, erect herb to 30 cm; submerged or partially floating. Leaves radical, heteromorphous, submerged and floating; submerged leaves oblong, floating leaves ovate-orbicular, 5–15 × 3–10 cm, chartaceous, base cuneate, margin entire, apex obtuse, glabrous. Flowers solitary, white with yellow spots, unisexual and/or bisexual; spathe elliptic-ovate, greenish, 6-winged, apex 2-lobed; peduncle 10–50 cm long, 3-angled. Flowers solitary, sessile, sepals 3, linear, persistent; petals 3, white with yellow spots, oblong, stamens 9–15, ovary oblong, beaked, 6–9-celled, ovules arranged on parietal placentation, styles 10–15, bifid. Fruits ellipsoid, glabrous; seeds many.

Common, throughout in water sources. Fl. & Fr.: Oct. -Mar. Kalasamudram (ATP), *KRKS & BSS* 39872.

VALLISNERIA L.

Vallisneria natans (Lour.) H. Hara, Jap. Bot. 49:136. 1974; FAP 3. 928. 1997. *Physkium natans* Lour., Fl. Cochinch. 663. 1970. *Vallisneria spiralis* auct. non L. 1753; FBI 5: 660. 1888 *p.p.*; FPM 3:1369. 1928.

Acaulescent, stoloniferous herb. Leaves radical, linear, 10–30 × 0.5–1cm, membranous, base narrow, margin denticulate, apex obtuse, glabrous. Flowers unisexual. Male flower: minute, many in spathe; spathe oblong. Female flower: solitary on long filiform spiral scape; sepals 3; petals absent. Ovary narrow. Fruit linear, to 2 cm long, many-seeded.

Occasional, in stagnant water. Fl. & Fr.: Sep. -Mar. Sanevaripalli (ATP), *KRKS* 40163.

ORCHIDACEAE

EULOPHIA R. Br.

Eulophia epidendraea (J. Koenig ex Retz.) C.E.C. Fischer, Fl. Madras 3: 1434. 1928. *Serapias epidendraea* J. Koenig ex Retz., Observ. Bot. 6: 65. 1791. *Eulophia virens* (R. Br.) Spreng., Syst. Veg. 3: 720. 1826; FBI 6: 1. 1894.

Local name: *Damma gadda.*

Perennial tuberous herb, rhizome green, rarely pseudobulbous, emerging above ground. Leaves narrow to 1 cm width, grass like, linear, to 30 cm long, glabrous, 1-ribbed, tip acute. Scape lateral, erect, to 1 m long, sheath oblong, to 5 mm long, connate at base, up to 2 mm, branched, axis glabrous, slender; bracts ovate, to 3 mm long, acute at apex. Flowers in raceme, 1.5–2 cm long, green; pedicel to 5 mm long. Sepals 3, spatulate, to 1.2 × 0.3 cm, mucronate-acuminate at apex, usually spreading, nerves red. Petals elliptic-oblong, spatulate, subequal, obtuse at apex, 0.8–1 × 0.2–0.3 cm, nerves red. Lip obovate, oblong, as long as petals, white, adnate to the foot of the column, nerves and crests red. Column 0.3 cm long, top oblique, entire, rarely lobed, sometimes winged; lateral lobes erect, embracing the column; mid-lobe curved, disk ridged with 5-crested nerves, base spurred; spur to 4 mm long, curved, blunt at apex, foot ciliate, red. Anther terminal, 2-celled; pollinia 2, attached by a strap to a diskoid gland. Ovary angular, 3-locular, to 1.7 cm long. Fruit oblong, to 3 cm long, furrowed, ends tapering. Seeds minute like dust, white.

Very rare, in dry deciduous forests near rock crevices. Fl.: Sept. – Dec. Kalasamudram (ATP), *KRKS* 41865.

Note: Fruit formation has not been found in the field, but occurs in garden by self pollination through hand. Flowering and fruiting throughout the year in green house.

AGAVACEAE

AGAVE L.

Agave americana L., Sp. Pl.323: 1753; FBI 6: 277. 1892; FPM 3:1505. 1934; FAP 3:928. 1997.

Perennial, stemless, scapigerous shrub to 2 m high. Leaves crowded at the base, lanceolate, 50-120 × 10-15 cm, coriaceous, succulent, glabrous, waxy, margin with very strong patent or recurved spines, apex with 3 cm spine. Panicle to 8 m, many branched, widely patent, flowers densely crowded, with unpleasant smell, perianth funnel-shaped, yellowish-green. Stamens exserted. Capsule oblong-clavate, beaked, black when dry. Seeds many, shining.

Occasional, in forest peripheries and common, in field bunds. Fl. & Fr.: Sep. –Jan. Ambakampalli (KDP), *KRKS & BSS* 40468.

Note: Bulbils produced on inflorescence. Fibre from leaves is very strong and used to make ropes.

AMARYLLIDACEAE

CRINUM L.

Crinum viviparum (Lam.) R. Ansari & V.J.Nair, J. Econ. Taxon.Bot. 11: 205. 1987 publi. 1988. *Amaryllis vivipara* Lam., Encycl. 1: 123. 1783. *Crinum defixum* Ker-Gawl., J. Sci. Arts (London) 3:105. 1817; FPM 3: 1504. 1931; FAP 3: 971. 1997.

Perennial erect herb, to 1 m; root stock bulbous, bulb ovoid, to 10 × 8 cm, cylindric. Leaves linear-lanceolate, 60–100 × 8–10 cm, coriaceous, scaberulous, semi succulent, margin entire, apex tapering. Scape to 50 cm long, flowers in umbels, white, 6–8, salver-shaped, 20 cm length, perianth lobes oblong-lanceolate. Stamens 6. Ovary tricarperllary, syncarpous. Capsules sub-globose; seeds 1-2, rugose.

Rare in shady and sandy localities. Fl. & Fr.: Jul. – Dec. Kalasamudram (ATP), *BSS & KRKS* 39191.

Note: New record for Andhra Pradesh and Eastern Ghats.

HYPOXIDACEAE

Key to Genera

1a. Leaves with distinct petiole .... **Molineria**

1b. Leaves with short or obscure petiole .... **Curculigo**

CURCULIGO Gaertn.

Curculigo orchioides Gaertn., Fruct. 1: 63. t. 16. f. 11. 1788; FPM 3: 1502. 1931; FAP 3: 976. 1997.

Perennial acaulescent herb; root stock stout to 30 cm long, tuberous covered with leaf bases. Leaves lanceolate, 10–25 × 1–1.5 cm, plicate, membranous, base sheathed, margin ciliate-entire, apex acute, glabrous or softly hairy. Flowers in axillary or simple racemes, distichous, bright yellow; lowest flower in the raceme bisexual, upper ones male. Stamens 6. Ovary 3-celled, stigmas 3. Capsules oblong, glabrescent. Seeds *ca.* 8, glossy, beaked.

Occasional, throughout in moist localities and in plains. Fl. & Fr.: All seasons. Kalasamudram (ATP), *KRKS* & *BSS* 39183.

MOLINERIA Colla

Molineria trichocarpa (Wight) N. P. Balakr., J. Bombay Nat. Hist. Soc. 63: 330. 1967. *Hypoxis trichocarpa* Wight, Icon. Pl. Ind.Orient. t. 2045. 1853. *Curculigo trichocarpa* (Wight) Bennet & Raizada, Indian J. Forest. 4: 68. 1981.

Perennial acaulescent herb; root stock stout, to 30 cm long, tuberous, covered with leaf bases. Leaves lanceolate, 30–45 × 2.5–3.5 cm, plicate, membranous, soft hairs scattered, entire, acute. Flowers in simple racemes, distichous, bright yellow; lowest flower in the raceme bisexual, upper ones male. Stamens 6. Ovary 3-celled; stigmas 3. Capsules oblong, glabrescent; seeds *ca.* 8, glossy, beaked.

Rare in moist localities under shade. Fl. & Fr.: All seasons. Batrepalli (ATP), *BSS* & *KRKS* 39603.

Note: New distributional record for Andhra Pradesh.

DIOSCOREACEAE

DIOSCOREA L.

Dioscorea pentaphylla L., Sp. Pl. 1032. 1753; FBI 6: 292. 1982; FPM 3:1511. 1934; FAP 3: 981. 1997.

Vine, to 5 m long; tubers oblong, stem slender, prickly towards the base, often bulbiferous in the leaf axils. Leaves 3–5-foliolate; terminal one elliptic ovate, 4–8 × 3–4 cm, membranous, glabrous above, sparsely pubescent beneath, entire, mucronate-acuminate; laterals inequilateral, 3–7 × 2–3 cm long, coriaceous, pubescent, entire, acute-mucronate. Flowers in spikes, male flowers pale greenish, solitary or in fascicles; female in axillary, pendulous spikes, pale green. Capsules oblong, deeply angled, glabrous; seeds 4–6, apically winged, black.

Occasional, throughout in dry deciduous forests. Fl. & Fr.: Feb. – Sep. Batrepalli (ATP), *BSS & KRKS* 39580.

LILIACEAE

Key to Genera

1a. Climbers .... (2)
1b. Herbs, not climbers .... (3)
2a. Leaves scaly; cladodes present .... **Asparagus**
2b. Leaves well developed; cladodes absent .... **Gloriosa**
3a. Leaves fleshy, margin spiny .... **Aloe**
3b. Leaves not fleshy, margin entire .... (4)
4a. Acaulescent herbs .... (5)
4b. Caulescent herbs .... **Iphigenia**
5a. Root stock with fasciculated tuberous roots .... **Chlorophytum**
5b. Root stock with bulb .... (6)
6a. Leaves linear, curved; scape erect .... **Drimia**
6b. Leaves not as above; scape ascending .... **Ledebouria**

ALOE L.

Aloe vera (L.) Burm. f., Fl. Ind. 83. 1768; FPM 3: 1520. 1934; FAP 3: 984. 1997. *A. perfoliata* L. var. *vera* L., Sp. Pl. 320. 1753.

Perennial, stoloniferous herb, to 50 cm. Leaves lanceolate, 20–40 × 4–6 cm, succulent, ensiform, margin prickled, apex acute, glabrous, aggregated. Flowers in simple scapigerous raceme, reddish-yellow, campanulate. Scape to 50 cm long; bracts persistent, flowers in racemes, perianth tube slightly curved. Stamens 6; filaments inserted in a pit in the connective. Ovary tricarpellary syncarpous, many ovules in each locule. Capsules oblong; seeds shining.

Occasional, throughout barren lands. Fl. & Fr.: Sep. – Dec. Batrepalli (ATP), *KRKS* & *BSS* 40187.

ASPARAGUS L.

1a. Cladodes 2–5 cm .... **A. laevissimus**
1b. Cladodes 0.5–1 cm .... **A. racemosus**

Asparagus laevissimus Steud. ex Baker, J. Linn. Soc 14: 623. 1875; FBI 6: 317. 1892; FPM 3: 1517. 1927; FAP 3: 984. 1997.

Perennial, subscandent climbing under shrub, to 1.5m long. Stems shiny, branches striate. Leaves with conical stout spurs; cladodes 3–6-nate, spreading, triquetrous. Flowers in racemes, pedicels short; bracts large, cymbiform. Tepals linear-oblong, obtuse, white. Stamen 6, as long as perianth. Style short, stout; stigmas 3. Fruit berry.

Rare in forests. Fl. & Fr.: Jun.–Sep. Batrepalli (ATP), *KRKS & BSS* 39579.

Asparagus racemosus Willd., Sp. Pl. 2: 152.1799; FPM 3: 1517. 1934; FAP 3: 984. 1997.

Vine, to 5 m; root-stock tuberous, fasciculated; branches angular, prickled. Leaves minute, slender, cladodes 3-6, slightly succulent, glabrous. Flowers in axillary racemes. Perianth lobes 6, oblong-obovate, white. Stamens 6, fused with base of perianth. Ovary obovoid, trilocular, syncarpous, 2 in each locule. Berries globose, red when ripe; seeds 3–6, globose or angled.

Common, throughout in shady areas. Fl. & Fr.: July – Oct. K.K. Kottala (KDP), *KRKS & KRS* 36427.

CHLOROPHYTUM Ker-Gawl.

1a. Scape erect, densely-flowered .... **C. tuberosum**

1b. Scape not erect, few-flowered .... **C. laxum**

Chlorophytum laxum R. Br., Prodr. 277. 1810; FBI 6: 336. 1892; FPM 3: 1526. 1934; FAP 3: 987. 1997.

Annual erect herb, to 30 cm; tubers hanging from the root tips. Leaves 6–12, linear-lanceolate, 20–28 × 0.2–0.4cm, membranous, base sheathed, margin entire, apex acute, glabrous, longer than the scape. Flowers in lax, simple or forked racemes; scape filiform, bracts ovate. Flowers distant, perianth oblong, obscurely 3-nerved, greenish-white. Stamens 6. Capsules broadly obcordate, triquetrous; seeds 1–4 in each lobe, black, minutely papillose.

Rare in shady localities of forest plains. Fl. & Fr.: May-Jun. Somulavandlapalli (ATP), *BSS & KRKS* 41907.

Chlorophytum tuberosum (Roxb.) Baker, J. Linn. Soc. Bot. 15: 332.1876; FBI 6: 334. 1892; FPM 3:1526. 1934; FAP 3:988. 1997. *Anthericum tuberosum* Roxb., Fl. Ind. 1: 149. 1832.

Annual erect herb, to 20 cm; roots cylindric, ellipsoid, tuberous, hanging. Leaves 6–12, lanceolate, 10–22 × 0.7–2.5 cm, membranous, base sheathed, margin entire, apex acuminate, glabrous, shorter than the scape, usually falcate, recurved. Scape to 25 cm, stout. Flowers in simple racemes, perianth lobes oblong, 1–1.3 × 0.3–0.5 cm, white. Stamens 6. Ovary tricarpellary, syncarpous, ovules 4 or more in each locule. Capsules oblong, triquetrous, glabrous. Seeds many, subquadrate, irregularly pitted, black.

Occasional, in shady localities of forests. Fl. & Fr.: Jun. – Nov. Near Rathnachalam Anjaneya Swamy Temple (ATP), *KRKS & BSS* 39670.

DRIMIA Jacq. ex Willd.

Drimia indica (Roxb.) Jessop, J.S. African Bot. 43: 272. 1977; FAP 3: 989. 1997. *Scilla indica* Roxb., Fl. Ind. 2: 147. 1832. *Urginea coromandeliana* (Roxb.) Hook. f., Fl. Brit. India 6: 347. 1892; FPM 3: 1527. 1934. *Scilla coromandeliana* Roxb., Fl. Ind. 2: 148. 1832.

Acaulescent herb, with tunicated bulb. Leaves linear, 15–20 × 0.3–0.5 cm, sub-bifarious, base sheathed, margin entire, apex acuminate, glabrous, appearing after flowering. Scape erect to 30 cm, stout at the base. Flowers in racemes; distant. Perianth lobes oblong-lanceolate, obtuse, dingy brown. Stamens 6, adnate at the base of perianth lobes. Capsules 3-celled, ellipsoid, tapering to both ends. Seeds 6–9 in each cell, black.

Common, throughout in open areas near forest margins. Fl. & Fr.: Mar. - Apr. Kalasamudram (ATP), *KRKS* 39903; Vempalli (KDP), *KRKS* & *BSS* 39087.

GLORIOSA L.

Gloriosa superba L., Sp. Pl. 305. 1753; FBI 6: 358. 1892; FPM 3: 1519. 1934; FAP 3: 990. 1997.

Local name: *Nagetigadda*.

Vine, to 2 m, branching; stem grooved; root stock tuberous, fleshy, jointed, white and brown, "L"-shaped. Leaves oblong-lanceolate, 5–9×1–1.5 cm, membranous, margin entire, apex tendrillar, glabrous. Flowers herkogamous, axillary, solitary or sub-corymbose; perianth lobes oblong-lanceolate, margins waved, yellowish at base, red towards apex. Stamens 6, spreading. Ovary oblong, tricarpellary, syncarpous, many ovules in each locule. Capsules ellipsoid, torulose; seeds many, compressed, warty.

Common, throughout in forests near bushes and in field edges. Fl. & Fr.: Aug. – Sep. Rachakuntapalle (KDP), *KRKS & KRS* 37915.

IPHIGENIA Kunth

Iphigenia indica (L.) A. Gray ex Kunth, Enum. Pl. 4: 213. 1843; FBI 6: 357. 1892; FPM 3: 1528. 1934; FAP 3: 991. 1997. *Melanthium indicum* L., Mant. Pl. 226. 1771.

Annual erect herb, to 30 cm high; corm small, tunicate, sub-globose, white. Leaves linear-lanceolate, 12–22 × 0.1–0.5 cm, upper leaves smaller than lower, membranous, margin entire, apex acuminate, glabrous. Flowers in terminal, erect raceme. Bracts subulate, perianth lobes linear-oblong, dark purple. Stamens 6, dark purple. Capsules oblong, glabrous. Seeds many, warty, brown.

Occasional, near marshy localities of forests. Fl. & Fr.: Jul. – Sep. Kalasamudram (ATP) *KRKS* 39959; Somulavaripalli (ATP), *KRKS & BSS* 41904.

LEDEBOURIA Roth

Ledebouria revoluta (L.f.) Jessop, J. S. African Bot. 36: 255. 1970. *Hyacinthus revolutus* L.f., Suppl. Pl. 204. 1782. *Scilla hyacinthina* (Roth) J.F.Macbr., Contr. Gray. Herb. 56: 14. 1918; FAP3: 992. 1997. *Ledebouria hyacinthina* Roth, Nov. Pl. Sp. 195. 1821 (illegitimate). *Scilla indica* (Wight) Baker, Saunders, Refug. Bot. 3(App.): 12. 1870; (non Roxb. 1832); FBI 6: 348. 1892; FPM 3: 1527. 1928. *Barnardia indica* Wight, Icon. Pl. Ind. Orient. t. 2041. 1853.

Annual bulbous herb; bulb ovate-globose. Leaves radical, oblong-lanceolate, 8–12×2–3cm, chartaceous, brown tinged, base sheathed, margin entire, apex acuminate, green or blotched with purple, glabrous. Scapes 1–3, racemes to 20cm, densely flowered. Perianth deeply lobed, greenish-purple. Stamens 6, adnate at the base of the perianth. Ovary tricarpellary, syncarpous, ovules 2 or more in each locule. Capsules globose. Seeds many.

Common, throughout in open areas. Fl. & Fr.: Jun. –Sep. Mabbuchinthalapalle cross (KDP), *KRKS* 37682.

SANSEVIERIACEAE

SANSEVIERIA Thunb. *nom. cons.*

Sansevieria roxburghiana Schult. & Schult. f., Syst. 7. 357. ff. 12 D & E 1829; FPM 3: 1520. 1934; FAP 3: 975. 1997.

Herb with very short stem; root stock creeping. Leaves about 8 or 9 in a tuft, 75–90 × 2–5 cm, towards the middle, suberect, rigid, pale green with transverse, dark green bands, thick, with spine-like tip. Flowers greenish-white tinged with violet, in fascicles of 3–6 on 30–60 cm long racemes, bracts membranous. Perianth-tube slender; lobes 6, narrow. Stamens 6, on the perianth tube; filaments filiform; anthers dorsifixed. Ovary 3-celled; ovules solitary in each cell, erect; style filiform; stigma simple. Fruit membranous, indehiscent. Seeds 1–3, globose, fleshy, ripening outside the pericarp.

Common, throughout in rocky shades and near bushes. Fl. & Fr.: Jun. – Aug. Kalasamudram (ATP), *KRKS* & *BSS* 40438.

PONTEDERIACEAE

MONOCHORIA Presl.

Monochoria vaginalis (Burm. f.) Presl, Relq. Haenk. 1:128. 1827; FBI 6: 363. 1892; FPM 3: 1529. 1934; FAP 3: 996. 1997. *Pontederia vaginalis* Burm. f., Fl. Ind. 80. 1768.

Aquatic erect herb, to 40 cm; root stock short, sub erect, spongy. Leaves variable, linear-ovate or ovate-cordate, 7–15×2–5cm, chartaceous, base sheathed, margin entire, acuminate, glabrous. Racemes inserted in the hollow sheath of the petiole. Flowers in solitary racemes, deep-blue with red spots. Stamens 6, inserted at the base of the perianth. Ovary oblong, 3-celled, many ovules in each locule. Capsules oblong, glabrous. Seeds many, ribbed.

Common, throughout in water sources and muddy places. Fl. & Fr.: Aug. – Mar. Kalasamudram (ATP), *KRKS* & *BSS* 39187; Batrepalle (ATP), *KRKS* 36967.

XYRIDACEAE

XYRIS L.

Xyris pauciflora Willd., Phytogr. 2. t. l. f. 1: 1974; FBI 6: 365. 1892; FPM 3: 1532. 1928: FAP 3: 997. 1997.

Erect, tufted, reed-like scapigerous herb, to 40 cm high. Leaves narrowly linear, 10 × 0.2 cm, papillose. Scape 15 cm long, slightly compressed, hardly winged, papillate, ribbed. Flower-heads sub-globose, bracts chestnut brown, apex dorsally keeled; median bract elliptic, mucronate; basal bract orbicular, margin membranous, apex triangular, papillate, spinulose. Sepals 3, median one broad, terminal curved. Petals 3, yellow, limbs obovate, emarginated. Stamens 3. Ovary oblong, 1-celled, ovules on parietal placentae. Fruit capsule.

Rare in moist locations. Fl. & Fr. Apr. – Jun. Talupula (ATP), *KRKS* & *BSS* 41910, 41918.

COMMELINACEAE

1a. Inflorescence terminal; flowers dark purple .... **Murdannia**

1a. Inflorescence axillary; flowers not as above .... (2)

2a. Inflorescence leaf-apposed .... **Commelina**

2b. Inflorescence not as above, pedunculate .... **Cyanotis**

COMMELINA L.

Key to Species

1a. Spathes ovate-lanceolate .... (2)

1b. Spathes funnel-shaped or cucullate .... (4)

2a. Seeds deeply grooved, petals yellow, peduncle 3 mm long .... **C. subulata**

2b. Seeds not grooved, petals blue, peduncle 10 mm long .... (3)

3a. Capsule 2-celled, constricted in the middle, sheaths more than 2.5 cm long .... **C. clavata**

3b. Capsule 3-celled, not constricted, sheaths less than 1.5 cm long .... **C. diffusa**

4a. Leaves ovate-oblong, cleistogamous flowers present .... **C. benghalensis**

4b. Leaves elliptic-lanceolate, cleistogamous flowers absent .... **C. maculata**

Commelina benghalensis L., Sp. Pl. 41. 1753; FBI 6: 370. 1892; FPM 3: 1539. 1931; FAP 3: 999. 1997.

Local name: *Enneddalaku*.

Annual ascending herb, to 1m high, rooting at lower nodes; root with cleistogamous flowers. Leaves ovate-oblong, 2–4.5 × 1–2.5 cm, membranous, base sheathed, margin entire, acute-obtuse, pubescent; sheaths to 2 cm, pubescent. Flowers in axillary cymes, upper cymes 2–3-flowered, lower one 1–2-flowered, spathes clustered, funnel-shaped. Sepals 3, pale green. Petals 3, linear-orbicular, blue. Ovary 3-celled, 2 cells 2-ovulate, one cell 1-ovulate. Capsules pyriform. Seeds 5, oblong, pitted.

Common, throughout in shady and marshy areas. Fl. & Fr.: Jul. -Nov. Thummalapalle (KDP), *KRKS, KRS,SSR & TP* 33669.

Commelina clavata C.B. Clarke, Commel, Cyrt. Bengal t. 5. 1874; FBI 6: 371. 1892; FPM 3: 1539. 1931; FAP 3: 1000. 1997.

Annual diffuse herb, to 1m high; branches spreading; rooting at nodes. Leaves ovate, elliptic-lanceolate or linear, 2–6 × 0.5–1 cm, membranous, base tapering, margin entire, apex acute-acuminate, glabrescent; sheaths apically hirsute. Flowers in cymes, spathes ovate-lanceolate, 2–3.5 cm. Sepals small. Petals lilac. Capsules oblong-quadrate, constricted in the middle. Seeds 4, black.

Occasional, in moist localities. Fl. & Fr.: Sep. – Feb. Thummalapalle (KDP), *SSR & TP* 33620.

Commelina diffusa Burm.f., Fl. Ind. 18, t. 7, f. 2. 1768; FAP 3: 1000. 1997. *Commelina nudiflora* sensu Hook.f., FBI 6: 369. 1892; non L. 1753; FPM 3: 1538. 1931.

Diffuse herb, to 40 cm high, stem glabrescent. Leaf sheath greenish, 1.5 cm long, ciliate. Lamina ovate-lanceolate, 4 – 6 × 0.5 – 0.8 cm, chartaceous, base subcordate, margin entire, apex acuminate. Spathes complicate, oblong-lanceolate, 3.5 cm long, glabrous, base cordate. Peduncle 3.5 cm long. Sepales 3, to 5 mm long, pale coloured. Petals blue, lyrate, lobes 3 mm across. Stamens 3, yellow, curved. Staminodes 3. Ovary 3-celled, 4 mm long. Capsule 3-celled, oblong, apex beaked, to 8 mm. Seeds tuberculate and reticulate to 3 mm.

Occasional, in marshy area of dry deciduous forest, plains merged with grass. Fl. & Fr.: Oct.-Dec. Batrepalli (ATP), *KRKS* 429 (HJCB).

Commelina maculata Edgew., Trans. Linn. Soc. 20: 89. 1846; FAP 3: 1003. 1997. *C. paludosa* Bl., Enum. Pl. Jav. 1: 2. 1827; FPM 3: 1539. 1931.

Annual, subscandent herb, to 1m high. Leaves lanceolate or elliptic-lanceolate, 6–15 × 1.5–3 cm, membranous, base tapering, entire, acute, pubescent or glabrous. Leaf sheaths to 4 cm, mouth bearded. Spathes solitary, funnel-shaped, glabrous, filled with glutinous liquid. Flowers in racemes. Petals blue or rarely white, large ones clawed. Capsules obovoid. Seeds 3, smooth.

Occasional, in moist and shady localities. Fl. & Fr.: Jun. -Oct. Batrepalli (ATP), *KRKS & BSS* 39616.

Commelina subulata Roth. Nov. Pl. Sp. 23. 1821; FAP 3: 1004. 1997; FBI 6: 369. 1892; FPM 3: 1538. 1931.

Diffuse or prostrate herb, to 60 cm long, stem glabrous, internodes 5 – 10 cm long. Leaves alternate, sheath to 1.5 cm with dark red veins, margin ciliate, lamina linear-lanceolate, 3–6 × 0.2–0.5 cm, chartaceous, glabrascent, base obtuse, margin entire, apex acute. Spathes subsessile, axillary, ovate, to 8 mm long, base cordate, margin ciliate, apex acute. Flowers light yellow, to 6 mm long. Peduncle to 3 mm long. Sepals unequal to 2.5 mm across, pale in color, curved. Corolla lyrate, 5 mm long, lobes elliptic-rounded, 3 mm across. Stamens 3, yellow, 4.5 mm long, staminodes 3. Ovary oblong, 3-lobed, to 1.5 mm long, style to 2.5 mm. Capsule oblong, to 5 mm long, 3-celled. Seeds ovoid, dark, 4-grooved, prominently punctate.

Very rare, in dry deciduous forest near moist locations and grass fields in high altitude. Fl. & Fr.: Nov. – Jan. Batrepalle forest (ATP), *KRKS* 428 (HJCB).

CYANOTIS D. Don *nom. cons.*

1a. Petals free .... **C. axillaris**

1b. Petals fused at the base .... (2)

2a. Plants cobwebby-woolly .... (3)

2b. Plants not cobwebby-woolly, nearly glabrous .... (4)

3a. Cymes densely aggregated .... **C. arachnoidea**

3b. Cymes lax .... **C. fasciculata**

4a. Leaves various, densely pubescent; roots tuberous .... **C. tuberosa**

4b. Leaves similar, sparsely pubescent or glabrous; roots not tuberous .... **C. cristata**

Cyanotis arachnoidea C.B. Clarke in DC., Monogr. Phan. 3: 250. 1881; FBI 6: 386. 1892; FPM 3:1550. 1934; FAP 3:1005. 1997.

Annual, prostrate or creeping herb, to 20 cm; often rooting at nodes. Leaves various, basal oblong-elliptic, 3–5 × 1–1.5 cm, woolly; upper elliptic-lanceolate, 2–7 × 0.3–1 cm, wooly. Flowers in terminal and axillary cymes, bracts exceeding cymes. Calyx pale yellow, bearded. Petals blue. Capsule oblong-lanceolate. Seeds terete, pitted.

Common, in moist localities on rocks etc. Fl. & Fr.: Aug. – Oct. Somulavaripalli (ATP), *KRKS* & *KRS* 37830.

Cyanotis axillaris (L.) D. Don ex Sweet, Hort. Brit. 430. 1826; FPM 3: 1550. 1931. *Tonningia axillaris* (L.) Kuntze, Revis. Gen. Pl. 2: 721. 1819; FAP 3: 1013. *Commelina axillaris* L., Sp. Pl. 42. 1753. *Amischophacelus axillaris* (L.) Rao & Kamm., J. Linn. Soc. Bot. 59. 306. 1966;

Local name: *Gorlaku.*

Annual erect or prostrate herb, to 1m high; branches spreading; rooting at nodes. Leaves narrow-lanceolate, 3–6 × 0.3–0.6 cm, succulent, plicate or flat, chartaceous, base obtuse, margin entire, apex acuminate, glabrous; sheaths pouched. Flowers in clusters at axils of leaf sheaths. Calyx lobes lanceolate. Petals ovate, violet-blue or white. Stamens 6, filaments blue with papillose hairs. Capsule ellipsoid. Seeds oblong, pitted.

Common, weed in cultivated fields and in moist localities. Fl. & Fr.: Jul. -Nov. Rachakuntapalle (KDP), *KRKS & KRS* 37626, 37901; Namalagundu (KDP), *KRKS* & *BSS* 40109.

Cyanotis cristata (L.) D. Don, Prodr. Fl. Nepal 46. 1825; FBI 6: 385. 1892; FPM 3:1549. 1934; FAP 3:1006. 1997. *Tradescantia cristata* L., Sp. Pl. 42. 1753.

Annual prostrate or creeping herb, to 30 cm; rooting at lower nodes. Leaves oblong-lanceolate, 2–4 × 0.5–1 cm, chartaceous, base cordate, margin entire, apex acute, sparsely pubescent. Flowers in scorpioid cymes, bracts exceeding cymes, bracteoles 5–8 pairs. Petals blue-purple. Capsules oblong, trigonous; seeds with 2 large pits on 2 sides.

Occasional, in moist rocky places. Fl. & Fr.: Jul. – Nov. Kalasamudram (ATP), *KRKS* & *BSS* 40134; Batrepalli (ATP), *KRKS* 243 (HJCB-E).

Cyanotis fasciculata (B. Heyne ex Roth) Schult. & Schult. f., in L. Syst. Veg. 7: 1152. 1830; FBI 6: 387. 1892; FPM 3: 1550. 1934; FAP 3: 1006. 1997. *Tradescantia fasciculata* B. Heyne ex Roth, Nov. Pl. Spec. 189. 1821.

Annual, decumbent herb, to 20 cm. Leaves linear-lanceolate, 3–9.5 × 0.7–1 cm, chartaceous, silky-cobwebby, base cordate, margin entire, apex subacute, woolly. Flowers in axillary or terminal cymes, cymes solitary or 1–2 in a cluster; bracts narrowly lanceolate. Calyx woolly, lobes lanceolate; petals oblong, pink. Stamens 6. Ovary ovate-oblong, style blue, bearded. Capsules oblong. Seeds 3–6, faintly rugose.

Common, throughout in marshy areas on rocks. Fl. & Fr.: Jun. – Dec. Batrepalli (ATP), *KRKS* 242 (HJCB-E).

Cyanotis tuberosa (Roxb.) Schult. & Schult. f., in L., Syst. Veg. 7: 1153.1830; FBI 6: 386. 1892; FPM 3: 1549. 1934; FAP 3: 1007. 1997. *Tradescantia tuberosa* Roxb., Pl. Coromandel. t. 108. 1799.

Perennial, sub-erect or prostrate herb, to 50 cm high, rooting at nodes; roots tuberous, fusiform, stem hirsute at base. Leaves radical and cauline reduced towards apex, basal oblong, upper linear-ensiform, 10–25 × 1–2 cm, chartaceous, glabrous or pubescent above, downy-sericeous below, base cordate-obtuse, margin ciliate, apex subacute. Flowers in axillary and terminal pedunculate cymes; bracts ovate, slightly exceeding cyme; bracteoles 10–15 pairs, 2-seriate. Calyx lobes linear, villous. Corolla lobes ovate, bluish-purple. Capsules oblong, upper half hairy, lower half glabrous. Seeds 4–6, obscurely rugose.

Common, throughout on slopes and plains. Fl. & Fr.: Aug.–Sep. Kalasamudram (ATP), *KRKS* 39691; K. K. Kottala (KDP), *KRKS* & *KRS* 36429.

MURDANNIA Royle *nom. cons.*

1. Leaves linear-lanceolate, 10 times longer than broad .... **M. nudiflora**
1. Leaves broadly linear-lanceolate, to 3 times longer than broad .... **M. spirata**

Murdannia nudiflora (L.) Brenan, Kew Bull. 7: 189. 1952; FAP 3: 1010. 1997. *Commelina nudiflora* L., Mant. 177. 1771 (non Sp. Pl. 41. 1753). *Aneilema nudiflorum* (L.) R. Br., Prodr. 271. 1810; FBI 6: 378. 1892; FPM 3: 1545. 1934.

Annual erect or decumbent herb, to 30 cm high; rooting at lower nodes. Leaves linear-lanceolate, 4–7 × 0.3–0.8 cm, chartaceous, base obtuse, margin entire, apex acuminate, glabrous. Flowers in cymes on terminal panicles. Sepals oblong, green. Petals sub orbicular, blue-purple. Stamens 2 perfect, 2–4 sterile, filaments of the fertile only or all bearded; ovary ovate, style white, stigma capitate. Capsules trigonous-subglobose, 3-celled. Seeds 2 in each cell, cuboid, deeply pitted, dark brown.

Common, throughout in moist localities. Fl. & Fr.: Jun. -Oct. Kalasamudram (ATP), *KRKS* & *KRS* 37690; Batrepalli (ATP), *BSS* & *KRKS* 40339, *KRKS* 240 (HJCB-E).

Murdannia spirata (L.) Brueckner in Engl., Nat. Pflanzenfam. (ed. 2) 15 a: 173. 1930. *Commelina spirata* L., Mant. Pl. 176. 1767. *Aneilema spiratum* (L.) R. Br., Prodr. 271. 1810; FBI 6: 377. 1892; FPM 3: 1545. 1931.

A small, erect or diffuse, annual herb, often rooting at nodes; branches 15-25 cm long, decumbent. Leaves sessile, the lower ones broadly linear-lanceolate, the upper ones oblong, cordate, all amplexicaul, glabrous, 1.5-3.5 × 0.6 cm, margin entire, apex acute. Flowers in dichotomously branched panicles; sepals 3, oblong; petals 3, blue, broadly oblong, obtuse; stamens 3 perfect, the anthers apiculate, blue, the filaments bearded, staminodes 3, shorter; ovary 3-celled, cells 3-many-ovulate. Capsules oblong, smooth, shining, trigonous, the cells 3-7-seeded; seeds 1-seriate, minute, cubical, white or straw-coloured.

Common, throughout. Fl. & Fr.: Aug. – Nov. Kalasamudram RF (ATP), *NY* 1168 (MH).

ARECACEAE

1a. Leaves compound, leaflets end with sharp spine .... **Phoenix**

1b. Leaves simple .... **Borassus**

BORASSUS L.

Borassus flabellifer L., Sp. Pl. 1187. 1753; FPM 3: 1562. 1931; FAP 3: 1015. 1997.

Tree with tall trunk, to 50 m high, dioecious. Leaves simple, palmately fan-shaped, plicate, margins multifid, apex acuminate, glabrous-glaucous; petioles stout, prickled. Spadices interfoliar, large, deciduous, branched. Male flowers small, clustered, mixed with scaly bracts, secund, 2-seriate; tepals 6, outer oblong, inner spatulate, sepaloid. Stamens 6, pistillodes of 3 bristles. Female flowers large, globose, perianth lobes reniform-convolute, fleshy, accrescent. Staminodes 6-9. Drupe subglobose, pyrenes 1-3, fibrous, yellow when ripe; pericarp thinly fleshy. Seed oblong, apex 3-lobed, testa adhering to the pyrene.

Occasional, in barren lands. Fl. & Fr.: Feb. -Apr.

PHOENIX L.

1a. Stems more than 10 m high .... **P. sylvestris**

1b. Stems less than 2 m high .... **P. loureiroi**

Phoenix loureiroi Kunth, Enum. Pl. 3: 257. 1841; FAP 3:1017. 1997. *P. humilis* Royle var. *pedunculata* Becc., Malesia 3:379 & 378.t. 44. ff. 13-15, 18-21, 25-27. 1890; FBI 6: 427. 1892; FPM 3:1560. 1931.

Arborescent trees, to 2 m high; densely covered with fallen petioles. Leaves to 2.5 m long; leaflets 10–20 × 0.5–1 cm, pliable, chartaceous, base narrow, margin entire, apex acuminate, spinous, glabrous. Flowers in spadix. Male spadix to 30 cm long; spikes in clusters, to 10 cm long. Calyx lobes triangular, petals oblong. Stamens 6. Female spadix to 50 cm long; spikes to 8 cm long. Drupe oblong, 1 cm long, black when ripe. Seed 1, grooved.

Rare in hillocks. Fl. & Fr.: Mar. – May. Chinnapalli (ATP), *KRKS & BSS* 40379.

Phoenix sylvestris (L.) Roxb., Fl. Ind. 3:787. 1832; FBI 6: 425. 1892; FPM 3:1559. 1931; FAP 3:1018. 1997. *Elate sylvestris* L., Sp. Pl. 1189. 1753.

Arborscent trees, to 10 m high; covered with fallen petioles. Crown thick, hemispherical. Leaves to 4 m long; leaflets 10–30 × 1–2 cm, chartaceous, rigid, base narrow, margin entire, apex acuminate, spinous, glabrous; spine strong. Flowers in spadix. Spadix to 1 m long; spathes as long as spadix, coriaceous. Male spadix to 30 cm long; spikes many in clusters, to 20 cm long. Calyx cup shaped; petals 3–4 times longer than calyx lobes, white. Stamens 6. Female spadix to 50 cm long, spikes to 20 cm long. Drupe oblong-ellipsoid, orange-yellow, 2 cm long, black when ripe. Seed 1, grooved, brown.

Common, throughout near streams. Fl. & Fr.: Jan. -Apr. Pulivendula (KDP), *KRKS* 42054.

TYPHACEAE

TYPHA L.

Typha domingensis Pers., Syn. Pl. 2: 532. 1807; Napper in Redhead & Polhill, Fl. Trop. E. Afr. 1. 1971; Cook in Tutin *et al.*, Fl. Eur. 276. 1980. *T. angustata* Bory & Chaub., Nouv. Fl. Pelop 4. 1838; FBI 6: 489. 1893; FPM 3: 1571. 1931.

Perennial, erect robust herb, to 3 m high. Leaves linear-lanceolate, 200 × 1–2 cm, erect, spongy, base sheathing, margin entire, apex acuminate, glabrous, exceeding the flowering stem. Flowers in cylindric spikes; spike to 50 cm long; male and female spikes separated with an interval. Male spike to 20 cm long, interruption up to 3 cm long. Stamens 2 or 3. Female spike to 20 cm long, 3 cm in diameter, gynophore white. Ovary linear. Fruits fusiform.

Common, throughout in streams. Fl. & Fr.: Feb. – Jul. Batrepalli (ATP), *KRKS* 37955.

ALISMATACEAE

LIMNOPHYTON Miq.

Limnophyton obtusifolium (L.) Miq., Fl. Ned. Ind. 3: 242. 1856; FBI 6: 560. 1893; FPM 3: 1595. 1931; FAP 3: 1032. 1997. *Sagittaria obtusifolia* L., Sp. Pl. 993. 1753.

Perennial, scapigerous herb to 80 cm high; root stock stout. Leaves broadly sagittate, 20–45×2–13cm, chartaceous, base cordate-sagittate, margin entire, apex obtuse-mucronate, glabrous. Flowers in panicles, branches whorls of 2–6, 10–15 flowers in each whorl. Flowers white, upper male, lower bisexual. Male flowers: sepals 3, oblong-elliptic, hyaline along the margins; petals 3, obovate, white. Stamens 6, yellow. Achenes in globose cluster. Seeds oblong.

Occasional, in stagnant water sources. Fl. & Fr.: Jan. –Mar. & Jul. – Sep. Kalasamudram *KRKS* 40132.

NAJADACEAE

NAJAS L.

Najas graminea Del., Descr. Egypt. Hist. Nat. 282. t. 50. f. 3. 1813; FBI 6: 569. 1893; FPM 3: 1604. 1931.

A submerged, aquatic herb, rooting from the base and lower nodes; stems or shoots terete, up to 50 cm long. Leaves in pseudowhorls, linear-subulate, obtuse or acute with a spinulosely dentate margin, 1-2.5 × 0.005-0.1 cm, sheaths spinulosely denticulate with triangular, acute or obtuse, entire or rarely shallowly notched, 0.2 – 0.5 cm long, auricles 0.4 – 0.8 cm long. Flowers solitary or 2 – 4 together. Male: Spathe absent. Perianth segments rounded, anther ellipsoid-ovoid or oblong. Female: Spathe absent; flowers solitary, axillary, subsessile. Perianth lobes closely appressed to ovary, ovary elliptic. Fruit ellipsoid-oblong, attenuate at the apex, 0.15-0.2 cm long, areoles minute, subquadrate or polygonal.

Occasional, in ponds. Fl. & Fr.: August – November. Kalasamudram (ATP), *NY* 12025.

APONOGETONACEAE

APONOGETON L. f. *nom. cons.*

Aponogeton natans (L.) Engl. & K. Krause, in Engl., Pflanzenr. 4.13.24:11. 1906; FPM 3: 1598. 1931; FAP 3: 1036. 1997. *Saururus nutans* L., Mant. Pl. 227. 1771.

Perennial, rhizomatous submerged or floating herb, to 30 cm; tubers oblong. Leaves mostly floating, oblong, 2–8 × 0.5–2 cm, base sub-cordate, margin entire, apex obtuse, membranous, glabrous. Flowers in dense spikes; peduncle to 30 cm long. Tepals 2, obovate, violet. Stamens 6, anthers blue-purple. Fruit oblong, beaked, glabrous. Seeds 2–8.

Common, throughout in stagnant water. Fl. & Fr.: Aug. – Feb. Batrepalli (ATP) *KRKS* 36967.

POTAMOGETONACEAE

POTAMOGETON L.

Potamogeton nodosus Poir. in Lam., Encycl. (Suppl. 4): 535. 1816. *P. indicus* Roxb., Fl. Ind. 1: 471. 1820; non Roth ex Roem. & Schult. 1818; FBI 6: 565. 1893; FPM 3: 1600. 1831.

An aquatic submerged herb; stem terete, branched, smooth. Leaves heterophyllous, 4-12 × 2-3.5 cm, upper ones floating, broadly ovate-elliptic, subacute-obtuse at the ends, lower ones submerged, lanceolate, 7-9-nerved, base attenuate-cuneate, margin entire, apex acute-acuminate; stipules free, membranous, keeled, apex tapering. Spikes axillary, leaf-opposed, 2-5 cm long; peduncle 6 cm. Flowers sessile. Tepals 4, suborbicular-ovate. Stamens 4. Ovaries 3, free, ovule 1. Drupelets obovoid, shortly beaked, 1-seeded.

Common in ponds and ditches. Fl. & Fr.: Jul. – Dec. Kalasamudram RF (ATP), *NY* 654 (MH).

ERIOCAULACEAE

ERIOCAULON L.

1a. Floral bracts glabrous .... **E. truncatum**

1b. Floral bracts villous or pubescent .... **E. quinquangulare**

Eriocaulon quinquangulare L., Sp. Pl. 87. 1753; FBI 6: 582. 1893; FPM 3: 1620. 1931; FAP 3: 1041. 1997.

Annual tufted herb, to 15 cm high. Leaves lanceolate, 3–8 × 0.2–0.5 cm, membranous, purple-red, glabrous. Flowers in globose heads, snow white; peduncles twisted; involucral bracts obovate, glabrous; floral bracts obovate, hyaline, ciliate. Male floret: Calyx lobes oblong, apex hairy, petals subequal, white. Stamens 6, dark colour. Female floret: Calyx oblanceolate, ciliate; petal lobes subequal, black gland at apex. Ovary stipitate. Seeds oblong-ellipsoid, smooth or striate.

Common, throughout in moist localities. Fl. & Fr.: Aug.-Dec. Talupula (ATP), *BSS & KRKS* 41912.

Eriocaulon truncatum Buch.-Ham. ex Mart., Pl. Asiat. Rar. 3: 29. 1832; FPM 3:1619. 1931; FAP 3: 1043. 1997.

Annual tufted herb, to 8 cm. Leaves oblong-lanceolate, 7 × 0.5 cm, membranous, acute, glabrous. Flowers in heads, peduncles shallowly 5-8-ribbed, pale straw-coloured. Involucral bracts oblong-obovate, glabrous, acute; floral bracts obovate, concave, glabrous. Male floret: sepals 2, obovate; petals 3, ovate, glabrous. Female floret: sepals 2, linear; petals oblanceolate, glabrous. Ovary trilobed. Seeds ellipsoid, longitudinally ribbed, brown.

Occasional, throughout in moist localities. Fl. & Fr.: Aug. -Dec. Kalasamudram (ATP) *BSS & KRKS* 40106.

CYPERACEAE

1a. Glumes on spikelets distichous .... (2)
1b. Glumes on spikelets spiral .... (5)
2a. Style 2-fid; nut bilaterally flattened .... **Pycreus**
2b. Style 3-fid, nut triangular or dorsiventrally flattened .... (3)
3a. Achenes trigonus or triquetrous .... (4)
3b. Achenes bilaterally flattened .... **Kyllinga**
4a. Glumes winged .... **Courtoisina**
4b. Glumes not winged .... **Cyperus**
5a. Hypogynous scales present .... **Lipocarpha**
5b. Hypogynous scales absent .... (6)
6a. Style base dilated and constricted or articulated above the nut .... (7)
6b. Style base not dilated, continuous with the nut .... (9)
7a. Hypogynous bristles present, nut beaked .... **Eleocharis**
7b. Hypogynous bristles absent, but not beaked .... (8)
8a. Style base persistent; if falling not leaving a tumour on the nut .... **Fimbristylis**
8b. Style deciduous leaving a tumour on the nut .... **Bulbostylis**
9a. Plants pubescent or ciliate, culms noded .... **Fuirena**
9b. Plants glabrous, culms not noded .... (10)
10a. Leaves well developed .... **Bolboschoenus**
10b. Leaved reduced to sheaths .... (11)
11a. Spikelets clustered on lateral or terminal heads .... **Schoenoplectiella**
11b. Spikelets many, stalked, in large terminal corymb .... **Actinoscirpus**

ACTINOSCIRPUS (Ohwi) Haines & Lye

Actinoscirpus grossus (L.f.) Goetgh. & D.A. Simpson, Kew Bull. 46: 171. 1991; Kukkonen in Rech.f., Fl. Iran 173: 15. 1998. *Scirpus grossus* L. f., Suppl. Pl. 104. 1782; FBI 6: 659. 1893; FPM 3: 1666. 1931. *Schoenoplectus grossus* (L.f.) Palla, Allg. Bot. Z. Syst. 17: Beibl. 3: 1911; FAP 3: 1107.1997.

Perennial, stoloniferous herb; stolons producing tubers. Culms to 200 cm high, sharply 3-angled, spongy, faces concave, septate-nodose, with a corm like hard enlargement at base. Leaves equaling the culm, 1-ribbed, flattish-plicate, septate-nodose, sheaths spongy, prominently septate-nodose, tightly surrounding the culm base. Inflorescence a large terminal corymb with primary, secondary and tertiary rays, rather dense, to 12 cm long. Involucral bracts 4, foliaceous, exceeding the inflorescence, the longest to 40 cm long. Spikelets solitary, ellipsoid, to 10 mm long, densely many-flowered, light brownish, rust-colured. Achenes broadly obovate, subcompressed-triangular, stramineous brown.

Rare species found in marshy fields. Fl. & Fr.: Aug. – Dec. Kalasamudram (ATP), *KRKS* 39139.

BOLBOSCHOENUS (Aschers.) Palla

BOlboschoenus maritimus (L.) Palla in Koch., Syn. Deutsch. Fl. ed. 3. 2532. 1905. subsp. **maritimus**. *Scirpus maritimus* L., Sp. Pl. 75. 1753; FBI 6: 658. 1893; FPM 3 : 1666. 1931. *Schoenoplectus maritimus* (L.) Lye, Blyttia 29: 145. 1971; FAP 3:1109.1997. *Scirpus tuberosus* Desf., Fl. Atlant. 1: 50. 1798; Maheshwari, Fl. Delhi 363. 1963; Verma & Veenachandra, Rec. Bot. Surv. India 21: 263. 1981; Karthikeyan *et al.*, Florae Indicae Enumeratio Monocotyledonae 71. 1989.

Perennial, erect, rhizomatous creeping herb with tubers, to 75 cm high; culms triquetrous. Leaves flat, linear, to 30 cm long, acuminate, ligulate, sheaths long. Inflorescence terminal umbel, rayed, unequal. Involucral bracts 3–5, leaflike; spikelets ovoid to cylindric, acute. Achenes obovate, trigonous, smooth, to 4 mm long.

Occasaional in moist wastelands, marshes, in beds of drying ponds and canals. Fl. & Fr.: Oct. – Mar. Kalasamudram (ATP), *KRKS* 39533; Nigidi forest (ATP), *KRKS* 212 (HJCB-E).

BULBOSTYLIS Kunth *nom. cons.*

Bulbostylis barbata (Rottb.) Kunth ex C.B. Clarke in Hook. f., Fl. Brit. India 6: 651. 1893; FAP 3: 1049. 1997. subsp. **barbata.** *Scirpus barbatus* Rottb., Progr. 27. 1772. *Stenophyllus barbatus* (Rottb.) T. Cooke, Fl. Pres. Bombay 2: 887. 1908; FPM 3: 1662. 1931.

Annual, densely tufted erect herb, to 30 cm high, culms slender, striate, leaved at base. Leaf blade filiform, 4 to 8 cm long, sheaths membranous, brownish. Inflorescence a terminal head of 2–20 sessile spikelets, 5–15 mm across, brownish. Involucral bracts 3, spreading, to 2 cm long, setaceous. Spikelets sessile, lanceolate, 3–8 mm long, angular, subdensely 7 to 15-flowered, brownish; glumes ovate, to 2 mm long with recurved short awn, membranous, yellowish-green and becoming rusty brown, prominently keeled, ciliolate; stamen 1; style base depressed-globular, style filiform, stigmas 3. Achenes obovate-orbicular, trigonous, pale rounded at apex.

Very common, species found everywhere. Usually seen in open dry, especially sandy areas, rocky surfaces, wastelands, road sides, and as a weed in cultivated lands. Fl. & Fr.: Aug. - Dec. Kalasamudram (ATP), *KRKS & BSS* 39988.

COURTOISINA Soják

Courtoisina cyperoides (Roxb.) Soják, Cas. Nár. Mus., Odd. Prir. 148: 193. 1979. *Cyperus pseudokyllingoides* Kuk., in Engler, Pflanzenreich IV. 20, (Ht.101): 501. 1936. *Indocourtoisia cyperoides* (Roxb.) Bennet & Raizada, Indian Forest. 107: 432. 1981; FAP 3: 1089. 1997. *Kyllinga cyperoides* Roxb., Fl. Ind. 1: 187. 1820. *Courtoisia cyperoides* (Roxb.) Nees, Linnaea, 9: 286. 1834; FBI 6: 625. 1893; FPM 3: 1645. 1931.

Annual, erect, tufted, non rhizomatous herb, to 30 cm high, culms trigonous, glabrous. Leaves few, basal, as long as the stem, gradually acuminate; sheaths 2.5 cm, yellow, membranous. Inflorescence compound, branches ending in spikes. Involucral bracts 6, lower one much exceeding the inflorescence, spreading. Spike globose, 6 mm long; spikelets many, 4 mm long, clustered, compressed, 1-flowered, broadly elliptic, golden-brown-coloured, shining, rhachilla slender, narrowly winged. Achenes trigonous, linear-oblong, 2.8 mm long, apiculate, golden brown, strongly puncticulate, about ¾ of the length of the glume.

Occasionally in rice fields as a weed and also found in wastelands. Fl. & Fr.: Nov. – Jan. Batrepalli (ATP), *KRKS* 37852; *KRKS* 218 (HJCB-E).

CYPERUS L.

1a. Style 2-fid .... **C. laevigatus**
1b. Style 3-fid .... (2)
2a. Rhachilla articulated, deciduous .... (3)
2b. Rhachilla not articulated, persistent .... (4)
3a. Base of the culms hardly enlarged .... **C. paniceus**
3b. Base of the culm bulbous .... **C. clarkei**
4a. Spikelets spicately disposed .... (5)
4b. Spikelets digitately disposed .... (10)
5a. Rhachilla of spikelets winged with base of the glumes .... (6)
5b. Rhachilla of spikelets without wings .... (9)
6a. Plants with creeping rhizome or stolons; rachilla of spikelets not or scarcely creeping .... (7)
6b. Plant with corm-like rhizome; rachilla of spikelets distinctly winged (-spikelets suberect, achenes triquetrous) .... **C. nutans**
7a. Leaves clothed at base with bladeless sheaths .... **C. pangorei**
7b. Leaves well developed .... (8)
8a. Glumes more than 2.5 mm long .... **C. rotundus**
8b. Glumes less than 2.5 mm long .... **C. stoloniferous**
9a. Spikelets more than 1 cm long .... **C. compressus**
9b. Spikelets less than 1cm long .... **C. iria**
10a. Glumes muticous; achenes obovate .... (11)
10b. Glumes awned or mucronate; achenes oblong .... (13)
11a. Inflorescence up to 15-spikelets in ray .... (12)
11b. Inflorescence many-spikelets in ray .... **C. difformis**
12a. Perennial; glumes closely arranged .... **C. haspan**
12b. Annual; glumes loosely arranged .... **C. tenuispica**

13a. Inflorescence umbell developed .... **C. castaneus**

13b. Inflorescence head-like.. **C. rubicundus**

Cyperus castaneus Willd., Sp. Pl. 1: 278. 1797; FBI 6: 598. 1893: FPM 3: 1639. 1931, FAP 3: 1059.1997

Annual, tufted, glabrous herb, to 20 cm high, culms slender, triquetrous, 10-leaved at base. Leaves 9 cm long; blade setaceous, weakly scabrid, reddish brown. Umbel simple, to 8 cm long, frequently congested in a single head. Involucral bracts 4, surpassing the inflorescence; rays 5, to 5 cm long. Spikelets stellately clustered in groups of 5–10, slenderly linear, to 10 mm long, flattened, 20–40-flowered, chestnut black; rhachilla striaght, wingless. Achenes oblong with parallel sides to oblong-obovoid, brown, puncticulate.

Common, in open wet lands, grassy place, rice fields and in swamps etc. Fl. & Fr.: Aug. – Dec. Batrepalli (ATP), *KRKS & SSR* 37861.

Cyperus clarkei T. Cooke, Fl. Bombay 2: 873. 1908. *Mariscus clarkei* (T. Cooke) T. Koyama, J. Jap. Bot. 51: 313. 1976. FAP 3 : 1094. 1997. *M. bulbosus* C.B.Clarke in Hook. f., Fl. Brit. India 6: 620. 1839. non Steud.; FPM 3: 1644. 1931.

Perennial, rhizomatous, stoloniferous herb; stolons to 6 cm long, the internodes to 5 mm long, clothed with pale brownish membranous scales, culms solitary, to 22 cm high, triquetrous, the base thickened, to form a lanceolate enlargement, to 4 cm long. Leaves 6 to a culm, shorter than the culm; blade 18 cm long, herbaceous, 1-costate, plicate with incurved margins, glaucous green; sheaths 4 cm long, membranous, dusky brown. Inflorescence simple, to 10 cm long; spikes cylindrical, subsessile, 2 cm long, obtuse. Achenes oblong-elliptic, trigonous, 1.5 mm long, maturing dark brown.

Occasional, in marshy places and along streams. Fl. & Fr.: Sep. – Dec. Rachakuntapalle (KDP), *KRKS & KRS* 37642

Cyperus compressus L., Sp. Pl. 46. 1753; FBI 6 : 605. 1893; FPM 3 : 1640; FAP 3:1059. 1997. *C. pectiniformis* Roem. & Schult., Mant. 2 : 128. 1824.

Annual, tufted herb. Culms patent, to 30 cm high, 3-sided. Leaves 4 to a culm; basal; blade linear, shorter than the culms, light green; sheaths membranus, pale brownish, striate. Inflorescence umbellate, to 10 cm long; involucral bracts 7, foliaceous, unequal, the longest one to 20 cm long; rays 8, patent, to 10 cm long. Spikes bearing to 10 spikelets on abbreviated axis. Spikelets lance-oblong, 1–2.5 × 0.2–0.3 long, *ca.* 20-flowered, compressd, greenish and stramineous at maturity; rhachilla flexuose. Achenes 1.8 mm long, broadly obovate, 3-sided, dark brown, shiny.

Common, weed of cultivated fields, wastelands, in open grass lands, along the sides of water courses. Fl. & Fr.: Sep. – Feb. Rachakuntapalle (KDP), *KRKS & SSR* 37858; Batrepalli (ATP), KRKS 342 (HJCB-E).

Cyperus difformis L., Cent. Pl. II. 6. 1756; FBI 6 : 598. 1893; FPM 3 : 1640. 1931; FAP 3 : 1061. 1997.

Annual, glabrous, tufted herb with many reddish fibrous roots; culms to 40 cm high, weak, triquetrous towards the apex. Leaves few, shorter than the stem, acuminate, flaccid, to 13 cm long; sheaths stramineous or brown. Spikes in a simple or compound or contracted into a head, to 1.5 cm across; involucral bracts 3, overtopping, longest to 20 cm long, leaflike; rays of the umbel 3-6, to 3 cm long. Spikelets many in dense heads, dusty or brown, oblong, 10-20-flowered, compressed, rhachilla not winged. Achenes ellipsoid, trigonous, yellow or brown.

Very common, found as a weed in rice fields and cultivated lands. Fl. &Fr.: Throughout the year. Near Muniswara temple, Kalasamudram (ATP), *KRKS & BSS* 39104.

Cyperus halpan L., Sp. Pl. 45. 1753 (*haspan*); FBI 6: 600. 1893; FPM 3: 1640. 1931; FAP 3: 1064. 1997. subsp. **halpan.**

Perennial, tufted, rhizomatous herb, culms to 50 cm high, acutely triquetrous, leaved at base. Leaves 2 to a culm, blade linear, to 15 cm long, herbaceous, light green; sheaths to 4 cm long, lower ones membranous, red purple. Umbel compound, dense to subloose, to 10 cm long. Involucral bracts 2, the lowest one surpassing the inflorescence, the longest one to 8 cm long; primary rays to 10, unequal, patent, up to 7 cm long; secondary rays to 2 cm long bearing 3-5 stellately spreading spikelets. Spikelets linear-oblong, to 10 mm long, subdensely to 28-flowered, chestnut brown, the wingless rhachilla hidden by densely disposd glumes. Achenes broadly obovate, trigonous, cream-yellow.

Common, weed in rice fields, abundant in wet and marshy areas. Fl. & Fr.: Aug. – Dec. Batrepalli (ATP), *BSS & KRKS* 40351.

Cyperus iria L., Sp. Pl. 45. 1753; FBI 6: 606; FPM 3: 1640. 1931; FAP 3: 1064. 1997. *C. parviflorus* Nees in Wight, Contr. Bot. India 87. 1834 excl. Syn. non Vahl, 1806. *C. iria* var. *parviflorus* (Nees) Miov. in Fl. Ind. Bot. 3: 270. 1859. *C. iria* var. *paniciformis* (Franch. & Sav.) C.B.Clarke in Hook. f. Fl. Brit. India 6: 607. 1893.

Annual, glabrous, erect, tufted, marshy herb; culms to 50 cm long, triquetrous, striate, rigid. Leaves flat or channeled, to 30 cm long, 6 mm wide, scaberulous, multistriate, acuminate, flaccid; sheaths purplish. Umbel decompound, 5-15 cm long; primary rays 5-7, to 11 cm long; secondary rays 3-5, to 8 cm long, bearing irregularly fascicled umbellules, formed of narrow interrupted spikes of 15-20, small, few-flowered spikelet. Rhachilla of the spike glabrous. Spikelets spicate, oblong, 8 × 1 mm, to 15-flowered. Achenes obovoid, triquetrous, black or brown.

Common, weed in wet and flooded paddy fields. Fl. & Fr.: Jul. – Dec. Kalasamudram (ATP), *KRKS* 39855; Batrepalli (ATP), *KRKS* 229 (HJCB-E).

Cyperus laevigatus L., Mant. Alt. 179. 1771. *Juncellus laevigatus* (L.) C.B.Clarke in Hook. f., Fl. Brit. India 6: 596. 1893; FPM 3: 1629. 1931; FAP 3 : 1090.1997.

Plate 20

A. *Cyperus iria*, B. *C. clarkei*, C. *Fimbristylis tetragona*, D. *Kyllinga bulbosa*, E. *Schoenoplectiella articulata*, F. *S. juncoides*

Perennial, tufted, rhizomatous herb, rhizome creeping horizontally with solitary distant stems, their bases enclosed by imbricate, shining, chestnut red scales. Culms to 30 cm high, round, fleshy. Leaf blade undeveloped. Spikelets 30 in one apparently lateral head, straw-coloured. Involucral bracts 2, the lower as though a continuation of stem, to 7 cm long, the other much shorter. Spikelets ovate, compressed but thick, straw-coloured dotted with brown spots. Achenes obovoid, brown, stipitate.

Occasional, in moist localities, marshes, sandy beds of rivers and muddy soil. Fl. & Fr.: Sep. – Mar. Kalasamudram (ATP), *KRKS* 39535, 202 (HJCB-E).

Cyperus nutans Vahl, Enum. Pl. Obs. 2: 363 1805; FAP 3: 1066. 1997; FBI 6: 697. 1893; FPM 3: 1640. 1931.

Perennial herb, rhizome short and corm-like. Stems solitary or few together, up to 1.2 m high, trigonous. Leaves few, shorter than the stem, linear, 6–15 mm wide, narrowed to an acute apex; sheaths reddish-brown towards base. Umbel compound to decompound; involucral bracts 5–7, lower 2 or 3 much longer than the inflorescence; primary rays 7–9, very unequal, up to 30 cm long. Spikes up to 4 cm long. Spikelets suberect, up to 15 × 1.5 mm, linear to linear-oblong, flattened, 13–15-flowered, rhachilla persistent. Glumes 2–2.5 mm long, oblong, obtuse, keeled, dark brown to yellowish-brown, 3-nerved with broad whitish-hyaline margins, nerves on both sides less stronger than midnerve. Achene *ca*. 1.5 mm long, trigonous, oblong, apiculate, brown.

Occasional, on the margins of ponds, ditches, rivers, rice fields and in wet places. Fl. & Fr.: Sep. – Dec. Batrepalli (ATP), *KRKS & SSR* 37850

Cyperus pangorei Rottb., Descr. Icon. Rar. Pl. 31. 7. f. 3. 1773; FAP 3: 1067. 1997; FPM 3: 1641. 1931. *Cyperus tegetum* Roxb., Fl. Ind. 1: 211. 1820; FBI 6: 613. 1893.

Perennial, glabrous herb; rhizome stout, creeping. Culms up to 1.2m high, sharply trigonous above. Leaves few, shorter than stem, with long purple sheaths. Umbel compound; rays many, up to 12 cm long; umbellules corymbose. Spikes numerous. Involucral bracts leaf-like, the lowest longer than the inflorescence. Spikelets spicate, linear, many-flowered, usually reddish-brown; rachilla narrowly to broadly winged. Glumes elliptic, obtuse, 3–5-nerved on the back, sides red-brown, nerveless. Stamens 3; styles shorter than achene; branches linear. Achene *ca*. 1.5 mm long, elliptic-oblong, trigonous, brown;

Common, in marshy areas near river banks, canals, along the sides of the ponds, in ditches etc. Fl. & Fr.: Aug. – Dec. Kalasamudram (ATP), *KRKS* 37960; Batrepalli(ATP), *KRKS* 39005.

Cyperus paniceus (Rottb.) Boeckeler, Linnaea 36: 381. 1870; FAP 3: 1097.1997. *Mariscus paniceus* (Rottb.) Vahl, Enum. Pl. 2: 373. 1806; FBI 6: 620. 1893; FPM 3: 1644. 1931. *Kyllinga panicea* Rottb., Descr Ic. Rar. 15, t. 4, f. 1. 1773.

Perennial, rhizomatous, stoloniferous herb; rhizome decumbent, knotty, 2 cm long, clothed with dark red brown fibres, emitting stolons; stolons covered with reddish-brown scales, soon disintegrating into brown fibres, culms to 40 cm high, trigonous. Leaves 4 to a culm; blade nearly equaling the culm, flattish-plicate; sheaths to 5 cm long, reddish-brown. Umbel of spikes or digitate-contracted, to 3 cm long, light glaucous-greenish. Involucral bracts 7, patent, surpassing the inflorescence, the longest to 25 cm long. Spikes cylindrical, to 1.5 cm long, densely bearing up to 30 spikelets. Spikelets horizontally divergent, lanceolate, whitish green, slightly recurved, 3 mm long, bearing 4 glumes, and 1-flowered. Achenes elliptic-oblong, trigonous, yellow brown.

Though rare it is sometimes seen growing along forest margins, in damp places preferably under shade. Fl. & Fr.: Jul. – Dec. Batrepalli (ATP), *BSS & KRKS* 39607.

Cyperus rotundus L., Sp. Pl. 45. 1753; FBI 6: 614. 1893; FPM 3: 1641. 1931; FAP 3: 1069.1997.

Perennial, glabrous, erect, stoloniferous herb; stolons bearing ovoid, tunicate, black, fragrant tubers; culms to 30 cm high, triquetrous at the top, tuberous at the base. Leaves several, narrowly linear, flat, 10–30 cm long, nerves prominent; sheaths brown. Umbel compound, 4–9 cm long. Involucral bracts 3–5, unequal, longest to 15 cm long; primary rays 4–6, to 6 cm long; secondary rays 3, bearing, short spikes of 4–10 slender, spreading, red-brown spikelets. Spikelets variable in length, linear, red-brown, compressed, spicate, to 3 cm long, pale or purplish, 15-30-flowered; rhachilla winged, pale. Achenes oblong, trigonous, greyish-black.

Very Common, weed in rice fields and cultivated lands. Fl. & Fr.: Oct. – Dec. Dorigallau (ATP), *KRKS & BSS* 40474.

Cyperus rubicundus Vahl, Enum. Pl. 2 : 308. 1805; Raynal, Adansonia (ser.2). 17 : 45. 1976; FAP 3: 1070. 1997. *C. teneriffae* Poir. in Lam., Encycl. 7: 245. 1806; FBI 6 : 601. 1893; FPM 3 : 1639. 1931.

Small caespitose annual sedge; stems tufted, reddish-brown 3–10 cm high, trigonous. Leaves, closely distichous, half or as long as stem. Inflorescence simple, capitate, to 1.5 cm; involucral bracts 3, unequal. Spikelets 5–7, digitately arranged, 2 cm across, stellate, purple, 15–40-flowered; rachilla persistent, corky, wingless; glumes broadly ovate, 7-nerved, margins hyaline, 3 × 2 mm. Nut stipitate, obovoid, trigonous, chestnut coloured, 0.8 mm.

Common, weed on road sides and wastelands. Fl. & Fr.:Aug. – Nov. Kalasamudram (ATP), *KRKS* 40078, 40444.

Cyperus stoloniferus Retz., Observ. Bot. 4: 10. 1786; FBI 6: 615. 1893; FPM 3: 1641. 1931.

Perennial, rhizomatous, stoloniferous herb; rhizome woody, creeping, forming ellipsoid or ovoid enlargement at base, clothed with dusky brown parallel fibres. Culms to 30 cm high, obtusely trignous, glaucous green, leaved at base. Leaves rigid, filiform, shorter than the culm; blades linear, to 4 mm wide, folded, recurved; sheaths pale, split into parallel brown fibres. Umbels simple and compact, to 5 cm long. Involucral bracts 3, foliaceous, lowest to 30 cm long, second briefly surpassing the inflorescence; primary rays 4, unequal, rigid, to 5 cm long. Spikes ovate, to 2 cm long with 3-6 spikelets in short axis. Spikelets linear-oblong, 10 mm long, to 18-flowered, slightly thickened with rather blunt edges, glaucous yellow and stained with sanguineous brown; rhachilla widely winged. Achenes ovate, oval, 3-sided, 1.5 mm long, maturing dark-brown.

Occsional near ponds and canals. Fl. & Fr.: May – Dec. Kalasamudram (ATP), *KRKS* 39140

Cyperus tenuispica Steud., Synops. Pl. Glumac. 2: 11. 1855; FPM 3: 1640. 1931.; FAP 3: 1071. 1997. *C. flavidus* sensu Clarke in FBI 6: 600. 1893.

Annual, slender, erect, tufted herb; culm to 20 cm high, triquetrous, finely ribbed. Leaves flat, linear, acute, to 15 cm long, umbel compound, to 7 cm long; involucral bracts 3, longest to 13 cm long, overtopping; primary rays 7–10, to 6 cm long; secondary rays 5–7, to 1 cm long, bearing heads of 5–10 stellately spreading minute spikelets. Spikelets narrowly linear, to 5 mm long, pale brown, *ca.* 20-flowered; rhachilla flexuous. Achenes globosely obovoid, trigonous, rounded, at first pale-yellow becoming white when ripe.

Occasional, in moist places. Fl. & Fr.: Nov. – Jan. Batrepalli (ATP), *KRKS* 42037, 222, 251 (HJCB-E).

ELEOCHARIS R. Br.

Key to Species

1a. Spikes 0.7 cm, elliptic, spikelet ovoid-ellipsoid *ca.* 8 mm long .... **E. congesta**

1b. Spikes 0.5 cm long, ovate, spikelet oblong-ovoid *ca.* 6 mm long .... **E. atropurpurea**

Eleocharis atropurpurea (Retz.) J. Presl & C. Presl, Reliq. Haenk. 1: 196. 1828; FBI 6: 627. 1893; FPM 3: 1648. 1931; FAP 3: 1073. 1997.

Annual tufted plant. Culms filiform, 2–15 cm high. Leaves reduced to sheaths. Spikelets ovoid-ellipsoid, 1.5–6 mm long, obtuse. Glumes ovate-oblong, *ca.* 1.3 mm long, obtuse, midrib green, sides membranous, purplish-brown; hypogynous bristles 4–7, shorter or as long as achene, white, retrorsely scabrid. Achene obovoid, bicovex, smooth, black, *ca.* 0.75 mm long; style 2-fid, style base white, triangular.

Occasional, in marshy places. Fl. & Fr.: Aug. – Nov. Nigidi forest: Batrepalli (ATP), *KRKS* 37849; Kalasamudram (ATP), *KRKS* 40098.

Eleocharis congesta D. Don, Prodr. Fl. Nepal 41. 1825; FBI 6: 630, 1893; FPM 3: 1648. 1931; FAP 3: 1073 & 1074.1997. subsp. **japonica** (Miq.) T. Koyama in Fl. Taiwan 5: 220. t. 1324. 1978 & in Dassan. & Fosb., Rev. Fl. Ceylon 5: 264. 1985.

Perennial, decumbent. Culms slender, 10–40 cm tall, with several longitudinal thin ribs. Leaf sheaths 2; cauline sheaths blood-red but later becoming pale blood-red apically, 3–8 cm, mouth truncate, apex mucronate to aristate. Spikelet oblong to narrowly oblong-ovoid, 5–8 × 3–4 mm, densely many-flowered, sometimes proliferous at spikelet base, apex obtuse to acute. Basal most glume empty, amplexicaul for whole spikelet base; fertile glumes blood-red but green medially, tightly imbricate, oblong, *ca.* 2.2 × 0.9 mm, margin very narrowly hyaline, apex subrounded. Perianth bristles 6, longer than nutlet, slightly densely and shortly retrorsely spinulose. Stigmas 2. Achenes black, 1 mm long.

Occasional, in rice fields. Fl. & Fr.: Oct – Jan. Kalasamudram (ATP), *KRKS* 39964.

FIMBRISTYLIS Vahl

1a. All glumes spirally arranged .... (2)
1b. Atleast lower glumes distichously arranged .... **F. ovata**
2a. Achenes sub-orbicular, obovate; stigmas 3 or 2 .... (3)
2b. Achenes oblong-cylindrical; stigmas 2 .... **F. tetragona**
3a. Stigmas 3; achenes trigonous .... (4)
3b. Stigmas 2; achenes biconvex: (occasionally 3 in *F. cymosa*) .... (7)
4a. Spikes elongated, 0.5–1.5 cm long .... (5)
4b. Spikes globose or subglobose, 0.5 cm along, spiklets few .... **F. littoralis**
5a. Perennials, with conspicuous rizhome, stem 40–70 cm high .... (6)
5b. Annuals, rhizome absent, stem < 30 cm high .... **F. tenera**
6a. Leaves surrounding culm bases bladed .... **F. complanata**
6b. All or upper few leaves surrounding culm bases blade less .... **F. quinquangularis**
7a. Inflorescence corymbose with more than 2 rays and 5 spikelets .... (8)
7b. Inflorescence head like or consisting of 1-4 solitary spikes only .... (11)
8a. Styles hardly flattened, glabrous .... **F. cymosa**
8b. Styles dorsiventrally flattened, fimbriate or long ciliate atleast on upper part .... (9)
9a. Glumes pubescent on upper half .... **F. sieberiana**
9b. Glumes glabrous .... (10)
10a. Spikelets 1-1.8 mm wide, oblong .... **F. bisumbellata**
10b. Spikelets 2-4 mm wide, ovate .... **F. dichotoma**
11a. Inflorescence head like with many spikelets; style glabrous .... **F. argentea**
11b. Inflorescence with 1-4 solitary spikes; styles flimbriate .... **F. schoenoides**

Fimbristylis argentea (Rottb.) Vahl, Enum. Pl. 2: 294. 1806; FBI 6: 640. 1893; FPM 3: 1659. 1931; FAP 3: 1079.1997. *Scirpus argenteus* Rottb., Descr. Pl. Rar. 27. 1777 & Descr. Icon. Rar. 51. t. 17. f. 6. 1773.

Annual, caespitose, glaucous herb, to 10 cm high; culms filiform, trigonous, striate. Leaves filiform, to 5 cm long; ligule 0; sheaths stramineous. Inflorescence capitate, 5–12 in a cluster. Involucral bracts 4, overtopping, longest to 3 cm long. Spikelets sessile, cylindric-oblong, terete, obtuse, many-flowered, 3 mm long. Achenes obovoid, compressed, biconvex, fulvous.

Occasional, on wet sandy grounds of open grasslands and rice fields. Fl. & Fr.: Sep. – Dec.Bat repalli (ATP), *KRKS* 37984, 37847.

Fimbristylis bisumbellata (Forssk.) Bubani, Dodecanthea 30. 1850; FPM 3: 1658. 1931; FAP 3: 1079.1997. *Scirpus bisumbellata* Forssk., Fl. Aegypt. – Arab. 1: 15. 1775. *Fimbristylis dichotoma* C.B.Clarke in Hook. f., Fl. Brit. India 6: 636. 1893 non Vahl, 1806.

Annual, erect, tufted herb; culms to 30 cm high, trigonous. Leaves flat, to 12 cm long, pubescent; ligule hairy; sheaths membranous. Umbels compound, terminal, 5 × 5 cm. Involucral bracts 3, longest to 5 cm long; primary rays 3–5, to 3 cm long; secondary rays 3–5. Spikelets solitary, oblong, 7 × 1.8 mm, many-flowered. Achenes obovoid, biconvex, creamy-yellow.

Common, in rice fields, wet places and marshy areas. Fl. & Fr.: Sep. – Dec. Kalasamudram (ATP), *KRKS* 40101

Fimbristylis complanata (Retz.) Link, Hort. Berol. 1: 292. 1827; FBI 6: 646. 1893; FPM 3: 1659. 1931; FAP 3: 1089.1997. *Scirpus complanatus* Retz., Observ. Bot. 5; 14. 1791.

Perennial, tufted, rhizomatous herb, to 40 cm high; culms densely caespitose, apex compressed, base trigonous. Leaves crowded at the base, 20 cm long, coriaceous, glaucous, ligule shortly hairy; sheaths compressed, keeled. Umbels decompound, effuse, 6 cm long; involucral bracts 2, not overtopping, to 6 cm long, largest leaf like; primary rays 3–5, to 3 cm long; secondary rays 3–4. Spikelet solitary, ovoid-oblong, to 8 mm long, acute, *ca.* 12-flowered, brown with distichously arranged glumes. Achenes obovoid, trigonous, stipitate and stramineous.

Occasional, in wet or moist places, on grassy slopes and in forest margins. Fl. & Fr.: Aug. – Apr. Kalasamudram (ATP), *KRKS* 39536

Fimbristylis cymosa R. Br., Prodr. Pl. Nov. Holl. 228. 1810. subsp. **spathacea** (Roth) T. Koyama, Micronesica 1: 83. 1964 & in Dassan. & Fosb., Rov. Handb. Fl. Ceylon 5: 301. 1985. *F. spathacea* Roth, Nov. Sp. Pl. 24. 1821; FBI 6: 640. 1893; FPM 3: 1659. 1931; FAP 3: 1080. 1997.

Perennial, erect, tufted, rhizomatous herb, to 40 cm high; culms densely leafy below, obtusely trigonous. Leaves densely crowded on the root stock, spreading and recurved, narrowly linear, to 15 cm long, rigidly coriaceous, margins recurved and scaberulous, sheaths short, coriaceous. Umbels compound, to 6 cm long; involucral bracts 3, filiform, to 2.5 cm long, leaf like, with broadly dilated base; primary rays 3–5, to 4 cm long; secondary rays 3, spikelet globose, to 3 mm long, pale brown, rhachilla winged. Achenes obovoid, biconvex, turbinate, compressed.

Rare in sandy areas near river beds and marshy low lands. Fl.& Fr.: Mar. – Oct. Sanevaripalli (ATP), *KRKS* 40155; Kalasamudram (ATP), *KRKS & TP* 40155.

Fimbristylis dichotoma (L.) Vahl, Enum. Pl. 2: 287. 1806; FBI 6: 635. 1893; FPM 3: 1658. 1931; FAP 3 : 1081. 1997. subsp. **glauca** (Vahl) T. Koyama, Bot. Mag. Tokyo 87: 324. 1974. *Scirpus dichotomus* L., Sp. Pl. 50. 1753. *Fimbristylis glauca* Vahl, Enum. Pl. Obs. 2: 288. 1805.

Annual, tufted herb in a small clump; culms to 40 cm high, trigonous, leaves at base. Leaves to 10, as long as the culms; blade erect, patent, linear, to 30 cm high, herbaceous, glaucous green beneath, whitish green on upper surface, several veined; ligule of a fringe of dense white hairs; sheaths cylindrical with rounded back, the dorsal side pale greenish, the hyaline ventral side lightly brown. Umbels compressed, bearing *ca.* 6 spikelets, lax to subdense, to 7 cm long. Involucral bracts to 6, the lowest one foliaceous, remaining setaceous; rays to 5, unequal, to 3 cm long, compressed, patent, terminated by 3 spikelets. Spikelets solitary, ovoid, to 7 mm long, terete, chestnut brown, densely many-flowered. Achenes obovate, biconvex, 1.8 mm long, excluding gynophore, rounded-truncate at mucronate apex, each side cancellated with 7–13 rows of transversely oblong cells.

Common, along the sides of water courses, open grassy places, sandy localities and as a weed in cultivated lands. Fl. & Fr.: Mar. – Dec. Kalasamudram (ATP), *KRKS* 39932, 37692, 40101.

Fimbristylis ferruginea (L.) Vahl, Enum. Pl. Obs. 2: 291. 1805; FAP 3:1086. 1997; subsp. **sieberiana** (Kunth) Lye, Nord. J. Bot. 2: 335. 1987. *F. sieberiana* Kunth Enum. Pl. 2: 237. 1837.

Perennial, tufted, rhizomatous herb, rhizome small, woody; culms to 45 cm high, basally scabrid. Leaves flat, to 40 cm long, margins inrolled, scabrid; ligule short, hairy; sheaths purple brown, hairy. Inflorescence simple, to 3 cm long. Involucral bracts 3-5, lowest overtopping, to 5 cm long; rays 2, to 2 cm long. Spikelets solitary, terete, to 1 cm long, obtuse. Achenes obovoid, biconvex, stramineous, smooth, stipitate.

Occasional, in open marshy places and in cultivated fields. Fl. & Fr.: Jul. – Dec. Batrepalli (ATP), *KRKS* 37860.

Fimbristylis littoralis Gaudich, Voy. Uranie 413. 1829. *F. miliacea* auct. non (L.) Vahl, FAP 3: 1083. 1997.

A tufted, annual, herbaceous, erect, non tuberous sedge, to 50 cm high; stem quadrangular below, triangular above. Leaves longer than the stem, narrow, tapering to a fine point; sheaths subdistichous, striate with scarious margins. Inflorescence a decompound umbel; primary rays 4-many, unequal, slender, up to 7 cm long; involucral bracts 2–3, filiform from a broad base, shorter than the umbel. Spikelets subglobose, reddish-brown, 2.5 – 5 mm long. Glumes *ca.* 1.2 mm long, ovate, obtuse, rounded on the back, membranous, brown. Stamens 1–3. Achene *ca.* 0.5 mm long, narrowly obovate, trigonous, surface yellow, verruculose.

Common, in moist places of deciduous forests and low elevations. Fl. & Fr.: Oct. – Feb. Kalasamudram (ATP), *KRKS* 40094; Batrepalli (ATP), *KRKS* 207 (HJCB-E).

Fimbristylis ovata (Burm. f.) Kern, Blumea 15: 126. 1967. & in Steenis, Fl. Males. 1. 7: 595. 1974. *Carex ovata* Burm. f., Fl. Indica 194. 1768. *Abildgaardia ovata* (Burm. f.) Kral, Sida 4: 72. 1971. *Cyperus monostachyos* L., Mant. 2: 180. 1771. *Abildgaardia monostachyos* (L.) Vahl, Enum. 2: 296. 1805. *Fimbristylis monostachyos* (L.) Hassk., Fl. Jav. Rav. 61. 1848, (*monostachya*); FBI 6: 649. 1893; FPM 3: 1660. 1931

Perennial, erect, caespitose and leafy herb, to 40 cm high; culms filiform, trigonous. Leaves crowded on the swollen base of the stem, filiform, to 20 cm long, scabrid; ligule 0; sheaths stramineous. Spikelets solitary on the stem. Involucral bracts 2, not overtopping, to 1 cm long. Spikelets compressed, ovate, acute, shining, 2 lower glumes empty and longer than the others, cuspidate, to 1 cm long; rhachilla winged. Achenes obovoid, trigonous, stipitate, stramenious.

Occasional, on wet grass lands, stream banks, in bunds of rice fields and in marshy areas. Fl. & Fr.: Dec. – Mar. Kalasamudram (ATP), *KRKS* 39881; Nigidi forest (ATP), *KRKS* 216 (HJCB-E).

Fimbristylis quinquangularis (Vahl) Kunth, Enum. Pl. 2: 229. 1807; FBI 6: 644. 1893; FPM 3: 1659; FAP 3:1085.1997 1931. *Scirpus quinquangularis* Vahl, Enum. Pl. 2: 279. 1806. *Fimbristylis miliacea* (L.) Vahl, Enum. Pl. 267. 1805; FBI 6: 644. 1893; FPM 3: 1660. 1931. *Scirpus miliaceus* L., Syst. Nat (ed. 10) 2: 868. 1759.

Annual, densely tufted, rhizomatous herb, to 70 cm high. Culms arising from axils of bladed leaves which first form a leaf sheath; pentagonous, clothed at base with 4 sheaths. Culm sheaths bladeless, subterete, pale green, the ventral side hyaline and light yellow brown, the orifice obliquely truncate, hyaline on margin, the uppermost sheath to 18 cm long, second to 10 cm. Leaf blade dorsiventrally flat, shorter than the culm, to 30 m long, herbaceous, whitish green and 5-nerved on upper surface, smooth except on upper margins where spinulose scabrid; sheaths to 7 cm long, pale green, yellow-brownish, ventral side split as culm grows. Corymbs compound, to 9 cm long; rays 7, patent, to 5 cm long, scabrous; secondary corymbs bearing to 30 spikelets. Involucral bracts 5, setaceous, much shorter than the rays. Spikelets solitary, ovoidal, to 5 cm long, rusty brown, *ca.* 40-flowered.

Occasional in wet places, grass lands, along canal banks and often in rice fields. Fl. & Fr.: Aug. – Nov. Batrepalli (ATP), *KRKS* 37856; Somulavaripalli (ATP), *KRKS* 40155.

Fimbristylis schoenoides (Retz.) Vahl, Enum. Pl. 2: 286. 1806; FBI 6: 634. 1893; FPM 3: 1658. 1931; FAP 3: 1085.1997. *Scirpus schoenoides* Retz., Observ. Bot. 5: 14. 1788.

Annual, caespitose, glaucous herb, to 70 cm high; culms filiform, obtusely angled, striate. Leaves few, blade filiform, shorter than the culm, to 16 cm long, ligule shortly hairy; sheaths pinkish-brown. Involucral bracts 3, stiff, filiform, not overtopping, to 1.5 cm long. Spikelets 3, ovoid-subglobose, to 1 × 0.4 cm, rhachilla strongly winged. Achenes compressed, obovoid, biconvex, to 2 mm long, stipitate, striate.

Abundant in sandy areas along canals and also along the borders of rice fields. Fl. & Fr.: Sep. – Dec. Batrepalli (ATP), *KRKS* 39001, 217 (HJCB-E).

Fimbristylis tenera Schult., Mant. 2: 57. 1825; FBI 6: 642. 1893; FPM 3: 1660. 193. *F. monticola* Hochst. ex Steud., Syn. Pl. Cyp. 3: 111. 1855.

Annual, slender, caespitose herb, to 25 cm high; culms filiform, obtusely trigonous, striate. Leaves erect, many at the base of the culm, shorter than the culm, filiform, glaucous-green, 15 cm long. Umbel subcompound, rays 4 with 2 spikelets on each. Involucral bracts very short, setiform. Spikelets to 1 cm long, oblong, acute, compressed, dark brown; rhachilla with deep pits. Achenes stipitate, globosely obovoid, obtusely trigonous, yellow or brown.

Usually seen in moist low lands, sandy soils, grass fields. Fl. & Fr.: Aug. – Nov.. Kalasamudram (ATP), *KRKS* 40135, 40093

Fimbristylis tetragona R. Br., Prodr. 226. 1810; FBI 6: 631. 1893; FPM 3: 1658. 1931.

Perennial, caespitose herb, to 35 cm tall; culms slender, ribbed, quadrangular. Leaves reduced to membranous sheaths, grey. Spikelet solitary, hemispherical, or sub-globose, lowest glume hardly brachiform, coriaceous, deciduous; rhachilla thickly studded with deep tetragonal pits with raised edges. Achenes oblong, cylindrical, biconvex, tapering towards the base, stipitate, fulvous grey, to 2 mm long.

Occasional, in wet or moist open places, rice fields. Fl. & Fr.: Jul. – Dec. Batrepalli (ATP), *KRKS* 37950.

FUIRENA Rottb.

Fuirena ciliaris (L.) Roxb., Fl. Ind. 1: 184. 1820; FAP 3:1088. 1997. *Scirpus ciliaris* L., Mant. Pl. 182. 1771. *Fuirena glomerata* Lam., Tabl. Encycl. 1 : 150. 1791; FBI 6 : 666. 1893; FPM 3 : 1669. 1931.

Annual, slender erect, leafy herb, to 30 cms tall; culms sub-cylindric below and trigonous above, 2 mm wide, sparsely hairy above, glabrous below. Leaves linear-lanceolate, acuminate, to 9 × 0.4 cm; sheaths long, closed striate, slightly pubescent. Inflorescence umbellate, 3–10 in a cluster in terminal and axillary spikelets in subcapitate heads, 1.8 × 1.8 cm. Involucral bracts 0. Spikelets oblong, obtuse, brown, squarrose, lowest glumes longest, empty ciliate, 0.8 × 0.3 cm, many-flowered.

Common, usually seen in wet open ground, river banks, margins of ponds and rice fields. Fl. & Fr.: Sep. – Mar. Kalasamudram (ATP), *KRKS* 39537; Batrepalli (ATP), *KRKS* 220 (HJCB); *KRKS* 39002.

KYLLINGA Rottb.

Key to Species

1a. Culms up to 90 cm high; rhizomatous, steloniferous .... **K. brevifolia**

1b. Culms less than 60 cm high; rhizomes short or absent, stems tufted .... (2)

2a. Culm to 15 cm high; involucral bracts 3; spikelets oblong; achenes oblong .... **K. bulbosa**

2b. Culm to 30 cm high; involucral bracts 4; spikelets elliptic; achenes elliptic-obovate .... **K. odorata**

Kyllinga brevifolia Rottb., Desc. Nov. Pl. 13. t. 4, f. 3. 1773; FBI 6: 588. 1893; FPM 3: 1624. 1931; FAP 3: 1091. *Cyperus brevifolius* (Rottb.) Hassk., Cat. Hort. Bogor. 24. 1844.

Perennial, stoloniferous, erect, rhizomatous, perennial herb, to 90 cm high; rhizome, creeping, slender, covered with brownish scales; culms arranged in a single row along the rhizome, 9–26 cm long, triquetrous. Leaves generally shorter than the stem, basal, 1.5–17 × 0.1–0.3 cm, narrowly linear, acuminate, scabrid on margins in the upper part; sheaths membranous, 1.5–5 cm long, purplish-brown, lower ones bladeless. Inflorescence a terminal, single, globose head, 5–10 mm long and wide, greenish; leafy bracts 3–4, very unequal in length, the lowest up to 9 cm long. Spikelets *ca.* 3 × 1 mm, oblong. Glumes 4, distichous, basal 2 sterile, small, 3rd glume fertile, *ca.* 2.5 × 1 mm, ovate-mucronate, 4th glume sterile, *ca.* 3 mm long, keeled. Stamens 1 or 2 stigmas 2. Nut 1–1.5 mm long, *ca.* 0.5 mm wide, oblong, apiculate, biconvex, brown.

Common, in moist sandy places. Fl. & Fr.: Jul. – Dec. Kalasamudram (ATP), *PVP* 2805.

Kyllinga bulbosa P. Beauv., Fl. d' Oware & Benin 1: 11. t. 8. f. 1. 1805; FAP 3: 1091. 1997. *K. triceps* Rottb., Descr. Icon. Rar. 1 & t. 4. f. 6. 1773. *nom. illegit.*, FBI 6: 587. 1893; FPM 3: 1623. 1931.

Perenial, rhizomatous herb; culms to 15 cm high, densely tufted with short erect rhizome, obtusely trigonous, the base forming an ovoidal, corm-like enlargement clothed with brown fibres. Leaves few; blade flattish-plicate, sheaths stout, pale and basal ones tinged with light brown, sheaths disintegrate leaving a tuft of fibres. Inflorescence a head with 3 sessile spikes, pale greenish, central spike ovoid-globose, to 8 mm long, obtuse, the lateral ones globose. Involucral bracts 3, eventually reflexed, the longest to 10 cm long. Spikelets oblong, pale greenish, 1-flowered. Achenes oblong, laterally flattered, brownish.

Common, species usually found in grass lands, lawns, and almost everywhere in the plains. Fl. & Fr.: Aug. – Dec. Kalasamudram (ATP), *KRKS* 39698.

Kyllinga odorata Vahl, Enum. Pl. 2: 382. 1906. subsp. **cylindrica** (Nees ex Wight) T. Koyama, Gard. Bull. Singapore 30. 161. 1977 & in Dassan. & Fosb., Rev. Handb. Fl. Ceylon 5: 244. 1985. *K. cylindrica* Nees ex Wight, Contr. Bot. Ind. 91. 1834; FBI 6: 588. 1893; FPM 3: 1624. 1931.

Perennial, tufted, rhizomatous herb; rhizome clothed with brown fibres; culms to 30 cm high, slender, triquetrous, hardly thickened at base. Leaves few, shorter than the culm; blades flattish, plicate; sheaths pale brownish, remain intact. Inflorescence a head with 3 spikes, whitish green; central spike cylindrical; lateral spikes elliptical, much shorter than the central spike, each densely bearing many spikelets. Involucral bracts 4, reflexed, the lowest to 10 cm long. Spikelets patent, elliptic. Achenes elliptic-obovate laterally flattened, maturing dark brown.

Common, in moist places and open grass lands. Fl. & Fr.: May – Oct. Thummalapalle (KDP), *KRKS & KRS* 36991.

LIPOCARPA R. Br. *nom. cons.*

Key to Species

1a. Spikelet bracts awned at apex, spikelet prophyll and glume absent .... **L. squarrosa**

1b. Spikelet bracts not awned, spikelet prophyll and glume present .... (2)

2a. Glumes whitish to silver; spikelet bracts almost twice as long as the achene .... **L. chinensis**

2b. Glumes dark purple; spikelet bracts almost as long as the nut or achene .... **L. gracilis**

Lipocarpha chinensis (Osbeck) Kern, Blumea, Suppl. 4: 167. 1958 & in Steenis, Fl. Males. 1, 7: 521. 1974; FAP 3:1093.1997. *Scirpus chinensis* Osbeck, Dagb. Osting. Rosa 220. 1757. *Lipocarpha argentea* (Vahl) R. Br. ex Nees, Linnaea 9: 287. 1835; FBI 6: 667. 1893; FPM 3: 1670. 1931.

Annual, caespitose herb with reddish brown fibrous roots, culms to 50 cm high, obtusely trigonous, whitish green. Leaves 4 to a culm; blade linear, rather soft and slightly spongy, whitish green; sheaths to 15 cm long, subloosely surrounding the culm, pale, the lower ones stained with reddish purplish colour. Inflorescence a head with 7 spikes. Involucral bracts 3, unequal, dilated at base, much surpassing the head, the longest up to 20 cm long. Spikes oblong-ellipsoidal, terete, to 8 mm long, rounded at apex, whitish, densely bearing numerous bracteoles. Achenes oblong, 1.5 mm long, trigonous, straight to weakly curved, straw-brown.

Common, in marshy areas, swamps and rice fields. Fl. & Fr.: May – Oct. Kalasamudram (ATP), *KRKS* 40107.

Lipocarpha gracilis (Rich. ex Pers.) Nees, Linnaea 9: 287. 1834. *Hypolytrum gracile* Rich. ex Pers. Syn. Pl. 1: 70. 1805. *Lipocarpha sphacelata* (Vahl) Kunth, Enum. Pl. 2: 267. 1837; FBI 6: 667. 1893; FAP 3: 1094.1997. *Hypaelyptum sphacelatum* Vahl,

Enum. Pl. 2: 283. 1806. *Lipocarpha triceps* Nees in Wight, Contrib. Bot. India 92. 1834; FPM 3: 1670. 1931.

Annual, erect, caespitose herb, to 20 cm high; culms triquetrous. Leaves basal, canaliculated, shorter than stem, to 14 cm long; ligule 0. Involucral bracts 3, foliar, over-topping, to 7 cm long. Inflorescence terminal, capitate. Spikelets 3-5 in a cluster, globose or ovoid, many-flowered. Achenes ellipsoid, beaked, fuscous, stoutly sptipitate and trigonous.

Occasional, in damp grassy places and open wet fields and margins of swamps. Fl. & Fr.: Aug. – Jan. Talupula (ATP), *KRKS & BSS* 41915; Kalasamudram (ATP), *KRKS* 40090; Batrepalli (ATP), *KRKS* 37857, *KRKS* 241 (HJCB-E).

Lipocarpha squarrosa (L.) Goetgh., Wageningen Agric. Univ. Pap. 89 (1): 71. 1989. *Scirpus squarrosus* L., Mant. Pl. 2: 181 1771; FBI 6: 663. 1893; FPM 3 : 1666. 1931. *Rikliella squarrosa* (L.) J.Raynal, Adansonia n.s., 13: 154 1973; FAP 3: 1105. 1997.

Stem filiform, up to 20 cm high. Leaves shorter than stem, filiform, acuminate, erect. Spikes 1–4, in a lateral cluster, ovoid, sessile, 3–5 mm long, awned. Involucral bracts 2–3, lower erect, filiform, up to 10 mm long. Spikelet bracts 0.5–0.75 mm long, narrowly obovate, suddenly narrowed into a curved awn, caducous, reddish-brown, midrib greenish. Spikelet prophyll and glume absent. Achene 0.4–0.5 mm long, trigonous, obovoid-ellipsoid, dark brown, reticulate.

Common, in marshy sand areas merged with other cypers and grasses. Fl. & Fr. : Nov. – Jan. Nigidi forest, Batrepalli (ATP), *KRKS* 337 (HJCB-E), 431 (HJCB).

PYCREUS P. Beauv.

Key to Species

1a. Glumes distinctly cuspidate, apex retuse..**P. pumilus**

1b. Glumes not cuspidate (obtuse to acute), apex not retuse ....(2)

2a. Achene ellipsoid or obovoid, glumes dark red-brown to black .... **P. flavidus**

2b. Achene narrowly oblong, glumes straw-coloured to reddish-brown... **P. polystachyos**

Pycreus flavidus (Retz.) T.Koyama, J. Jap. Bot. 51: 316 1976; FAP 3: 1099. 1997. *Cyperus flavidus* Retz. Observ. Bot. 5: 13 1788. *Pycreus globosus* (All.) Rchb., Fl. Germ. Excurs. 140(10) 1831; FPM 3: 1627. 1931. *Cyperus globosus* All., Fl. Pedem. 1: 49 1785. *Pycreus capillaris* (J.Koenig ex Roxb.) Nees ex C.B.Clarke in Hook.f., Fl. Brit. India 6: 591 1893. *Cyperus capillaris* J. Koenig ex Roxb., Fl. Ind. 1: 194. 1820.

Annual, tufted plant. Stems up to 60 cm high. Leaves shorter than the stem, filiform, stiff. Umbel simple or compound; rays 3–5, subsessile, up to 5 cm long. Involucral bracts 2–4, up to 20 cm long. Spikelets loosely spicate in two rows, linear, tapering into an acute apex, 3–18 × 2–3 mm, 20–50-flowered, dark brown to black. Rachilla straight. Glumes *ca.* 2.5 mm long, obovate-oblong, obtuse, keel greenish-yellow, sides brown with hyaline margins. Achene 0.8 mm long, obovoid, apiculate, brownish, finely granular, pitted.

Occasional, in marshy areas, merged with grass. Fl. & Fr.: Aug. – Oct. Kalasamudram (ATP), *KRKS* 40091.

Pycreus polystachyos (Rottb.) P.Beauv., Fl. De Oware & Benin 2: 48, t. 86, f. 2. 1807; FBI 6: 592. 1893; FAP 3: 1100.1997.var. **polystachyos.** *Cyperus polystachyos* Rottb., Descr. Ic. Rar. Nov. Pl. 39. t. 11, f. 1. 1773. *Pycreus odoratus* Urban, Symb. Antill 2: 164. 1900; FPM 3: 1627. 1931.

Perennial erect, tufted rhizomatous herb, to 50 cm high; rhizome covered with purple scales; culms trigonous, smooth. Leaves shorter than the culms, flat, linear, gradually acuminate, scabrid on margins in the upper part; sheaths purple brown, 5 cm long. Inflorescence simple, reduced to a single head, 3 cm long and wide, rays sometimes present with spikelets spicately arranged. Involucral bracts 8, lower 5 much exceeding the inflorescence, others short; spikelets strongly compressed, yellow to straw-coloured, 20 mm long, linear-lanceolate, rhachilla flexuous, narrowly winged. Achenes oblong, biconvex, 1.2 mm long, dark brown, sharply apiculate.

Common, in marshy areas, rice fields and along water streams. Fl. & Fr.: Sep. – Dec. Batrepalli (ATP), *KRKS* 37951.

Pycreus pumilus (L.) Nees ex C.B. Clarke in Hook. f., Fl. Brit. India 6: 591. 1893; FPM 3: 1627. 1931; FAP 3: 1101.1997. *Cyperus pumilus* L., Cent. Pl. 2: 6. 1756. *Pycreus nitens* Nees, Nov. Act. Acad. Caes. Leopcarol. Nat. Cur. 19, Suppl. 1: 53. 1843; FBI 6: 591. 1893.

Key to Subspecies

1a. Inflorescence rays well developed ....subsp. **membranaceus**

1b. Inflorescence congested, head-like ....subsp. **pumilus**

Pycreus pumilus (L.) Nees ex C.B. Clarke subsp. **pumilus.**

Annual, densely tufted herb, to 15 cm high, culms slender, triquetrous. Leaves few, surpassing the culm, blade narrowly linear, soft, flat, light green, sheaths pale. Involucral bracts 4, patent, lower 3 much surpassing the inflorescence, the longest 15 cm long; primary rays 4, patent, up to 3 cm long, slender. Spikes subglobose, densely bearing many spikelets, the short rhachis up to 3 mm long. Spikelets spreading, oblong, 7 mm long, flattened, *ca.* 30-flowered, light greenish, rhachilla straight, wingless. Achenes elliptic, laterally flattened, lenticular, maturing brownish, contracted to truncate apex.

Occasional, in the drying beds and marshy fields. Fl. & Fr.: Aug. – Nov. Batrepalli (ATP), *KRKS* 37848, 37846

Pycreus pumilus (L.) Nees ex C.B. Clarke subsp. **membranaceus** (Vahl) T.Koyama, Gard. Bull. Singapore 30: 151. 1977. *Cyperus membranaceus* Vahl, Enum. Pl. 2: 330. 1806.

Culms densely tufted, 5–16 cm, smooth. Leaves shorter than culms or overtopping them, flat or channeled, acuminate, to 0.2 cm wide; sheaths pale brown. Inflorescence compound, to 5 cm, involucral bracts 3–5, unequal, overtopping, to 12 cm; spikes ovoid, 5–18-spikeletted, to 3 cm; spikelets oblong, compressed, subacute, 10–40-flowered, 0.4–1 cm; glumes ovate to elliptic, to 0.15 cm, membranaceous. Nut obovoid, biconvex, apiculate, stalked, brownish.

Rare in marshy lands of foot hills. Fl. & Fr.: Aug. – Dec. Kalasamudram (ATP), *KRKS* 40078.

SCHOENOPLECTIELLA Lye

1a. Spiklets 3–6 in a pseudolateral head .... **S. juncoides**

1b. Spiklets more than 10 in a pseudolateral head .... (2)

2a. Culms flowering in the upper half, not septate; glumes mucronate .... **S. lateriflora**

2b. Culms flowering in the lower half, septate when dry; glumes not mucronate .... (3)

3a. Culms robust; sheaths lax; achenes with transverse wavy lines .... **S. articulata**

3b. Culms slender; sheaths close; achenes with tranverse undulate ridge .... **S. senegalensis**

Schoenoplectiella articulata (L.) Lye, Lidia 6: 20: 2003; FAP 3: 1111. 1997. *Scirpus articulatus* L., Sp. Pl. 47. 1753; FBI 6: 656. 1893; FPM 3: 1666. 1931. *Schoenoplectus articulata* (L.) Palla, Bot. Jahrb. Syst. 10: 299. 1888; FAP 3: 1107.1997.

Perennial, short living, robust herb, to 40 cm high; culms terete; 0.8 cm wide below, transversely septate at intervals of 1 cm, hollow, smooth, glabrous deeply green and slightly shiny when alive, clothed at base with 3 bladeless, stramineous-brown, lax sheaths; upper 2 basal sheaths cylindrical, to 12 cm long, herbaceous, pale green and later becoming yellow-brownish, finely many-nerved, lower 1 sheath reduced and scale like. Inflorescence a pseudolateral head, globose or spherical, 4 cm in diameter, located at the midway portion of the (culm) stem or much lower down near the apex of upper-most sheath (depending upon the depth of the water at the locality as the inflorescence appear immediately above the water surface) densely bearing several to over 40 spikelets, light green. Involucral bract 1, stem like, to 40 cm long, terete, septate. Spikelets cylindrical or oblong, to 14 × 4 mm, pink-fulvous, acuminate. Achenes obovid, triquetrous, with transverse wavy lines and apiculate.

Common, species usually found in marshy places, shallow waters and also along the margins of ponds. Fl. & Fr.: Aug. – Dec. Kalasamudram (ATP), *KRKS* 39105.

Schoenoplectiella juncoides (Roxb.) Lye, Lidia 6 (1): 25. 2003. *Scirpus juncoides* Roxb. Hort. Bengal. 81. 1814; Fl. Ind., ed. Carey & Wall, 1: 218. 1820; Fl.Ind., ed. Carey, 1: 216. 1832; FPM 3: 1666. 1931. *Schoenoplectus juncoides* (Roxb.) Palla, Bot, Jahrb. Syst. 10: 229. 1889; FAP 3: 1108. 1997.

Annual, erect, caespitose herb without conspicuous rhizome, to 30 cm high; culms slender, terete, obtusely several-angled, light green, clothed at base with few sheaths only. Sheaths 3, the lower ones scale like, brownish, the upper one to 5 cm long, pale green, obliquely truncate. Inflorescence a pseudolateral head with 3-6 spikelets; involucral bracts to 13 cm long, rather suddenly subacute at apex, dilated at base. Spikelets oblong to ovoid-oblong, terete, 3 × 8 mm, straw-coloured, acute, densely many-flowered. Achenes broadly obovoid, unequally biconvex, 2 mm long, compressed, faintly transversely undulate, apiculalte.

Common, in the margins of rice fields, river banks, swamps and marshy localities. Fl. & Fr.: Sep. – Dec. Batrepalli (ATP), *KRKS* 37949.

Schoenoplectiella lateriflora (J.F.Gmel.) Lye, Lidia 6: 25. 2003. *Scirpus lateriflorus* J.F.Gmel., Syst. Nat. 2: 127 1791. *Scirpus supinus* var. *lateriflorus* (J.F.Gmel.) T.Koyama, J. Fac. Sci. Univ. Tokyo, Sect. 3, Bot. 7: 302 1958. *Schoenoplectus lateriflorus* (J.F.Gmel.) Lye, Bot. Not. 124: 290. 1971; FAP 3: 1108. 1997. *Scirpus supinus* Roxb., Fl. Ind. 1: 219. 1820 auct. non L. 1753. *S. supinus* var. *uninodis* sensu Clarke. in Hook. f., FBI 6: 656. 1893.

Tufted annual herb. Culm up to 30 cm high, terete. Leaves reduced to sheaths or with a very short erect blade. Sheaths 5 cm long, membranous, pale green, the orifice obliquely truncate with a 2 mm long mucro, frequently the upper most sheaths born 3 cm above the base of the culm. Inflorescence pseudolateral, head like. Involucral bract 1, stem-like, 10 cm long, shorter than the stem. Spikelets *ca.* 15, in a lateral cluster, all sessile or a few peduncled, ovoid, 4–6 mm long. Glumes 2–2.5 mm long, ovate, apiculate, strongly keeled above, pale brown. Hypogynous bristles absent. Anthers 3; styles slender, branches 3. Achene 1–1.2 mm long, globosely obovoid, trigonous, black, sharply transversely ridged, black.

Common, in wet grassy places, margins of ponds, rice fields. Fl. & Fr.: Aug. – Dec. Batrepalli (ATP), *KRKS* 37859.

Schoenoplectiella senegalensis (Steud.) Lye Lidia 6: 27. 2003. *Schoenoplectus senegalensis* (Steud.) Palla Bot. Jahrb. Syst. 10(4): 299. 1888; FAP 3: 1110. 1997. *Isolepis senegalensis* Steud., Syn. Pl. Glumac. 2: 96. 1855. *Scirpus jacobi* C.E.C.Fisch., Bull. Misc. Inform. Kew 1931: 103 1931; FPM 3: 1666. 1931.

Perennial, erect, caespitose herb to 30 cm high; culms slender, subterete, transversely septate. Sheaths close, obliquely truncate. Inflorescence pseudolateral, capitate. Involucral bracts greater than 20 cm long, stem like. Spikelets *ca.* 20 in a cluster, ovoid-subglobose, 8 mm long, fulvous. Achenes obovoid, triquetrous, to 2 mm long, undulate, sharply angled, apiculate.

Common, in marshy places. Fl. & Fr.: Aug. – Dec. Kalasamudram (ATP), *NY* 1180.

POACEAE

Key to Sub-families

1a. Spikelets 2-flowered; upper hermaphrodite, lower male or barren; falling entire at maturity, dorsally compressed .... **Panicoideae**

1b. Spikelets 1 or more than 2-flowered; all alike; breaking up at maturity, leaving more or less persistent glumes or if falling entire, then not 2-flowered; laterally compressed................................ **Pooideae**

Key to the Tribes of Subfamily Panicoideae

1a. Spikelets usually paired or in threes, with one sessile and the other pedicelled; glumes as long as the spikelets, more or less rigid; upper lemma usually awned or reduced to an awn...... **Andropogoneae**

1b. Spikelets solitary or paired, more or less similar; glumes usually membranous, lower smaller or suppressed; upper lemma usually without or shortly awned... **Paniceae**

Key to Genera of Tribe Andropogoneae

1a. Spikelets of each pair usually similar .... (2)

1b. Spikelets in each pair dissimilar, the sessile bisexual, pedicelled male or barren .... (3)

2a. Inflorescence panicle; spikelets enclosed in white silky hairs........... **Saccharum**

2b. Inflorescence raceme; spikelets not as above .... **Lophopogon**

3a. Racemces solitary; joints of rachis and pedicel of pedicelled spikelets swollen, 3-angled, rounded or flattened.. (4)

3b. Racemes in pairs, digitate, fascicled or paniculate; joints of the rachis and the pedicel narrow, seldom thickened upwards, occasionally with a translucent longitudinal groove .... (8)

4a. Spikelets awned .... (5)

4b. Spikelts unawned... (7)

5a. All spikelets heterogamous .... (6)

5b. Lowest pairs of spikelets homogamous .... **Heteropogon**

6a. Sessile spikelet pitted; perennials.. **Dichanthium** (*p.p*)

6b. Sessile spikelets not pitted; annuals .... **Schizachyrium**

7a. Sessile spikelet globose, lower glume densely pitted.................... **Hackelochloa**

7b. Sessile spikelet and lower glume not as above.................................. **Rottboellia**

8a. Spikelets in threes .... **Chrysopogon**

8b. Spikelets in pairs .... (9)

9a. Racemes spathate.. (10)

9b. Racemes espathate .... (14)

10a. Involucral spikelets present.. (11)

10b. Involucral spikelets absent .... (12)

11a. Involucral spikelets pedicellate, deciduous .... **Iseilema**

11b. Involucral spikelets sessile, persistent .... **Themeda**

12a. Basal joints of racemes bulbous; spathes boat shaped............................ **Apluda**

12b. Basal joints of racemes and spathes not as above .... (13)

13a. Aromatic grasses; racemes in pairs often collected into huge panicles .... **Cymbopogon**

13b. Non aromatic grasses; racemes binate or rarely digitate, not collected into huge panicles.. **Andropogon**

14a. Upper lemma well developed; awn arises from the sinus............. **Ischaemum**

14b. Upper lemma reduced to a translucent base of awn; awn arises from the apex of the lemma .... (15)

15a. Joints and pedicels with a translucent longitudinal furrow .... **Bothriochloa**

15b. Joints and pedicels without a translucent furrow............... **Dichanthium** (*p.p*)

Key to Genera of Tribe Paniceae

1a. Inflorescence of 1-3 racemes, each consisting of a broad, flat and disarticulating rachis, bearing on the under surface of each joint shortly peduncled clusters of 1-6 spikelets mixed with small scale-like glumes .... **Trachys**

1b. Inflorescence and spikelets not as above... (2)

2a. Spikelets finely awned or mucronate from notched tips of the upper glume and lower lemma .... **Melinis**

2b. Spikelets not awned or if awned, awns from the entire tips of upper glumes and lower lemmas (*Echinochloa*) or from the tips of both the glumes (*Oplismenus*); upper lemma mostly crustaceous, rarely chartaceous .... (3)

3a. Spikelets falling singly, not subtended by bristles, or if present, then the bristles persisting after spikelets have fallen.. (4)

3b. Spikelets with an involucre of bristles; falling with or without the bristles .... (15)

4a. Inflorescence more or less open panicles or contracted and spike like panicles .... (5)

4b. Inflorescence one-sided spikes or spike like racemes; spikes or racemes digitate or scattered, rarely solitary .... (7)

5a. Spikelets subtended by bristles.. **Setaria**

5b. Spikelets not subtended by bristles.. (6)

6a. Spikelets arranged in open panicles..**Panicum**

6b. Spikelets arranged in cylindrical spike like panicles........................ **Sacciolepis**

7a. Upper lemma more or less crustaceous or coriaceous ....(8)

7b. Upper lemma thinly cartilaginous ....(14)

8a. Lower glume reduced to annular callus at base of spikelet .... **Eriochloa**

8b. Lower glume not as above...(9)

9a. Glumes as long as spikelets; well developed and awned.............. **Oplismenus**

9b. Lower glume usually reduced or absent; mucronate, not awned ....(10)

10a. Lower glume (when present) turned away from rachis; back of upper lemma facing it..(11)

10b. Lower glume turned towards rachis, back of upper lemma turned away from it.. **Brachiaria**

11a. Lower glume absent ....**Paspalum**

11b. Lower glume present ....(12)

12a. Ligule absent.. **Echinochloa**

12b. Ligule present..(13)

13a. Upper lemma acute or mucronate.. **Paspalidium**

13b. Upper lemma not as above...**Urochloa**

14a. Spikelets awned... **Alloteropsis**

14b. Spikelets awnless ....**Digitaria**

15a. Involucral bristles rigid or spiny; united at base into a cup .... **Cenchrus**

15b. Involucral bristles plumose; free at base ....**Pennisetum**

Key to the Tribes of Sub-family Pooideae

1a. Culms woody, arborescent; lodicules 3..................**Bambuseae**(*Dendrocalamus*)

1b. Culms otherwise; lodicules 2 ....(2)

2a. Spikelets arranged in solitary, digitate or scattered spikes, spike like racemes .... **Chlorideae**

2b. Spikelets arranged in open or contracted or spike like panicles in racemes or spikes...(3)

3a. Spikelets with one fertile floret; with or without one or two male or barren florets below it ....(4)

3b. Spikelet with 2 or more fertile florets or if with one, then with sterile reduced florets above it ....(9)

4a. Glumes minute, scale like; fertile lemma and palea similar in texture .... **Oryzeae**

4b. Glumes well developed; fertile lemma and palea dissimilar ....(5)

5a. Inflorescence of solitary spike .... **Perotideae**(*Perotis*)

5b. Inflorescence otherwise .... (6)

6a. Spikelets awned .... (7)

6b. Spikelets unawned .... (8)

7a. Awn trifid .... **Aristideae**(*Aristida*)

7b. Awn single, not trifid .... **Arundinellae** (*Arundinella*)

8a Upper glume prickled .... **Zoysieae** (*Tragus*)

8b. Upper glume smooth .... **Sporoboleae** (*Sporobolus*)

9a. Spikelets 4-many-flowered .... **Eragrostideae**

9b. Spikelets 2-3-flowered .... (10)

10a. Culms robust; found near streams and river beds; lemmas 1-awned .... **Arundineae**

10b. Culms not robust; found in dry lands; lemmas 9-awned .... **Pappophoreae** (*Enneapogon*)

Key to the Genera of Tribe Arundineae

1a. Panicle contracted; lemmas hairy all over the back, rhachilla glabrous .... **Arundo**

1b. Panicle effuse; lemmas glabrous; rhachilla villous .... **Phragmites**

Key to the Genera of Tribe Chlorideae

1a. Inflorescence terminal, spicate .... (2)

1b. Inflorescence of digitate spikes or spike like racemces .... (4)

2a. Inflorescence of spaced clusters of awned spikelets on a simple rachis .... **Melanocenchris**

2b. Inflorescence a simple spike or spikes of awned or awnless spikelets .... (3)

3a. Culms more than 50 cm height; spikelets not sunk in the rachis .... **Enteropogon**

3b. Culms not more than 10 cm height; spikelets sunk in the rachis .... **Oropetium**

4a. Spikelets awned, with one or more sterile florets above the fertile floret .... (5)

4b. Spikelets unawned, without sterile florets .... **Cynodon**

5a. Spikes 1-3; fertile florets 2 or more .... **Tetrapogon**

5b. Spikes more than 4; fertile floret one .... **Chloris**

Key to Genera of Tribe Eragrostideae

1a. Lemmas entire, obtuse, acute or acuminate at the apex .... (2)

1b. Lemmas usually emarginated or 2-4-lobed or toothed at the apex .... (6)

2a. Inflorescence long pyramidal spike; glumes aristate or aristate-acuminate .... **Dinebra**

2b. Inflorescence and glumes otherwise .... (3)

3a. Inflorescence of open or contracted panicles or solitary secund spikes .... (4)

3b. Inflorescence of digitate or racemosely arranged spikes or spike-like racemes .... (5)

4a. Inflorescence a solitary secund spike .... **Eragrostiella**

4b. Inflorescence of open or contracted panicle .... **Eragrostis**

5a. Culms flattened; axis and branches of the inflorescence ending in a spikelet .... **Eleusine**

5b. Culms terete; axis of spikes terminating in a sharp point .... **Dactyloctenium**

6a. Spikelets in panicles or in digitate or racemosely arranged spike like racemes .... **Leptochloa**

6b. Spikelets in terminal solitary secund spikes .... (7)

7a. Lower glume of sessile spikelet when young with two tufts of hair, upper glume with one tuft; ligule membranous .... **Tripogon**

7b. Lower glume of sessile spikelet without tufts of hair; ligule a row of hairs .... **Sehima**

Key to Genera Tribe Oryzeae

1a. Fertile lemma supported by two empty scales; tip of pedicel with 2 lips .... **Oryza**

1b. Fertile lemma solitary; tip of pedicel without lips; spikelets imbricate .... **Leersia**

ALLOTEROPSIS Presl

Alloteropsis cimicina (L.) Stapf, Prain, Fl. Trop. Afr. 9: 487.1919; FPM 3: 1766. 1934; FAP 3: 1127.1997. *Milium cimicinum* L., Mant. Alt. 184. 1771.

Annual, culms erect, tufted, to 80 cm high. Leaf-sheaths covered with tubercle based hairs. Ligules fimbriate. Leaf-blades ovate-lanceolate, 2–10 × 1–1.5 cm, margins tubercle based hairy, acute. Racemes 4–10, digitate, 5–15 cm. Spikelets broadly elliptic, 2–4 × 2 mm, solitary. Lower glume ovate-lanceolate, 2 × 1 mm, acuminate, 3-nerved. Upper glume broadly ovate, 2–4 × 2 mm, ciliate, acuminate, 5-nerved. Lower lemma ovate, 1.5–3 × 1 mm, acuminate, 5-nerved. Lower palea elliptic, 1.5–2 × 0.5 mm, hyaline, bilobed at apex. Stamens 3. Lodicules truncate, 0.5 mm. Upper lemma elliptic, 2–3 × 1 mm, aristate, margin ciliate. Upper palea ovate, 2–3 × 1 mm, 2–keeled, chartaceous, papillose. Grains ovate.

Common, as undergrowth in forests and in open dry habitats. Fl. & Fr.: Jun. – Mar. Kalasamudram (ATP), *KRKS* & *BSS* 37905.

ANDROPOGON L.

Andropogon pumilus Roxb., Fl. Ind. 1: 277. 1820; FPM 3: 1753. 1934; FAP 3: 1128.1997.

Annual, culms erect, tufted, to 1 m high; often geniculate. Leaf-sheaths glabrous. Ligules truncate, fimbriate. Leaf-blades linear-lanceolate, 2–7 × 0.5 cm, glabrous, acuminate. Racemes 2, digitate, paired in axillary clusters, spathaceous. Spikelets in pairs, sessile bisexual and pediceled staminate or neuter. Sessile spikelet oblong, 3–4 × 1.5 mm. Lower glume linear-lanceolate, 3–4 × 1 mm, chartaceous, acuminate. Upper glume boat-shaped, 3–4 × 1 mm, 3-nerved, aristate. Lower lemma oblong, 2–2.5 × 1 mm, hyaline. Upper lemma linear, 2–2.5 × 1 mm, hyaline base of the awn; awn to 2 cm long. Palea lanceolate, 1–2 × 1 mm, hyaline. Stamens 3. Grains ovoid. Pedicelled spikelet akin to sessile spikelet.

Common, in plains, field edges and usually in dry habitats. Fl. & Fr.:Aug.–Jan. Batrepalli (ATP), *KRKS* & *BSS* 39601; Batrepalli (ATP), *KRKS* HJCB-E-345.

APLUDA L.

Apluda mutica L., Sp. Pl. 82. 1753; FPM 3: 1750. 1934; FAP 3: 1128. 1997.

Perennial, culms erect, tufted, to 2 m high; often rooting at lower nodes. Leaf-blades linear-lanceolate, 5–25 × 0.5–1.5 cm, usually petiolate, glabrous or villous on veins. Panicles terminal, to 90 cm long. Racemes solitary, spatheolate. Spikelets 3, one sessile, 2 pedicelled. Sessile spikelets oblong, 4–10 mm, laterally compressed, awned. Lower glume elliptic, 5–8 × 1.5 mm, chartaceous, acute, 11–14-nerved. Upper glume broadly–ovate, 5–6 × 1.3 mm, chartaceous, 13-nerved. Lower lemma oblong-lanceolate, 3–5 × 0.4–1 mm, hyaline, 3–5-nerved. Lower palea akin to its lemma, 2-nerved. Stamens 3. Upper lemma lanceolate, 7–8 × 2 mm, hyaline, 3-nerved, awned. Upper palea ovate-acute, 2–2.5 × 1 mm, hyaline. Stamens 3. Grains oblong. Pedicelled spikelet lanceolate, 4–6 × 1.5 mm, sterile. Lower glume lanceolate, 3.5–4 × 1 mm, chartaceous, 13-nerved. Upper glume ovate–lanceolate, 4 × 2 mm, chartaceous, 6–11-nerved. Florets more or less akin to that of sessile spikelet.

Common, in plains, wastelands, forests and moist localities. Fl. & Fr.:Jun. – Dec. Thummalapalle (KDP), *KRKS* & *KRS* 36977.

ARISTIDA L.

Key to Species

1a. Awns unequal; laterals much reduced .... **A. redacta**

1b. Awns subequal; laterals well–developed, but slightly shorter than the central .... (2)

2a. Panicles effuse, as broad as long, culms often creeping .... **A. hystrix**

2b. Panicles and culms not as above .... (3)

3a. Panicles subcylindric, glossy .... **A. mutabilis**

3b. Panicles not subcylindric, dull green .... (4)

4a. Awns of the lemma to 1.5 cm; perennial .... **A. setacea**

4b. Awns of the lemma more than 1.5 cm; annuals .... (5)

5a. Culms more than 30 cm long; leaves folded; panicles flabelliform .... **A. funiculata**

5b. Culms, leaves and panicles not as above .... **A. adscensionis**

Aristida adscensionis L., Sp. Pl. 82. 1753; var. **adscensionis;** Bor, Grass. India 407. 1960; FAP 3: 1130. 1997. *A. depressa* Retz., Observ. Bot. 4: 22. 1786; FPM 3: 1809. 1934.

Local Name: *Ooba gaddi.*

Annual or perennial; culms erect, tufted, to 30 cm high. Leaf-sheaths glabrous. Ligules membranous. Leaf-blades linear, 5–10 × 0.2 cm, glabrous, acuminate. Panicles contracted, to 15 cm long. Spikelets linear-lanceolate, 4–8 mm long, purplish; callus bearded. Lower glume lanceolate, 5 × 1 mm, chartaceous, 1-nerved. Upper glume oblong-lanceolate, 5–8 × 1 mm, chartaceous, 1-nerved. Lemma linear-oblong, 6–7 × 1 mm, chartaceous, compressed, scabrid on the keels, 3-awned, awns to 40 mm. Palea oblong, 2 × 1 mm, hyaline. Stamens 3. Grains linear.

Gregarious on dry sandy soil of the forests, wastelands and dry lands. Fl. & Fr.: Feb. – Sept. Kalasamudram (ATP), *KRKS* & *BSS* 40448.

Aristida funiculata Trin. & Rupr., Sp. Gram. Stip. 159.1842; FPM 3: 1809.1934; FAP 3: 1131. 1997.

Annual, culms erect or ascending, tufted, to 70 cm high. Leaf-sheaths glabrous. Ligules membranous. Leaf-blades linear, 5–15 × 0.3 cm, folded, glabrous. Panicles linear, to 15 cm long. Spikelets linear-lanceolate, 8–15 × 2 mm; callus pungent. Lower glume linear, 10–12 × 1 mm, chartaceous, hyaline, aristate, 1-nerved. Upper glume linear-lanceolate, 10–15 × 1 mm, chartaceous, aristate, 1-nerved. Lemma linear-lanceolate, 9–11 × 1 mm, convolute, chartatceous, 3-awned, awn to 20 mm long. Palea oblong, 3 × 1 mm, hyaline. Stamens 3. Grains linear.

Common, in dry wastelands, cultivated fields, hedges and forest plains. Fl. & Fr.: Sept. – Dec. Batrepalli (ATP), *KRKS* & *BSS* 39608.

Aristida hystrix L. f., Suppl. Pl. 113. 1781; FPM 3: 1809. 1934; FAP 3: 1131. 1997.

Perennial, culms ascending or creeping, tufted, to 80 cm high. Leaf-sheaths glabrous, pilose at the mouth. Ligule membranous, soft hairy. Leaf-blades linear, 2–8 × 0.1–0.3 cm, chartaceous, glabrous. Panicles lax, 8–25 × 5–10 cm. Spikelets lanceolate-oblong, 7–15 × 1.5 mm, purple; callus pungent. Lower glume oblong-lanceolate, 5–10 × 1–1.5 mm, chartaceous, aristate, 1-nerved. Upper glume elliptic-lanceolate, 8–15 × 1.5–2 mm, chartaceous, aristate, 1-nerved. Lemma linear, 7–15 × 1 mm, chartaceous, 3-awned, awn 40 mm. Palea narrowly oblong, 5 × 1 mm, hyaline. Stamens 3. Grains linear.

Common, in dry habitats of wastelands, forest plains and cultivated fields. Fl. & Fr.: Sept. – Dec. Kalasamudram (ATP), *KRKS* & *KRS* 36497.

Aristida mutabilis Trin. & Rupr., Sp. Gram. Stip. 150. 1842; FPM 3: 1809. 1934; FAP 3: 1132. 1997.

Annual, culms erect or ascending, tufted, to 40 cm. Leaf-sheaths glabrous. Ligule membranous. Leaf-blades linear, 5–15 × 0.5 cm, convolute, glabrous. Panicles subcylindric, to 15 cm. Spikelets linear-lanceolate, 8 × 1 mm; callus pungent. Lower glume oblong-lanceolate, 4.5 × 0.5 mm, chartaceous, aristate, 1-nerved. Upper glume oblong-lanceolate, 7 × 1 mm, chartaceous, aristate, 1-nerved. Lemma linear, 4 × 0.7 mm, chartaceous, 3-awned; awn 15 mm. Palea oblong, 1.3 × 0.3 mm, hyaline. Stamens 3. Grains linear.

Rare in plains of dry habitats. Fl. & Fr.: Apr. – Sept. Batrepalli (ATP), *KRKS & BSS* 39602.

Aristida redacta Stapf, Kew Bull. 85. 1892; FPM 3: 1809. 1934; FAP 3: 1132. 1997.

Perennial, culms creeping, tufted, branched, to 80 cm high. Leaf-sheaths glabrous. Ligule ciliate, membranous. Leaf-blades linear-lanceolate, 5–15 × 0.1–0.2 cm, acuminate, hairy. Panicles effuse, to 40 cm long. Spikelets linear-lanceolate, 8–10 × 1 mm; callus pungent. Lower glume linear, chartaceous, aristate, 1-nerved. Upper glume linear-lanceolate, 9–10 × 1 mm, chartaceous, aristate, 1-nerved. Lemma linear-lanceolate, 9–10 × 1 mm, chartaceous, 1-nerved, 3-awned, middle awn to 40 mm. Palea oblong, 3 × 1 mm, hyaline. Stamens 3. Grains linear–oblong.

Rare in dry habitats of hillocks. Fl. & Fr.: Sept. – Mar. Bakkannagaripalli (KDP), *KRKS & BSS* 41944.

Aristida setacea Retz., Observ. Bot. 4: 22. 1786; FPM 3: 1809. 1934; FAP 3: 1132. 1997.

Local Name: *Nakkusa gaddi.*

Perennial, culms erect, tufted, to 2 m high, base woody. Leaf-sheaths glabrous. Ligules membranous. Leaf-blades linear-lanceolate, 10–30 × 0.2–0.5 cm, glabrous. Panicles flabelliform, contracted, to 30 cm. Spikelets linear-lanceolate, 10–15 mm; callus pungent. Lower glume oblong-lanceolate, 10–12 × 0.5–1 mm, chartaceous, aristate, 1-nerved. Upper glume oblong-lanceolate, 10–15 × 1.5 mm, chartaceous, aristate, 1-nerved. Lemma linear or lanceolate, 9–12 × 1 mm, convolute, 3-awned, awn 15 mm long. Palea small, 2 × 1 mm, hyaline. Stamens 3. Grains linear.

Common, in forest plains, open dry, rocky places and near streams. Fl & Fr.:All seasons. Mabbuchintalapalle (KDP), *KRKS & KRS* 33666; *KRKS & BSS* 39156; Talupula (ATP), *KRKS* 224 HJCB-E; Batrepalli (ATP), *KRKS* 232 HJCB-E.

ARUNDINELLA Raddi

Key to Species

1a. Spikelets 6–7 mm; perennials .... **A. setosa**

1b. Spikelets 5–6 mm; annuals .... **A. nervosa**

Arundinella nervosa (Roxb.) Nees ex Hook. & Arn., Bot. Beechey Voy. 237.1836; FPM 3: 1801.1934.

Annual, culms erect, tufted, to 80 cm high. Leaf-sheaths covered with tubercle based hairs. Ligules fimbriate. Leaf-blades linear-lanceolate, 4–8 × 0.5–1 cm, covered with tubercle based hairs. Panicles to 15 cm. Spikelets elliptic-lanceolate, 4–5 × 1 mm. Lower glume oblong-lanceolate, 4 × 1.3 mm, chartaceous, 3-nerved. Upper glume ovate-lanceolate, 4.5 × 1.3 mm, chartaceous, 5-nerved. Lower lemma elliptic or oblong, 2.5–3 × 1 mm, chartaceous, 5-nerved. Lower palea lanceolate, 2.5 × 0.8 mm, hyaline, 2-nerved. Stamens 3. Upper lemma oblong-elliptic, 2 × 1.5 mm, sub-coriaceous, notched, awned with 2 setae, 1.5 mm on either side; awn 5 mm. Upper palea lanceolate, 1.5 × 0.5 mm, hyaline, 2-nerved. Stamens 3. Grains ovate.

Occasional, in forest slopes, grass lands usually in moist localities. Fl. & Fr.: Oct. – Jan. Chinnapalli (ATP), *KRKS* 40372.

Arundinella setosa Trin., Gram. Panic. 63.1826; var. **setosa**. FAP 3: 1137.1997. *A. mutica* Nees ex Steud., Syn. Pl. Glum. 1: 116.1854; FPM 3: 1801.1934.

Perennial, culms erect, tufted, to 1.5 m high, woody at base. Leaf-sheaths hirsute, tubercled based hairy. Ligules membranous. Leaf-blades linear-lanceolate, 8–20 × 0.5–0.8 cm, villous. Panicles effuse, to 20 cm. Spikelets ovate-lanceolate, 6–8 × 3 mm. Lower glume oblong, 4.5–5 × 1 mm, cymbiform, chartaceous, 3-nerved, sparsely hairy, awned. Upper glume elliptic-lanceolate, 5–6 × 1.5 mm, chartaceous, 5-nerved, sparsely hairy. Lower lemma elliptic-lanceolate, 4.5–5.5 × 1 mm, chartaceous, 7-nerved. Lower palea ovate, 4–4.5 × 1 mm, hyaline, 2-nerved. Upper lemma oblong-lanceolate, 4–5 × 1 mm, awn to 15 mm long, setae 2, 1.5 mm long, scaberulous. Upper palea oblong-lanceolate, 4 × 1 mm, hyaline, 2-nerved. Stamens 3. Grains elliptic.

Common, in forest plains, near streams and especially in saline moist soils. Fl. & Fr.: Jul. – Dec. Batrepalli (ATP), *KRKS* & *BSS* 40338; Batrepalli (ATP), *KRKS* 232 (HJCB-E).

ARUNDO L.

Arundo donax L., Sp. Pl. 1: 81. 1753; FPM 3: 1805. 1934; FAP 3: 1139. 1997.

Perennial, rhizomatous, culms erect, robust, woody below, to 4 m high. Leaf-sheaths glabrous. Ligules long hairy. Leaf-blades linear-lanceolate, 20–40 × 2–4 cm, glabrous. Panicles to 40 cm long, large, decompound, erect, thyrsiform. Spikelets laterally compressed, oblong, 10–15 mm, 3-flowered, rhachilla glabrous. Lower glume lanceolate, 10–12 × 2 mm, acuminate, 3-nerved. Lemmas 2–4, lanceolate, 10–15 × 2 mm, chartaceous, 3-nerved, lower half with long-silky hairs. Palea oblong-lanceolate, 5 × 1.5 mm, hyaline, 2-nerved. Stamens 3. Ovary oblong.

Occasional, along stream beds and riverbeds. Fl. & Fr.: Jul. – Dec. Gandi temple (KDP), *KRKS* & *BSS* 39069.

BOTHRIOCHLOA Kuntze.

Key to Species

1a. Lower glume of the sessile spikelet with two pits .... **B. pertusa**

1b. Lower glume of the sessile spikelets without pits ....**B. pseudoischaemum**

Bothriochloa pertusa (L.) A.Camus, Ann. Soc. Linn. Lyon, 1930, n.s. 76: 164. 1931. *Dichanthium pertusum* (L.) W.D. Clayton, Kew Bull. 32: 4. 1977; FAP 3: 1174. 1997. *Andropogon pertusus* (L.) Willd., Sp. Pl. 4: 922. 1806; FBI 7: 173. 1896. *Amphilophis pertusus* (L.) Nash ex Stapf, Agric. News W. Ind. 15: 179. 1916; FPM 3: 1731. 1934.

A perennial, stoloniferous grass. Culms up to 50 cm high, ascending; nodes sparsely hairy. Leaf-sheaths 2–6 cm long, glabrous; ligule membranous. Leaf-blades 3–10 × 0.2–0.4 cm, linear, base truncate, apex acuminate, glabrous. Racemes 2–6, 5 cm long, slender, digitate, silky hairy. Spikelets 6 mm, binate, pitted. Sessile spikelet 4 mm, oblong-lanceolate, pitted; callus bearded. Lower glume elliptic-oblong, bearded below, dorsally pitted, subchartaceous. Upper glume ovate-lanceolate, finely pointed at the tip. Lower lemma lanceolate, nerveless, hyaline. Upper lemma reduced to a slender awn. Stamens 3. Grain oblong. Pedicelled spikelets 4 mm, male or barren, similar to sessile spikelet, but awnless.

Common, in plains and foot hills of the deciduous forests. Fl. & Fr.: Jul. – Nov. Mabbuchintalapalle (KDP), *KRKS* 206 (HJCB-E); Batrepalli (ATP), *KRKS* 236 (HJCB-E), Thummalaplle (KDP), *KRKS* 263 (HJCB-E).

Bothriochloa pseudoischaemum (Nees ex Steud.) Henr., Blumea 3: 457. 1940. *Andropogon pseudoischaemum* Nees ex Steud., Syb. Pl. Glum. 1: 380. 1854; FBI 7: 172. 1896. *Amphilophis pseudoischaemum* (Steud.) C.E.C. Fischer, Fl. Madras 3: 1731. 1934. *Dichanthium pseudoischaemum* (Nees ex Steud.) S.K.Jain & Deshp., Bull. Bot. Surv. India 20: 134. 1979; FAP 3: 1174. 1997.

A perennial, tufted, herbaceous grass. Culms to 2 m, erect, 3–6 mm wide, hollow, glabrous, braching from the base; internodes 10–18 cm; nodes bearded. Leaf-sheaths to 13 cm, glabrous, hispid at mouth. Ligule membranous, to 4 mm, shortly fimbriate. Leaf-blades oblong, to 25 cm, scaberulous, base rounded, apex setaceous. Racemes to 10 cm long; rachis straight, 2.5 mm, pilose along margins. Spikelets 1 cm, 2-nate. Sessile spikelet: to 5 mm, elliptic-lanceolate, bisexual. Lower glumes to 4.5 mm, flat at the back, 5-7-nerved, not pitted, 2-keeled, ciliate, apex 2-denate; upper glumes 5 mm, elliptic-lanceolate, rigid, 3-nerved, ciliate, apex aristate, to 0.5 mm long; lower lemma to 4 mm, hyaline, palea 0; upper lemma to 1.5 mm; awn to 2 cm. Stamens 3. Grains ellipsoid, compressed, 1.5 mm. Pedicelled spikelet to 5 mm as sessile spikelet, but awnless.

Common, along rocky streams and water courses. Fl. & Fr.: Aug. – Nov. Batrepalli (ATP), *KRKS* 37985; Nigidi forest (ATP), *KRKS* 39935.

BRACHIARIA Griseb.

Key to Species

1a. Panicle linear; lower glume minute, scale like .... **B. eruciformis**

1b. Panicle lanceolate or oblong; lower glume up to half as long as the spikelets.. (2)

2a. Racemes two or rarely 3–4.. **B. distachya**

2b. Racemes more than two... (3)

3a. Spikelets 1.2–2 mm long .... **B. reptans**

3b. Spikelets 3–4.5 mm or more long.. **B. ramosa**

Brachiaria distachya (L.) Stapf, in Prain, Fl. Trop. Afr. 9: 565. 1919; FPM 3: 1769.1934; FAP 3: 1141. 1997. *Panicum distachyon* L., Mant. Pl. 183. 1771.

Annual, culms erect or ascending, to 50 cm, rooting at nodes. Leaf-sheaths hairy. Ligule fimbriate. Leaf-blades elliptic-lanceolate, 1–5 × 0.2–0.5 cm, slightly pubescent. Racemes 2 or rarely 3, hairy below the rhachis. Spikelet ovate-elliptic, 1.5–2.5 × 1.5 mm, solitary. Lower glume ovate-obovate, 1.2 × 1.3 mm, membranous, 7-nerved. Upper glume broadly ovate, 1.5–2.5 × 1–1.5 mm, chartaceous, 7-nerved. Lower lemma akin to the upper glume, 5-nerved. Palea oblong, 1.5–2 × 0.5 mm, hyaline, 2-nerved. Upper lemma ovate, 2–2.5 × 1–1.5 mm, crustaceous, rugulose, 5-nerved. Palea ovate-oblong, 2–2.3 × 0.3–1 mm, crustaceous, rugulose, 2-nerved. Stamens 3. Grains elliptic.

Occasional, in plains near moist areas. Fl. & Fr.: Nov. – Apr. Vemula (KDP), *KRKS* & *BSS* 41951.

Brachiaria eruciformis (Sm.) Griseb., in Ledeb., Fl. Ross. 4: 469. 1853; FPM 3: 1769. 1934; FAP 3: 1142. 1997. *Panicum eruciforme* Sm., in Sibth. & Sm., Fl. Graeca 1: 44. t. 59. 1806.

Annual, culms erect or ascending, to 40 cm high, rooting at lower nodes. Leaf-sheaths tubercled based hairy. Ligule a tuft of hairs. Leaf-blades elliptic-lanceolate, 2–5 × 0.5 cm, glabrous. Panicles sub-secund, racemosly arranged spikes, 4–5 cm. Spikelets ovate-oblong, 2.5 × 1 mm, pubescent. Lower glume ovate-lanceolate, 0.3 × 0.2 mm, membranous, minute, scale like. Upper glume broadly ovate, 2 × 1 mm, membranous, 5-nerved, hairy. Lower lemma lanceolate-ovate, 2 × 0.8 mm, chartaceous, pubescent, 3-nerved. Lower palea truncate, 1.5 × 0.7 mm, hyaline, 2-nerved. Upper lemma ovate, 1.5 × 0.9 mm, crustaceous, 3-nerved. Upper palea broadly elliptic, 1.4 × 0.7 mm, crustaceous, rugulose, 2-nerved. Stamens 3. Grains ovoid.

Common, in irrigated fields and moist places in plains. Fl. & Fr.: Jul. – Dec. Velpula (KDP), *KRKS* & *KRS* 37679.

Brachiaria ramosa (L.) Stapf. in Prain, Fl. Trop. Afr. 9: 542. 1919; FPM 3: 1770. 1934; FAP 3: 1143. 1997. *Panicum ramosum* L., Mant. Pl. 1: 29. 1767.

Annual or perennial, culms erect or decumbent, to 50 cm high, rooting at lower nodes. Leaf-sheaths glabrous. Ligules hairy. Leaf-blades linear-lanceolate, 5–15 × 1–2 cm, pubescent. Panicles lax, to 15 cm. Spikelets broadly ovate–elliptic, 2–3 × 1–1.5 mm. Lower glume obovate, 1–1.5 × 3 mm, chartaceous, 7-nerved. Upper glume broadly ovate, 2.5–3 × 2 mm, chartaceous, 7-nerved. Lower lemma ovate, 2.5–3 × 2 mm, chartaceous, 5-nerved. Palea lanceolate-oblong, 2 × 0.5 mm, hyaline, 2-nerved. Upper lemma broadly ovate, 2–2.5 × 1.5–2 mm, crustaceous, rugose, 5-nerved. Palea ovate, 2 × 1 mm, crustaceous, 2-nerved. Stamens 3. Grains oblong.

Common, in plains, wastelands and in moist localities. Fl. & Fr.: Feb. – Oct. Kalasamudram (ATP), *KRKS* 36926; Rachakuntapalle (KDP), *KRKS* & *BSS* 39842.

Brachiaria reptans (L.) Gard. & C. E. Hubb., in Hook., Icon. Pl. Sub tab. 3363. 1938; FAP 3: 1144. 1997. *Panicum reptans* L., Syst. Nat. 870.1759. *Urochloa reptans* (L.) Stapf in Prain, Fl. Trop. Afr. 9: 610. 1920; FPM 3: 175. 1934.

Annual, culms erect or creeping, to 50 cm high, rooting at nodes. Leaf-sheaths glabrous, margins ciliate. Ligules hairy. Leaf-blades lanceolate, 2–5 × 0.5–1 cm, glabrous or sparsely hairy. Panicles to 5 cm, racemes 5–10, alternate. Spikelets elliptic, 1.5–2 × 1 mm, acute. Lower glume truncate or obovate, 0.5–1 × 0.5 mm, chartaceous, 5-nerved. Upper glume ovate, 1.5–2 × 1 mm, chartaceous, 5-nerved. Lower lemma ovate, 1.5–2 × 1 mm, chartaceous, 7-nerved. Lower lemma ovate, 1.5–2 × 1 mm, chartaceous, 7-nerved. Lower palea ovate-elliptic, 1.5–2 × 1 mm, hyaline, 2-nerved. Upper lemma ovate, 1–1.5 × 1 mm, crustaceous, rugulose. Upper palea ovate, 1–1.2 × 1 mm, 2-nerved, crustaceous, rugulose. Stamens 3. Grains obovate.

Common, in dry and moist localities of wastelands. Fl. & Fr.: Jul. – Dec. Kalasamudram (ATP), *KRKS & KRS* 33678, *KRKS* & *BSS* 39155.

CENCHRUS L.

Key to Species

1a. Bristles or spines of the involucres antrorsely scabrid.....................................(2)

1b. Bristles or spines of the involucres retrorsely scabrid........................ **C. biflorus**

2a. Bristles connate at the base only... **C. ciliaris**

2b. Bristles connate into a cup 1–3 mm long. – inner bristles rigid ...... **C. setigerus**

Cenchrus biflorus Roxb., Fl. Ind. 1: 238. 1820; FAP 3: 1145. 1997. *C. barbatus* Schum., Beskr. Guin Pl. 43.1827; FPM 3: 1749.1934.

Annual, culms erect or ascending, tufted, to 70 cm high. Leaf-sheaths glabrous, ciliate at mouth. Ligules ring of hairs. Leaf-blades linear lanceolate, 5–10 × 0.5–0.8 cm, sparsely hairy, acuminate. Racemes solitary, cylindric, 5–12 cm long; involucral bristles retrosely barbed. Spikelets subglobose, 3–5 mm. Lower glume absent. Upper glume oblong-lanceolate, 3–4.5 × 2 mm, membranous, 3-nerved. Lower lemma lanceolate, 4–4.5 × 1.5 mm, membranous, 5-nerved. Lower palea ovate-lanceolate, 3–4 × 1 mm, hyaline, male or empty. Upper lemma oblong-lanceolate, 3.5–4 × 1 mm, membranous, scabrid on nerves. Upper palea akin to its lemma. Stamens 3. Grains oblong.

Occasional weed of dry land agricultural fields, especially edges. Fl. & Fr.: Aug. – Dec. Vemula (KDP), *KRKS* & *BSS* 41950.

Cenchrus ciliaris L. Mant. Pl. 2: 302. 1771; FAP 3: 1146. 1997; FPM 3: 1749. 1934. *Pennisetum cenchroides* Rich. in Pers. Syn. Pl. 1: 72. 1805, *nom. superfl;* FBI 7: 88. 1896.

Perennial, culms erect, to 50 cm high, much branched. Leaf-sheaths 4 – 6 cm long, keeled, glabrous, ciliate near mouth. Ligule a pubescent ring. Leaf-blades 7–15 × 0.2–0.5 cm, linear-lanceolate, base rounded, apex acuminate, glabrous-sparsely tuberculate pilose. Racemes 6 – 10 cm long, solitary, cylindric. Involucral bristles antrorsely barbellate, base of the involucres small, elliptic, naked; outer ones filiform, inner bristles not very stout or thickened at the base, densely ciliate on the margins, flattened, connate below into a ring. Spikelets 6 mm, oblong-lanceolate, 1 – 3 in each involucral. Lower and upper glumes similar, ovate, hyaline, 2.8-3.3 × 1 mm, 1-nerved. Lower lemma ovate, 2 × 1.5 mm, hyaline. Upper lemma ovate-lanceolate, 4.5-5.5 × 1 mm, acuminate, 5-nerved. Lower palea hyaline, male. Upper palea lanceolate, bisexual. Stamens 3. Grain oblong.

Occasional in plains and foot hills of the forests, also roadsides. Fl. & Fr.: Sep. – Jan. Agriculture fields, Pulivendula to Kadapa roadside (KDP), *KRKS* 221 (HJCB-E).

Cenchrus setigerus Vahl, Enum. Pl. Obs. 2: 395. 1806.*C. biflorus* sensu FBI 7: 89. 1896; FAP 3: 1145. 1997 *p.p.*non Roxb.

Perennial, culms erect or ascending, to 60 cm high. Leaf-sheaths keeled Leaf-blades linear, 15–30 × 0.4–0.6 cm, adaxial surface pilose with long scattered hairs. Ligule 0.5 mm, membranous. Inflorescence rather stiff, 4–12 × 0.6–0.7 cm, burrs overlapping by about half their length, rachis scabrous-puberulous. Burrs broadly oblong with rounded base, 0.3–0.7 cm, subsessile; inner spines connate to half the their length forming a tough cupule, flattened free tips narrowly triangular, erect, antrorsely scaberulous with a broad green groove on outer side, inner side shortly ciliate; outer spines very short or almost suppressed, reduced to bristles around periphery of cupule. Spikelets 1–4 in burr, 3.5–5 mm; lower glume half the spikelet length; upper glume smaller as spikelet length. Grain units are inserted along a zig-zag axis, each bur-like fascicle comprising a single spikelet or cluster of 2 or 3 spikelets.

Common in agriculture fields. Fl. & Fr.: Oct. – Dec. Pulivendula to Kadapa roadside (KDP), *KRKS* 201 (HJCB-E).

Note: A hardy and drought-tolerant grass, *Cenchrus setigerus* is a valuable fodder in dry areas, though it is tussocky and lacks bulk. It is quite palatable and readily accepted by stock. It makes useful hay in areas that can be mown though the yield is not high.

CHLORIS Swartz.

Key to Species

1a. Perfect florets two .... **C. bournei**

1b. Perfect floret one .... (2)

2a. Culms puberulous below the inflorescence .... **C. quinquesetica**

2b. Culms glabrous below the inflorescence .... **C. barbata**

Chloris barbata Sw., Fl. Ind. Occ. 1: 200.1797; FPM 3: 1838.1934; FAP 3: 1148. 1997.

Annual or perennial, culms erect or ascending, tufted, to 50 cm high. Leaf-sheaths glabrous. Ligules truncate, membranous. Leaf-blades linear-lanceolate, 5–10 × 0.3–0.5 cm, glabrous, acuminate. Spikes 4–15, digitate. Spikelets obovate, 2–3 mm long, 1–3-flowered. Lower glume lanceolate, 1.5–2 × 1 mm, membranous, 1-nerved, scabrid on the keels. Upper glume lanceolate, 2–3 × 1 mm, membranous, 1-nerved, scabrid on the keels. Fertile lemmas ovate-elliptic, 1.5–3 × 1 mm, acute, awned, 3-nerved, scabrid. Empty Lemmas 1–2, obovate, 1 mm, coriaceous, 3-nerved, awned. Paleas elliptic, 2–2.5 × 1 mm, hyaline, lanceolate, 2-nerved. Stamens 3. Grains sub-trigonous.

Common, in plains, usually in moist areas. Fl. & Fr.: Apr. – Dec. Kalasamudram (ATP), *KRKS* & *BSS* 39549.

Chloris bournei Rang. & Tad., J. Indian Bot. 2: 189. 1921; FPM 3: 1838. 1934; FAP 3: 1148. 1997.

Perennial, culms erect or decumbent, to 1.2 m high. Leaf-sheaths glabrous. Ligules a ciliate rim. Leaf-blades linear-lanceolate, 20 × 0.4 cm, glabrous, acuminate. Spikes 4–8, digitate. Spikelets ovate, 4.1 × 2.5 mm. Lower glume lanceolate, 3 × 0.8 mm, chartaceous, 1-nerved, acute, scabrid. Upper glume lanceolate, 4 × 0.9 mm, chartaceous, 1-nerved, acute, scabrid. First fertile lemma elliptic–ovate, 4.2 × 2 mm, 3-nerved, chartaceous, acute, scabrid, awned. Second fertile lemma ovate, 3.2 × 1.3 mm, chartaceous, 3-nerved, scabrid, awned. Third fertile lemma ovate, 3 × 1.3 mm, chartaceous, 3-nerved, scabrid, awned. First sterile lemma ovate, 2 × 1.3 mm, chartaceous, 3-nerved, awned. Second sterile lemma ovate, 1.5 × 0.8 mm, chartaceous, 3-nerved, awned. Paleas 3, ovate, 3.2 × 1.3 mm, hyaline, 2-nerved, scabrid, acute. Stamens 3. Grains ovoid.

Rare along agricultural field margins especially in black cotton soils. Fl. & Fr.: Sept. – Jan. Way to Gollalaguduru (KDP), *KRKS & BSS* 41962.

Chloris quinquesetica Bhide, J. Proc. Asiat. Soc. Beng. (n.s.) 8: 311. 1912; FAP 3: 1150. 1997.

Perennial, culms erect or ascending, to 50 cm high; rooting at lower nodes. Leaf-sheaths glabrous. Ligules fimbriate, membranous. Leaf-blades linear-lanceolate, 5–15 × 0.5 cm, glabrous, acuminate. Spikes 4–6, digitate. Spikelets obovate, 2–4 mm, 1–4–flowered. Lower glume oblong-lanceolate, 2–2.5 × 1 mm, membranous,

1-nerved, scabrid on the keels. Upper glume elliptic-oblong or lanceolate, 2–3 × 1 mm, membranous, 1-nerved, scabrid on the keels. Fertile lemmas elliptic-ovate, 2–3.5 × 1.5 mm, acute, 3-nerved, awned, lateral nerves bearded. Empty lemmas broadly ovate, 2–3 × 2 mm, truncate, coriaceous, 3-nerved, awned. Paleas elliptic, 2–3 × 1.5 mm, hyaline, 2-nerved. Stamens 3. Grains trigonous.

Occasional, in plains and wastelands, especially in saline soils. Fl. & Fr.: Aug.–Dec. Peddarangapuram (KDP), *KRKS* & *BSS* 40487.

CHRYSOPOGON Trin. *nom. cons.*

1a. Pedicels half the length of the sessile spikelets or longer .... **C. asper**

1b. Pedicels shorter than half the length of the sessile spikelet............................. (2)

2a. Leaves glabrous.. **C. fulvus**

2b. Leaves puberulous.. **C. aucheri**

Chrysopogon asper B. Heyne ex Blatt. & McCann, Bombay. Grass. 68. 1935; FPM 3: 1738. 1934; FAP 3: 1153. 1997. *Andropogon asper* B. Heyne ex Hook. f., Fl. Brit. India 7: 189. 1896.

Perennial, culms erect, robust, tufted, to 120 cm high. Leaf-sheaths with tubercle-based bristles. Ligules fimbriate, membranous. Leaf-blades linear-lanceolate, 5–35 × 0.5–1.5 cm, glabrous, margin serrulate. Panicles to 10–20 cm, branches whorled. Sessile spikelets oblong-lanceolate, 4–6 mm long, awned. Lower glume linear–oblong, 5–6 × 1 mm, chartaceous, 3–5-nerved. Upper glume oblong–ovate, 5–6 × 1.5 mm, awned, chartaceous, 7-nerved. Lower lemma linear-lanceolate, 3–5 × 1 mm, hyaline, acute, 2-nerved, margins ciliate. Upper lemma oblong-lanceolate, 2.5–4 × 1 mm, hyaline, awned, 2-nerved. Stamens 3. Grains oblong. Pedicelled spikelets linear-lanceolate, 8–13 mm, awned. Lower glume lanceolate, 8–12 × 1.5–2 mm, chartaceous, awned, 7-nerved, keels ciliate. Upper glume lanceolate, 10–12 × 1.5–2.5 mm, acute, 3-nerved, ciliate. Lower lemma lanceolate, 8–10 × 1–1.5 mm, hyaline, 2-nerved, margins ciliate. Upper lemma linear-oblong, 4.5–6 × 1 mm, hyaline, 1-nerved. Stamens 3.

Occasional, in forests usually in slopes. Fl. & Fr.: Aug. – Dec. Rachakuntapalle (KDP), *KRKS & BSS* 40176; Kalasamudram (ATP), *KRKS* & *BSS* 40321.

Chrysopogon aucheri (Boiss.) Stapf, Kew Bull. 1907: 21. 1907. *Andropogon aucheri* Boiss., Diagn. Pl. Orient. 5: 77. 1844.

Perennial, culms erect or ascending, densely tufted, to 50 cm high, basal sheaths silky. Leaf-sheaths compressed, puberulous, basal one distinct with tubercle based hairs. Ligules fimbriate. Leaf-blades linear-lanceolate, 12–20 × 0.6–0.8 cm, puberulous, margins with tubercle based hairs near the base, acuminate. Panicle lax, 5–10 × 4 cm. Sessile spikelets elliptic, 6–7 × 2–2.5 mm, laterally compressed. Lower glume linear, 4–5 × 1 mm, chartaceous, obscurely 6-nerved. Upper glume oblong, 5 × 2.5 mm, chartaceous, 1-nerved, awned. Lower lemma linear-obtuse, 3.5 × 1 mm, hyaline, ciliate above. Upper lemma linear, 4 × 1.5 mm, hyaline, modified as dilated base of the awn. Stamens 3. Grains linear. Pedicelled spikelets linear-lanceolate,

6.5 × 1.2 mm. Lower glume lanceolate, 6 × 1.5 mm, chartaceous, 8-nerved, awned. Upper glume lanceolate, 6 × 1 mm, chartaceous, 3-nerved, ciliate above. Lower lemma lanceolate, 5 × 0.7 mm, hyaline, 3-nerved, shortly ciliate on the margins. Upper lemma lanceolate, 3 × 0.3 mm, hyaline, 1-nerved. Stamens 3.

Rare on rock slopes in dry forests. Fl. & Fr.: Jun. – Jan. Chinnapalli (ATP), *KRKS & BSS* 40378.

Chrysopogon fulvus (Spreng.) Chiov., Fl. Somala 1: 327. 1929; FAP 3:1153. 1997. *Pollina fulva* Spreng., Pl. Minus Cognit. Pug. 2: 10. 1815. *Chrysopogon montanus* Trin. ex Spreng., Neue Entdeck. 2: 93. 1821; FPM 3: 1739. 1934.

Perennial, culms erect, tufted, to 70 cm high. Leaf-sheaths glabrous, ciliate at mouth. Ligules membranous. Leaf-blades linear-lanceolate, 5–15 × 0.3–0.5 cm, glabrous, acuminate, margins sparsely scaberulous. Panicles ovate, 3–6 cm, branches whorled. Sessile spikelets elliptic-lanceolate, 5–6 × 2 mm, callus bearded. Lower glume linear-lanceolate, 4–4.5 × 1.5 mm, chartaceous, margins ciliate. Upper glume boat-shaped, 4.5–5 × 1.5 mm, chartaceous, densely hispid on back, awned. Lower lemma elliptic-lanceolate, 3–3.5 × 1 mm, membranous, ciliate. Upper lemma elliptic-lanceolate, 2.5–3.5 × 1 mm, membranous, 3-nerved, awned. Palea oblong, 3 × 1 mm, hyaline. Stamens 3. Grains linear. Pedicelled spikelets elliptic-lanceolate, 5–6 mm, male, pedicel villous. Lower glume oblong, 3.5–4 × 1 mm, chartaceous, 7-nerved, awned. Upper glume elliptic-lanceolate, 3–4 × 1 mm, chartaceous, 3-nerved. Lower lemma ovate, 3–3.5 × 1 mm, chartaceous, 2-nerved, ciliate. Upper lemma elliptic, 2–3 × 1 mm, hyaline. Palea lanceolate, 2–3 × 1 mm, hyaline.

Common, in plains, hill slopes of the forests, wastelands and grass lands. Fl. & Fr.: Aug. – Dec. Rachakuntapalle (KDP), *KRKS* & *SSR* 33629.

CYMBOPOGON Spreng.

1a. Lower glume of the sessile spikelet with a deep slot-like depressions occupying most of its length .... **C. gidarba**

1b. Lower glume of the sessile spikelet without a slot, either flat, convex, shallowly concave or with a deep narrow slit in the lower half .... (2)

2a. Lower glume of the sessile spikelet with a narrow slit-like groove in the lower half which appears as a rib on the inner surface .... **C. caesius**

2b. Lower glume of the sessile spikelet flat, slightly convex, shallowly concave, wrinkled or with 1–3 very shallow depressions .... **C. coloratus**

Cymbopogon caesius (Nees ex Hook. & Arn.) Stapf, Kew Bull. 341. 360. 1960; FPM 3: 1757. 1934; FAP 3: 1160. 1997. *Andropogon caesius* Nees ex Hook. & Arn., Bot. Beech. Voy. 244. 1838.

Perennial, culms erect, tufted, to 2.5 m high. Leaf-sheaths glabrous. Ligules ovate, membranous. Leaf-blades linear, 10–25 × 0.5–1 cm, glabrous, acuminate. Panicles contracted, to 20 cm, spathes 3 cm, spatheoles 2 cm long. Racemes paired, 1.5 cm, joints villous. Sessile spikelets oblong, 2.5–4 × 0.5–0.8 mm. Lower glume oblong, 3–3.8 × 1 mm, coriaceous, 3–5-nerved, acute, keels winged. Upper glume elliptic-lanceolate, 3.5–4 × 1 mm, coriaceous, keels winged, margins ciliate. Lower lemma elliptic-oblong, 2.5–3 × 1 mm, hyaline, 2-nerved, margins ciliate. Upper lemma linear, 2 × 0.5 mm, hyaline, 1-nerved, awned. Stamens 3. Grains oblong. Pedicelled spikelets elliptic, 2.5–4 × 1 mm, male. Lower glume oblong, 3.5–4 × 1.5 mm, coriaceous, 12-nerved, keels scaberulous. Upper glume elliptic, 3–4 × 1 mm, hyaline, 3-nerved, ciliate. Lower lemma oblong, 3.5–4 × 1 mm, hyaline, 2–3-nerved, margins ciliate. Stamens 3.

Common, in forest peripheries and field edges. Fl. & Fr.: Jul. – Dec. Chinnakudala (KDP), *KRKS* & *BSS* 40466; Kalasamudram (ATP), *KRKS* 40334; Thummalapalle (KDP), *KRKS* & *SSR* 33652.

Cymbopogon coloratus (Hook.f.) Stapf, Kew Bull. 321. 1906; FPM 3: 1756. 1934; FAP 3: 1161. 1997. *Andropogon nardus* L. var. *coloratus* Hook. f., Fl. Brit. India 7:206. 1896.

Perennial, culms erect, tufted, to 2 m high. Leaf-sheath base tomentose. Ligules ovate, membranous. Leaf-blades linear-lanceolate, 15–25 × 0.5–1 cm, chartaceous, glabrous. Panicles 10–40 cm, spathes 2 cm long; spatheoles 1.5 cm. Racemes paired, 1.5 cm, joints villous. Sessile spikelets oblong, 4–6 × 0.8 mm. Lower glume elliptic-lanceolate, 4–5 × 1 mm, chartaceous, 2–dentate, 3–5-nerved. Upper glume oblong-lanceolate, 3–4 × 1 mm, membranous, keels ciliate. Lower lemma lanceolate, 2–3 × 1 mm, hyaline, keels ciliate. Upper lemma lanceolate, 2–3 × 1 mm, hyaline, bifid, awned, ciliate. Stamens 3. Grains linear. Pedicelled spikelets linear-lanceolate, 4.5 × 1 mm, male. Lower glume elliptic-lanceolate, 3–4 × 1 mm, chartaceous, keeled. Upper glume lanceolate-oblong, 3.5–4 × 1 mm, chartaceous, keeled. Lemmas oblong-lanceolate, 2.5–3.5 × 1 mm, hyaline. Stamens 3.

Common, in dry localities especially on hill slopes. Fl. & Fr.: Jul. – Dec. Rachakuntapalle (KDP), *KRKS* & *SSR* 33622; Kalasamudram (ATP), *KRKS* 264 (HJCB-E).

Cymbopogon gidarba (Buch.-Ham. ex Steud.) Haines, Bot. Bihar and Orissa Pt. 5: 48. 1924; FAP 3: 1162. 1997; FPM 3: 1757. 1934. *Andropogon gidarba* Buch. Ham. Num. List. no. 8797:1828., FBI 7: 208. 1896.

An aromatic, tufted, perennial grass. Culms 50–100 cm high. Leaf-sheaths terete, glabrous. Ligule short, truncate. Leaf-blades linear, 4–40 × 0.4 – 0.6 cm, flat or folded, abaxial glabrous, adaxial scabrid-puberulous, sometimes with long scattered hairs, margin ciliate at base, entire towards apex, apex acuminate, glabrous. Panicle narrow, spathe slender; spatheoles linear, tightly convolute, 2–3 cm branches simple. Racemes pinkish grey, *ca.* 2 cm; rachis internodes and pedicels stoutly cuneate, margins shortly but densely ciliate or woolly, back pubescent or subglabrous; pedicel of homogamous pair and lowest internode swollen. Spikelets 6 mm long, 2-nate, joints and pedicels shortly villous. Sessile spikelet pale or dark green, bisexual. Lower glume deeply concave in lower half, channels puberulous with a prominent swelling at base, keels rounded alongside channel, sharp and scabrid above, wingless, 2–3-veined between keels; upper glume cymbiform, 1-nerved, acute. Lower lemma lanceolate, 2-nerved; upper lemma linear, bidentate,

A. *Cymbopogon coloratus*, B. *Phragmites karka*, C. *Ischaemum afrum*, D. *Leersia hexandra*, E. *Oropetium thomaeum*, F. *Themeda triandra*

ciliate, 0.7–0.9 cm long awned. Stamens 3, stigmas plumose. Pedicelled spikelet 3–4 mm; longer than sessile spikelet, lanceolate, acute.

Rare in plains and near rocks in deciduous forests. Fl. & Fr.: Oct. – Jan. Batrepalli, Nigidi forest (ATP), *KRKS* 237 (HJCB-E).

CYNODON Rich. *nom. cons.*

Cynodon dactylon (L.) Pers., Syn. Pl. 1: 85. 1805; FPM 3: 1835. 1934; FAP 3: 1165. 1997. *Panicum dactylon* L., Sp. Pl. ed. 1: 58. 1753.

Perennial, culms erect or geniculate, slender, stoloniferous, rhizomatous, to 50 cm high. Leaf-sheaths glabrous. Ligules fimbriate, membranous. Leaf-blades linear, 5–10 cm long, glabrous, scaberulous on the margins. Racemes 4-6, digitate. Spikelets elliptic-lanceolate, 2–2.5 × 1 mm. Lower glume lanceolate, 1.5 × 0.5 mm, chartaceous, 1-nerved. Upper glume lanceolate, 1.5–2 × 0.5 mm, chartaceous, 1-nerved. Lemma firm, 2–2.5 × 1 mm, boat-shaped, pubescent with clavate hairs, 3-nerved. Palea oblong, 2–2.5 × 1 mm, hyaline, 2-nerved. Stamens 3. Grains oblong.

Common, in plains, fields and margins of the forests in moist localities. Fl. & Fr.: Apr.–Nov. Thummalapalle (KDP), *KRKS* & *KRS* 33693.

DACTYLOCTENIUM Willd.

1a. Spikes 2–6; stolons present .... **D. aegyptium**

1b. Spikes one or rarely 2; stolons absent .... **D. aristatum**

Dactyloctenium aegyptium(L.) P. Beauv., Ess. Agrost. Expl. Pl. 15. 1812; FPM 3: 1840. 1934; FAP 3: 1167. 1997. *Cynosurus aegyptius* L., Sp. Pl. 72. 1753.

Annual, culms ascending or erect, stoloniferous, to 60 cm high, rooting at lower nodes. Leaf-sheaths glabrous. Ligules membranous. Leaf-blades linear-lanceolate, 8–20 × 0.5 cm, glabrous, sparsely ciliate on the margins. Spikes 2-6, digitate. Spikelets oblong, 3–4 × 2 mm, 3-5-flowered. Lower glume ovate, 2–2.5 × 1 mm, boat-shaped, chartaceous, 1-nerved, keels scabrid. Upper glume obovate, 2 × 1 mm, chartaceous, 1-nerved, keels scabrid. Lemma ovate-lanceolate, 2–2.5 × 1.5 mm, chartaceous, 3-nerved. Palea lanceolate, 2 × 1 mm, hyaline, 2-nerved. Stamens 3. Grains obovate, transversely rugose.

Common, in plains, cultivated fields and forest margins in moist localities. Fl. & Fr.: Jun. – Sept. Thummalapalle (KDP), *KRKS* & *KRS* 36412.

Dactyloctenium aristatum Link, Hort. Berol. 1: 59. 1827; FAP 3: 1167. 1997.

Annual, culms erect or geniculate, tufted, to 8 cm high; rooting at lower nodes. Leaf-sheaths hairy. Ligules ovate, membranous. Leaf-blades linear-lanceolate, 2–3 × 0.5 cm, acuminate, glabrous. Spike solitary, 1-2 cm long; rhachis slender. Spikelets ovate, 3.5 × 3.5 mm, 2-3-flowered. Lower glume ovate, 2.2 × 0.7 mm, chartaceous, 1-nerved, keels scabrid. Upper glume ovate-elliptic, 2.5 × 0.8 mm, chartaceous, keels scabrid, aristate. Lemma ovate-lanceolate, 4 × 1.5 mm, membranous, 3-nerved. Palea ovate-lanceolate, 3 × 0.8 mm, hyaline. Stamens 3. Grains globose, rugose.

Rare in plains and forest margins in moist areas. Fl. & Fr.: Jun.– Nov. Somulavaripalli (ATP), *KRKS & BSS* 40180.

DENDROCALAMUS Nees

Dendrocalamus strictus Nees, Linnaea 9: 476. 1834; FPM 3: 1858. 1934; FAP 3: 1168. 1997. *Bambusa stricta* Roxb., Pl. Coromondel t. 80. 1798.

Perennial, culms erect, robust, densely tufted, to 15 m high; branched root stock, nodes swollen. Culm sheaths glabrous, auricle short, ciliate. Ligules narrow, serrate. Leaf-blades linear-lanceolate, 6–15 × 1–2 cm, stiff, petiolate, glabrous above, softly hairy benath, scabrid on margins. Panicles at large globular heads. Spikelets spiny, hairy, 1-3-flowered. Glumes ovate, spiny, many-nerved. Lemmas ovate, tipped with a sharp spine, surrounded by tufts of hairs. Paleas ovate or obovate, emarginate. Stamens 6. Grains ovoid–sub globose, hairy.

Rare in forest slopes. Fl. & Fr.: Very infrequent. Thummalapalle (KDP), *KRKS & KRS* 36410.

DICHANTHIUM Willemet.

1a. Racemes solitary.. **D. foveolatum**

1b. Racemes more than one .... **D. annulatum**

Dichanthium annulatum (Forssk.) Stapf in Prain, Fl. Trop. Afr. 9: 178. 1917; FPM 3: 1741. 1934; FAP 3: 1170. 1997. *Andropogon annulatus* Forssk., Fl. Aegypt. Arab. 173. 1775.

Annual or perennial, culms erect or geniculately ascending, to 80 cm high; woody at base. Leaf-sheaths scabrid. Ligules membranous. Leaf-blades linear-lanceolate, 8–30 × 0.5–1 cm, membranous, glaucous, midrib prominent. Racemes 2-10, each 3-7 cm long, arranged on short primary axis. Spikelets binate, sessile and pedicelled. Sessile spikelets ovate-lanceolate, 3–4.5 × 1.5 mm, callus hairy. Lower glume ovate-oblong, 2–3.5 × 1 mm, chartaceous, 7-nerved. Upper glume lanceolate, 2.5–3.5 × 1 mm, chartaceous, 3-nerved. Lower lemma elliptic-lanceolate, 2–2.5 × 1 mm, hyaline. Upper lemma linear, 2 × 1 mm, hyaline, awned. Stamens 3. Grains elliptic. Pedicelled spikelets male or sterile, more or less similar to sessile.

Common, in plains and in edges of irrigated fields. Fl. & Fr.: July–Nov. Thummalapalle (KDP), *KRKS & KRS* 33683.

Dichanthium foveolatum (Del.) Roberty, Boissiera 9: 170. 1960. *Andropogon foveolatus* Del., Fl. d' Egypte 16, t. 8, f. 2. 1812. *Eremopogon foveolatus* (Del.) Stapf in Prain, Fl. Trop. Afr. 9: 183. 1917; FAP 3: 1201.1997.

Perennial, culms erect or ascending, tufted, to 100 cm high. Leaf-sheaths scabrid or glabrous. Ligules fimbriate. Leaf-blades linear-lanceolate, 4–6 × 0.3 cm, glabrous. Racemes solitary, spiciform, 2-6 cm, spatheate. Spikelets binate, sessile and pedicelled. Sessile spikelets oblong-lanceolate, 4–6 × 2–4 mm, hermaphrodite, callus bearded. Lower glume ovate-oblong, 5–6 × 2 mm, chartaceous, pitted, 3-5-nerved. Upper glume ovate-lanceolate, 5–6 × 2 mm, chartaceous, 3-nerved. Lower lemma linear-lanceolate, 3 × 1.5 mm, chartaceous. Upper lemma narrow, hyaline, awned. Stamens 3. Grains obovoid. Pedicelled spikelets akin to sessile, awnless. Lemmas absent, sometimes hyaline.

Common, in forest margins, wastelands and plains. Fl. & Fr.: Aug. – Dec. Kalasamudram (ATP), *KRKS* 36901.

DIGITARIA Heiser ex Fabricius

1a. Inflorescence effuse panicle .... **D. tomentosa**

1b. Inflorescence spike like racemes, digitate .... (2)

2a. Spikelets glabrous or hairy; if so hairs never verrucose .... **D. bicornis**

2b. Spikelets hairy; hairs verrucose. .... (3)

3a. Annuals; racemes 2 or rarely 4; spikelets solitary, 1.5–2 mm........ **D. longiflora**

3b. Perennials; racemes 3–15; spikelets paired, 3–4 mm .... **D. ciliaris**

Digitaria bicornis (Lam.) Roem. & Schult. ex Loud., Hort. Brit. 24: n.1578. 1830; FAP 3: 1177. 1997. *Paspalum bicorne* Lam., Tab. Encycl. Meth. Bot. 1: 176. 1791.

Annual, culms erect, tufted, to 70 cm high. Leaf-sheaths glabrous or sparsely hairy. Ligules membranous. Leaf-blades linear-lanceolate, 5–30 × 0.3–0.5 mm, sparsely hairy or glabrous. Racemes 2-8, each 5-10 cm long. Spikelets in pairs, heteromorphous. Sessile spikelet glabrous or pubescent, elliptic-oblong. Lower glume scaly or absent. Upper glume lanceolate, 1.5 × 0.5 mm, chartaceous, 3-nerved, silky pubescent. Lower lemma oblong-lanceolate, 3.2 × 1.3 mm, membranous, 7-nerved, pubescent. Pedicelled spikelet pubescent, pectinate. Upper lemma elliptic-lanceolate, 3 × 1 mm, chartaceous, 3-nerved. Palea elliptic-lanceolate, 2.8 × 1.7 mm, hyaline, 2-nerved. Stamens 3. Grains oblong.

Common, in plains, cultivated fields usually in dry habitats. Fl. & Fr.: Mar. – Oct. Kalasamudram (ATP), *KRKS* & *BSS* 40126; Batrepalli (ATP), *KRKS* 205 (HJCB-E).

Digitaria ciliaris (Retz.) Koeler, Descr. Gram. 27. 1802. subsp. **ciliaris**; FAP 3: 1177. 1997. *Panicum ciliare* Retz., Observ. Bot. 4: 16. 1786. *Digitaria marginata* Link, Enum. Hort. Berol. 1: 102. 1821; FPM 3: 1764. 1934.

Annual, culms erect or ascending, tufted, to 90 cm high; rooting at lower nodes. Leaf-sheaths with tubercle-based hairs or glabrous. Ligules membranous, truncate. Leaf-blades linear-lanceolate, 5–25 × 0.5–1 cm, villous or glabrous. Racemes 3-15, sub-digitate; rhachis serrate. Spikelets binate, homomorphous, sessile and pedicelled. Spikelets elliptic, 3–3.5 × 1 mm. Lower glume scaly, 0.8 × 0.5 mm, membranous, hairy. Upper glume oblong-lanceolate, 2.8 × 1 mm, chartaceous, pubescent, 3-nerved. Lower lemma elliptic-lanceolate, 2.8 × 1 mm, chartaceous, pubescent, 7-nerved. Upper lemma elliptic-lanceolate, 2.5 × 1 mm, chartaceous, acuminate. Palea elliptic-lanceolate, 2.5 × 0.8 mm, hyaline, acuminate, 2-nerved. Stamens 3. Grains linear-oblong.

Common, in cultivated fields especially dry lands and irrigated fields. Fl. & Fr.: Jun. – Apr. Kalasamudram (ATP), *KRKS* & *BSS* 40125.

Digitaria longiflora (Retz.) Pers., Syn. Pl. 1: 85. 1805; FPM 3: 1765. 1934; FAP 3: 1178. 1997. *Paspalum longiflorum* Retz., Observ. Bot. 4: 15. 1786.

Annual, culms erect or creeping, slender, to 50 cm high; rooting at nodes. Leaf-sheaths glabrous or sparsely hairy. Ligules membranous. Leaf-blades elliptic-lanceolate or oblong-lanceolate, 1–10 × 0.2–1 cm, glabrous, acute. Racemes 2–3, rarely 4, digitate; rhachis flat, winged. Spikelets solitary, ovate–elliptic, 1.5 × 0.5 mm, hairy. Lower glume absent. Upper glume elliptic–ovate, 1.5 × 1 mm, chartaceous, 7-9-nerved, pubescent. Lower lemma akin to the upper glume. Upper lemma ovate-elliptic, 1.4 × 0.9 mm, crustaceous, acute. Palea broadly elliptic, 1.3 × 0.6 mm, hyaline, acute, 2-nerved. Stamens 3. Grains ovate.

Common, in plains, open grass lands and forests. Fl. & Fr.:All seasons. Kalasamudram (ATP), *KRKS* & *KRS* 36494, 40096, 40119; Kalasamudram (ATP), *KRKS* 204 (HJCB-E).

Digitaria tomentosa (J.Koenig ex Rottler) Henrard, Blumea 1: 100. 1934; FAP 3: 1180. 1997. *Milium tomentosum* J.Koenig ex Rottler, Neue Schriften Ges. Naturf. Freunde Berlin 4: 220. 1803. *Panicum subeglume* Trin., Mem. Acad. Imp. Sci. St. Petsreb. Ser. 6, Sci. Math. 3: 292. 1834; FPM 3: 1783. 1934.

Annual, culms erect or geniculate, to 70 cm high. Leaf-sheaths glabrous. Ligules membranous. Leaf-blades oblong-lanceolate, 5–20 × 1–1.5 cm, glabrous or hairy. Panicles effuse, 10-30 cm long; rhachis elongate, pilose at branch axils; pedicels long, terete. Spikelets solitary, oblong, 2.5 × 0.7 mm, pubescent. Glumes minute. Lower glume truncate, 0.1 × 0.1 mm. Upper glume oblong, 2.5 × 0.7 mm, acute, pubescent. Lower lemma and lower palea absent. Upper lemma elliptic-oblong, 2 × 0.6 mm, coriaceous, acute. Upper palea similar to upper lemma. Stamens 3. Grains oblong.

Common near rocky shades in forest slopes. Fl. & Fr.: Jun – Dec. Batrepalli (ATP), *KRKS* & *BSS* 37855.

DINEBRA Jacq.

Dinebra retroflexa (Vahl) Panz., Denksehr., Acad. Wiss. Munchen. 4: 270. t. 12. 1814; FPM 3: 1841. 1934; FAP 3: 1182. 1997. *Cynosurus retroflexus* Vahl, Symb. Bot. 2: 20. 1791.

Annual, culms creeping or ascending, tufted, to 100 cm high. Leaf-sheaths glabrous. Ligules membranous. Leaf-blades linear-lanceolate, 5–10 × 0.5 cm, glabrous. Inflorescence 10–20 cm long, elliptic-oblong to pyramidal, spikes 2–5 cm long, stift. Spikelets 1-3-flowered, alternate, 5–7 × 2–3 mm. Glumes persistent. Lower glume elliptic-lanceolate, 5–7 × 2 mm, coriaceous, 1-nerved, awned. Upper glume elliptic-lanceolate, 6–7 × 2 mm, coriaceous, 1-nerved, awned or cuspidate. Lemmas narrowly ovate, 2.5 × 1.5 mm, hyaline, 1-nerved. Paleas ovate, 2–3 × 1 mm, hyaline, sparsely ciliate on the keels. Stamens 3. Grains ovoid, trigonous.

Common, weed in cultivated fields especially in dry land irrigated fields. Fl. & Fr.: Aug. – Sept. Chinnapalli (ATP), *KRKS* & *KRS* 37809.

ECHINOCHLOA P. Beauv.

1a. Ligule present .... **E. stagnina**
1b. Ligule absent .... (2)
2a. Spikelets not awned .... **E. colona**
2b. Spikelets distinctly awned .... **E. crus-galli**

Echinochloa colona (L.) Link, Hort. Berol. 2: 209. 1833. FPM 3: 1776. 1934; FAP 3: 1184. 1997. *Panicum colonum* L., Syst. Nat. (ed.10) 810. 1759.

Annual, culms erect, tufted, to 70 cm high. Leaf-sheaths glabrous. Ligules absent. Leaf-blades linear-lanceolate, 4–20 × 0.2–0.5 cm, glabrous, acuminate, margin undulate. Panicles spiciform, racemes 3–10 cm long. Spikelets ovate-lanceolate, 2-4 mm. Lower glume ovate, 2–2.5 × 1.5 mm, chartaceous, 5-nerved. Upper glume boat-shaped, 2.5–3.5 × 2 mm, chartaceous, 7-nerved. Lower lemma broadly ovate, 2.5–3.5 × 2 mm, cuspidate, 7-nerved. Lower palea oblong, 2–2.5 × 1 mm, hyaline, 2-nerved. Upper lemma elliptic, 2.5–3 × 1 mm, chartaceous, 5-nerved. Upper palea oblong, 1–2 × 1 mm, crustaceous, 3-nerved. Stamens 3. Grains broadly elliptic.

Common, in cultivated fields and moist localities. Fl. & Fr.: All seasons. Kalasamudram (ATP), *KRKS* 33665, 40122.

Echinochloa crus-galli (L.) P. Beauv., Ess. Agrost. 53: 161.1812; FPM 3: 1777.1934; FAP 3: 1184.1997. *Panicum crus-galli* L., Sp. Pl. 56.1753.

Annual, culms erect, tufted, to 150 cm high. Leaf-sheaths glabrous. Ligules absent. Leaf-blades lanceolate, 5–25 × 0.5–1 cm, glabrous. Panicles congested, racemes 4–12 cm long. Spikelets ovate, 4-6 mm, awned. Lower glume ovate-cuspidate, 1.5–2.5 × 2 mm, chartaceous, 5-nerved. Upper glume oblong-lanceolate, 3–4 × 2 mm, chartaceous, 5-nerved. Lower lemma ovate-oblong, 3.5–4 × 2 mm, awned, 5-nerved, scabrid. Lower palea oblong–elliptic, 3–3.5 × 1 mm, hyaline. Upper lemma oblong, 3.5–4.5 × 2 mm, chartaceous, acuminate, 3-nerved. Upper palea oblong, 2.5–3.5 × 2 mm, crustaceous, 3-nerved. Stamens 3. Grains ellipsoid.

Common, in rice fields and occasional in marshy localities of plains and forests. Fl. & Fr.: Apr.– Dec. N.S.P.Kottala (ATP), *KRKS* & *BSS* 40481.

Echinochloa stagnina (Retz.) P. Beauv., Ess. Agrost. 53. 161 & 171. 1812; Fischer, Fl. Madras 3: 1777. 1934; Bor, Grass. India 311. 1960. *Panicum stagninum* Retz., Observ. Bot. 5: 17. 1789. *P. crusgalli* sensu Hook. f., Fl. Brit. India 7: 30. 1896 *p.p.* non L. 1753.

Annual, culms erect, stoloniferous; up to 50 cm high, rooting at the nodes; nodes glabrous. Leaf-sheaths 5-10 cm long, glabrous. Ligule a fringe of stiff hairs. Leaf-blades 4-15 × 0.4-0.8 cm, linear-lanceolate, apex acuminate, midrib prominent. Spikelike panicle 5-10 cm long, racemes interrupted; rachis angular, scabrid. Spikelets ovate, 5-6 mm, crowded, acutely acuminate, awned, secund, pubescent. Lower glume broadly ovate, 1.5–2.8 × 2 mm, pubescent. Upper glume broadly ovate, 3.5–4.5 × 2 mm, acuminate, hairy at apex and margins. Lower lemma ovate, 3.5-4

× 2 mm, hairy and scabrid at apex, awned. Palea oblong-ovate, 3 × 1 mm, hyaline, male. Upper lemma ovate, 3 × 2mm, chartaceous, acute. Palea similar to upper lemma, bisexual. Stamens 3. Grains broadly ellipsoid.

Occasional, in rice fields and marshy areas. Fl. & Fr.: Throughout the year. Kalasamudram RF (ATP), *NY* 1172 (MH).

ELEUSINE Gaertn.

Eleusine indica (L.) Gaertn., Fruct. 1: 8. 1789; FPM 3: 1839. 1934; FAP 3: 1186. 1997. *Cynosurus indicus* L., Sp. ed. 1.72. 1753.

Annual or perennial, culms erect, tufted, to 80 cm high. Leaf-sheaths glabrous. Ligules membranous. Leaf-blades linear or linear-lanceolate, 5–30 × 0.5–0.8 cm, glabrous. Spikes digitate or sub–digitate, 5–15 cm long. Spikelets oblong-lanceolate, 3–5.5 × 2 mm, 2-5-flowered, secund. Lower glume oblong-lanceolate, 2–2.5 × 1 mm, chartaceous, 1–3-nerved, acute, scabrid on the keels. Upper glume ovate-oblong, 2–3 × 1–1.5 mm, chartaceous, 5-7-nerved. Lemmas ovate-oblong, 2.5–3 × 1–2 mm, chartaceous, 3-nerved. Paleas elliptic-oblong, 2–2.5 × 1 mm, hyaline, 2-nerved. Stamens 3. Grains ovate.

Occasional, in open moist places. Fl. & Fr.: Jul. – Dec. Batrepalli (ATP), *KRKS & BSS* 40361.

ENNEAPOGON Desv. ex P. Beauv.

Enneapogon persicus Boiss., Dign. Pl. Orient. 5: 71. 1844. *E. schimperianus* (Hoscht. ex A. Rich.) Renvoize, Kew Bull. 22. 400. 1968; FAP 3: 1187. 1997. *Pappophorum schimperianum* Hochst. ex A. Rich., Tent. Fl. Abyss. 2: 403. 1854. *Enneapogon elegans* (Nees ex Steud.) Stapf, Bull. Misc. Inform. 1907. 224. 1907; FPM 3: 1841. 1934. *Pappophorum elegans* Nees ex Steud., Syn. Pl. Glumac. 1: 199. 1854.

Perennial, culms erect or ascending, tufted, to 50 cm high. Leaf-sheaths puberulous. Ligules membranous. Leaf-blades linear-lanceolate, 2–4 × 0.5 cm, acute, glabrous. Panicles contracted, to 10 cm long. Spikelets 3-flowered. Glumes persistent. Lower glume ovate-lanceolate, 4.3 × 1.2 mm, chartaceous, 7-9-nerved, pubescent. Upper glume ovate-lanceolate, 6 × 1.3 mm, chartaceous, 5-7-nerved, pubescent. Lemma ovate-oblong, 2 × 2 mm, chartaceous, 9-nerved, 9-awned. Palea elliptic-ovate, 2.7 × 1.4 mm, hyaline, 2-toothed, apex ciliate. Stamens 3. Grains ovoid.

Occasional, in foot hills and scrub jungles in and around bushes. Fl. & Fr.: Oct. – Dec. Kalasamudram (ATP), *KRKS & BSS* 39547, *PVP* 2607.

ENTEROPOGON Desv. ex P. Beauv.

Enteropogon monostachyos (Vahl) K. Schum. ex Engl., Abh. Preuss. Akad. Wiss. 17. 1894; FPM 3: 1834. 1934; FAP 3: 1188. 1997.

Perennial, culms erect, slender, to 100 cm high. Leaf-sheaths glabrous. Ligule membranous, apex hairy. Leaf-blades linear, 10–30 × 0.3–0.5 cm, involute, acuminate, glabrous. Spikes solitary, 10–15 cm long; rhachis flat or trigonous. Spikelets oblong, 6.5–7 × 0.5 mm. Lower glume oblong-lanceolate, 2–2.5 × 1 mm, hyaline, 1-nerved, acuminate. Upper glume lanceolate, 6–7 × 0.5–0.8 mm, hyaline, 1-nerved, acuminate.

Lower lemma oblong, 5–6 × 0.8 mm, chartaceous, bidentate, awned. Lower palea oblong, 4–6 × 0.8 mm, hyaline, scabrid. Upper lemma lanceolate, 5 × 0.8 mm, chartaceous. Upper palea lanceolate, 5 × 0.8 mm, hyaline. Stamens 3. Grains ellipsoid.

Occasional, in forest plains near bushes. Fl. & Fr.: Jul. – Aug. Rachakuntapalle (KDP), *KRKS* & *BSS* 37924; Batrepalli (ATP), *KRKS* 235 (HJCB-E).

ERAGROSTIELLA Bor

1a. Leaves from the basal sheaths long; reaching the base of the inflorescence .... **E. walkeri**

1b. Leaves from the basal sheaths much shorter; never reaching the base of the inflorescence .... (2)

2a. Keels at the palea broadly-winged .... **E. bifaria**

2b. Keels at the palea narrowly-winged .... **E. brachyphylla**

Eragrostiella bifaria (Vahl) Bor, Indian Forest. 66: 270. 1940; FAP 3: 1189. 1997. *Poa bifaria* Vahl, Sym. Bot. 2: 19. 1791. *Erogrostis bifaria* (Vahl) Wight ex Steud., Nom. Bot. Ed. 2.1: 562. 1840; FPM 3: 1828. 1934.

Annual or perennial, culms erect, tufted, to 90 cm high. Leaf-sheaths glabrous. Ligules membranous or fimbriate. Leaf-blades linear or acicular, 3–20 × 0.2 cm, often rolled, acute, glabrous. Inflorescence solitary raceme, 5–25 cm long. Spikelets slightly compressed, ovate-lanceolate, 4–5 × 2–3 mm, 5-30-flowered, grey. Lower glume oblong-lanceolate, 2.5–3 × 1 mm, chartaceous, 1-nerved, keel scabrid. Upper glume ovate-lanceolate, 2–2.5 × 1 mm, chartaceous, 1-nerved, keels scabrid. Lemmas broadly ovate, 2–3.5 × 1–2 mm, chartaceous, 3-nerved. Paleas elliptic-ovate, 2–2.5 × 1 mm, hyaline, keels winged, ciliate. Stamens 3. Grains globose.

Common, in plains, grass lands and margins of the forests. Fl. & Fr.: Oct. – Dec. Rachakuntapalle (KDP), *KRKS* & *SSR* 33626; Batrepalli (ATP), *KRKS* 234, 231 (HJCB-E).

Eragrostiella brachyphylla (Stapf) Bor, Indian Forest. 66: 270. 1940; FAP 3: 1189. 1997. *Eragrostis brachyphylla* Stapf in Hook f., Fl. Brit. India 7: 327. 1896; FPM 3: 1828. 1934.

Perennial, culms erect, tufted, slender, to 50 cm high. Leaf-sheaths glabrous. Ligules ciliate rim. Leaf-blades linear, 2–8 × 0.2 cm, acute, glabrous. Inflorescence solitary raceme, 10–20 cm long. Spikelets ovate-oblong, 2–4 mm, 5-18-flowered. Lower glume linear-lanceolate, 2–2.5 × 0.5 mm, chartaceous, acute, 1-nerved. Upper glume lanceolate, 2–3 × 1.5 mm, chartaceous, acute, 1-nerved. Lemmas broadly ovate, 2–3.5 × 1–2 mm, chartaceous, 3-nerved, margin hyaline. Palea obovate, 2–2.5 × 1 mm, hyaline, keels ciliate, margin inflexed. Stamens 3. Grains ellipsoid.

Rare in forest plains, hill slopes and wastelands. Fl. & Fr.: Jul. – Nov. Batrepalli (ATP), *KRKS* & *BSS* 39576; Vempalle (KDP), *KRKS 208* (HJCB-E).

Eragrostiella walkeri (Stapf) Bor, Indian Forest. 66: 270. 1940; FAP 3: 1190. 1997. *Eragrostis walkeri* Stapf, in Trim., Fl. Ceyl. 5: 298.1900; FPM 3: 1828. 1934.

Perennial, culms erect, tufted, slender, to 70 cm high. Leaf-sheaths glabrous. Ligules membranous. Leaf-blades filiform, 5–15 × 0.1 cm, convolute, glabrous, margin scabrous. Inflorescence solitary raceme, 20–30 cm long. Spikelets ovate, 2–6 × 1–2 mm, secund, 2-3-flowered at base, 6-8-flowered at apex. Lower glume linear-lanceolate, 2–2.5 × 0.5–0.6 mm, chartaceous, acute, 1-nerved. Upper glume boat-shaped, 2.5–3 × 1.5 mm, chartaceous, keeled, acute, 1-nerved. Lemmas broadly ovate, 2–5 × 1.5 mm, chartaceous, margins hyaline, 3-nerved. Paleas spatheolate, 1.5–2 × 0.5 mm, hyaline, keels ciliate, 2-nerved. Stamens 3. Grains ellipsoid.

Common, in wastelands, forest plains and open slopes. Fl. & Fr.: Jul. – Dec. Kalasamudram (ATP), *KRKS* & *BSS* 39696.

ERAGROSTIS Wolf

Key to Species

1a. Spikelets breaking up from above downwards; rhachis fragile .... (2)

1b. Spikelets breaking up from below upwards; rhachis tough .... (7)

2a. Lemmas ciliate on the margins.-lemma margins with a few cilia below .... **E. coarctata**

2b. Lemma not cilate on the margins .... (3)

3a. Keels of the palea more or less ciliate .... (4)

3b. Keels of the palea scabrid or smooth, not ciliate .... (6)

4a. Panicles spiciform, compact and dense.- Lemmas 1.25 mm long........ **E. riparia**

4b. Panicles effuse .... (5)

5a. Culms and leaves viscous .... **E. viscosa**

5b. Culms and leaves not viscous .... **E. amabilis**

6a. Panicles thyrisiform or very broadly oblong with numerous capillary branches. - 25–60 cm long, 35 cm wide; lemmas 1.5 mm long, markedly truncate .... **E. aspera**

6b. Panicles oblong or linear, not thyrsiform, not more than 15 cm wide .... **E. japonica**

7a. Annuals .... (8)

7b. Perennials .... (12)

8a. Lemmas up to 1.5 mm long .... (9)

8b. Lemmas over 1.5 mm long .... (11)

9a. Plants glandular .... (10)

9b. Plants eglandular. – lemmas 1-1.25 mm long .... **E. gangetica**

10a. Lowest branches of the panicle fascicled or subwhorled, with pitted glands just above and below the lowest node; glandular band below culm nodes.. **E. pilosa**

10b. Lowest branches not whorled.-leaf-sheaths bearded in the throat.. **E. minor**

11a. Spikelets 2 mm or more wide. –ovate or ovate-oblong................ **E. unioloides**

11b. Spikelets 0.75 – 2 mm wide. –spikelets many flowered, seated on the pedicels.. **E. tremula**

12a. Spikelets 1.75 – 2.5 mm wide; paleas deciduous with the lemmas. – inflorescence a more or less contracted panicle of shortly pedicelled spikelets.. **E. atrovirens**

12b. Spikelets 1 – 1.25 mm wide; paleas persistent.. (13)

13a. Panicle contracted; spikelets crowded, straight; lemmas 1.25 mm long, 1 mm wide, oblong-acute .... **E. nutans**

13b. Panicle widely spreading; spikelets loose, slightly curved; lemmas 1 mm long, 1.5 mm wide, almost orbicular........................... **E. papposa**

Eragrostis amabilis (L.) Wight & Arn., Cat. Indian Pl. 2: 105. 1834. *Poa amabilis* L., Sp. Pl. 68. 1753. *Eragrostis tenella* (L.) P. Beauv. ex Roem. & Schult., Syst. Veg. 2: 576. 1871; FAP 3: 1199. 1997. *Poa tenella* L., Sp. Pl. 69. 1753. *Eragrostis plumosa* (Retz.) Link, Hort. Berol. 1: 192. 1827; FPM 3: 1826. 1934.

Annual, culms erect, tufted, slender, to 40 cm high. Leaf-sheaths hairy at mouth. Ligules membranous, fimbriate. Leaf-blades linear, 4–10 × 0.2–0.3 cm, glabrous, acuminate. Panicles contracted or lax, 5–15 cm long. Spikelets oblong-ovate, 3–4 × 1.5 mm, 5-9-flowered. Lower glume lanceolate, 0–3–1 × 0.5 mm, chartaceous, 1-nerved. Upper glume ovate-lanceolate, 0.5–1 × 0.5 mm, chartaceous, 1-nerved. Lemmas oblong-ovate, 1–1.5 × 0.5–0.7 mm, chartaceous, 3-nerved. Paleas lanceolate, 1 × 0.5 mm, hyaline, 2-nerved, keels ciliate. Stamens 3. Grains ovate–elliptic.

Common, in wastelands, slopes and plains. Fl. & Fr.: Apr. – Nov. Thummalapalle (KDP), *KRKS* & *KRS* 33677.

Eragrostis aspera (Jacq.) Nees, Fl. Afr. Austr. 911. 408. 1841; FPM 3: 1825. 1934; FAP 3: 1191. 1997. *Poa aspera* Jacq., Hort. Bot. Vindob 332.t.56. 1776.

Annual, culms erect, tufted, to 80 cm high. Leaf-sheaths glabrous. Ligules ring of hairs. Leaf-blades linear-lanceolate, 5–20 × 0.5 cm, acuminate, glabrous. Panicles lax, 10–30 cm long, grey. Spikelets ovate-lanceolate, 3–7 × 1–2 mm, 4-15-flowered. Lower glume oblong, 1–1.5 × 1 mm, chartaceous, 1-nerved. Upper glume oblong, 1.5–2 × 1.5 mm, chartaceous, 1-nerved. Lemmas broadly ovate, 2–3 × 1–2 mm, chartaceous, 3-nerved. Paleas ovate-lanceolate, 2 × 1 mm, hyaline, keels scabrid, 2-nerved. Stamens 3. Grains ovate-lanceolate.

Occasional, in dry land agriculture especially groundnut fields. Fl. & Fr.: Jul. – Nov. Nuthnakalava (ATP), *KRKS* & *BSS* 41928.

Eragrostis atrovirens (Desf.) Trin. ex Steud., Nom. Bot. Ed. 2.1: 562. 1840; FAP 3: 1192. 1997. *Poa atrovirens* Desf., Fl. Atlant. 1: 73.t.14. 1798.

Perennial, culms erect, tufted, to 150 cm high. Leaf-sheaths glabrous. Ligules membranous. Leaf-blades linear-lanceolate, 5–25 × 0.2–0.5 cm, glabrous, acuminate. Panicles contracted, often lax, 10–20 cm long. Spikelets ovate-lanceolate, 2–3 × 2.5 mm, 3-30-flowered, grey. Lower glume lanceolate, 2 × 1 mm, chartaceous, 1-nerved. Upper glume ovate-lanceolate, 2–2.5 × 1.5 mm, chartaceous, 1-nerved. Lemmas ovate, 2–2.5 × 1.5 mm, chartaceous, 3-nerved. Paleas lanceolate, 1–2 × 0.5 mm, hyaline, keels ciliate. Stamens 3. Grains ellipsoid.

Common, in stream banks and moist localities. Fl. & Fr.: Jul. – Feb. Near Rathnachalam Anjaneyaswamy temple (KDP), *KRKS* & *BSS* 39667.

Eragrostis coarctata Stapf in Hook.f., Fl. Brit. India 7: 313. 1896; FPM 3: 1825. 1934; FAP 3: 1193. 1997.

Perennial, tufted grass. Culms up to 60 cm tall, erect or geniculately ascending at base; nodes glabrous. Leaf-sheaths 4–10 cm, ciliate at mouth; ligule minutely hairy; Leaf-blades 4–12 × 0.2–0.5 cm, linear-lanceolate, flat, base rounded, apex acuminate, glabrous. Panicles compact, up to 10 cm long, sub-clavate, dense; branches short, villous at the nodes. Spikelets 3 mm, oblong, compressed, breaking up from apex. Glumes subequal, lanceolate, 1-nerved, keeled; lemma broadly ovate, 3-nerved, sparsely ciliate below, apex acute; palea narrow, keels ciliolate. Stamens 3, stigmas plumose. Grains ovoid.

Occasional in grasslands and near streams. Fl. & Fr.: Nov. – Jun. Kalasamudram (ATP), *NY* 1166.

Eragrostis gangetica (Roxb.) Steud., Syn. Pl. Glumac. 1(3): 266. 1854; FAP 3: 1195. 1997; FPM 3: 1826. 1937. *Poa gangetica* Roxb. Fl. Ind. 1: 341. 1820. *Eragrostis stenophylla* Hochst. ex Miq., Verh. Konink.-Nederl. Inst. 3: 4, 39. 1851 *p.p.*; FBI 7: 318. 1896.

Annual, culms erect, tufted slender, to 50 cm high. Leaf-sheaths glabrous, mouth bearded. Ligule a ciliate rim. Leaf-blade linear-lanceolate, 5–10 × 0.5 cm, glabrous, margin entire, apex finely acuminate. Panicle ovate-oblong, 4–15 cm long, lax, grey. Spikelets ovate-oblong or oblong-lanceolate, 3–5 × 0.5–1.25 mm, slate-grey, crowded, 5–20-flowered. Lower glume ovate, 1 × 0.5 mm, chartaceous, 1-nerved. Upper glume oblong-lanceolate, 1-1.5 × 1 mm, chartaceous, 1-nerved. Lemmas ovate-lanceolate, 1-2 × 1 mm, chartaceous, 3-nerved. Paleas elliptic-lanceolate, 1 × 0.6 mm, 2-nerved, keels ciliate. Stamens 3. Grains orbicular.

Common, in moist places in cultivated fields, plains and foot hills of the forests. Fl. & Fr.: Sep. – May. Batrepalli (ATP), *KRKS* & *BSS* 40342.

Eragrostis japonica (Thunb.) Trin., Mem. Acad. Imp. Sci. St.-Petersbourg Ser. 6. Sci. Math. 1: 405. 1830. *Poa japonica* Thunb., Fl. Jap. 51. 1784. *Eragrostis diarrhena* (Schult. & Schult.f.) Steud., Syn. Pl. Glum. 1: 266. 1854; FAP 3: 1194. 1997. *Poa diarrhena* Schult. & Schult.f., Mant. 3: 616. 1827.

Perennial, culms erect, tufted, to 180 cm high. Leaf-sheaths glabrous. Ligules membranous. Leaf-blades linear-lanceolate, 5–25 × 0.3–0.5 cm, glabrous, acuminate. Panicles contracted, interrupted, branches dense, 10–35 cm long. Spikelets ovate, 2–4 × 2 mm, 6-15-flowered. Lower glume lanceolate, 2 × 1 mm, chartaceous, 1-nerved. Upper glume oblong-lanceolate, 2–2.5 × 1.5 mm, chartaceous, 1-nerved. Lemmas broadly ovate, 2–3 × 1.5 mm, chartaceous, 3-nerved. Paleas oblong, 2 × 1 mm, hyaline, 2-nerved, keels scaberulous. Stamens 3. Grains ovate.

Common, in muddy stagnant water, stream beds and paddy fields. Fl. & Fr.: Jan.–Dec. Gollalaguduru (KDP), *KRKS* & *BSS* 41974.

Eragrostis minor Host, Gram. Austr. 4: 15. 1809; FBI 7: 327. 1896; FAP 3: 1196. 1997. *E. poaeoides* P. Beauv. Ess. Agrost. 162. 1812; FPM 3: 1827. 1934.

Annuals, culms erect, tufted, 30–60 cm high. Leaf-sheaths glabrous, ciliate at mouth. Ligule a pubescent ridge. Leaf-blades lanceolate, 10-12 × 0.5-1 cm, glabrous, usually with a row of warty glands along the margin. Panicle lax, 4–10 cm long, stiffly branched with short pedicels, usually with glands on pedicels and branchlets. Spikelets oblong-linear, 3-9 × 1-2 mm, leaden grey, 6–16-flowered. Glumes subequal, ovate, boat-shaped, 1-1.7 mm long, 1-3-nerved, often glandular on the keel, acute. Lemmas broadly ovate, 1.5-2 × 1 mm, chartaceous, often glandular on the keel, 3-nerved. Palea spathulate, 1.2 × 1mm, scabrid on the keels, persistent. Stamens 3. Grains broadly oblong.

Rare in agriculture fields and wastelands. Fl. & Fr.: Jul. – Dec. Kalasamudram (ATP), *NY* 393.

Eragrostis nutans (Retz.) Nees ex Steud., Nom. Bot. ed. 2. 563. 1840; FPM 3: 1826. 1934. *Poa nutans* Retz., Observ. Bot. 4:19.1786. *Eragrostis elegantula* Nees ex Steud., Syn. Pl. Glum. 1: 266.1854; FBI 7: 318. 1896.

Perennials, culms erect, tufted, to 90 cm high, slender. Leaf-sheaths glabrous. Ligules hairy. Leaf-blades 6-15 × 0.3-0.5 cm, linear-lanceolate, apex acuminate. Panicles 5-10 cm, contracted, black or grey colour. Spikelets oblong-lanceolate, 6-15-flowered. Lower glume ovate, 0.4 × 0.2 mm, chartaceous, 1-nerved. Upper glume oblong, 0.6 × 0.2 mm, chartaceous, 1-nerved. Lemmas ovate, 1-1.2 × 1 mm, chartaceous, 3-nerved. Palea, obovate, 2-nerved, ciliate on the keels. Stamens 3. Grains ellipsoid.

Occasional in moist habitats. Fl. & Fr.: Jun. – Apr. Kalasamudram RF (ATP), *NY* 1184; Gollalaguduru (KDP), *KRKS* & *BSS* 41966 (HJCB).

Eragrostis papposa (Desf. ex Roem. & Schult.) Steud., Nomencl. Bot. ed. 2, 1: 564. 1840; FAP 3: 1197. 1997. *Megastachya papposa* Roem. & Schult., Syst. Veg. 2: 585. 1817.

Perennial, culms erect or ascending, to 40 cm high. Leaf-sheaths glabrous, ciliate near the mouth. Leaf-blades flat or inrolled, linear, 2–15 × 0.1–0.3 cm, stiff, glaucous, often forming a compact cushion. Panicle ovate, to 20 cm long, lax, open, with or without glands on the slender pedicels, glabrous. Spikelets linear, 3.5–11 × 1 mm,

5–23-flowered, grey-green to purplish green. Glumes unequal; the lower a lanceolate, hyaline scale-like, 0.3–1 mm long; the upper ovate and 0.8–1.3 mm long. Lemmas broadly elliptic to rotund, 1.1–1.7 mm long, appressed to the rhachilla, obtuse. Palea persistent, obtuse, minutely scaberulous on the keels, persistent. Anthers 3, 0.1–0.2 mm long. Grains oblong-ellipsoid, dorsally grooved.

Occasional, along water sources in deciduous forests and plains. Fl. & Fr.: Apr. – Oct. Thummalapalle (KDP), *KRKS* 37845.

Eragrostis pilosa (L.) P. Beauv., Ess. Agrost. 71: 162.175. 1812; FPM 3: 1827. 1934; FAP 3: 1198. 1997. *Poa pilosa* L., Sp. Pl. 68. 1753.

Annual, culms erect, tufted, slender, to 120 cm high. Leaf-sheaths glabrous. Ligules membranous. Leaf-blades linear-lanceolate, 5–25 × 0.2–0.5 cm, glabrous, acuminate. Panicles lax, 10–40 cm long, filiform, grey or black. Spikelets ovate-lanceolate, 3–4 × 1.5 mm, 4-10-flowered. Lower glume ovate-lanceolate, 0.6 × 0.4 mm, chartaceous, 1-nerved. Upper glume lanceolate, 1 × 0.5 mm, chartaceous, 1-nerved. Lemmas oblong-lanceolate, 1–1.5 × 1 mm, chartaceous, 3-nerved. Paleas ovate, 1–1.5 × 0.4–1 mm, 2-nerved, keels scabrid. Stamens 3. Grains ovate.

Occasional, in forest margins and plains in marshy localities. Fl. & Fr.: Aug.– Dec. Kalasamudram (ATP), *KRKS* & *BSS* 39108.

Eragrostis riparia (Willd.) Nees, Agrost. Bras. 512. 1829; FPM 3: 1825. 1934; FAP 3: 1198. 1997. *Poa riparia* Willd., Ges., Natur. Freunde Berlin, Neue Schrift. 4: 185. 1803.

Perennial, culms erect or ascending, tufted, to 40 cm high. Leaf-sheaths glabrous, mouth hairy. Ligules membranous. Leaf-blades linear-lanceolate, 4–8 × 0.2–0.4 cm, hairy, acute. Panicles spiciform, dense, 3–8 cm long, rarely grey. Spikelets ovate-oblong, 2 × 0.5–1 mm, 5–10–flowered. Lower glume ovate-linear, 1–1.5 × 1 mm, chartaceous, 1-nerved. Upper glume oblong, 2 × 0.5–1 mm, chartaceous, 1–nerved. Lemmas oblong, 1–2 × 1–1.5 mm, chartaceous, 3-nerved. Paleas oblong, 1.5 × 1 mm, hyaline, 2-nerved, keels ciliate. Stamens 3. Grains elliptic.

Common, on sandy soils near streams and in dry cultivated fields. Fl. & Fr.: May – Nov. Batrepalli (ATP), *KRKS* 37986.

Eragrostis tremula Hochst. ex Steud., Syn. Pl. Glum. 1: 269. 1854; FBI 7: 320.1896; FPM 3: 1827. 1934.

Annuals or perennials, culms erect, tufted, to 90 cm high. Leaf-sheaths bearded at mouth; ligules fimbriate. Leaf-blades 5-15 × 0.2-0.5 cm, linear, apex acuminate, glbrous. Panicles up to 20 cm long, effuse, filiform; spikelets oblong, 10-15×1-2 mm, 15-50-flowered. Lower glume oblong, 1 × 0.5 mm, chartaceous, 1-nerved; upper glume oblong, 0.5-1 × 1mm, chartaceous 1-nerved. Lemmas ovate-oblong, 1.5 × 1 mm, chartaceous, 3-nerved. Paleas elliptic-oblong, 1.2 × 1 mm, 2-nerved, keels scabrid. Stamens 3. Grains globose.

Occasional, in forest plains. Fl. & Fr.: Jun. - Nov. Namalagundu (KDP), *KRKS* & *BSS* 40147; Kalasamudram RF (ATP), *NY* 393, *BR* & *AMR* 24239.

Eragrostis unioloides (Retz.) Nees ex Steud., Syn. Pl. Glum. 1: 264. 1854; FPM 3: 1826. 1934; FAP 3: 1200. 1997. *Poa unioloides* Retz., Observ. Bot. 5: 19. 1789.

Annual or perennial, culms erect or ascending, tufted, to 50 cm high. Leaf-sheaths glabrous. Ligules membranous. Leaf-blades linear–oblong, 4–15 × 0.3 cm, glabrous, acuminate. Panicles lax, 5–15 cm long, purplish. Spikelets ovate, 6–10 × 2–4.5 mm, 10-30-flowered. Lower glume ovate-oblong, 0.5–1 × 0.5 mm, chartaceous, 1-nerved. Upper glume ovate-oblong, 1.5 × 0.7 mm, chartaceous, 1-nerved. Lemmas broadly ovate, 1–1.5 × 1.5 mm, chartaceous, 3-nerved. Paleas oblong-elliptic, 2 × 1 mm, hyaline, 2-nerved, keels scabrid. Stamens 3. Grains obovoid.

Common, in streams, marshy areas and in moist habitats. Fl. & Fr.:Jul. – Apr. Namalagundu (KDP), *KRKS* & *BSS* 40175.

Eragrostis viscosa (Retz.) Trin., Mem., Acad. Imp. Sci. St.-Petersbourg. Ser. 6. Sci. Math. 1: 397. 1830; FPM 3: 1826. 1934; FAP 3: 1200.1997. *Poa viscosa* Retz., Observ. Bot. 4: 20. 1786.

Annual, culms erect or ascending, tufted, viscous below, to 50 cm high. Leaf-sheaths glabrous, viscous. Ligules membranous. Leaf-blades linear, 4–10 × 0.3 cm, convolute, glabrous, acuminate. Panicles narrowly cylindric, glandular, 5–15 cm long. Spikelets oblong-lanceolate, 2–4 ´1.5 mm, 4-20-flowered. Lower glume linear, 0.3–0.8 × 0.4 mm, chartaceous, 1-nerved. Upper glume oblong, 0.5–1 × 0.8 mm, chartaceous, 1-nerved. Lemmas ovate, 1–1.5 × 0.8 mm, chartaceous, 3-nerved. Paleas linear, 1 × 0.3 mm, hyaline, ciliate on the keels, 2-nerved. Stamens 3. Grains ovoid.

Common, in wastelands and plains. Fl. & Fr.: Apr.–Nov. Batrepalli (ATP), *KRKS* & *BSS* 39605.

ERIOCHLOA Kunth

Eriochloa procera (Retz.) C.E. Hubb., Kew Bull. 256. 1930; FPM 3: 1767. 1934; FAP 3: 1202. 1997. *Agrostis procera* Retz., Observ. Bot. 4: 19. 1786.

Perennial, culms erect, slender, tufted, to 150 cm high. Leaf-sheaths glabrous. Ligules membranous. Leaf-blades linear-lanceolate, 8–35 × 0.3–1 cm, glabrous, acuminate. Panicles lax, 5–20 cm long, grey. Spikelets ovate-lanceolate, 3–4 × 2 mm. Lower glume obscure. Upper glume oblong-lanceolate, 3–4 × 1.5 mm, chartaceous, 5-nerved, pilose. Lower lemma obovate, 2–3.5 × 1–1.5 mm, chartaceous, 5-nerved, pilose. Palea oblong, 2–3 × 1 mm, crustaceous, 2-nerved. Upper lemma ovate, 1.5–2.5 × 1.3 mm, crustaceous, mucronate, 3-5-nerved. Upper palea oblong, 2–4 × 1 mm, crustaceous, 2-nerved. Stamens 3. Grains oblong.

Frequent in streams and in wetlands. Fl. & Fr.: Mar. – Oct. Kalasamudram (ATP), *KRKS* & *KRS* 36470.

HACKELOCHLOA Kuntze

Hackelochloa granularis (L.) Kuntze, Revis. Gen. Pl. 776. 1891; FPM 3: 1758. 1934. *Mnesithea granularis* (L.) Koning & Sosef, Blumea 31: 295. 1986; FAP 3: 1220. *Cenchrus granularis* L., Mant. 2. App. 575. 1771.

Annual, culms erect, branched, tufted, to 120 cm high. Leaf-sheaths tubercle based hairy. Ligules membranous. Leaf-blades lanceolate, 5–10 × 0.5–1 cm, hirsute with tubercle based hairs. Racemes solitary, 2–5 cm long, cylindrical. Spikelets binate. Sessile spikelets globose, 2 × 2 mm. Lower glume globose, 1.5–2 × 1 mm, crustaceous, 7-9-nerved, pitted. Upper glume broadly ovate, 1–1.5 × 1 mm, chartaceous, 3-nerved. Lower lemma ovate, obtuse, hyaline. Upper lemma broadly ovate, hyaline. Palea ovate, hyaline. Stamens 3. Grains ovoid. Pedicelled spikelets narrowly globose, 2–3 mm. Lower glume ovate-lanceolate, 1–2.5 × 1 mm, chartaceous, 7-nerved, keels ciliate. Upper glume boat-shaped, 1.5–2 × 1 mm, chartaceous, 7-nerved, keels winged, ciliate. Lower lemma ovate, 1–2 × 1 mm, hyaline. Upper lemma ovate, 1.5–2 × 1 mm, hyaline. Palea oblong, 1.5 × 0.8 mm, hyaline. Stamens 3. Grains orbicular.

Common, in forest margins and groundnut fields. Fl. & Fr.: Jun. – Nov. Kottala (KDP), *KRKS* & *KRS* 37667.

HETEROPOGON Pers.

1a. Racemes glabrous .... **H. fischerianus**

1b. Racemes hairy .... **H. contortus**

Heteropogon contortus (L.) P. Beauv. ex Roem. & Schult., Syst. Veg. 2: 836. 1817. FPM 3: 1743. 1934; FAP 3: 1206. 1997. *Andropogon contortus* L., Sp. Pl. 1045. 1753.

Perennial, culms erect or geniculate, tufted, to 100 cm high. Leaf-sheaths compressed, sparsely hairy towards collar. Ligules membranous. Leaf-blades lanceolate, 5–25 × 0.2–0.5 cm, scaberulous. Racemes solitary, 3–8 cm long, awned, twisted. Spikelets binate, lower homogamous and upper heterogamous. Homogamous sessile spikelets linear, 4–7 mm. Lower glume lanceolate, 4–6 × 1.5–2 mm, chartaceous, many-nerved, tubercle based hairs on dorsal side. Upper glume lanceolate, 4–5 × 1.5 mm, chartaceous, 3-nerved, ciliate. Lower lemma lanceolate, 4–6 × 1–1.5 mm, chartaceous, 3-nerved, ciliate. Upper lemma linear-lanceolate, 4–6 × 1–1.5 mm, chartaceous, 3-nerved, margin ciliate. Stamens 3. Homogamous pedicelled spikelets akin to sessile spikelets. Heterogamous sessile spikelets linear, 4.5–7.5 mm, awned. Lower glume lanceolate, 4–7 × 1.5 mm, chartaceous, 9-nerved. Upper glume linear, 5-6 × 1.2 mm, chartaceous, keels pubescent. Lower lemma oblong, 4-5 × 1.5 mm, hyaline, 1-nerved. Upper lemma, hyaline, 4-5 × 1 mm, awned, awn hairy. Grains oblong. Heterogamous pedicelled spikelets linear, 5–7 mm. Glumes not winged. Florets akin to homogamous pedicelled spikelets.

Common, in wastelands, plains and forests. Fl. & Fr.: June – Nov. Vemula (KDP), *KRKS* & *BSS* 41956; Batrepalli (ATP), *KRKS* 233 (HJCB-E).

Heteropogon fischerianus Bor, Kew Bull. 1951. 170: 1951. *Heteropogon contortus* var. *distichous* Fischer, Fl. Madras. 3: 1743. 1934.

Perennial, culms erect, tufted, to 100 cm high. Leaf-sheaths glabrous. Ligules ciliate rim. Leaf-blades linear, 5–15 × 0.2 cm, slightly hairy at the base, distichous, glabrous. Racemes solitary, 2–5 cm long, glabrous. Spikelets binate, lower homogamous, upper heterogamous. Lower sessile spikelet, 5.5–6 × 1.5 mm, male. Lower glume lanceolate, 5.3 × 1.5 mm, crustaceous, glabrous, winged, 9-nerved. Upper glume lanceolate, 5.5–6 × 1.5 mm, glabrous, membranous, 3-nerved, winged. Lower lemma lanceolate, 5 × 1 mm, chartaceous, hairy. Upper lemma lanceolate, 4.5 × 1 mm, chartaceous, hairy. Stamens 3. Pedicelled spikelet 7 × 1.5 mm, pedicel 1.5 mm, male. Lower glume lanceolate, 5.2–5.5 × 1.5 mm, crustaceous, 9-nerved, winged one side. Upper glume lanceolate, 5.6–5.8 × 1.8 mm, membranous, 3-nerved, winged upper side. Lower lemma lanceolate, 3.8–4 × 1 mm, hyaline. Stamens 3. Upper sessile spikelet 8–8.5 × 1 mm, bisexual, awned. Lower glume oblong-lanceolate, 6.2–6.5 × 1 mm, brown hispid, oblong, crustaceous. Upper glume lanceolate, 6–6.2 × 1 mm, centre coriaceous glabrous. Lemma reduced to a hyaline base of the awn, awn 5.7 cm. Grains oblong.

Occasional, on slopes and plains of forest. Fl. & Fr.: Jul. – Dec. Bhumaiahgari palle (KDP), *KRKS* & *KRS* 37610; Batrepalli (ATP), *KRKS* 256 (HJCB-E).

ISCHAEMUM L.

Ischaemum afrum (J.F.Gmel.) Dandy, Fl. Pl. Sudan 3: 476. 1956. *Andropogon afer* J.F.Gmel., Syst. Nat. 166. 1791. *Ischaemum pilosum* (Klein ex Willd.) Wight, Madras Lit. Sci. J. 3: 138. 1835; FPM 3: 1723. 1934; FAP 3: 1211. 1997. *Andropogon pilosus* Klein ex Willd., Sp. Pl. 4:920. 1806.

Perennial, culms erect, rhizomatous, tufted, to 1.5 m high. Leaf-sheaths glabrous. Ligules membranous. Leaf-blades linear-lanceolate, 5–25 × 0.5–0.8 cm, glabrous. Racemes 4–6, sub digitate, 5–15 cm long; joints and pedicels subclavate, compressed, brownish-pilose, callus bearded. Sessile spikelets lanceolate, 5–6 × 3.5 mm. Lower glume lanceolate, 5–6 × 2.5 mm, chartaceous, scabrid below, 7-nerved. Upper glume oblong-lanceolate, 4.5–5 × 2 mm, chartaceous, scabrid, 7-nerved. Lower lemma oblong-lanceolate, 4.4–5 × 1.5 mm, ciliate. Lower palea lanceolate, 4 ×1 mm, hyaline. Stamens 3. Upper lemma oblong, 4.5 × 1mm, 2–cleft, ciliate, awned. Upper palea lanceolate, hyaline. Stamens 3. Grains linear. Pedicelled spikelets akin to sessile spikelets, upper lemma with a slender arista.

Occasional, in black soils in forests margins and plains. Fl. & Fr.: Sept. – Dec. Chinnakudala (KDP), *KRKS* & *BSS* 40464.

ISEILEMA Andersson

1a. Pedicels of involucral spikelets as broad as long .... **I. anthephoroides**

1b. Pedicels of involucral spikelets longer than broad at apex .... (2)

2a. Tubercles absent on spatheoles or spikelets .... **I. laxum**

2b. Tubercles present on spatheoles and margins of involucral spikelets .... **I. prostratum**

Iseilema anthephoroides Hack., in DC., Monogr. Phan. 6: 683. 1889; FPM 3: 1748. 1934; FAP 3: 1213. 1997.

Annual or perennial, culms erect, tufted, to 1m high. Leaf-sheaths margins scabrid. Ligules membranous. Leaf-blades linear-lanceolate, 5–2 × 0.5–0.8 cm, glabrous, acuminate or acute. Panicles 3–8 cm long, spatheoles cymbiform, glabrous. Involucral spikelets oblong-lanceolate, 3–4 × 2 mm. Lower glume oblong-lanceolate, 3–4 × 1.5 mm, chartaceous, 2-nerved, glabrous. Upper glume oblong, 3–3.5 × 1.5 mm, chartaceous 1-nerved, glabrous. Lemmas obovate–oblong, 3 × 1 mm, hyaline, 3-nerved. Stamens 3. Sessile spikelets linear or oblong lanceolate, 4–5 × 2.5 mm. Lower glume ovate, 4–5 × 2 mm, scabrid, 3-4-nerved. Upper glume oblong, 3–4.5 × 1.5 mm, chartaceous, scabrid, 3-nerved. Lower lemma oblong-lanceolate, 3–3.5 × 1 mm, hyaline, 2-nerved, ciliate. Upper lemma reduced to membranous base of the awn, awn 10–15 mm, slender. Stamens 3. Grains ellipsoid. Pedicel spikelets akin to involucral spikelets.

Common, in forest margins, streams and plains. Fl. & Fr.: Jun. – Dec. Bhumaiahgari palle (KDP), *KRKS* & *KRS* 37609.

Iseilema laxum Hack., in DC., Monogr. Phan. 6: 682. 1889; FPM 3: 1748. 1934; FAP 3: 1213. 1997.

Perennial, culms erect, slender, tufted, to 90 cm high. Leaf-sheaths scabrid or glabrous. Ligules membranous. Leaf-blades lanceolate, 4–15 × 0.3–0.5 cm, glabrous. Panicles 2–13 cm long, spatheoles cymbiform. Involucral spikelets oblong-lanceolate, 3–3.5 × 2 mm. Lower glume linear-oblong, 3–3.5 × 1–2 mm, chartaceous, scabrid, 3-nerved. Upper glume oblong, 3 × 1.5 mm, chartaceous, 3-nerved. Lemma oblong, 2.5 × 1 mm, hyaline, 3-nerved. Stamens 3. Sessile spikelets elliptic-lanceolate, 3–3.5 × 1.5 mm. Lower glume oblong, 3 × 1.5 mm, chartaceous, 1–3-nerved, ciliate. Upper glume lanceolate, 2–2.5 × 1 mm, chartaceous, 3-nerved, scabrid. Lower lemma oblong-lanceolate, 2.5 × 1 mm, hyaline, 2-nerved. Upper lemma reduced to membranous base of the awn, awn 5–10 mm. Stamens 3. Grains oblong. Pedicelled spikelets akin to involucral spikelets.

Common, in moist places, wastelands and forests plains. Fl. & Fr.: July – Feb. Kalasamudram (ATP), *KRKS* & *BSS* 40457.

Iseilema prostratum (L.) Anderson, Nov. Act. Soc. Sci. Upsal. Ser. 3.2: 251. 1856; FPM 3: 1747. 1934; FAP 3: 1214. 1997. *Andropogon prostratus* L., Mant. All. 304. 1771.

Perennial, culms erect or ascending, tufted, to 100 cm high. Leaf-sheaths glabrous. Ligules membranous. Leaf-blades lanceolate or linear-lanceolate, 5–10 × 0.3–0.5 cm, glabrous, margins with tubercled based hairs. Panicles lax, 5–20 cm long, rhachis nodes bearded, spatheoles cymbiform. Involucral spikelets oblong, whorled, 3–4 × 2–3 mm. Lower glume oblong, 3–4 × 1.5–2 mm, chartaceous, 5-nerved, pilose. Upper glume lanceolate, 3–3.5 × 1–1.5 mm, chartaceous, 1-nerved, glabrous. Lemmas oblong, 2.5–3 × 1 mm, hyaline, 2-nerved. Stamens 3. Sessile spikelets lanceolate, 4–5 × 1.5–2 mm. Lower glume oblong-ovate, 3–4 × 1.5 mm, chartaceous, 9-nerved. Upper glume elliptic-lanceolate, 3–3.5 × 1 mm, chartaceous, 3-nerved. Lower lemma lanceolate, 2–3 × 1 mm, hyaline, 2-nerved. Upper lemma reduced to membranous base of the awn, awn 5–8 mm. Stamens 3. Grains oblong. Pedicelled spikelets akin to involucral spikelets.

Common, in moist localities of plains, Wastelands and dried ponds. Fl. & Fr.: Sept. – June. Somulavaripalli (ATP), *KRKS* & *KRS* 37835; Batrepalli (ATP), *KRKS* 335 (HJCB-E).

LEERSIA Solander ex Swartz. *nom. cons.*

Leersia hexandra Swartz., Prodr. Veg. Ind. Occ. 21. 1788; FPM 3: 1845. 1934; FAP 3: 1215. 1997.

Perennial or annual, aquatic, culms erect or ascending, to 100 cm high. Leaf-sheaths glabrous. Ligules membranous. Leaf-blades linear-lanceolate, 5–20 × 0.2–0.3 cm, glabrous. Panicles lax, 5–20 cm long, branches flexuous. Spikelets ovate-oblong, 4–5 × 1–1.5 mm, laterally compressed, yellow. Glumes obscure or reduced. Lemma oblong, 4 × 1 mm, coriaceous, 5-nerved, scabrid along the keels and nerves. Palea linear-oblong, 3.5 × 1 mm, coriaceous, 3-nerved, ciliate on the keels. Stamens 6. Grains ovate.

Rare in stagnant water bodies in forests. Fl. & Fr.: Jul. – Oct. Near Rathnachalam Anjaneyaswamy temple (KDP), *KRKS* & *BSS* 40385.

LEPTOCHLOA P. Beauv.

1a. Lemmas 3-lobed; racemes loosely spiculate .... **L. fusca**

1b. Lemmas not lobed; racemes densely spiculate................................. **L. chinensis**

Leptochloa chinensis (L.) Nees, Syll. Ratisb. 1: 4. 1824; FPM 3: 1820. 1934; FAP 3: 1215. 1997. *Poa chinensis* L., Sp. Pl. 69. 1753.

Annual, culms erect or ascending, tufted, to 150 cm high. Leaf-sheaths keeled, glabrous. Ligules membranous. Leaf-blades linear or linear-lanceolate, 5–30 × 0.3–0.5 cm, glabrous. Panicles open, 5–15 cm long. Spikelets ovate-oblong, 3–5 × 1–1.5 mm, 3-7–lowered. Lower glume lanceolate, 1–2 × 0.5–0.8 mm, chartaceous, 1-nerved. Upper glume ovate-lanceolate, 1.3 × 0.8 mm, chartaceous, 1-nerved. Lemmas ovate, 1–1.5 × 1 mm, chartaceous, 3-nerved. Paleas ovate-oblong, 1.5 × 1 mm, 2-nerved, keels ciliate. Stamens 3. Grains oblong.

Occasional, in moist places and muddy water sources. Fl. & Fr.: Jul. – Dec. Nallacheruvupalli (KDP), *KRKS* & *BSS* 41981.

Leptochloa fusca (L.) Kunth, Revis. Gramin. 1: 91. 1829. *Diplachne fusca* (L.) P. Beauv., Ess. Agrost. 80 & 163. 1812; FPM 3: 1829. 1934; FAP 3: 1183. 1997. *Festuca fusca* L., Sp. Pl. (ed.2) 109. 1762.

Perennial, culms erect, tufted, to 120 cm high, rooting at lower nodes. Leaf-sheaths glabrous. Ligules membranous. Leaf-blades linear-lanceolate, 5–10 × 0.2–0.4 cm, glabrous. Panicles densely spikelate, 5–25 cm long. Spikelets oblong-lanceolate, 5–6 mm, 4-8-flowered, grey. Lower glume oblong, 1–2 × 1 mm, chartaceous, 1-nerved glabrous. Upper glume oblong-lanceolate, 2–2.5 × 1 mm, chartaceous, 1-nerved glabrous. Lemmas oblong, 3–4 × 1.5 mm, chartaceous, 3-lobed, aristate, scabrid. Paleas oblong, 2–3 × 1 mm, hyaline, 2-nerved, keels scabrid.

Rare in muddy water bodies and moist localities of stream beds. Fl. & Fr.: Oct. – Apr. N.S.P. Kottala (ATP), *KRKS* & *BSS* 40482.

LOPHOPOGON Hackel

Lophopogon tridentatus (Roxb.) Hack, in Engl. & Prantl, Naturl. Pflanzenf. 2.2: 26.1887; FPM 3: 1724.1934; FAP 3: 1216.1997. *Andropogon tridentatus* Roxb., Fl. Ind. 1: 261.1820.

Annual, culms erect or ascending, tufted, to 30 cm high, base white silky villous. Leaf-sheaths glabrous. Ligules fimbriate. Leaf-blades linear-lanceolate, 3–5 × 0.2–0.4 cm, glabrous. Racemes binate, solitary, covered with sheaths. Spikelets oblong, 3–5 mm long, 1-2-flowered. Sessile spikelets oblong, compressed, 3.5–4.5 × 2–3 mm. Lower glume oblong-lanceolate, 3–4 × 2 mm, chartaceous, 5-nerved, 3-toothed, tuft of rufous hairs in the middle, margins ciliate. Upper glume oblong, 3.5–4 × 2 mm, chartaceous, margins with rufous hairs, aristate. Lower lemma ovate-oblong, 3–3.5 × 1–1.5 mm, 1-nerved. Lower palea lanceolate, 3 × 1 mm, hyaline, male or neuter. Upper lemma lanceolate, 3–3.5 × 1 mm, hyaline, 2-cleft, awned from sinus. Upper palea lanceolate, 3 × 1 mm, hyaline, 2-nerved. Stamens 3. Grains linear. Pedicelled spikelet akin to sessile spikelets.

Occasional, in forests margins, wastelands and plains. Fl. & Fr.: Jun – Nov. Batrepalli (ATP), *KRKS* & *BSS* 40331; Batrepalli (ATP), *KRKS* 238 (HJCB-E).

MELANOCENCHRIS Nees

Melanocenchris jacquemontii Jaub. & Spach., Ill. Pl. Or. 4: 36. 1851; FAP 3: 1218. 1997. *M. royleana* Nees, Proc. Linn. Soc. 1: 95. 1841 nomen. FPM 3: 1831. 1934.

Annual or perennial, culms tufted, erect or decumbent, to 15 cm high. Leaf-sheaths hairy. Ligules fimbriate. Leaf-blades lanceolate or oblong-lanceolate, 2–5 × 0.2–0.4 cm, glabrous, margins tubercled based hairy. Spikes filiform, 4–8 cm long. Spikelets sessile, ovate-oblong, 5–6 × 3 mm, 2-flowered. Lower glume lanceolate, 3–5 × 3 mm, coriaceous, ciliate on margins, awned. Upper glume oblong-lanceolate, 3.5–5 × 3–3.5 mm, coriaceous, ciliate on margins, awned. Lower lemma broadly ovate, 4–5 × 3 mm, membranous, cuspidate, 3-aristate. Lower palea ovate, 3.5–4 × 2–2.5 mm, membranous, 2-cuspidate. Stamens 3. Grains oblong. Upper lemma ovate-oblong, 3–4.5 × 2–2.5 mm, membranous, keels ciliate, awned. Upper palea ovate-oblong, hyaline, 2-cuspidate. Stamens 3.

Common, in wastelands, dry fields and forests margins. Fl. & Fr.: Jun – Nov. Rachakuntapalle (KDP), *KRKS* & *KRS* 37644.

MELINIS P.Beauv.

Melinis repens (Willd.) Zizka, Biblioth. Bot. 38: 55. 1988. *Saccharum repens* Willd., Sp. Pl. 1. 322.1798. *Rhychelytrum repens* (Willd.) C.E. Hubb., Bull. Misc. Inform. Kew110. 1934; FAP 3: 1246. 1997. *R. villosum* (Parl.) Chiov., Amuar., R. Inst. Bot. Roma. 8: 310. 1908; FPM 3: 1791. 1934.

Perennial, culms erect or geniculate, tufted, to 60 cm high. Leaf-sheaths tubercle-based hairy. Ligules hairy. Leaf-blades linear-lanceolate, 5–10 × 0.2–0.3 cm, pubescent. Panicle open. Spikelets ovate, 4–6 × 2–3 mm, villous, orange at maturity. Lower glume ovate-oblong, 1–2 × 0.4–0.6 mm, chartaceous, 1-nerved, pilose. Upper glume ovate-lanceolate, 2.5–3.5 × 1–2.5 mm, aristate, sub–coriaceous, 5-nerved. Lower palea elliptic, 3–4 × 1–1.3 mm, hyaline, 2-nerved. Upper lemma ovate-oblong, 2.5–3 × 2 mm, coriaceous, 5-nerved. Upper palea elliptic, 2–2.5 × 1 mm, coriaceous. Stamens 3. Grains obovoid.

Rare in marshy areas and grass lands. Fl. & Fr.: Apr. – Dec. Namalagundu (KDP), *KRKS & BSS* 39661.

OPLISMENUS P. Beauv.

Oplismenus composites (L.) P. Beauv., Ess. Agrost. 54: 168.169. 1812; FPM 3: 1778. 1934; FAP 3: 1223. 1997. *Panicum compositum* L., Sp. Pl. ed. 1: 57. 1753.

Perennial, culms erect or creeping, to 100 cm high; rooting at lower nodes. Leaf-sheaths ciliate along the margins. Ligules fimbriate. Leaf-blades elliptic-lanceolate, 3–8 × 0.5–1.5 cm, hispid or villous. Racemes 10–20 cm long, rhachis glabrous or villous. Spikelets elliptic, 3–5 × 2 mm, paired. Lower glume oblong-lanceolate, 3–4 × 1.5 mm, chartaceous, awned. Upper glume ovate-lanceolate, 3–3.5 × 1.5 mm, chartaceous, aristate. Lower lemma ovate-oblong, 3–4 × 1.5–2 mm, chartaceous, 7-nerved. Lower palea oblong, 1–2.5 × 1 mm, hyaline, 2-nerved. Upper lemma elliptic-lanceolate, 2.5–3.5 × 1.5 mm, coriaceous, 7-nerved. Upper palea elliptic, 2–2.5 × 1 mm, subcoriaceous, 2-nerved. Stamens 3. Grains linear.

Occasional, in moist and shady localities of the forests near streams. Fl. & Fr.: Oct. – Feb. Batrepalli (ATP), *KRKS* & *BSS* 40350.

OROPETIUM Trin.

Oropetium thomaeum (L.f.) Trin., Fund. Agrost. 98.t.3. 1820; FPM 3: 1830. 1934; FAP 3: 1224. 1997. *Nardus thomaea* L.f., Suppl. 105. 1781.

Annual, culms erect, tufted, to 8 cm high. Leaf-sheaths small, compressed, glabrous. Ligules membranous. Leaf-blades linear, filiform, 1–3 × 0.1–0.2 cm, glabrous, ciliate on margins. Spikes solitary, slender, terminal, 3–5 cm long. Spikelets sessile, solitary, lanceolate, 1–2.5 mm, sunk in the rhachis, 1-4-flowered. Lower glume minute. Upper glume linear-lanceolate, 1–2 mm, chartaceous, 1-nerved. Lemmas ovate-oblong, 1.5 × 1 mm, hyaline, 1-3-nerved. Paleas lanceolate, 1.5 × 1 mm, 2-nerved. Stamens 3. Grains oblong.

Common, in dry localities of the plains, wastelands and forests. Fl. & Fr.: Jul. – Nov. Batrepalli (ATP), *KRKS* & *BSS* 39610, *KRKS* 239 (HJCB-E).

ORYZA L.

Oryza sativa L., Sp. Pl. 1: 333. 1753; FPM 3: 1814; FAP 3: 1226.1997.

Annual, culms erect, tufted, to 1m high. Leaf-sheaths glabrous. Ligule membranous, 2-partite. Leaf-blades linear-lanceolate, 8–20 × 0.3–0.6 mm, glabrous. Panicles lax, compound, to 30 cm long. Spikelets linear-oblong, 5 × 1.5 mm. Glumes oblong, minute, 1-nerved. Lemma boat-shaped, 5 × 1.5 mm, coriaceous, 3-nerved. Palea linear, 5 × 1.5 mm, coriaceous, scabrid, 3-nerved. Stamens 6. Grains oblong.

Cultivated for its grains. Fl. & Fr.: Sept. – Dec. Namalagundu (KDP), *KRKS* & *BSS* 40166.

PANICUM L.

Key to Species

1a. Culm erect to 90 cm high; spikelets ovate-lanceolate; upper glume ovate-lanceolate; upper lemma crustaceous .... **P. psilopodium**

1b. Culm ascending to 120 cm high; spikelets elliptic-lanceolate; upper glume ovate or elliptic-lanceolate; upper lemma ovate–oblong .... (2)

2a. Lower lemma epaleate, elliptic-lanceolate; grains oblong........... **P. paludosum**

2b. Lower lemma oblong-lanceolate; grain ovate ....**P. curviflorum**

Panicum curviflorum Hornem, Hort. Bot. Hafn. Suppl. 116. 1819. *P. trypheron* Schult., Syst. Veg 2: Mant. 244. 1824; FPM 3: 1783. 1934; FAP 3: 1231. 1997.

Annual, culms erect or geniculate, tufted, to 120 cm long. Leaf-sheaths 3.5 – 7 cm, ciliate-glabrous. Ligules membranous or tuft of hairs to 4 mm. Leaf-blades linear-lanceolate, 5–20 × 0.3–0.5 cm, hispid, base truncate, margins ciliate, apex acuminate. Panicles open, 10–30 cm long, branches alternate, pedicel 2.7 mm long. Spikelets elliptic-lanceolate, 2.5–3 × 2 mm, widely gaping. Lower glume ovate, 2–2.5 × 1 mm, cuspidate, membranous, 5-nerved. Upper glume ovate-lanceolate, 2–3 × 1.5 mm, membranous, 7-nerved. Lower lemma oblong-lanceolate, 2–3 × 2 mm, membranous, 9-nerved. Lower palea oblong, 1–2 × 0.5 mm, hyaline, 2-nerved. Upper lemma oblong-lanceolate, 2–3 × 2 mm, subcoriaceous, 9-nerved. Upper palea elliptic-oblong, 1–2 × 1 mm, subcoriaceous. Stamens 3. Grains ovate.

Common, in forests, plains and in grasslands and fallow lands. Fl. & Fr.: Jun. – Nov. Kalasamudram (ATP), *KRKS & BSS* 40097; Batrepalli (ATP), *KRKS* 230 (HJCB-E).

Panicum paludosum Roxb., Fl. Ind. 1:310. 1820; FPM 3: 1783. 1934; FAP 3: 1229. 1997.

Perennial, aquatic, culms erect or ascending, tufted, to 120 cm high; rooting at lower nodes. Leaf-sheaths glabrous. Ligules membranous. Leaf-blades linear-lanceolate, 5–20 × 0.3–0.5 cm, glabrous. Panicles effuse, 8–15 cm long. Spikelets elliptic-lanceolate, 2.5–3.5 × 1–2 mm, acute. Lower glume ovate-orbicular, 1 × 1.3 mm, membranous, 5-nerved. Upper glume elliptic-lanceolate, 2.5–3 × 1.5 mm, chartaceous, 7-9-nerved. Lower lemma epaleate, elliptic-lanceolate, 3–4 × 1.2 mm, chartaceous, 9-11-nerved. Upper lemma ovate–oblong, 2–2.5 × 1 mm, chartaceous, 7-nerved. Upper palea oblong–elliptic, 1.5 × 1 mm, coriaceous. Stamens 3. Grains oblong.

Occasional, near streams and in marshy areas. Fl. & Fr.: Jul. – Dec. Idupulapaya (KDP),*KRKS* & *BSS* 40304; Talupula (ATP), *KRKS* 219 (HJCB-E).

Panicum psilopodium Trin., Gram. Panic. 217. 1826; FBI 7: 46. 1896; FPM 3: 1782. 1934.

A perennial tufted grass; culms up to 50 cm high; nodes glabrous. Leaf-sheaths 4-8 cm long, glabrous; ligule a fimbriate membrane. Leaf-blades 4-10 × 0.3-0.9 cm, elliptic-lanceolate, base cordate, apex acuminate, glabrous. Panicle up to 12 cm long, ovate-oblong, effuse; branches numerous, alternate. Spikelets 3 mm, elliptic-oblong, deciduous. Lower glume less than half the length of the spikelet, ovate, obscurely 3-nerved. Upper glume broadly ovate, acute, 9-nerved, margins incurved. Lower lemma broadly ovate, 10-nerved. Upper lemma oblong, acute. Palea oblong, obtuse, scabrid on margin, bisexual. Stamens 3. Grains ellipsoid.

Common, in shady and wet or marshy areas. Fl. & Fr.: Jul. – Sep. Kalasamudram RF (ATP), *NY* 1171, *PVP* 2611.

PASPALIDIUM Stapf

1a. Spikes not overlapping the nodes of the inflorescence; spikelets to 6; annuals .... **P. flavidum**

1b. Spikes overlapping the nodes of the inflorescence; spikelets to 12; perennials .... **P. punctatum**

Paspalidium flavidum (Retz.) A. Camus, in Lecomte, Fl. Gen. De1' Indo–Chine 7: 419. 1922; FPM 3: 1744. 1934; FAP 3: 1232. 1997. *Panicum flavidum* Retz., Observ. Bot. 4: 15. 1786.

Annual, culms erect or geniculate, slender, to 80 cm long. Leaf-sheaths glabrous. Ligules fringed. Leaf-blades linear-lanceolate, 5–20 × 0.3–0.8 cm, glabrous. Inflorescence spiciform, 10–20 cm long, spikes 2–10, alternate. Spikelets gibbous, ovate, 2.5–3.5 × 2.5 mm. Lower glume orbicular, 0.4–1 × 1.5 mm, chartaceous, 3-nerved. Upper glume broadly–ovate, 2.5–3 × 2 mm, chartaceous, 9-nerved. Lower lemma ovate, 2.5–3 × 2 mm, chartaceous, 5-nerved. Lower palea oblong, 2–2.5 × 1 mm, hyaline, 2-nerved. Upper lemma oblong-elliptic, 2–2.5 × 1 mm, crustaceous, 5-7-nerved. Upper palea elliptic, 1.5 × 1 mm, sub coriaceous. Stamens 3. Grains oblong.

Occasional, in marshy areas in shady places and grasslands. Fl. & Fr.: Jun. – Nov. Kalasamudram (ATP), *KRKS* & *BSS* 37910.

Paspalidium punctatum (Burm.) A.Camus, Fl. Indo-Chine 7: 419. 1922; FAP 3: 1233. 1997; FPM 3: 1774. 1934. *Panicum punctatum* Burm.f., Fl. Indica 26. 1768; FBI 7: 29. 1896

Perennial, culms creeping or ascending, to 80 cm high. Leaf-sheaths glabrous. Ligules membranous. Leaf-blades linear-lanceolate, 5–15 × 0.5 cm, glabrous. Inflorescence spiciform, to 15 cm long, spikes many, alternate. Spikelets ovoid, 2 mm long. Lower glume orbicular, 1-1.5 × 1mm, chartaceous. Upper glume ovate, 2.5-3.5 × 1-2mm, chartaceous, 5-7-nerved. Lower lemma ovate-oblong, 3-3.5 × 1.5mm, membranous, 5-7-nerved. Palea ovate, 2.5-3 × 1.5mm, hyaline, 2-nerved. Stamens

3. Upper lemma broadly ovate, 3-3.5 × 1.5mm, coriaceous, mucronate. Upper palea akin to its lemma. Stamens 3. Grains ellipsoid.

Occasional, in marshy areas and near water bodies. Fl. & Fr.: Oct. – July. Dorigallu (ATP), *KRKS & BSS* 40473.

PASPALUM L.

Paspalum scrobiculatum L., Mant. Pl. 1: 29.1767; FPM 3: 1772.1934; FAP 3: 1235.1997.

Perennial, culms erect, tufted, to 90 cm high, rooting at lower nodes, rhizomatous. Leaf-sheaths sparsely hairy or glabrous. Ligules membranous. Leaf-blades linear-lanceolate, 8–20 × 0.3–0.5 cm, glabrous. Racemes subdigitate, 2-6. Spikelets in 2 rows, obovate or orbicular, 2.5–3 × 2–2.5 mm. Lower glume absent. Upper glume broadly ovate, 2–2.5 × 2–2.2 mm, membranous, 3-nerved. Lower lemma ovate, 2–2.5 × 2–2.2 mm, membranous, 5-nerved. Upper lemma ovate, 2.2 × 2 mm, coriaceous. Palea ovate–obovate, 1.5–2 × 1.5–2 mm, coriaceous, margins inflexed. Stamens 3. Grains ovate.

Occasional, in wet lands. Fl. & Fr.: Apr. – Jan. Kalasamudram (ATP), *KRKS & BSS* 40112.

PENNISETUM Richard ex Persoon

1a. Lower lemma often 3-lobed .... **P. pedicillatum**

1b. Lower lemma not 3-lobed .... **P. hohenackeri**

Pennisetum hohenackeri Hochst. ex Steud., Syn. Pl. Glum. 1: 103. 1854; FPM 3: 1792.1934; FAP 3: 1237. 1997.

Perennial, culms erect, tufted, to 2 m high. Leaf-sheaths glabrous. Ligules hairy. Leaf-blades linear, convolute, 10–40 × 0.3–0.5 cm, sparsely hairy. Panicles spiciform, 5–20 cm long. Spikelets elliptic-lanceolate, 5–7 × 2–2.5 mm, bristles unequal, to 2 cm. Lower glume ovate-suborbicular, 1–1.5 × 1 mm, membranous, 3-nerved. Upper glume oblong-lanceolate, 3–4.5 × 1–1.5 mm, chartaceous, 5-nerved. Lower lemma oblong-lanceolate, 5–7.5 × 2 mm, chartaceous, 9-nerved. Upper lemma ovate-lanceolate, 5–7 × 2 mm, chartaceous, 2-nerved. Stamens 3. Grains oblong.

Rare along stream banks. Fl. & Fr.: Sept. – Feb. Naganupula (ATP), *KRKS & BSS* 40493.

Pennisetum pedicellatum Trin., Mem. Acad. Imp. Sci. St. Petersbourg ser 6. Sci. Math. 3: 184. 1834; FPM 3: 1792. 1934; FAP 3: 1237. 1997.

Annual, culms erect, tufted, to 100 cm high. Leaf-sheaths glabrous. Ligules membranous. Leaf-blades linear-lanceolate, 15–30 × 0.2–0.5 cm, glabrous. Panicles spiciform, 5–10 cm long, involucral bristles densely plumose. Spikelets 3, two sessile, one pediceled, lanceolate, 3–5 × 2 mm. Lower glume oblong, 1.5 × 0.5 mm, hyaline, 3-nerved, silky. Upper glume ovate-lanceolate, 2 × 1.5 mm, hyaline, 5-nerved, silky. Lower lemma ovate, 3.5 × 1.8 mm, hyaline, 5-nerved, 3-lobed at apex. Palea small, hyaline. Stamens 3. Upper lemma elliptic, 3.5 × 1.5 mm, coriaceous, 5-nerved. Upper palea elliptic, 3 × 1 mm, hyaline. Stamens 3. Grains oblong.

Rare near marshy areas. Fl. & Fr.:S ept. – Feb. Rachakuntapalle (KDP), *KRKS* 39042.

PEROTIS Aiton

Perotis indica (L.) Kuntze, Revis. Gen. Pl. 2: 787. 1891; FPM 3: 1814. 1934; FAP 3: 1239. 1997. *Anthoxanthum indicum* L., Sp. Pl. ed 1: 28. 1753.

Annual, culms erect or geniculate, tufted, to 30 cm high. Leaf-sheaths glabrous. Ligules membranous. Leaf-blades ovate-lanceolate, 2–5 × 0.5–1 cm, glabrous, margins scabrid. Racemes spike like, 5–15 cm long, purpulish. Spikelets linear, sub sessile, 2.1 mm. Glumes linear-lanceolate, subequal, 2 × 1 mm, chartaceous, 1-nerved, keels ciliate, awned. Lemmas linear-lanceolate, 2 × 1 mm, chartaceous, 1-nerved. Palea linear-lanceolate, 1.5 × 1 mm, hyaline, 2-nerved. Stamens 3. Grains linear.

Common, in dry land agriculture, wastelands and in forests plains especially sandy soils. Fl. & Fr.: All seasons. Kalasamudram (ATP), *KRKS* & *KRS* 36464; Rachakuntapalle (KDP), *KRKS* 200 (HJCB-E).

PHRAGMITES Trin.

Phragmites karka (Retz.) Trin. ex Steud., Nomencl. Bot. ed. 2.2: 324. 1841; FPM 3: 1807. 1934; FAP 3: 1240. 1997. *Arundo karka* Retz., Observ. Bot. 4: 21.1 786.

Perennial, stoloniferous, culms erect, tufted, to 3 m high. Leaf-sheaths glabrous. Ligules narrow rim. Leaf-blades linear-lanceolate, 10–50 × 0.5–2 cm, glabrous or scabrid. Panicles pyramidal, to 40 cm long, woolly. Spikelets oblong-lanceolate, 7–12 × 2–3 mm, 3-5-flowered; rhachilla silky villous. Lower glume oblong-lanceolate, 2–4 × 1 mm, chartaceous, 3-nerved. Upper glume linear, 2–4 × 1 mm, chartaceous, 3-nerved. Lower lemma elliptic, 1.5–3 × 1 mm, chartaceous, empty. Upper lemmas ovate-lanceolate, 5–10 × 1 mm, chartaceous, 3-nerved. Paleas oblong-lanceolate, 3–4 × 1 mm, hyaline, 2-nerved. Stamens 3. Grains oblong.

Occasional, on river banks, near streams and ponds. Fl. & Fr.: Apr. – Dec. Idupulapaya (KDP), *KRKS* & *BSS* 41988; Gandi river (KDP), *KRKS* 223 (HJCB-E).

ROTTBOELLIA L. f.

Rottboellia cochinchinensis (Lour.) W.D. Clayton, Kew Bull. 35: 817. 1981; FAP 3: 1246. 1997. *Stegosia cochinchinensis* Lour., Fl. Cochinch. 1: 51. 1790. *Rottboellia exaltata* L. f., Nov. Gram. Gen. 40. t.1. 1779; FPM 3: 1759. 1934.

Perennial, culms erect, tufted, to 2 m high; supported by stilt roots. Leaf-sheaths basally tubercle-–based bristly, upper glabrous. Ligules membranous. Leaf-blades linear-lanceolate, 5–30 × 1–1.5 cm, glabrous. Racemes terete, 5–10 cm long, terminating into narrow end, joints hollow, rachis fragile. Sessile spikelets ovate, 2.5–3 × 2–3 mm. Lower glume broadly ovate, 2.5–3 × 3 mm, coriaceous or crustaceous, 13-nerved, close at the cavity. Upper glume boat-shaped, 2–3 × 3 mm, coriaceous, muticous, winged, 13-nerved. Lower lemma ovate-lanceolate, 2–3 × 1.5 mm, sub coriaceous. Lower palea ovate–lanceolate, 2.5–3 × 1.5 mm, hyaline.

Stamens 3. Upper lemma boat-shaped, 2–3 × 1.5 mm, hyaline. Upper palea ovate–lanceolate, 2–3 × 1.5 mm, hyaline. Stamens 3. Grains ellipsoid. Pedicelled spikelets ovate–lanceolate, 3–4 × 1–2 mm. Lower glume oblong, 3–4 × 1 mm, chartaceous, 13-nerved. Upper glume boat-shaped, 2.5–3 × 1 mm, chartaceous, 7-nerved. Lemmas ovate-oblong, hyaline. Palea ovate–oblong, hyaline, male or empty.

Common, in groundnut fields, wastelands and hill slopes. Fl. & Fr.: Jun. – Apr. Batrepalli (ATP), *KRKS* & *BSS* 37987.

SACCHARUM L.

Saccharum spontaneum L., Mant. Alt. 183. 1771; FPM 3: 1709. 1934; FAP 3: 1247. 1997.

Perennial, stoloniferous, culms erect, tufted, to 3 m high. Leaf-sheaths glabrous. Ligules membranous. Leaf-blades linear-lanceolate, 20–75 × 1–1.5 cm, margins serrulate, glaucous. Panicles lax, to 50 cm long, silky villous, joints long villous. Lower glume lanceolate, 3.5–4 × 1.2 mm, chartaceous, hyaline, 3-nerved, ciliate. Upper glume ovate-lanceolate, 3.5–4 × 1–1.5 mm, chartaceous, hyaline, 1-nerved, ciliate. Lower lemma ovate, 2–2.5 × 1 mm, hyaline, 3-nerved margins ciliate. Upper lemma linear-lanceolate, 2–2.5 × 0.3 mm, hyaline. Lodicules 2, 2-horned, ciliate. Stamens 3. Grains oblong. Pedicelled spikelets akin to sessile ones.

Common, in rivers, stream beds and all water courses especially in sandy soils. Fl. & Fr.: Jun. – Dec. Gandi River (KDP), *KRKS* & *BSS* 39068.

SACCIOLEPIS Nash

Sacciolepis indica (L.) A. Chase, Proc. Biol. Soc. Wash. 21: 8. 1908; FPM 3: 1787. 1934; FAP 3: 1248. 1997. *Aira indica* L., Sp. Pl. 1753. in errata.

Annual, culms erect or ascending, slender, tufted, to 50 cm high. Leaf-sheaths glabrous. Ligules membranous. Leaf-blades linear-lanceolate, 5–10 × 0.5 cm, glabrous. Panicles spiciform, 1–5 cm long. Spikelets elliptic-lanceolate, 2–3.5 × 2 mm. Lower glume ovate, 1 × 0.5 mm, chartaceous, 3-nerved. Upper glume oblong–ovate, 1–2.5 × 1 mm, chartaceous, 9-nerved. Lower lemma ovate, 1–2.5 × 1 mm, chartaceous, 9-nerved. Palea scale like or absent. Upper lemma ovate-oblong, 1 × 0.5 mm, coriaceous. Upper palea oblong, 1 × 0.5 mm, coriaceous. Stamens 3. Grains linear.

Occasional, in moist localities and wet lands. Fl. & Fr.: Apr. – Dec. Talupula (ATP), *KRKS* & *BSS* 41913.

SCHIZACHYRIUM Nees

Schizachyrium brevifolium (Sw.) Nees ex Buse, Pl. Jungh. 359. 1854; FPM 3: 1752. 1934; FAP 3: 1250.1997. *Andropogon brevifolius* Sw., Prodr. Veg. Ind. Occ. 26. 1788.

Annual, culms erect or ascending, tufted, to 35 cm high. Leaf-sheaths glabrous. Ligules membranous. Leaf-blades oblong-obtuse, 2–5 × 0.2–0.4 cm, glabrous. Racemes 1–3 cm long, spatheolate. Spikelets binate, fragile. Sessile spikelets linear-lanceolate, 2–3 × 2 mm, awned, callus bearded. Lower glume oblong-lanceolate, 1–2.5 × 0.2–0.5 mm, chartaceous, 5-nerved, bifid at apex. Upper glume linear, 2–3

× 1 mm, chartaceous, 1-nerved, bidentate. Lower lemma elliptic-lanceolate, 1–2 × 0.5 mm, hyaline, 2-nerved; epaleate. Upper lemma deeply 2-lobed, 1.5–2 × 0.5 mm, hyaline, awn 5–10 mm long. Upper palea minute. Stamens 3. Grains oblong. Pedicelled spikelets reduced to minute-arista.

Rare on soil present on rocky slopes, hill slopes and grass lands. Fl. & Fr.: Oct. – Jan. Rachakuntapalle (KDP), *KRKS* 41811.

SEHIMA Forssk.

Sehima nervosum (Rottl.) Stapf, Prain Fl. Trop. Afr. 9: 36. 1917; FAP 3: 1250. 1997. *Andropogon nervosus* Rottl. Neue Schriften Ges. Naturf. Freunde Berlin 4: 218. 1803. *Ischaemum laxum* R.Br., Prodr. Fl. Nov. Holl. 205 1810.

Perennial, culms erect, tufted, up to 1 m high. Leaf-sheaths 5–9 cm long, glabrous or pilose. Ligule membranous. Leaf-blades linear, 5–30 × 0.3–0.5 cm, coriaceous, apex acuminate, glabrous. Racemes terminal, fragile to 5 cm long, exserted from the sheath. Spikelets lanceolate, 12 mm long, laterally compressed, 2-nate. Sessile spikelet 3 mm, keels with a transculent longitudinal depression in the center. Lower glume linear-lanceolate, 5-7 × 1 mm, chartaceous, margins winged, grooved, 7-nerved. Upper glume boat-shaped, 6 × 2 mm, hyaline, apex scabrid, 5-nerved, awn to 7 mm long. Lower lemma 6 × 1 mm, hyaline, lanceolate, 2-nerved, 2-keeled. Lower palea lanceolate, 5 × 1 mm, hyaline, apex scabrid. Upper lemma ovate-oblong, 5-6 × 1 mm, hyaline, 3-nerved, awned, awn to 25 mm. Upper palea linear-lanceolate, 3.5 × 0.5 mm, hyaline, 2-nerved. Stamens 3. Grains oblong. Pedicelled spikelet lanceolate, 7-8 × 1.5mm, unawned. Lower glume lanceolate, 7 × 1.5mm, chartaceous, winged on the keels, ciliate. Upper glume lanceolate, 6 × 1.5 mm, chartaceous, 3-nerved, villous. Lemmas akin to those of sessile spikelets.

Common, in plains and open forests. Fl. & Fr.: Aug. – Dec. K.K. Kottala (KDP), *KRKS* 37668

SETARIA P. Beauv.

1a. Involucral bristles retrorsely barbed. – Panicle spiciform .... **S. verticillata**

1b. Involucral bristles antrorsely barbed .... (2)

2a. Inflorescense cylindrical false spike.- spikes to 4 cm long, spikelets compact .... **S. pumila**

2b. Inflorescence a narrow panicle, often lobed below .... **S. intermedia**

Setaria intermedia Roem. & Schult., Syst. Veg. 2: 489. 1817; FPM 3: 1789. 1934; FAP 3: 1252. 1997. *Setaria tomentosa* (Roxb.) Kunth., Rev. Gram. 1: 47. 1829. *Panicum tomentosum* Roxb., Fl. Ind. 1: 303.1820.

Annual, culms erect or decumbent, tufted, to 70 cm high. Leaf-sheaths hairy. Ligules hairy. Leaf-blades linear-lanceolate, 5–20 × 0.4–1 cm, hairy. Panicles spiciform, 5–10 cm long, interrupted. Spikelets broadly-ovate, 1–3 × 2 mm. Lower glume orbicular, 1 × 0.5 mm, chartaceous, 3-nerved. Upper glume broadly–ovate, 1–1.5 × 1 mm, chartaceous, 5-nerved. Lower lemma ovate–lanceolate, 1–2.5 × 1.5 mm, chartaceous, 5-nerved. Lower palea ovate-orbicular, 1.5 × 1 mm, hyaline, 2-nerved.

Upper lemma broadly ovate, 1–2 × 1 mm, crustaceous, transversely rugose, 5-nerved. Upper palea elliptic, 1–2 × 1 mm, crustaceous, rugose. Stamens 3. Grains ellipsoid.

Common, in wastelands, field edges and shady places. Fl. & Fr.: Aug. – Oct. Mabbuchintalapalle (KDP), *KRKS* & *KRS* 36998.

Setaria pumila (Poir.) Roem. & Schult., Syst. Veg. 2: 891. 1817; FAP 3: 1254. 1997. *Panicum pumilum* Poir., Encycl. Meth. Bot. Suppl. 4: 273. 1816. *Setaria pallidefusca* (Schum.) Stapf & C.E. Hubb., Kew Bull. 1930: 25. 1930; FPM 3: 1739. *Panicum pallidefuscum* Schum., Beskr. Guin. Pl. 58. 1827.

Annual, culms erect or decumbent, tufted, to 50 cm high. Leaf-sheaths glabrous, ciliate at mouth. Ligules hairy. Leaf-blades linear-lanceolate, 5–15 × 0.5–1 cm, sparsely hairy. Panicles spiciform, 5–10 cm long, brownish-yellow, bristly. Spikelets ovate, gibbous, 1–3 × 2 mm. Lower glume ovate, 1–1.5 × 0.2–0.4 mm, chartaceous, 3-nerved. Upper glume ovate, 1.5–2 × 1–1.5 mm, chartaceous, 5-nerved. Lower lemma elliptic-ovate, 1–2.5 × 1–1.5 mm, chartaceous, 7-nerved. Lower palea elliptic, 1.5 × 1 mm, hyaline, 2-nerved. Stamens 3. Upper lemma boat-shaped, 2–3 × 2 mm, rugose, 5-nerved. Upper palea elliptic-oblong, 1.5 × 1 mm, crustaceous, rugose. Stamens 3. Grains ovoid.

Common, in plains, periphery of the forests, in and around the shrubs in scrub. Fl. & Fr.: Apr. – Nov. Bhumaiahgari palle (KDP), *KRKS* & *KRS* 37633.

Setaria verticillata (L.) P. Beauv., Ess. Agrostogr. 51. 1812; FAP 3: 1255. 1997; FBI 7: 80. 1896; FPM 3: 1789. 1934. *Panicum verticillatum* L. Sp. Pl. ed. 2 82 1762.

Annual, culms erect, tufted, up to 1 m high. Leaf-sheaths 3–7 cm long, glabrescent. Ligule a fringe of hairs. Leaf-blades lanceolate, 3–10 × 0.3–1 cm, flat or folded, tuberculate-pubescent, apex acuminate. Panicles spiciform, to 10 cm long, involucral bristles retrorsely barbed, 6–10 mm long. Spikelets oblong-ovate, 2 mm long. Lower glume orbicular, 1.5 × 1 mm, chartaceous, 3-nerved. Upper glume ovate, 2 × 1 mm, chartaceous, 3-nerved; lower lemma oblong-ovate, 2 × 1 mm, chartaceous, 5-nerved. Lower palea elliptic-linear; 1.8 × 1 mm, hyaline, 2-nerved, male. Upper lemma ovate-oblong, 1.6 × 1 mm, crustaceous, transversely rugose. Upper palea similar to its lemma. Stamens 3. Grain ellipsoid.

Occasional, in scrub jungle, plains and near cultivated fields especially shaded places. Fl. & Fr.: Apr. – Nov. Bhumaiahgari palle (KDP), *KRKS* 36998.

SPOROBOLUS R. Br.

1a. Panicle effuse or pyramidal .... (2)

1b. Panicle contracted .... (3)

2a. Culms to 15 cm; leaf margins with tubercled based hairs; upper glume as long as the lemma .... **S. coromandelianus**

2b. Culms more than 15 cm; leaf margins not as above; upper glume shorter than the lemma .... **S. wallichii**

3a. Annual, stoloniferous grass; lemma linear-lanceolate .... **S. virginicus**

3b. Perennial or annual, non stoloniferous grass; lemma oblong–ovate .... **S. indicus**

Sporobolus coromandelianus (Retz.) Kunth, Revis. Gram. 1: 68. 1829; FPM 3: 1817. 1934; FAP 3: 1258. 1997. *Agrostis coromandeliana* Retz., Observ. Bot. 4: 19. 1786.

Perennial, culms erect or decumbent, tufted, to 10 cm high. Leaf-sheaths glabrous. Ligules membranous. Leaf-blades linear-lanceolate, 1–3 × 0.2–0.4 cm, sparsely tubercle based hairy, margins spinulosely toothed. Panicles pyramidal, 3–8 cm long, decompound, whorled. Spikelets ovate, 1–2 m long. Lower glume oblong, 0.5 × 0.3 mm, membranous. Upper glume ovate, 1 × 0.5 mm, chartaceous, 1-nerved. Lemma ovate, 1–1.5 × 0.5 mm, chartaceous, 1-nerved. Palea elliptic, 1 × 0.5 mm, hyaline, 2-nerved. Stamens 3. Grains ellipsoid.

Common, in plains, wastelands in dry habitats especially saline soils. Fl. & Fr.: Aug. – Dec. Dorigallu (ATP), *KRKS* & *BSS* 40470; Gollalaguduru (KDP), *KRKS* & *BSS* 41979.

Sporobolus indicus (L.) R. Br., Prodr.Fl. Nov. Holl. 170. 1810. *Agrostis indica* L., Sp. Pl. 63. 1753.

Key to Varieties

1a. Spikelets obovoid; lower glume lanceolate...................................... var. **diander**

1b. Spikelets obovoid; lower glume ovate-oblong .... var. **fertilis**

Sporobolus indicus (L.) R. Br. var. **diander** (Retz.) Jovet & Guedes, Taxon 22: 163. 1973; FAP 3: 1259. 1997. *Agrostis diandra* Retz., Observ. Bot. 5: 19. 1789. *Sporobolus diander* (Retz.) P. Beauv., Ess. Agrost. 26.147.178. 1812; FPM 3: 1817. 1934.

Annual or perennial, culms erect, tufted, to 100 cm high. Leaf-sheaths glabrous. Ligules fimbriate, membranous. Leaf-blades linear-lanceolate, 10–25 × 0.3–0.5 cm, glabrous. Panicles effuse, to 30 cm long, filiform. Spikelets obovoid-lanceolate, 1–2 mm. Lower glume lanceolate, 0.4 × 0.2 mm, truncate, chartaceous, 1-nerved. Upper glume ovate, 1 × 1 mm, chartaceous, 1-nerved. Lemma lanceolate, 1–2 × 1 mm, chartaceous, 1-nerved. Palea oblong–elliptic, 1.3 × 1 mm, hyaline, 2-nerved. Stamens 2. Grains obovoid.

Rare in moist and shady localities of the forests especially in sandy soils. Fl. & Fr.: Jul. – Dec. Kalasamudram (ATP), *KRKS* & *BSS* 40344.

Sporobolus indicus (L.) R. Br., var. **fertilis** (Steud.) Jovet & Guedes, Bull. Centre. Et. Rech. Sc. Biarritz. 7: 50. 1968; FAP 3: 1259. 1997. *Agrostis fertilis* Steud., Syn. Pl. Glumac. 1: 170. 1854. *Sporobolus indicus* sensu Hook. f., Fl. Brit. India 7: 247. 1896; FPM 3: 1817. 1934.

Perennial, culms erect, tufted, to 100 cm high. Leaf-sheaths glabrous. Ligules membranous. Leaf-blades linear-lanceolate, 15–30 × 0.5 cm, glabrous. Panicles spiciform, to 30 cm long. Spikelets obovoid, 1–2 × 1 mm. Lower glume ovate-oblong, 1 × 0.5 mm, membranous, 1-nerved. Upper glume ovate, 1.5 × 1 mm, chartaceous, 1-nerved. Lemma oblong-ovate, 2 × 1 mm, chartaceous, 1-nerved. Palea oblong-elliptic, 1.5 × 1 mm, hyaline, 2-nerved. Stamens 3. Grains ovate–oblong.

Occasional, in forests plains, especially in sandy soils. Fl. & Fr.: Sept. – Dec., Thimmamma Marrimanu (ATP,) *KRKS* & *BSS* 41937.

Sporobolus virginicus (L.) Kunth, Revis. Gramin. 1: 67. 1829. *Agrostis virginica* L. Sp. Pl. 1: 63. 1753. *Sporobolus tremulus* (Willd.) Kunth, Rev. Gram. 1: 67. 1829; FPM 3: 1817. 1934; FAP 3: 1261. 1997. *Agrostis tremula* Willd., Sp. Pl. 1: 372. 1798.

Annual, stoloniferous, culms erect, to 40 cm high. Leaf-sheaths glabrous. Ligules hairy. Leaf-blades linear-lanceolate, 1–2 × 0.2 cm, glabrous. Panicles subspiciform, interrupted. Spikelets ellipsoid, 3 mm long. Lower glume lanceolate, 1 × 0.5 mm, chartaceous, 1-nerved. Upper glume linear-lanceolate, 1.5 × 1 mm, chartaceous, glabrous, 1-nerved. Lemma oblong-lanceolate, 1.5 × 1 mm, chartaceous, glabrous, 3-nerved. Palea ovate, 1 × 0.5 mm, hyaline. Stamens 3. Grains ellipsoid.

Occasional, in forests in moist places. Fl. & Fr.: Jul. – Dec. Batrepalli (ATP), *KRKS & BSS* 39575.

Sporobolus wallichii Munro ex Thwaites., J. Bot. 27: 171. 1889; FPM 3: 1817. 1934; FAP 3: 1262. 1997.

Perennial, culms erect, slender, tufted, to 90 cm high. Leaf-sheaths glabrous. Ligules membranous. Leaf-blades linear, 10–30 × 0.2–0.4 cm, convolute, glabrous. Panicles effuse, 10–40 cm long, flexuous. Spikelets oblong-lanceolate, 1–2.5 × 0.5 mm, chartaceous, 1-nerved. Lemma elliptic, 1–2 × 1 mm, chartaceous, 1-nerved. Palea oblong, 1 × 1 mm, hyaline, 2-nerved. Stamens 3. Grains oblong.

Occasional, in shady and moist localities of forests. Fl. & Fr.: July – Dec. Batrepalli (ATP), *KRKS* & *BSS* 40347.

TETRAPOGON Desf.

Tetrapogon tenellus (Roxb.) Chiov., Annuario Reale Ist. Bot. Roma 8: 352. 1908; FAP 3: 1262. 1997. *Chloris tenella* Roxb., Fl. Ind. 1: 330. 1820; FPM 3: 1838. 1934.

Annual, culms erect or ascending, tufted, to 30 cm high. Leaf-sheaths glabrous. Ligules fimbriate, membranous. Leaf-blades linear-lanceolate, 5–10 × 0.2–0.5 cm, glabrous. Racemes solitary or rarely 2, 3–5 cm long. Spikelets cuneate, 5–6 × 3 mm, 2-4-awned, 4-6-flowered. Lower glume ovate-lanceolate, 3–5 × 2–2.5 mm, chartaceous, 1-nerved. Upper glume broadly-ovate, 3–5 × 2–3 mm, chartaceous, 3-nerved, awned, bifid. Lowest lemma 3–4 × 3 mm, chartaceous, 3-nerved, ciliate, awn 5–10 mm long; next lemma longer, 4th very short; empty lemmas 2, muticous. Paleas elliptic-oblong, 2–3 × 2 mm, hyaline, 2-nerved, ciliate. Stamens 3. Grains linear-oblong.

Common, in plains, wastelands, field edges and margins of the forests. Fl. & Fr.: Jul. – Dec. Mabbuchintalapalle (KDP), *KRKS* & *KRS* 36975; *KRKS* 209 (HJCB-E).

THEMEDA Forssk.

Themeda triandra Forssk., Fl. Aegypt. Arab. 123 & 178. 1775; FPM 3: 1746. 1934; FAP 3: 1265. 1997.

Perennial, culms erect, tufted to 1.5 m high. Leaf-sheaths glabrous. Ligules membranous. Leaf-blades linear-lanceolate, 10–30 × 0.5 cm, scabrid, margins scaberulous. Panicles spatheolate, 10–20 cm; racemes 3 cm long. Involucral spikelets oblong-lanceolate, 5.5 × 1.5 mm, inserted at same level. Lower glume ovate–oblong, 4.5 × 1 mm, coriaceous, sparsely tubercled based hairs. Upper glume linear-lanceolate, 5.5 × 1 mm, coriaceous, glabrous. Lower lemma oblong-linear, 4.5 × 1 mm, hyaline. Upper lemma linear, 3–3.5 × 1 mm, hyaline. Stamens 3. Sessile spikelets solitary, linear, 6.5 × 2 mm. Lower glume oblong, 6 × 1.5 mm, crustaceous, scabrid. Upper glume oblong–obtuse, 5.5 × 1 mm, keeled. Lower lemma oblong-lanceolate, 4.5 × 1 mm, hyaline, apex ciliate. Upper lemma hyaline base of the awn, awn 5–6 cm. Stamens 3. Grains linear. Pedicelled spikelets akin to the involucral spikelets.

Rare in forests along stream beds. Fl. & Fr.: Sept. – Feb. Batrepalli (ATP), *KRKS* & *BSS* 40337.

TRACHYS Pers.

Trachys muricata (L.) Pers. ex Trin., Mem., Acad. Imp. Sci. St.-Petersbourg. Ser. 6. Sci. Math. 3: 189. 1834; FPM 3: 1813. 1934; FAP 3: 1267. 1997. *Cenchrus muricatus* L., Mart. 2: 302. 1771.

Annual, culms slender, erect or decumbent, to 40 cm high; rooting at lower nodes. Leaf-sheaths villous, loose. Ligules membranous. Leaf-blades elliptic-lanceolate, 3–8 × 0.5–1 cm, hispid. Racemes 2, rarely 3, to 6 cm long. Spikelets clustered on a flat rhachis, 5–6 × 3 mm. Lower glume lanceolate, 2.5–3 × 0.5 mm, chartaceous, 1-nerved. Upper glume lanceolate, 3–4.5 × 0.4 mm, chartaceous, 1-nerved. Lower lemma broadly-ovate, 4–5 × 2.5 mm, coriaceous, many-nerved. Upper lemma lanceolate, 4–4.5 × 2 mm, coriaceous, 3-nerved. Palea lanceolate, 4–4.5 × 1.5 mm, hyaline. Stamens 3. Grains ovate.

Common, in Groundnut fields. Fl. & Fr.: Jul. – Dec. Thummalapalle (KDP), *KRKS & KRS* 36978.

TRAGUS Hall.

Tragus mongolorum Ohwi, Acta Phytotax. Geobot. 10: 268. 1941. *T. roxburghii* Panigrahi, Kew Bull. 29: 495.1974; FAP 3: 1267.1997. *T. racemosus* Hook. f., FBI 7: 97. 1896 non Scop. 1777. *T. biflorus* auct. non Schult. 1824; FPM 3: 1813.1934.

Perennial, culms ascending or decumbent or erect, to 10 cm high. Leaf-sheaths glabrous. Ligules fimbriate, membranous. Leaf-blades ovate-lanceolate, 2–4 × 0.3–0.5 cm, margin ciliate, glabrous. Racemes spiciform, 3–5 cm. Spikelets ovate-lanceolate, 2–3 × 1.5 mm, 1–flowered. Lower glume minute or absent. Upper glume lanceolate, 1.5–2 × 1 mm, chartaceous, prickled, apex hooked. Lemma elliptic-lanceolate, 1.5–2

× 1 mm, chartaceous, glabrous. Palea lanceolate, 1.5 × 0.5 mm, hyaline. Stamens 3. Grains oblong.

Common, in wastelands and forest peripheries. Fl. & Fr.: Aug. – Dec. Kalasamudram (ATP), *KRKS* & *BSS* 39151.

TRIPOGON Roem. & Schult.

1a. Lemmas 2-lobed at apex; leaves with white matted hairs....... **T. purpurascens**

1b. Lemmas 4-lobed at apex; leaves pilose... **T. bromoides**

Tripogon bromoides Roem. & Schult., Syst. Veg. 2: 600. 1817; FPM 3: 1834. 1934; FAP 3: 1268. 1997.

Perennial, culms erect, tufted to 40 cm high. Leaf-sheaths glabrous. Ligules ciliate. Leaf-blades linear-lanceolate, 5–20 × 0.2–0.4 cm, hairy. Spikes to 25 cm long. Spikelets lanceolate, 4–8-flowered. Lower glume oblong-lanceolate, 2.5–3 × 1 mm, chartaceous, 1-nerved. Upper glume lanceolate–ovate, 3–4 × 1.5 mm, 2-lobed, 1-nerved. Lemmas ovate-lanceolate, 6.5 × 1.5 mm, chartaceous, 4-lobed, 3-nerved, 3-awned. Paleas elliptic-oblong, 2–3 × 1 mm, hyaline, 2-nerved. Stamens 3. Grains oblong.

Occasional, on rocky hill slopes. Fl. & Fr.: Aug. – Feb. Rachakuntapalle (KDP), *KRKS* 41812.

Tripogon purpurascens Duthie, Ann., Roy. Bot. Gard. Cal. 9: 75.1901.

Perennial, culms erect, tufted, slender, to 15 cm high. Leaf-sheaths glabrous. Ligules ciliate membranous. Leaf-blades filiform, convolute, 2–5 × 0.1 cm, tip obtuse with purple colour, densely hairy near the base. Inflorescence terminal, contracted. Spike 5–10 cm, rhachis zigzag, ciliate. Spikelets lanceolate, 4–5 × 1–2 mm, alternate, 2–3-flowered. Lower glume small, 1.5 × 0.5 mm, membranous, triangular, 1-nerved, slightly awned. Upper glume lanceolate, 4–4.5 × 1 mm, membranous, 1-nerved, central thickened, margin hyaline, upper part ciliate. Lemmas 3, fertile, bisexual, 3-nerved, cleft, 2-lobed, median nerve slightly awned in the cleft, margin hyaline. Paleas 3, 2-lobed at apex, except third palea. First lemma 2.5 × 1.3 mm. Palea 2 × 1 mm. Grains linear. Second lemma 2.3 × 1.2 mm. Palea 1.5 × 1 mm. Third lemma 1.5 × 1 mm. Palea 1 × 0.5 mm. Stamens 2–3.

Occasional, in moist areas, dried ponds and marshy areas of forests. Fl. & Fr.: July–Nov. Kalasamudram (ATP), *KRKS* & *BSS* 40458.

UROCHLOA P. Beauv.

Urochloa panicoides P. Beauv., Ess. Agrost. 53. t. 11. F. 1. 1812; FPM 3: 1775. 1934; FAP 3: 1270. 1997.

Key to Varieties

1a. Lower lemma with a fringe of bristles..var. **velutina**

1b. Lower lemma without a fringe of bristles......................................var. **pubescens**

Urochloa panicoides P. Beauv. var. **velutina** (Henrard) Bor. Grass. India 372. 1960.

Annual, culms erect or ascending, tufted, to 50 cm high, rooting at lower nodes. Leaf-sheaths tubercle based hairy. Ligules ciliate. Leaf-blades linear-lanceolate, 5–10 × 0.5–1 cm, tubercle based hairy, margin ciliate. Racemes 3–4. Spikelets solitary or paired, ovate-elliptic, 4–5 × 2 mm. Lower glume ovate-orbicular, 1 × 1 mm, 3-nerved. Upper glume broadly ovate, 3–4.5 × 2–2.5 mm, chartaceous, 7-nerved, ciliate. Lower lemma broadly-ovate, 3–4.5 × 1.5–2 mm, chartaceous, 3-nerved. Lower palea ovate, 3–4 × 2 mm, hyaline, 2-nerved. Stamens 3. Upper lemma ovate, 2.5–3 × 2 mm, crustaceous, transversely rugose, mucronate. Upper palea ovate, 2–2.5 × 1 mm, crustaceous, rugose. Stamens 3. Grains ovate.

Common, in forests margins, cultivated fields and plains especially in shady localities. Fl. & Fr.:Jun. – Dec. Batrepalli (ATP), *KRKS* & *BSS* 40066.

Urochloa panicoides P. Beauv. var. **pubescens** (Kunth.) Bor, Grass. India 372. 1960.

Annual, culms erect or geniculate, stoloniferous, up to 70 cm high, rooting at lower nodes. Leaf-sheath 2–5 cm, loose, glabrous, keeled; ligule ciliate rim. Leaf-blades 4–12 × 0.5-1 cm, lanceolate, margin tuberculate hairy, apex acute, glabrous. Racemes spiciform, of 4–9, up to 5 cm long; rachis serrate, flat, scabrid. Spikelets ovate-orbicular, 4 × 1 mm, hairy, 1-nerved. Lower glume ovate, 1 × 0.8 mm, membranous, 5-nerved. Upper glume ovate, acute, membranous, 7 nerved. Lower lemma with fringe of bristles, ovate, 3.5 × 2 mm, membranous, 3-nerved. Lower plea ovate-lanceolate, 3 × 1.5 mm, hyaline, 2-nerved. Upper lemma ovate-obovate, 3 × 2 mm, crustaceous, 3-nerved, rugose. Upper palea akin to its lemma. Stamens 3. Grains ovate, compressed.

Occasional, on road sides. Fl. & Fr.: Jun. – Dec. Sanevaripalli (ATP), *KRKS* 40164.

7

Pteridophytes

Key to Families

1a. Leaves microphyllous with a single vein and no petiole (fern allies)............ (2)

1b. Leaves megaphyllous with open dichotomous venation .... (3)

2a. Aquatic; rosette herbs, stem corm form; leaves grass like, 20 cm long .... **Isoetaceae**

2b. Terrestrial; stem prostrate, branched; leaves linear to 2 mm long .... **Selaginellaceae**

3a. Plants free floating .... (4)

3b. Plants terrestrial .... (5)

4a. Leaves pinnately arranged, sporangia on narrow leaflets with reflexed edges .... **Parkeriaceae**

4b. Leaves simple, fronds 4-foliolate, sporangia in specialized sporocarps .... **Marsileaceae**

5a. Fronds comprising a sterile lamina and fertile spike diverging from a common, basal stalk .... **Ophioglossaceae**

5b. Sterile and fertile fronds borne separately on the rhizome or all mature fronds fertile

6a. Fronds branching in a series of forks with dormant apex in each fork .... **Gleicheniaceae**

6b. Fronds not branching .... (7)

7a. Sori on upper surface of simple leaf covering more or less the entire lower surface of the fertile leaves .... **Hemionitidaceae**

7b. Sori on upper surface of leaflets of pinnate leaf not occurred by entire lower surface .... **(8)**

8a. Pinnae articulated at the base .... **Oleandraceae**

8b. Pinnae not articulated at the base .... (9)

9a. Sori on near margin protected by reflexed margin .... (10)

9b. Sori not protected by reflexed margin .... **Thelypteridaceae**

10a. Sori in veins which are produced on the reflexed margins .... **Adiantaceae**

10b. Sori on leaf, not on margin which are lacking veins .... (11)

11a. Fronds repeatedly dichotomous with linear segment........ **Actiniopteridaceae**

11b. Fronds pinnately or palmately divided .... **Sinopteridaceae**

SELAGINELLACEAE

1a. Leaves ovate-lanceolate, distant, arranged in four rows .... **Bryodesma**

1b. Leaves linear, dense, spirally arranged .... **Selaginella**

BRYODESMA Sojak

Bryodesma wightii (Hieron) Sojak, Preslia 64(2): 155. 1992. *Selaginella wightii* Hieron, Hedw. 39 (26): 319. 1900; Pullaiah *et al.*, Pterid. Andhra Pradesh 29. 2003.

Plants form tuft of cushion-like patches. Stem prostrate, profusely branched, completely covered by leaves. Leaves dense, spirally arranged, linear, 2 mm long, apex long acuminate, margin ciliated. Cones borne on ultimate branches, fertile branches quadrangular, up to 8 × 1 mm; Sporophylls uniform, spirally arranged, ovate, up to 2 mm long, apex acuminate, margin dentate, midvein distinct; microsporangia and megasporangia borne on the same spike.

Occasional, at 422 – 462 m alt., grows on open rocky crevices on the hill slopes.

Somulavaripalli (ATP), *KRKS* 41942.

SELAGINELLA P.Beauv.

Selaginella bryopteris (L.) Baker, J. Bot. 22: 376. 1884. *Lycopodium bryopteris* L., Sp. Pl. 2: 1103. 1753.

Local name: *Mrithi sanjeevini.*

Stem 5–30 cm, sub-erect to erect, simple in the basal lower third but profusely branched above; branches flabellate; rhizophores confined to the base, distant, ovate-lanceolate, denticulate, long-acuminate at tip, heteromorphic on the branches, contiguous, ascending; lateral leaves ovate, oblique at base, acute to acuminate at apex, midrib short, obscure, inner half-leaf dilated, thin, white-transluscent, imbricate, denticulate, outer half-leaf thick, green; axillary leaves more or less similar to the lateral leaves; median leaves small, ovate, oblique at base, acute to acuminate at apex, entire to minutely denticulate. Strobili rare due to prevalence of vegetative propagation, 3 – 5 × 1 – 2.5 mm, single at the apex of the branchlets; sporophylls uniform, ovate, entire to minutely denticulate, acuminate. Megasporos 200 – 270 µm, dull yellow, verrucoid; microspores 15 – 25 µm, yellow, granulose.

Common, in foot hills of dry deciduous forest near moist locations and rocky crevices. Nigidi forest (ATP), *PVP* 2894.

ISOETACEAE

CALAMARIA Boehm.

Calamaria coromandeliana (Lf.) Kuntze, Revis. Gen. Pl. 2: 828. 1891. *Isoetes coromandelina* L.f., Suppl. Pl. 447. 1781; Dixit & Vohra, Dict. Pterid. India. 23. 1984; Pullaiah *et al.*, Pterid. Andhra Pradesh 31. 2003.

Corm subterranean, broadly triangular, up to 4 × 6 mm, fleshy, hard, bearing long fibrous roots densely all over the corm. Leaves tufted, erect, linear, grass like, up to 20 × 0.2 cm, at the base, progressively tapered towards acicular apex, dark green, herbaceous, glabrous. Sporangia borne at the base of the leaves, heterosporangiate; megasporangia common, reniform, connected with ligule at the apex, sporangial wall thin with horizontal trabeculae on the inner walls, microsporangia similar to megasporangia; megaspores granular, whitish; microspores whitish when dry, powdery, greyish when wet.

Occasional, at 424 – 854 m alt., grows in water, marshy places near ponds of dry deciduous forest.

Talupula (ATP), *KRKS* 41909.

OPHIOGLOSSACEAE

OPHIOGLOSSUM L.

1a. Sterile blade linear, grass- like .... **O. lusitanicum** subsp. **coriaceum**

1b. Sterile blade spatulate, ovate-elliptic .... **O. nudicaule**

Ophioglossum lusitanicum L., Sp Pl. 2: 1063. 1753. subsp. **coriaceum** (A. Cunn.) R.T. Clausen, Mem.Torrey Bot. Cl. 19(2): 161. 1938. *O. coriaceum* A. Cunn., Companion Bot. Mag. 2: 361. 1836. *O. gramineum* Willd., Nov. Acad. Erf., 2: 18, t. 1. 1802; Beddome, Handb. Ferns. Br. India. Suppl. 108. 1892; Pullaiah *et al.*, Pterid. Andhra Pradesh 38. 2003.

A tiny fern, with 3 cm tall, un-branched stem, rhizome cylindrical, 2 mm thick, fronds 1-3, nearly all of which are fertile; stipe 1–2 cm long, erect, naked and greenish when fresh. Sterile blade linear, up to 15 × 1 mm, naked, simple, entire, lamina thin, soft, green, glossy and glabrous on both surfaces. Fertile segment distinctly stalked, with stalk markedly longer than stipe, up to 1 cm, slender and with the sporangia bearing region up to 15 mm long with 9 pairs, dorsiventral. Spores globose-tetrahedral.

Rare, at 621 – 626 m alt., grows in marshy areas in dry deciduous forest.

Kalasamudram (ATP), *KRKS* 41874.

Ophioglossum nudicaule L.f., Suppl. PI. 443. 1781; Beddome, Handb. Ferns Br. India 464. 1883; Pullaiah *et al.*, Pterid. Andhra Pradesh 39. 2003.

A tiny fern, erect, up to 9 cm tall, un-branched, rhizome small, cylindrical, 5 mm thick. Fronds seasonal, usually only 1 or 2 at a time, stipe up to 2 – 3.5 cm long; sterile leaves borne on stalk, 3–8 mm long; blades spatulate, ovate-elliptic, base cuneate, apex obtuse; veins obscure; four or five veins passing up through the stalk of the blade; fertile stalk up to 10 cm long; spike solitary, up to 6 cm long with a sterile, acuminate apex and a strobilar region which is dorsiventral; sporangia up to 25 mm long, with 20 pairs. Spores whitish.

Very rare, at 621 – 626 m alt., occurs gregarioulsy amidst grasses in marshy locations of dry deciduous forest.

Kalasamudram (ATP), *KRKS* 41873.

ACTINIOPTERIDACEAE

ACTINIOPTERIS Link

Actiniopteris radiata (Sw.) Link, Fil. Spec. 80. 1841; Dixit & Vohra, Dict. Pterid. India, 1.1984; Pullaiah *et al.*, Pterid. Andhra Pradesh 58. 2003. *Asplenium radiatum* Sw., Schrad. J. Bot. 1800(2): 50. 1801.

Rhizome suberect, 1 cm thick, hard, densely covered by scales; scales lanceolate, acuminate, 5 × 1 mm, entire. Stipes numerous, tufted, fertile ones up to 15 cm, sterile ones up to 8 cm, abaxially deeply grooved, adaxially rounded, edges of the groove reflexed, pale brown at the base, pale green above; sparsely covered by pale brown, thin transparent scales. Laminae flabellate, up to 3 × 5 cm, repeatedly, dichotomously divided into six times; costa and veins indistinct; lamina pale green; texture coriaceous. Sporangia borne in intramarginal grooves throughout, protected by the reflexed margin of the segments.

Common, at 342 – 426 m alt., terrestrial or lithophytic fern grows as large colony among rocky crevices in fully exposed dry places.

Kalasamudram (ATP), *KRKS* 39679; Rachakuntapalle (KDP), *SSR* & *KRKS* 33646.

SINOPTERIDACEAE

CHEILANTHES Sw.

Cheilanthes mysorensis Wall. ex Beddome, Ferns S. India, t. 190. 1863; Handb, Ferns Br. India t. 46. 1883; Pullaiah *et al.*, Pterid. Andhra Pradesh 62. 2003.

Rhizome short, erect, usually unbranched, 1–1.5 cm long, 1 cm thick. Stipes tufted, erect, scaly at the base, 4 cm long. Lamina elongate-oblanceolate, 10–18 × 2.5–3 cm, compact, 2-pinnate to 3-pinnatifid, narrowing towards both ends; pinnae 1–1.4 cm apart, sub-opposite in the basal half and alternate in the anterior half, sessile, basal pinnae gradually reduced to auricles, middle and upper pinnae patent and slightly ascending, largest up to 2×0.7 cm, linear-deltoid. Sori protected by reflexed marginal flaps, whitish, sub-orbicular.

Common, at 502 – 537 m alt., lithophytic or terrestrial fern growing on fully exposed dry places along the slopes, rocky crevices.

Batrepalli (ATP), *KRKS* 37862.

PARKERIACEAE

CERATOPTERIS Brongn.

Ceratopteris thalictroides (L.) Brongn., Bull. Sci. Soc. Philom. Paris 186. 1821; Beddome, Ferns Br. India, t. 75. 1864; Pullaiah *et al.*, Pterid. Andhra Pradesh 67. 2003. *Acrostichum thalictroides* L., Sp. PI. 2: 1070. 1753.

Aquatic marshy plant with short, erect, unbranched rhizome, dimorphic, fibrous roots arise from the abaxial side, apex covered by scales. Fronds in rosettes. Lamina varying much in size and form, dimorphic; sterile lamina bipinnate or tripinnate, up to 15×10 cm, ovate, acute, cuneate; primary pinnae 4–5-pairs, alternate, stalked, up to 6 × 3 cm, secondary pinnae about 3–4- pairs, alternate, shortly stalked, ovate, up to 3 × 2 cm. Fertile lamina ovate, up to 25 × 10 cm, tripinnate, ultimate segment needle like, acute, margin reflexed covering the lower surface on which two rows of larger sporangia are borne.

Very rare, at 529 – 702 m alt., grows in hill regions at marshy places in dry deciduous forest.

Ratnachalam Anjaneyaswamy temple (KDP), *KRKS* & *BSS* 40056; Rachakuntapalli (KDP), *KRKS* 41817.

HEMIONITIDACEAE

HEMIONITIS L.

Hemionitis arifolia (Burm.) T. Moore, Index Fil., 114.1859; Beddome, Handb. Ferns Br. India, 413. 1883; Pullaiah *et al.*, Pterid. Andhra Pradesh 69. 2003. *Asplenium arifolium* Burm., Fl. Ind. 231. 1768. *Hemionitis cordata* Hook. & Grev., Ic. Fil. t. 64. 1828; Beddome, Ferns S. India t. 53. 1864.

Rhizome erect or suberect, unbranched, 1–2 cm long, 2–4 mm thick. Stipes tufted, sterile ones up to 19 cm long, fertile up to 31 cm long, 1.5 mm thick, faintly grooved on the abaxial surface, dark, glossy. Lamina dimorphous, simple; sterile fronds up to 8 × 5.7 cm, deeply cordate, margin entire, apex rounded; fertile lamina sagittate, apex obtuse or acute, base forming an inverted 'V' shape; texture coriaceous; costule dark and raised below, grooved and concolorous with the lamina above, veins obscure, anastomosing closely, areoles seen as depressions in dry fronds. Sori continuous along the veins, filling the entire surface of lamina when mature, intermixed with copious hairs and scales.

Occasional, at 562 – 628 m alt., grows in relatively dry localities and on shaded steep slopes in dry deciduous forest.

Batrepalli (ATP), *KRKS* 39645.

ADIANTACEAE

ADIANTUM L.

Adiantum incisum Forssk., Fl. Aegypt. 187. 1775; Pullaiah *et al.*, Pterid. Andhra Pradesh 75. 2003.

Rhizome erect, 1 × 0.6 cm, covered by tuft of wiry roots and linear paleae which are lanceolate, acuminate, entire, uniformly thickened, up to 6 × 1 mm. Stipes 4–8 cm long, 1 cm thick, fronds simply pinnate, up to 44 × 4 cm, linear-lanceolate, distal nonleafy, apex rooting, base slightly narrower than middle part; pinnae dimidiate, largest up to 2 × 0.8 cm, lower edge straight, upper margin descending towards the apex, inner margin slightly concave, upper and outer edges lobed to half way to the centre; lobes serrate. Sori linear, borne on the under surface of the reflexed flaps of lobes; spores trilete, exine granulated.

Common, at 378 – 506 m alt., on slopes, rock-crevices of dry deciduous forest.

Kalasamudram (ATP), *KRKS* 39680; Chinnapalli (ATP), *KRKS* 37997.

OLEANDRACEAE

NEPHROLEPIS Schott.

Nephrolepis cordifolia (L.) Presl., Tent. Pterid., 79. 1836; Beddome, Handb. Ferns Br. India, 282. 1883; Pullaiah *et al.*, Pterid. Andhra Pradesh 91. 2003. *Polypodium cordifolium* L., Sp. PI. 2: 1089. 1753. *Nephrolepis tuberosa* Presl. in Beddome, Ferns S. India t. 78. 1864.

Rhizome erect, up to 0.5 cm thick, with tuft of fronds and dense wiry, tuber-bearing roots. Stipes erect, up to 12 cm long, 2 mm thick, densely scaly throughout, distally grooved. Lamina pinnate, 40 × 6.5 cm, linear, oblong-lanceolate, slightly narrowing above the middle, apex obtuse; pinnae jointed to rachis, numerous, sessile, closely set, edges crenate or serrate, apex obtuse; costules dark, distinct; veins indistinct, branched once at base. Pinnae herbaceous, light green, glabrous above and below. Sori seated on vein-ends, reniform.

Occasional, at 554 – 568 m alt., terrestrial, rock crevices in moist places, near waterfalls in dry deciduous forest.

Namalagundu (KDP), *KRKS* & *BSS* 39653.

GLEICHENIACEAE

DICRANOPTERIS Bernh.

Dicranopteris linearis (Burm.f.) Underwood, Bull. Torr. Bot. CI. 34: 249. 1907; Pullaiah *et al.*, Pterid. Andhra Pradesh 96. 2003. *Polypodium lineare* Burm.f., Fl. Ind. 265. t. 67. 1768. *Gleichenia linearis* (Burm.f.) C.B.Clarke, Ferns North India, 428. 1879; Beddome Handb. Ferns Br. India 4. 1883.

Rhizome long creeping, up to 5 mm thick, densely hairy. Stipes scattered, up to 25 × 0.1 cm, grey brown, glabrous. Lamina 25 × 20 cm, primary branches two or three pairs, about 10 cm apart, forked one or two times; secondary branches unequal, accessory branches borne at the fork, stipules lobed; ultimate leafy branch up to 10 × 2 cm, oblanceolate, acuminate, pinnatifid up to 1–2 mm to the costa, leaf segments 2.5 × 0.4 cm, oblong, slightly narrowed towards notched apex. Sori sub median on the veinlets, consisting of up to 8 sporangia per sorus.

Occasional, at 554 – 568 m alt., terrestrial, near and on rocky crevices in dry deciduous forest.

Namalagundu (KDP), *KRKS* & *BSS* 39652.

THELYPTERIDACEAE

CHRISTELLA Ker-Gawl.

Christella dentata (Forssk.) Brownsey & Jeremy, British Fern Gaz. 10: 338. 1973; Pullaiah *et al.*, Pterid. Andhra Pradesh 104. 2003. *Polypodium dentatum* Forssk., Fl. Aegypt. Arab. 185. 1775. *Nephrodium molle* Desv. in Bedd., Handb. Ferns Br. India 277. 1876. *Cyclosorus dentatus* (Forssk.) Ching, Bull. Fan. Mem. Inst. Biol. Bor. 8: 26. 1938; Beddome, Handb. Ferns Br. India 67. 1974.

Rhizome short-creeping, 5 mm thick, densely scaly. Scales lanceolate, acuminate, up to 6 × 1 mm. Stipes 15 cm long, 3 mm thick; lamina pinnate, oblong-lanceolate, up to 30 × 20 cm, narrowing towards apex, pinnae about 20 pairs, with sub-opposite, reduced basal pinnae, rest of the pinnae alternate, apex acuminate, base broadly cuneate, basal ones sub-sessile, upper ones sessile, long hairs borne on costa and costules below, dense long hairs on costa, short hairs on veins and rarely between veins above. Sori median on veins; indusia with short hairs.

Occasional, at 554 – 568 m alt., shaded stream banks and on rock crevices in loamy soil in dry deciduous forest.

Namalagundu (KDP), *KRKS* & *BSS* 39654.

MARSILEACEAE

MARSILEA L.

Marsilea minuta L., Mant. 308. 1771; Sledge, Bot. J. Linn. Soc. 84: 22. 1982; Pullaiah *et al.,* Pterid. Andhra Pradesh 157. 2003.

Rhizome long creeping, branched, subterranean, green, semi aquatic or aquatic plant, dark brown in terrestrials, covered by soft slender hairs all over the plant parts; roots from below the stem in the nodal region. Leaves quadrifoliolate, leaflets sessile, arranged at the tip of the stipe in clover leaf model, obovate, up to 2 × 1.5 cm, base cuneate, lateral margin above and below, flabellately branched. Sporocarps 2–4 in each cluster, peduncle 5 × 1 mm; sporocarps adnate to the peduncle laterally and perpendicularly. Microsporangia enclosed in the same sporocarp and covered by gelatinous layer.

Common, 360 – 391 m alt., in shallow ponds, ditches at the boundaries of paddy fields.

Kalasamudram (ATP), *KRKS* 39960.

8

Bryophytes

Key to Families

1a. Plant body foliose; usually midrib present; rhizoids multicellular; peristome present; elaters absent .... **Musci (2)**

1b. Plant body thalloid or foliose; without midrib; rhizoids unicellular; operculum and peristome absent; elaters or pseudoelaters present .... **Hepaticae (Ricciaceae)**

2a. Sporophyte immersed, without seta, capsule sunken at the tip of the gametophyte .... **Archidiaceae**

2b. Sporophyte exposed, with prominent seta, capsule elevated from the gametophyte .... (3)

3a. Leaves distichous, with characteristic sheathing (or) vaginate lamina (apparently split on the inner edge and sheathing each other and the stem) .... **Fissidentaceae**

3b. Leaves not distichous, without splitting (sheathing lamina) .... (4)

4a. Peristome when present single (haplolepidae) .... (5)

4b. Peristome usually double (diplolepidae) .... (6)

5a. Leaves mostly elongated and broad, leaf cells isodiametric from the middle to upwards .... **Pottiaceae**

5b. Leaves mostly elongate to subulate, cells usually rectangular to linear .... **Dicranaceae**

6a. Leaves ovate, orbicular or lanceolate .... (7)

6b. Leaves elongate subulate, usually margin dentate or serrate by means of paired papillae .... **Bartramiaceae**

7a. Rhizoids ends warty papillose; leaf margin not bordered .... **Splachnaceae**

7b. Rhizoids simple; leaf margins bordered by one or two rows of narrow elongated cells .... **Bryaceae**

RICCIACEAE

RICCIA L.

Key to Species

1a. Plants aquatic and terrestrial; air cavities present in thallus .... **R. fluitans**

1b. Plants only terrestrial on moist or dry soil, never aquatic; air cavities absent in thallus .... (2)

2a. Plants dioecious .... **R. discolor**

2b. Plants monoecious .... **R. billiardieri**

Riccia billardieri Mont. & Nees ex Gottsche *et al.*, Hepat. 602.1846; Bapna & Kachroo, J. Indian Bot. Soc. 54: 219-224.1975; Sandhya Rani *et al.*, Bryo. Andhra Pradesh 63. 2014.

Monoecious. Thallus yellowish green, robust, bluish green, overlapping, forming dense patches, spreading on open areas, 1–2 times dichotomously branched, 22 × 12 mm, lobes ovate to oblong with prominent groove at apex and shallow behind, apical notch distinct.

Occasional, at 570 – 639 m alt., grows on moist soil in exposed places in dry deciduous forests.Batrepalli (ATP), *KRKS* 36970, 37989, 40330.

Riccia discolor Lehm. & Lindenb., Nov. Stirp. Pug.4: 1. 1832; Bapna & Kachroo, Hepatic. India 2: 451. 2000; Sandhya Rani *et al.*, Bryo. Andhra Pradesh 64. 2014.

Plants dioecious, overlapping in dense patches, rarely forming irregular rosettes. Thalli green, forked twice, lobes oblong, 5–6×3–4 mm in male plants and 6–14 × 3–4 mm in female plants, margins purple and entire often bend downwards, ventral surface bears simple and tuberculate rhizoids; ventral scales small, semilunar, overlapping, dark pink, projecting beyond the thallus margin.

Very common, at 650 – 719 m alt., on moist soil and rocks associated with other *Riccia* sps. in dry deciduous and scrub forests. Batrepalli (ATP), *KRKS* & *BSS* 37989, *KRKS* 39084.

Riccia fluitans L., Sp. Pl. 1139. 1753; Bapna & Kachroo, J. Indian Bot. Soc. 54: 221.1975 & Hepatic. India 2: 452. 2000; M.C.Nair *et al.*, Bryophyt. Wayanad 39. 2005; Sandhya Rani *et al*. Bryo. Andhra Pradesh 66.2014.

Monoecious. Thallus terrestrial, yellowish green, 30–50 × 1.3 mm, several times dichotomously branched, segments divergent, linear, apex emarginate or truncate, air chambered. Rhizoids numerous, ventral scales small, confined to below the apex, air cavities conspicuous. Spores brownish-yellow transluscent.

Rare, at 565 – 576 m alt., grows in marshy places near streams and water falls in dry deciduous forests. Batrepalli (ATP), *KRKS* 36970.

Note: It generally occurs in two forms, free floating form with very narrow lobes and terrestrial form with some what thicker and broader lobes.

ARCHIDIACEAE

ARCHIDIUM Brid.

Archidium birmannicum Mitt. ex Dix., J. Indian Bot., 2: 175.1921; Gangulee, Moss. E. India1: 172. 1971; R.S. Chopra, Taxon. Indian Mosses 11. 1975; Sandhya Rani *et al*. Bryo. Andhra Pradesh 105.2014.

Densely caespitose, olive-green to yellow-green, lax, fertile plants up to 4.3 mm long, sometimes longer in sterile plants. Fertile plants often showing persistent protonema at base. Stem branching with bud-like fertile shoots. Leaves erect on sterile stems, patent on fertile shoots. Stems leaves about 1 × 0.2–0.3 mm broad at base, lanceolate to narrow lanceolate, acuminate, percurrent, margin entire, leaf cells elongated, thin walled and transparent, perichaetial leaves longer ±1.4 mm long and 0.25 mm wide at base, narrow lanceolate, spirally encircling the sporophyte like a flower. Sporophyte is typical. Capsule green when young, dark-brown when mature, spores spherical.

Rare, at 618 – 867 m alt., grows on black loamy soil in dry deciduous forests. Batrepalli (ATP), *KRKS* & *BSS* 40364, 40366.

DICRANACEAE

DICRANELLA (Mull. Hal.) Schimp.

Dicranella amplexans (Mitt.) A. Jaeger, Ber. Thatigk. St. Gallischen Naturwiss. Ges., 1870-71: 376. 1872; Gangulee, Moss. E. India 1: 261. 1971; Sandhya Rani *et al*. Bryo. Andhra Pradesh 114. 2014. *Leptotrichum amplexans* Mitt., Musc. Ind. Or. 9. 1859.

Plants green, usually tufted, branched. Stem about 6–7 mm height. Leaves laxly arranged, 1.15 × 0.2 mm, tapering into a long subula from a broad concave sheathing, base elliptical, costa prominent, percurrent, ending just below the apex, margin entire, slight denticulation at apex.

Common, at 414 – 576 m alt., on rocks, soils, slopes of moist valleys in dry deciduous and scrub forests. Rachakuntapalli (KDP), *KRKS* 40330; Rekkamanu (ATP), *KRKS* 37925; Kalasamudram (ATP), *KRKS* & *BSS* 39178.

FISSIDENTACEAE

FISSIDENS Hedw.

Key to Species

1a. Leaf showing limbidium of narrow elongated cells of all leaves .... **F. splachnobryoides**

1b. Leaf not showing limbidium .... **F. involutus**

Fissidens involutus Wilson ex Mitt., J. Proc. Linn. Soc. Bot. Suppl. 138. 1859. subsp. **involutus;** Gangulee, Moss. E. India 1: 547. 1971; Sandhya Rani *et al.* Bryo. Andhra Pradesh 128. 2014.

Plant yellowish green or brown, 1–16 × 0.2 mm with leaves. Leaves lax, closer on young shoots, curled when dry, ovate-lanceolate, acuminate at apex, 1.8 × 0.05 mm, sheathing lamina equal, 1.3 × 0.03 mm, leaf cells quadrate to round, chlorophyllous, 0.01 × 0.01 mm, costa prominent, percurrent, margin denticulate at apex by projecting cells, cell walls thicker than usual, limbidium absent.

Occasional, at 560 – 624 m alt., on moist soil near streams, along cement bridges associated with other mosses in moist and dry deciduous forests. Somulavaripalli (ATP), *KRKS* 37839; Batrepalli (ATP), *KRKS* 39083.

Fissidens splachnobryoides Broth., Fl. Schutzgeb. Sudsee: 81. 1900; Gangulee, Moss. E. India 1: 462. 1971; Sandhya Rani *et al.* Bryo. Andhra Pradesh 129. 2014.

Plant yellowish green. Shoot 7–10 × 2 mm with 12–16 pairs of leaves. Leaves lax and small at base, crowded and long at apex, oblong-lanceolate, widely acuminate, up to 1.7 × 0.4 mm, sheathing lamina up to 1 × 0.2 mm, limbidium with 1 to 3 rows of narrow, elongated cells, up to 0.2 × 0.01 mm, costa prominent, occupy nearly ¾ of lamina ending far below tip.

Common, at 587 – 628 m alt., on moist soil in dry deciduous forests. Somulavaripalli (ATP), *KRKS* 37839; Kalasamudram (ATP), *KRKS* 39131.

POTTIACEAE

Key to the Genera

1a. Peristome absent .... (2)

1b. Peristome present .... (3)

2a. Shoot arise from long horizontal radiculous part, leaf spatulate, margin denticulate at apex .... **Hyophila**

2b. Shoot with out horizontal radiculous part, leaf linear lanceolate, margin involute, undulate at apex .... **Hymenostomum**

3a. Upper laminar cells highly multipapillose .... **Barbula**

3b. Upper laminar cells few papillate, leaf cells mostly transparent or transluscent........ **Hydrogonium**

BARBULA Hedw.

Barbula indica (Hook.) Spreng., Nomencl. Bot. 2: 72. 1824. *Tortula indica* Hook., Musci Exotici, 2: 135. 1819; Sandhya Rani *et al.* Bryo. Andhra Pradesh 136.2014. *Semibarbula orientalis* (F.Weber.) Wijk. & Marg., Taxon 8: 75. 1959; Gangulee, Moss. E. India 1: 717. 1972. *Trichostomum orientale* F. Weber, Arch. Syst. Naturgesch., 1: 129. 1804.

Plants yellowish green, in dense tufts 8–10 mm height, unbranched, stem lower part brown, upper part green. Leaves spirally arranged on main stem, lower leaves sparsely arranged, upper leaves in the form of a comal tuft, narrow-lanceolate, *ca.*1.5 × 0.2 mm, base rounded, tip acute, margin entire at base, costa prominent, short excurrent, light green, papillose at back. Perichaetial leaf small. Antheridial bunch at apex, single antheridia 0.3 × 0.06 mm with multicellular uniseriate paraphyses, 0.4 mm long. Gemmas sac like, multicellular, 0.1 × 0.04 mm.

Very common, at 405 – 468 m alt., grow in dense tufts on rock over soil, along stream side, rock crevices (calciphilous and terricolous) of water falls in black loamy soils in dry deciduous forests. Rachakuntapalli (KDP), *KRKS* & *GK* 33567; *KRKS* 40329; K. K. Kottala (KDP), *KRKS* & *GK* 33572, 33573.

HYDROGONIUM (Mull. Hal.) A.Jaeger

Hydrogonium javanicum (Dozy & Molk) Hilap., Beih. Bot. Centralbl 50 (2) : 632.1933; Sandhya Rani *et al.* Bryo. Andhra Pradesh 142.2014. *H.consanguineum* (Thw. & Mitt.) Hilp., Beih. Bot. Centrabl., 50(2): 626. 1933; Gangulee, Moss. E. India 1: 729. 1972. *Tortula consanguinea* Thw. & Mitt., J. Linn. Soc., 13: 300. 1873.

Plants green, tufted, slender. Stem reddish brown, erect, unbranched *ca.* 6 mm long. Leaves erect, appressed to stem when dry, distant below, forming comal tuft above, curled when dry, lanceolate, broad at base, slightly narrowing towards tip, bluntly pointed, margin highly papillate, costa brown, prominent, percurrent. Sporophyte from the emerging tip of the plant, seta *ca.* 7mm long, brown, smooth. Capsule erect, cylindrical, *ca.* 1 mm. peristome teeth arise from inside walls of the capsule, slender, long, rough, reddish brown.

Very common, at 547 – 653 m alt., grows on moist soil of dry deciduous and scrub forests. Somulavaripalli (ATP), *KRKS* 37838; Rekkamanu (ATP), *KRKS* 37925; Thimmammamarimanu (ATP), *KRKS* 41945.

HYOPHILA Brid.

Key to the Species

1a. Costa excurrent, leaf marginal cells not denticulate at apex .... **H. nymaniana**

1b. Costa percurrent, leaf marginal cells denticulate at apex .... **H. involuta**

Hyophila involuta (Hook.) A. Jaeger, Ber. S. Gall. Naturw. Ges. 1871–72: 356. 1873; Gangulee, Moss. E. India 1: 681. 1972; Sandhya Rani *et al.*, Bryo. Andhra Pradesh 145. 2014. *Gymnostomum involutum* Hook., Musc. Exot. 2: 154. 1819.

Plants in dense tufts, erect, green, unbranched or rarely branched, shoot erect, *ca.* 1 cm long, arise from long, horizontal, radiculous part, rhizoids reddish. Lower leaves sparsely arranged, upper leaves spreading in rosettes, leaves inrolled when dry, ligulate to spatulate, 3 × 0.8 mm, apex broadly pointed, margin denticulate near apex, entire below, costa prominent, reddish, percurrent. Sporophyte apical, seta long, erect, brown, *ca.*1 cm long, capsule cylindrical, up to 2 mm long, peristome teeth absent.

Very common, at 385 – 423 m alt., on reddish black soils, black soils, on rock grooves, rocks near streams, black loamy bauxite soil in open dry deciduous and scrub forests. K.K. Kottala (KDP), *KRKS* 37670; Bhumaiahgaripalli (KDP), *KRKS* & *GK* 33567.

Hyophila nymaniana (M. Fleisch.) M. Menzel, Willdenowia 22: 198. 1992. *Glyphomitrium nymanianum* Fleish., Musci Fl. Buitenz., 1:372.1904; Sandhya Rani *et al.* Bryo. Andhra Pradesh 147.2014. *Hyophila rosea* Williams, Bull. N.Y. Bot. Gard., 8: 341. 1914; Gangulee, Moss. E. India 1: 685. 1972. *H. comosa* Dix. &Varde, Arch. Bot., 1: 166.1927; Gangulee, Moss. E. India 1: 687. 1972.

Plant green – greenish yellow, usually unbranched, sometimes branched, up to 1.5 cm tall. Leaves in interrupted rosettes, erect, spreading, inrolled and curled when dry, spatulate, narrow base, *ca.* 2 × 0.65 mm, tip broadly acuminate, costa prominent, broad, slightly excurrent, brown. Capsule cylindrical, erect, 2 mm long, peristome teeth absent.

Occasional, at 570 – 938 m alt., on black loamy soils near streams in dry deciduous forests. Batrepalli (ATP), *KRKS* 39083, *KRKS* & *BSS* 40363.

HYMENOSTOMUM R.Br.

Hymenostomum edentulum (Mitt.) Besch., Bull. Soc. Bot. France, 34: 95. 1887; Gangulee, Moss. E. India 1: 673. 1972; R.S. Chopra, Tax. Indian Mosses 128. 1975. *Weisia edentula* Mitt. in Musc. Ind. Or.: 27. 1859; Sandhya Rani *et al.* Bryo. Andhra Pradesh 149. 2014.

Plant yellowish green, stem brown, branched or unbranched, *ca.* 3–6 mm long. Leaves canaliculated, densely arranged on stem, large towards apex of the stem, up to 2.1 × 0.4 mm spreading, crisped when dry, linear–lanceolate, sheathing base, tip narrow, pointed, leaf margin involute, undulate at apex, entire at base, costa prominent, strong, brown, excurrent. Sporophyte emerging from top, seta erect, light reddish brown *ca.* 5 mm height, capsule reddish brown, erect, slightly bent at apex.

Very common, at 423 – 624 m alt., on black soils and red soils near streams, moist rocks, soil cuttings termite mounds in dry deciduous and scrub forests.

Thummalapalli (KDP), *KRKS* & *GK* 33566; K. K. Kottala (KDP), *KRKS* & *GK* 33570, 33571, 33572, 33573, 33574; Batrepalli (ATP), *KRKS* & *BSS* 40364; Kalasamudram (ATP), *KRKS* 40054.

SPLACHNACEAE

SPLACHNOBRYUM Müller Hal.

Splachnobryum assamicum Dixon, J. Bombay Nat. Hist. Soc. 39: 778. 1937; Gangulee, Moss E. India 2: 866. 1974; Sandhya Rani *et al.*, Bryo. Andhra Pradesh 153.2014. *S. synoicum* H. Rob., J. Hattori Bot. Lab. 27: 127. 1964; Gangulee, Moss E. India 2: 870. 1974.

Plant synoicious, slender, stem radiculose at the base, olive green, laxly caespitose, single shoot about 2.7 to 5 mm long. Leaves laxly imbricate, lower leaves small, upper leaves large, oblong-lanceolate to ligulate, 0.7 × 0.2 mm, broad at base, apex round, margin entire or slightly projecting, costa prominent, ending below the tip.

Rare, at 543 – 583 m alt., on red rocky soils in dry deciduous forests.

Kalasamudram (ATP), *KRKS* 40054; Batrepalli (ATP), *KRKS* 40160.

BRYACEAE

1a. Capsule erect or shortly inclined; peristome process mostly rudimentary........**Brachymenium**

1b. Capsule mostly inclined to pendulous; peristome processes and cilia normal and well developed........(2)

2a. Leaf margin flat, without border........**Gemmabryum**

2b. Leaf margin flat or revolute, with distinct or indistinct borders........(3)

3a. Leaves spirally twisted when dry........**Ptychostomum**

3b. Leaves not spirally twisted........**Bryum**

BRACHYMENIUM Schwägr.

1. Plants up to 5 mm high; leaves 1 × 0.3 mm........**B. bryoides**

1. Plants 5-8 mm high; leaves 1.5 × 0.6 mm........**B. indicum**

Brachymenium bryoides Hook. ex Schwägr., Sp. Musc. Frond. Suppl. 2(1): 134. 1824; Gangulee, Moss. E. India 2: 940. 1974; Sandhya Rani *et al.*, Bryo. Andhra Pradesh 155. 2014

Plant green, stem up to 5 mm high, branched with subfloral innovation. Leaves dense, ovate-lanceolate, 1.5 × 0.6 mm, tip acuminate, basal cells rectangular, and become quadrate towards margin, costa strong, excurrent in a long denticulate arista, margin flat and entire but poorly denticulate at apex, perichaetial leaves small. Sporophyte terminal, seta long, red, *ca.* 1.4 mm long, capsule erect or inclined, brown, *ca.* 1 mm long and 0.3 mm in diameter, narrow mouthed.

Occasional, at 262 – 462 m alt., grows on moist soil in dry deciduous forests. Rachakuntapalli (KDP), *KRKS* 40330; Gandi (KDP), *KRKS* & *BSS* 41992.

Brachymenium indicum (Dozy & Molk.) Bosh & Sande Lac., Bryol. Jav., 1: 141. 1860; Gangulee, Moss. E. India 2: 942. 1974; Sandhya Rani *et al*. Bryo. Andhra Pradesh 63. 2014. *Bryum indicum* Dozy & Molk., Musci Fr. Ined. Archip. Indici, 1: 22. 1845.

Plants green, slightly glossy, densely tufted, 5–8 mm height, lower leaves small, distinctly arranged, upper leaves large, densely arranged, imbricate, 1 × 0.3 mm at base, ovate-lanceolate, tip acuminate, leaf cells rhomboidal- elongated rhomboidal, 0.05 × 0.04 mm at apex, gradually becoming shorter and wider below, subrectangular, 0.06 × 0.02 mm, costa strong, excurrent as a denticulate arista, margin entire showing poor denticulations at apex.

Occasional, at 414 – 547 m alt., grows on red soil, in open dry deciduous forests. Rekkamanu (ATP), *KRKS* 37925, Rachakuntapalli (KDP), *KRKS* 40329.

BRYUM Hedw.

1a. Costa percurrent .... (2)

1b. Costa excurrent .... (3)

2a. Plants pale green to silvery white, leaves spreading on stem, 2.5 × 1mm .... **B. argenteum**

2b. Plants reddish green to green, leaves compact, 1-1.5 × 0.4 – 0.5 mm........ **B. cellulare**

3a. Leaf margin usually flat. .... **B. mildenaum**

3b. Leaf margin more or less revolute all along .... **B. paradoxum**

Bryum argenteum Hedw., Sp. Musc. Frond, 181. 1801. var. **lanatum** (P. Beauv.) Hamp., Linnaea, 13: 44. 1839; Ber S. Gall., Bryol. Eur., 4: 148. 1839; Gangulee, Moss. E. India 2 : 970.1974; Sandhya Rani *et al*., Bryo. Andhra Pradesh 161.2014. *Mnium lanatum* P.Beauv., Prodr. Aetheogam: 75. 1805.

Plants small, densely tufted, glossy. Stem short, reddish brown, *ca.* 7 mm long, erect, julaceous branches. Leaves densely arranged on the stem, broadly ovate, concave, short acuminate, 1.5 × 0.5 mm at base, margin entire, costa percurrent, *ca.* 0.4 mm long with hyaline arista.

Common, at 546 – 624 m alt., exhibits both the terricolous and saxicolous habitat. On cement bridges, Teak plantation, on forest rocks in blackish red soils in moist deciduous forests. Batrepalli (ATP), *KRKS & BSS* 37988, 40365, 40366; Kalasamudram (ATP), *KRKS* 40054.

Bryum cellulare Hook., Sp. Musc. Suppl. 3(1): 214a. 1827; Gangulee, Moss. E. India 2: 969. 1974. Sandhya Rani *et al*., Bryo. Andhra Pradesh 166. 2014

Plants slender, reddish green, unbranched or rarely branched. Stem erect, up to 1.2 cm long, reddish. Leaves sparsely arranged, alternate, ovate-lanceolate, 2.5×1 mm, concave, apical leaf hook like, curled when dry, apex acute, margin entire or involute, costa reddish, percurrrent or ending one cell below the apex.

Occasional, at 264 – 570 m alt., on rocks, red soil, moist shady and wet localities in dry deciduous forests. Batrepalli (ATP), *KRKS* & *BSS* 40367; Naganupula (KDP), *KRKS* & *BSS* 40496.

Bryum mildenaum Jur., Verh. Zool.-Bot. Ges. Wien 12: 967. 1862. *B. porphyroneuron* C. Muell., Bot. Zeitung (Berlin), 11: 22. 1853. var. **erythrinum** (Mitt.) Fleisch. in Broth. in Nat. Pfl., 1(3): 590. 1904; Gangulee, Moss. E. India 2: 1008. 1974; Sandhya Rani *et al.*, Bryo. Andhra Pradesh 170. 2014.

Plants brown below, yellowish green above, loosely tufted, 1–1.3 cm height, branched from the base or unbranched. Leaves arranged on main stem, lower leaves small, red, sparsely arranged, upper leaves larger, up to 2 × 0.5 mm, arranged in tufts, leaves appressed to the stem and slightly curled when dry, elliptic-lanceolate, acuminate, margin entire, slightly denticulate at apex, marginal cells slightly narrow forming indistinct border, costa excurrent, *ca.* 0.2 mm, prominent, reddish, with denticulate arista.

Occasional, at 414 – 902 m alt., on black loamy soil in dry deciduous forests. Rachakuntapalli (KDP), *GK* & *KRKS* 33568, 33569; Batrepalli (ATP), *KRKS* 39083.

Bryum paradoxum Schwägr., Sp. Musc. Frond., Suppl. 3 (1): 224a.1827; Gangulee, Moss. E. India 2: 1012. 1974.

Plants dioecious small, *ca.* 10 mm high, yellowish green above, blackish below, in loose tufts. Stem often branched by innovations. Leaves appressed when dry, lanceolate or oblong-lanceolate, acuminate, lower leaves *ca.* 1.7 mm long; upper leaves *ca.* 2.8 mm long, indistinctly concave; margins often narrowly recurved, upper margins serrulate, costae excurrent, ending in short awns, reddish brown at base. Setae *ca.* 10–20 µm long, curved, reddish brown; capsules oblong-pyriform to club-shaped, pendulous, reddish brown; neck slightly shorter than urn. Spores 10–17 µm in diameter, mostly smooth.

Rare, at 415 – 945 m alt., on soil or thin soil over rocks in dry deciduous forests. Rachakuntapalli (KDP), *KRKS* 40329; Batrepalli (ATP), *KRKS* & *BSS* 40362; Rahakuntapalli (KDP), *KRKS* 40330.

GEMMABRYUM J.R.Spence & H.P.Ramsay

Gemmabryum apicuatum (Schwägr) J.R. Spence & H.P. Ramsay, Phytologia 87: 65. 2005. *Bryum plumosum* Dozy & Molk., Ann. Sc. Nat. Bot. ser. 3,3 (2): 301. 1844; Gangulee, Moss. E. India 2: 1010. 1974. Sandhya Rani *et al.*, Bryo. Andhra Pradesh 170. 2014

Plants tufted, slender, yellowish green, shoot 6 mm long. Stem erect, branched or with sub floral innovations. Leaves scattered at base, densely crowded at apex, 1.4–2.5 mm long, ovate-lanceolate, acuminate, 1.4×0.5 mm, margin flat, entire, fine denticulate towards apex, border indistinct, costa excurrent in a short arista.

Rare, at 414 – 432 m alt., grows on the soils near waterfalls, moist earth cuttings of dry deciduous forests.

PTYCHOSTOMUM J.R.Spence & H.P.Ramsay

Ptychostomum capillare (Hedw.) D.T. Holyoak & N. Pedersen, J. Bryo. 29: 119. 2007. *Bryum capillare* L. ex Hedw., Sp. Musc. Frond. 182. 1801; Bruehl, Rec. Bot. Surv. India 13(1): 53. 1931; Gangulee, Moss. E. India 2: 987. 1974. Sandhya Rani *et al.*, Bryo. Andhra Pradesh 163. 2014

Plants densely tufted yellowish green above, reddish brown below, 3–8 mm, dense tufts, unbranched, rarely branched. Stem reddish. Leaves spirally arranged on stem, sparsely at below, tufts at apex, ovate, acuminate, concave, 2.4 × 0.7 mm, margin entire, rarely denticulate at apex, costa prominent, reddish brown, excurrent, spinulose aristate, up to 0.2 mm long. Sporophyte apical, seta erect, *ca.* 1.8 cm long, red, capsule horizontal, 0.2 × 0.1 mm, reddish brown, up to 0.2 mm long, 0.1 mm wide.

Very common, at 356 – 902 m alt., reddish black soil, black loamy soil, red soil, rocks and bark of trees in dry deciduous forests. Batrepalli (ATP), *KRKS* & *BSS* 37988; *KRKS* 39085; Bhumaiahgaripalli (KDP), *KRKS* & *GK* 33567; Rachakuntapalli (KDP), *KRKS* & *GK* 33568, 33569; Vemula (KDP), *KRKS* & *GK* 33575.

BARTRAMIACEAE

PHILONITIS Bridel

Philonotis hastata (Duby) Wijk & Margad., Taxon 8: 74. 1959; Gangulee, Moss. E. India 2: 1127.1974; R.S.Chopra, Taxon. Indian Moss. 252. 1975; Sandhya Rani *et al.*, Bryo. Andhra Pradesh 174.2014. *Hypnum hastatum* Duby in Moritzi, Syst. Verz. 132. 1846.

Plants yellowish green-brown, many branched by subapical innovations, stem soft, delicate, filiform unbranched, *ca.* 5–9 mm long, leaves lax, curled and appressed to stem when dry, lanceolate, shortly acuminate or blunt slightly, rounded at base, 0.4 × 0.1 mm, margin serrulate by upturned tips or mamillae of the cells, from just above the base. Costa prominent, percurrent, 0.31 mm long, papillose at the back, marginal teeth in two rows.

Occasional, at 546 – 583 m alt., on rocks, rock crevices near waterfalls and moist place in dry deciduous forests. Kalasamudram (ATP), *KRKS* 39131, 39178; Batrepalli (ATP), *KRKS* & *BSS* 37988.

References

Ahmedullah, M. (2000). Endemism in Indian flora; In Flora of India Introductory Vol. II.BSI. Kolkata.

Ahmedullah, M. and M.P. Nayer (1987). *Endemic Plants of Indian Region.* Vol.1 (Peninsular India) Botanical Survey of India, Calcatta.

Balakrishnan, N. P. and T. Chakrabarthy (2007). The Family Euphorbiaceae in India. Bishen Singh Mahendra Pal Singh, Dehra Dun.

Balakrishnan, N. P., T. Chakrabarty, M. Sanjappa, P. Lakshminarsimhan and P. Singh (2012). *Flora of India, Loranthaceae – Daphniphyllaceae.* Vol. 23. Botanical Survey of India.

Beddome, R. H. (1869-74). *The Flora of Sylvatica for Southern India* (3 vols), Madras.

Binojkumar, M. S. and N. P. Balakrishnan (2010). *The genus Euphorbia L.* (*Euphorbiaceae*) *in India: A Taxonomic Revision.*Bishen Singh Mahendra Pal Singh, Dehra Dun.

Blasco, F. (1971). Orophytes of south India and Himalyas. *J. Indian Bot. Soc.* 50: 377 – 381.

Bor, N. L. (1960) *The Grasses of Burma. Ceylon, India and Pakistan.* Pergamon Press. Oxford.

Bose, T. K., S. K. Bhattacharjee, P. Das and U. C. Bhaskar (1980). *Orchids of India*. Naya Prokash, Calcutta.

Brandis, D. (1906). *Indian trees*. Archibald Constable and Co, London. (repr. Ed. 1998. Bishen Singh Mahendra Pal Singh, Dehra Dun).

Chakrabarthy, T. and M. Gangopadhyay (1996a). The genus. *Acacia* P. Miller (Leguminosae: Mimosoideae) in India. *J. Econ. Tox. Bot.*, 20: 499-533.

Chakrabarthy, T. and M. Gangopadhyay (1996b). The genus *Albizia* Durazz. (Leguminosae: Mimosoideae) in India. *J. Econ. Tox. Bot.*, 20: 581-597.

Champion, H. G. and S. K. Seth (1968). *A revised survey of the forest types of India*. Government of India, New Delhi.

Chatterjee, D. (1940). Studies on the endemic flora of India and Burma. *J. Roy. Asiat. Soc. Bengal* (*Sci.*) 5: 19-16.

Chaudhuri, A.B. (1967). Studies on the Meliaceae of India and neighbouring countries. *Bull. Bot. Soc. Bengal.*, 21: 1-24.

Chennaiah, E. (1997) *Flora of Kalasamudram Reserved forest*. M.Phil. Dissertation. S. K. Universtiy, Anantapur.

Elliot, W. (1859). *Flora Andhrica*. Madras.

FAO (2000). *Global Forest Resources Assessment*. United Nations, Rome, Italy.

Gamble, J. S. (1881). *A manual of Indian timbers*. (repr. ed. 2001) International Book Distributors, Dehra Dun.

Gamble, J. S. and C. E. C. Fischer (1915-35). *Flora of the Presidency of Madras*. London.

Hajra, P. K. and V. Mudgal (1997). Plant diversity Hot spots in India – An overview. Botanical Survey of India. Calcutta.

Hajra, P. K., R. R. Rao, D. K. Singh and B. P. Uniyal (1995a). *Flora of India* Volume-12 & 13. *Asteraceae*. Botanical Survey of India. Calcutta.

Hajra, P. K., V. J. Nair and P. Daniel (1997). *Flora of India Volume-4 Malpighiaceae-Dichapetalaceae*. Botanical Survey of India. Calcutta.

Hooker, J. D. (1872-1897). *Flora of British India* (ed.) (7 vols.) London.

IUCN, (2010). Species Survival Commission, *IUCN Red List Categories, Version 8.1.*, www.intranet.iucn.org/webfiles/doc/SSC/RedList/RedListGuidelines.pdf.

Jain, S. K. and R. R. Rao (1977). *Field and Herbarium Methods*. New Delhi.

Janarthanam, M. K. & A.K. Henry (1992). *Bladderworts of India*. Botanical Survey of India. Coimbatore.

Karthikeyan, S., S. K. Jam, M. P. Nayar and M. Sanjappa (1989). *Florae Indicae Enumeration Monocotyledonae*. Fl. Ind. Ser. 4. Botanical survey of India. Calcutta.

Karuppusamy, K., A. Ugraiah and T. Pullaiah (2013). *Caralluma* (*sensu lato*) *Antiobesity plants*. Regency publications, New Delhi.

Kumar, S. V. and M. Bhanja (1992). Forestry seed manual of Andhra Pradesh. Research & Development Circle, Andhra Pradesh Forest Department, Hyderabad, India.

Madhava Chetty, K. and K. N. Rao (1990). Endemic plants of Tirumala hills in Chittoor district of Andhra Pradesh. *Vegetos* 3(1): 12 - 15.

Madhusudhana Rao, A. (1989). *Floristic studies on the flora of Kadapa district, A.P.* Ph.D, Thesis. S. V. University, Tirupati.

Matthew, K. M. (1981-1983). *The Flora of Tamilnadu Carnatic.* Vol 3. The Rapinat Herbarium, Tiruchirapalli.

Matthew, K. M. (1982). *Illustrations on the Flora of Tamilnadu Carnatic.* The Rapinat Herbarium, Tiruchirapalli.

Mc Neely, J.A., K.R.Miller, W.V. Reid, R.A. Mittermeier and T.B. Werner (1990). Conserving the World's Biological Diversity. IUCN, WRI, CI, WWF-US and The World Bank.

Mishra, S. (2007). *Orchids of India*, Bishen Singh & Mahendra Pal Singh, Dehra Dun.

Nayar, M. P. (1980a). Endemism and patterns of distribution of endemic genera (Angiosperms). *J. Econ. Tax. Bot.*1: 99- 110.

Nayar, M. P. (1980b). Endemic flora of peninsular India and its significance. *Bull. Bot. Surv. India* 22: 12 – 23.

Nayar, M. P. (1996). *'Hot Spots' of endemic plants of India, Nepal and Bhutan.* Tropical Botanical Garden and Research Institute, Palode, Trivandrum.

Nayar, M. P. and A. R. K. Sastry (1987). *Red data book of Indian Plants.*vol. 1. Botanical Survey of India, Calcutta.

NRSA (2007). Biodiversity characterization at landscape level in Eastern Ghats and East coast using satellite remote sensing and Geographical Information Systems. National Remote Sensing Agency. Department of Space & Department of Biotechnology, GOI.

Phani, P.R.C. (2011) Geological Excursion to Eparchaean Unconformity at Namalagundu, Anantapur District, Andhra Pradesh, India. e-Journal Earth Science India 1–8. (http://www.earthscienceindia.info/pdfupload/esi_geosites_1.pdf)

Pullaiah, T. (1997). *Flora of Andhra Pradesh, India.* Vol. 3. Scientific Publishers, Jodhpur.

Pullaiah, T. (2008). Biodiversity and Conservation of Eastern Ghats. In: *Forest Biodiversity*. 1: 96-99.

Pullaiah, T. and E. Chennaiah (1997). *Flora of Andhra Pradesh, India.* Vol. 1. Scientific Publishers, Jodhpur.

Pullaiah, T. and S. Karuppusamy (2008). *Flora of Andhra Pradesh, India.* Vol. 5. Scientific Publishers, Jodhpur.

Pullaiah, T., S. Karuppusamy and S. Sandhya Rani (2008). The district floras of Andhra Pradesh- an overview. *Proc. A. P. Akademi of Sciences* 12(1&2): 34-47.

Pullaiah, T. and D. A. Moulali (1997). *Flora of Andhra Pradesh, India.* Vol. 2. Scientific Publishers, Jodhpur.

Pullaiah, T. and D. Muralidhara Rao (2002). *Flora of Eastern Ghats Hill Ranges of Southeast India.* Vol. 1(*Ranunculaceae-Moringaceae*). Regency Publications, New Delhi.

Pullaiah, T. and S. Sandhya Rani (1999). *Trees of Andhra Pradesh, India.* Regency Publications. New Delhi.

Pullaiah, T., S. Sandhya Rani and S. Karuppusamy (2011). *Flora of Eastern Ghats Hill Ranges of Southeast India* Vol. 4 (*Stylidaceae – Plantaginaceae*). Regency Publications, New Delhi.

Pullaiah, T. and K. Sri Rama Murthy (2001). *Flora of Eastern Ghats Hill Ranges of Southeast India.* Vol.2 (*Leguminosae-Fabaceae*). Regency Publications, New Delhi.

Pullaiah, T., K. Sri Rama Murthy and S. Karuppusamy (2007). *Flora of Eastern Ghats Hill Ranges of Southeast India.* Vol. 3. (*Rosaceae-Asteraceae*). Regency Publications New Delhi.

Pullaiah, T. and P. Surya Prakash Babu (1998). *Flora of Andhra Pradesh, India.*Vol. 4. *Illustrations.* Scientific Publishers, Jodhpur.

Pullaiah, T. and N. Yesoda (1989). *Flora of Anantapur district, Andhra Pradesh.* Bishen Singh and Mahendra Pal Singh, Dehradun.

Raja Kullayiswamy, K., D. Ramanjaneyulu, S. Sandhyarani and T, Pullaiah (2013a). *Utricularia australis* R. Br. (Lentibularaceae) a new distributional record from Eastern Ghats, India. *J. Indian Bot. Soc.* 93 (1&2): 101-103.

Raja Kullayiswamy, K., S. Sandhyarani and T. Pullaiah (2013b). *Euphorbia clarkeana* Hook. f.- New distributional Record from Eastern Ghats. *J. Econ. Taxon. Bot.* 37 (2): 305-307.

Raja Kullayiswamy, K., S. Sandhyarani and T. Pullaiah (2013c). *Brachystelma ciliatum* (Apocynaceae): A new record for Eastern Ghates of Andhra Pradesh, India. *Nelumbo* 55: 191-195.

Raju V. S., C. Sudhakar Reddy, K. N. Reddy, K. Seshagiri Rao and B. Bahad (2008). Orchid wealth of Andhra Pradesh, India. *Proc. A.P. Academy of Sciences*12 (1&2): 180-192.

Rao, B.R.P., M. V. Suresh Babu, A. Madhusudhana Reddy, S. Sunitha, A. Narayanaswamy, G. Lakshminarayana and M. Ahmedullah (2011). Conservation status of *Hildegardia populifolia* (Roxb.) Schott & Endl. (Malvaceae: Sterculioideae: Sterculieae), an endemic of southern peninsular India. *J. Threatened Taxa* 3(8): 2018-2022.

Rao, C. K. (1972). Angiosperm genera endemic to the Indian floristic regions and its neighboring areas. *Indian Forester* 98: 560 – 566.

Roxburgh, W. (1795 - 1810). *Flora Indica.* 3 Vols. Calcutta.

Sandhya Rani, S. and T. Pullaiah (2002). A taxonomic survey of trees in Eastern Ghats. In: *Proc. of the Nat. Seminar on Conservation of Eastern Ghats. EPTRI-ENVIS.* pp. 5-15.

Sanjappa, M. (1991). *Legumes of India.* Bishen singh Mahendra pal Singh, Dehra Dun.

Santapau, H. (1958). *Botanical collectoris Manual.* Bot. Surv. India. Calcutta.

Satyanarayana, P. and M. Sanjappa (1998). Endemic legumes of Eastern Ghats. In: *Proc. Nat. seminar on Conservaton of Eastern Ghats.* 501-503.

Seshagiri Rao, R. (1998). Vegetation and valuable plant resources of the Eastern Ghats. *Proc. National. Seminar on Conservation of Eastern Ghats. EPTRI-ENVIS.* pp. 59-87.

Shali Saheb, T. (2008). *Medicinal Plant resources and conservation in Nallamalais, Andhra Pradesh.* Ph. D., Thesis. S. K. University, Anantapuram, A.P.

Sharma B. D. and M. Sanjappa (1993). *Flora of India.* Volume 3. *Portulcaceae – Ixonanthaceae.* Botanical Survey of India. Calcutta.

Sharma B. D. and N. P. Balakrishanan (1993). *Flora of India.* Volume 2. *Papaveraceae-Caryophyllaceae.* Botanical Survey of India. Calcutta.

Sharma B. D., N. P. Balakrishanan, R.R. Rao and P.K. Hajra (1993). *Flora of India.* Volume 1. *Ranunculaceae-Barclayaceae.* Botanical Survey of India. Calcutta.

Sharma, J. R., V. Mudgal and P. K. Hajra (1997). Floristic diversity-review, scope and perspectives. In: *Floristic diversity and conservation strategies in India.* Mudgal, V and P.K. Hajra (eds.) Botanical Survey of India.vol. I. pp. 1-45.

Singh, J.S. (2002). The Biodiversity Crisis: A multifaceted review. *Curr. Sci.* 82: 638-647.

Singh, N. P., J. N. Vohra, P. K. Hajra and D. K. Singh (2000). *Flora of India,* Volume-5, *Olacaceae-Connaraceae.* Botanical Survey of India. Calcutta.

Singh, V. (2001a). *Monograph on Indian Leucas R. Br. (Dronapushpi) Lamiaceae.* Scientific publishers Jodhpur.

Singh, V. (2001b). *Monograph on Indian subtribe Cassiinae (Caesalpiniaceae).* Scientific publishers, Jodhpur.

Sivaraj, N., B. Narayani and M. C. Kiran (2006). Medicinal Plant Wealth of Eastern Ghats with special reference to indigenous knowledge systems. *J. Swamy. Bot. Club.,* 23: 165-172.

Sudhakar Reddy, C., K. N. Reddy, E. N. Murthy and V. S. Raju (2009). Trees Wealth of Eastern Ghats of Andhra Pradesh, India: An updated checklist. *Check List* 5(2): 173-194.

Sudhakar Reddy, C., K. N. Reddy, Vatsavaya S. Raju (2008). Supplement of Flora of Andhra Pradesh, India. Deep Publications, New Delhi.

Suresh Babu, M. V. (2011). Plant biodiversity assessment of Seshachalam hill ranges, Andhra Pradesh, India. Thesis submitted to Sri Krishnadevaraya University, Anantapuram

Venkatappa, N., T. Pullaiah, K. Sri Ramamurthy and S. Sandhya Rani (1998). Convolvulaceae in Eastern Ghats. In: Proc. Nat. Seminar on Conservation of Eastern Ghats. p.p. 106-113.

Wight, R. (1838 - 1853). *Icones Plantarum Indiae Orientalis*. J. B. Pharoah, Madras.

World Conservation Monitoring Centre (1998). In: IUCN 2011. IUCN Red List of Threatened Species. Version 2011.1. <www.iucnredlist.org>.

BRACHYSTELMA R. Br. *nom. cons.*

1a. Climbing herb; pedicel glabrous; calyx lobes shorter than the corolla tube and glabrous; corolla tube urceolate, 7 – 9 mm long .... **B. nigidianum**

1b. Erect herb; pedicel pubescent; calyx lobes longer than the corolla tube and pubescent; corolla tube ovate, 1 mm long .... **B. ciliatum**

Brachystelma nigidianum Raja Kullayisw. *et al.*, Kew Bull. 71: 1. 2016. *Brachystelma volubile* auct. non Hook. f.; Raja Kullayiswamy *et al.* Rheedea 22: 107. 2012.

Perennial climbing herb, to 1.5 m long. Tuber discoid-subglobose, 1 or 2 subterranean, brown, *ca.* 2.8 × 3.5 cm. Stem herbaceous, terete, glabrous, sometimes branched. Leaves linear-lanceolate, 5 – 16 × 0.2 – 0.5 cm, narrowed into 3 – 5 mm long petiole, acute at base, acuminate at apex, entire at margin, scabrous above, hirsute beneath on raised midrib, margin ciliate, caducous, leaves reduced towards apex on stem, sessile *ca.* 3 mm long, scabrous. Petiole scabrous, canaliculate with brown colleter (gland) at base. Inflorescence extraaxillary, 2- or 3-flowered. Peduncles 5 – 6 mm long, glabrous; bracts lanceolate, glabrous, *ca.* 4 mm long; bracteoles lanceolate, glabrous *ca.* 3 mm long. Pedicels *ca.* 9 mm long, glabrous, purplish-green. Calyx linear-lanceolate, subulate, 9 – 10 mm long, glabrous, purplish-green Corolla 3.3 – 3.5 cm long, *ca.* 4.4 cm across (when open); corolla tube *ca.* 9 mm long, urceolate or urn-shaped, base, middle and mouth of corolla tube *ca.* 9 mm, *ca.* 7 mm and *ca.* 11 mm wide respectively, glabrous, greenish with brown spots without, and slightly ribbed, pubescent with fine hairs, dark brown with light yellow lines within; lobes deltoid at base (two times wider than long) tapering towards apex, *ca.* 2.6 × 0.3 mm, cohering at apex when young and free when mature, acuminate at apex, glabrous, greenish outside, reddish maroon, densely pubescent within, scattered pilose along margin (hairs *ca.* 2 mm long), margin slightly outcurved, yellow striations at base of lobes. Corona *ca.* 2 mm high, *ca.* 4 mm across; outer corona cupular, reddish maroon, *ca.* 2 mm across, pubescent, 5-lobed, each lobe slightly bifid; inner corona 5-lobed, dorsiventrally flattened, c. 1 mm long, basally fused with outer ring, lobes 5, free, oblong-ovate, yellow with maroon tinge at middle, apex obtuse, overlapping anthers with pollen chambers below. Pollinaria *ca.* 0.5 × 0.5 mm across; pollinia ovoid, *ca.* 0.4 × 0.3 mm, yellow, basally attached by hallow hyaline caudicles to a brown corpuscle, pollinia with germination crests towards apex more than half as long as pollinium. Ovary c. 2 mm long, style short. Follicles in pairs, slightly fusiform, 5.5 – 7.5 cm × 2.5 – 4 mm, tapering, connate at apex, slender, smooth, glabrous, pale green. Seeds flattened, ovate, *ca.* 5 × 2.5 mm, light brown, outer margins thine, comose; coma 9 – 14 mm, white silky.

Rare in Nigidi forest in Anantapuramu district. Fl. & Fr.: Oct.- Dec. Nigidi forest (ATP), *KRKS* 41853 (holotype SKU; isotypes BSID, CAL, K, MH).

Index to FAMILIES, Genera and Species

Italicized names indicate synonymy

Rottboellia L.f. 499
cochinchinensis (Lour.) W.D.Clayton 499
exaltata L.f. 499

Index to the Local Names

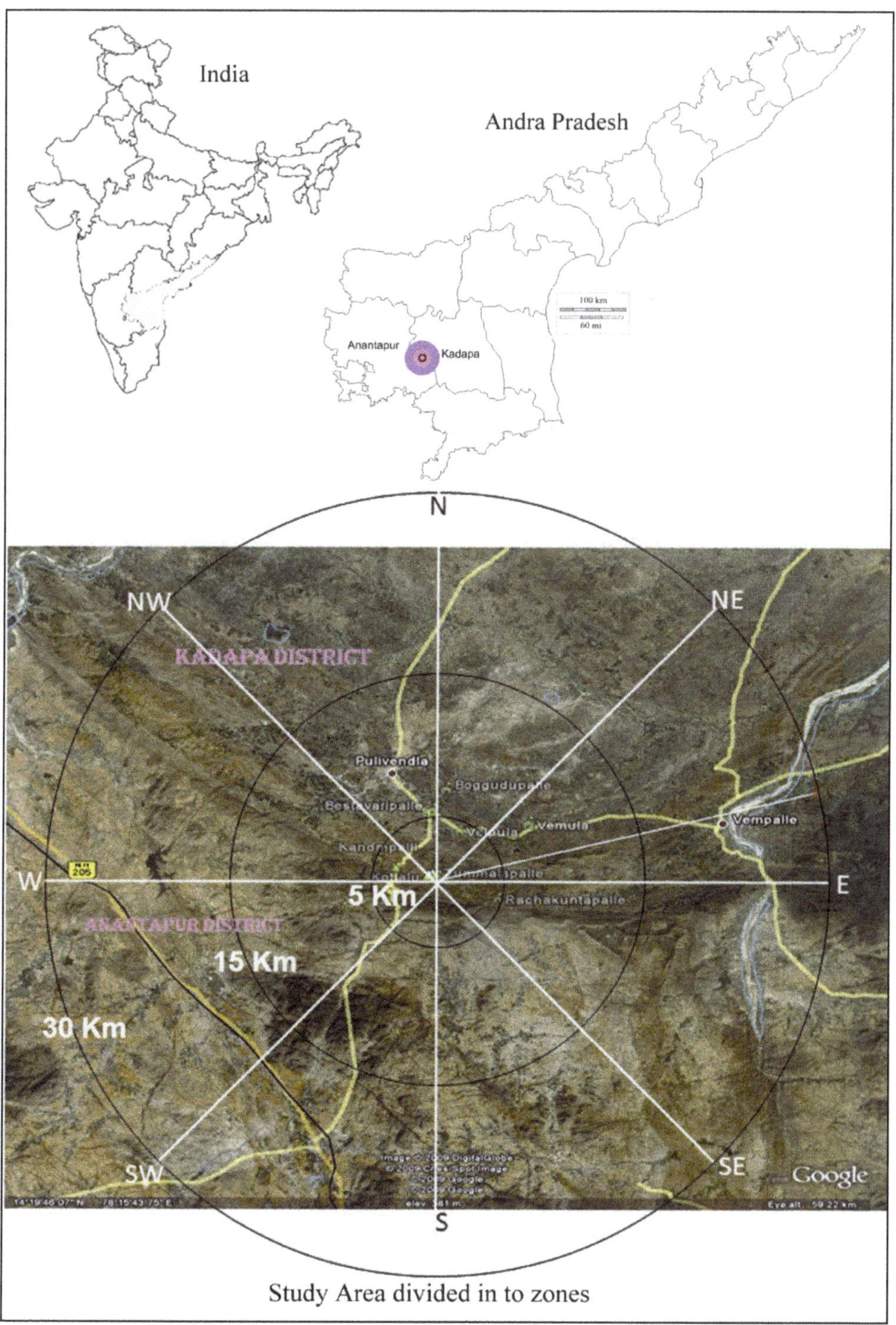

Study Area divided in to zones

p. 7

Vegetation

A. Dry deciduous forest, Batrepalli

B. Dry deciduous forest, Idupulapaya

C. Scrub forest, Thummalapalle

D. Ground nut crop, Kalasamudram

E. Bengal gram crop, Peddakudala

F. Thummalapalle village

p. 11

Water bodies

Stream at Malakavemula (Buffer zone II)

Papagni rever at Gandi (Buffer zone II)

Tank at Batrepalle (Buffer zone I)

Tank at Thalupula (Buffer zone I)

Yogi Vemana reservoir at Mudigubba (Buffer zone II)

Plate 1

A. *Stephania japonica*, B. *Nymphaea nouchali*, C. *Nymphaea rubra*, D. *Capparis sepiaria*, E. *Maerua apetala*, F. *Hybanthus vatsavayae*

Plate 2

A. *Polygala arvensis*, B. *Polygala wightiana*, C. *Polycarpaea aurea*, D. *Portulaca pilosa*, E. *Portulaca tuberosa*, F. *Abutilon pannosum*

p. 59

Plate 3

A. *Decaschistia rufa*, B. *Hildegardia populifolia*, C. *Melhania hamiltoniana*, D. *Grewia orientalis*, E. *Biophytum reinwardtii*, F. *Naringi crenulata*

Plate 4

A. *Ochna obtusata* var. *gamblei*, B. *Gymnosporia heyneana,* C. *Ampelocissus tomenosa*, D. *Cardiospermum corindum*, E. *Moringa concanesis,* F. *Alysicarpus heyneanus*

Plate 5

A. *Crotalaria angulata*, B. *C. laburnifolia*, C. *Galactia tenuiflora*, D. *Indigofera oblongifolia*, E. *Ormocarpum sennoides*, F. *Pterocarpus santalinus*

p. 123

Plate 6

A. *Rhynchosia rufescens*, B. *R. ravii*, C. *Zornia diphylla*, D. *Senna holosericea*, E. *Albizia thompsonii*, F. *Neptunia triquetra*

p. 149

Plate 7

A. *Mimosa polyancistra*, B. *Terminalia coriacea*, C. *Ammannia octandra*, D. *Rotala densiflora*, E. *R. fimbriata*, F. *R. illecebroides*,

p. 180

Plate 8

A. *Rotala rosea*, B. *Cucumis prophetarum*, C. *Trianthem triquetra*, D. *Mollugo disticha*, E. *Dentella repens*, F. *Gardenia latifolia*,

p. 193

Plate 9

A. *Knoxia wightiana*, B. *Morinda pubescens* var. *stenophylla*, C. *Amberboa ramosa*, D. *Cynthillium albicans*, E. *Epaltes divaricata*, F. *Blepharispermum subsessile*

p. 215

Plate 10

A. *Carissa carandas*, B. *Brachystelma ciliatum*, C. *Caralluma adscendens* var. *attenuata*, D. *C. adscendens* var. *fimbriata*, E. *Ceropegia bulbosa* var. *lusii*, F. *Ceropegia pullaiahii*

p. 249

Plate 11

A. *Ceropegia spiralis*, B. *Cynachum sarcomedium*, C. *Cryptolepis buchanani*, D. *Decalepis hamiltonii*, E. *Tylophora fasciculata*, F. *Hoppea dichotoma*

p. 263

Plate 12

A. *Heliotropium subulatum*, B. *Trichodesma sedgewickianum*, C. *Cordia monoica*, D. *Argyreia kleiniana*, E. *Rivea ornata*, F. *Xenostegia tridentata* subsp. *hastata*

p. 281

Plate 13

A. *Datura stramonium*, B. *Glossostigma diandra*, C. *Limnophila indica*, D. *Lindernia oppositifolia*, E. *Utricularia australis*, F. *Henckelia incana*

p. 301

Plate 14

A. *Andrograpis paniculata*, B. *Barleria acuminata*, C. *B. stocksii*, D. *Justicia vahlii*, E. *Lepidagathis subarmata*, F. *Vitex pinnata*

p. 322

A. *Basilicum polystachyon*, B. *Leucas lavandulifolia*, C. *L. mathewiana*, D. *Orthosiphon rubicundus*, E. *Achyranthes aspera* var. *sicula*, F. *Allmania longipedunculata*

p. 350

A. *Nothosaerva brachiata*, B. *Dendrophthoe falcata* var. *pubescens*, C. *Scurrula parasitica*, D. *Viscum articulatum*, E. *Acalypha fruticosa*, F. *Croton scabiosus*

p. 374

Plate 17

A. *Acalypha lanceolata*, B. *Euphorbia clarkeana*, C. *E. fusiformis*, D. *E. kadapensis*, E. *E. longistyla*, F. *E. prostrata*

p. 387

Plate 18

A. *Euphorbia tortilis*, B. *Phyllanthus indofischeri*, C. *P. rheedei*, D. *P. rotundifolius*, E. *Ficus tsjakela*, F. *Eulophia epidendraea*

p. 405

A. *Ceratophyllum demersum*, B. *Molineria trichocarpa*, C. *Commelina diffusa*, D. *C. subulata*, E. *Najas graminea*, F. *Eriocaulon quinquangulare*

p. 423

A. *Cyperus iria*, B. *C. clarkei*, C. *Fimbristylis tetragona*, D. *Kyllinga bulbosa*, E. *Schoenoplectiella articulates*, F. *S. juncoides*

p. 444

A. *Cymbopogon coloratus*, B. *Phragmites karka*, C. *Ischaemum afrum*, D. *Leersia hexandra*, E. *Oropetium thomaeum*, F. *Themeda triandra*

p. 476

www.ingramcontent.com/pod-product-compliance
Ingram Content Group UK Ltd.
Pitfield, Milton Keynes, MK11 3LW, UK
UKHW021322190726
13845UKWH00003B/27